W9-CCA-286

| | | | | metals | | | | nonmetals | | | metalloids | | | noble gases |

										VIII
										2 **He** 4.00
				III	**IV**	**V**	**VI**	**VII**	
				5 **B** 10.81	6 **C** 12.01	7 **N** 14.01	8 **O** 16.00	9 **F** 19.00	10 **Ne** 20.18
				13 **Al** 26.98	14 **Si** 28.09	15 **P** 30.97	16 **S** 32.06	17 **Cl** 35.45	18 **Ar** 39.95
28 **Ni** 58.69	29 **Cu** 63.55	30 **Zn** 65.38	31 **Ga** 69.72	32 **Ge** 72.59	33 **As** 74.92	34 **Se** 78.96	35 **Br** 79.90	36 **Kr** 83.80
46 **Pd** 106.42	47 **Ag** 107.87	48 **Cd** 112.41	49 **In** 114.82	50 **Sn** 118.69	51 **Sb** 121.75	52 **Te** 127.60	53 **I** 126.90	54 **Xe** 131.29
78 **Pt** 195.08	79 **Au** 196.97	80 **Hg** 200.59	81 **Tl** 204.38	82 **Pb** 207.2	83 **Bi** 208.98	84 **Po** (209)	85 **At** (210)	86 **Rn** (222)

63 **Eu** 151.96	64 **Gd** 157.25	65 **Tb** 158.92	66 **Dy** 162.50	67 **Ho** 164.93	68 **Er** 167.26	69 **Tm** 168.93	70 **Yb** 173.04	71 **Lu** 174.97

95 **Am** (243)	96 **Cm** (247)	97 **Bk** (247)	98 **Cf** (251)	99 **Es** (252)	100 **Fm** (257)	101 **Md** (258)	102 **No** (259)	103 **Lr** (260)

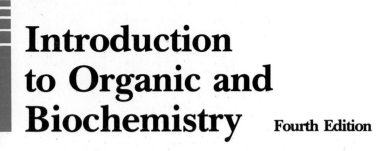

Introduction to Organic and Biochemistry

Fourth Edition

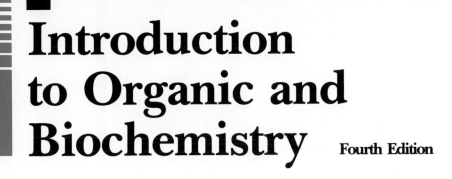

Introduction to Organic and Biochemistry

Fourth Edition

William H. Brown
Beloit College

Brooks/Cole Publishing Company
Monterey, California

Brooks/Cole Publishing Company
A Division of Wadsworth, Inc.

Printed in the United States of America
10 9 8 7 6 5 4 3 2 1

Library of Congress Cataloging-in-Publication Data

Brown, William Henry, [date]
  Introduction to organic and biochemistry.

  Includes index.
  1. Chemistry, Organic. 2. Biological chemistry.
I. Title.
QD253.B73 1987   547   86-21539
ISBN 0-534-07386-7

Sponsoring Editor: *Sue Ewing*
Editorial Assistant: *Lorraine McCloud*
Production Editor: *Phyllis Larimore*
Manuscript Editor: *Janet Wright*
Interior and Cover Design: *Sharon L. Kinghan*
Cover Photo: © *Manfred Kage, Peter Arnold, Inc.*
Art Coordinators: *Judith Macdonald and Sue C. Howard*
Interior Illustration: *John Foster*
Photo Editor and Researcher: *Judy Blamer*
Typesetting: *Syntax International, Ltd., Singapore*
Cover Printing: *Phoenix Color Corporation, Long Island City, New York*
Printing and Binding: *R. R. Donnelley & Sons Co., Crawfordsville, Indiana*

Photo Credits

Page **107B:** Chevron Corporation; **191C:** United States Department of Agriculture; **214B:** Upjohn
Corporation; **247F:** United States Army, Natick Research and Development Center; **386B:** SEM
Photomicrographs taken by Professor Marion I. Barnhart, Wayne State University.

Preface

With this book I intend to provide an introduction to organic and biochemistry for those students working toward careers in the life sciences. These students usually do not plan to become professional chemists, but they do need to understand how living systems depend on chemistry. In this survey of organic and biochemistry I have tried to reveal this relationship and to provide the necessary foundations in organic and biochemistry for further study.

This text is divided into three major sections: organic chemistry, biomolecules, and metabolism.

Text Organization

Organic Chemistry: Chapters 1–10

The first part of the book begins with the Lewis model of covalent bonding in organic compounds and introduces the concepts of structural and functional group isomerism. In Chapter 1 I also introduce the hybridization of atomic orbitals and covalent bond formation by the overlap of atomic orbitals. In Chapters 2–8 I present the chemistry of specific functional groups: saturated hydrocarbons, unsaturated hydrocarbons, alcohols and phenols, amines, aldehydes and ketones, carboxylic acids, and functional derivatives of carboxylic acids. These chapters are similarly organized to include structure, nomenclature, physical properties, and reactions. Chapter 9 covers pH and acid-base buffers with particular emphasis on the bicarbonate and phosphate buffers. The part on organic chemistry concludes in Chapter 10 with a discussion of chirality, a major theme that connects the study of organic chemistry and biochemistry.

My goal in revising the organic chapters has been to tighten them as much as possible and to make them flow smoothly and directly into the following sections on biomolecules and metabolism. Further, I have made two major revisions in chapter organization. First, the chapter on amines has been moved forward to Chapter 5, so that discussion of these molecules follows immediately the discussion of alcohols, ethers, and phenols in Chapter 4.

Second, discussion of chirality has been moved to the end of the part on organic chemistry where it now serves as a transition from organic molecules to biomolecules.

Biomolecules: Chapters 11–15

The major classes of biomolecules are introduced in Chapters 11–15. In Chapter 11 I discuss the chemistry of carbohydrates, and in Chapter 12, the chemistry of lipids. Amino acids and proteins are presented in Chapter 13 and enzymes in Chapter 14. This part concludes with Chapter 15 on the structure and function of nucleic acids.

Among the major revisions in these chapters are addition of Michaelis-Menton kinetics and a more quantitative treatment of enzyme kinetics, expanded discussion of allosteric regulation of enzyme activity, and new material on the use of enzymes in the health sciences.

Metabolism: Chapters 16–19

In Chapter 16 I introduce metabolism and bioenergetics and clearly delineate the several stages in the oxidation of foodstuffs and the generation of ATP. A discussion of the metabolism of carbohydrates, fatty acids, and amino acids follows in Chapters 17–19. Throughout these chapters I point out that the metabolism of these foodstuff molecules is interrelated and precisely regulated.

Among major revisions in this group of chapters is a much more detailed discussion of oxidative phosphorylation, including the chemiosmotic theory.

Features

Several features of this edition make it an especially effective teaching tool.

Example problems with step-by-step solutions appear in each chapter, and each example is followed by a similar problem for the student to solve.

End-of-chapter problems are grouped according to chapter section. This feature ensures a balanced and representative group of problems for each section of the text.

Key terms and concepts are listed at the end of each chapter. The section reference following each term or concept directs the student to the place in the text where it is defined and used.

Eleven mini-essays are included. Five of them are new to this edition: *Alkaloids, Biogenic Amines and Emotions, Nylon and Dacron, Clinical Enzymology,* and *The Penicillins.* The six retained from the third edition have been revised and updated. Mini-essays have several purposes: they bridge the gap between the study of chemistry and the projected vocational areas of life science students; they demonstrate some of the creative excitement inherent in chemistry; and finally, they offer a glimpse of the human involvement in research and development.

Extensive use of graphics, all of it newly designed and redrawn specifically for this edition, enhances the visual appeal and pedagogical effectiveness of the text.

Supplements

In addition to the text, I have prepared the following supplemental materials.

Student Study Guide   Contains complete solutions to all text problems and a detailed guide, using a step-by-step learning approach, to the important concepts and terms used in the text.

Instructor's Manual   Contains suggestions for course organization and scheduling.

Acknowledgments

I would like to acknowledge the help I have received from Bruce Thrasher of Willard Grant Press on three previous editions of this text and from Sue Ewing of Brooks/Cole on this edition. And I thank the very capable staff of Brooks/Cole, coordinated by Phyllis Larimore, for leading me so gently through the intricacies of design, copyediting, and production. Certainly I thank users of the first three editions, and the following reviewers of this edition for their valuable contributions: Thomas Brownlee, Virginia Polytechnic Institute; Jenny Caruthers, University of Colorado; Charles Horn, Mesa Community College; Ted Huiatt, Iowa State University; Sally Jacobs, University of Maine; Richard Luibrand, California State University—Hayward; Fran Mathews, California State University—Fullerton; Darwin Mayfield, California State University—Long Beach; Atilla Tuncay, Indiana University Northwest; and Ann Willbrand, University of Georgia. Most of all I acknowledge with profound thanks the patience and understanding of my wife Hazel and my children, all of whom I fear were sometimes given short shrift as deadlines came and went.

William H. Brown

Contents

Chapter 15   Nucleic Acids and the Synthesis of Proteins   418

Chapter 16   Flow of Energy in the Biological World   451

Introduction to Organic and Biochemistry

Fourth Edition

1 Organic Chemistry: The Compounds of Carbon

Organic chemistry is the study of the compounds of carbon. While the term *organic* reminds us that many compounds of carbon are of either plant or animal origin, by no means is that the limit of organic chemistry. Certainly, organic chemistry is the study of the natural medicines such as penicillin, cortisone, and streptomycin; but it also includes the study of novocaine, the sulfa drugs, aspirin, and other man-made medicines. Organic chemistry is the study of natural textile fibers such as cotton, silk, and wool. It is also the study of man-made textile fibers such as nylon, Dacron, Orlon, and rayon. Organic chemistry is the study of Saran, Teflon, Styrofoam, polyethylene, and other man-made polymers used to manufacture the films and molded plastics with which we are so familiar today. The list could go on and on.

Perhaps the most remarkable feature of organic chemistry is that it comprises the chemistry of carbon and of only a few other elements—chiefly hydrogen, nitrogen, and oxygen. Let us begin with a description of how atoms of these elements combine by sharing electron pairs to form molecules.

1.1 Electronic Structure of Atoms

You are already familiar from a previous course in chemistry with the fundamentals of the electronic structure of atoms. Briefly, an atom contains a small, dense nucleus in which neutrons and positively charged protons are packed. The nucleus is surrounded by a much larger, extranuclear space containing negatively charged electrons. Electrons are concentrated about the nucleus in regions of space called principal energy levels, identified by the **principal quantum numbers** 1, 2, 3, and so on. Each principal energy level can contain up to $2n^2$ electrons, where n is the number of the principal energy level. Thus, the first energy level can contain 2 electrons, the second 8 electrons, the third 18 electrons, the fourth 32 electrons, and so on. For our study of organic chemistry and biochemistry, we need to consider only elements of the first three principal energy levels (rows 1–3 of the periodic table) because these are the elements from which most organic compounds are built.

Each **principal energy level** is subdivided into regions of space called orbitals; each **orbital** can hold two electrons. The first principal energy level contains a

1

single orbital, called a $1s$ orbital. The second principal energy level contains one s orbital and three p orbitals; these orbitals are designated $2s$, $2p_x$, $2p_y$, and $2p_z$. The third principal energy level contains one $3s$ orbital, three $3p$ orbitals, and five $3d$ orbitals. The first three principal energy levels, the maximum number of electrons each can hold, and the orbitals of each are summarized in Table 1.1. Shown in Table 1.2 are electron configurations for the first 18 elements of the periodic table.

Table 1.1
Orbitals of the first three principal energy levels.

Principal Energy Level	Maximum Number of Electrons	Orbitals
$n = 1$	$2n^2 = 2$	$1s$
$n = 2$	$2n^2 = 8$	$2s, 2p_x, 2p_y, 2p_z$
$n = 3$	$2n^2 = 18$	$3s, 3p_x, 3p_y, 3p_z$ plus five $3d$ orbitals

Table 1.2
Electron configurations for the first 18 elements.

Element	Atomic Number	Orbitals $1s$	$2s$	$2p_x$	$2p_y$	$2p_z$	$3s$	$3p_x$	$3p_y$	$3p_z$
H	1	1								
He	2	2								
Li	3	2	1							
Be	4	2	2							
B	5	2	2	1						
C	6	2	2	1	1					
N	7	2	2	1	1	1				
O	8	2	2	2	1	1				
F	9	2	2	2	2	1				
Ne	10	2	2	2	2	2				
Na	11	2	2	2	2	2	1			
Mg	12	2	2	2	2	2	2			
Al	13	2	2	2	2	2	2	1		
Si	14	2	2	2	2	2	2	1	1	
P	15	2	2	2	2	2	2	1	1	1
S	16	2	2	2	2	2	2	2	1	1
Cl	17	2	2	2	2	2	2	2	2	1
Ar	18	2	2	2	2	2	2	2	2	2

When discussing the properties of an element, we often focus on the outermost orbitals of the element, because the electrons in these orbitals are the ones involved in the formation of chemical bonds and in chemical reactions. To show the outermost electrons of an atom, we commonly use a representation called a Lewis structure, after the American chemist G. N. Lewis, who devised this notation. A **Lewis structure** shows the symbol of the element surrounded by a number of dots equal to the number of electrons in the outer shell of that element. In a Lewis structure, the atomic symbol represents the nucleus and all completely filled inner shells. Outer-shell electrons are called **valence electrons,** and the energy level in which they are found is called the valence shell. Table 1.3 shows Lewis structures for the first 18 elements of the periodic table. The noble gases helium, neon, and argon have filled valence shells; the valence shell of helium is filled with two electrons, and those of neon and argon are filled with eight electrons. The valence shells of all other elements shown in Table 1.3 are only partially filled.

Table 1.3
Lewis structures for the first 18 elements of the periodic table.

IA	IIA	IIIA	IVA	VA	VIA	VIIA	VIIIA
H·							He:
Li·	Be:	B:	·C:	·N:	:O:	:F:	:Ne:
Na·	Mg:	Al:	·Si:	·P:	:S:	:Cl:	:Ar:

1.2 The Lewis Model of Bonding

A. Formation of Ions

In 1916, G. N. Lewis devised a beautifully simple model that unified many of the observations about chemical reactions of the elements. He pointed out that the chemical inertness of the noble gases indicates a high degree of stability of the electron configurations of these elements: helium with an outer shell of two electrons; neon with shells of two and eight electrons; and argon with shells of two, eight, and eight electrons. The tendency of atoms to react in ways to achieve an outer shell of eight valence electrons is particularly common among elements of Groups I–VII and is given the special name **octet rule**.

Example 1.1

Show how sodium, in forming Na^+, follows the octet rule.

☐ *Solution*

The electron configuration of sodium is $1s^2 2s^2 2p^6 3s^1$.
The electron configuration of Na^+ is $1s^2 2s^2 2p^6$.
The electron configuration of Ne is $1s^2 2s^2 2p^6$.
By losing the single electron in its $3s$ orbital, sodium achieves a complete octet of electrons in its outermost (valence) shell and an electron configuration resembling that of neon, the noble gas nearest it in the periodic table.

Problem 1.1

Show that formation of the following ions obeys the octet rule:

a. Sulfur forms sulfide ion, S^{2-} **b.** Magnesium forms Mg^{2+}.

B. Formation of Chemical Bonds

According to Lewis's model, atoms bond together in such a way that each atom participating in a chemical bond acquires a completed outer-shell-electron configuration resembling that of the noble gas nearest it in the periodic table. Atoms acquire completed outer shells in two ways.

1. An atom may lose or gain enough electrons to acquire a completely filled outer shell. An atom that gains electrons becomes a negatively charged ion (an **anion**); an atom that loses electrons becomes a positively charged ion (a **cation**). A chemical bond between a positively charged ion and a negatively charged ion is called an ionic bond.
2. An atom may share electrons with another atom or atoms to complete its outer shell. A chemical bond formed by sharing electrons is called a covalent bond.

C. Electronegativity and Chemical Bonds

One way of estimating the degree of ionic or covalent character in a chemical bond is to compare the electronegativities of the atoms involved. **Electronegativity** measures the force of attraction an atom has for the electrons it shares in a chemical bond with another atom. The most widely used scale of electronegativities was devised by Linus Pauling in the 1930s; it is based on bond energies of diatomic molecules. On this scale, fluorine, the most electronegative element, is assigned an electronegativity of 4.0 and all other elements are assigned values relative to fluorine. Figure 1.1 shows electronegativities of common elements.

Electronegativity is a periodic property of the elements. It increases from left to right in a horizontal row of the periodic table, that is, from the most metallic element in a row to the least metallic. It decreases from top to bottom within a vertical column; that is, it decreases with increasing atomic number. You should remember that the values given in Figure 1.1 only approximate the re-

Increasing electronegativity →

	I	II											III	IV	V	VI	VII
1	H 2.1																
2	Li 1.0	Be 1.5											B 2.0	C 2.5	N 3.0	O 3.5	F 4.0
3	Na 0.9	Mg 1.2				Transition elements							Al 1.5	Si 1.8	P 2.1	S 2.5	Cl 3.0
4	K 0.8	Ca 1.0	Sc 1.4	Ti 1.5	V 1.6	Cr 1.7	Mn 1.6	Fe 1.8	Co 1.9	Ni 1.9	Cu 2.0	Zn 1.6	Ga 1.8	Ge 2.0	As 2.2	Se 2.6	Br 2.8
5	Rb 0.8	Sr 1.0	Y 1.2	Zr 1.3	Nb 1.6	Mo 2.2	Tc —	Ru 2.2	Rh 2.3	Pd 2.2	Ag 1.9	Cd 1.7	In 1.8	Sn 1.8	Sb 2.0	Te 2.1	I 2.5
6	Cs 0.79	Ba 0.9						Pt 2.3	Au 2.5	Hg 2.0	Tl 2.0	Pb 2.3	Bi 2.0	Po —			

Increasing electronegativity ↑

Figure 1.1 Electronegativities of common elements (Pauling scale).

lative tendencies of atoms in a chemical bond to attract electrons. The electronegativity of a particular element depends not only on its position in the periodic table but also on its oxidation state; the higher the oxidation state of an element, the greater its electronegativity. The electronegativity of Cu(I) in Cu_2O, for example, is 1.80, whereas the electronegativity of Cu(II) in CuO is 2.0. In spite of these variations, electronegativity is still a useful guide to the distribution of electrons in a chemical bond.

1. Ionic Bonds

An **ionic bond** is formed by the transfer of electrons from an atom of low electronegativity to one of high electronegativity. The more electronegative atom gains electrons and becomes an anion; the less electronegative atom loses electrons and becomes a cation. An example of an ionic bond is that formed between sodium, one of the least electronegative elements, and fluorine, the most electronegative element.

$$\text{Na}\cdot + \;:\!\ddot{\text{F}}\!: \longrightarrow \text{Na}^+ \quad :\!\ddot{\ddot{\text{F}}}\!:^-$$

This electron is transferred to fluorine

We say, as an approximation, that an ionic bond is formed when the difference in electronegativity between two elements is 1.7 or greater. The difference in electronegativity between fluorine and sodium, for example, is $4.0 - 0.9$, or 3.1.

2. Covalent Bonds

A **covalent bond** is formed between two atoms of smaller differences in electronegativity. The simplest example of a covalent bond is that in a hydrogen molecule. When two hydrogen atoms combine, the single electrons from each combine to form an electron pair. According to the Lewis model, the pair of electrons in this covalent bond is shared by both hydrogen atoms, and at the same time, fills the valence shell of each hydrogen.

$$\text{H}\cdot + \cdot\text{H} \longrightarrow \text{H}\!:\!\text{H}$$

The Lewis model accounts for the stability of covalent bonds in the following way. In forming a covalent bond, an electron pair occupies the region between two nuclei and shields one positively charged nucleus from the repulsive force of the other positively charged nucleus. At the same time, the electron pair attracts both nuclei. An electron pair in the space between two nuclei fixes the internuclear distance within very narrow limits. The distance between nuclei of two atoms in a molecule or an ion is called the bond length.

3. Percentage of Ionic Character of a Covalent Bond

While all covalent bonds involve sharing of electrons, they differ widely in degree of sharing. Consider, for example, a covalent bond between carbon and hydrogen. Because these atoms have almost identical electronegativities, electrons in a C—H bond are shared almost equally. Covalent bonds in which electrons are shared equally or almost equally are called nonpolar covalent bonds. We say, as an approximation, that a nonpolar bond is formed when the difference in electronegativity between two bonded elements is 0.4 or less.

Another situation arises when the difference in electronegativity between two atoms joined by a covalent bond is greater than 0.4. Covalent bonds in which sharing of electrons is unequal are called polar covalent bonds. We refer to them also as **covalent bonds with partial ionic character**. On the Pauling scale, for example, the difference in electronegativity between hydrogen and chlorine is 0.9. The H—Cl bond is covalent but the sharing of electrons is not equal. Electrons of the bond are attracted more toward chlorine than to hydrogen; chlorine has a greater proportion of the shared electrons around it, and consequently, chlorine has a partial negative charge. A partial negative charge is indicated by the symbol $\delta-$. Hydrogen has a smaller proportion of the shared electrons, and consequently, it has a partial positive charge, indicated by the symbol $\delta+$. The distribution of electrons in a nonpolar covalent bond, a polar covalent bond, and an ionic bond is illustrated in Figure 1.2.

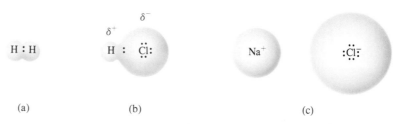

(a)   (b)   (c)

Figure 1.2 Sharing of electrons in (a) a nonpolar covalent bond, (b) a polar covalent bond, and (c) transfer of electrons in an ionic bond. These drawing show relative sizes of the atoms and ions involved in each bond.

As we shall see throughout this textbook, the degree of polarity (partial ionic character) of covalent bonds is one of the most important factors in determining the physical and chemical properties of organic molecules. For this reason, chemists have sought ways to calculate the partial ionic character (polarity) of covalent bonds. The simplest way is based on the following equation, where E_a is the electronegativity of the more electronegative atom, and E_b of the less electronegative atom:

$$\text{Percentage ionic character} = \frac{E_a - E_b}{E_a} \times 100$$

Example 1.2

Calculate the percentage of ionic character of the following covalent bonds. For each, show which atom bears the partial positive charge, and which the partial negative charge.

a. C—O   **b.** C—H   **c.** H—O

☐ *Solution*   **a.** The electronegativity of oxygen is 3.5 and of carbon 2.5.

$$\text{Percentage ionic character} = \frac{3.5 - 2.5}{3.5} \times 100 = 29\% \qquad \overset{\delta+}{C}\!-\!\overset{\delta-}{O}$$

A C—O bond has 29% ionic character. Oxygen, the more electronegative atom, bears a partial negative charge, and carbon, the less electronegative atom, bears a partial positive charge.

   b. The electronegativity of carbon is 2.5, and of hydrogen is 2.1. A C—H bond has approximately 16% ionic character. Carbon, the more electronegative element, bears a partial negative charge and hydrogen, the less electronegative element, bears a partial positive charge.

$$\text{Percentage ionic character} = \frac{2.5 - 2.1}{2.5} \times 100 = 16\% \qquad \overset{\delta-}{C}\!-\!\overset{\delta+}{H}$$

   c. An H—O bond has approximately 40% ionic character; oxygen bears a partial negative charge, and hydogen bears a partial positive charge.

$$\text{Percentage ionic character} = \frac{3.5 - 2.1}{3.5} \times 100 = 40\% \qquad \overset{\delta+}{H}\!-\!\overset{\delta-}{O}$$

Problem 1.2   Calculate the percentage of ionic character of the following covalent bonds. For each, show which atom bears the partial negative charge and which the partial positive charge.

a. N—H   **b.** C—Mg   **c.** B—H

D. Lewis Structures of Covalent Molecules and Ions

The ability to write Lewis structures for covalent molecules and ions is a fundamental skill for the study of organic chemistry and biochemistry. The following guidelines will help develop this skill:

1. Determine the number of valence electrons in the molecule or ion. To do this, add the number of valence electrons each atom contributes. For ions, add one electron for each negative charge on the ion; subtract one electron for each positive charge on the ion.

2. Determine the order of attachment of atoms in the molecule or ion. Except for the simplest molecules and ions, this arrangement must be given in advance; that is, it must be determined experimentally.

3. Connect the atoms with single bonds. Then arrange the remaining electrons in pairs so that each atom in the molecule or ion has a complete outer shell. Each hydrogen atom must be surrounded by two electrons. Each atom of carbon, oxygen, nitrogen, and halogen must be surrounded by eight electrons, as the octet rule requires.

4. A pair of electrons involved in a covalent bond (bonding electrons) is shown as a dash; an unshared (nonbonding) pair of electrons is shown as a pair of dots.

5. In a single bond, two atoms share one pair of electrons. In a double bond, they share two pairs of electrons; and in a triple bond, they share three pairs of electrons.

Table 1.4 shows Lewis structures, molecular formulas, and names for several molecules. After the molecular formula of each is given the number of valence electrons it contains. Notice that in these molecules, each hydrogen atom is surrounded by two valence electrons, and each atom of carbon, nitrogen, oxygen, and chlorine is surrounded by eight valence electrons.

Table 1.4
Lewis structures for several small molecules.

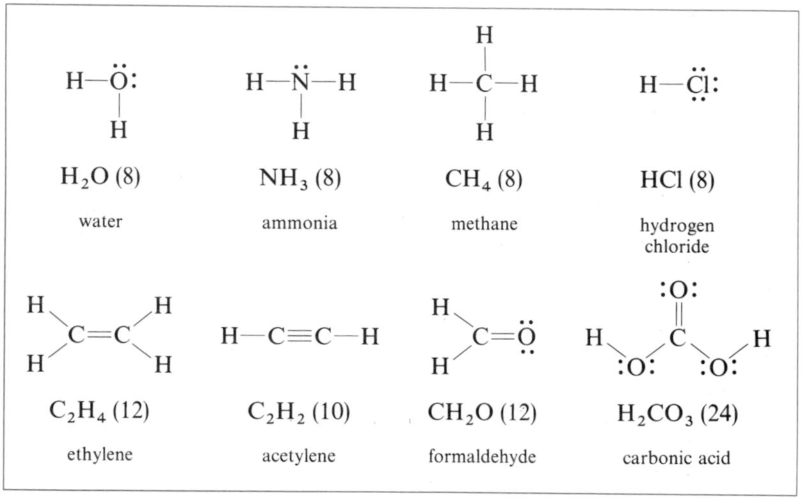

Example 1.3

Draw Lewis structures, showing all valence electrons, for the following molecules:

a. CO_2 **b.** CH_4O **c.** CH_3Cl

☐ *Solution*

a. A Lewis structure for carbon dioxide must show 16 valence electrons—12 from the two oxygens and 4 from carbon:

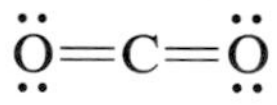

$$\ddot{O}=C=\ddot{O} \qquad \text{(16 valence electrons)}$$

With four shared pairs of electrons and four unshared pairs, the Lewis structure shows the required 16 valence electrons. Furthermore, each atom of carbon and oxygen has a complete octet.

b. A Lewis structure for CH_4O must show 14 valence electrons—4 from carbon, 4 from the four hydrogens, and 6 from oxygen:

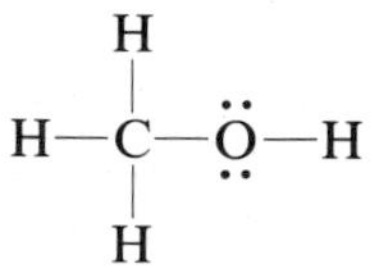

This structure shows five single bonds and two unshared pairs of electrons and has the correct number of valence electrons. Each atom of hydrogen is surrounded by 2 electrons, and each atom of carbon and oxygen is surrounded by 8 electrons.

c. A Lewis structure for CH_3Cl must show 14 valence electrons:

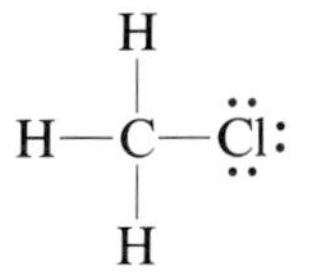

$$H\text{---}\overset{\displaystyle H}{\underset{\displaystyle H}{\overset{|}{\underset{|}{C}}}}\text{---}\overset{\cdot\cdot}{\underset{\cdot\cdot}{Cl}}\colon \qquad \text{(14 valence electrons)}$$

This structure shows four single bonds and three unshared pairs of electrons and thus has the correct number of valence electrons. Each atom has a complete valence shell.

Problem 1.3

Draw Lewis structures, showing all valence electrons, for the following covalent molecules:

a. C_2H_6 **b.** CS_2 **c.** HCN

E. The Covalence of Carbon, Hydrogen, Nitrogen, and Oxygen

In treating organic compounds, we shall deal almost exclusively with just four atoms: carbon, hydrogen, nitrogen, and oxygen. Lewis structures for these four elements are given in Table 1.3. The single valence electron of hydrogen belongs to the first principal energy level; this shell is completely filled with two electrons. Thus, hydrogen can form only one covalent bond with another element; hydrogen has a valence of 1.

Carbon, with four valence electrons, needs four additional electrons to complete its octet; carbon has a valence of 4. The valence of carbon can be satisfied by appropriate combinations of single, double, or triple bonds, as illustrated by methane, ethane, ethene, and ethyne:

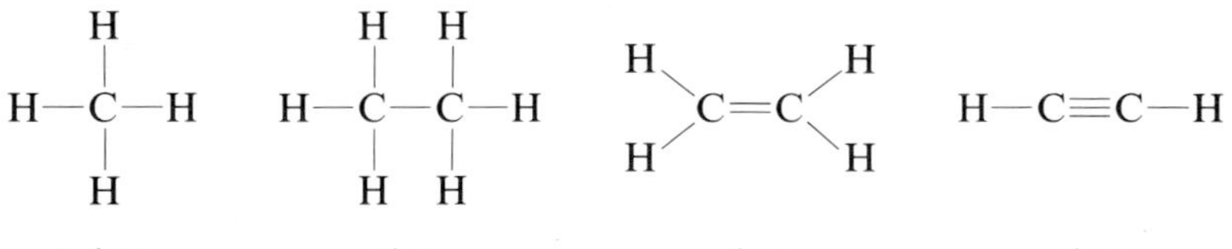

| methane | ethane | ethene
(ethylene) | ethyne
(acetylene) |

Nitrogen, with five valence electrons, needs three additional electrons to complete its octet; nitrogen has a valence of 3. This valence can be satisfied by appropriate combinations of single, double, or triple bonds, as illustrated by ammonia, nitrous acid, hydrogen cyanide, and nitrogen:

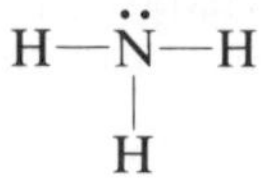

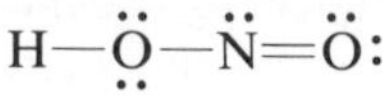

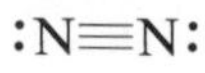

$$H—\overset{\displaystyle ..}{N}—H \qquad H—\overset{\displaystyle ..}{\underset{\displaystyle ..}{O}}—\overset{\displaystyle ..}{N}{=}\overset{\displaystyle ..}{O}: \qquad H—C{\equiv}N: \qquad :N{\equiv}N:$$

$$\underset{\displaystyle H}{|}$$

ammonia   nitrous   hydrogen   nitrogen
                       acid   cyanide

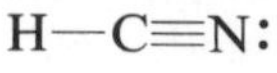

Oxygen, with six valence electrons, needs two additional electrons to complete its octet; oxygen has a valence of 2. This valence can be satisfied by either two single bonds, as in water and methanol, or by one double bond, as in methanal (formaldehyde) and acetic acid:

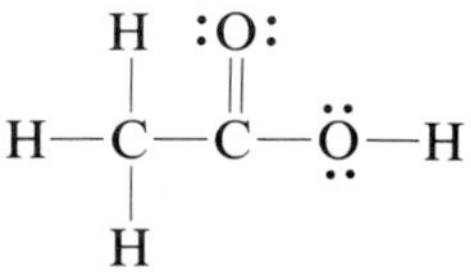

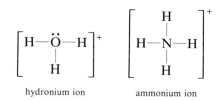

water 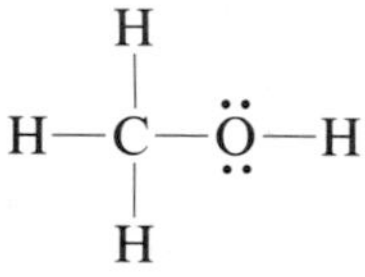   methanol 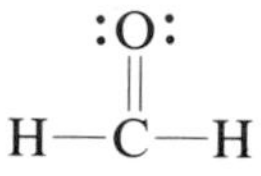   methanal   acetic acid
                                             (formaldehyde)

After examining these molecules, we can make the following generalizations. For stable, uncharged molecules containing atoms of carbon, nitrogen, oxygen, and hydrogen:

1. Carbon forms four covalent bonds.
2. Nitrogen forms three covalent bonds and has one unshared pair of electrons.
3. Oxygen forms two covalent bonds and has two unshared pairs of electrons.
4. Hydrogen forms only one covalent bond.

F. Formal Charge

Following are structural formulas for the hydronium and ammonium ions:

$$\left[H—\overset{\displaystyle ..}{O}—H \atop \underset{\displaystyle H}{|} \right]^{+} \qquad \left[{\overset{\displaystyle H}{|} \atop H—N—H} \atop \underset{\displaystyle H}{|} \right]^{+}$$

hydronium ion   ammonium ion

It is important that you be able to determine which atom or atoms in these and other polyatomic ions bear the positive or the negative charge. The charge on an atom in an ion or a molecule is called its **formal charge.** To derive formal charge:

1. Write a correct Lewis structure for the molecule or ion.
2. Assign to each atom all its unshared (nonbonding) electrons and half its shared (bonding) electrons.

3. Compare this number with the number of valence electrons in the neutral, unbonded atom. If the number of electrons assigned a bonded atom is less than the number assigned to the unbonded atom, then the atom has a formal positive charge. On the other hand, if the number of electrons assigned to a bonded atom is greater than the number assigned to the unbonded atom, then the atom has a formal negative charge.

Example 1.4

Draw Lewis structures for the following ions and show which atom in each bears the formal charge:

a. H_3O^+ **b.** NH_4^+ **c.** HCO_3^-

☐ *Solution*

a. The Lewis structure for the hydronium ion must show 8 valence electrons; 3 from the three hydrogens, 6 from oxygen minus 1 for the single positive charge. An unbonded oxygen atom has 6 valence electrons. The oxygen atom in this ion is assigned 5 electrons—the 2 unshared electrons and 1 from each shared pair. Therefore, in the hydronium ion, oxygen has a formal charge of $+1$ $(6 - 5 = +1)$.

Assigned 5 valence electrons; formal charge of +1

$$H - \overset{\cdot\cdot}{\underset{|}{\overset{+}{O}}} - H$$
$$H$$

hydronium ion

b. The Lewis structure for the ammonium ion must show 8 valence electrons. An unbonded nitrogen atom has 5 valence electrons. Nitrogen is assigned 4 valence electrons and therefore has a formal charge of $+1$ $(5 - 4 = +1)$.

Assigned 4 valence electrons; formal charge +1

$$H$$
$$|$$
$$H - \overset{+}{N} - H$$
$$|$$
$$H$$

ammonium ion

c. The Lewis structure for the bicarbonate ion must show 24 valence electrons: 1 from hydrogen, 4 from carbon, and 18 from the three oxygens, plus 1 for the single negative charge. In the Lewis structure, carbon is assigned 4 valence electrons and therefore has no formal charge $(4 - 4 = 0)$. Two oxygens are assigned 6 valence electrons and have no formal charges $(6 - 6 = 0)$. The third is assigned 7 valence electrons and has a formal charge of -1 $(6 - 7 = -1)$.

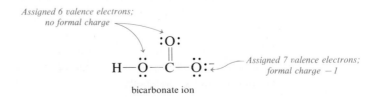

Assigned 6 valence electrons; no formal charge

Assigned 7 valence electrons; formal charge −1

bicarbonate ion

■
Problem 1.4

Draw Lewis structures for the following ions and state which atom in each bears the formal charge:

a. CH_3^+ **b.** CH_3^- **c.** OH^-

1.3 Bond Angles and Shapes of Molecules

In Section 1.2, we used a shared pair of electrons as the fundamental unit of covalent bonding; and we drew Lewis structures for several small molecules and ions containing various combinations of single, double, and triple bonds. We can predict bond angles in these and other covalent molecules and ions in a straightforward way using the **valence-shell electron-pair repulsion (VSEPR) model.** According to the VSEPR model, an atom is surrounded by an outer shell of valence electrons. These valence electrons may be involved in the formation of single, double, or triple bonds, or may be unshared. Each of these combinations creates a negatively charged region of space, and because like charges repel each other, these regions of electron density around an atom will be arranged so that the regions are as far apart as possible.

Let us use the valence-shell electron-pair repulsion model to predict the shape of methane, CH_4. The Lewis structure for CH_4 shows a carbon atom surrounded by four separate regions of electron density, each containing a pair of electrons forming a bond to a hydrogen atom. According to the VSEPR model, these four regions radiate from carbon so that they are as far from each other as possible. This occurs when the angle between any two pairs of electrons is 109.5°. Therefore, all H—C—H bond angles in methane are predicted to be 109.5°, and the shape of the molecule is predicted to be tetrahedral. The H—C—H bond angles in methane have been measured experimentally and found to be 109.5°. Thus, the bond angles and shape of methane predicted by the VSEPR model are identical to what has been observed. Figure 1.3 shows a Lewis structure for methane, the tetrahedral arrangement of the four regions of electron density around carbon, and a space-filling model.

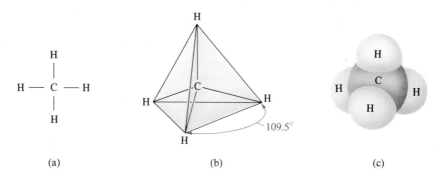

(a)	(b)	(c)

Figure 1.3 The shape of the methane molecule, CH_4: (a) Lewis structure; (b) the three-dimensional shape; (c) a space-filling model.

We can predict the shape of the ammonia molecule similarly. The Lewis structure of NH_3 shows nitrogen surrounded by four regions of electron density. Three regions contain single pairs of electrons forming covalent bonds with hydrogen atoms. The fourth region contains an unshared pair of electrons. These four regions of electron density are arranged in a tetrahedral manner around the central nitrogen atom, as shown in Figure 1.4. According to the VSEPR model, the four regions of electron density around nitrogen are arranged in a tetrahedral manner and the predicted H—N—H bond angles in ammonia are 109.5°. The observed bond angles are 107.3°. This small difference between the predicted and observed angles can be explained by proposing that the unshared pair of electrons on nitrogen repels adjacent electron pairs more strongly than bonding pairs of electrons do.

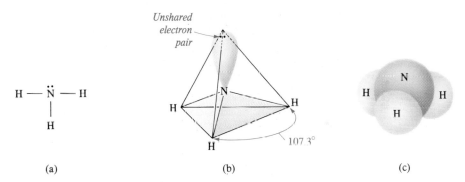

(a)   (b)   (c)

Figure 1.4 The shape of the ammonia molecule, NH_3: (a) Lewis structure; (b) the three-dimensional shape; (c) a space-filling model (unshared pair of electrons not shown).

Figure 1.5 shows a Lewis structure, the three-dimensional shape, and a space-filling model for a water molecule. In H_2O, oxygen is surrounded by four separate regions of electron density. Two regions contain shared pairs of electrons used to form covalent bonds with hydrogens; the remaining two regions contain unshared pairs of electrons. According to the VSEPR model, the four

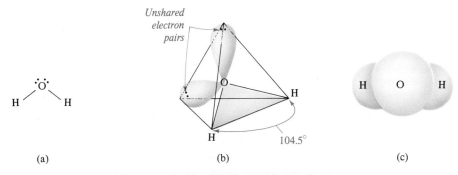

(a)   (b)   (c)

Figure 1.5 The shape of the water molecule, H_2O: (a) Lewis structure; (b) the three-dimensional shape; (c) a space-filling model (unshared pairs of electrons not shown).

regions of electron density around oxygen are arranged in a tetrahedral manner and the predicted H—O—H bond angle is 109.5°. Experimental measurements show that the actual bond angle is 104.5°, a value smaller than predicted. This difference between the predicted and observed bond angles can be explained by proposing, as we did for NH_3, that unshared pairs of electrons repel more strongly than shared pairs do. Note that the distortion from 109.5° is greater in H_2O, which has two unshared pairs of electrons, than in NH_3, which has only one unshared pair.

A general prediction emerges from this discussion of the shapes of CH_4, NH_3, and H_2O. For four separate regions of electron density around an atom, the VSEPR model predicts a tetrahedral distribution of electron density and bond angles of approximately 109.5°.

In many of the molecules we will encounter in organic and biochemistry, an atom is surrounded by three regions of electron density. Shown in Figure 1.6 are Lewis structures for formaldehyde and ethylene molecules.

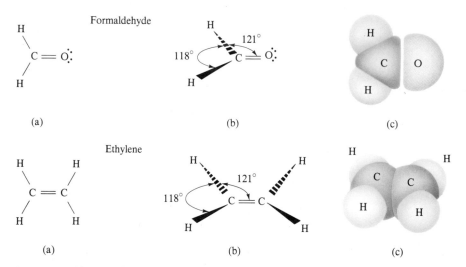

Figure 1.6 Shapes of formaldehyde and ethylene molecules: (a) Lewis structures; (b) planar arrangements of three regions of electron density around carbon atoms; (c) space-filling models. In three figures, ◄ represents a bond projecting in front of the plane of the paper and ⅲⅲⅲ represents a bond projecting behind the plane of the paper.

According to the VSEPR model, a double bond is treated as a single region of electron density. In formaldehyde, carbon is surrounded by three regions of electron density. Two regions contain shared pairs of electrons, forming single bonds to hydrogen atoms; the third region contains two pairs of electrons, forming a double bond to oxygen. In ethylene, each carbon atom is also surrounded by three regions of electron density. Two contain single pairs of electrons, and the third contains two pairs of electrons. Three regions of electron density about an atom are farthest apart if they are all in the same plane at angles of 120° to each other. Thus, the predicted H—C—H and H—C—O bond angles in formaldehyde are 120°; the predicted H—C—H and H—C—C bond angles in ethylene are also 120°. Such an arrangement of atoms is described as trigonal planar.

In still other types of molecules, a central atom is surrounded by only two regions of electron density. Shown in Figure 1.7 are Lewis structures and space-filling models of carbon dioxide, hydrogen cyanide, and acetylene. In carbon dioxide, carbon is surrounded by two regions of electron density; each contains two shared pairs of electrons and forms a double bond to an oxygen atom. In acetylene, each carbon is also surrounded by two regions of electron density. One contains a single pair of electrons and forms a single bond to a hydrogen atom, whereas the other contains three pairs of electrons and forms a triple bond to a carbon atom. In each case, the two regions of electron density are farthest apart if they form a straight line through the central atom and create an angle of 180°. Both carbon dioxide and acetylene and linear molecules.

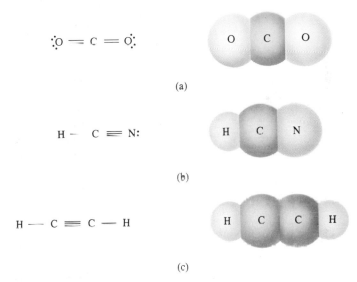

(a)

(b)

(c)

Figure 1.7 Shapes of (a) carbon dioxide, (b) hydrogen cyanide, and (c) acetylene.

Predictions of the valence-shell electron-pair repulsion model are summarized in Table 1.5.

Predict all bond angles in the following molecules and ions:

Example 1.5   **a.** CH_3Cl **b.** HCN   **c.** CH_2=CHCl

☐ *Solution*   **a.** The Lewis structure for CH_3Cl shows carbon surrounded by four separate regions of electron density. Predict, therefore, that the distribution of electron pairs is tetrahedral, all bond angles are 109.5°, and the shape of CH_3Cl is tetrahedral. The actual H—C—H bond angle is 110°.

Table 1.5
Predicted
molecular shapes
(VSEPR model).

Regions of Electron Density around an Atom	Arrangement in Space	Predicted Bond Angles	Examples
4	tetrahedral	109.5°	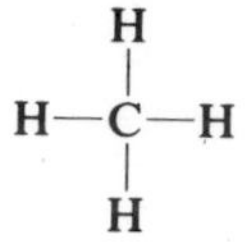 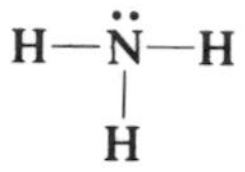 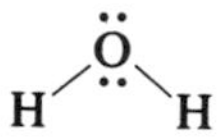 $H-\overset{\overset{\displaystyle H}{\vert}}{\underset{\underset{\displaystyle H}{\vert}}{C}}-H$ $H-\overset{\cdot\cdot}{\underset{\underset{\displaystyle H}{\vert}}{N}}-H$ $\overset{\cdot\cdot}{\underset{\displaystyle H}{O}}_{H}$
3	trigonal planar	120°	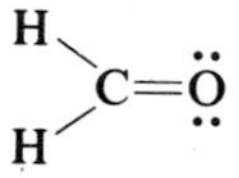 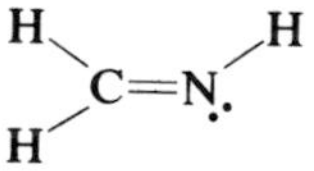 $\overset{\displaystyle H}{\underset{\displaystyle H}{>}}C=\overset{\cdot\cdot}{\underset{\cdot\cdot}{O}}$ $\overset{\displaystyle H}{\underset{\displaystyle H}{>}}C=\overset{\displaystyle N}{\underset{\cdot\cdot}{}}\overset{\displaystyle H}{}$
2	linear	180°	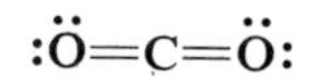 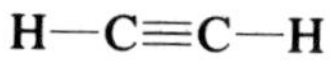 $:\overset{\cdot\cdot}{O}=C=\overset{\cdot\cdot}{O}:$ $H-C\equiv C-H$

$H\overset{\overset{\displaystyle H}{\vert}}{\underset{\underset{\displaystyle H}{}}{\overset{\cdot}{C}}}\overset{\cdot\cdot}{Cl}:$

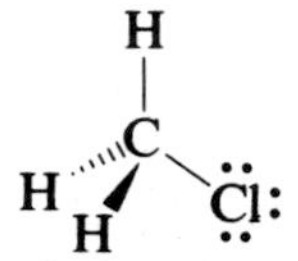

all bond angles
109.5°
(predicted)

b. In the Lewis structure of HCN, carbon is surrounded by two regions of electron density. Therefore, predict 180° for the H—C—N bond angle and the shape of HCN as linear.

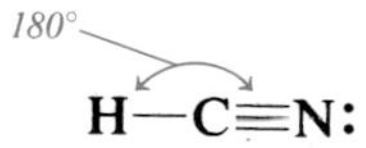

180°
$H-C\equiv N:$

c. The Lewis structure of CH_2=CHCl shows each carbon surrounded by three regions of electron density. Predict, therefore, that all bond angles are 120°.

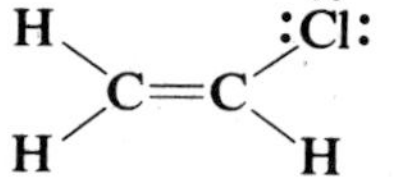

$\overset{\displaystyle H}{\underset{\displaystyle H}{>}}C=C\overset{\overset{\cdot\cdot}{Cl}:}{\underset{\displaystyle H}{}}$

Problem 1.5

Predict all bond angles for the following molecules:

a. CH_3OH **b.** CH_2Cl_2 **c.** CH_3NH_2

1.4 Functional Groups

Carbon combines with other atoms (C, H, N, O, S, halogens, and so on) to form characteristic structural units called **functional groups**. Functional groups are important for three reasons. First, they are the units by which we divide organic molecules into classes. Second, they are sites of chemical reaction; a particular functional group, in whatever compound it is found, undergoes the same types of chemical reactions. Third, functional groups are a basis for naming organic compounds.

A. Alcohols and Ethers

The characteristic structural feature of an **alcohol** is an atom of oxygen bonded to one carbon atom and one hydrogen atom. We say that an alcohol contains an —OH, or **hydroxyl group**. The characteristic structural feature of an **ether** is an atom of oxygen bonded to two carbon atoms.

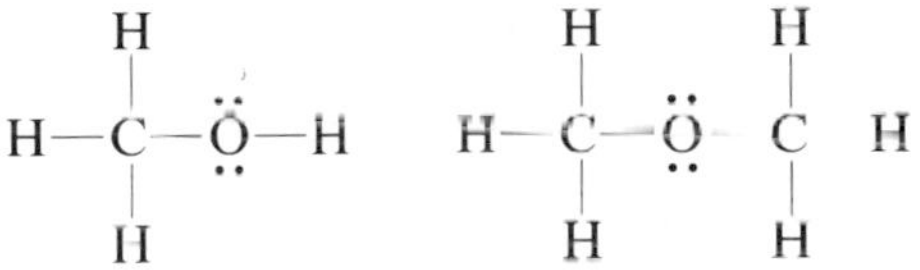

Characteristic structural feature: an alcohol — an ether

We can write formulas for this alcohol and ether in a more abbreviated form by using what are called **condensed structural formulas**. In a condensed structural formula, CH_3—indicates a carbon with three attached hydrogens, —CH_2— indicates a carbon with two attached hydrogens, and —CH— indicates a carbon with one attached hydrogen. Following are Lewis structures and condensed structural formulas for the alcohol and ether of the molecular formula C_2H_6O.

Lewis Structures

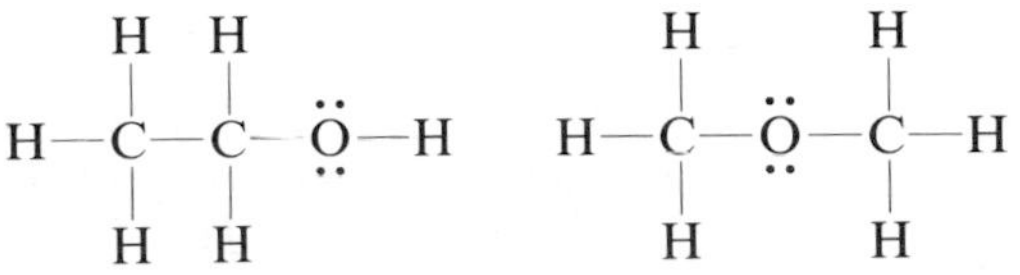

Condensed Structural Formulas

$$CH_3CH_2OH \qquad CH_3OCH_3$$

Example 1.6

There are two alcohols of molecular formula C_3H_8O. Draw Lewis structures and condensed structural formulas for each.

☐ *Solution*

The characteristic structural feature of an alcohol is an atom of oxygen bonded to one carbon and one hydrogen atom:

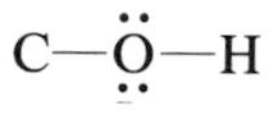

$$C—\overset{..}{\underset{..}{O}}—H$$

The molecular formula contains three carbon atoms. These can be bonded together in a chain with the —OH (hydroxyl) group attached to the end carbon of the chain or attached to the middle carbon of the chain:

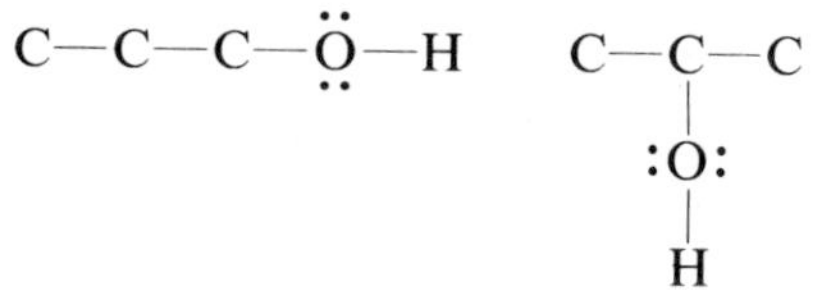

$$C—C—C—\overset{..}{\underset{..}{O}}—H \qquad C—C—C$$
$$\underset{\underset{H}{|}}{\underset{:O:}{|}}$$

Finally, add seven hydrogens to satisfy the tetravalence of carbon and give the correct molecular formula:

Lewis Structures

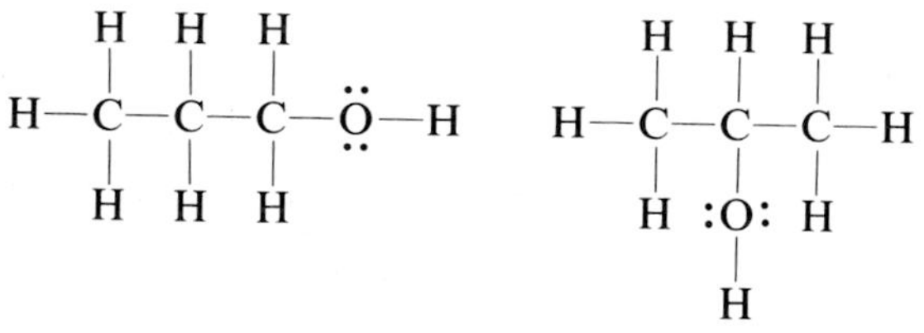

Condensed Structural Formulas

$$CH_3CH_2CH_2OH \qquad\qquad CH_3CHCH_3$$
$$\underset{OH}{|}$$

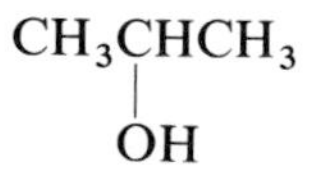

Problem 1.6

There is one ether of molecular formula C_3H_8O. Draw a Lewis structure and a condensed structural formula for this compound.

B. Aldehydes and Ketones

Both aldehydes and ketones contain a **carbonyl group**, **C=O**, the most important functional group in organic chemistry. The characteristic structural feature of an **aldehyde** is a carbonyl group bonded to a hydrogen atom. Formaldehyde is the only aldehyde that contains two hydrogen atoms bonded to a carbonyl group. All other aldehydes have one carbon and one hydrogen bonded to a carbonyl group. The characteristic structural feature of a **ketone** is a carbonyl group bonded to two carbon atoms.

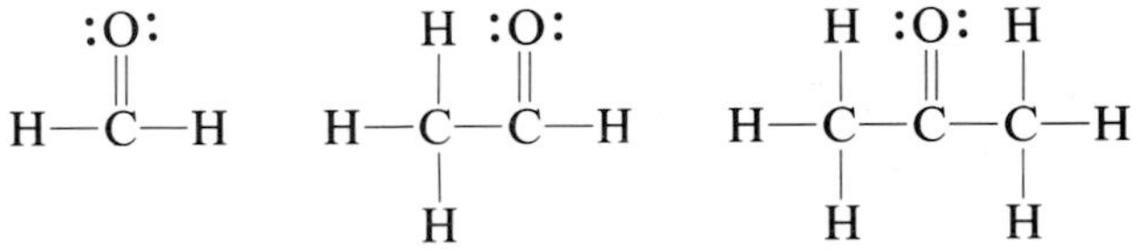

$$
\begin{array}{ccc}
\text{:O:} & \text{H :O:} & \text{H :O: H} \\
\| & \| & \| \\
\text{H--C--H} & \text{H--C--C--H} & \text{H--C--C--C--H} \\
 & | & |\ \ \ \ \ | \\
 & \text{H} & \text{H}\ \ \ \ \ \text{H}
\end{array}
$$

| an aldehyde (formaldehyde) | an aldehyde (acetaldehyde) | a ketone (acetone) |

Characteristic structural feature:

$$
\begin{array}{ccc}
\text{:O:} & \text{:O:} & \text{:O:} \\
\| & \| & \| \\
\text{--C--H} & \text{C--C--H} & \text{C--C--C}
\end{array}
$$

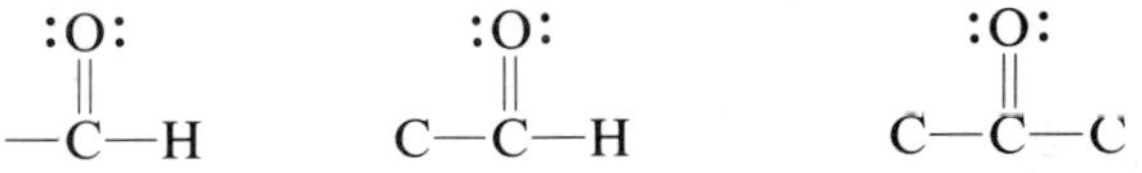

Example 1.7

Draw Lewis structures and condensed structural formulas for the two aldehydes of molecular formula C_4H_8O.

Solution

First draw the characteristic structural feature of the aldehyde group and then add the remaining carbons. These may be attached in two different ways.

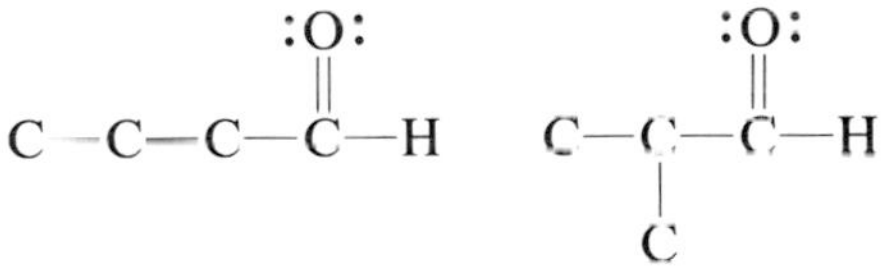

$$
\begin{array}{cc}
\text{:O:} & \text{:O:} \\
\| & \| \\
\text{C--C--C--C--H} & \text{C--C--C--H} \\
 & | \\
 & \text{C}
\end{array}
$$

Finally add seven hydrogens to complete the tetravalence of carbon and give the correct molecular formula. The aldehyde group may be written as above, or alternatively, it may be written —CHO.

Lewis Structures

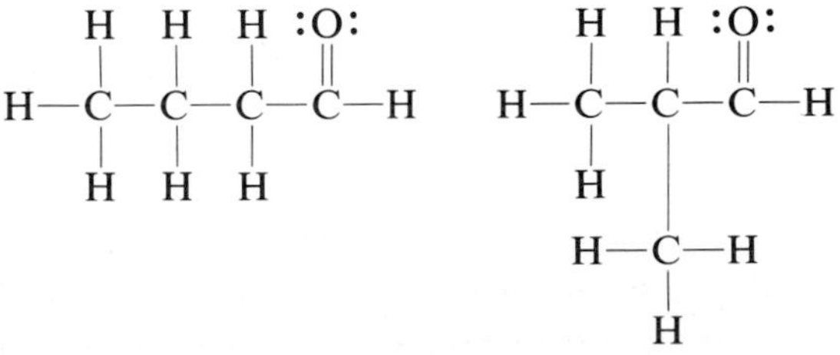

$$
\begin{array}{cc}
\text{H H H :O:} & \text{H H :O:} \\
|\ \ \ |\ \ \ |\ \ \ \| & |\ \ \ |\ \ \ \| \\
\text{H--C--C--C--C--H} & \text{H--C--C--C--H} \\
|\ \ \ |\ \ \ | & |\ \ \ | \\
\text{H H H} & \text{H} \\
 & \text{H--C--H} \\
 & | \\
 & \text{H}
\end{array}
$$

Condensed Structural Formulas

$$\underset{\text{CH}_3\text{CH}_2\text{CH}_2\overset{\displaystyle\overset{\text{O}}{\|}}{\text{C}}\text{H}}{} \qquad \underset{\underset{\text{CH}_3}{|}}{\text{CH}_3\overset{\displaystyle\overset{\text{O}}{\|}}{\text{C}}\text{HCH}}$$

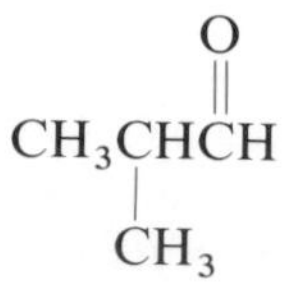

or   or

$$\text{CH}_3\text{CH}_2\text{CH}_2\text{CHO} \qquad (\text{CH}_3)_2\text{CHCHO}$$

Problem 1.7   Draw Lewis structures and condensed structural formulas for the three ketones of molecular formula $C_5H_{10}O$.

C. Carboxylic Acids

The characteristic structural feature of a **carboxylic acid** is a carboxyl (carbonyl + hydroxyl) group. The **carboxyl group** may be written in any of the following ways, all of which are equivalent.

A Carboxyl Group

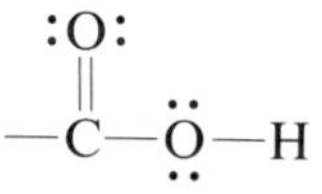

$$\underset{}{-\overset{\displaystyle\overset{:\text{O}:}{\|}}{\text{C}}-\overset{..}{\underset{..}{\text{O}}}-\text{H}} \qquad \text{or} \qquad -\text{COOH} \qquad \text{or} \qquad -\text{CO}_2\text{H}$$

A Carboxylic Acid

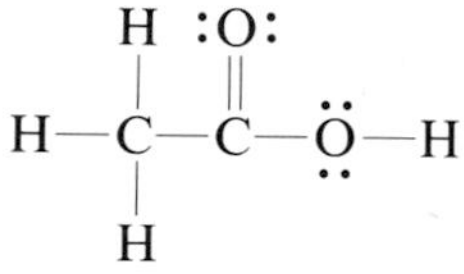

$$\text{H}-\overset{\displaystyle\overset{\text{H}}{|}}{\underset{\displaystyle\underset{\text{H}}{|}}{\text{C}}}-\overset{\displaystyle\overset{:\text{O}:}{\|}}{\text{C}}-\overset{..}{\underset{..}{\text{O}}}-\text{H} \qquad \text{or} \qquad \text{CH}_3\text{COOH} \qquad \text{or} \qquad \text{CH}_3\text{CO}_2\text{H}$$

Example 1.8   Draw a Lewis structure and condensed structural formula for the single carboxylic acid of molecular formula $C_3H_6O_2$.

☐ *Solution*   **Lewis Structure**

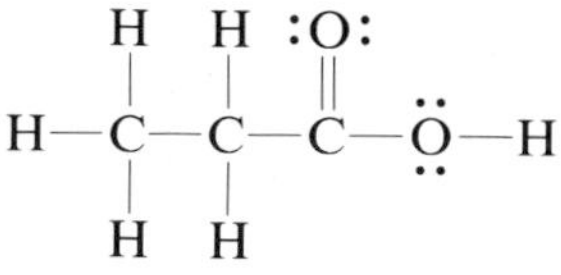

$$\text{H}-\overset{\displaystyle\overset{\text{H}}{|}}{\underset{\displaystyle\underset{\text{H}}{|}}{\text{C}}}-\overset{\displaystyle\overset{\text{H}}{|}}{\underset{\displaystyle\underset{\text{H}}{|}}{\text{C}}}-\overset{\displaystyle\overset{:\text{O}:}{\|}}{\text{C}}-\overset{..}{\underset{..}{\text{O}}}-\text{H}$$

Condensed Structural Formula

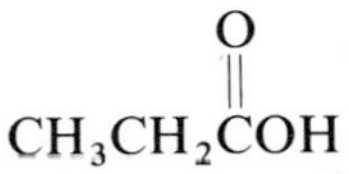

$$CH_3CH_2\overset{\overset{\displaystyle O}{\|}}{C}OH \quad or \quad CH_3CH_2COOH \quad or \quad CH_3CH_2CO_2H$$

Problem 1.8 Draw Lewis structures and condensed structural formulas for the two carboxylic acids of molecular formula $C_4H_8O_2$.

1.5 Structural Isomerism

For most molecular formulas it is possible to draw more than one structural formula. For example, the following compounds have the same molecular formula, C_2H_6O, but different structural formulas and different functional groups; the first is an alcohol, the second an ether.

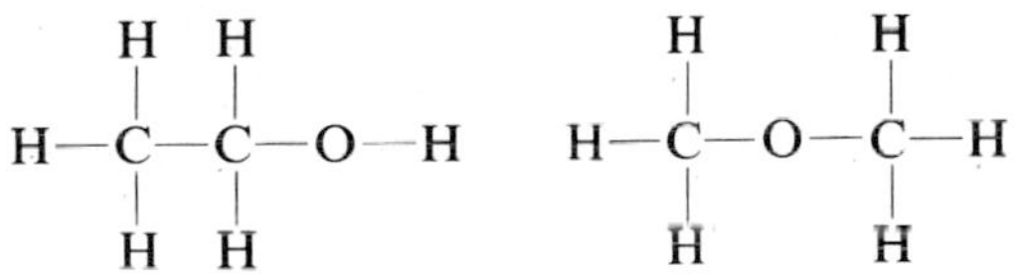

H H   H   H
| | | |
H—C—C—O—H   H—C—O—C—H
| | | |
H H   H   H

     an alcohol   an ether

Similarly, the following compounds have the same molecular formula, C_3H_6O, and the same functional group (a carbonyl group). The first is an aldehyde, however, and the second is a ketone.

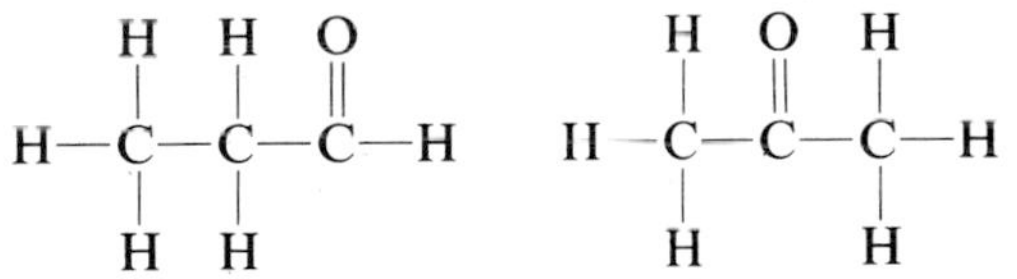

H H O   H O H
| | ‖ | ‖ |
H—C—C—C—H   H—C—C—C—H
| | | |
H H   H   H

    an aldehyde   a ketone

The following alcohols have the same molecular formula but different structural formulas:

$$CH_3—CH_2—CH_2—OH \qquad CH_3—\underset{\underset{\displaystyle OH}{|}}{CH}—CH_3$$

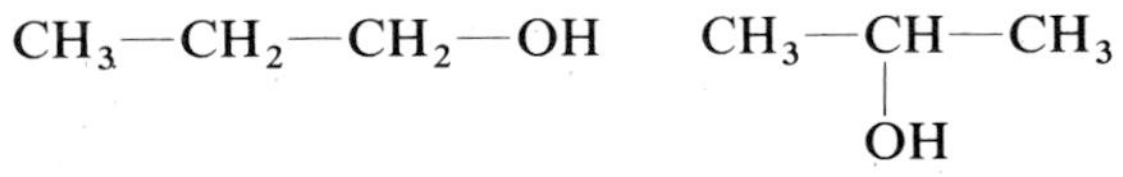

Compounds that have the same molecular formula but different structural formulas (different orders of attachment of atoms) are called **structural isomers**. To determine whether two or more compounds are structural isomers, first write the molecular formula of each compound and then compare them. All those that have the same molecular formula but different orders of attachment of atoms are structural isomers.

Example 1.9

Divide the following into groups of structural isomers.

$$\textbf{a. } CH_3-CH_2-\overset{\overset{\textstyle O}{\|}}{C}-OH \qquad\qquad \textbf{b. } CH_3-O-CH_2-\overset{\overset{\textstyle O}{\|}}{C}-H$$

$$\textbf{c. } CH_2{=}CH-CH_2-O-CH_3 \qquad \textbf{d. } CH_3-\underset{\underset{\textstyle CH_3}{|}}{CH}-\overset{\overset{\textstyle O}{\|}}{C}-H$$

$$\textbf{e. } CH_3-\overset{\overset{\textstyle O}{\|}}{C}-CH_2-CH_2-CH_2-OH \qquad \textbf{f. } CH_3-\overset{\overset{\textstyle O}{\|}}{C}-CH_2-OH$$

☐ *Solution*

Compounds (a), (b), and (f) have the same molecular formula, $C_3H_6O_2$, but different structural formulas and are structural isomers. Compounds (c) and (d) have the same molecular formula, C_4H_8O, but different structural formulas and are also structural isomers. There are no structural isomers in this problem for compound (e).

Problem 1.9

Divide the following into groups of structural isomers.

$$\textbf{a. } CH_2{=}CH-O-CH{=}CH_2 \qquad \textbf{b. } HC{\equiv}C-\overset{\overset{\textstyle O}{\|}}{C}-CH_3$$

$$\textbf{c. } CH_3-CH_2-O-C{\equiv}CH \qquad \textbf{d. } CH_3-CH{=}CH-\overset{\overset{\textstyle O}{\|}}{C}-H$$

1.6 Need for Another Model of Covalent Bonding

As much as the Lewis and valence-shell electron-pair repulsion models (Section 1.3) have helped us understand covalent bonding and the geometry of organic molecules, they leave many important questions unanswered. The most important is the relation between molecular structure and chemical reactivity. For example, a carbon-carbon single bond is different in chemical reactivity from a carbon-carbon double bond. Most carbon-carbon single bonds are unreactive, but carbon-carbon double bonds react with a great variety of reactants. The Lewis model of bonding gives us no way to account for these differences. Therefore let us turn to a new model of bonding, namely, formation of covalent bonds by the overlap of atomic orbitals.

1.7 Covalent Bond Formation by the Overlap of Atomic Orbitals

According to modern bonding theory, formation of a covalent bond between two atoms amounts to bringing them together in such a way that an atomic orbital of one atom overlaps an atomic orbital of the other. In forming the covalent bond in H_2, for example, two hydrogen atoms approach each other so their 1s atomic orbitals overlap (Figure 1.8). The new orbital formed by the overlap of two atomic orbitals encompasses both hydrogen nuclei and is called a **molecular orbital**. Like an atomic orbital, a molecular orbital can accommodate two electrons. In the covalent bond illustrated in Figure 1.8, the orbital overlap, and therefore the electron density in the resulting molecular orbital, are concentrated about the axis joining the two nuclei. A covalent bond in which orbital overlap of the bond is concentrated along the axis joining the two nuclei is called a **sigma (σ) bond**.

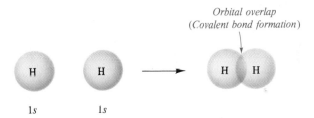

Figure 1.8 Formation of a covalent bond by overlap of two 1s atomic orbitals.

Carbon, nitrogen, and oxygen form covalent bonds using atomic orbitals of the second principal energy level. The second principal energy level consists of four atomic orbitals: a single 2s orbital and three 2p orbitals (Section 1.1). The 2p orbitals are designated $2p_x$, $2p_y$, and $2p_z$ and are oriented along the x axis, y axis, and z axis respectively. Figure 1.9 shows these atomic orbitals.

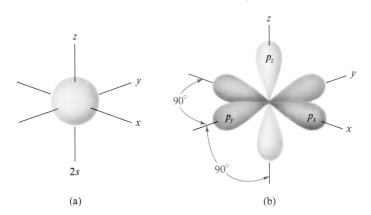

Figure 1.9 Atomic orbitals of the second principal energy level. (a) Shape of the 2s orbital; (b) orientations in space of the three 2p orbitals relative to each other.

The three $2p$ orbitals are at 90° angles to each other (Figure 1.9), and if atoms of carbon, nitrogen, or oxygen used these orbitals to form covalent bonds, bond angles around each would be approximately 90°. Bond angles of 90° are not observed in organic compounds, however (Section 1.3). What we find instead are bond angles of approximately 109.5°, 120°, or 180°. To account for these observed bond angles, Linus Pauling proposed that atomic orbitals may combine to form new orbitals, which then interact to form bonds with the angles we do observe. The combination of atomic orbitals is called **hybridization**, and the new atomic orbitals formed are called hybrid atomic orbitals, or more simply, hybrid orbitals.

A. sp^3 Hybrid Orbitals

Combination of one $2s$ orbital and three $2p$ orbitals produces four equivalent orbitals called sp^3 **hybrid orbitals** (Figure 1.10). The four sp^3 hybrid orbitals are directed toward the corners of a regular tetrahedron, and sp^3 hybridization produces bond angles of approximately 109.5°.

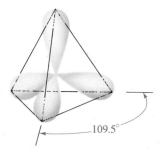

109.5°

Figure 1.10 sp^3 hybrid orbitals.

In Section 1.2, we described the covalent bonding in CH_4, NH_3, and H_2O by the Lewis model. Now let us consider the bonding in these molecules in terms of the overlap of atomic orbitals. To bond to four other atoms, carbon uses sp^3 hybrid orbitals. Carbon has four valence electrons, and one electron is placed in each sp^3 orbital. Each partially filled sp^3 orbital then overlaps a partially filled $1s$ orbital of hydrogen to form a sigma bond, and hydrogen atoms occupy the corners of a regular tetrahedron (Figure 1.11). In bonding with three other atoms, the five valence electrons of nitrogen are distributed so that one sp^3 is filled with a pair of electrons and the other three sp^3 orbitals have one electron each. Overlap of these partially filled sp^3 orbitals by $1s$ orbitals of hydrogen gives the NH_3 molecule (Figure 1.11). In bonding with two other atoms, the six valence electrons of oxygen are distributed so that two sp^3 orbitals are filled and the remaining two have one electron each. Each partially filled sp^3 orbital overlaps a $1s$ orbital of hydrogen, and hydrogen atoms occupy two corners of a regular tetrahedron. The remaining two corners of the tetrahedron are occupied by unshared pairs of electrons (Figure 1.11).

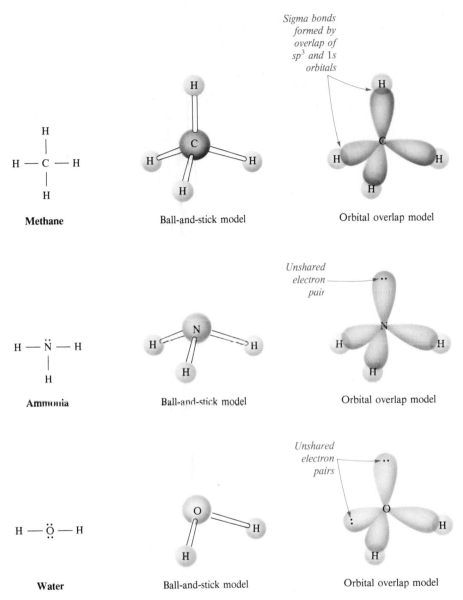

Figure 1.11 Lewis structures, ball-and-stick models, and orbital overlap diagrams for molecules of methane, CH_4, ammonia, NH_3, and water, H_2O.

B. sp^2 Hybrid Orbitals

Combination of one $2s$ orbital and two $2p$ orbitals produces three equivalent orbitals called sp^2 hybrid orbitals. The three sp^2 hybrid orbitals lie in a plane and are directed toward the corners of an equilateral triangle; the angle between sp^2 orbitals is 120° [Figure 1.12(a)]. Figure 1.12(b) shows three equivalent sp^2 hybrid orbitals along with the remaining unhybridized $2p$ orbital.

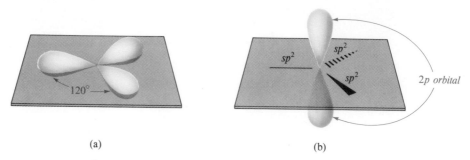

(a)
(b)

Figure 1.12 sp^2 Hybrid orbitals. (a) Three sp^2 orbitals in a plane with angles of 120° between them; (b) the unhybridized $2p$ orbital is perpendicular to the plane created by the three sp^2 hybrid orbitals.

The sp^2 hybrid orbitals are used by second-row elements to form double bonds. Consider ethylene, C_2H_4 [see the Lewis structure in Figure 1.13(a)]. A sigma bond between carbons is formed by overlap of sp^2 orbitals along a common axis [Figure 1.13(b)]. Each carbon also forms sigma bonds to two hydrogens. The remaining $2p$ orbitals on adjacent carbon atoms lie parallel to each other and overlap to form a bond in which electron density is concentrated above and below the axis of the two nuclei. A bond formed by the overlap of parallel p orbitals is called a **pi (π) bond**. Because of the lesser degree of overlap of orbitals forming pi bonds compared with those forming sigma bonds, pi bonds are generally weaker than sigma bonds.

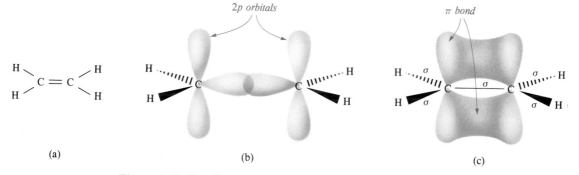

(a)
(b)
(c)

Figure 1.13 Covalent bond formation in ethylene. (a) Lewis structure; (b) a sigma bond between carbon atoms is formed by overlap of sp^2 orbitals (the unhybridized $2p$ orbitals are shown uncombined); (c) the overlap of parallel $2p$ orbitals forms a pi bond.

All double bonds can be described in the same manner as we have described a carbon-carbon double bond. In formaldehyde, which is the simplest organic molecule containing a carbon-oxygen double bond, carbon forms sigma bonds to two hydrogens by overlap of sp^2 orbitals of carbon and $1s$ orbitals of hydrogen. Carbon and oxygen are joined by a sigma bond formed by overlap of sp^2 orbitals. Figure 1.14 shows the Lewis structure of formaldehyde, the sigma bond framework, and overlap of parallel $2p$ orbitals to form a pi bond. Similarly, a carbon-nitrogen double bond is a combination of one sigma bond and one pi bond.

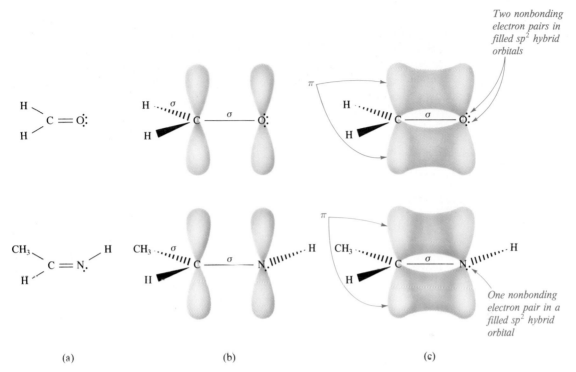

Figure 1.14 Carbon-oxygen and carbon-nitrogen double bonds. (a) Lewis structures of formaldehyde, CH_2=O, and methylene imine, CH_2=NH; (b) the sigma bond framework and nonoverlapping $2p$ orbitals; (c) overlap of parallel $2p$ orbitals to form a pi bond.

C. *sp* Hybrid Orbitals

Combination of one $2s$ orbital and one $2p$ orbital produces two equivalent **sp hybrid orbitals** that lie at an angle of 180° with the nucleus. The unhybridized $2p$ orbitals lie in planes perpendicular to each other and perpendicular to the plane of the *sp* hybrid orbitals. In Figure 1.15, *sp* hybrid orbitals are shown on the *x* axis and unhybridized $2p$ orbitals on the *y* axis and the *z* axis.

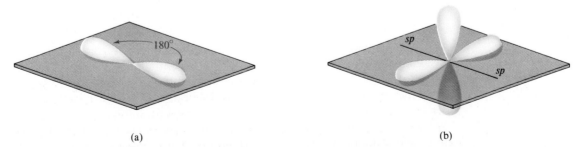

Figure 1.15 *sp* Hybrid orbitals. (a) Two *sp* hybrid orbitals; (b) the two remaining $2p$ orbitals are perpendicular to each other and to the linear *sp* orbitals.

Figure 1.16 shows Lewis structures and orbital overlap pictures for acetylene and hydrogen cyanide. A carbon-carbon triple bond consists of one sigma bond formed by overlap of *sp* hybrid orbitals and two pi bonds. One pi bond is formed by overlap of $2p_y$ orbitals and the second by overlap of $2p_z$ orbitals. Similarly, a carbon-nitrogen triple bond also consists of one sigma bond and two pi bonds.

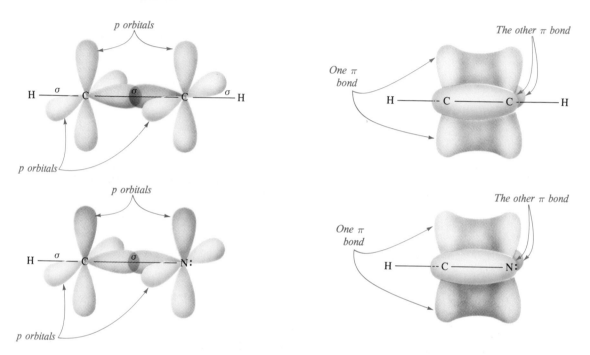

Figure 1.16 Covalent bonding in acetylene and hydrogen cyanide. (a) The sigma bond framework, shown along with nonoverlapping $2p$ orbitals; (b) formation of two pi bonds by the overlap of two sets of $2p$ orbitals.

If you compare the predictions of bond angles in the small molecules we have studied in this chapter, you will see that the valence-shell electron-pair repulsion model (Section 1.3) and the orbital overlap model give equally good predictions. The orbital overlap model, however, gives us a more useful understanding of double and triple bonds. A double bond is not a combination of two identical bonds. Rather, it consists of one sigma bond and one pi bond. The distribution of the bonding electrons in a pi bond is quite different from distribution in a sigma bond. Similarly, a triple bond is not a combination of three identical bonds. Rather, it consists of one sigma bond and two pi bonds. The relations among the number of groups bonded to carbon, orbital hybridization, and types of bonds involved are summarized in Table 1.6.

We will use the orbital overlap picture of covalent bonding in later chapters to help us understand why compounds containing double and triple bonds have different chemical properties from compounds containing only single bonds.

Table 1.6
Covalent bonding of carbon.

Groups Bonded to Carbon	Orbital Hybridization	Types of Bonds Involved	Example
4	sp^3	four sigma bonds	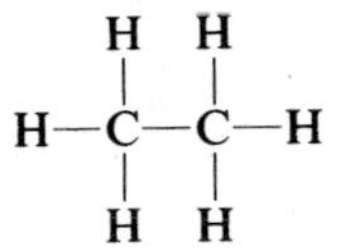 H—C—C—H (ethane structure with H H above and H H below)
3	sp^2	three sigma bonds and one pi bond	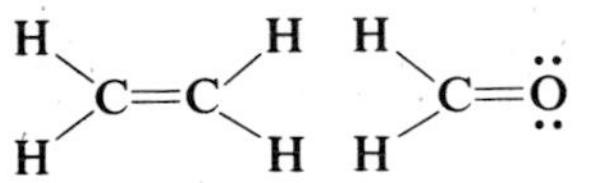 C=C and C=Ö structures
2	sp	two sigma bonds and two pi bonds	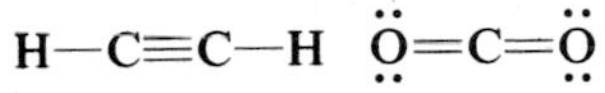 H—C≡C—H Ö=C=Ö

Example 1.10

Describe the bonding in these molecules in terms of the atomic orbitals involved, and predict all bond angles.

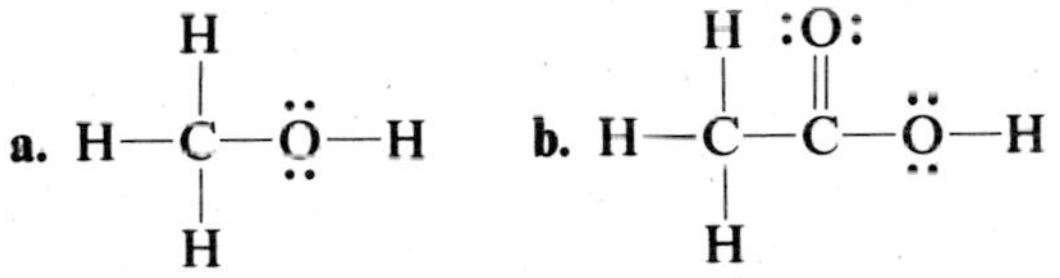

a. H—C—Ö—H (with H above and H below the C)   b. H—C—C—Ö—H (with H and :O: above, H below)

Solution

The problem here is how to show clearly and concisely on a structural formula (1) the hybridization of each atom, (2) the atomic orbitals involved in each covalent bond, and (3) all bond angles. One way to do this is in three separate diagrams, as follows: Labels on the first diagram point to atoms and show the hybridization of each atom. Labels on the second diagram point to bonds and show the type of bond, either sigma or pi. Labels on the third diagram point to atoms and show predicted bond angles about each atom.

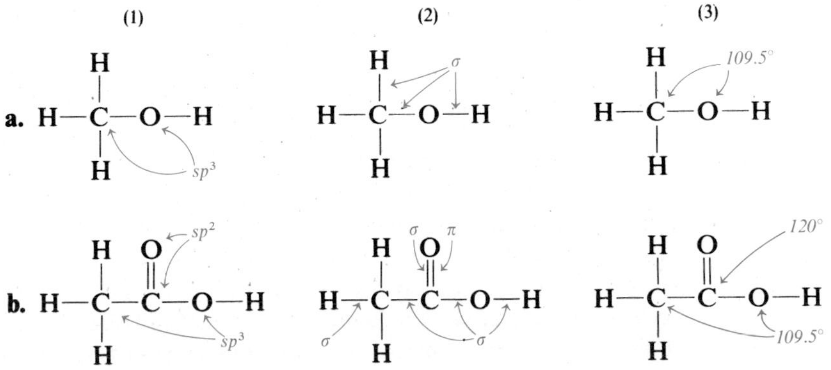

(1)   (2)   (3)

a. H—C—O—H with sp^3 labels; second diagram with σ labels; third diagram with 109.5° labels

b. H—C—C—O—H with sp^2 and sp^3 labels; second diagram with σ and π labels; third diagram with 120° and 109.5° labels

Problem 1.10

Describe the bonding in the following molecules in terms of atomic orbitals involved and predict all bond angles.

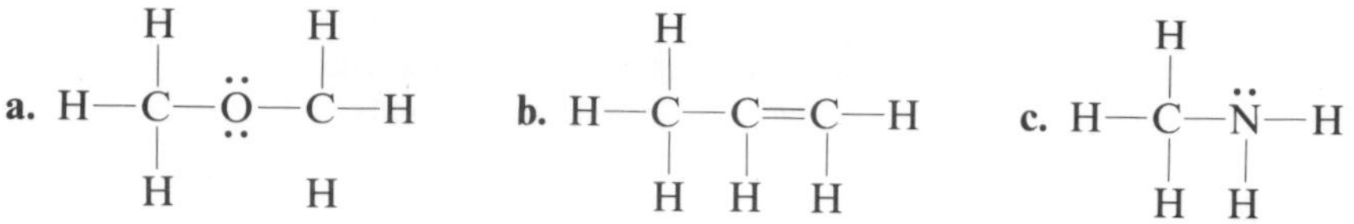

a.
$$\begin{array}{c} \;\;HH \\ || \\ H-C-\overset{..}{\underset{..}{O}}-C-H \\ || \\ \;\;HH \end{array}$$

b.
$$\begin{array}{c} H \\ | \\ H-C-C=C-H \\ |\;\;\;|\;\;| \\ H\;\;H\;\;H \end{array}$$

c.
$$\begin{array}{c} H \\ | \\ H-C-\overset{..}{N}-H \\ |\;\;\;| \\ H\;\;H \end{array}$$

Key Terms and Concepts

alcohol (1.4A)

aldehyde (1.4B)

anion (1.2B)

condensed structural formula (1.4A)

carbonyl group (1.4B)

carboxyl group (1.4C)

carboxylic acid (1.4C)

cation (1.2B)

covalent bond (1.2C)

covalent bond with partial ionic character (1.2C)

electronegativity (1.2C)

ether (1.4A)

formal charge (1.2F)

functional group (1.4)

sp^2 hybrid orbital (1.7C)

sp^3 hybrid orbital (1.7B)

sp hybrid orbital (1.7A)

hydridization (1.7)

hydroxyl group (1.4A)

ionic bond (1.2C)

ketone (1.4B)

Lewis structure (1.1, 1.2D)

molecular orbital (1.7)

octet rule (1.2)

orbital (1.1)

organic chemistry (introduction)

pi bond (1.7B)

principal energy level (1.1)

principal quantum number (1.1)

sigma bond (1.7)

structural isomerism (1.5)

valence electron (1.1)

valence-shell electron-pair repulsion (VSEPR) model (1.3)

Problems

Lewis structures of covalent molecules and ions (Section 1.2)

1.11 What is the relationship between the Lewis structures of H, C, N, O, and F and the valence of these elements?

1.12 Write Lewis structures for the following atoms and state how many electrons each must gain, lose, or share to achieve the same electron configuration as that of the noble gas nearest it in the periodic table.

a. carbon   b. nitrogen   c. sulfur

d. oxygen   e. hydrogen   f. chlorine

g. phosphorus   h. bromine   i. iodine

1.13 Write Lewis structures for these molecules. Be certain to show all valence electrons. In some parts of this problem where the order of attachment of atoms may not be obvious to you, the name of the functional group present in the molecule is given to help you arrive at the correct order of attachment of atoms.

a. H_2O_2 b. N_2H_4 c. CH_3OH
d. CH_3SH e. CH_3NH_2 f. CH_3Cl
g. CH_3OCH_3 h. C_2H_6 i. C_2H_4
j. C_2H_2 k. CO_2 l. H_2CO_3
m. CH_2O n. CH_3CHO o. CH_3COCH_3
        (an aldehyde)   (a ketone)

p. HCO_2H q. CH_3CO_2H
   (a carboxylic   (a carboxylic
   acid)   acid)

1.14 Write Lewis structures for the following ions. Be certain to show all valence electrons and all formal charges.

a. CH_3O^- b. H_3O^+ c. $CH_3NH_3^+$
d. Cl^- e. HCO_3^- f. CO_3^{2-}
g. $CH_3CO_2^-$ h. HCO_2^-

1.15 Following are Lewis structures for several ions. All valence electrons on each atom are shown. Assign formal charges to each structure as appropriate.

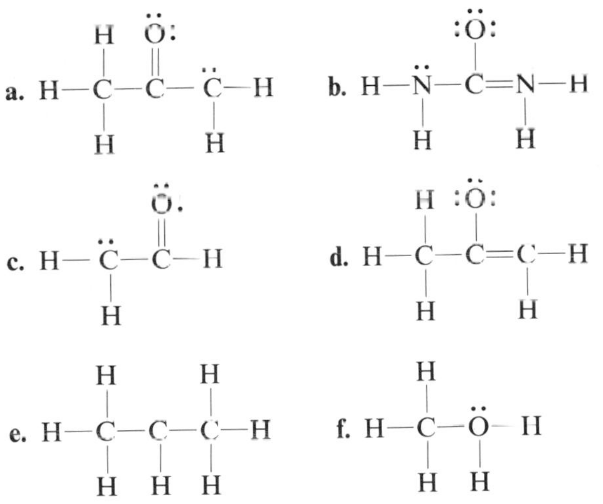

1.16 Following the rule that each atom of carbon, oxygen, nitrogen, and the halogens reacts to achieve a complete outer shell of eight valence electrons, add unshared pairs of electrons as necessary to complete the valence shells of these molecules and ions. Then assign formal charges as appropriate.

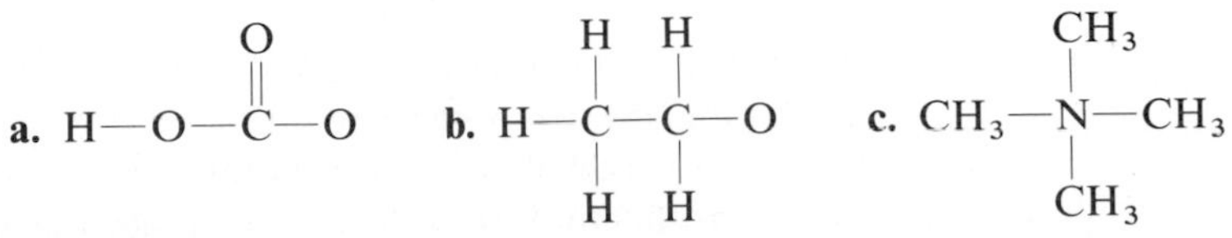

$$\textbf{d.} \quad H-\overset{\overset{\displaystyle H}{|}}{C}-\overset{\overset{\displaystyle H}{|}}{\underset{\underset{\displaystyle H}{|}}{C}} \qquad\qquad \textbf{e.} \quad H-\overset{\overset{\displaystyle H}{|}}{\underset{\underset{\displaystyle H}{|}}{C}}-\overset{\overset{\displaystyle O}{\|}}{C}-O$$

$$\textbf{f.} \quad H-\overset{\overset{\displaystyle H}{|}}{\underset{\underset{\displaystyle H}{|}}{N}}-\overset{\overset{\displaystyle H}{|}}{\underset{\underset{\displaystyle H}{|}}{C}}-\overset{\overset{\displaystyle O}{\|}}{C}-O$$

1.17 Following are compounds containing both ionic and covalent bonds. Draw Lewis structures for each. Show covalent bonds by dashes and ionic bonds by positive and negative charges on the appropriate atoms.
- **a.** NaOH
- **b.** NH_4Cl
- **c.** $NaHCO_3$
- **d.** Na_2CO_3
- **e.** HCO_2Na
- **f.** CH_3CO_2Na

1.18 Arrange the single covalent bonds within each set in order of increasing partial ionic character.
- **a.** C—H, O—H, N—H
- **b.** C—H, C—O, C—N
- **c.** C—H, C—Cl, C—I
- **d.** C—S, C—O, C—F

1.19 For each polar covalent bond in these molecules, label the partially positive atom with δ^+ and the partially negative atom with δ^-.

a. $H-\underset{\underset{\displaystyle H}{|}}{N}-H$ **b.** $H-\overset{\overset{\displaystyle H}{|}}{\underset{\underset{\displaystyle H}{|}}{C}}-\overset{\overset{\displaystyle O}{\|}}{C}-O-H$

c. $H-\overset{\overset{\displaystyle H}{|}}{\underset{\underset{\displaystyle H}{|}}{C}}-S-H$ **d.** $H-\overset{\overset{\displaystyle H}{|}}{\underset{\underset{\displaystyle H}{|}}{C}}-\overset{\overset{\displaystyle O}{\|}}{C}-\overset{\overset{\displaystyle H}{|}}{\underset{\underset{\displaystyle H}{|}}{C}}-H$

e. $H-\overset{\overset{\displaystyle H}{|}}{\underset{\underset{\displaystyle H}{|}}{C}}-\overset{\overset{\displaystyle H}{|}}{\underset{\underset{\displaystyle H}{|}}{C}}-O-H$ **f.** $H-\overset{\overset{\displaystyle H}{|}}{\underset{\underset{\displaystyle H}{|}}{C}}-\overset{\overset{\displaystyle H}{|}}{\underset{\underset{\displaystyle H}{|}}{C}}-F$

g. $H-\overset{\overset{\displaystyle H}{|}}{\underset{\underset{\displaystyle H}{|}}{C}}-O-\overset{\overset{\displaystyle H}{|}}{\underset{\underset{\displaystyle H}{|}}{C}}-H$ **h.** $\underset{H}{\overset{Cl}{\diagdown}}C=C\underset{\diagdown H}{\overset{\diagup H}{}}$

1.20 An organometallic compound is one containing a carbon-metal bond. Following are two organometallics, both of which are probably familiar to you because of particular hazards they pose to human health and the environment. For each, state whether the carbon-metal bond is pure covalent, polar covalent, or ionic. (*Hint:* Review the section of electro-negativity and its relation to chemical bonding.)

$$CH_2{-}CH_3$$
a. 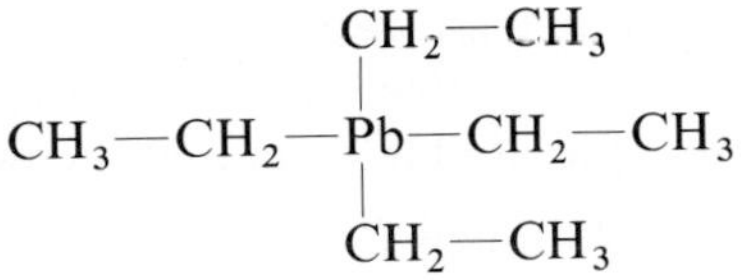 $CH_3{-}CH_2{-}Pb{-}CH_2{-}CH_3$ **b.** $CH_3{-}Hg{-}CH_3$
$$CH_2{-}CH_3$$

tetraethyl lead
(the octane-improving
additive in leaded
gasoline)

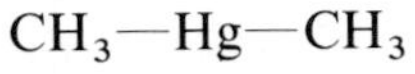

dimethyl mercury
(anaerobic bacteria in ocean
and lake sediments can convert
mercuric ion, Hg(II), to dimethyl
mercury, a substance very toxic
to higher organisms)

*Bond angles
and shapes of
molecules
(Section 1.3)*

1.21 Explain how the valence-shell electron-pair repulsion model is used to predict bond angles.

1.22 Following are Lewis structures for several molecules. Use the valence-shell electron-pair repulsion model to predict bond angles about each circled atom.

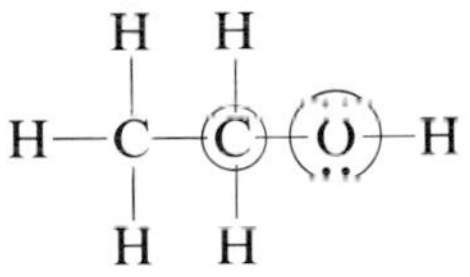

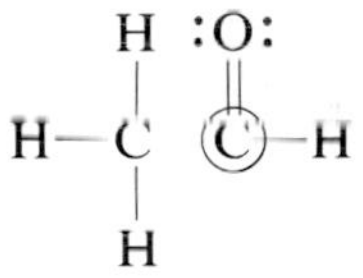

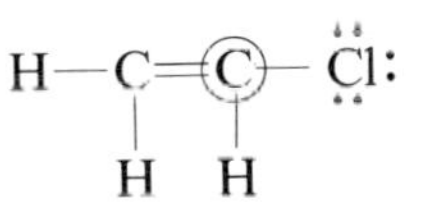

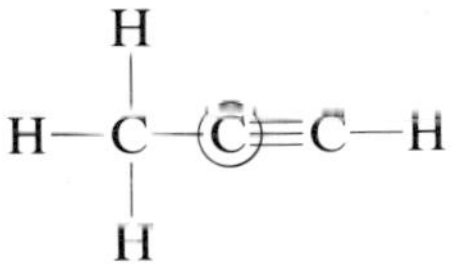

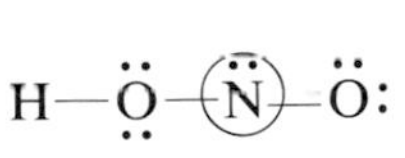

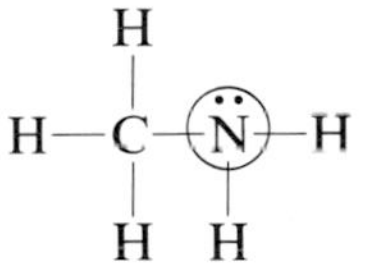

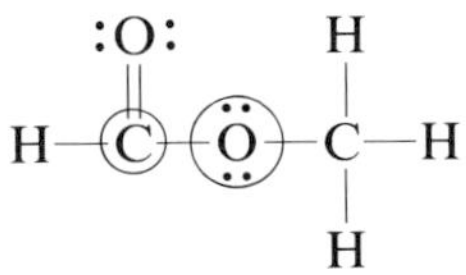

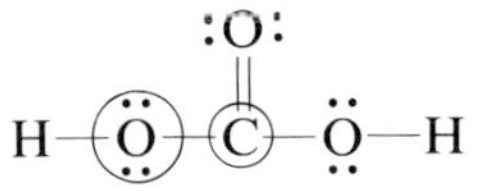

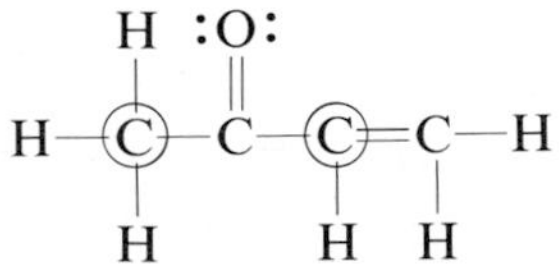

*Functional
groups
(Section 1.4)*

1.23 Draw Lewis structures for the following functional groups. Be certain to show all valence electrons on each.
a. carbonyl group   **b.** carboxyl group   **c.** hydroxyl group

1.24 Draw condensed structural formulas for all compounds of molecular formula C_4H_8O that contain the following functional groups:
a. a ketone (there is only one)
b. an aldehyde (there are two)
c. a carbon-carbon double bond and an ether (there are four)
d. a carbon-carbon double bond and an alcohol (there are eight)

1.25 Draw structural formulas for:
a. the eight alcohols of molecular formula $C_5H_{12}O$
b. the six ethers of molecular formula $C_5H_{12}O$
c. the eight aldehydes of molecular formula $C_6H_{12}O$
d. the six ketones of molecular formula $C_6H_{12}O$
e. the eight carboxylic acids of molecular formula $C_6H_{12}O_2$

Structural
isomerism
(Section 1.5)

1.26 Which of the following are true about structural isomers?
a. They have the same molecular formula.
b. They have the same molecular weight.
c. They have the same order of attachment of atoms.
d. They have the same physical properties.

1.27 Are the following pairs of molecules structural isomers or are they identical?

a.

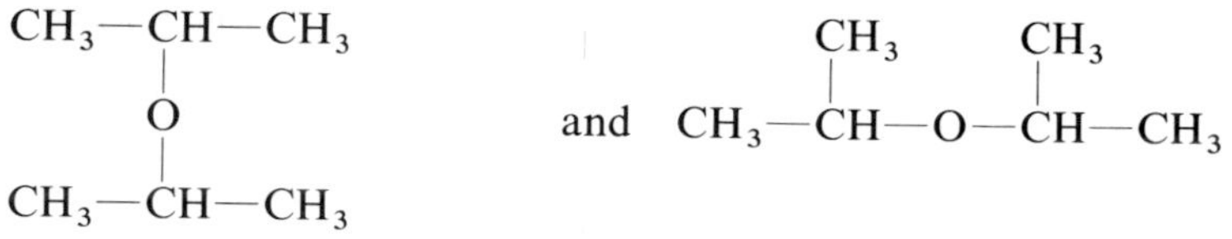

$$CH_3-\underset{\underset{\underset{CH_3-CH-CH_3}{|}}{O}}{CH}-CH_3 \quad \text{and} \quad CH_3-\underset{\underset{CH_3}{|}}{CH}-O-\underset{\underset{CH_3}{|}}{CH}-CH_3$$

b. $CH_2{=}CH-CH_2-CH_3$ and $CH_3-CH{=}CH-CH_3$

c. $CH_3-O-\overset{\overset{O}{\|}}{C}-H$ and $CH_3-\overset{\overset{O}{\|}}{C}-O-H$

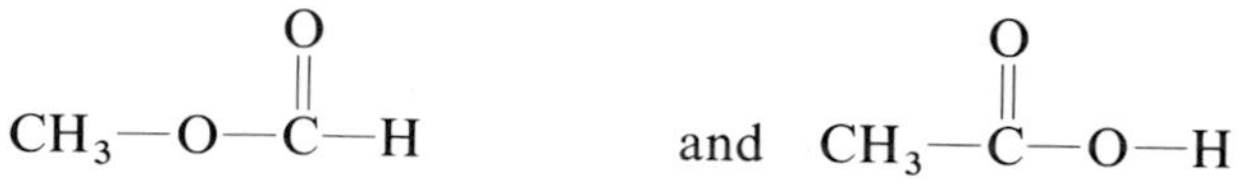

d. $HO-CH_2-\underset{\underset{CH_3}{|}}{CH}-OH$ and $CH_3-\underset{\underset{CH_2-OH}{|}}{CH}-OH$

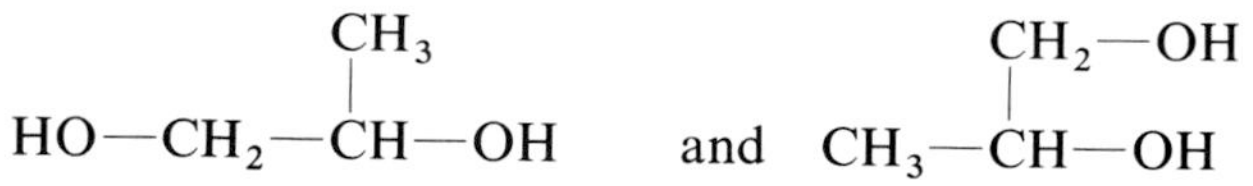

e. $H-\overset{\overset{O}{\|}}{C}-CH_2-\underset{\underset{CH_3}{|}}{CH}-CH_3$ and $\underset{CH_3}{\overset{CH_3}{>}}CH-CH_2-\overset{\overset{O}{\|}}{C}-H$

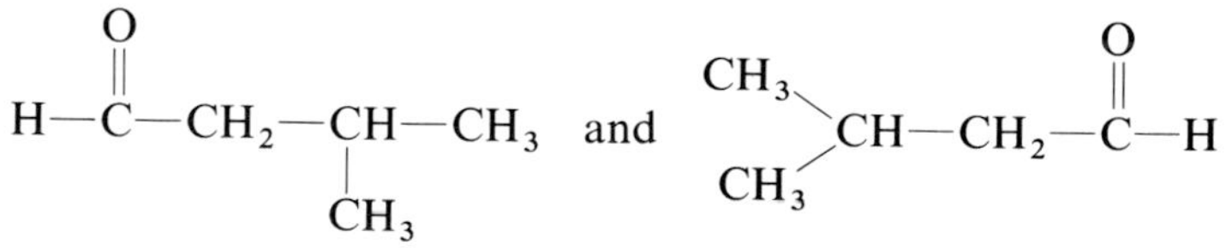

f. $CH_3-\underset{\underset{\underset{CH_3}{|}}{CH_2-CH-CH_3}}{CH}-CH_2-CH_3$ and

$CH_3-\underset{\underset{CH_3}{|}}{CH}-CH_2-CH_2-\underset{\underset{CH_3}{|}}{CH}-CH_3$

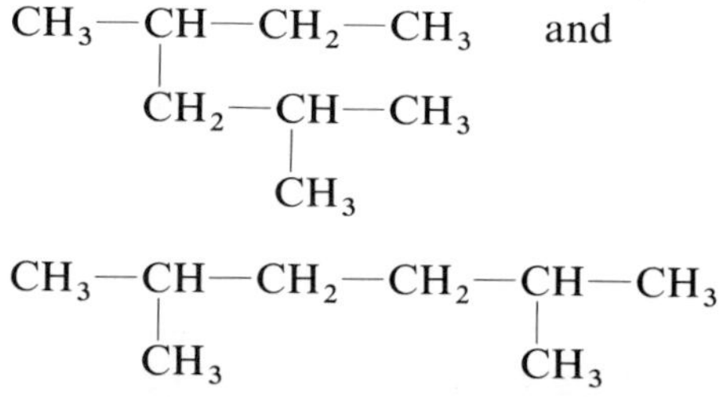

Covalent bond
formation by
the overlap
of atomic
orbitals
(Section 1.7)

1.28 Following are Lewis structures for several molecules. State the orbital hybridization of each circled atom.

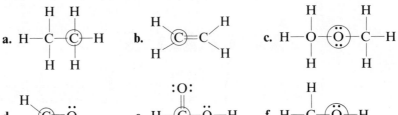

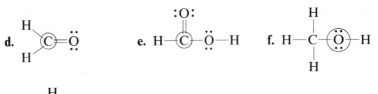

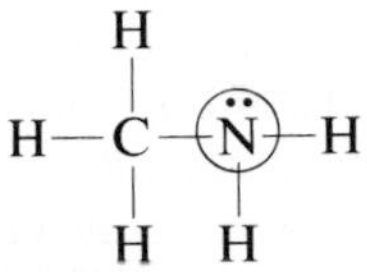

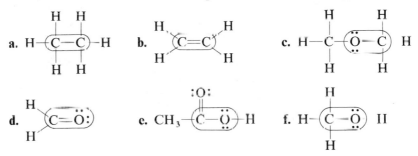

1.29 Following are Lewis structures for several molecules. Describe each circled bond in terms of the overlap of atomic orbitals.

Alkanes
and Cycloalkanes

Compounds containing only carbon and hydrogen are called **hydrocarbons**. If the hydrocarbon contains only single bonds, it is called a **saturated hydrocarbon**, or **alkane**. If a hydrocarbon contains one or more double or triple bonds, it is called an **unsaturated hydrocarbon**. In this chapter we shall discuss saturated hydrocarbons, the simplest class of organic compounds.

2.1 Structure of Alkanes

Methane, CH_4, and ethane, C_2H_6, are the first members of the alkane family. Shown in Figure 2.1 are Lewis structure and ball-and-stick models for these molecules. The shape of methane is tetrahedral, and all H—C—H bond angles are 109.5° (Section 1.3). Each of the carbon atoms in ethane is also tetrahedral, and all bond angles are approximately 109.5°. The next members of the series are propane, C_3H_8; butane, C_4H_{10}; and pentane, C_5H_{12}.

Lewis Structures

$$\begin{array}{ccc}
& \text{H} \quad \text{H} \quad \text{H} & \\
\text{H}-\text{C}-\text{C}-\text{C}-\text{H} & \\
& \text{H} \quad \text{H} \quad \text{H} &
\end{array}$$

$$\begin{array}{cccc}
\text{H} \quad \text{H} \quad \text{H} \quad \text{H} \\
\text{H}-\text{C}-\text{C}-\text{C}-\text{C}-\text{H} \\
\text{H} \quad \text{H} \quad \text{H} \quad \text{H}
\end{array}$$

$$\begin{array}{ccccc}
\text{H} \quad \text{H} \quad \text{H} \quad \text{H} \quad \text{H} \\
\text{H}-\text{C}-\text{C}-\text{C}-\text{C}-\text{C}-\text{H} \\
\text{H} \quad \text{H} \quad \text{H} \quad \text{H} \quad \text{H}
\end{array}$$

Condensed Structural Formulas

$$CH_3CH_2CH_3 \qquad CH_3CH_2CH_2CH_3 \qquad CH_3CH_2CH_2CH_2CH_3$$

        propane   butane   pentane

Condensed structural formulas for these and higher alkanes can be written in an even more abbreviated form. For example, the structural formula of pentane contains three —CH_2— (**methylene**) groups in the middle of the chain. They can be grouped together and the structural formula written $CH_3(CH_2)_3CH_3$. Names, molecular formulas, and condensed structural formulas of the first twenty alkanes are given in Table 2.1.

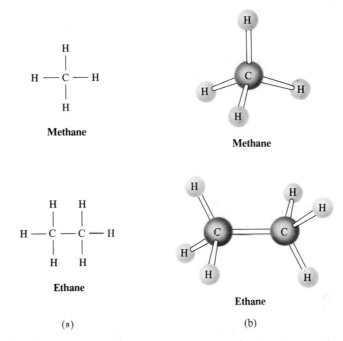

Methane

Methane

Ethane

Ethane

(a)

(b)

Figure 2.1 Methane and ethane. (a) Lewis structures; (b) ball-and-stick models.

Table 2.1 Names, molecular formulas, and condensed structural formulas for the first twenty alkanes.

Name	Molecular Formula	Condensed Structural Formula	Name	Molecular Formula	Condensed Structural Formula
methane	CH_4	CH_4	undecane	$C_{11}H_{24}$	$CH_3(CH_2)_9CH_3$
ethane	C_2H_6	CH_3CH_3	dodecane	$C_{12}H_{26}$	$CH_3(CH_2)_{10}CH_3$
propane	C_3H_8	$CH_3CH_2CH_3$	tridecane	$C_{13}H_{28}$	$CH_3(CH_2)_{11}CH_3$
butane	C_4H_{10}	$CH_3(CH_2)_2CH_3$	tetradecane	$C_{14}H_{30}$	$CH_3(CH_2)_{12}CH_3$
pentane	C_5H_{12}	$CH_3(CH_2)_3CH_3$	pentadecane	$C_{15}H_{32}$	$CH_3(CH_2)_{13}CH_3$
hexane	C_6H_{14}	$CH_3(CH_2)_4CH_3$	hexadecane	$C_{16}H_{34}$	$CH_3(CH_2)_{14}CH_3$
heptane	C_7H_{16}	$CH_3(CH_2)_5CH_3$	heptadecane	$C_{17}H_{36}$	$CH_3(CH_2)_{15}CH_3$
octane	C_8H_{18}	$CH_3(CH_2)_6CH_3$	octadecane	$C_{18}H_{38}$	$CH_3(CH_2)_{16}CH_3$
nonane	C_9H_{20}	$CH_3(CH_2)_7CH_3$	nonadecane	$C_{19}H_{40}$	$CH_3(CH_2)_{17}CH_3$
decane	$C_{10}H_{22}$	$CH_3(CH_2)_8CH_3$	eicosane	$C_{20}H_{42}$	$CH_3(CH_2)_{18}CH_3$

2.2 Structural Isomerism in Alkanes

Two or more compounds that have the same molecular formula but different orders of attachment of atoms are called **structural isomers** (Section 1.5). For the molecular formulas CH_4, C_2H_6, and C_3H_8, there is only one possible order of attachment of atoms, and therefore methane, ethane, and propane have no structural isomers. For the molecular formula C_4H_{10}, two orders of attachment of atoms are possible. In one, the four carbon atoms are attached in a chain; in the other, they are attached three in a chain with the fourth carbon as a branch on the three-carbon chain. These isomeric alkanes are named butane and 2-methylpropane:

$$CH_3-CH_2-CH_2-CH_3 \qquad CH_3-\overset{\displaystyle |}{\underset{\displaystyle CH_3}{CH}}-CH_3$$

butane
bp −0.5°C

2-methylpropane
bp −11.2°C

Butane and 2-methylpropane are structural isomers; they have the same molecular formula but different orders of attachment of their atoms. Structural isomers are different compounds and have different physical and chemical properties. Notice that the boiling points of butane and 2-methylpropane differ by over 10°C.

There are three structural isomers of molecular formula C_5H_{12}, five structural isomers of C_6H_{14}, eighteen of C_8H_{18}, and seventy-five of $C_{10}H_{22}$. It should be obvious that for even a small number of carbon and hydrogen atoms, a very large number of structural isomers is possible. In fact, the potential for structural and functional group individuality from just the basic building blocks of carbon, hydrogen, nitrogen, and oxygen is practically limitless.

Example 2.1

Identify these pairs as formulas of identical compounds or as formulas of structural isomers.

a. $CH_3-CH_2-CH_2-CH_2-CH_2-CH_3$ and

$$CH_3-CH_2-\overset{\displaystyle |}{\underset{\displaystyle CH_2-CH_2-CH_3}{CH_2}}$$

b. $CH_3-CH_2-\overset{\displaystyle |}{\underset{\displaystyle CH_3}{CH}}-CH_2-CH_2-CH_3$ and

$$CH_3-CH_2-CH_2-\overset{\displaystyle |}{\underset{\displaystyle \underset{\displaystyle CH_3}{\overset{\displaystyle |}{CH_2}}}{CH}}-CH_3$$

c. $CH_3-\overset{\displaystyle \overset{\displaystyle CH_3}{\displaystyle |}}{\underset{\displaystyle CH_3}{CH}}-CH_2-\overset{\displaystyle |}{\underset{\displaystyle CH_3}{CH}}$ and $CH_3-CH_2-\overset{\displaystyle \overset{\displaystyle CH_3}{\displaystyle |}}{CH}-\overset{\displaystyle |}{\underset{\displaystyle CH_3}{CH}}-CH_3$

Solution

To determine whether these formulas are identical or represent structural isomers, find the longest chain of carbon atoms and number it from the end nearest the first branch. Note that in finding the longest chain, it makes no difference whether the chain is drawn straight or bent. As structural formulas are drawn in this problem, there is no attempt to show three-dimensional shapes. After you have found the longest carbon chain and numbered it from the correct end, compare the lengths of each chain and the size and locations of any branches.

a. $\overset{1}{C}H_3-\overset{2}{C}H_2-\overset{3}{C}H_2-\overset{4}{C}H_2-\overset{5}{C}H_2-\overset{6}{C}H_3$ and

$\overset{1}{C}H_3-\overset{2}{C}H_2-\overset{3}{C}H_2$
$\quad\quad\quad\quad\;\;|$
$\quad\quad\quad\underset{4}{C}H_2-\underset{5}{C}H_2-\underset{6}{C}H_3$

Each formula has an unbranched chain of six carbons; they are identical and represent the same compound.

b. $\overset{1}{C}H_3-\overset{2}{C}H_2-\overset{3}{C}H-\overset{4}{C}H_2-\overset{5}{C}H_2-\overset{6}{C}H_3$ and
$\quad\quad\quad\quad\quad\;|$
$\quad\quad\quad\quad\;\;CH_3$

$\overset{6}{C}H_3-\overset{5}{C}H_2-\overset{4}{C}H_2-\overset{3}{C}H-CH_3$
$\quad\quad\quad\quad\quad\quad\quad|$
$\quad\quad\quad\quad\quad\quad\;\;\underset{2}{C}H_2$
$\quad\quad\quad\quad\quad\quad\quad|$
$\quad\quad\quad\quad\quad\quad\;\;\underset{1}{C}H_3$

Each has a chain of six carbons with a CH_3- group on the third carbon atom of the chain; they are identical and represent the same compound.

$\quad\quad\quad\quad\quad\quad\quad\quad\overset{5}{C}H_3\quad\quad\quad\quad\quad\quad\quad\quad\quad\overset{1}{C}H_3$
$\quad\quad\quad\quad\quad\quad\quad\quad\;\;|\quad\quad\quad\quad\quad\quad\quad\quad\quad\;\;|$
c. $\overset{1}{C}H_3-\overset{2}{C}H-\overset{3}{C}H_2-\overset{4}{C}H$ and $\overset{5}{C}H_3-\overset{4}{C}H_2-\overset{3}{C}H-\overset{2}{C}H-CH_3$
$\quad\quad\quad|\quad\quad\quad\quad\;\;|\quad\quad\quad\quad\quad\quad\quad\quad\quad|$
$\quad\quad\;\;CH_3\quad\quad\;\;CH_3\quad\quad\quad\quad\quad\quad\;\;CH_3$

Each has chains of five carbons with two CH_3- branches. While the branches are identical, they are at different locations on the chains. Therefore, these formulas represent structural isomers.

Problem 2.1

Identify the pairs as identical or as formulas of structural isomers.

$\quad\quad\quad\quad\quad CH_3$
$\quad\quad\quad\quad\quad\;\;|$
$\quad\quad\quad\quad\quad CH_2$
$\quad\quad\quad\quad\quad\;\;|\quad\quad\quad\quad\quad\quad\quad\quad CH_3\quad\quad CH_3$
$\quad\quad\quad\quad\quad\quad\quad\quad\quad\quad\quad\quad\quad\quad|\quad\quad\quad\;|$
a. $CH_3-CH-CH-CH_3$ and $CH_3-CH_2-CH-CH_2-CH-CH_3$
$\quad\quad\quad\quad\quad\quad\quad|$
$\quad\quad\quad\quad\quad CH_2-CH_3$

$\quad\quad\quad\quad CH_3\quad\quad\quad\quad\quad\quad\quad\quad\quad CH_3\;\;CH_3$
$\quad\quad\quad\quad\;\;|\quad\quad\quad\quad\quad\quad\quad\quad\quad\;|\quad\;\;|$
b. $CH_3-CH-CH-CH_3$ and $CH_3-CH-CH-CH_2-CH_3$
$\quad\quad\quad\quad\quad\quad\;|$
$\quad\quad\quad\quad\quad CH_2-CH_3$

Example 2.2

Draw structural formulas for the five structural isomers of molecular formula C_6H_{14}.

☐ *Solution*

In solving problems of this type, you should devise a strategy and then follow it. One strategy for this example is the following. First, draw the structural isomer with all six carbons in an unbranched chain. Then, draw all structural isomers with five carbons in a chain and one carbon as a branch on the chain. Finally, draw all structural isomers with four carbons in a chain and two carbons as branches.

Six Carbons in an Unbranched Chain

$$\overset{1}{C}H_3-\overset{2}{C}H_2-\overset{3}{C}H_2-\overset{4}{C}H_2-\overset{5}{C}H_2-\overset{6}{C}H_3$$

Five Carbons in a Chain; One Carbon as a Branch

$$\overset{1}{C}H_3-\overset{2}{C}H-\overset{3}{C}H_2-\overset{4}{C}H_2-\overset{5}{C}H_3 \qquad \overset{1}{C}H_3-\overset{2}{C}H_2-\overset{3}{C}H-\overset{4}{C}H_2-\overset{5}{C}H_3$$
$$\qquad | \qquad\qquad\qquad\qquad\qquad | $$
$$\qquad CH_3 \qquad\qquad\qquad\qquad\qquad CH_3$$

Four Carbons in a Chain; Two Carbons as Branches

$$\begin{array}{ccc} & CH_3 & \\ & | & \\ CH_3-&C&-CH_2-CH_3 \\ & | & \\ & CH_3 & \end{array} \qquad \begin{array}{ccc} CH_3 & & \\ | & & \\ CH_3-CH&-&CH-CH_3 \\ & & | \\ & & CH_3 \end{array}$$

Problem 2.2

Draw structural formulas for the three structural isomers of molecular formula C_5H_{12}.

2.3 Nomenclature of Alkanes

A. The IUPAC System

Ideally every organic compound should have a name that clearly describes its structure and from which a structural formula can be drawn. For this purpose, chemists throughout the world have accepted a set of rules established by the International Union of Pure and Applied Chemistry (IUPAC). This system is known as the **IUPAC system**, or alternatively, the Geneva system, because the first meetings of the IUPAC were held in Geneva, Switzerland. The IUPAC names of alkanes with an unbranched chain of carbon atoms consist of two parts: (1) a prefix that indicates the number of carbon atoms in the chain; and (2) the ending *-ane*, to show that the compound is an alkane. Prefixes used to

Table 2.2
Prefixes used in the IUPAC system to indicate one to twenty carbon atoms in a chain.

Prefix	Number of Carbon Atoms	Prefix	Number of Carbon Atoms
meth	1	undec	11
eth	2	dodec	12
prop	3	tridec	13
but	4	tetradec	14
pent	5	pentadec	15
hex	6	hexadec	16
hept	7	heptadec	17
oct	8	octadec	18
non	9	nonadec	19
dec	10	eicos	20

show the presence of from one to twenty carbon atoms are given in Table 2.2. The first four prefixes listed in Table 2.2 were chosen by the International Union of Pure and Applied Chemistry because they were well established in the language of organic chemistry. In fact, they were well established even before there were hints of the structural theory underlying the discipline. For example, the prefix *but-* appears in the name butyric acid, a compound of four carbon atoms present in butter fat (Latin *butyrum*, butter). Roots to show five or more carbons are derived from Greek or Latin roots. Names, molecular formulas, and condensed structural formulas for the first twenty alkanes are given in Table 2.1.

The IUPAC names of substituted alkanes consist of a parent name, which indicates the longest chain of carbon atoms in the compound, and substituent names, which indicate the groups attached to the parent chain.

$$CH_3—CH_2—CH_2—CH—CH_2—CH_2—CH_2—CH_3$$
$$\overset{|}{CH_3}$$

Parent

Substituent

A substituent group derived from an alkane is called an **alkyl group**. The symbol R— is commonly used to show the presence of an alkyl group. Alkyl groups are named by dropping the *-ane* from the name of the parent alkane and adding the suffix *-yl*. For example, the alkyl substituent $CH_3CH_2—$ is named ethyl.

$$\begin{matrix} & H & H \\ & | & | \\ H— & C— & C—H \\ & | & | \\ & H & H \end{matrix} \qquad \begin{matrix} & H & H \\ & | & | \\ H— & C— & C— \\ & | & | \\ & H & H \end{matrix}$$

ethane
(parent hydrocarbon)

ethyl group
(an alkyl group)

Names and structural formulas for eleven of the most common alkyl groups are given in Table 2.3.

Table 2.3 Common alkyl groups.

IUPAC Name	Condensed Structural Formula	IUPAC Name	Condensed Structural Formula
methyl	—CH_3		
ethyl	—CH_2—CH_3	tert-butyl	$-\overset{\displaystyle CH_3}{\underset{\displaystyle CH_3}{\overset{\displaystyle \mid}{\underset{\displaystyle \mid}{C}}}}-CH_3$
propyl	—CH_2—CH_2—CH_3		
isopropyl	—$\underset{\displaystyle \underset{\displaystyle CH_3}{\mid}}{CH}$—$CH_3$	pentyl	—CH_2—CH_2—CH_2—CH_2—CH_3
butyl	—CH_2—CH_2—CH_2—CH_3	isopentyl	—CH_2—CH_2—$\underset{\displaystyle \underset{\displaystyle CH_3}{\mid}}{CH}$—$CH_3$
isobutyl	—CH_2—$\underset{\displaystyle \underset{\displaystyle CH_3}{\mid}}{CH}$—$CH_3$		
sec-butyl	—$\underset{\displaystyle \underset{\displaystyle CH_3}{\mid}}{CH}$—$CH_2$—$CH_3$	neopentyl	—CH_2—$\overset{\displaystyle \overset{\displaystyle CH_3}{\mid}}{\underset{\displaystyle \underset{\displaystyle CH_3}{\mid}}{C}}$—$CH_3$

Following are the rules of the IUPAC system for naming alkanes.

1. The general name of a saturated hydrocarbon is *alkane*.
2. For branched-chain hydrocarbons, the hydrocarbon derived from the longest chain of carbon atoms is taken as the parent chain and the IUPAC name is derived from that of the parent chain.
3. Groups attached to the parent chain are called substituents. Each substituent is given a name and a number. The number shows the carbon atom of the parent chain to which the substituent is attached.
4. If the same substituent occurs more than once, the number of each carbon of the parent chain on which the substituent occurs is given. In addition, the number of times the substituent group occurs is indicated by a prefix *di-*, *tri-*, *tetra-*, *penta-*, *hexa-*, and so on.
5. If there is one substituent, number the parent chain from the end that gives it the lower number. If there are two or more substituents, number the parent chain from the end that gives the lower number to the substituent encountered first.
6. If there are two or more different substituents, list them in alphabetical order.

Example 2.3

Give IUPAC names for these compounds.

a. CH_3—$\underset{\displaystyle \underset{\displaystyle CH_3}{\mid}}{CH}$—$CH_2$—$CH_3$

b. CH_3—$\underset{\displaystyle \underset{\displaystyle CH_3}{\mid}}{CH}$—$CH_2$—$\underset{\displaystyle \underset{\displaystyle CH_2—CH_3}{\mid}}{CH}$—$CH_2$—$CH_3$

$$CH_3-CH_2-CH_2-\overset{\displaystyle CH_3}{\underset{\displaystyle CH_3}{\overset{\displaystyle |}{\underset{\displaystyle |}{C}}}}-CH_3$$

c.

Solution   **a.** There are four carbon atoms in the longest chain, and therefore the name of the parent chain is butane (Rule 2). The butane chain must be numbered so that the single methyl group is on carbon 2 of the chain (Rule 5). The correct name of this alkane is 2-methylbutane.

$$\overset{1}{CH_3}-\overset{2}{\underset{\displaystyle CH_3}{\overset{\displaystyle |}{CH}}}-\overset{3}{CH_2}-\overset{4}{CH_3}$$

2-methylbutane

b. The longest chain contains six carbons, and therefore the parent chain is a hexane (Rule 2). There are two alkyl substituents: a methyl group and an ethyl group. The hexane chain must be numbered so that the substituent encountered first (the methyl group) is on carbon 2 of the chain (Rule 5). The ethyl and methyl substituents are listed in alphabetical order (Rule 6) to give the name 4-ethyl-2-methylhexane.

$$\overset{1}{CH_3}\quad\overset{2}{\underset{\displaystyle CH_3}{\overset{\displaystyle |}{CH}}}\quad\overset{3}{CH_2}\quad\overset{4}{\underset{\displaystyle CH_2-CH_3}{\overset{\displaystyle |}{CH}}}\quad\overset{5}{CH_2}-\overset{6}{CH_3}$$

4-ethyl-2-methylhexane

c. The longest chain contains five carbon atoms and therefore the parent chain is a pentane (Rule 2). The pentane chain must be numbered so that the substituents are on carbon 2 of the chain (Rule 5). There are two substituents and each must have a name and a number (Rule 4). Because the substituents are identical, they are grouped together using the prefix *di-* (Rule 4). The IUPAC name is 2,2-dimethylpentane.

$$\overset{5}{CH_3}-\overset{4}{CH_2}-\overset{3}{CH_2}-\overset{2}{\underset{\displaystyle CH_3}{\overset{\displaystyle \overset{1}{CH_3}}{\overset{\displaystyle |}{\underset{\displaystyle |}{C}}}}}-CH_3$$

2,2-dimethylpentane

Problem 2.3   Name these alkanes by the IUPAC system.

a.

$$CH_3-\overset{\displaystyle \overset{CH_3}{|}}{CH}-CH_2-CH_2-\overset{\displaystyle \overset{CH_3}{|}}{CH}-CH-CH_3$$
$$\underset{\displaystyle CH_2-CH_3}{\underset{\displaystyle |}{CH_2}}$$

$$\text{CH}_2\text{—CH}_2\text{—CH}_3$$

b. $\text{CH}_3\text{—CH}_2\text{—CH}_2\text{—}\overset{\displaystyle |}{\underset{\displaystyle |}{\text{C}}}\text{—CH}_2\text{—CH}_2\text{—CH}_3$

$$\text{CH—CH}_3$$

$$\text{CH}_3$$

B. Common Names

In spite of the precision of the IUPAC system, routine communication in organic chemistry still relies on a combination of trivial, semisystematic, and systematic names. The reasons for this are rooted in both convenience and historical development.

In the older, semisystematic nomenclature, the total number of carbon atoms in an alkane, regardless of their arrangement, determines the name. The first three alkanes are methane, ethane, and propane. All alkanes of formula C_4H_{10} are called butanes, all alkanes of formula C_5H_{12} are called pentanes, and those of formula C_6H_{14} are called hexanes. For alkanes beyond propane, **normal**, or **n-** is used to indicate that all carbons are joined in a continuous chain, and **iso-** is used to indicate that one end of an otherwise continuous chain terminates in a $(CH_3)_2CH$— group. The first compounds of molecular formula C_5H_{12} to be discovered and named were pentane and its isomer isopentane. Subsequently, another compound of molecular formula C_5H_{12} was discovered, and because it was a "new" pentane (at least it was new to those who first discovered it), this isomer was named neopentane. Following are examples of common names.

$$\text{CH}_3\text{—CH}_2\text{—CH}_2\text{—CH}_3 \qquad \text{CH}_3\text{—}\overset{\displaystyle \text{CH}_3}{\overset{\displaystyle |}{\text{CH}}}\text{—CH}_3$$

n-butane   isobutane

$$\text{CH}_3\text{—CH}_2\text{—CH}_2\text{—CH}_2\text{—CH}_3 \qquad \text{CH}_3\text{—}\overset{\displaystyle \text{CH}_3}{\overset{\displaystyle |}{\text{CH}}}\text{—CH}_2\text{—CH}_3$$

n-pentane   isopentane

$$\text{CH}_3\text{—}\overset{\displaystyle \text{CH}_3}{\underset{\displaystyle \text{CH}_3}{\overset{\displaystyle |}{\underset{\displaystyle |}{\text{C}}}}}\text{—CH}_3 \qquad \text{CH}_3\text{—CH}_2\text{—CH}_2\text{—CH}_2\text{—CH}_2\text{—CH}_3$$

neopentane   *n*-hexane

$$\text{CH}_3\text{—}\overset{\displaystyle \text{CH}_3}{\overset{\displaystyle |}{\text{CH}}}\text{—CH}_2\text{—CH}_2\text{—CH}_3 \qquad \text{CH}_3\text{—}\overset{\displaystyle \text{CH}_3}{\underset{\displaystyle \text{CH}_3}{\overset{\displaystyle |}{\underset{\displaystyle |}{\text{C}}}}}\text{—CH}_2\text{—CH}_3$$

isohexane   neohexane

This system of common names has no good way of handling other branching patterns, and for more complex alkanes it is necessary to use the more flexible, IUPAC system of nomenclature.

We will concentrate on IUPAC names. However, we will also use common names, especially when the common name is used almost exclusively in the everyday discussions of chemists. Where both IUPAC and common names are given in this textbook, we will always give the IUPAC name first, followed by the common name in parentheses. In this way, you should have no doubt about which name is which.

2.4 Cycloalkanes

So far we have considered only chains (branched and unbranched) of carbon atoms. A molecule that contains carbon atoms joined to form a ring is called a cyclic hydrocarbon. Further, when all carbons of the ring are saturated, the molecule is called a **cycloalkane**.

Cycloalkanes of ring size from three to over thirty are found in nature; and in principle, there is no limit to ring size. Five-membered rings (cyclopentanes) and six-membered rings (cyclohexanes) are especially abundant in nature and therefore have received special attention.

Figure 2.2 shows structural formulas of cyclopropane, cyclobutane, cyclopentane, and cyclohexane. For convenience, organic chemists usually do not write out structural formulas for cycloalkanes showing all carbons and hydrogens. Rather, the rings are represented by regular polygons with the same number of sides. For example, cyclopropane is represented by a triangle and cyclohexane by a hexagon.

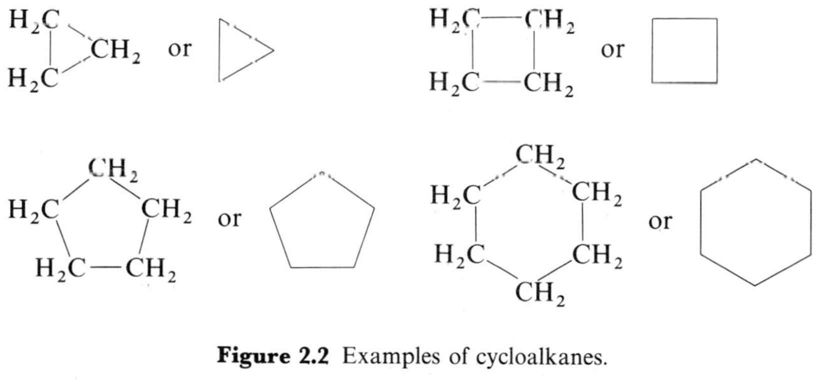

Figure 2.2 Examples of cycloalkanes.

To name cycloalkanes, prefix the name of the corresponding open-chain hydrocarbon with *cyclo-* and name each substituent on the ring. If there is only a single substituent on the cycloalkane ring, there is no need to give it a number. If there are two or more substituents, each substituent must be given a number to indicate its location on the ring.

Example 2.4

Name these cycloalkanes.

a. 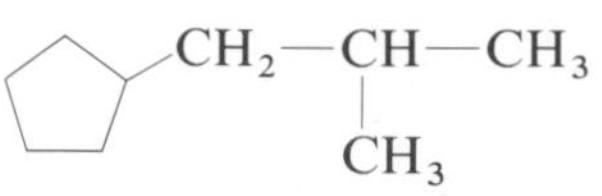 (cyclopentane ring)—CH_2—CH—CH_3 with CH_3

b. 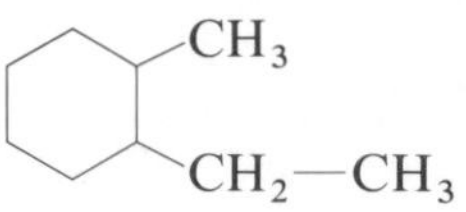 (cyclohexane ring) with CH_3 and CH_2—CH_3

☐ *Solution*

a. The ring contains five atoms and is a cyclopentane. Because only one substituent, an isobutyl group, is on the ring, there is no need to number the atoms of the ring. The IUPAC name is isobutylcyclopentane.

b. The ring is a cyclohexane. Number the atoms of the ring, beginning with the substitutent of lowest alphabetical order, in this case, *ethyl*. The IUPAC name of this cycloalkane is 1-ethyl-2-methylcyclohexane.

Problem 2.4

Name these cycloalkanes.

a. 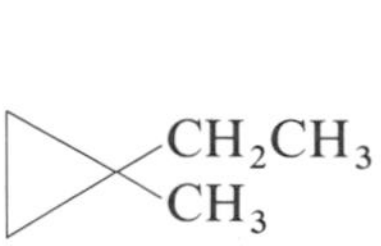 (cyclopropane ring) with CH_2CH_3 and CH_3

b. 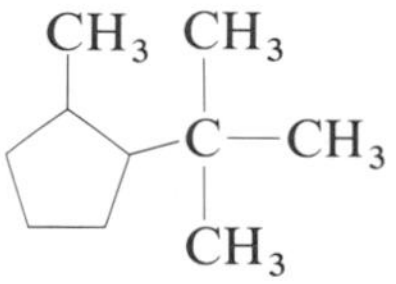 (cyclopentane ring) with CH_3 CH_3, C—CH_3, CH_3

The use of carbon bonds to close a ring means that cycloalkanes contains two fewer hydrogen atoms than an alkane of the same number of carbon atoms. Compare, for example, the molecular formulas of cyclopropane, C_3H_6, and propane, C_3H_8; or of cyclohexane, C_6H_{12}, and hexane, C_6H_{14}. The general formula of a cycloalkane is C_nH_{2n}.

2.5 The IUPAC System—A General System of Nomenclature

The naming of alkanes and cycloalkanes (Sections 2.3 and 2.4) illustrated the application of the IUPAC system of nomenclature to two specific classes of organic compounds. Now, let us describe the general approach of the IUPAC system. The name assigned to any compound with a chain of carbon atoms consists of three parts: a prefix, an infix, and a suffix. Each part provides specific information about the structural formula of the compound.

1. The prefix tells the number of carbon atoms in the parent chain. Examples are:

Prefix	Number of Carbon Atoms
but-	4
pent-	5
hex-	6

2. The infix tells the nature of the carbon-carbon bonds in the parent chain. Examples are:

Infix	Nature of Carbon-Carbon Bonds
-an-	all single bonds
-en-	one or more double bonds
-yn-	one or more triple bonds

3. The suffix tells the class of compound to which the substance belongs. Examples are:

Suffix	Class of Compound
-e	hydrocarbon
-ol	alcohol
-al	aldehyde
-one	ketone
-oic acid	carboxylic acid

Example 2.5

Following are IUPAC names and structural formulas for several compounds. Divide each name into a prefix, an infix, and a suffix and specify the information about the structural formula that is contained in each.

a. $CH_2=CH-CH_3$

propene

b. CH_3-CH_2-OH

ethanol

c. $CH_3-CH_2-\overset{\displaystyle O}{\overset{\|}{C}}-CH_3$

butanone

d. $CH_3-CH_2-CH_2-CH_2-\overset{\displaystyle O}{\overset{\|}{C}}-OH$

pentanoic acid

e.

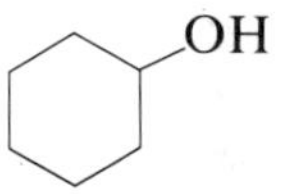

cyclohexanol

☐ *Solution*

One carbon-carbon double bond

a. →prop-en-e

Three carbon atoms   *A hydrocarbon*

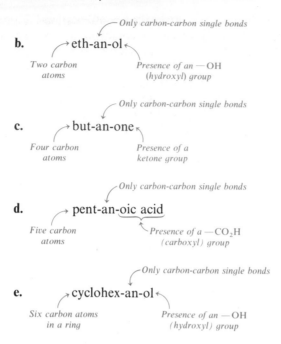

b. *Only carbon-carbon single bonds* → eth-an-ol ← *Presence of an —OH (hydroxyl) group* / *Two carbon atoms*

c. *Only carbon-carbon single bonds* → but-an-one ← *Presence of a ketone group* / *Four carbon atoms*

d. *Only carbon-carbon single bonds* → pent-an-oic acid ← *Presence of a —CO₂H (carboxyl) group* / *Five carbon atoms*

e. *Only carbon-carbon single bonds* → cyclohex-an-ol ← *Presence of an —OH (hydroxyl) group* / *Six carbon atoms in a ring*

Problem 2.5

Combine the proper prefix, infix, and suffix and write IUPAC names for the following.

a. $CH_2{=}CH_2$

b. $CH_3{-}C{\equiv}CH$

c. $CH_3{-}CH_2{-}CH_2{-}\overset{\overset{\textstyle O}{\|}}{C}{-}OH$

d. $CH_3{-}CH_2{-}CH_2{-}\overset{\overset{\textstyle O}{\|}}{C}{-}H$

e.

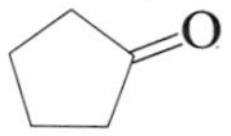

f.

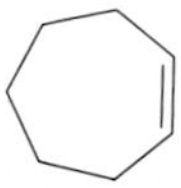

2.6 Conformation of Alkanes and Cycloalkanes

A. Alkanes

Structural formulas are useful to show the order of attachment of atoms. However, they do not show actual three-dimensional shapes. As chemists try to understand more about how structure and the chemical and physical properties of molecules are related, it becomes increasingly important to understand more about the three-dimensional shapes of molecules.

Alkanes of two or more carbons can be twisted into a number of different three-dimensional arrangements by rotation about a carbon-carbon bond or bonds. The different three-dimensional arrangements of atoms that result by

rotation about single bonds are called **conformations**. Figure 2.3(a) shows a
ball-and-stick model of a staggered conformation of ethane. In a **staggered
conformation**, all C—H bonds on adjacent carbons are as far apart as possible.
Figure 2.3(b) is a Newman projection of this conformation of ethane. In the
Newman projection shown, the molecule is viewed along the axis of the C—C
bond. The three hydrogens nearer your eye are shown on lines extending from
the center of the circle at angles of 120°. The three hydrogens of the carbon farther from your eye are shown on lines extending from the circumference of the
circle. Remember that bond angles about each carbon are 109.5° and not 120°,
as this Newman projection might suggest.

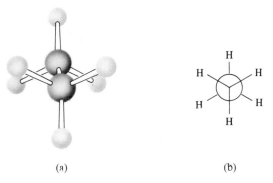

(a)   (b)

Figure 2.3 A staggered conformation of ethane. (a) Ball-and-stick model; (b) a Newman
projection.

Figures 2.4(a) and 2.4(b) show ball-and-stick models for an eclipsed conformation of ethane viewed from two different perspectives. In an **eclipsed conformation**, C—H bonds on adjacent conformations are as close together as
possible. An ethane molecule can be twisted into an infinite number of conformations between the staggered and eclipsed conformations. Yet any given
ethane molecule spends most of its time in the staggered conformation, because
in this conformation, C—H bonds on one carbon are as far apart as possible

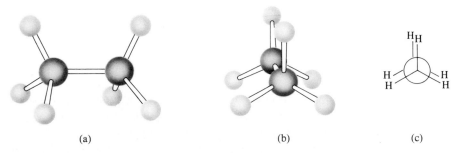

(a)   (b)   (c)

Figure 2.4 An eclipsed conformation of ethane. (a, b) Ball-and-stick models; (c) a
Newman projection.

from C—H bonds on the adjacent carbon. For this reason, the staggered conformation is the most stable or preferred conformation.

Example 2.6

Draw Newman projections for two staggered conformations of butane. Consider only conformations along the bond between carbons 2 and 3 of the butane chain. Which of these is the more stable? Which is the less stable?

☐ *Solution*

The condensed structural formula of butane is

$$\overset{1}{CH_3}-\overset{2}{CH_2}-\overset{3}{CH_2}-\overset{4}{CH_3}$$

First view the molecule along the bond between carbons 2 and 3. Then to see the possible conformations asked for, hold carbon 2 in place and rotate carbon 3 about the single bond between the two carbons. Staggered conformation (a) is the more stable because in it the two —CH_3 groups are as far apart as possible; staggered conformation (b) is less stable because the two —CH_3 groups are closer together.

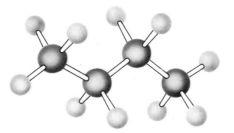

Ball-and-stick model of a staggered
conformation of butane

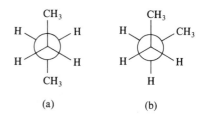

Newman projections of two staggered
conformations of butane

Problem 2.6

Draw two eclipsed conformations for butane. Consider only conformations along the bond between carbons 2 and 3. Which of these is the more stable eclipsed conformation? Which is the less stable eclipsed conformation?

B. Cycloalkanes

Figure 2.5 shows the shapes of cyclopropane, cyclobutane, and cyclopentane. For all practical purposes, these molecules are planar.

Cyclohexane and all larger rings exist in nonplanar, or puckered, conformations. Shown in Figure 2.6(b) is a ball-and-stick model of the most stable puckered conformation of cyclohexane. This conformation is called a **chair** because of its resemblance to a beach chair with a back rest, a seat, and a leg

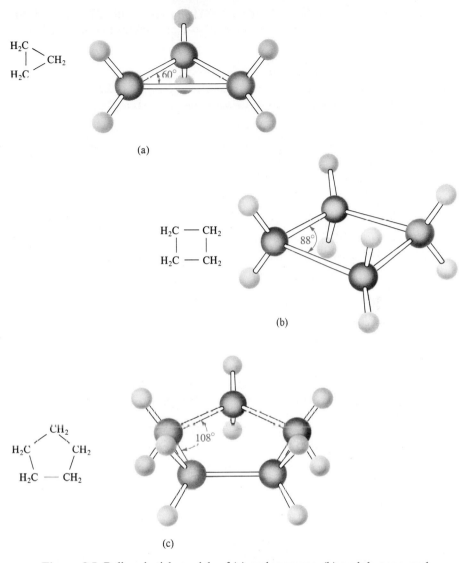

Figure 2.5 Ball-and-stick models of (a) cyclopropane, (b) cyclobutane, and (c) cyclopentane.

rest. All C—C—C bond angles in a chair conformation are approximately 109.5°.

Hydrogens in a chair conformation are in two different geometrical positions (Figure 2.6c). Six, called equatorial hydrogens, project straight out from the ring and are roughly parallel to the plane of the ring. The other six, called axial hydrogens, are perpendicular to the plane of the ring.

If a hydrogen of cyclohexane is replaced by a methyl group or other substituent, the group can occupy either an axial or an equatorial position.

CH_3 — Axial

H

H — *Equatorial*

CH_3

2.7 Configurational Isomerism in Cycloalkanes

All cycloalkanes with different substituents on two or more carbons of the ring show a type of isomerism called **configurational isomerism**. Configurational isomers have the same molecular formula, the same order of attachment of atoms, and an arrangement of atoms that cannot be interchanged by rotation about sigma bonds under ordinary conditions. By way of comparison, conformations such as those of ethane, propane, and butane can be interconverted easily by rotation about single bonds. The term **cis-trans isomerism**, or its alternative, **geometric isomerism**, is applied to the type of configurational isomerism that depends on the arrangement of substituent groups, either in a cyclic structure (as we shall see in this chapter), or on a double bond (Chapter 13).

Configurational isomerism in cyclic structures can be illustrated by models of 1,2-dimethylcyclopentane. In the following drawings, the cyclopentane ring is shown as a planar pentagon viewed through the plane of the ring. Carbon-carbon bonds of the ring projecting forward are shown as heavy lines. When the ring is viewed from this perspective, substituents attached to the ring project above and below its plane. In one configurational isomer of 1,2-dimethyl-cyclopentane, the methyl groups are on the same side of the ring; in the other, they are on opposite sides of the ring. The prefix *cis-* (Latin, on the same side) is used to indicate that the substituents are on the same side of the ring; the prefix *trans-* (Latin, across) is used to indicate that they are on opposite sides of the ring. In each isomer, the configuration of the methyl groups is fixed, and

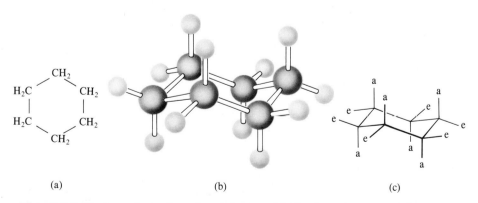

(a)   (b)   (c)

Figure 2.6 Chair conformation of cyclohexane. (a) Condensed structural formula, (b) ball-and-stick model, and (c) three-dimensional drawing with axial and equatorial positions labeled by "a" and "e," respectively.

no amount of twisting about carbon-carbon bonds can convert the cis isomer to the trans isomer or vice versa.

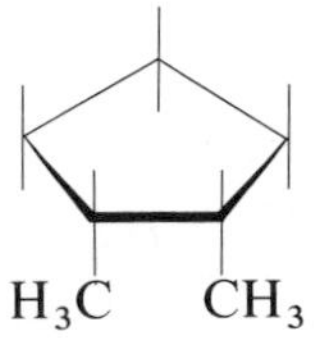

cis-1,2-dimethylcyclopentane 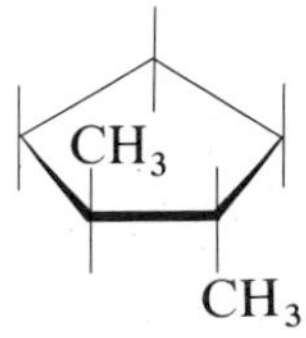   trans-1,2-dimethylcyclopentane

Cyclopropane, cyclobutane, and cyclopentane are accurately represented by planar drawings. Cyclohexanes and all larger rings are nonplanar, however, and therefore drawing them is more difficult. Cyclohexane, for example, is best represented as a nonplanar chair conformation. Fortunately, for the purpose of deciding how many cis-trans isomers are possible for a given substituted cyclo- alkane, it is adequate to draw the ring as a planar polygon, as is done below for cyclohexane. There are two cis-trans isomers for 1,4-dimethylcyclohexane.

trans-1,4-dimethylcyclohexane   cis-1,4-dimethylcyclohexane

Example 2.7

Following are several cycloalkanes of molecular formula C_6H_{12}. State which show cis-trans isomerism, and for each that does, draw the cis and trans isomers.

a. 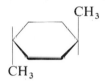 **b.** 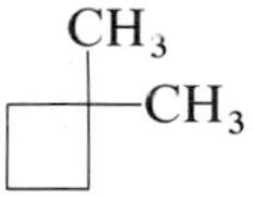 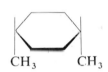 **c.**

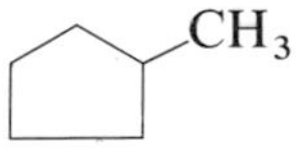

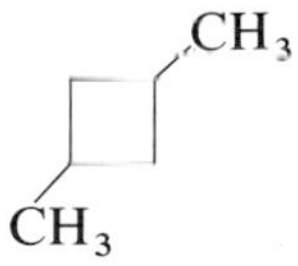

Solution

a. Because methylcyclopentane has only one substituent on the ring, it does not show cis-trans isomerism.

b. 1,1-dimethylcyclobutane does not show cis-trans isomerism. There is only one possible arrangement for the two methyl groups on the ring; they must be trans to each other.

c. 1,3-dimethylcyclobutane shows cis-trans isomerism.

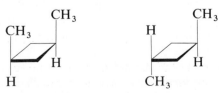

cis-1,3-dimethylcyclobutane   trans-1,3-dimethylcyclobutane

Problem 2.7

Following are several cycloalkanes of molecular formula C_7H_{14}. State which show cis-trans isomerism and for each that does, draw the cis and trans isomers.

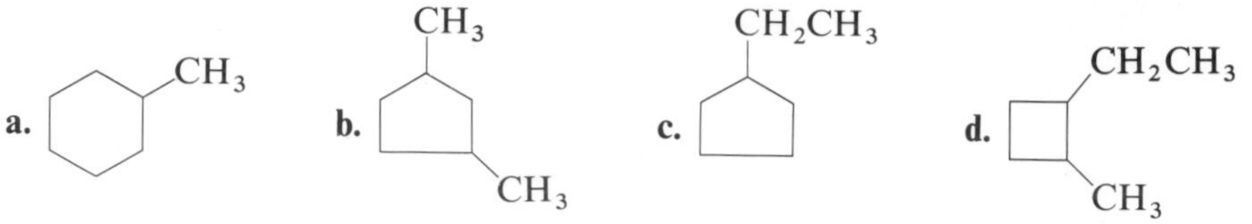

a. (cyclohexane with CH_3)   b. (cyclopentane with CH_3 and CH_3)   c. (cyclopentane with CH_2CH_3)   d. (cyclobutane with CH_2CH_3 and CH_3)

2.8 Physical Properties of Alkanes and Cycloalkanes

You are already familiar with the physical properties of some alkanes and cycloalkanes from your everyday experiences. The low-molecular-weight alkanes, such as methane (marsh gas), ethane, propane, and butane, are gases at room temperature and atmospheric pressure. Higher-molecular-weight alkanes, such as those in gasoline and kerosene, are liquids. Very high-molecular-weight alkanes, such as those found in paraffin wax, are solids. Melting points, boiling points, and densities of the first ten alkanes are listed in Table 2.4.

Table 2.4
Physical properties of some alkanes.

Name	Condensed Structural Formula	mp (°C)	bp (°C)	Density of Liquid (g/mL at 0°C)
methane	CH_4	-182	-164	(a gas)
ethane	CH_3CH_3	-183	-88	(a gas)
propane	$CH_3CH_2CH_3$	-190	-42	(a gas)
butane	$CH_3(CH_2)_2CH_3$	-138	0	(a gas)
pentane	$CH_3(CH_2)_3CH_3$	-130	36	0.626
hexane	$CH_3(CH_2)_4CH_3$	-95	69	0.659
heptane	$CH_3(CH_2)_5CH_3$	-90	98	0.684
octane	$CH_3(CH_2)_6CH_3$	-57	126	0.703
nonane	$CH_3(CH_2)_7CH_3$	-51	151	0.718
decane	$CH_3(CH_2)_8CH_3$	-30	174	0.730

Methane, the lowest-molecular-weight alkane, is a gas at room temperature and atmospheric pressure. It can be converted to a liquid if cooled to $-164°C$ and to a solid if cooled to $-182°C$. That methane (or any other compound, for that matter) can exist as a liquid or a solid depends on the existence of **intermolecular forces** of attraction between particles of each pure compound. Although the forces of attraction between particles are all electrostatic, they vary widely in relative strength. The main types of intermolecular forces we deal with in this book are listed in Table 2.5. Note that both ion-ion and ion-dipole forces involve ions rather than molecules, and in this regard, the term *intermolecular* is not strictly accurate for these forces. Nonetheless, they are

Table 2.5
Summary of types of intermolecular forces.

Type of Interaction	Where Found	Example	Relative Strength
ion-ion	ionic solids	$Na^+\text{------}Cl^-$	strongest type of interaction
ion-dipole	solutions of ionic solids	$Na^+\text{------}\overset{\delta-}{O}\diagup^H_{\diagdown H}$	
dipole-dipole	hydrogen bonding	$\overset{H}{\diagdown}O\text{---}\overset{\delta+}{H}\text{------}\overset{\delta-}{O}\overset{\diagup H}{\diagdown H}$	
dispersion	nonpolar molecules	$CH_4\text{------}CH_4$	weakest type of interaction

grouped with dipole-dipole interactions and dispersion forces for purposes of comparison.

Let us use these concepts of the nature of intermolecular forces to examine the relation between the physical properties of alkanes and their molecular structure. Alkanes are nonpolar compounds and the only forces of attraction between them are dispersion forces. Because interactions between molecules of alkanes are so weak, the alkanes have lower boiling points than almost any other type of compound of the same molecular weight. As the number of atoms and molecular weight of an alkane increase, the strength of dispersion forces per molecule also increases. Therefore, the boiling points of alkanes increase as molecular weight increases (compare Table 2.4).

Melting points of alkanes also increase with increasing molecular weight. However, the increase is not as regular as that observed for boiling points. The average density of the alkanes listed in Table 2.4 is about 0.7 g/mL; the density of higher-molecular-weight alkanes is about 0.8 g/mL. All liquid alkanes are less dense than water (1.0 g/mL).

An observation you have certainly heard is "oil and water do not mix." For example, gasoline and crude oil do not dissolve in water. We refer to these and other water-insoluble compounds as hydrophobic (water-hating). Alkanes do dissolve in nonpolar liquids such as carbon tetrachloride, CCl_4, carbon disulfide, CS_2, and of course other hydrocarbons.

Alkanes that are structural isomers are different compounds and have different physical and chemical properties. Table 2.6 lists the boiling points, melting points, and densities of the five structural isomers of molecular formula C_6H_{14}. The boiling point of each of the branched-chain isomers of C_6H_{14} is lower than that of hexane itself, and the more branching there is, the lower the boiling point. These differences in boiling point are related to molecular shape in the following way. The only forces of attraction between alkane molecules are dispersion forces. As branching increases, the shape of an alkane molecule becomes more compact and its surface area decreases. As surface area decreases, contact between adjacent molecules decreases, the strength of dispersion forces

Table 2.6
Physical
properties of
the isomeric
alkanes of
molecular formula
C_6H_{14}.

Name	bp (°C)	mp (°C)	Density (g/mL)
hexane	68.7	−95	0.659
2-methylpentane	60.3	−154	0.653
3-methylpentane	63.3	−118	0.664
2,3-dimethylbutane	58.0	−129	0.661
2,2-dimethylbutane	49.7	−98	0.649

decreases, and boiling points also decrease. For any group of alkane structural isomers, it is usually observed that the least branched isomer has the highest boiling point and the most branched isomer has the lowest boiling point. Figure 2.7 shows structural formulas and ball-and-stick models of hexane (a

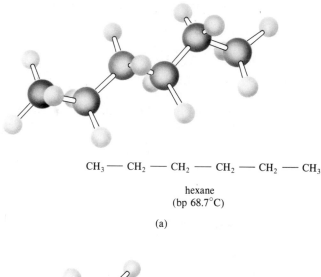

$$CH_3 \!-\! CH_2 \!-\! CH_2 \!-\! CH_2 \!-\! CH_2 \!-\! CH_3$$

hexane
(bp 68.7°C)

(a)

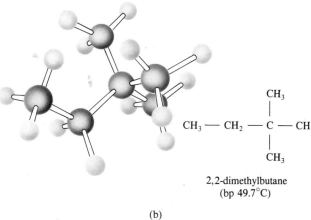

$$CH_3 \!-\! CH_2 \!-\! \underset{\underset{CH_3}{|}}{\overset{\overset{CH_3}{|}}{C}} \!-\! CH_3$$

2,2-dimethylbutane
(bp 49.7°C)

(b)

Figure 2.7 (a) Ball-and-stick models of hexane (fully staggered conformation), and (b) 2,2-dimethylbutane.

fully staggered conformation) and its most highly branched structural isomer, 2,2-dimethylbutane.

Example 2.8

Arrange the following in order of increasing boiling point.

a. $CH_3CH_2CH_2CH_3$ $CH_3CH_2CH_2CH_2CH_2CH_3$ $CH_3(CH_2)_8CH_3$

b. $CH_3(CH_2)_6CH_3$
$$CH_3\overset{\overset{\displaystyle CH_3}{|}}{C}CH_2\overset{\overset{\displaystyle CH_3}{|}}{C}HCH_3$$
$$\underset{\displaystyle CH_3}{|}$$
$$CH_3\overset{}{C}H(CH_2)_4CH_3$$
$$\underset{\displaystyle CH_3}{|}$$

Solution

a. All are unbranched alkanes. As the number of carbon atoms in the chain increases, dispersion forces between molecules increase and boiling points increase. Predict that decane has the highest boiling point and that butane has the lowest.

$CH_3CH_2CH_2CH_3$ $CH_3CH_2CH_2CH_2CH_2CH_3$ $CH_3(CH_2)_8CH_3$

butane   hexane   decane
(bp −0.5°C)   (bp 69°C)   (bp 174°C)

b. These three alkanes have the same molecular formula, C_8H_{18}, and are structural isomers of each other. Their relative boiling points depend on the degree of branching. Predict that 2,2,4-trimethylpentane, the most highly branched isomer, has the lowest boiling point, and octane, the unbranched isomer, the highest boiling point.

$$CH_3\overset{\overset{\displaystyle CH_3}{|}}{\underset{\underset{\displaystyle CH_3}{|}}{C}}CH_2\overset{\overset{\displaystyle CH_3}{|}}{C}HCH_3 \qquad CH_3\overset{\overset{\displaystyle CH_3}{|}}{C}HCH_2CH_2CH_2CH_2CH_3 \qquad CH_3(CH_3)_6CH_3$$

2,2,4-trimethylpentane   2-methylheptane   octane
(bp 99°C)   (bp 118°C)   (bp 126°C)

Problem 2.8

Arrange the following in order of increasing boiling point.

a. 2-methylbutane, 2,2-dimethylpropane, pentane
b. 3,3-dimethylheptane, 2,2,4-trimethylpentane, nonane

2.9 Reactions of Alkanes

Alkanes and cycloalkanes are unreactive to most reagents, a behavior consistent with the facts that they are nonpolar compounds and composed entirely of strong sigma bonds. However, they do react under certain conditions with oxygen and with halogens.

A. Oxidation

By far the most economically important reaction of alkanes is their **oxidation (combustion)** by O_2 to form carbon dioxide and water. Oxidation of saturated hydrocarbons is the basis for their use as energy sources for heat (natural gas, LPG, and fuel oil) and power (gasoline, diesel fuel, and aviation fuel). Following are balanced equations for complete oxidation of methane, the major component of natural gas, and 2,2,4-trimethylpentane, a component of gasoline. Also given is the heat of combustion, ΔH^0, for oxidation of one mole of each compound at 25°C, to give gaseous carbon dioxide and liquid water.

$$CH_4 + 2O_2 \longrightarrow CO_2 + 2H_2O \qquad \Delta H^0 = -212 \text{ kcal/mol}$$

methane

$$\underset{\substack{| \\ CH_3}}{\overset{\substack{CH_3 \quad CH_3 \\ | \qquad |}}{CH_3CCH_2CHCH_3}} + \frac{25}{2}O_2 \longrightarrow 8CO_2 + 9H_2O \qquad \Delta H^0 = -1304 \text{ kcal/mol}$$

2,2,4-trimethylpentane
(isooctane)

B. Halogenation

If a mixture of methane and chlorine gas is kept in the dark at room temperature, no detectable change occurs. If, however, the mixture is heated or exposed to visible or ultraviolet light, a reaction begins almost at once with the evolution of heat. The products are chloromethane (methyl choride) and hydrogen chloride. What occurs is a substitution reaction, in this case substitution of a hydrogen atom in methane by a chlorine atom and the production of an equivalent amount of hydrogen chloride.

$$\underset{\substack{| \\ H}}{\overset{\substack{H \\ |}}{H-C-H}} + Cl-Cl \xrightarrow[\text{heat}]{\text{light or}} \underset{\substack{| \\ H}}{\overset{\substack{H \\ |}}{H-C-Cl}} + H-Cl$$

methane   chloromethane
(methyl chloride)

Substitution is defined as a reaction in which an atom or group of atoms in a compound is replaced by another atom or group of atoms.

The IUPAC names for haloalkanes are derived according to the rules we have already used for naming alkanes. The parent chain is numbered from the direction that gives the substituent encountered first (whether it is halogen or an alkyl group) the lowest number. Halogen substituents are indicated by the prefixes *fluoro-*, *chloro-*, *bromo-*, and *iodo-*. In common names, the carbon chain is named followed by the word for the name of the halide, as in *methyl chloride*.

If chloromethane is allowed to react with more chlorine, further chlorination produces a mixture of dichloromethane (methylene chloride), trichloromethane (chloroform), and tetrachloromethane (carbon tetrachloride):

$$CH_3Cl + Cl_2 \longrightarrow CH_2Cl_2 + HCl$$

dichloromethane
(methylene chloride)

$$CH_2Cl_2 + Cl_2 \longrightarrow CHCl_3 + HCl$$

trichloromethane
(chloroform)

$$CHCl_3 + Cl_2 \longrightarrow CCl_4 + HCl$$

tetrachloromethane
(carbon tetrachloride)

Reaction of ethane with chlorine gives chloroethane (ethyl chloride); reaction with bromine gives bromoethane (ethyl bromide).

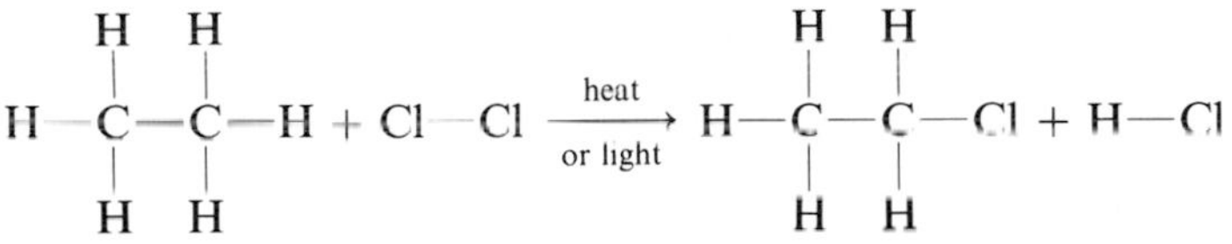

ethane

chloroethane
(ethyl chloride)

Reaction of propane with bromine gives two isomeric bromopropanes:

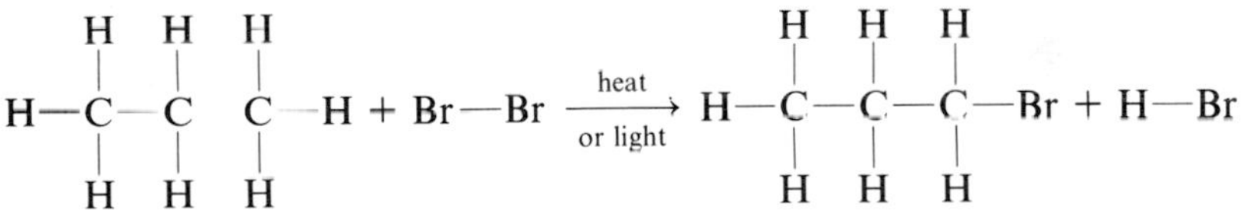

propane

1-bromopropane
(*n*-propyl bromide)

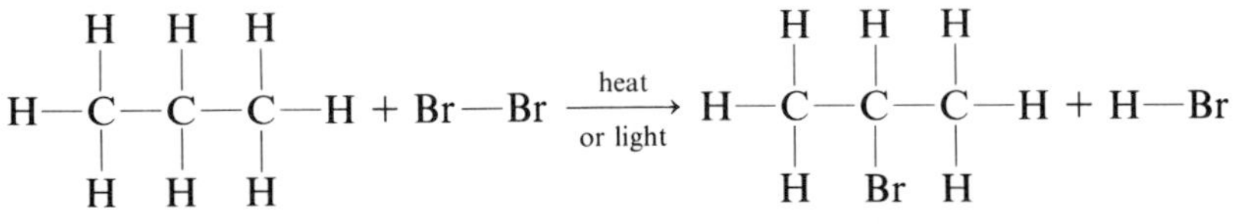

propane

2-bromopropane
(isopropyl bromide)

These are different reactions and are written separately. However, it is more common to show all products formed in one equation. The single equation below means that reaction of one mole of propane with one mole of bromine produces one mole of HBr and one mole of C_3H_7Br. Of the C_3H_7Br, approximately 8% is 1-bromopropane and the remaining 92% is 2-bromopropane.

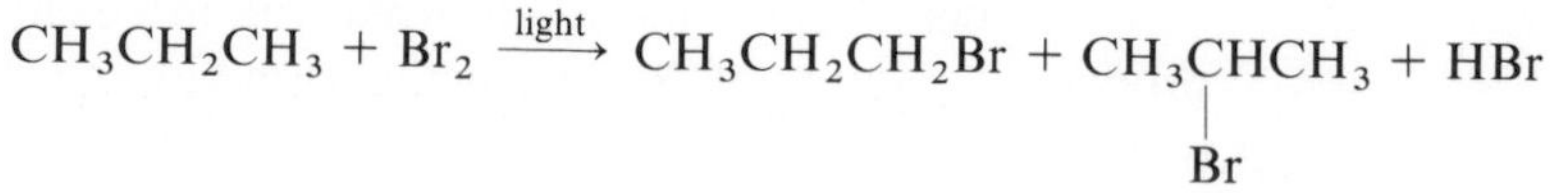

$$\text{CH}_3\text{CH}_2\text{CH}_3 + \text{Br}_2 \xrightarrow{\text{light}} \text{CH}_3\text{CH}_2\text{CH}_2\text{Br} + \underset{\underset{\text{Br}}{|}}{\text{CH}_3\text{CHCH}_3} + \text{HBr}$$

propane   1-bromopropane   2-bromopropane
(8%)   (92%)

Example 2.9

Draw structural formulas for all monohalogenation products formed in the following reactions. Give each product an IUPAC name and where possible, also a common name.

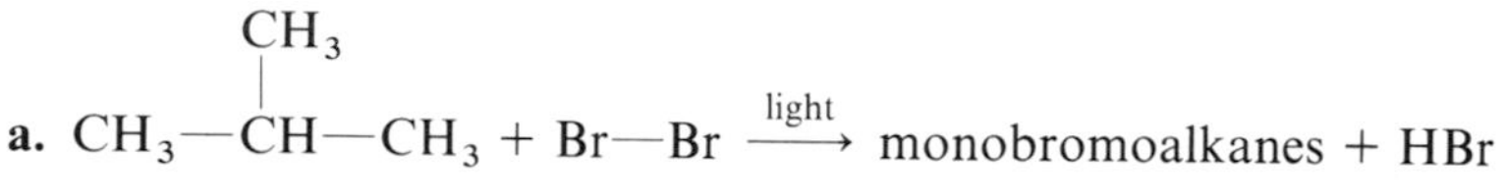

a. $\text{CH}_3\text{—}\underset{\underset{\text{CH}_3}{|}}{\text{CH}}\text{—CH}_3 + \text{Br—Br} \xrightarrow{\text{light}}$ monobromoalkanes + HBr

2-methylpropane
(isobutane)

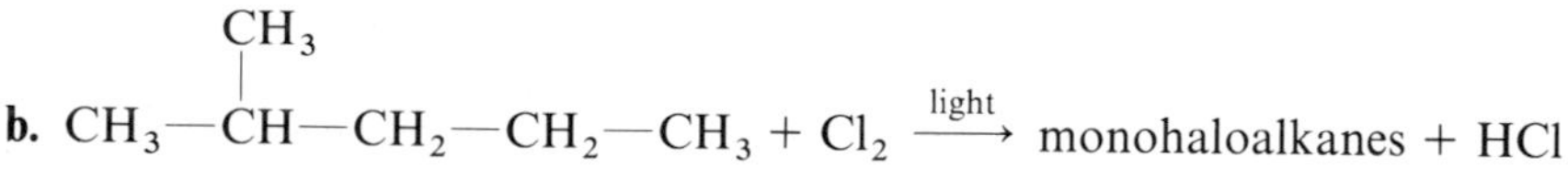

b. $\text{CH}_3\text{—}\underset{\underset{\text{CH}_3}{|}}{\text{CH}}\text{—CH}_2\text{—CH}_2\text{—CH}_3 + \text{Cl}_2 \xrightarrow{\text{light}}$ monohaloalkanes + HCl

2-methylpentane
(isohexane)

☐ Solution

a. As we did when drawing all possible structural isomers of a given molecular formula, it is best to devise a system and then follow it. The most direct way is to start at one end of the carbon chain and substitute —Br for —H. Then do the same thing on each carbon until you come to the other end of the chain. There are only two monobromination products from 2-methylpropane. The IUPAC names are given first and then common names in parentheses.

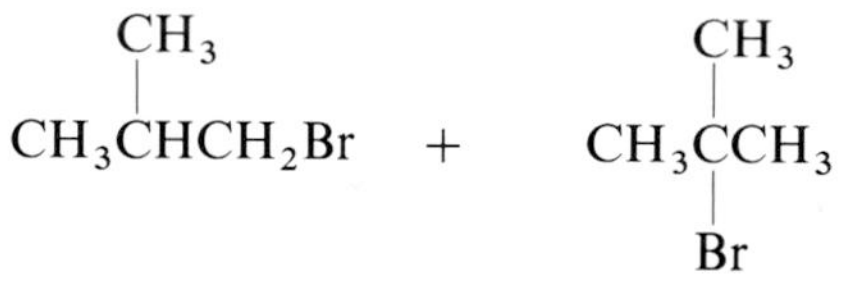

$\underset{\underset{\text{CH}_3}{|}}{\text{CH}_3\text{CHCH}_2\text{Br}}$ + $\underset{\underset{\text{Br}}{|}}{\overset{\overset{\text{CH}_3}{|}}{\text{CH}_3\text{CCH}_3}}$

1-bromo-2-methylpropane   2-bromo-2-methylpropane
(isobutyl bromide)   (*tert*-butyl bromide)

b. There are five monohaloalkanes possible from 2-methylpentane. The IUPAC names for each are given. No convenient common names exist for these compounds.

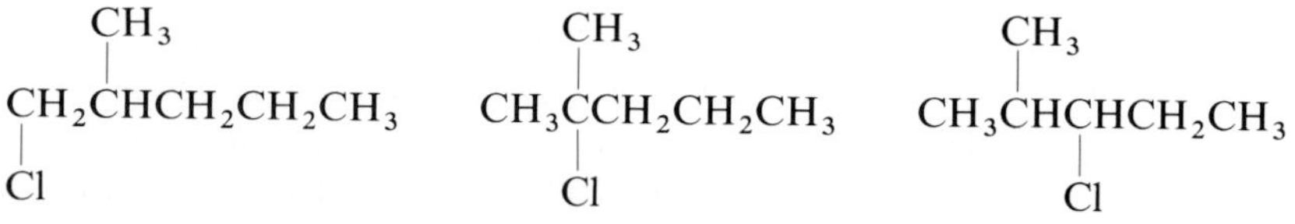

$\underset{\underset{\text{Cl}}{|}}{\overset{\overset{\text{CH}_3}{|}}{\text{CH}_2\text{CHCH}_2\text{CH}_2\text{CH}_3}}$ $\underset{\underset{\text{Cl}}{|}}{\overset{\overset{\text{CH}_3}{|}}{\text{CH}_3\text{CCH}_2\text{CH}_2\text{CH}_3}}$ $\underset{\underset{\text{Cl}}{|}}{\overset{\overset{\text{CH}_3}{|}}{\text{CH}_3\text{CHCHCH}_2\text{CH}_3}}$

1-chloro-2-methylpentane   2-chloro-2-methylpentane   3-chloro-2-methylpentane

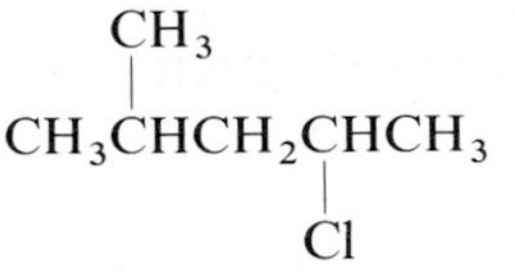

$$CH_3$$
$$|$$
$$CH_3CHCH_2CHCH_3$$
$$|$$
$$Cl$$

2-chloro-4-methylpentane

$$CH_3$$
$$|$$
$$CH_3CHCH_2CH_2CH_2$$
$$|$$
$$Cl$$

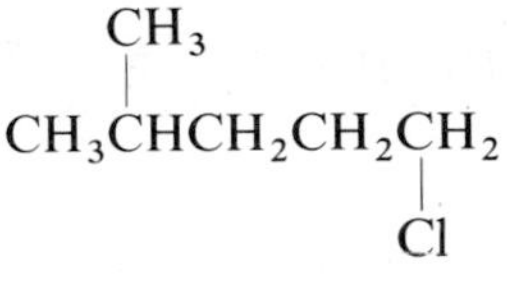

1-chloro-4-methylpentane

■

Problem 2.9

Draw structural formulas for all monohalogenation products of the following reactions. Give each product an IUPAC name and where possible, a common name.

$$CH_3$$
$$|$$
a. 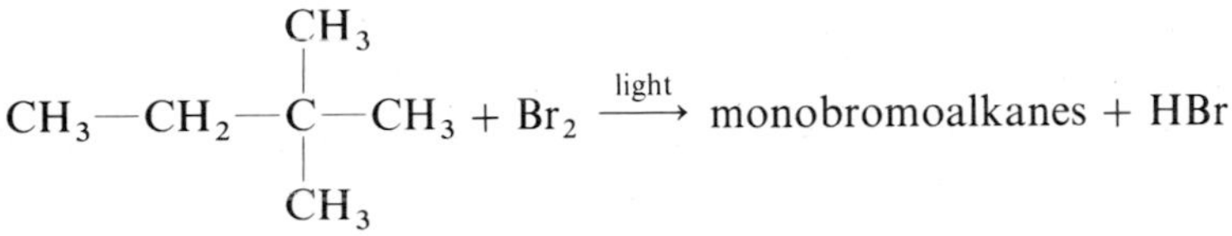 $CH_3—CH_2—C—CH_3 + Br_2 \xrightarrow{\text{light}}$ monobromoalkanes + HBr
$$|$$
$$CH_3$$

b. 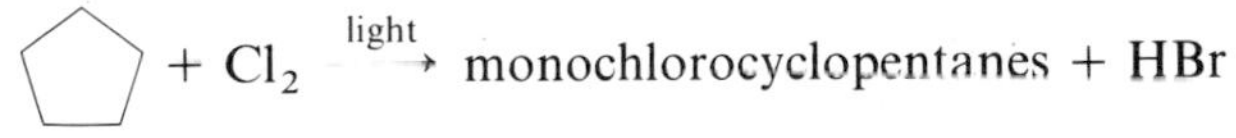 ⬠ $+ Cl_2 \xrightarrow{\text{light}}$ monochlorocyclopentanes + HBr

■

Many halogenated hydrocarbons, because of their physical and chemical properties, have found wide commercial use as solvents, refrigerants, dry-cleaning agents, local and inhalation anesthetics, and insecticides. Of the halo-alkanes, the one most widely used as a solvent today is dichloromethane (methylene chloride)

Chloroethane (ethyl chloride) is used as a fast-acting topical anesthetic. This chloroalkane owes its anesthetic property more to its physical properties than to its chemical properties. Chloroethane boils at 12°C and, unless under pressure, is a gas at room temperature. When sprayed on the skin, it evaporates and cools the skin surface and underlying nerve endings. Skin and underlying nerve endings become anesthetized when skin temperature drops to about 15°C. Halothane is a widely used inhalation anesthetic.

$$F \quad H$$
$$| \quad |$$
$$F—C—C—Br$$
$$| \quad |$$
$$F \quad Cl$$

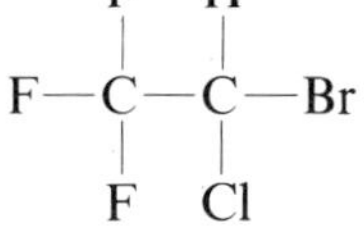

2-bromo-2-chloro-1,1,1-trifluoroethane
(Halothane)

Of all the fluoroalkanes, those manufactured under the trade name Freon are the most widely used. **Freons** were developed in a search for new refrigerants, compounds that would be nontoxic, nonflammable, odorless, and noncorrosive. In 1930, General Motors Corp. announced the discovery of just such a compound, dichlorodifluoromethane, which was marketed under the trade name Freon-12. Freons are a class of compounds manufactured by reacting a chlorinated hydrocarbon with hydrofluoric acid in the presence of an antimony pentafluoride or pentachloride catalyst:

$$CCl_4 + HF \xrightarrow{\text{SbF}_5} CCl_3F + HCl$$
Freon-11

$$CCl_3F + HF \xrightarrow{\text{SbF}_5} CCl_2F_2 + HCl$$
Freon-12

By 1974, U.S. production of Freons had grown to more than 1.1 billion pounds annually, almost one-half of world production. Worldwide production of these compounds is now at an all-time high, reflecting their use in refrigeration and air-conditioning systems, in aerosols (outside of the United States), and as solvents.

Concern about the environmental effect of Freons arose in 1974 when Drs. Sherwood Rowland and Marion Molina proposed that when Freons are used in aerosols, they escape through the lower atmosphere to the stratosphere. There they absorb ultraviolet radiation from the sun, decompose, and set up a chemical reaction that may also lead to destruction of the stratospheric ozone layer that shields the earth from excess ultraviolet radiation. An increase in ultraviolet radiation reaching the earth may lead to destruction of certain crops and agricultural species, and to increased skin cancer in sensitive individuals. Controversy continues over the potential for ozone depletion and its effect on the environment. In the meantime, both government and the chemical industry in the United States have taken steps to limit sharply the use of Freons.

2.10 Sources of Alkanes

A. Natural Gas

The two main sources of alkanes throughout the world are natural gas and petroleum. **Natural gas** consists of approximately 80% methane, 10% ethane, and a mixture of other relatively low-boiling alkanes—chiefly propane, butane, and 2-methylpropane (isobutane).

B. Petroleum

Petroleum is a liquid mixture of literally thousands of compounds, most of them hydrocarbons, formed from the decomposition of marine plants and animals. Petroleum and petroleum-derived products fuel automobiles, aircraft, and trains. They provide most of the greases and lubricants required for the machinery of our highly industrialized society. Furthermore, petroleum, along with natural gas, provides close to 90% of the organic raw materials for the manufacture of synthetic fibers, plastics, detergents, drugs, dyes, and a multitude of other products.

The task of the petroleum refinery industry is to produce usable products, with a minimum of waste, from the thousands of different hydrocarbons in this liquid mixture. The various physical and chemical processes for this purpose fall into two broad categories: separation processes, which separate the complex mixture into various fractions, and conversion processes, which alter the molecular structure of the hydrocarbon components themselves.

The fundamental separation in refining petroleum is distillation. Practically all crude oil that enters a refinery goes to distillation units, where it is heated to temperatures as high as 370°C to 425°C and separated into fractions. Each fraction contains a mixture of hydrocarbons that boils within a particular range. Following are the common names associated with several of these fractions along with the principal uses of each.

1. Gases boiling below 20°C are taken off at the top of the distillation column. This fraction is a mixture of low-molecular-weight hydrocarbons, predominantly propane, butane, and 2-methylpropane, substances that can be liquefied under pressure at room temperature. The liquefied mixture, known as **liquefied petroleum gas (LPG)**, can be stored and shipped in metal tanks and is a convenient source of gaseous fuel for home heating and cooking.

2. Naphthas, bp 20–200°C, are a mixture of C_4 to C_{10} alkanes and cycloalkanes. Naphthas also contain some aromatic hydrocarbons such as benzene, toluene, and xylene (Section 3.9). The light naphtha fraction, bp 20–150°C, is the source of what is known as straight-run gasoline and averages approximately 25% of crude petroleum. In a sense, naphthas are the most valuable distillation fractions because they are useful not only as fuel but also as sources of raw materials for the organic chemical industry.

3. Kerosene, bp 175–275°C, is a mixture of C_9 to C_{15} hydrocarbons.

4. Gas oil, bp 200–400°C, is a mixture of C_{15} to C_{25} hydrocarbons. Diesel fuel is obtained from this fraction.

5. Lubrication oil and heavy fuel oil distill from the column at temperatures over 350°C.

6. Asphalt is the name given to the black, tarry residue remaining after removal of the other volatile fractions.

Gasoline is a complex mixture of C_4 to C_{10} hydrocarbons. The quality of gasoline as a fuel for internal combustion engines is expressed by **octane number**, or antiknock index. When an engine is running normally, the air-fuel mixture is ignited by a spark plug and burns smoothly as the flame moves outward from the plug, building up pressure that forces the piston down during the compression stroke. Engine knocking occurs when a portion of the air-fuel mixture explodes prematurely (usually as a result of heat developed during compression), and independently of ignition by the spark plug. The basic procedure for measuring the antiknock quality of a gasoline was established in 1929. Two compounds were selected as reference fuels. One of these, 2,2,4-trimethylpentane (isooctane), has very good antiknock properties and was assigned an octane number of 100. The other, heptane, has very poor antiknock properties and was assigned an octane number of 0. The octane rating of a particular gasoline is that percentage of isooctane in a mixture of isooctane and heptane that has equivalent knock properties. For example, 2-methylhexane has the same knock properties as a mixture of 42% isooctane and 58% heptane; therefore, the octane rating of 2-methylhexane is 42. Octane itself has an octane rating of -20, which means that it produces even more engine knocking than heptane.

The antiknock properties of tetraethyl lead, $(CH_3CH_2)_4Pb$, and related compounds were discovered by Thomas Midgley in the 1930s. Gasoline so treated is known as ethyl, or **leaded gasoline**. Addition of 3 g of tetraethyl lead to one gallon of gasoline raises the octane rating by 15 to 20 units. At one time, regular gasoline contained 2.4 to 3.2 g of tetraethyl lead per gallon and premium gasoline contained up to 4.2 g per gallon.

The use of leaded gasoline in the United States reached a peak in 1970, when approximately 99% of all gasoline contained lead additives. In that year, almost 280,000 tons of lead was consumed in the manufacture of tetraethyl lead. This tremendous quantity of lead was emitted directly into the atmosphere, much of it as very small particles, or aerosols. Because lead is toxic, there has been great concern over the long-term effects of spewing this waste into the environment. One result of this concern has been legislation designed to restrict the use of lead additives. There are also serious research efforts to discover substances that will increase octane ratings but not simultaneously pollute our air with dangerous byproducts. One of the most promising additives is *tert*-butyl methyl ether (Chapter 4).

A second way of increasing the octane rating of hydrocarbons derived from petroleum is catalytic cracking. Catalytic cracking accomplishes two things. First, hydrocarbons of high molecular weight are broken or cracked into hydrocarbons of smaller molecular weight, thus increasing the yield of gasoline from crude oil. Second, carbon skeletons of the hydrocarbons themselves are rearranged and made more highly branched. Branched-chain hydrocarbons have higher octane numbers than linear hydrocarbons (compare, for example, the octane ratings of octane and its structural isomer 2,2,4-trimethylpentane). The first commercial catalytic cracking plant was built in 1937 by Sun Oil. By 1976, the United States produced over 150 million gallons of feedstock a day.

Catalytic re-forming supplements catalytic cracking as a means of converting low-octane components to higher octane fuels. Catalytic re-forming, for example, converts hexane to cyclohexane and then to benzene, a high-octane unsaturated hydrocarbon:

$$CH_3CH_2CH_2CH_2CH_2CH_3 \xrightarrow[-H_2]{\text{catalyst}} \bigcirc \xrightarrow[-3H_2]{\text{catalyst}} \bigcirc$$

hexane   cyclohexane   benzene

The first catalytic re-forming process came into use in 1940 and used a silica-molybdena catalyst in the presence of hydrogen gas. In 1949, Universal Oil Products introduced a platinum catalyst. This process, called Platforming, is extremely effective and remains the dominant catalytic re-forming process in use today. The petroleum industry now uses catalytic re-forming to treat more than

120 million gallons of feedstock per day, or close to 25% of the crude oil that enters refineries.

Key Terms and Concepts

alkane (introduction)
alkyl group (2.3A)
chair conformation (2.6B)
cis-trans isomerism in cycloalkanes (2.7)
combustion of alkanes (2.9A)
configurational isomerism (2.7)
conformation (2.6A)
cycloalkane (2.4)
eclipsed conformation (2.6A)
Freons (2.9B)
gasoline (2.10B)
geometric isomerisms (2.7)
halogenation of alkanes (2.9B)
hydrocarbons (introduction)
intermolecular forces (2.8)
iso- (2.3B)
IUPAC system of nomenclature (2.3A and 2.5)

leaded gasoline (2.10B)
liquefied petroleum gas, LPG (2.10B)
methylene group (2.1)
n- (2.3B)
naphthas (2.10B)
natural gas (2.10A)
Newman projection (2.6A)
normal (2.3B)
octane number (2.10B)
oxidation of alkanes (2.9A)
petroleum (2.10B)
saturated hydrocarbon (introduction)
staggered conformation (2.6A)
structural isomers (2.2)
unsaturated hydrocarbon (introduction)

Key Reactions

1. Oxidation/combustion of alkanes with oxygen to give carbon dioxide and water (Section 2.9A).
2. Halogenation of alkanes with bromine or chlorine to give bromo- or chloro-alkanes (Section 2.9B).

Problems

Structural isomerism in alkanes and cycloalkanes (Sections 2.2 and 2.4)

2.10   Are the following molecules pairs of structural isomers or are they identical?

a. 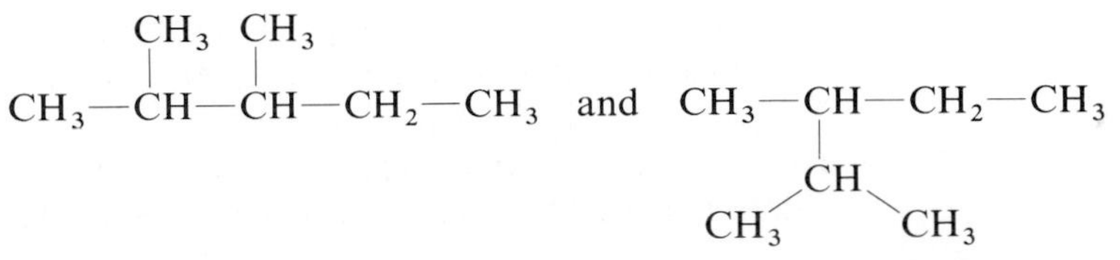 $CH_3-CH-CH-CH_2-CH_3$ (with CH_3 and CH_3 substituents) and $CH_3-CH-CH_2-CH_3$ (with CH bearing CH_3 and CH_3)

b. 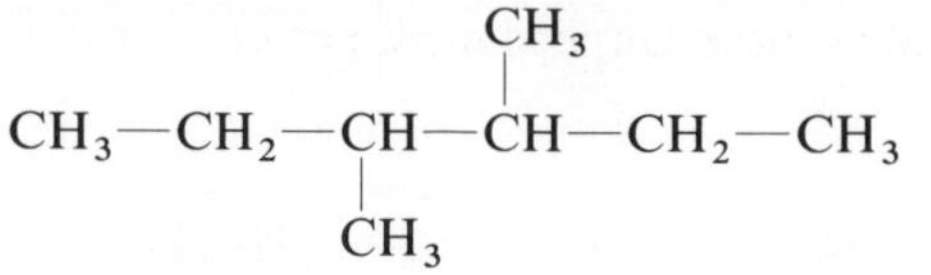
$$CH_3-CH_2-\underset{\underset{CH_3}{|}}{\overset{\overset{CH_3}{|}}{CH}}-CH-CH_2-CH_3 \quad \text{and}$$

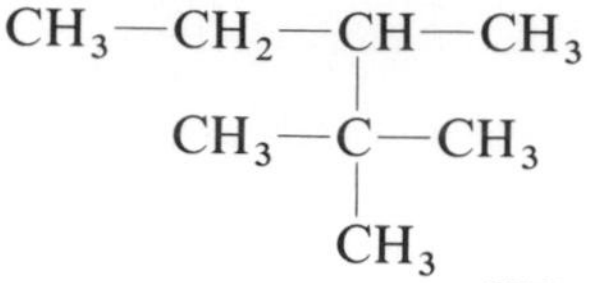

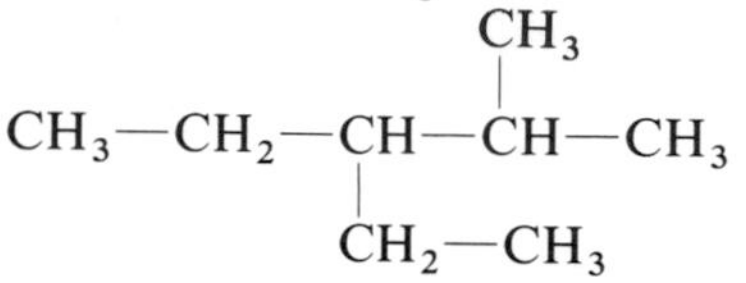

$$CH_3-CH_2-\underset{\underset{CH_3-\underset{\underset{CH_3}{|}}{\overset{|}{C}}-CH_3}{|}}{CH}-CH_3$$

c.
$$CH_3-CH_2-\underset{\underset{CH_2-CH_3}{|}}{\overset{\overset{CH_3}{|}}{CH}}-CH-CH_3 \quad \text{and}$$

$$\underset{CH_3}{\overset{CH_3}{\diagdown}}CH-CH\underset{\diagdown CH_2-CH_3}{\overset{\diagup CH_2-CH_3}{}}$$

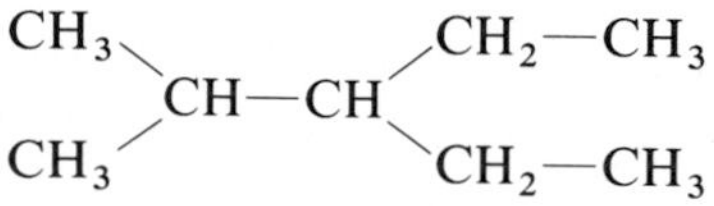

2.11 Which of the following are identical compounds and which are structural isomers?

a. 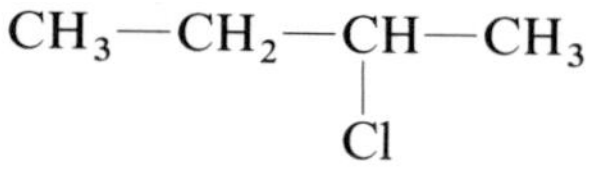
$$CH_3-CH_2-\underset{\underset{Cl}{|}}{CH}-CH_3$$

b. 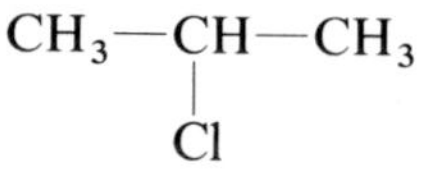
$$CH_3-\underset{\underset{Cl}{|}}{CH}-CH_3$$

c. 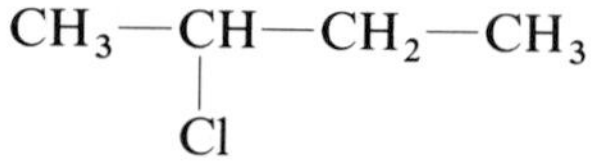
$$CH_3-\underset{\underset{Cl}{|}}{CH}-CH_2-CH_3$$

d. 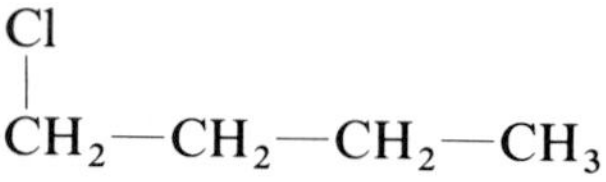
$$\underset{\underset{}{}}{\overset{\overset{Cl}{|}}{CH_2}}-CH_2-CH_2-CH_3$$

e. 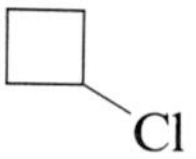
with Cl

f. 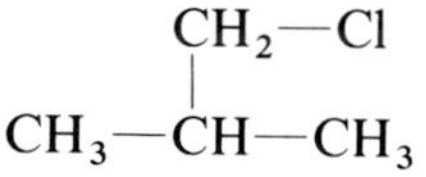
$$CH_3-\underset{\underset{}{}}{\overset{\overset{CH_2-Cl}{|}}{CH}}-CH_3$$

g. $CH_3-CH_2-CH_2-CH_2-Cl$

h. 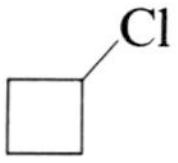
with Cl

i. 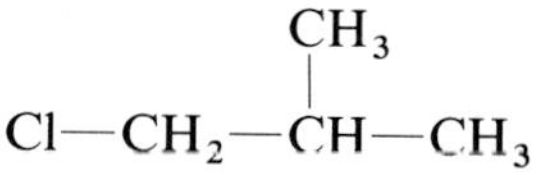
$$Cl-CH_2-\underset{\underset{}{}}{\overset{\overset{CH_3}{|}}{CH}}-CH_3$$

j. 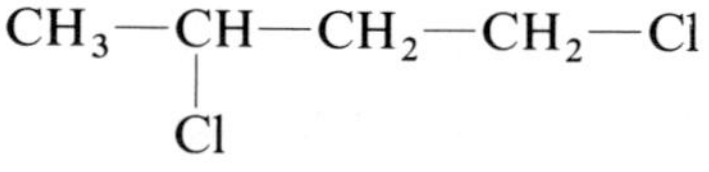
$$CH_3-\underset{\underset{Cl}{|}}{CH}-CH_2-CH_2-Cl$$

k. 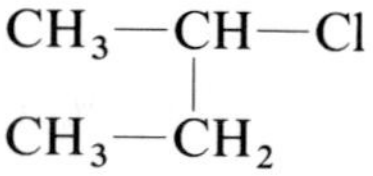
$$CH_3-\underset{\underset{CH_3-CH_2}{|}}{CH}-Cl$$

l. 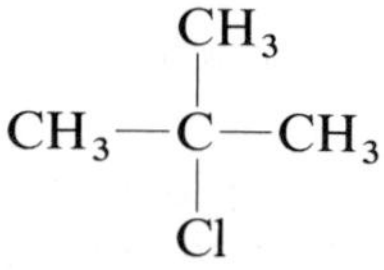
$$CH_3-\underset{\underset{Cl}{|}}{\overset{\overset{CH_3}{|}}{C}}-CH_3$$

Names and structural formulas of alkanes and cycloalkanes (Sections 2.3 and 2.4)

2.12 Name these alkyl groups.

a. CH_3—

b. CH_3CH_2—

$$CH_3$$
c. 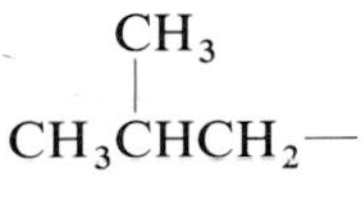 CH_3CHCH_2—

d. $CH_3CH_2CH_2$—

$$CH_3$$
e. 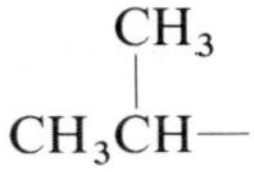 CH_3CH—

$$CH_2$$
f. 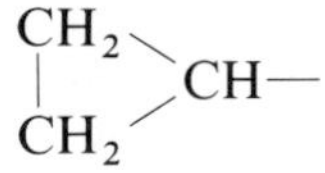 $\underset{CH_2}{\overset{CH_2}{|}}CH$—

$$CH_3$$
g. 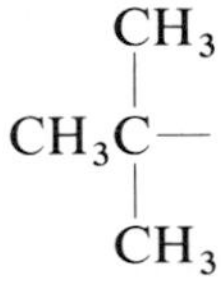 CH_3C—
$$CH_3$$

h. $CH_3CH_2CH_2CH_2$—

i. 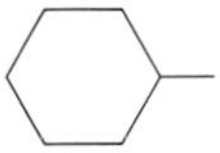 ⬡—

2.13 Write IUPAC names for these structural formulas.

a. 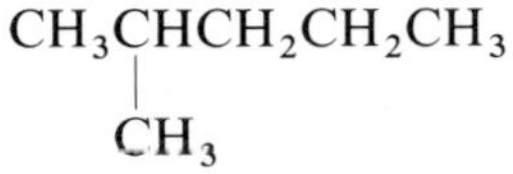 $CH_3CHCH_2CH_2CH_3$
$\quad\quad |$
$\quad\quad CH_3$

b. 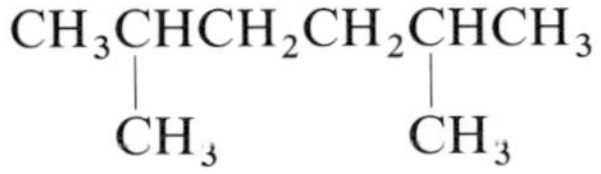 $CH_3CHCH_2CH_2CHCH_3$
$\quad\quad |\quad\quad\quad\quad |$
$\quad\quad CH_3\quad\quad CH_3$

c. 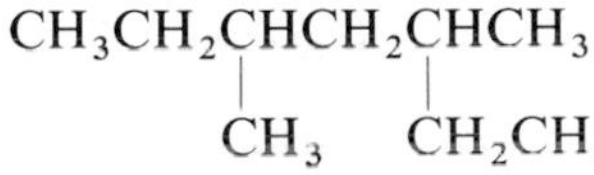 $CH_3CH_2CHCH_2CHCH_3$
$\quad\quad\quad\quad |\quad\quad\quad |$
$\quad\quad\quad\quad CH_3\quad CH_2CH_3$

d. 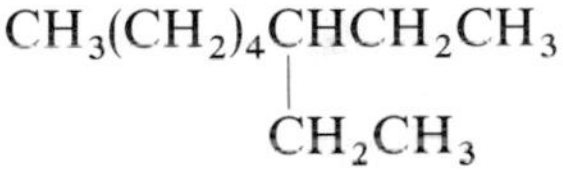 $CH_3(CH_2)_4CHCH_2CH_3$
$\quad\quad\quad\quad\quad\quad |$
$\quad\quad\quad\quad\quad\quad CH_2CH_3$

e. 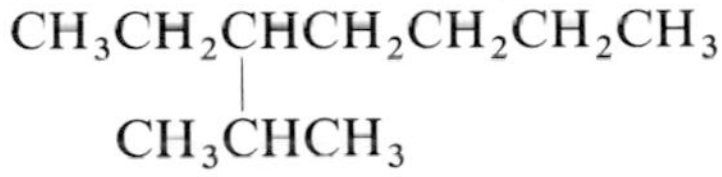 $CH_3CH_2CHCH_2CH_2CH_2CH_3$
$\quad\quad\quad\quad |$
$\quad\quad\quad\quad CH_3CHCH_3$

f. 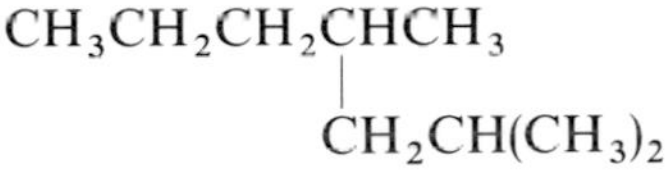 $CH_3CH_2CH_2CHCH_3$
$\quad\quad\quad\quad\quad |$
$\quad\quad\quad\quad\quad CH_2CH(CH_3)_2$

g. $CH_3(CH_2)_8CH_3$

h. $(CH_3)_2CHCH_2CH_2C(CH_3)_3$

i. 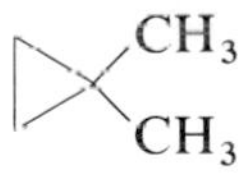 ◁$\overset{CH_3}{\underset{CH_3}{<}}$

j. 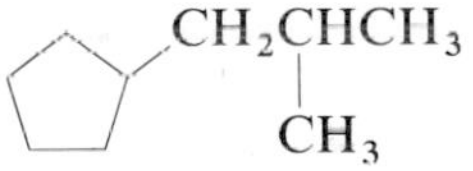 ⬠$\overset{CH_2CHCH_3}{\underset{CH_3}{}}$

k. 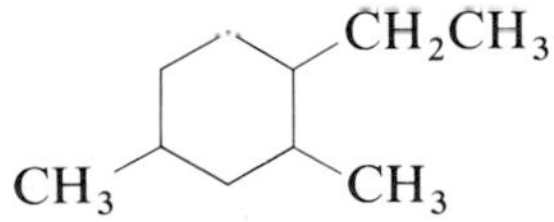 ⬡$\underset{CH_3\quad\quad CH_3}{\overset{CH_2CH_3}{}}$

l. 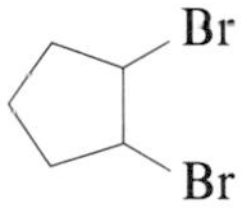 ⬠$\overset{Br}{\underset{Br}{}}$

2.14 Write structural formulas for these compounds.

a. 2,2,4-trimethylhexane

b. 1,1,2-trichlorobutane

c. 2,2-dimethylpropane

d. 3-ethyl-2,4,5-trimethyloctane

e. 2-bromo-2,4,6-trimethyloctane

f. 5-butyl-2,2-dimethylnonane

g. 4-isopropyloctane

h. 3,3-dimethylpentane

i. 1,1,1-trichloroethane

j. *trans*-1,3-dimethylcyclopentane

k. *cis*-1,2-diethylcyclobutane

l. 1,1-dichlorocycloheptane

2.15 Name and draw structural formulas for all isomeric alkanes of molecular formula C_6H_{14}.

2.16 Name and draw structural formulas for all isomeric alkanes of molecular formula C_7H_{16}.

2.17 Explain why each of the following is an incorrect IUPAC name. Write a correct IUPAC name for the intended compound.

a. 1,3-dimethylbutane
b. 4-methylpentane
c. 2,2-diethylbutane
d. 2-ethyl-3-methylpentane
e. 4,4-dimethylhexane
f. 2-propylpentane
g. 2,2-diethylheptane
h. 5-butyloctane
i. 2,2-dimethylcyclopropane
j. 2-*sec*-butyloctane
k. 4-isopentylheptane
l. 1,3-dimethyl-6-ethylcyclohexane

The IUPAC system of nomenclature (Section 2.5)

2.18 For each of these IUPAC names, draw the corresponding structural formula.

a. 3-pentanone   b. 2,2-dimethyl-3-pentanone   c. 2-butanone
d. ethanoic acid   e. hexanoic acid   f. propanoic acid
g. propanal   h. 1-propanol   i. 2-propanol
j. cyclopentene   k. cyclopentanol   l. cyclopentanone
m. cyclopropanol   n. ethene   o. ethanol
p. ethanal   q. decanoic acid   r. propanone

Conformations of alkanes and cycloalkanes (Section 2.6)

2.19 Given 1-bromo-2-chloroethane, draw structural formulas for:
a. a conformation in which Br— and Cl— are eclipsed by each other.
b. a conformation in which Br— and Cl— are eclipsed by hydrogens.
c. two different staggered conformations.

Cis-trans isomerism in cycloalkanes (Section 2.7)

2.20 There are four cis-trans isomers of 2-isopropyl-5-methylcyclohexanol.

2-isopropyl-5-methylcyclohexanol

a. In the cis-trans isomer found in nature and named *menthol*, the isopropyl group is trans to the hydroxyl group and the methyl group is cis to the hydroxyl group. Using a planar hexagon representation for the cyclohexane ring, draw a structural formula for menthol.
b. Also using a planar hexagon representation for the cyclohexane ring, draw structural formulas for the other three cis-trans isomers of 2-isopropyl-5-methylcyclohexanol.

2.21 The substance 1,2,3,4,5,6-hexachlorocyclohexane shows cis-trans isomerism. A crude mixture of the isomers is sold as the insecticide benzene hexachloride (BHC), under the trade names Lindane and Gammexane. The insecticidal properties of the mixture arise from one isomer known as the gamma isomer, which is *cis*-1,2,4,5-*trans*-3,6-hexachlorocyclohexane.
a. Draw a structural formula for 1,2,3,4,5,6-hexachlorocyclohexane, disregarding for the moment the existence of cis-trans isomerism. What is the molecular formula of this compound?
b. Using a planar hexagon representation for the cyclohexane ring, draw a structural formula for the gamma isomer.

*Physical
properties
(Section 2.8)*

2.22 In Problem 2.16, you drew structural formulas for all isomeric alkanes of molecular formula C_7H_{16}. Predict which of these isomers has the lowest boiling point, and which the highest.

2.23 What unbranched alkane has about the same boiling point as water? (Refer to Table 2.4 on the physical properties of alkanes.) Calculate the molecular weight of this alkane and compare it with the molecular weight of water.

*Reactions of
alkanes
(Section 2.9)*

2.24 What is the major component of natural gas? of bottled gas, or LPG?

2.25 Complete and balance these combustion reactions. Assume that each hydrocarbon is converted completely to carbon dioxide and water.
  a. propane + O_2 $\longrightarrow$ **b.** octane + O_2 $\longrightarrow$
  c. cyclohexane + O_2 $\longrightarrow$ **d.** 3-methylpentane + O_2 $\longrightarrow$

2.26 Following are heats of combustion per mole for methane, propane, and 2,2,4-trimethylpentane. On a gram-for-gram basis, which hydrocarbon is the best source of heat energy?

Hydrocarbon	Major Component of	ΔH^0 (kcal/mol)
CH_4	natural gas	-212
$CH_3CH_2CH_3$	LPG	-531
$\begin{matrix} & CH_3 \ \ CH_3 \\ & \vert \quad \ \vert \\ CH_3CCH_2CHCH_3 \\ & \vert \\ & CH_3 \end{matrix}$	gasoline	-1304

2.27 Name and draw structural formulas for all possible monohalogenation products that might be formed in these reactions.

  a. $\bigcirc$ + Cl_2 $\xrightarrow{\text{light}}$

  b. $CH_3{-}CH_2{-}CH_2{-}\overset{\overset{\displaystyle CH_3}{\vert}}{CH}{-}CH_3$ + Cl_2 $\xrightarrow{\text{light}}$

  c. $CH_3{-}\overset{\overset{\displaystyle CH_3}{\vert}}{CH}{-}\underset{\underset{\displaystyle CH_3}{\vert}}{CH}{-}CH_3$ + Br_2 $\xrightarrow{\text{light}}$

  d. $\begin{matrix} CH_2 \\ \vert \quad \diagdown \\ \ \quad \ \ CH_2 \\ \vert \quad \diagup \\ CH_2 \end{matrix}$ + Br_2 $\xrightarrow{\text{light}}$

2.28 There are three isomeric alkanes of molecular formula C_5H_{12}. When isomer A is reacted with chlorine gas at 300°C, the result is a mixture of four monochlorination products. Under the same conditions with isomer B, a mixture of three monochlorination products results, and with isomer C, only one monochlorination product is formed. From this information, assign structural formulas to isomers A, B, and C.

2.29   Consult a handbook of chemistry or other reference work, to find out the densities of dichloromethane (methylene chloride), trichloromethane (chloroform), and tetrachloromethane (carbon tetrachloride). Which are more dense than water? Which are less dense?

3

Unsaturated Hydrocarbons

Unsaturated hydrocarbons are hydrocarbons that contain one or more carbon-carbon double or triple bonds. There are three classes of unsaturated hydrocarbons: alkenes, alkynes, and aromatic hydrocarbons. **Alkenes** contain one or more carbon-carbon double bonds, and alkynes contain one or more carbon-carbon triple bonds. The structural formulas of ethene, the simplest alkene, and ethyne, the simplest alkyne, are

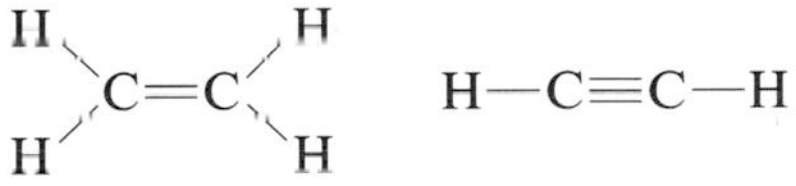

$$
\begin{array}{cc}
\text{H} \diagdown \quad \diagup \text{H} & \\
\quad \text{C}=\text{C} & \text{H}-\text{C}\equiv\text{C}-\text{H} \\
\text{H} \diagup \quad \diagdown \text{H} & \\
\end{array}
$$

ethene
(an alkene)

ethyne
(an alkyne)

Because alkenes and alkynes have fewer hydrogen atoms than alkanes with the same number of carbons atoms, they are commonly referred to as unsaturated hydrocarbons; they are unsaturated with respect to hydrogen atoms. Alkynes are not widely distributed in the biological world, and therefore, we will not study them further.

The third class of unsaturated hydrocarbons are the **aromatic hydrocarbons**. The Lewis structure of benzene, the simplest aromatic hydrocarbon, is

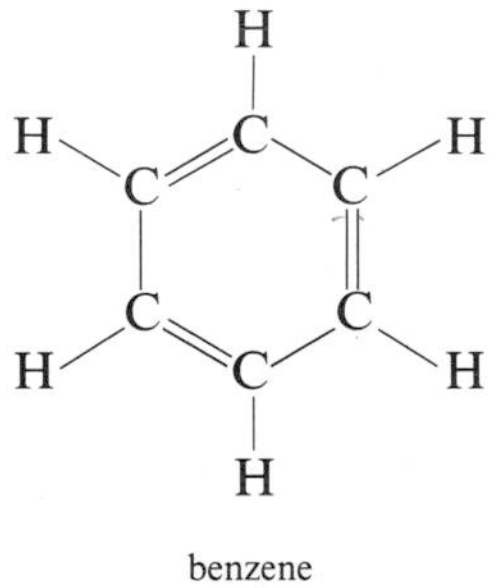

$$
\begin{array}{c}
\text{H} \\
| \\
\text{C} \\
\diagup \quad \diagdown \\
\text{H}-\text{C} \qquad \text{C}-\text{H} \\
\| \qquad | \\
\text{H}-\text{C} \qquad \text{C}-\text{H} \\
\diagdown \quad \diagup \\
\text{C} \\
| \\
\text{H}
\end{array}
$$

benzene
(an aromatic hydrocarbon)

Benzene and other aromatic hydrocarbons have chemical properties quite different from those of alkenes and alkynes. However, the structural feature that relates alkenes, alkynes, and aromatic hydrocarbons is the presence of one or more pi bonds. This chapter is a study of carbon-carbon pi bonds.

3.1 Structure and Bonding in Alkenes

We have examined the formation of carbon-carbon double bonds in terms of the overlap of atomic orbitals (Section 1.7). To bond with three other atoms, carbon uses sp^2 hybrid orbitals formed by combination of one $2s$ orbital and two $2p$ orbitals. The three sp^2 orbitals lie in a plane at angles of 120° to each other. The remaining $2p$ orbital of carbon is not hybridized and lies perpendicular to the plane created by the three sp^2 orbitals. A carbon-carbon double bond consists of one sigma bond formed by the overlap of sp^2 hybrid orbitals of adjacent carbon atoms and one pi bond formed by the overlap of unhybridized orbitals (Figure 3.1).

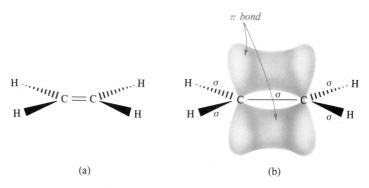

(a)   (b)

Figure 3.1   Covalent bonding in ethene. (a) Lewis structure; (b) orbital overlap model, showing sigma and pi bonds.

Using the orbital overlap model of a carbon-carbon double bond, we would predict a value of 120° for the H—C—C bond angles in ethene. The observed angle is 121.7°, a value close to that predicted. In substituted alkenes, deviations from the predicted angle of 120° may be larger. The C—C=C bond angle in propene, for example, is 124.7°.

3.2 Nomenclature of Alkenes

A. IUPAC Names

The IUPAC names of alkenes are formed by changing the -an- infix of the parent alkane to -en- (Section 2.5). Hence, CH_2=CH_2 is named ethene, and CH_3—CH=CH_2 is named propene.

$$CH_2{=}CH_2 \qquad CH_3{-}CH{=}CH_2$$

ethene　　　　　propene

Ethene and propene can contain a double bond in only one position. In higher alkenes where there are isomers that differ in location of the double bond, a numbering system must be used. According to the IUPAC system, the longest carbon chain that contains the double bond is numbered to give the carbon atoms of the double bond the lowest possible numbers. The location of the double bond is indicated by the number of the first carbon of the double bond. Branched or substituted alkenes are named like alkanes. Carbon atoms are numbered, substituent groups are located and named, the double bond is located, and the main chain is named.

$$\overset{6}{C}H_3{-}\overset{5}{C}H_2{-}\overset{4}{C}H_2{-}\overset{3}{C}H{=}\overset{2}{C}H{-}\overset{1}{C}H_3 \qquad \overset{6}{C}H_3{-}\overset{5}{C}H_2{-}\overset{4}{C}H{-}\overset{3}{C}H{=}\overset{2}{C}H{-}\overset{1}{C}H_3$$

$$\underset{CH_3}{\big|}$$

2-hexene　　　　　　　　　　4-methyl-2-hexene

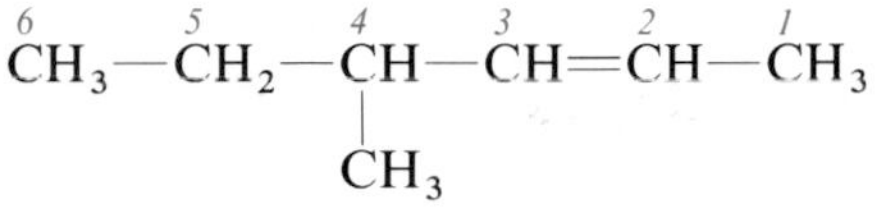

In the following alkene, there is a chain of five carbon atoms. However, because the longest chain that contains the double bond has only four carbons, the parent compound is butane and the IUPAC name of this alkene is 2-ethyl-3-methyl-1-butene.

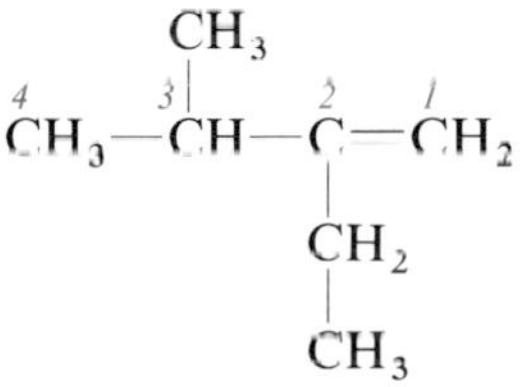

$$\overset{4}{C}H_3{-}\overset{3}{C}H{-}\overset{2}{C}{=}\overset{1}{C}H_2$$

with CH_3 on C-3 and $CH_2{-}CH_3$ on C-2

2-ethyl-3-methyl-1-butene

When cycloalkenes are named, the carbon atoms of the ring double bond are numbered 1 and 2 in the direction that gives the substituent encountered first the smallest number. If there are two or more different substituents, they are listed in alphabetical order.

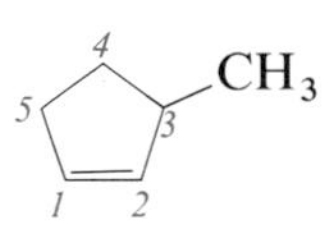

3-methylcyclopentene 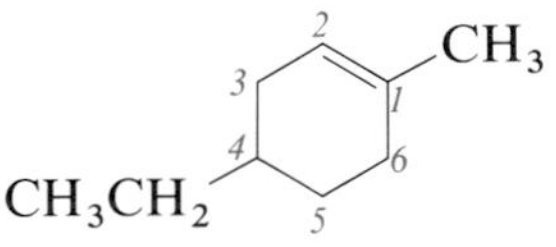　　　　　4-ethyl-1-methylcyclohexene

Alkenes that contain more than one double bond are called alkadienes, alkatrienes, or more simply dienes, trienes, and so on.

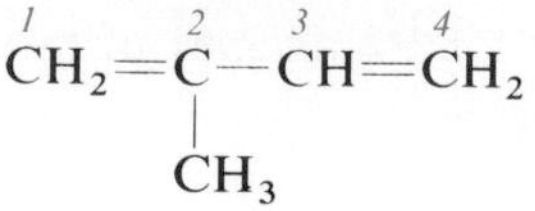

$$\overset{1}{C}H_2 = \overset{2}{C} - \overset{3}{C}H = \overset{4}{C}H_2 \qquad \overset{1}{C}H_2 = \overset{2}{C}H - \overset{3}{C}H_2 - \overset{4}{C}H = \overset{5}{C}H_2$$
$$\underset{CH_3}{|}$$

2-methyl-1,3-butadiene   1,4-pentadiene
(isoprene)

B. Common Names

Many alkenes, particularly those of low molecular weight, are known almost exclusively by their common names.

$$CH_2 = CH_2 \qquad CH_3 - CH = CH_2 \qquad CH_3 - \overset{CH_3}{\underset{|}{C}} = CH_2 \qquad CH_2 = CH - CH = CH_2$$

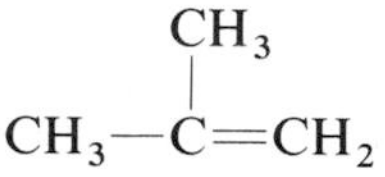

IUPAC name:	ethene	propene	2-methylpropene	1,3-butadiene
Common name:	ethylene	propylene	isobutylene	butadiene

Further, the common names **methylene**, **vinyl**, and **allyl** are often used to show the presence of alkenyl groups, as shown in Table 3.1. The name methylene is also used to show the presence of a $-CH_2-$ group, as illustrated in the following examples:

Methylene groups

$$CH_3(CH_2)_4CH_3 \qquad Cl-CH_2-Cl$$

hexane   dichloromethane
(methylene chloride)

Table 3.1 Names for common alkenyl groups.

Alkenyl Group	Group Name	Examples	
$CH_2=$	methylene		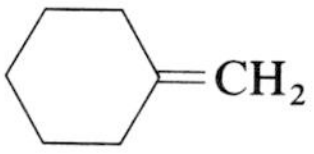 $\bigcirc\!\!=CH_2$ methylene cyclohexane
$CH_2=CH-$	vinyl	$CH_2=CH-Cl$ vinyl chloride	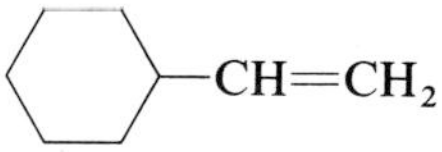 $\bigcirc\!-CH=CH_2$ vinyl cyclohexane
$CH_2=CH-CH_2-$	allyl	$CH_2=CH-CH_2-Cl$ allyl chloride	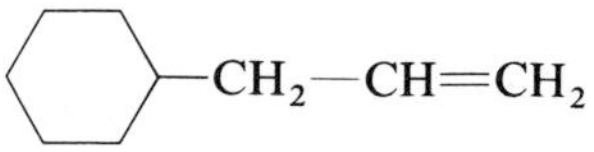 $\bigcirc\!-CH_2-CH=CH_2$ allyl cyclohexane

Example 3.1

Draw structural formulas for all alkenes of molecular formula C_5H_{10}. Give each an IUPAC name.

☐ *Solution*

Approach this type of problem systematically. First draw the carbon skeletons possible for molecules of five carbon atoms. There are three such skeletons.

$$\text{a. } C{-}C{-}C{-}C{-}C \qquad \text{b. } C{-}\underset{\underset{C}{|}}{C}{-}C{-}C \qquad \text{c. } C{-}\underset{\underset{C}{|}}{\overset{\overset{C}{|}}{C}}{-}C$$

Because skeleton (c) cannot contain a carbon-carbon double bond (the central carbon atom already has four bonds to it), we need consider only skeletons (a) and (b) when drawing structural isomers for alkenes of molecular formula C_5H_{10}. Next, locate the double bond between carbon atoms along the chain. For carbon skeleton (a), there are two possible locations for the double bond, and for carbon skeleton (b), three possible locations.

For (a): $C{=}C{-}C{-}C{-}C$ and $C{-}C{=}C{-}C{-}C$

For (b): $C{=}\underset{\underset{C}{|}}{C}\ C{-}C$ and $C{-}\underset{\underset{C}{|}}{C}{=}C{-}C$ and $C{-}C{-}\underset{\underset{C}{|}}{C}{=}C$

Finally, add hydrogens to complete the tetravalence of carbon and give the correct molecular formula. To derive IUPAC names for these alkenes, number the carbon chain from the direction that gives the carbons of the double bond the lowest numbers and then name and give a number to each substituent. Common names are given below for several of these alkenes. Note, however, that none is given for 2-pentene. You might be tempted to call it ethylmethylethylene, but that name is ambiguous. It does not tell you whether the ethyl and methyl groups are on the same carbon (as in 2-methyl-1-butene) or on adjacent carbons (as in 2-pentene) of the double bond.

$$\overset{1}{C}H_2{=}\overset{2}{C}H{-}\overset{3}{C}H_2{-}\overset{4}{C}H_2{-}\overset{5}{C}H_3 \qquad \overset{1}{C}H_3{-}\overset{2}{C}H{=}\overset{3}{C}H{-}\overset{4}{C}H_2\ \overset{5}{C}H_3$$

1-pentene
(*n*-propylethylene)   2-pentene

$$\overset{1}{C}H_2{=}\overset{2}{\underset{\underset{CH_3}{|}}{C}}{-}\overset{3}{C}H_2{-}\overset{4}{C}H_3 \qquad \overset{1}{C}H_3{-}\overset{2}{\underset{\underset{CH_3}{|}}{C}}{=}\overset{3}{C}H{-}\overset{4}{C}H_3 \qquad \overset{4}{C}H_3{-}\overset{3}{\underset{\underset{CH_3}{|}}{C}H}{-}\overset{2}{C}H{=}\overset{1}{C}H_2$$

2-methyl-1-butene   2-methyl-2-butene   3-methyl-1-butene
                           (trimethylethylene)   (isopropylethylene)

Problem 3.1

Draw structural formulas for all alkenes of molecular formula C_6H_{12} that have the following carbon skeletons. Give each alkene an IUPAC name and where possible a common name.

$$
\text{a. } C-\overset{\overset{\displaystyle C}{|}}{C}-C-C-C \qquad \text{b. } C-\overset{\overset{\displaystyle C}{|}}{C}-\overset{\overset{\displaystyle C}{|}}{C}-C \qquad \text{c. } C-\overset{\overset{\displaystyle C}{|}}{\underset{\underset{\displaystyle C}{|}}{C}}-C-C
$$

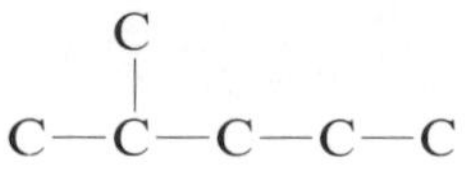

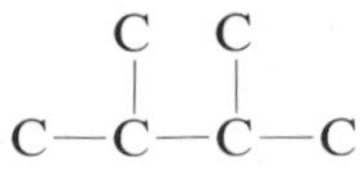

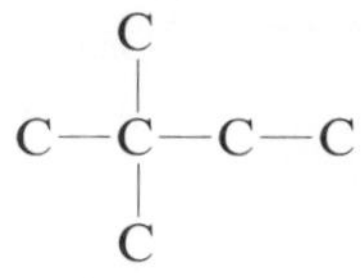

3.3 Cis-Trans Isomerism in Alkenes

A. The Existence of Cis-Trans Isomers

Configurational isomers have the same molecular formula, the same order of attachment of atoms, and an arrangement of atoms that is fixed in space because of restricted rotation about one or more bonds within the molecule. **Cis-trans isomerism** is one type of configurational isomerism. In cycloalkanes, cis-trans isomerism depends on the arrangement of substituent groups on a ring (Section 2.7); in alkenes, cis-trans isomerism depends on the arrangement of groups on a double bond. The structural feature that makes possible the existence of cis-trans isomerism in alkenes is restricted rotation about the two carbons of a double bond (Figure 3.2).

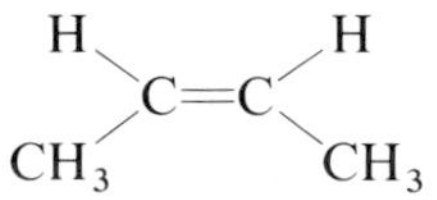

cis-2-butene
(mp − 139°C;
bp 4°C)

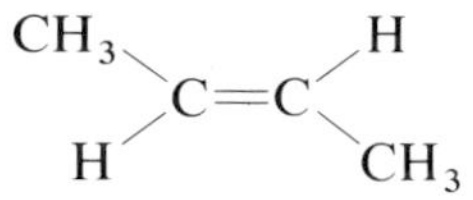

trans-2-butene
(mp − 106°C;
bp 1°C)

Figure 3.2 The cis and trans isomers of 2-butene.

Ethene, propene, 1-butene, and 2-methylpropene have only one possible arrangement of groups about the double bond. In 2-butene there are two possible arrangements. In one arrangement, the two methyl groups are on the same side of the double bond; this isomer is called *cis*-2-butene. In the other arrangement, the two methyl groups are on opposite sides of the double bond; this isomer is called *trans*-2-butene. Shown in Figure 3.2 are Lewis structures for these configurational isomers. Cis and trans isomers are different compounds and have different physical and chemical properties. The cis and trans isomers of 2-butene differ in melting points by 33°C and in boiling points by 3°C.

Example 3.2

Which of the following alkenes show cis-trans isomerism? For each that does, draw structural formulas for both isomers.

a. $CH_2{=}CH-CH_2-CH_2-CH_3$ b. $CH_3-CH{=}CH-CH_2-CH_3$

c. 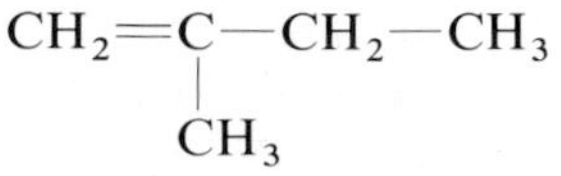 $CH_2{=}\overset{\displaystyle }{\underset{\underset{\displaystyle CH_3}{|}}{C}}-CH_2-CH_3$

☐ *Solution*

a. 1-Pentene. Begin by drawing a carbon-carbon double bond to show the surrounding bond angles of 120°:

$$\diagup C = C \diagdown$$

Next, complete the structural formula, showing all four groups attached to the double bond:

$$\begin{array}{c} H \diagdown \quad \diagup H \\ C = C \\ H \diagup \quad \diagdown CH_2CH_2CH_3 \end{array}$$

To determine whether this molecule shows cis-trans isomerism, exchange positions of the —H and —CH$_2$CH$_2$CH$_3$ groups and compare the two structural formulas to see whether they are identical or whether they represent cis and trans isomers.

$$\begin{array}{c} H \diagdown \quad \diagup H \\ C = C \\ H \diagup \quad \diagdown CH_2CH_2CH_3 \end{array} \qquad \begin{array}{c} H \diagdown \quad \diagup CH_2CH_2CH_3 \\ C = C \\ H \diagup \quad \diagdown H \end{array}$$

In this case, they are identical. To see this, imagine that you pick up either one and turn it over as you would turn your hand from palm down to palm up. If you do this correctly, you will see that one structural formula fits exactly on top of the other; that is, it is superimposable on the other. These structural formulas are identical, and 1-pentene does not show cis-trans isomerism.

b. 2-Pentene. Draw structural formulas for possible cis-trans isomers as you did in part (a).

$$\begin{array}{c} CH_3 \diagdown \quad \diagup H \\ C = C \\ H \diagup \quad \diagdown CH_2CH_3 \end{array} \qquad \begin{array}{c} CH_3 \diagdown \quad \diagup CH_2CH_3 \\ C = C \\ H \diagup \quad \diagdown H \end{array}$$

trans-2-pentene   *cis*-2-pentene

The four groups attached to the double bond in the structural formula on the left have a different orientation from that in the structural formula on the right. Therefore, the drawings represent cis-trans isomers.

c. 2-Methyl-1-butene. The structural formulas below are superimposable, and therefore this alkene does not show cis-trans isomerism.

$$\begin{array}{c} H \diagdown \quad \diagup CH_3 \\ C = C \\ H \diagup \quad \diagdown CH_2CH_3 \end{array} \qquad \begin{array}{c} H \diagdown \quad \diagup CH_2CH_3 \\ C = C \\ H \diagup \quad \diagdown CH_3 \end{array}$$

From this example and the preceding discussion, you should realize that an alkene shows cis-trans isomerism only if each of the carbon atoms of the double bond has two different groups attached to it. In 2-pentene, the different groups on the first carbon of the double bond are —H and —CH$_3$; on the second carbon of the double bond, they are —H and —CH$_2$CH$_3$. In 2-butene, the different groups on each carbon of the double bond are —H and —CH$_3$.

Problem 3.2

Which of the following alkenes show cis-trans isomerism? For each that does, draw structural formulas for the isomers.

a. 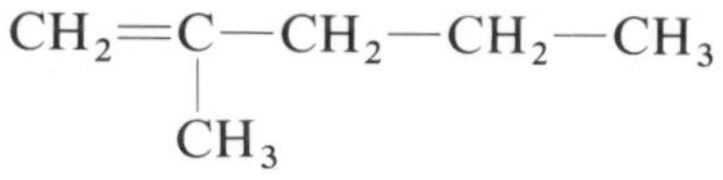 $CH_2{=}C{-}CH_2{-}CH_2{-}CH_3$
   $\quad\quad\;\; |$
   $\quad\quad\; CH_3$

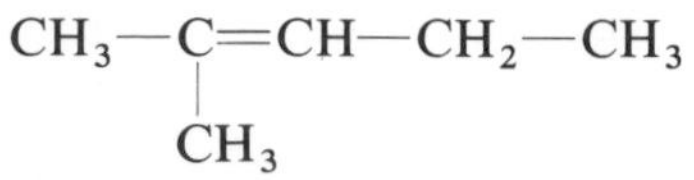

b. $CH_3{-}C{=}CH{-}CH_2{-}CH_3$
   $\quad\quad\;\; |$
   $\quad\quad\; CH_3$

c. $CH_3{-}CH{-}CH{=}CH{-}CH_3$
   $\quad\quad\quad |$
   $\quad\quad\; CH_3$

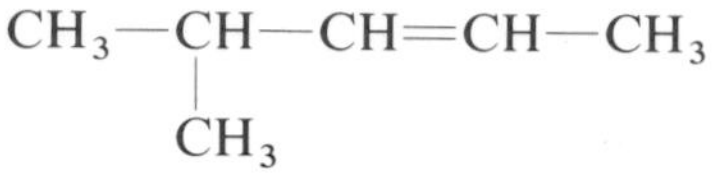

d. 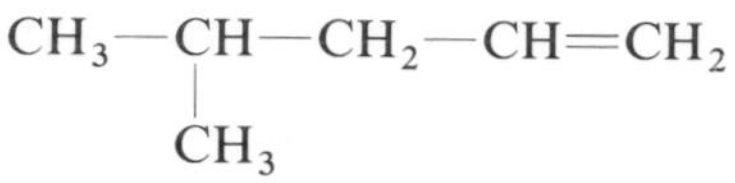 $CH_3{-}CH{-}CH_2{-}CH{=}CH_2$
   $\quad\quad\quad |$
   $\quad\quad\; CH_3$

B. Designating Configuration of Alkenes

The most common method for specifying configuration in alkenes uses the prefixes *cis* and *trans*. There is no doubt whatsoever which configurational isomers are intended by the names *cis*-2-butene and *trans*-3-hexene.

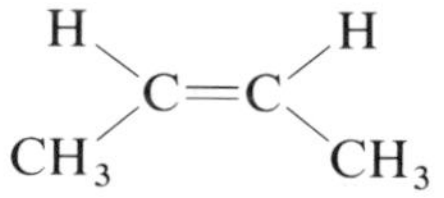

cis-2-butene 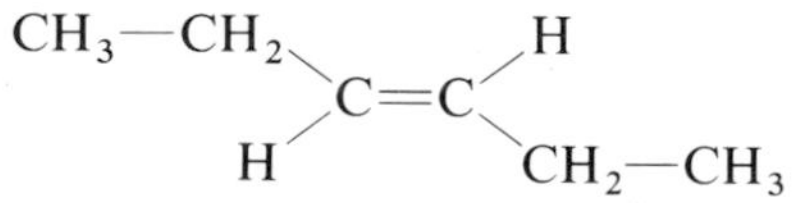   *trans*-3-hexene

For alkenes with more complex structural formulas, it is the orientation of the main carbon chain that determines cis and trans. Following is a structural formula for one configurational isomer of 3,4-dimethyl-2-pentene

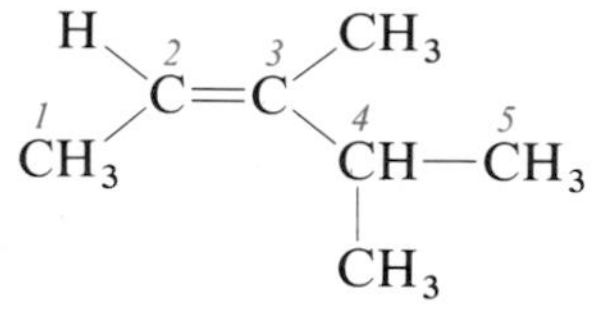

3,4-dimethyl-2-pentene

Should it be named cis because the carbon atoms of the parent chain (atoms 1-2-3-4-5) are on the same side of the double bond, or should it be named trans because the two methyl groups are on opposite sides? According to IUPAC rules, it should be named cis because of the cis orientation of the main carbon chain about the double bond.

C. Configurational Isomerism in Cycloalkenes

Configurational isomerism is possible about the carbon-carbon double bond in cycloalkenes only when the ring is large enough to accommodate a trans double bond. In cyclohexene and cycloheptene, ring formation is possible only if the carbon atoms attached directly to the double bond (carbons 3 and 6 of the cyclohexene ring, carbons 3 and 7 of the cycloheptene ring) are cis to each other (Figure 3.3). A trans configuration is not possible in these and smaller cycloalkenes.

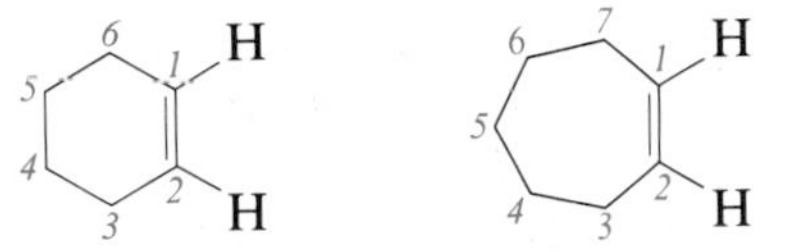

to form cyclohexene,
carbon atoms 3 and 6
must be cis to each
other

to form cycloheptene,
carbons atoms 3 and 7
must be cis to each
other

Figure 3.3 Configurational isomerism is not possible in cyclohexene and cycloheptene.

Only in cyclooctene and higher cycloalkenes is the number of carbon atoms involved in ring formation large enough so there is a possibility for configurational isomerism.

D. Configurational Isomerism in Dienes, Trienes, and Polyenes

Thus far we have considered cis-trans isomerism in compounds containing only one carbon-carbon double bond. Next let us consider compounds with two or more carbon-carbon double bonds. In 1,3-heptadiene, carbon 1 has two identical groups and there is no possibility for configurational isomerism about the first double bond. Each carbon of the double bond between carbons 3 and 4 has two different groups attached to it, and configurational isomerism is possible about this double bond.

$$\overset{1}{C}H_2=\overset{2}{C}H-\overset{3}{C}H=\overset{4}{C}H-\overset{5}{C}H_2-\overset{6}{C}H_2-\overset{7}{C}H_3$$

Each carbon atom of
this double bond has
two different groups
attached to it

1,3-heptadiene

An example of a biologically important molecule for which there are numerous cis-trans isomers is vitamin A. There is no possibility for cis-trans isomerism about the cyclohexene double bond, because the ring contains only six atoms. There are four carbon-carbon double bonds in the chain of atoms attached to the substituted cyclohexene ring, and each has the potential for cis-trans isomerism. There are $2 \times 2 \times 2 \times 2$, or 16, possible cis-trans isomers for vitamin A.

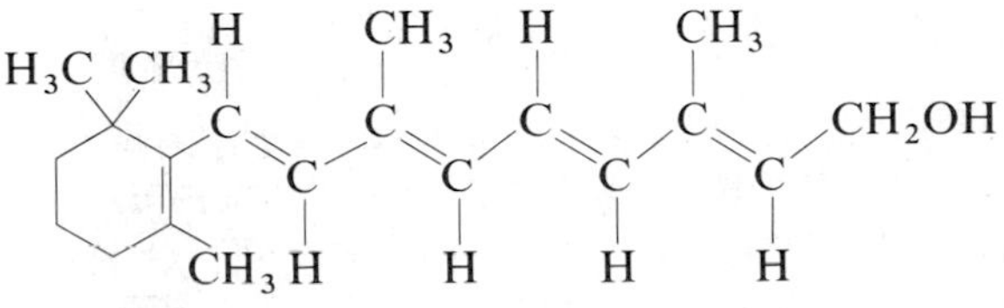

vitamin A

3.4 Reactions of Alkenes—An Overview

In contrast to alkanes, alkenes react with a variety of compounds. One characteristic reaction of alkenes takes place at the carbon-carbon double bond in such a way that the pi bond is broken and in its place are formed sigma bonds to two new atoms or groups of atoms. Such reactions are called addition reactions.

$$>\!\!C\!=\!\!C\!\!< \; + \; A\!-\!B \; \longrightarrow \; -\!\!\underset{\underset{A}{|}}{C}\!-\!\underset{\underset{B}{|}}{C}\!-$$

In addition reactions, one sigma bond (A—B) and one pi bond are broken and two new sigma bonds (C—A and C—B) are formed. Consequently, addition reactions to double bonds are almost always energetically favorable because there is net conversion of one pi bond to a sigma bond.

Each addition reaction is given a special name that describes the particular type of addition. Table 3.2 gives examples of four addition reactions.

Table 3.2
Four alkene addition reactions and the descriptive name or names associated with each.

Addition Reaction		Descriptive Name
$>\!\!C\!=\!\!C\!\!< \; + \; H\!-\!H \xrightarrow{\text{addition of hydrogen}} -\!\underset{\underset{H}{\|}}{C}\!-\!\underset{\underset{H}{\|}}{C}\!-$		hydrogenation; reduction
$>\!\!C\!=\!\!C\!\!< \; + \; Br\!-\!Br \xrightarrow{\text{addition of bromine}} -\!\underset{\underset{Br}{\|}}{C}\!-\!\underset{\underset{Br}{\|}}{C}\!-$		bromination; halogenation
$>\!\!C\!=\!\!C\!\!< \; + \; H\!-\!Br \xrightarrow{\text{addition of hydrobromic acid}} -\!\underset{\underset{H}{\|}}{C}\!-\!\underset{\underset{Br}{\|}}{C}\!-$		hydrobromination; hydrohalogenation
$>\!\!C\!=\!\!C\!\!< \; + \; H\!-\!OH \xrightarrow{\text{addition of water}} -\!\underset{\underset{H}{\|}}{C}\!-\!\underset{\underset{OH}{\|}}{C}\!-$		hydration

A second characteristic reaction of alkenes is conversion to alkanes by addition of two hydrogen atoms. Reaction of an alkene with hydrogen is addition, but because it is also reduction we will treat it separately. A balanced half-reaction for the conversion of an alkene to an alkane shows that this transformation is a two electron reduction.

Balanced Half-Reaction

$$\text{C=C} + 2H^+ + 2e^- \longrightarrow -\overset{|}{\underset{H}{C}}-\overset{|}{\underset{H}{C}}-$$

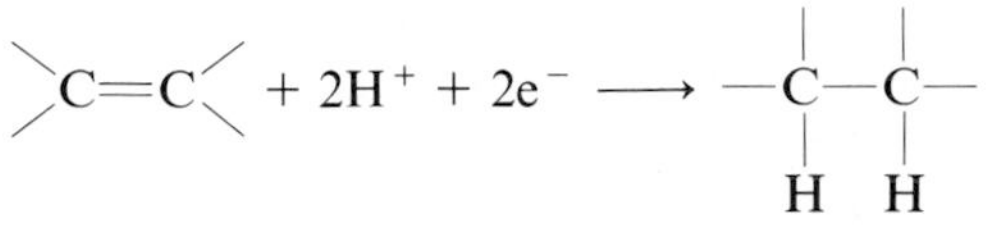

an alkene   an alkane

A third characteristic reaction of alkenes is oxidation. **Oxidation** almost invariably involves addition of oxygen, in some reactions without cleavage of the carbon skeleton, and in other reactions with cleavage, as illustrated by the following balanced half-reactions. The first is a two-electron oxidation and does not break the carbon skeleton. The second is a four-electron oxidation and does involve cleavage of the carbon-carbon double bond.

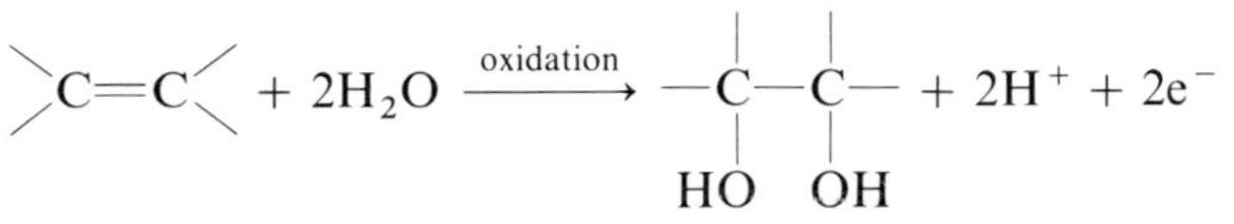

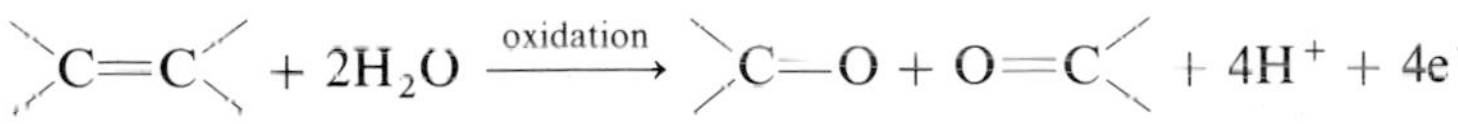

$$\text{C=C} + 2H_2O \xrightarrow{\text{oxidation}} -\overset{|}{\underset{HO}{C}}-\overset{|}{\underset{OH}{C}}- + 2H^+ + 2e^-$$

$$\text{C=C} + 2H_2O \xrightarrow{\text{oxidation}} \text{C—O} + \text{O=C} + 4H^+ + 4e^-$$

We will now study these characteristic alkene reactions in considerable detail, including what is known about how each occurs, or in the terminology of organic chemists, what is known about the mechanism of each reaction.

3.5 Addition to Alkenes

A. Addition of Bromine and Chlorine: Halogenation

Bromine, Br_2, and chlorine, Cl_2, add readily to alkenes to form single covalent carbon-halogen bonds (C—Br or C—Cl) on adjacent carbons. Fluorine also adds to alkenes, but because its addition is fast and not easily controlled, adding fluorine to alkenes is not a general laboratory procedure. Iodine is so unreactive that it does not add to alkenes. Halogenation with bromine or chlorine is generally carried out either with the pure reagents or by mixing them in CCl_4 or some other inert solvent.

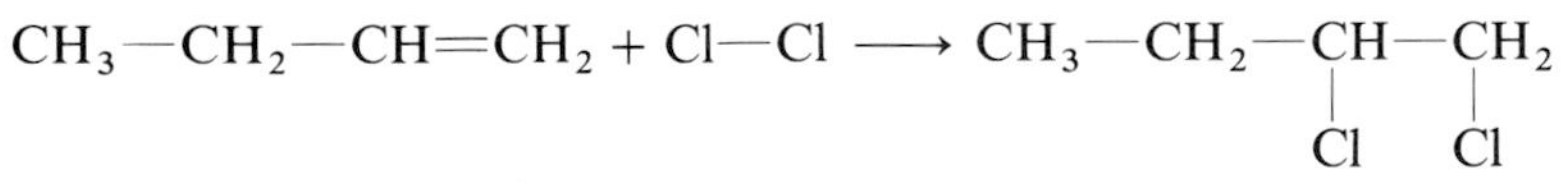

$$CH_3-CH_2-CH=CH_2 + Cl-Cl \longrightarrow CH_3-CH_2-\overset{|}{\underset{Cl}{CH}}-\overset{|}{\underset{Cl}{CH_2}}$$

1-butene   1,2-dichlorobutane

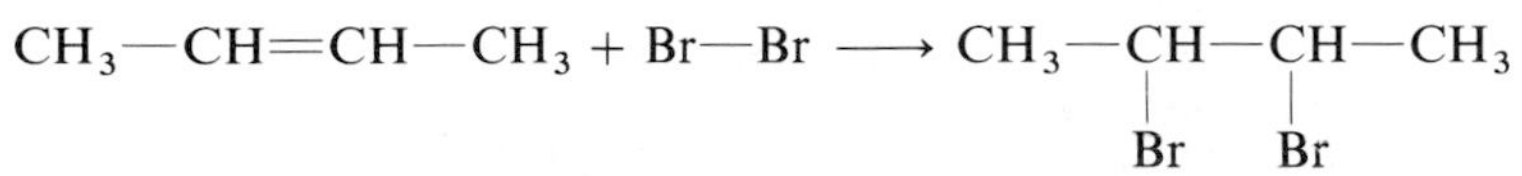

$$CH_3-CH=CH-CH_3 + Br-Br \longrightarrow CH_3-\overset{|}{\underset{Br}{CH}}-\overset{|}{\underset{Br}{CH}}-CH_3$$

2-butene   2,3-dibromobutane cyclohexene   1,2-dibromocyclohexane

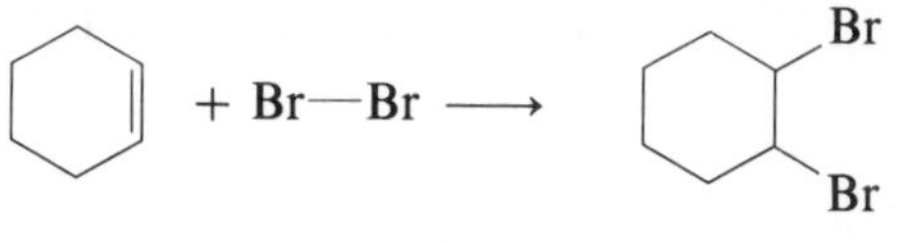

Alkenes are different from alkanes in reacting with Br_2 and Cl_2. Recall (Section 2.9B) that chlorine and bromine do not react with alkanes unless the halogen-alkane mixture is exposed to ultraviolet or visible light, or heated to temperatures of 250–400°C. The reaction that then occurs is substitution of halogen for hydrogen and formation of an equivalent amount of HCl or HBr. Halogenation of most alkanes invariably gives a complex mixture of products. In contrast, chlorine and bromine react with alkenes at room temperature by addition of halogen atoms to the two carbon atoms of the double bond, with the formation of two new carbon-halogen bonds.

Example 3.3

Name and draw structural formulas for the products of the following halogenation reactions:

a. $CH_3C{=}CHCH_3 + Br_2 \longrightarrow$ (with CH_3 above)

b. 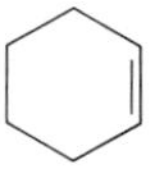 $+ Cl_2 \longrightarrow$

☐ *Solution*

a. $CH_3{-}\underset{\underset{Br}{|}}{\overset{\overset{CH_3}{|}}{C}}{-}\underset{\underset{Br}{|}}{CH}{-}CH_3$

b.

2,3-dibromo-2-methylbutane   1,2-dichlorocyclohexane

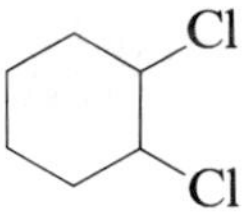

Problem 3.3

Name and draw structural formulas for the alkene that reacts with bromine, to give

a. $CH_3\underset{\underset{Br}{|}}{\overset{\overset{CH_3}{|}}{CH}}\underset{\underset{Br}{|}}{CH}CH_2$

b.

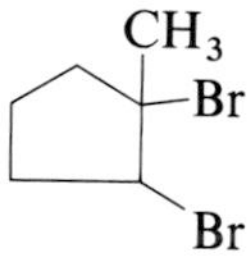

Reaction with bromine is a particularly useful qualitative test for the presence of an alkene. A solution of bromine in carbon tetrachloride is red, whereas alkenes and dibromoalkanes are usually colorless. If a few drops of bromine in carbon tetrachloride is added to an alkene, the red color of the test solution is discharged.

B. Addition of Hydrogen Halides: Hydrohalogenation

Dry HF, HCl, HBr, and HI add to alkenes to give haloalkanes. These additions may be carried out either with the pure reagents or in the presence of a polar solvent such as acetic acid. Addition of HCl to ethene gives chloroethane (ethyl chloride):

$$CH_2{=}CH_2 + H{-}Cl \longrightarrow \underset{\underset{H \qquad Cl}{|\qquad\;\,|}}{CH_2{-}CH_2}$$

ethene   chloroethane
(ethyl chloride)

Addition of HCl to propene gives two products, 2-chloropropane (isopropyl chloride) and 1-chloropropane (*n*-propyl chloride) depending on which part of the reagent goes to which carbon of the double bond:

$$CH_3{-}CH{=}CH_2 + H{-}Cl \longrightarrow \underset{\underset{Cl\;\;\;H}{|\quad\;\,|}}{CH_3{-}CH{-}CH_2} + \underset{\underset{H\;\;\;Cl}{|\quad\;\,|}}{CH_3{-}CH{-}CH_2}$$

propene   2-chloropropane   1-chloropropane
(major product)

Of the two possible products, only 2-chloropropane is formed. This selective pattern of addition was noted by Vladimir Markovnikov, who made the generalization that in additions of H—X to alkenes, hydrogen adds to the carbon of the double bond that has the greater number of hydrogens already attached to it. It is important to remember that while **Markovnikov's rule** provides a way to predict the major product of an alkene addition reaction, it does not explain why one product predominates over other possible products.

Example 3.4

Name and draw structural formulas for the products of these alkene addition reactions. Use Markovnikov's rule to predict which is the major product.

a. $CH_3 \underset{\underset{}{|}}{\overset{\overset{CH_3}{|}}{\,C}}{=}CH_2 + HI \longrightarrow$ **b.** [cyclopentene with CH$_3$] $+ HCl \longrightarrow$

Solution

a. HI adds to 2-methylpropene (isobutylene) to form two possible products:

$$CH_3{-}\underset{\underset{}{|}}{\overset{\overset{CH_3}{|}}{\,C}}{=}CH_2 + HI \longrightarrow CH_3{-}\underset{\underset{I\;\;\;H}{|\quad\;|}}{\overset{\overset{CH_3}{|}}{\,C}}{-}CH_2 + CH_3{-}\underset{\underset{H\;\;\;I}{|\quad\;|}}{\overset{\overset{CH_3}{|}}{\,C}}{-}CH_2$$

2-methylpropene   2-iodo-2-methyl-   1-iodo-2-methyl-
(isobutylene)   propane   propane
(*t*-butyl iodide)   (isobutyl iodide)
(major product)   (minor product)

In forming 2-iodo-2-methylpropane (*tert*-butyl iodide), hydrogen adds to the carbon of the double bond bearing two hydrogens; in forming 1-iodo-2-methylpropane (isobutyl iodide), hydrogen adds to the carbon of the double bond bearing no hydrogens. Markovnikov's rule predicts that 2-iodo-2-methylpropane is the major product.

b. Addition of H—Cl to 1-methylcyclopentene forms two products:

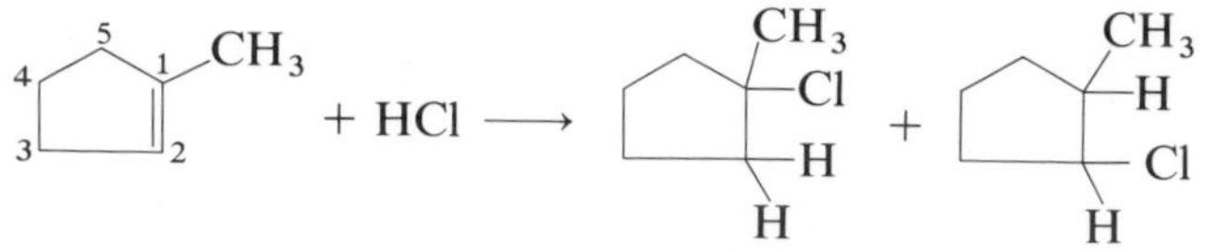

1-methylcyclo-
pentene

1-chloro-1-methyl-
cyclopentane
(major product)

1-chloro-2-methyl-
cyclopentane
(minor product)

Carbon 1 of this cycloalkene contains no hydrogen atoms and carbon 2 contains one hydrogen. Using Markovnikov's rule, predict that hydrogen adds to carbon 2 and that 1-chloro-1-methylcyclopentane is the major product.

Problem 3.4

Name and draw structural formulas for the two possible products of the following alkene addition reactions. Use Markovnikov's rule to predict which is the major product.

$$\text{CH}_3$$
$$|$$
a. $\text{CH}_3\text{—CH}{=}\text{C—CH}_3 + \text{HI} \longrightarrow$ **b.** $+ \text{HI} \longrightarrow$

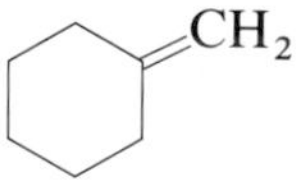

C. Addition of Water: Hydration

In the presence of an acid catalyst, most commonly concentrated sulfuric acid, water adds to alkenes to give alcohols. Addition of water to an alkene is called **hydration**. In simple alkenes, —H adds to the carbon of the double bond with the greater number of hydrogens and —OH adds to the carbon with the fewer hydrogens. Thus, H—OH adds to alkenes in accordance with Markovnikov's rule.

$$\text{CH}_3\text{—CH}{=}\text{CH}_2 + \text{H—OH} \xrightarrow{\text{H}^+} \text{CH}_3\text{—CH—CH}_2$$
$$\qquad\qquad\qquad\qquad\qquad\qquad\quad | \quad\;\; |$$
$$\qquad\qquad\qquad\qquad\qquad\qquad\;\; \text{HO} \quad \text{H}$$

propene   2-propanol

$$\qquad\quad\text{CH}_3 \qquad\qquad\qquad\qquad\qquad\qquad\;\; \text{CH}_3$$
$$\qquad\quad | \qquad\qquad\qquad\qquad\qquad\qquad\qquad\;\; |$$
$$\text{CH}_3\text{—C}{=}\text{CH}_2 + \text{H—OH} \xrightarrow{\text{H}^+} \text{CH}_3\text{—C—CH}_2$$
$$\qquad\qquad\qquad\qquad\qquad\qquad\qquad\qquad\;\; | \quad\;\; |$$
$$\qquad\qquad\qquad\qquad\qquad\qquad\qquad\;\; \text{HO} \quad \text{H}$$

2-methylpropene   2-methyl-2-
propanol

Example 3.5

Draw structural formulas for the major products of these hydration reactions:

a. $\text{CH}_3\text{—CH}_2\text{—CH}{=}\text{CH}_2 + \text{H}_2\text{O} \xrightarrow{\text{H}^+}$ **b.** $+ \text{H}_2\text{O} \xrightarrow{\text{H}^+}$

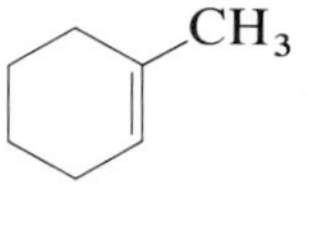

Solution **a.** $CH_3-CH_2-CH=CH_2 + H_2O \xrightarrow{H^+} CH_3-CH_2-\underset{\underset{\displaystyle OH}{|}}{CH}-CH_3$

2-butanol

b. 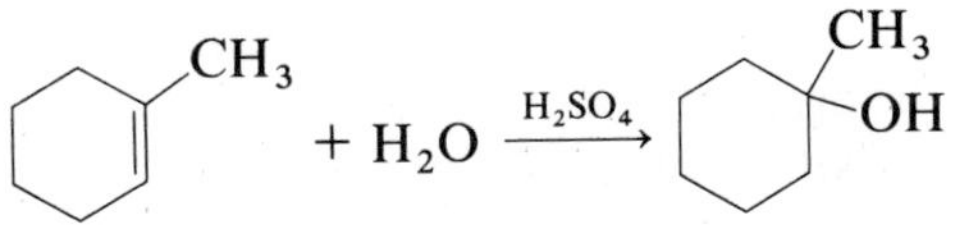 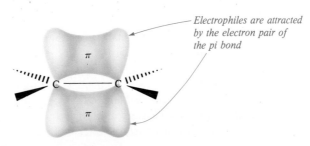 $+ H_2O \xrightarrow{H_2SO_4}$

1-methylcyclo-
hexanol

Problem 3.5

Draw structural formulas for the major products of these hydration reactions:

a. $CH_3-\underset{\underset{\displaystyle CH_3}{|}}{C}=CH-CH_3 + H_2O \xrightarrow{H^+}$

b. $CH_2=\underset{\underset{\displaystyle CH_3}{|}}{C}-CH_2-CH_3 + H_2O \xrightarrow{H^+}$

D. Mechanism of Addition to Alkenes

It became apparent to chemists studying the addition of HX, Br_2, and Cl_2 to alkenes that initial attack on a carbon-carbon double bond is by an electrophile. An **electrophile** is any molecule or ion that can accept a pair of electrons and in the process form a new covalent bond. Following are Lewis structures for two electrophiles:

$$H^+ \qquad\qquad :\overset{..}{\underset{..}{Br}}^+$$

hydrogen   bromonium
ion   ion

The susceptibility of a double bond to attack by electrophiles is consistent with our electronic formulation; this pictures the double bond as composed of one sigma bond, in which electron density is located on the bond axis, and one pi bond, in which electron density is located in two lobes, one above and the other below the plane created by the sigma bond framework.

Electrophiles are attracted by the electron pair of the pi bond

1. Use of Curved Arrows in Organic Reaction Mechanisms

To show the flow of electrons in the making and breaking of bonds during a chemical reaction, chemists have adopted a symbol called a **curved arrow**. A curved arrow is used to show the flow of electron pairs *from where to where*. *From where* is shown by the tail of the arrow and may be (1) from an atom to an adjacent bond or (2) from a bond to an adjacent atom. *To where* is shown by the head of the arrow. Further, curved arrows are always read in a specific way, namely from left to right. When a curved arrow or curved arrows are shown on the left side of an equation, then the result of that flow of electrons is always shown in the structure or structures on the right, as illustrated by the flow of electrons during reaction of hydroxide ion with a proton to form a water molecule. In this illustration, the curved arrow shows that a pair of unshared electrons in the valence shell of oxygen is used to form a new covalent bond with hydrogen:

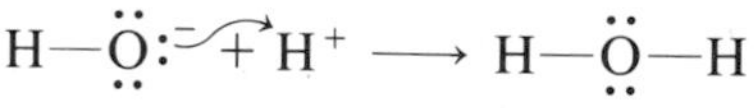

$$H - \ddot{\underset{\cdot\cdot}{O}} :\frown + H^+ \longrightarrow H - \ddot{\underset{\cdot\cdot}{O}} - H$$

In a sense, the curved arrow is nothing more than a bookkeeping device for keeping track of electron pairs. Do not be misled by its simplicity. Being able to follow the flow of electrons is fundamental to understanding the reactions of organic molecules and ions. Therefore, it is extremely important that you learn how to use this symbol properly.

2. Formation of Carbocation Intermediates

Addition of H—X to an alkene can be accounted for by a two-step mechanism. Addition begins by interaction of HX with the electron pair of the pi bond and formation of a new C—H bond, as illustrated by the reaction of HBr and 2-butene. The curved arrows on the left side of the equation show a combination of two bond-breaking steps and one bond-making step. First, the pi bond of the alkene is broken and the electron pair used instead to make a bond with the hydrogen atom of HBr. Second, the sigma bond in HBr is broken and the electron pair is given entirely to bromine, forming a bromide ion.

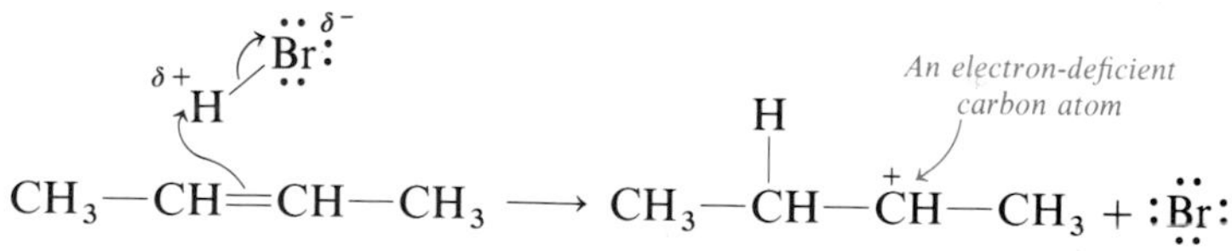

$$CH_3 - CH \overset{\overset{\displaystyle \delta^+ \,\, H}{|}}{=\!=} CH - CH_3 \longrightarrow CH_3 - \overset{\overset{\displaystyle H}{|}}{CH} - \overset{+}{CH} - CH_3 + :\ddot{\underset{\cdot\cdot}{Br}}:^-$$

An electron-deficient carbon atom

This step leaves one carbon atom with only six electrons in its valence shell. A carbon with six electrons in its valence shell bears a positive charge and is called a **carbocation** (carbon-containing cation). Carbocations are also called carbonium ions by analogy with the ammonium, NH_4^+, and hydronium, H_3O^+, ions. We will use the term *carbocation* throughout the text. You should, however, be aware of the term *carbonium ion*, since it is still used.

Because a carbocation contains an electron-deficient, positively charged carbon atom, it is an unstable intermediate. In reaction of H—X with an alkene, the carbocation intermediate reacts rapidly with halide ion to form a new C—X bond. The curved arrow on the left side of the equation shows that a pair of electrons on bromide ion is used to make a new C—Br bond:

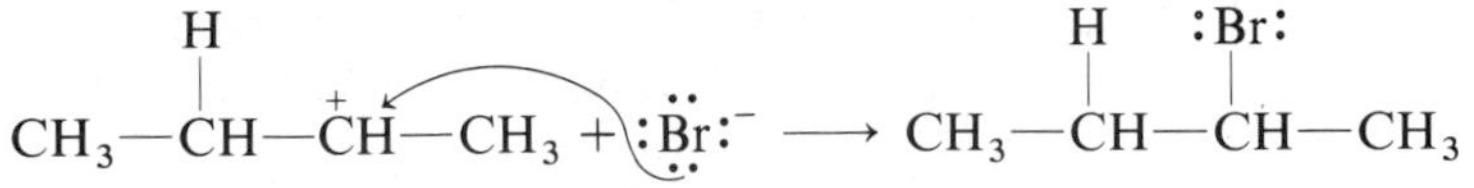

$$
\underset{\substack{| \\ \mathrm{CH_3-CH-\overset{+}{CH}-CH_3}}}{\overset{\mathrm{H}}{}} + :\!\overset{..}{\underset{..}{Br}}:^- \longrightarrow \underset{\substack{| \qquad | \\ \mathrm{CH_3-CH-CH-CH_3}}}{\overset{\mathrm{H} \quad :\overset{..}{Br}:}{}}
$$

3. Relative Ease of Formation of Carbocations: An Explanation for Markovnikov's Rule

Carbocations are classified as primary (1°), secondary (2°), or tertiary (3°), depending on the number of alkyl groups bonded to the carbon bearing the positive charge. There is much experimental evidence that a tertiary carbocation is formed more easily than a secondary carbocation, which is in turn formed more easily than a primary carbocation.

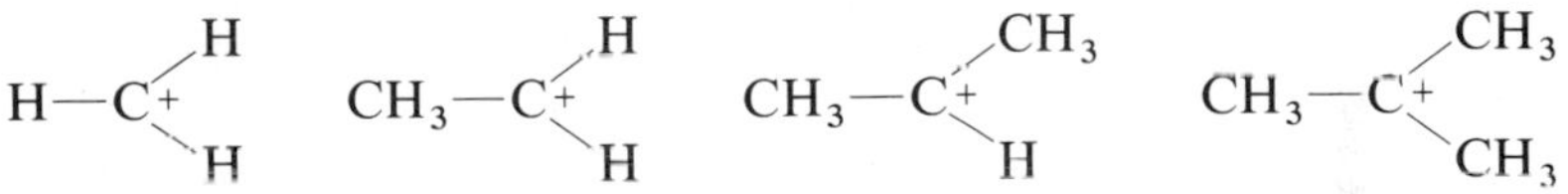

| methyl carbocation (methyl) | ethyl carbocation (1°) | isopropyl carbocation (2°) | tert-butyl carbocation (3°) |

Example 3.6

Label these carbocations primary, secondary, or tertiary, and arrange them in order of increasing ease of formation:

a. $\underset{\substack{| \qquad | \\ \quad \mathrm{CH_3} }}{\underset{\substack{| \\ \mathrm{CH_3-\overset{+}{CH}-C-CH_3}}}{\overset{\mathrm{CH_3}}{}}}$ b. $\underset{\substack{| \\ \mathrm{CH_3}}}{\underset{\substack{\mathrm{CH_3-\overset{+}{C}-CH-CH_3}}}{\overset{\mathrm{CH_3}}{}}}$

c. $\underset{\substack{| \\ \mathrm{CH_3}}}{\overset{\mathrm{CH_3}}{\underset{}{\mathrm{CH_3 \quad C-CH_2-\overset{+}{CH_2}}}}}$

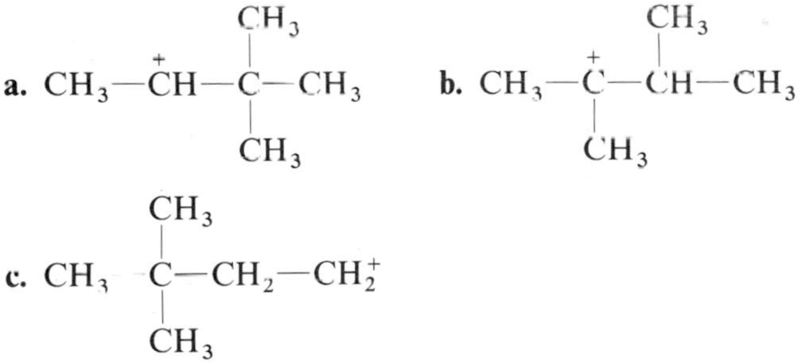

Solution

(a) is a secondary carbocation, (b) is tertiary and (c) is primary. In order of increasing ease of formation they are c, a, b.

Problem 3.6

Label these carbocations primary, secondary, or tertiary and arrange them in order of increasing ease of formation:

a.   b.   c.

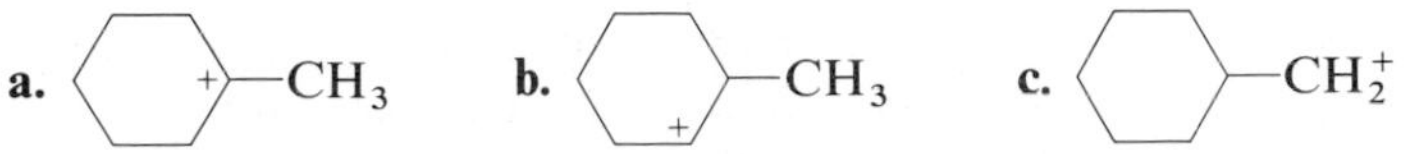

To account for the fact that HX adds to alkenes to give a major product and a minor product and for the observations generalized in Markovnikov's rule, the carbocation mechanism proposes that reaction of H—X and an alkene can give two different carbocation intermediates depending on how H⁺ adds to the double bond. This is illustrated for the reaction of HBr with propene:

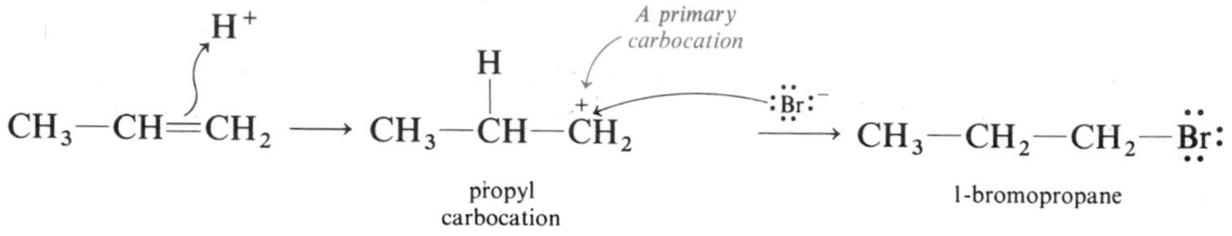

propyl carbocation — 1-bromopropane

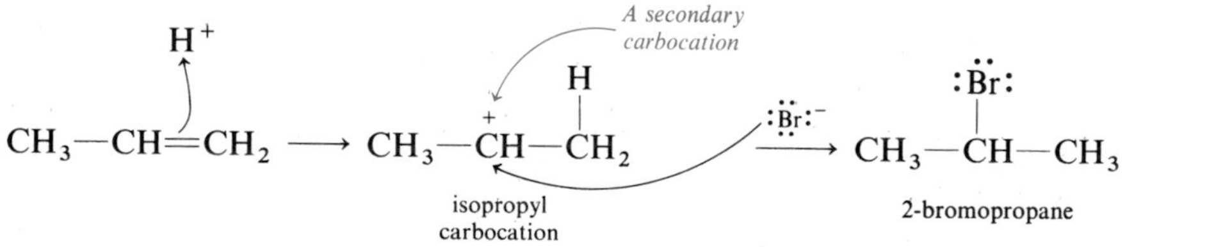

isopropyl carbocation — 2-bromopropane

The isopropyl carbocation is a secondary carbocation and formed more easily than the propyl carbocation (a primary carbocation). Therefore, the major product of reaction of propene with HBr is 2-bromopropane, formed by reaction of the isopropyl carbocation with bromide ion.

Similarly, in the reaction of HBr with 2-methylpropene (isobutylene), adding H⁺ to the carbon-carbon double bond gives either an isobutyl carbocation (a primary carbocation) or a *tert*-butyl carbocation (a tertiary carbocation).

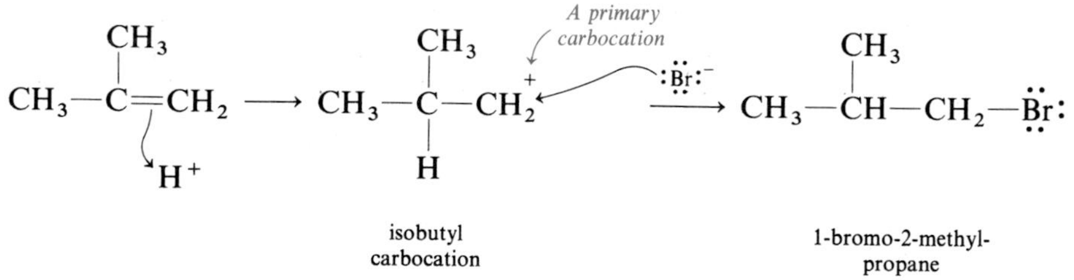

isobutyl carbocation — 1-bromo-2-methyl-propane

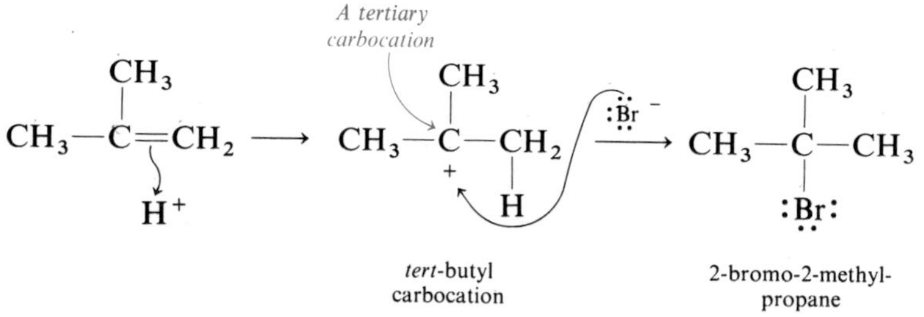

tert-butyl carbocation — 2-bromo-2-methyl-propane

The *tert*-butyl carbocation (a tertiary carbocation) is formed more easily than the isobutyl carbocation (a primary carbocation). Therefore, the major product formed in the reaction of 2-methylpropene with HBr is 2-bromo-2-methylpropane (*tert*-butyl bromide)—formed by the reaction of the *tert*-butyl carbocation with bromide ion.

The mechanism for acid-catalyzed hydration of alkenes is similar to what we have already proposed for addition of HCl and HBr to alkenes and is illustrated by conversion of propene to 2-propanol. In step 1, a proton (an electrophile) reacts with the electron pair of the pi bond to form a carbocation (also an electrophile). In the reaction of propene with H^+, the secondary isopropyl carbocation forms more easily than the alternative primary propyl carbocation, and only the isopropyl carbocation is shown in the following mechanism. This intermediate then completes its valence shell in step 2 by forming a new covalent bond with an unshared pair of electrons of the oxygen atom of H_2O. Finally, loss of a proton in step 3 causes formation of an alcohol and regeneration of a proton.

Step 1

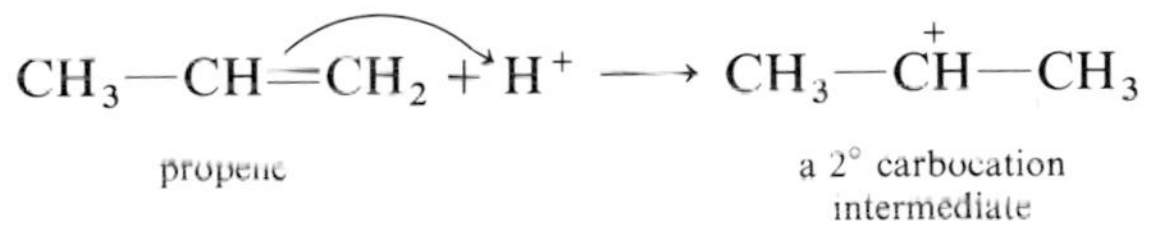

$$CH_3-CH{=}CH_2 + H^+ \longrightarrow CH_3-\overset{+}{C}H-CH_3$$

propene   a 2° carbocation intermediate

Step 2

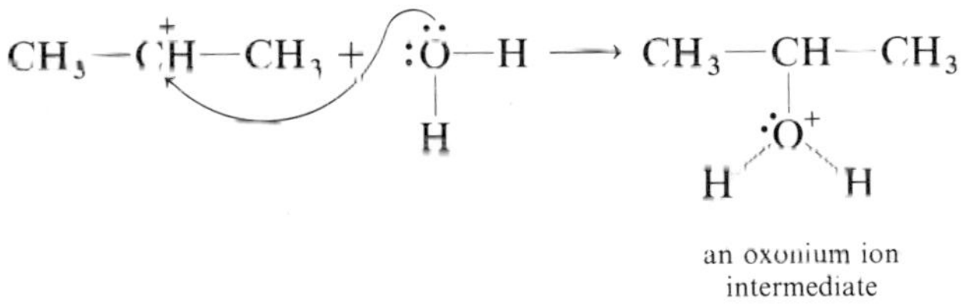

$$CH_3-\overset{+}{C}H-CH_3 + :\overset{\cdot\cdot}{O}-H \longrightarrow CH_3-CH-CH_3$$

$$\underset{H}{|} \qquad\qquad \underset{H\quad\;\;H}{\overset{|}{:\overset{+}{O}}}$$

an oxonium ion intermediate

Step 3

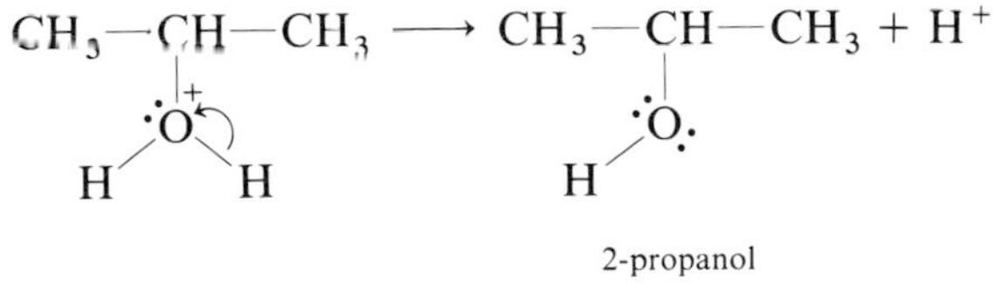

$$CH_3-CH-CH_3 \longrightarrow CH_3-CH-CH_3 + H^+$$

$$\underset{H\quad\;H}{\overset{|}{:\overset{+}{O}}} \qquad\qquad \underset{H}{\overset{|}{:\overset{\cdot\cdot}{O}}}$$

2-propanol

3.6 Reduction of Alkenes

Virtually all alkenes, no matter what the nature of the substituents on the double bond, react quantitatively with molecular hydrogen, H_2, in the presence of a metal catalyst. The reaction is one of addition of hydrogen to the double bond, and in the process, the alkene is converted to an alkane:

$$CH_2{=}CH_2 + H{-}H \xrightarrow{\text{metal catalyst}} \underset{\substack{| \\ H}}{CH_2}{-}\underset{\substack{| \\ H}}{CH_2}$$

ethylene   ethane

$$\bighexagon\!\!| \;+\; H{-}H \xrightarrow{\text{metal catalyst}} \bighexagon$$

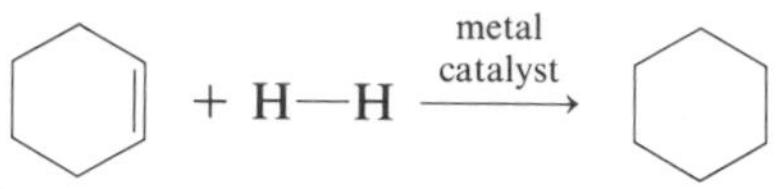

cyclohexene   cyclohexane

A balanced half-reaction for the conversion of an alkene to an alkane shows that this transformation is a two-electron reduction:

Balanced Half-Reaction

$$\overset{\diagdown}{\underset{\diagup}{C}}{=}\overset{\diagup}{\underset{\diagdown}{C}} + 2H^+ + 2e^- \longrightarrow \underset{\substack{| \\ H}}{-C}\,\underset{\substack{| \\ H}}{C-}$$

Because conversion of an alkene to an alkane involves reduction by hydrogen in the presence of a catalyst, the process is called catalytic reduction, or alternatively, catalytic hydrogenation.

The most common pattern in **catalytic reduction of an alkene** is cis addition of the two hydrogen atoms. For example, catalytic reduction of 1,2-dimethylcyclopentene gives the cis isomer in preference to the trans isomer:

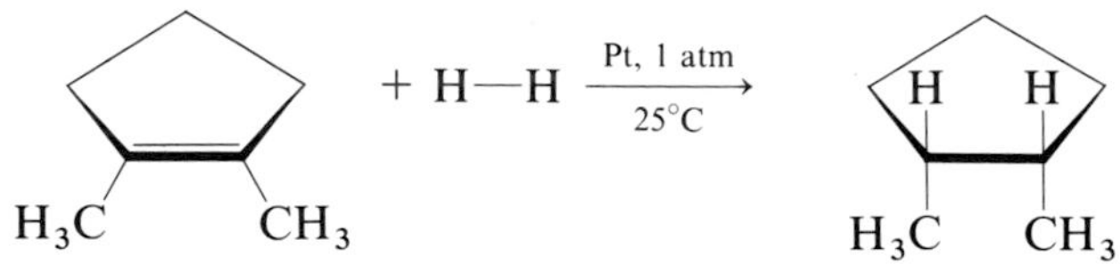

$$\text{(structure)} + H{-}H \xrightarrow[25°C]{Pt,\ 1\ atm} \text{(structure)}$$

1,2-dimethylcyclopentene   *cis*-1,2-dimethylcyclopentane

Reduction of an alkene to an alkane is immeasurably slow when no catalyst is present. It occurs rapidly, however, in the presence of transition metal catalysts such as platinum, palladium, and nickel. Separate experiments have shown that transition metals near the center of the periodic table adsorb large quantities of hydrogen onto their surfaces. During adsorption of H_2, the covalent bond between hydrogen atoms is weakened and hydrogen-metal bonds are formed. The exact nature of these bonds is not well understood. Similarly, alkenes are also adsorbed on metal surfaces and their pi bonds partially broken, with simultaneous formation of carbon-metal bonds. If both hydrogen and alkene are positioned properly on the metal surface, hydrogen atoms become attached to carbon atoms and the reduced alkane is desorbed (Figure 3.4).

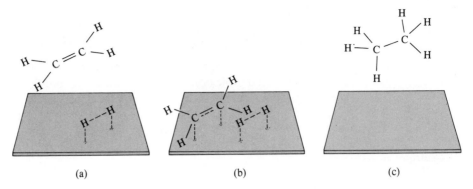

Figure 3.4 Addition of hydrogen to ethene, involving a metal cataylst.

3.7 Oxidation of Alkenes

A. Formation of Glycols

Reaction of an alkene with $KMnO_4$ in dilute alkaline solution oxidizes the alkene to a **glycol**, a compound with hydroxyl groups on adjacent carbons. In this two-electron oxidation (as we demonstrated by a balanced-half reaction in Section 3.4), permanganate is reduced to manganese dioxide, MnO_2, which precipitates as a brown solid:

$$3CH_3-CH=CH_2 + 2KMnO_4 + 4H_2O \longrightarrow 3CH_3-CH-CH_2 + 2MnO_2 + 2KOH$$
$$\qquad\qquad\qquad\qquad\qquad\qquad\qquad\qquad\qquad\qquad\qquad\quad HO \quad\; OH$$

| propene | potassium permanganate | 1,2-propanediol (a glycol) | manganese dioxide |

To form glycols in high yield, it is necessary to control reaction conditions very carefully. Most important, the reaction medium must be kept basic, generally between pH 11 and 12.

Reaction with permanganate is the basis for a qualitative test for alkenes. An aqueous solution of potassium permanganate is deep purple. When permanganate solution reacts with an alkene, the purple color of permanganate disappears and a brown precipitate of MnO_2 appears. Disappearance of the purple color coupled with appearance of a brown precipitate is evidence for the presence of an alkene. This test is not completely specific for alkenes, however, since several other functional groups also reduce permanganate to manganese dioxide.

B. Cleavage of Carbon-Carbon Double Bonds by Potassium Permanganate and Potassium Dichromate

Both potassium permanganate and potassium dichromate can be used to cleave a carbon-carbon double bond and form two C=O bonds in its place. Oxidation is carried out at elevated temperature in the prescence of sulfuric acid. If a carbon of a double bond has one attached hydrogen, it is oxidized to a carboxylic acid; if it has no attached hydrogens, it is oxidized to a ketone.

This carbon oxidized to a ketone

This carbon oxidized to a carboxylic acid

$$CH_3-\overset{\overset{\displaystyle CH_3}{|}}{C}=CH-CH_2-CH_3 + MnO_4^- \longrightarrow CH_3-\overset{\overset{\displaystyle CH_3}{|}}{C}=O + HO-\overset{\overset{\displaystyle O}{||}}{C}-CH_2-CH_3 + Mn^{2+}$$

2-methyl-2-pentene   a ketone   a carboxylic acid

$$\bigcirc + Cr_2O_7^{2-} \longrightarrow HO-\overset{\overset{\displaystyle O}{||}}{C}-CH_2-CH_2-CH_2-CH_2-\overset{\overset{\displaystyle O}{||}}{C}-OH + Cr^{3+}$$

cyclohexene   a dicarboxylic acid

Example 3.7

Draw structural formulas for the products formed when the following alkenes are oxidized by potassium dichromate. Name the new functional group formed in each oxidation.

a. $CH_3CH_2CH{=}\overset{\overset{\displaystyle CH_3}{|}}{C}HCHCH_3$ b. [cyclohexene with CH_3 substituent]

☐ *Solution*

a. $CH_3CH_2\overset{\overset{\displaystyle O}{||}}{C}OH + HO\overset{\overset{\displaystyle O}{||}}{C}\overset{\overset{\displaystyle OCH_3}{|}}{C}HCH_3$ b. $CH_3\overset{\overset{\displaystyle O}{||}}{C}CH_2CH_2CH_2CH_2\overset{\overset{\displaystyle O}{||}}{C}OH$

a carboxylic   a carboxylic   a ketone   a carboxylic
acid   acid   acid

Problem 3.7

Draw a structural formula for the alkene of the given molecular formula whose oxidation by potassium permanganate yields the products shown.

a. $C_9H_{16} \xrightarrow{\text{oxidation}} \bigcirc{=}O + CH_3CH_2\overset{\overset{\displaystyle O}{||}}{C}OH$

b. $C_7H_{12} \xrightarrow{\text{oxidation}} CH_3\overset{\overset{\displaystyle O}{||}}{C}CH_2CH_2CH_2\overset{\overset{\displaystyle O}{||}}{C}CH_3$

3.8 Polymerization of Alkenes

From the perspective of the chemical industry, the single most important reaction of alkenes is **polymerization**, the building together of many small units known as **monomers** (Greek, *mono + meros*, single parts) into very large, high-molecular-weight **polymers** (Greek, *poly + meros*, many parts). In *addition polymerization* monomer units are joined together without loss of atoms. An example of addition polymerization is the formation of polyethylene from ethylene. The following equation shows three molecules of ethylene. Curved arrows are used to show how the pi bond in each molecule is broken and the pair of electrons used to form a new sigma bond to an adjacent molecule.

$$CH_2\!=\!CH_2 + CH_2\!=\!CH_2 + CH_2\!=\!CH_2 \xrightarrow{\text{catalyst}}$$

ethylene
(monomer)

$$-CH_2-CH_2-CH_2-CH_2-CH_2-CH_2-$$

Monomer units

polyethylene
(polymer)

For a more complete picture of addition polymerization, you must also imagine other molecules to both left and right of the three ethylene units shown. In practice, hundreds of monomer units polymerize, and molecular weights of polyethylene molecules produced range from 50,000 to over 1,500,000.

Polymerization reactions are usually written in the following way, where n is a very large number, typically several thousand:

$$nCH_2\!=\!CH_2 \xrightarrow{\text{catalyst}} -(CH_2-CH_2)_n-$$

Repeating monomer unit

Propylene can also be polymerized, to give polypropylene, with methyl groups repeating regularly on every other carbon atom of the polymer chain:

$$nCH_3-CH\!=\!CH_2 \xrightarrow{\text{catalyst}} \overset{\displaystyle CH_3}{\underset{\displaystyle}{-(CH-CH_2)_n-}}$$

propylene   polypropylene

Table 3.3 lists several important polymers derived from ethylene and substituted ethylenes, along with their common names and most important uses.

The tetrafluoroethylene polymers were discovered accidentally in 1938 by Du Pont chemists. One morning a cylinder of tetrafluoroethylene appeared to be empty (no gas escaped when the valve was open) and yet the weight of the

Table 3.3
Polymers derived from ethylene and substituted ethylenes.

Monomer Formula	Common Name	Polymer Names and Common Uses
$CH_2{=}CH_2$	ethylene	polyethylene, Polythene unbreakable containers and packaging materials
$CH_2{=}CHCH_3$	propylene	polypropylene, Herculon fibers for carpeting and clothing
$CH_2{=}CHCl$	vinyl chloride	polyvinyl chloride, PVC tubing
$CH_2{=}CCl_2$	1,1-dichloro-ethylene	Saran food wrapping
$CH_2{=}CHCN$	acrylonitrile	polyacrylonitrile, Orlon acrylics and acrylates
$CF_2{=}CF_2$	tetrafluoro-ethylene	polytetrafluoroethylene, Teflon nonstick coatings
$CH_2{=}CHC_6H_5$	styrene	polystyrene, Styrofoam insulating materials
$CH_2{=}CCO_2CH_3$ $\quad\;\;\mid$ $\quad\;\;CH_3$	methyl methacrylate	polymethyl methacrylate, Lucite, Plexiglas, glass substitutes
$CH_2{=}CHCO_2CH_3$	methyl acrylate	polymethyl acrylate, acrylates, latex paints
$CH_2{=}CHOCOCH_3$	vinyl acetate	polyvinyl acetate adhesives

cylinder indicated it was full. The cylinder was opened and inside was found a waxy solid, the forerunner of Teflon.

$$n CF_2{=}CF_2 \xrightarrow{\text{polymerization}} {-}(CF_2{-}CF_2)_n{-}$$

tetrafluoro-ethylene polytetrafluoro-ethylene (Teflon)

Polytetrafluoroethylene proved to have unusual properties: extraordinary chemical inertness, outstanding heat resistance, a very high melting point, and unusual frictional properties. In 1948, Du Pont built the first commercial Teflon plant, and the product was used to make gaskets, bearings for automobiles, nonstick equipment for candy manufacturers and commercial bakers, and numerous other items. Teflon became a household word in 1961, with the introduction of nonstick frying-pans in the U.S. market.

The years since the 1930s have seen extensive research and development in polymer chemistry and physics; and an almost explosive growth in plastics, coatings, and rubber technology has created a worldwide multibillion dollar industry. A few basic characteristics account for this phenomenal growth. First, the raw materials for plastics are derived mainly from petroleum and natural gas. With the development of efficient refining technology, the raw materials for the synthesis of polymers have become generally cheap and plentiful. Second, within broad limits, scientists have learned how to tailor polymers to the requirements of the end use. Third, many plastics can be fabricated more cheaply than competing materials. For example, plastics technology created the water-based (latex) paints, which have revolutionized the coatings industry; plastic films and foams have done the same for the packaging industry. The list could go on and on as we think of the manufactured items that surround us in our daily lives.

3.9 Aromatic Hydrocarbons

Benzene is a colorless liquid with a boiling point of 80°C. Michael Faraday first isolated it in 1825 from the oily liquid that collected in the illuminating gas lines of London. Its molecular formula, C_6H_6, suggested a high degree of unsaturation. Remember that a saturated alkane of six carbons has the molecular formula C_6H_{14}, and a saturated cycloalkane, the molecular formula C_6H_{12}. Considering this high degree of unsaturation, you might expect benzene to undergo many of the same reactions as alkenes. Surprisingly, benzene does not undergo characteristic alkene reactions. For example, it does not react with bromine, chlorine, hydrogen chloride, hydrogen bromide, or other reagents that usually add to carbon carbon double bonds. It is not oxidized by potassium permanganate or potassium dichromate under conditions in which alkenes are oxidized to ketones and carboxylic acids. When benzene does react, it typically does so by substitution, in which a hydrogen atom is replaced by another atom or group of atoms. For example, benzene reacts with bromine in the presence of iron(III) bromide to form bromobenzene and hydrogen bromide:

$$C_6H_6 + Br_2 \xrightarrow{\text{FeBr}_3} C_6H_5Br + HBr$$

benzene   bromobenzene

The terms *aromatic* and *aromatic compound* have been used to classify benzene and its derivatives because many of them have distinctive odors. It has become clear, however, that a classification for these compounds should be based not on aroma but on structure and chemical reactivity. The term *aromatic* is still used today but rather than refer to aroma, it refers to the unusual chemical properties of these compounds. They do not undergo typical alkene addition and oxidation-reduction reactions; when they do react, they do so by substitution.

A. The Structure of Benzene

The six carbon atoms of benzene form a regular hexagon, with bond angles of 120°. One hydrogen atom is bonded to each carbon. Two Lewis structures, Ia and Ib, can be drawn for this arrangement of atoms. They differ only in the arrangement of double bonds within the ring.

Ia 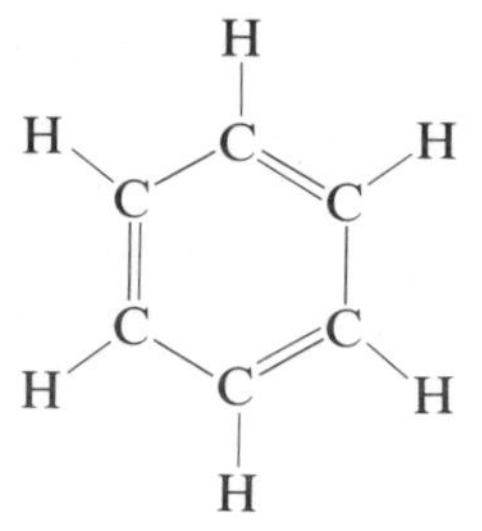   Ib

Lewis structures Ia and Ib do account for the fact that benzene is a cyclic, unsaturated hydrocarbon. They do not, however, explain why benzene does not
undergo typical alkene addition reactions, oxidations, or reductions. If benzene
contains three double bonds then, chemists asked, why doesn't it show reactions
typical of alkenes? Why, for example, doesn't benzene add three moles of bromine to form 1,2,3,4,5,6-hexabromocyclohexane?

The first adequate description of the structure and unusual chemical properties of benzene was provided by the resonance theory proposed by Linus
Pauling. When a molecule can be represented as two or more contributing
structures that differ only in position of valence electrons, the actual molecule
is best represented as a resonance hybrid. The two principal contributing structures for the benzene resonance hybrid are

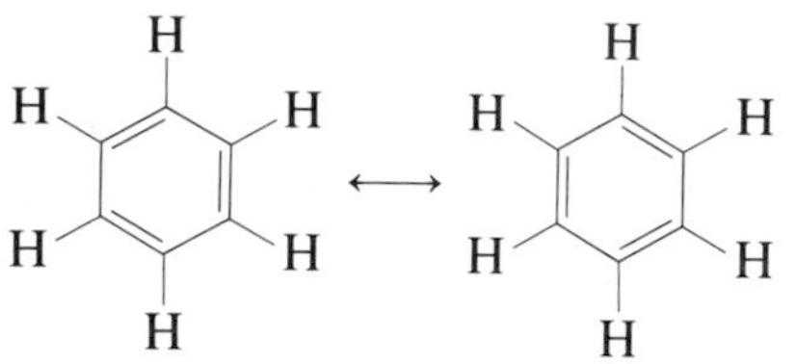

One of the consequences of resonance is a marked increase in stability of the
hybrid compared with any one of the contributing structures. Resonance stabilization is particularly large and important in benzene and other aromatic hydrocarbons. Because of their stability, benzene and other aromatic hydrocarbons
do not undergo typical alkene reactions.

B. Nomenclature of Aromatic Hydrocarbons

In the IUPAC system, monosubstituted alkylbenzenes are named as derivatives
of benzene; for example, ethylbenzene. The IUPAC system also retains certain
common names for several of the simpler monosubstituted benzenes. Examples
are toluene (rather than methylbenzene), cumene (rather than isopropylbenzene),
and styrene (rather than vinylbenzene).

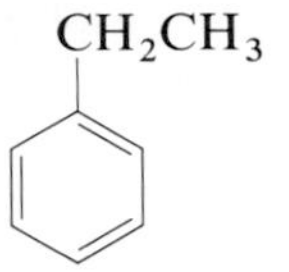

CH_2CH_3

ethylbenzene

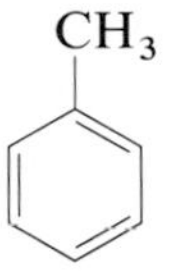

CH_3

toluene
(methylbenzene)

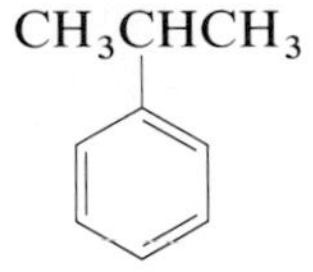

CH_3CHCH_3

cumene
(isopropylbenzene)

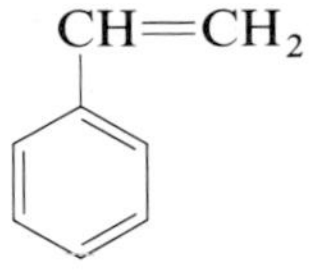

$CH=CH_2$

styrene
(vinylbenzene)

When there are two substituents on a benzene ring, three structural isomers are possible. The substituents may be located by numbering the atoms of the ring. Alternatively, the relative location of two substituents may be indicated by the prefixes *ortho*, *meta*, or *para*.

1,2- is equivalent to ortho
1,3- is equivalent to meta
1,4- is equivalent to para

These equivalent ways to locate two substituents are illustrated below by the three isomeric bromotoluenes.

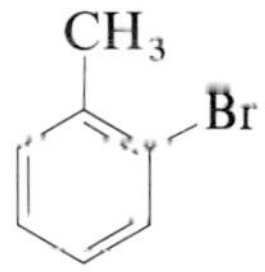

CH_3
Br

2-bromotoluene
(*o*-bromotoluene)

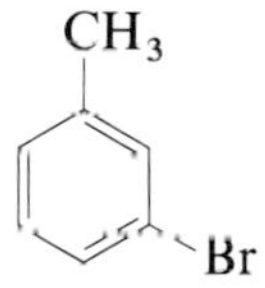

CH_3
Br

3-bromotoluene
(*m*-bromotoluene)

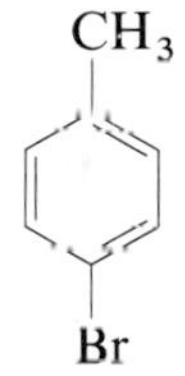

CH_3

Br

4-bromotoluene
(*p*-bromotoluene)

The IUPAC system retains the common name *xylene* for the three dimethylbenzenes.

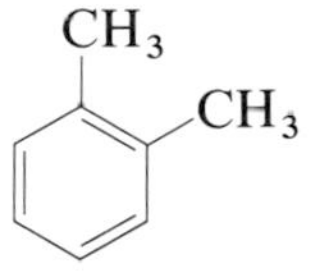

CH_3
CH_3

o-xylene

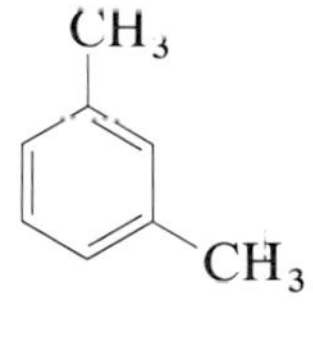

CH_3

CH_3

m-xylene

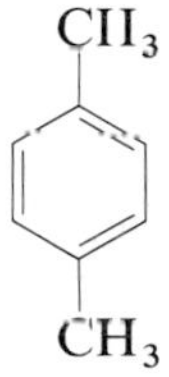

CH_3

CH_3

p-xylene

With three or more substituents, a numbering system must be used.

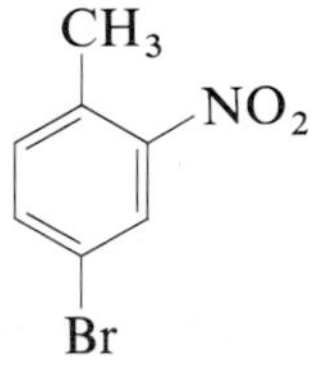

CH_3
NO_2

Br

4-bromo-2-nitrotoluene

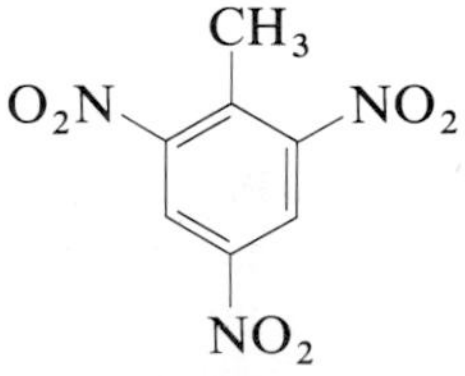

CH_3
O_2N NO_2

NO_2

2,4,6-trinitrotoluene
(TNT)

In more complex molecules, the benzene ring is often named as a substituent on a parent chain. In this case, the group C_6H_5— is called a phenyl group.

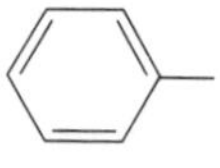

phenyl group 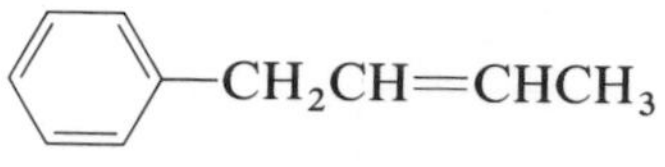   1-phenyl-2-butene 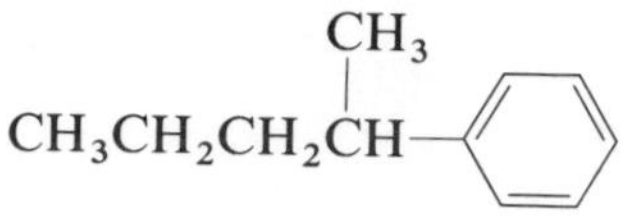   2-phenylpentane

Closely related to benzene are numerous aromatic hydrocarbons having two or more six-membered rings joined together.

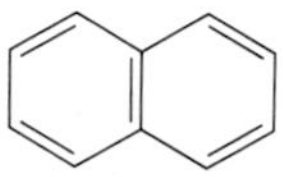

naphthalene   anthracene   phenanthrene
(mp 80°C)   (mp 217°C)   (mp 99°C)

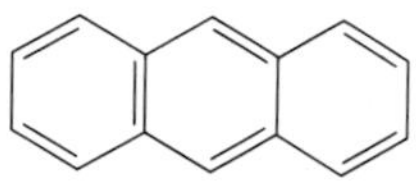

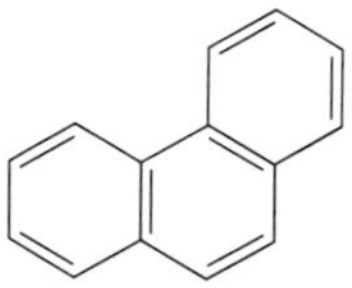

C. Reactions of Aromatic Hydrocarbons

By far the most characteristic reaction of aromatic compounds is substitution at a ring carbon. Some groups that can be introduced directly on the ring are the halogens (except fluorine); the nitro group, NO_2; the sulfonic acid group, SO_3H; and the alkyl group, R. The reaction of benzene under appropriate conditions is represented for each of these substitution reactions.

Chlorination

$$C_6H_6 + Cl_2 \xrightarrow{FeCl_3} C_6H_5Cl + HCl$$

chlorobenzene

Bromination

$$C_6H_6 + Br_2 \xrightarrow{FeBr_3} C_6H_5Br + HBr$$

bromobenzene

Nitration

$$C_6H_6 + HNO_3 \xrightarrow{H_2SO_4} C_6H_5NO_2 + H_2O$$

nitrobenzene

Sulfonation

$$C_6H_6 + H_2SO_4 \xrightarrow{SO_3} C_6H_5SO_3H + H_2O$$

benzenesulfonic
acid

Alkylation

$$C_6H_6 + CH_3CH_2Cl \xrightarrow{AlCl_3} C_6H_5CH_2CH_3 + HCl$$

ethylbenzene

Key Terms and Concepts

alkene (introduction)
allyl group (3.2B)
aromatic hydrocarbon
  (introduction and 3.9)
bonding in alkenes (3.1)
carbocation (3.5D2)
catalytic reduction of alkenes (12.6)
cis-trans isomerism in alkenes (3.3A)
curved arrows, use of (3.5D1)
electrophile (3.5D)
glycol (3.7A)
halogenation of alkenes (3.5A)
hydration of alkenes (3.5C)
hydrohalogenation of alkenes (3.5B)

Markovnikov's rule (3.5B)
methylene group (3.2B)
monomer (3.8)
nomenclature of alkenes
  (3.2A and 3.2B)
oxidation of alkenes (3.7)
polymer (3.8)
polymerization of alkenes (3.8)
reduction of alkenes (3.6)
unsaturated hydrocarbon
  (introduction)
vinyl group (3.2B)

Key Reactions of Alkenes

1. Addition of Br_2 and Cl_2: halogenation (Section 3.5A).
2. Addition of HCl, HBr, and HI: hydrohalogenation (Section 3.5B).
3. Addition of H_2O: hydration (Section 3.5C).
4. Addition of H_2: reduction (Section 3.6).
5. Oxidation to glycols by $KMnO_4$ (Section 3.7A).
6. Oxidation to ketones/carboxylic acids by $KMnO_4$ or $K_2Cr_2O_7$ (Section 3.7B).
7. Addition polymerization (Section 3.8).

Problems

*Structure
of alkenes
(Section 3.1)*

3.8 Predict all bond angles about each circled carbon atom. To make these predictions, use the valence-shell electron-pair repulsion model. (Review Section 1.3.)

a.

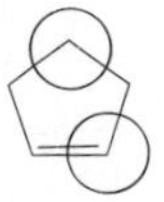

b. $CH_3 - ⓒH = CH - CH_2 - ⓒH_3$

c.  (structure) CH_2-OH d. 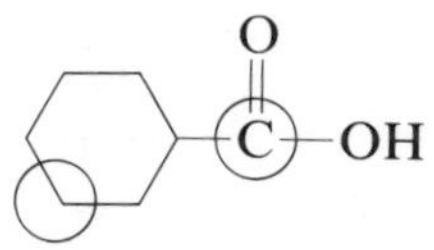 (structure) $\overset{\overset{O}{\parallel}}{C}-OH$

e. 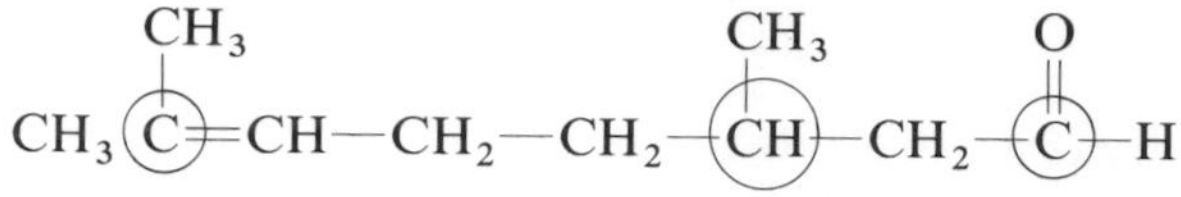 $CH_3\overset{\overset{\displaystyle CH_3}{|}}{C}=CH-CH_2-CH_2-\overset{\overset{\displaystyle CH_3}{|}}{(CH)}-CH_2-\overset{\overset{O}{\parallel}}{C}-H$

3.9 For each circled carbon atom in Problem 3.8, identify which atomic orbitals are used to form each sigma bond and which to form each pi bond.

Nomenclature
of alkenes
(Section 3.2)

3.10 Name these compounds:

a. 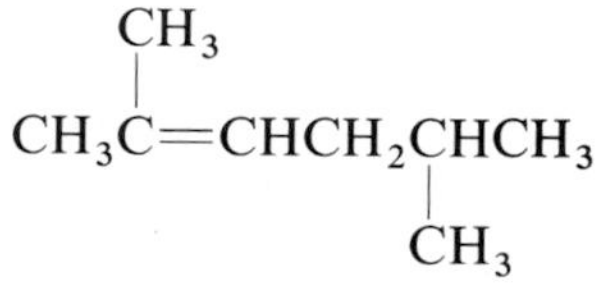 $CH_3\overset{\overset{\displaystyle CH_3}{|}}{C}=CHCH_2\overset{\underset{\displaystyle CH_3}{|}}{CHCH_3}$ b. $CH_2=\overset{\overset{\displaystyle CH_3}{|}}{C}CH_2CH_3$

c. $CH_2=\overset{\overset{\displaystyle CH_3}{|}}{C}CH=CH_2$ d. $ClCH=CHCl$

e. 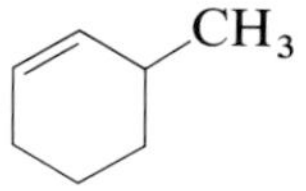 (structure with CH_3)   f. 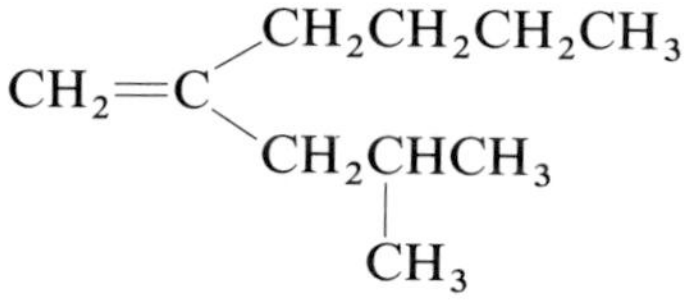 $CH_2=C\overset{\displaystyle \diagup CH_2CH_2CH_2CH_3}{\diagdown CH_2\overset{\underset{\displaystyle CH_3}{|}}{CHCH_3}}$

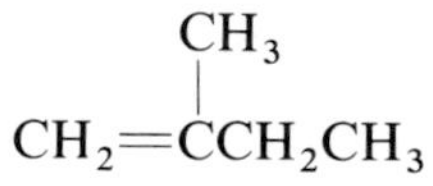

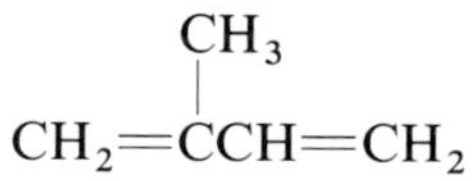

g. 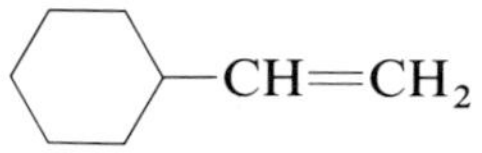 (cyclohexyl)$-CH=CH_2$ h. 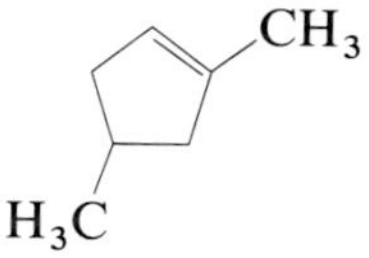 (structure with CH_3 and H_3C)

i. $(CH_3)_2CHCH=C(CH_3)_2$ j. $(CH_3)_3CCH_2CH=CH_2$

k. $CH_2=CHCH=CH_2$ l. $CH_2=CHCl$

m. 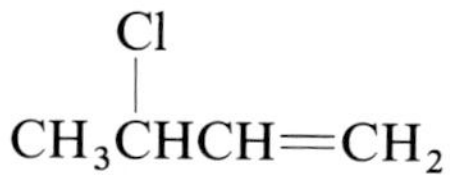 $CH_3\overset{\overset{\displaystyle Cl}{|}}{CHCH}=CH_2$ n. 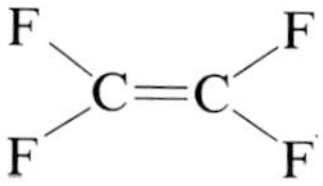 $\overset{\displaystyle F}{\underset{\displaystyle F}{\diagup}}C=C\overset{\displaystyle F}{\underset{\displaystyle F}{\diagdown}}$

o. 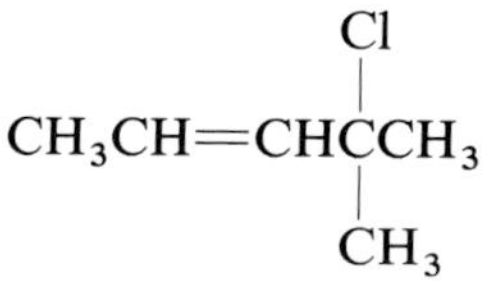 $CH_3CH=CH\overset{\overset{\displaystyle Cl}{|}}{\underset{\underset{\displaystyle CH_3}{|}}{C}}CH_3$ p. $ClCH_2CH=CHCH_2Cl$

3.11 Draw structural formulas for these compounds.
  a. 2-methyl-3-hexene
  b. 2-methyl-2-hexene

c. 3,3-dimethyl-1-butene
d. 3-ethyl-3-methyl-1-pentene
e. 2,3-dimethyl-2-butene
f. 1-pentene
g. 2-pentene
h. 1-chloropropene
i. 2-chloropropene
j. 3-chloro-3-methylcyclohexene
k. 1-isopropyl-4-methylcyclohexene
l. 1-phenylcyclohexene
m. 3-hexene
n. 5-isopropyl-3-octene
o. 3-phenyl-1-butene
p. tetrachloroethylene

Cis-trans
isomerism in
alkenes
(Section 3.3)

3.12   Which of the molecules in Problem 3.11 show cis-trans isomerism? For each that does, draw both cis and trans isomers.

3.13   Which of these molecules show cis-trans isomerism?

a. 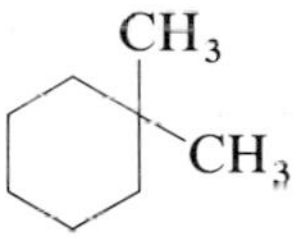   b. 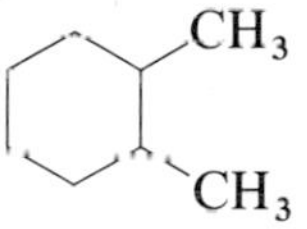   c.

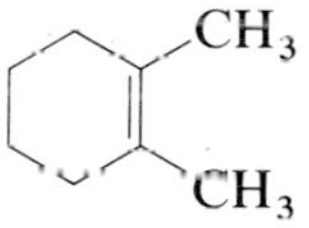

d. ClCH=CHCl

e. $CH_3(CH_2)_3CH=CH(CH_2)_7\overset{\displaystyle O}{\overset{\|}{C}}OH$

f. $HO\overset{\displaystyle O}{\overset{\|}{C}}CH=CH\overset{\displaystyle O}{\overset{\|}{C}}OH$

3.14   Draw structural formulas for all compounds of molecular formula C_5H_{10} that are:
a. alkenes that do not show cis-trans isomerism.
b. alkenes that do show cis-trans isomerism.
c. cycloalkanes that do not show cis-trans isomerism.
d. cycloalkanes that do show cis-trans isomerism.

3.15   Draw structural formulas for the four isomeric chloropropenes, C_3H_5Cl.

Reactions of
alkenes;
addition
(Sections
3.4–3.5)

3.16   Draw structural formulas for the major product of these alkene addition reactions:

a. $CH_3\overset{\displaystyle CH_3}{\overset{|}{C}}=CHCH_3 + H_2O \xrightarrow{H_2SO_4}$

b. $CH_3\overset{\displaystyle CH_3}{\overset{|}{C}}=CHCH_3 + HBr \longrightarrow$

$$\text{c. } CH_3\overset{\overset{\displaystyle CH_3}{|}}{C}{=}CHCH_3 + Br_2 \longrightarrow$$

$$\text{d. } CH_3\overset{\overset{\displaystyle CH_3}{|}}{C}{=}CHCH_3 + H_2 \xrightarrow{Pt}$$

e. 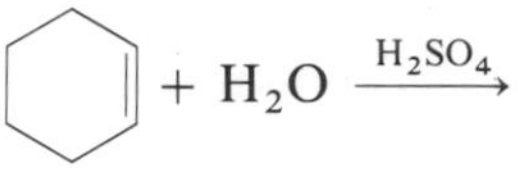 ⬡ $+ H_2O \xrightarrow{H_2SO_4}$

$$\text{f. } CH_2{=}\overset{\overset{\displaystyle CH_3}{|}}{C}CH_2CH_3 + H_2O \xrightarrow{H_2SO_4}$$

g. 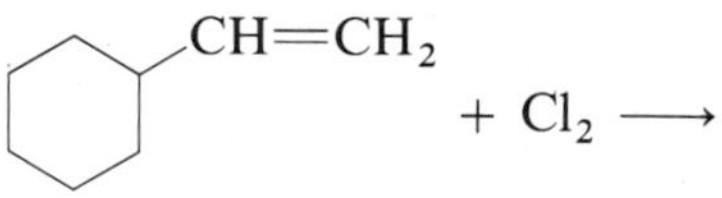 ⬡–CH=CH$_2$ $+ Cl_2 \longrightarrow$

$$\text{h. } \overset{\displaystyle CH_3}{\underset{\displaystyle H}{}}{C}{=}C\overset{\displaystyle CH_2CH_3}{\underset{\displaystyle H}{}} + HCl \longrightarrow$$

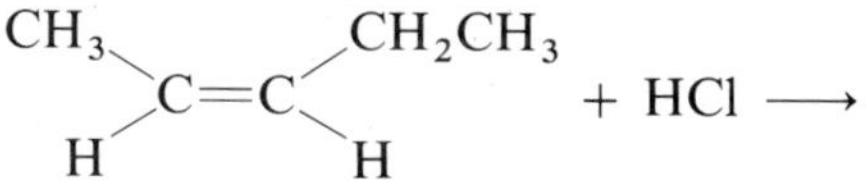

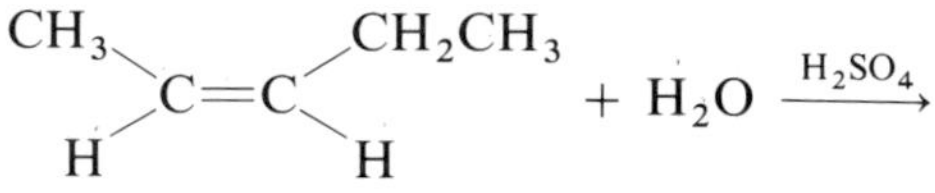

$$\text{i. } \overset{\displaystyle CH_3}{\underset{\displaystyle H}{}}{C}{=}C\overset{\displaystyle CH_2CH_3}{\underset{\displaystyle H}{}} + H_2O \xrightarrow{H_2SO_4}$$

$$\text{j. } \overset{\displaystyle CH_3}{\underset{\displaystyle H}{}}{C}{=}C\overset{\displaystyle H}{\underset{\displaystyle CH_2CH_3}{}} + H_2O \xrightarrow{H_2SO_4}$$

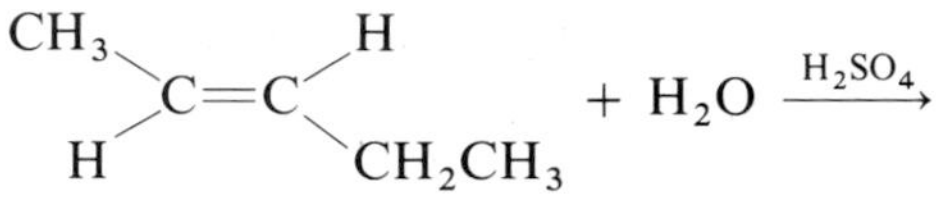

$$\text{k. } CH_2{=}\overset{\overset{\displaystyle CH_3}{|}}{C}CH{=}CH_2 + 2H_2 \xrightarrow{Pt}$$

$$\text{l. } CH_3\overset{\overset{\displaystyle H_3C}{|}}{C}{=}\overset{\overset{\displaystyle CH_3}{|}}{C}CH_3 + Br_2 \longrightarrow$$

m. 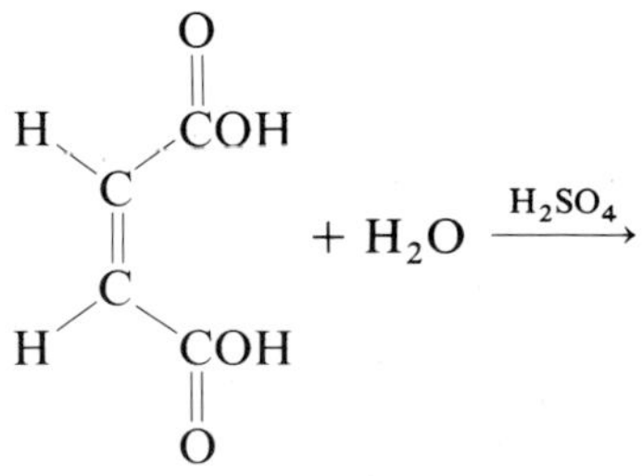
$$\overset{\displaystyle H}{\underset{\displaystyle H}{}}{C}{=}C\overset{\displaystyle \overset{O}{\overset{||}{C}}OH}{\underset{\displaystyle \underset{O}{\underset{||}{C}}OH}{}} + H_2O \xrightarrow{H_2SO_4}$$

3.17   Draw the structural formula for an alkene or alkenes of molecular formula C_5H_{10} that will react to give the indicated compound as the major product. Note that in several parts of this problem (for example, part a), more than one alkene will give the same compound as the major product.

$$\text{a. } C_5H_{10} + H_2O \xrightarrow{H_2SO_4} CH_3\overset{\displaystyle CH_3}{\underset{\displaystyle OH}{C}}CH_2CH_3$$

$$\text{b. } C_5H_{10} + Br_2 \longrightarrow CH_3\overset{\displaystyle CH_3}{CH}\underset{\displaystyle Br\ \ Br}{CHCH_2}$$

$$\text{c. } C_5H_{10} + H_2 \xrightarrow{Pt} CH_3CH_2CH_2CH_2CH_3$$

$$\text{d. } C_5H_{10} + H_2O \xrightarrow{H_2SO_4} CH_3\underset{\displaystyle OH}{CH}CH_2CH_2CH_3$$

$$\text{e. } C_5H_{10} + HCl \longrightarrow CH_3\overset{\displaystyle CH_3}{\underset{\displaystyle Cl}{C}}CH_2CH_3$$

3.18   Draw structural formulas for the carbocations formed by the reaction of H⁺ with the following alkenes. Where two different carbocations are possible, state which is more stable.

$$\text{a. } CH_3CH_2\overset{\displaystyle CH_3}{C}{=}CHCH_3 + H^+ \longrightarrow$$

$$\text{b. } CH_3CH_2CH{=}CHCH_3 + H^+ \longrightarrow$$

c. 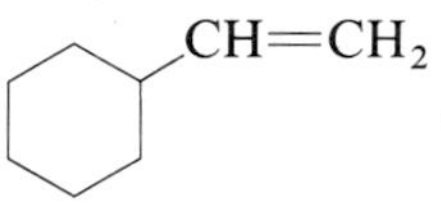   (cyclohexane ring)—CH=CH₂   + H⁺ ⟶

d. 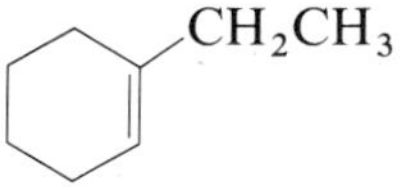   (cyclohexene ring)—CH₂CH₃   + H⁺ ⟶

3.19   Write a reaction mechanism for the following alkene addition reactions. For each mechanism, identify all electrophiles and reactive intermediates.

$$\text{a. } CH_3\overset{\displaystyle CH_3}{C}{=}CH_2 + HCl \longrightarrow CH_3\overset{\displaystyle CH_3}{\underset{\displaystyle Cl}{C}}CH_3$$

$$\underset{\substack{|\\ \text{CH}_3}}{\text{CH}_3\overset{\text{CH}_3}{\text{C}}=\text{CH}_2} + \text{H}_2\text{O} \xrightarrow{\text{H}_2\text{SO}_4} \text{CH}_3\underset{\substack{|\\ \text{OH}}}{\overset{\substack{\text{CH}_3\\|}}{\text{C}}}\text{CH}_3$$

b.

3.20   Terpin hydrate is prepared commercially by the addition of two moles water to limonene in the presence of dilute sulfuric acid. Limonene is found in lemon, orange, caraway, dill, bergamot, and some other oils. Terpin hydrate is used medicinally as an expectorant for coughs. It may be given as a mixture of terpin hydrate and codenine. Propose a structure for terpin hydrate and a reasonable mechanism to account for the formation of the product you have predicted.

$$+ 2\text{H}_2\text{O} \xrightarrow[\text{H}_2\text{SO}_4]{\text{dilute}} \text{C}_{10}\text{H}_{20}\text{O}_2$$

terpin hydrate

limonene

3.21   Reaction of 2-methylpropene with methanol in the presence of concentrate H_2SO_4 yields a product of molecular formula $\text{C}_5\text{H}_{12}\text{O}$.

$$\underset{\substack{|\\ \text{CH}_3}}{\text{CH}_3\overset{\text{CH}_3}{\text{C}}=\text{CH}_2} + \text{CH}_3\text{OH} \xrightarrow{\text{H}_2\text{SO}_4} \text{C}_5\text{H}_{12}\text{O}$$

a. Propose a structural formula for $\text{C}_5\text{H}_{12}\text{O}$.
b. Propose a reaction mechanism for the formation of this product.

3.22   Show how you could distinguish between the members of the following pairs of compounds by a simple chemical test. In each case, tell what test you would perform and what you would expect to observe, and write an equation for each positive test.
a. cyclohexane and 1-hexene
b. 1-hexene and 2-chlorohexane
c. 2,3-dimethyl-2-butene and 1,1-dimethylcyclopentane

Reactions of alkenes; oxidation (Section 3.7)

3.23   Define oxidation.

3.24   Show by writing a balanced half-reaction that these reactions are oxidations:

a. $$\underset{\text{CH}_3}{\overset{\text{CH}_3}{>}}\text{C}=\text{C}\underset{\text{CH}_3}{\overset{\text{CH}_3}{<}} \longrightarrow \underset{\text{CH}_3}{\overset{\text{CH}_3}{>}}\text{C}=\text{O} + \text{O}=\text{C}\underset{\text{CH}_3}{\overset{\text{CH}_3}{<}}$$

b. $\text{CH}_3\text{CH}_2\text{CH}=\text{CHCH}_2\text{CH}_3 \longrightarrow 2\text{CH}_3\text{CH}_2\overset{\overset{\text{O}}{\|}}{\text{C}}\text{OH}$

c. 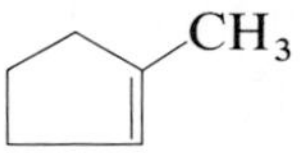 [cyclopentene with CH$_3$ structure] $\longrightarrow$ $CH_3\overset{O}{\overset{\|}{C}}CH_2CH_2CH_2\overset{O}{\overset{\|}{C}}OH$

3.25 Draw structural formulas for the products of oxidation of these alkenes:

a. $CH_3CH_2CH=\overset{\overset{\displaystyle CH_3}{|}}{C}CH_2CH_3$ $\xrightarrow{\text{oxidation}}$

b. 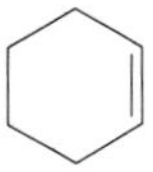 [cyclohexene structure] $\xrightarrow{\text{oxidation}}$

c. 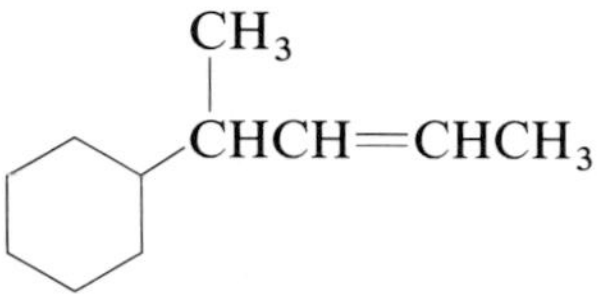 [cyclohexane with $\overset{\overset{\displaystyle CH_3}{|}}{C}HCH=CHCH_3$ substituent] $\xrightarrow{\text{oxidation}}$

d. $CH_3CH_2CH=CHCH_2CH_2\overset{O}{\overset{\|}{C}}OH$ $\xrightarrow{\text{oxidation}}$

3.26 Draw the structural formula for an alkene of given molecular formula that can be oxidized by hot potassium permanganate to give the indicated products.

a. C_6H_{12} $\xrightarrow{\text{oxidation}}$ $CH_3CH_2\overset{O}{\overset{\|}{C}}OH + CH_3\overset{O}{\overset{\|}{C}}CH_3$

b. C_6H_{12} $\xrightarrow{\text{oxidation}}$ $CH_3\overset{O}{\overset{\|}{C}}OH + CH_3CH_2CH_2\overset{O}{\overset{\|}{C}}OH$

c. C_7H_{12} $\xrightarrow{\text{oxidation}}$ $CH_3\overset{O}{\overset{\|}{C}}CH_2CH_2CH_2\overset{O}{\overset{\|}{C}}CH_3$

d. C_8H_{14} $\xrightarrow{\text{oxidation}}$ $HO\overset{O}{\overset{\|}{C}}CH_2\underset{\underset{\displaystyle CH_3}{|}}{\overset{\overset{\displaystyle CH_3}{|}}{C}}CH_2CH_2\overset{O}{\overset{\|}{C}}OH$

e. C_9H_{16} $\xrightarrow{\text{oxidation}}$ [cyclohexane ring]$=O + CH_3\overset{O}{\overset{\|}{C}}CH_3$

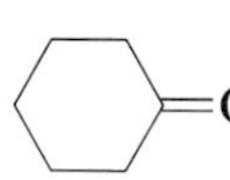

f. C_9H_{16} $\xrightarrow{\text{oxidation}}$ [cyclohexane ring]$-\overset{O}{\overset{\|}{C}}OH + CH_3\overset{O}{\overset{\|}{C}}OH$

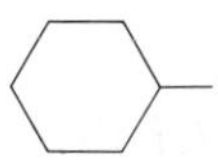

Reactions of alkenes; polymerization (Section 3.8)

3.27   Following is a structural formula for a section of polypropylene derived from three propylene monomers.

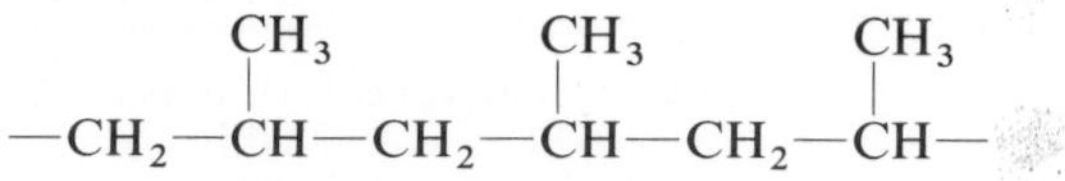

$$-CH_2-\underset{\underset{CH_3}{|}}{CH}-CH_2-\underset{\underset{CH_3}{|}}{CH}-CH_2-\underset{\underset{CH_3}{|}}{CH}-$$

a section of polypropylene

Draw structural formulas for comparable three-unit sections of:
a. polyvinyl chloride   **b.** Saran
c. Teflon   **d.** Plexiglas

3.28   Natural rubber is a polymer derived from the monomer 2-methyl-1,3-butadiene (isoprene). The repeating unit in this polymer is

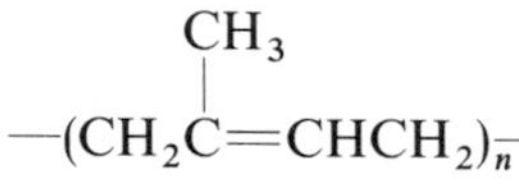

$$-(CH_2\underset{\underset{CH_3}{|}}{C}=CHCH_2)_n-$$

natural rubber

a. Draw the structural formula for a section of natural rubber, showing three repeating monomer units.
b. Draw the structural formula of the product of oxidation and cleavage of the carbon-carbon double bond in natural rubber, and name the two new functional groups in the product.
c. The smog prevalent in Los Angeles contains oxidizing agents. How could you account for the fact that this type of smog attacks natural rubber (automobile tires, and the like) but does not affect polyethylene or polyvinyl chloride?

Aromatic hydrocarbons (Section 3.9)

3.29   For each circled atom, predict bond angles formed by the attached atoms. Use the valence-shell electron-pair repulsion model to make these predictions.

a. 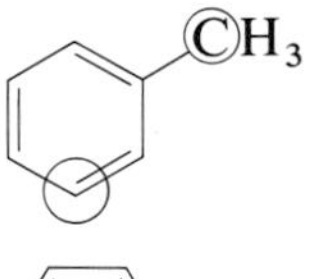   (ring)—ⒸH_3

b. 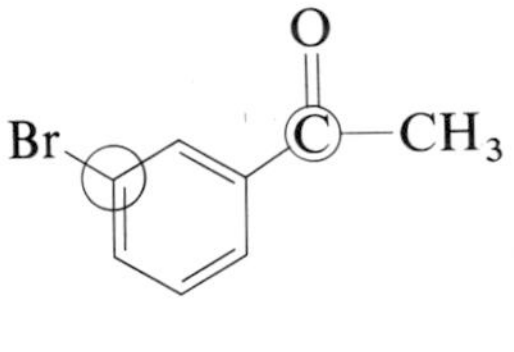   Br—(ring)—$\overset{\overset{O}{\|}}{C}$—$CH_3$

c.   (ring)—ⒸH=CH_2

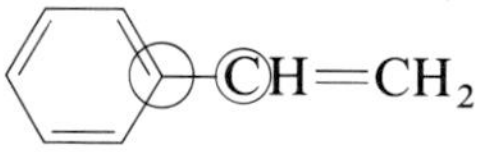

3.30   For each circled atom in Problem 3.29, identify which atomic orbitals are used to form each sigma bond and which to form each pi bond.

3.31   Name the compounds.

a. 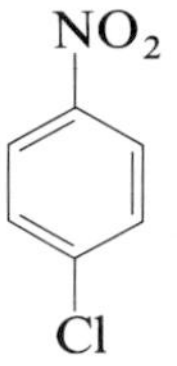 NO_2 (ring) Cl

b. 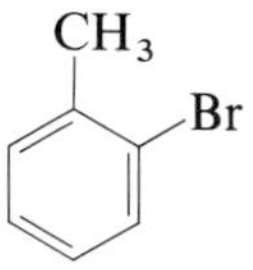 CH_3 (ring) Br

c. 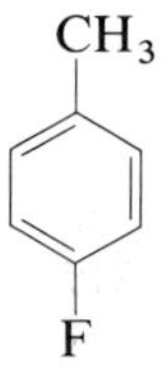 CH_3 (ring) F

d. 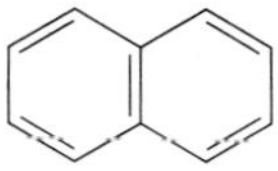 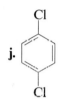   **e.**

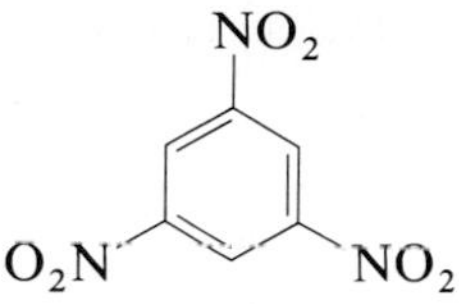

CH₃CHCH=CHCH₃

f.

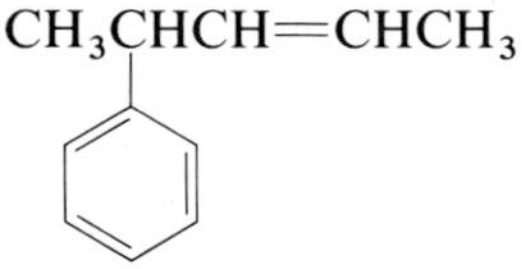

CH=CH₂   CH₃CHCH₂CH=CH₂   CH₃

g.   **h.** 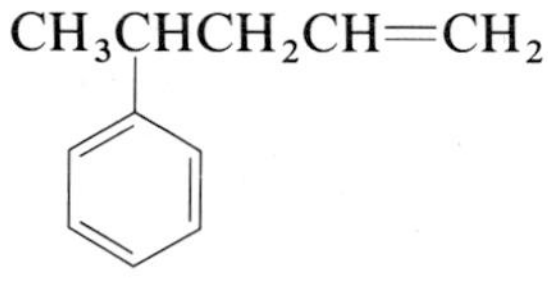   **i.**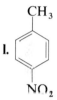
CH₃

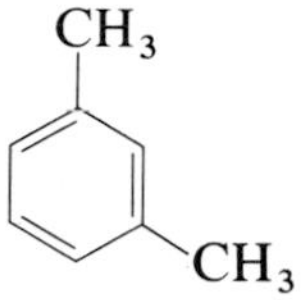

Cl   CH₂CH₂CH₃   CH₃

j.   **k.** 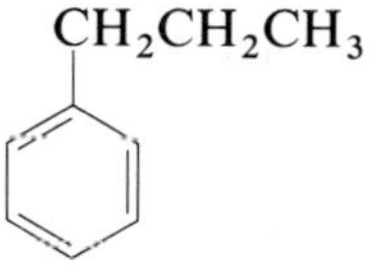   **l.**

Cl   NO₂

3.32 Draw structural formulas for these molecules:
   a. *m*-dibromobenzene   **b.** 2,4,6-trinitrotoluene (TNT)
   c. *p*-chloroiodobenzene   **d.** 2-ethyl-4-isopropyltoluene
   e. *p*-xylene   **f.** 2-ethylnaphthalene
   g. *p*-diiodobenzene   **h.** 2-phenyl-2-pentene
   i. phenanthrene   **j.** anthracene
   k. isopropylbenzene (cumene)

Ethylene

The U.S. chemical industry produces more ethylene, on a pound-per-pound basis, than any other organic chemical. Reports on this and other key chemicals can be found regularly in the weekly publication *Chemical & Engineering News* (*C&EN*). A report on ethylene and its accompanying graph (Figure I-1) follows.

Ethylene

How it is made: thermal (steam) cracking of hydrocarbons ranging from natural gas–derived ethane to oil-derived gas oil (fuel oil).

Major derivatives: polyethylenes 45%, ethylene oxide 20%, vinyl chloride 15%, styrene 10%.

Major end uses: fabricated plastics 65%, antifreeze 10%, fibers 5%, solvents 5%.

Commercial value: $4.60 billion total production in 1985.

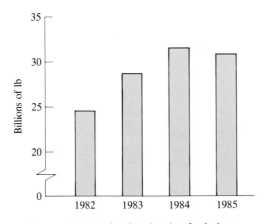

Figure I-1 Production levels of ethylene.

Of the estimated 250 billion pounds of organic chemicals produced each year in the United States, approximately 100 billion pounds are derived from ethylene. Clearly in terms of its volume and the volume of the chemicals derived from it, ethylene is the organic chemical industry's most important building block. Our focus is on how this vital starting material is produced, its principal end uses, and consumer products derived from it.

First, how do we obtain ethylene? More than 90% of all organic chemicals used by the chemical industry are derived from petroleum and natural gas. However, ethylene is not found in either of these resources. If we do not find ethylene in nature, then how do we make it from the raw materials available to us? The answer is not an easy one and it differs from one part of the world to another depending on availability of raw materials and economic demand. As we shall see presently, how we make ethylene today may be different from how we will be forced to make it in the future.

Ethylene is produced by cracking of hydrocarbons. In the United States where there are vast reserves of natural gas, the main process for ethylene production has been thermal cracking, often in the presence of steam, of the small quantities of ethane, propane, and butane that can be recovered from natural gas. (Recall from Section 2.10A that natural gas is approximately 80% methane and 10% ethane.) For this reason, ethylene-generating plants constructed in the past have been concentrated near sites of natural gas reserves. In the United States, for example, many are located on the Gulf Coast of Texas and Louisiana.

When the thermal (steam) cracking of ethane is written as a balanced equation, it seems simple enough:

$$CH_3—CH_3 \xrightarrow{\text{thermal cracking}} CH_2{=}CH_2 \ + \ H_2$$

Actually the reaction is complicated, and several other substances are produced along with ethylene. For example, for every billion pounds of ethylene produced from ethane, there are also

Figure II-2 Catalytic-cracking facilities, such as this refinery at Port Arthur, Texas, account for more than 80% of total ethylene production in the United States.

obtained 36 million pounds of propylene and 35 million pounds of butadiene as co-products. Although these quantities of starting materials and products may seem enormous to you, they reflect the scale on which the U.S. chemical industry operates. Although other low-molecular-weight alkanes can be cracked to give ethylene, thermal cracking of ethane gives the highest-percentage and highest-purity ethylene. Shown in Figure I-2 is an ethylene-cracking plant in Port Arthur, Texas.

In the United States, natural gas is currently the main source of the raw materials for manufacturing ethylene. Approximately 10% of the natural gas consumed each year is used for this purpose. In Europe and Japan, however, supplies of natural gas are much more limited. As a result, those countries depend almost entirely on catalytic cracking of petroleum-derived naphtha for their ethylene.

Now that we know how ethylene is made, let us turn to the second question: How do we use it? Each year ethylene is the starting material for the synthesis of almost 100 billion pounds of chemicals and polymers. As you can see from Table I-1, its major derivatives are polyethylene (45%), ethylene oxide and ethylene glycol (20%), vinyl chloride (15%), and styrene (10%).

We will concentrate on just one important derivative of ethylene, namely, the fabricated polyethylene plastics, which account for approx-

Table I-1 Principal derivatives and end uses of ethylene.

Principal Derivatives of Ethylene	Structural Formula	1985 Production (Billions of lb)	Major End Uses
polyethylene	$—(CH_2—CH_2)_n—$	12.0	fabricated plastics
ethylene oxide/ ethylene glycol	$CH_2—CH_2$, $CH_2—CH_2$ $\diagdown_O\diagup$ $OH \quad OH$	9.17	antifreeze, polyester textile fibers, solvents
vinyl chloride	$CH_2{=}CHCl$	6.72	polyvinyl chloride-fabricated plastics
styrene	$\langle\bigcirc\rangle—CH{=}CH_2$	6.61	polystyrene, fabricated plastics, and synthetic rubbers

imately 45% of all ethylene used in this country. The first commercial process for ethylene polymerization used peroxide catalysts at temperatures of 500°C and pressures of 1000 atm, producing a polymer known as low-density polyethylene (LDPE). Low-density polyethylene is a soft, tough plastic. It has a density between 0.91 and 0.94 g/cm³ and a melting point of about 115°C. Because LDPE's melting point is only slightly above 100°C, it is not used for products that will be exposed to boiling water. Low-density polyethylene is about 50–60% crystalline. Although polymers do not crystallize in the conventional sense, they often have regions where their chains are precisely ordered relative to each other and interact by noncovalent forces. Such regions are called crystallites. When we say that low-density polyethylene is 50–60% crystalline, we mean that this percentage is composed of crystallites.

The main end use of low-density polyethylene is film for packaging such consumer items as baked goods, vegetables, and other produce; for coatings for cardboard and paper; and perhaps most important, for trash bags.

An alternative method of ethylene polymerization uses catalysts composed of titanium chloride and organoaluminum compounds. With catalysts of this type, ethylene can be polymerized under conditions as low as 60°C and 20 atm pressure. Polyethylene produced in this manner has a density of 0.96 g/cm³ and is called high-density polyethylene (HDPE). High-density polyethylene has a higher degree of crystallinity (90%) than LDPE and a higher melting point (135°C). It is best described as a hard, tough plastic. The physical properties and cost of HDPE relative to other materials make it ideal for the production of plastic bottles, lids, caps, and so on. It is also molded into housewares such as mixing bowls and refrigerator and freezer containers.

Approximately 45% of all high-density polyethylene used in the United States is blow-molded, mostly into containers. The process of blow-molding an HDPE bottle is illustrated in Figure I-3.

Approximately 65% of all low-density polyethylene is used to manufacture films. These LDPE films are fabricated by a variation of the

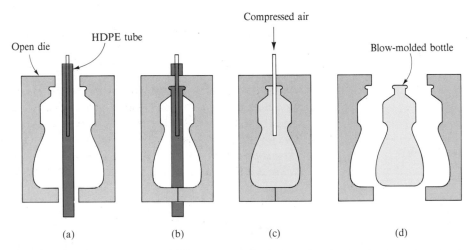

Figure I-3 Blow-molding a high-density polyethylene bottle. (a) A short length of HDPE tubing is inserted into an open die, (b) the die is closed, sealing the bottom of the tube, and the unit is heated, (c) compressed air is forced into the warm polyethylene/die assembly and the tubing is blown up to take the shape of the mold, (d) the die is opened and there is the bottle!

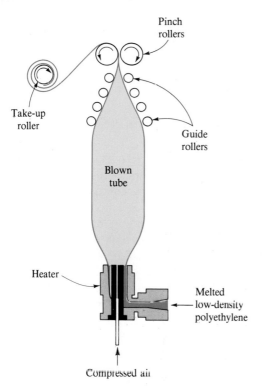

Take-up
roller

Pinch
rollers

Guide
rollers

Blown
tube

Heater

Melted
low-density
polyethylene

Compressed air

Figure I-4 Fabrication of LDPE film.

blow-molding technique illustrated in Figure I-4. A tube of LDPE, along with a jet of compressed air, is forced through an opening and blown into a giant, thin-walled bubble. The film is then cooled and taken up onto a roller. This double-walled film can be slit down the side to give LDPE film, or it can be sealed at points along its length to give LDPE trash bags.

Clearly our dependence on polyethylene is enormous, and it will continue to be so since our society is so industrialized.

Sources

American Chemical Society. 1973. *Chemistry in the Economy.* Washington, D.C.

Bilmeyer, F. W., Jr. 1971. *Textbook of Polymer Science,* 2d ed. New York: Wiley

Fernelius, Y. C., Wittcoff, H. A., and Varnerid, R. E., eds. 1979. Ethylene: The Organic Chemical Industry's Most Important Building Block. *J. Chem. Ed.* 56:385–387.

Webber, D. C&EN's Top 50 Chemical Products. *Chem. & Eng. News,* 21 April 1986.

Wittcoff, H. A., and Reuben, B. G. 1980. *Industrial Organic Chemicals in Perspective.* New York. Wiley.

Terpenes

A wide variety of substances in the plant and animal world contain one or more carbon-carbon double bonds. In this essay, we will focus on one group of natural alkenes, the terpene hydrocarbons. The characteristic structural feature of a terpene is a carbon skeleton that can be divided into two or more units that are identical with the carbon skeleton of isoprene. This generalization is known as the isoprene rule. In discussing terpenes and the isoprene rule, it is common to refer to the head and tail of an isoprene unit. The tail of an isoprene unit is the carbon atom farther from the methyl branch.

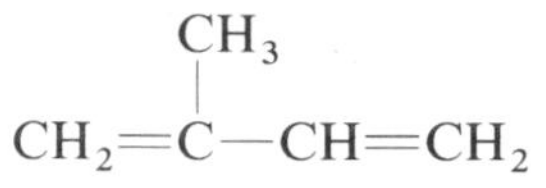

$$CH_2{=}\underset{\underset{CH_3}{|}}{C}{-}CH{=}CH_2$$

isoprene

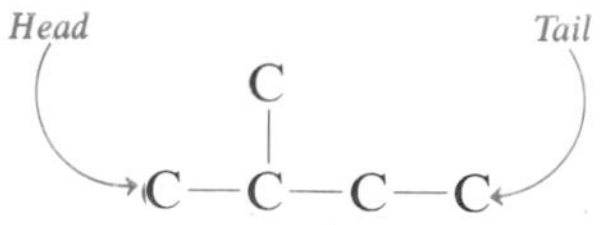

an isoprene unit

There are several important reasons for looking at this group of organic compounds. First, the number of terpenes found in bacteria, plants, and animals is staggering. Second, terpenes provide a glimpse at the wondrous diversity that nature generates from even a relatively simple carbon skeleton. Third, terpenes illustrate an important principle of the molecular logic of living systems: In building what might seem to be complex molecules, living systems piece together small subunits to produce complex but logically designed skeletal frameworks. In this mini-essay, we will show how to identify the skeletal framework of terpenes.

Probably the terpenes most familiar to you, at least by odor, are components of the so-called essential oils obtained by steam distillation or ether extraction of various parts of plants. Essential oils contain relatively low molecular weight substances that are largely responsible for characteristic plant fragrances. Many essential oil, particularly those from flowers, are used in perfumes.

An example of a terpene obtained from an essential oil is myrcene, $C_{10}H_{16}$, obtained from bayberry wax and from oils of bay and verbena. Its parent chain of eight carbon atoms contains three double bonds and two one-carbon branches [Figure II-1(a)]. Figure II-1(b) shows only the carbon skeleton of myrcene. As you can see from the position of the dashed lines in Figure II-1(b), myrcene can be divided into two isoprene units linked head to tail. Head-to-tail linkages of isoprene units are vastly more common in nature than the alternative head-to-head or tail-to-tail patterns.

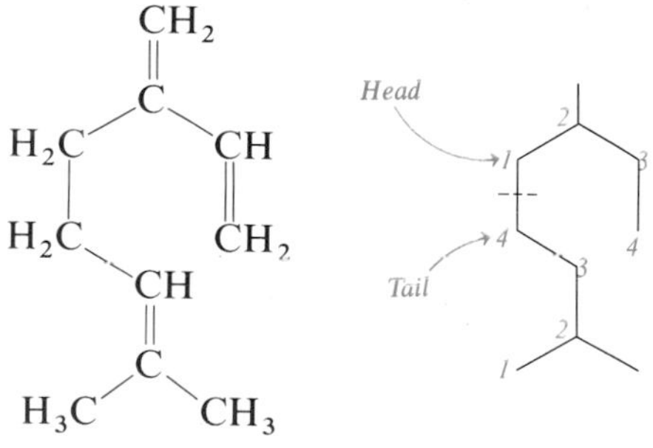

(a)   (b)

Figure II-1 (a) The structure and (b) carbon skeleton of myrcene, a terpene of two isoprene units (10 carbon atoms).

geraniol
(roses and other flowers)

an aggregating pheromone of bark beetles

limonene
(oil of lemon and orange)

menthol
(peppermint)

α-pinene
(turpentine)

camphor

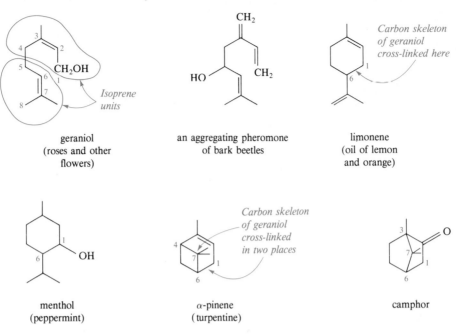

Figure II-2 Several terpenes of two isoprene units (10 carbon atoms).

Figure II-2 shows structural formulas for six more terpenes. Geraniol and the aggregating pheromone of bark beetles of the Ips family (see the mini-essay "Pheromones") have the same carbon skeleton as myrcene but different locations of carbon-carbon double bonds. In addition, each has an —OH group. In the last four terpenes shown in Figure II-2, the framework of carbon atoms present in myrcene, geraniol, and the bark beetle pheromone is cross-linked to form cyclic structures. To help you identify the points of cross-linkage and ring formation, the carbon atoms of the geraniol skeleton are numbered 1 through 8. Bond formation between carbon atoms 1 and 6 of the geraniol skeleton gives the carbon skeletons of limonene and menthol; and formation of bonds between carbons 1,6 and 4,7 gives the carbon skeleton of alpha-pinene; and between 1,6 and 3,7, the carbon skeleton of camphor.

Shown in Figure II-3 are structural formulas for several terpenes of 15 carbon atoms. For reference, the carbon atoms of the parent chain of farnesol are numbered 1 through 12. Bond formation between carbon atoms 1 and 6 of this skeleton gives the carbon skeleton of zingi-

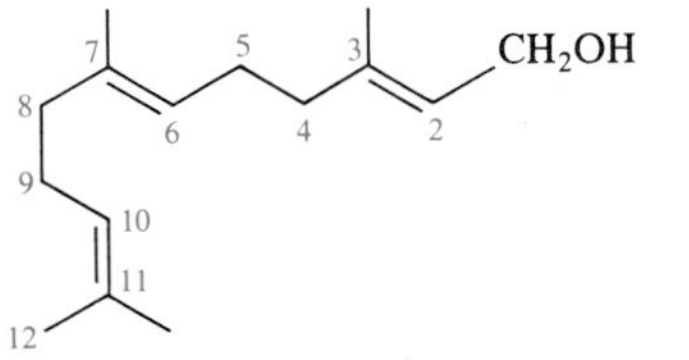

farnesol
(lily of the valley)

β-selinene
(celery)

caryophyllene
(cloves)

zingiberene
(ginger)

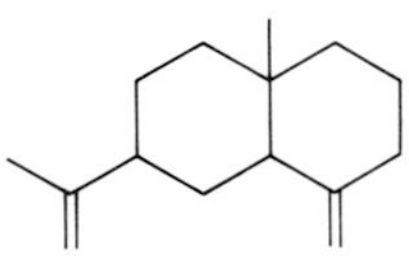

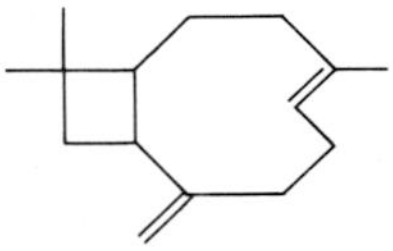

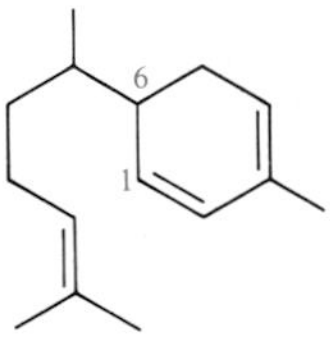

Figure II-3 Several terpenes of three isoprene units (15 carbon atoms).

berene. You might try to discover for yourself what patterns of cross-linking give the carbon skeletons of beta-selinene and caryophyllene.

Shown in Figure II-4 are structural formulas for vitamin A, a terpene of four isoprene units, and beta-carotene, a terpene of eight isoprene units. The four isoprene units of vitamin A are linked head to tail and cross-linked at one point to form a six-membered ring. The function of vitamin A is discussed in Section 12.6A. Beta-carotene can be divided into two 20-carbon terpenes, each identical to the carbon skeleton of vitamin A, and then these two 20-carbon units joined tail to tail. The function of beta-carotene is also discussed in Section 12.6A.

We have presented only a few of the terpenes that abound in nature, but these examples should be enough to suggest to you their widespread distribution in living systems, the biological individuality that plants and animals achieve through their synthesis, and the structural pattern (the isoprene rule) that underlies this apparent diversity in structural formula. In the future, when you encounter molecules of 10, 15, 20, and more carbon atoms derived from living systems, you might study their structural formulas to see whether they are terpenes.

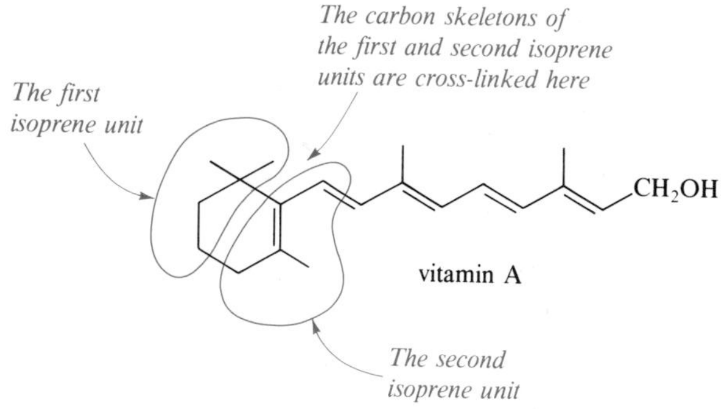

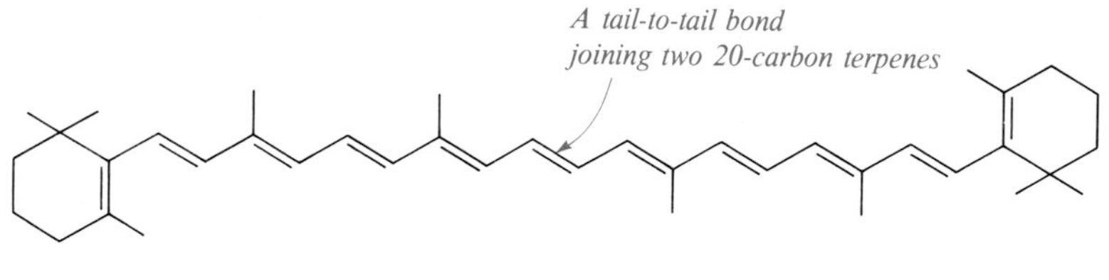

Figure II-4   Vitamin A, a terpene of 20-carbon atoms (4 isoprene units) and beta-carotene, a terpene of 40-carbon atoms (8 isoprene units).

4

Alcohols, Ethers, Phenols, and Thiols

In this chapter, we will cover the physical and chemical properties of three classes of oxygen-containing compounds and one class of sulfur-containing compounds. Alcohols and phenols both contain an —OH group. The difference between them is that in an alcohol, —OH is bonded to an alkyl group, whereas in a phenol, it is bonded to a benzene ring. A thiol is like an alcohol in structure except that —OH is replaced by —SH:

CH_3CH_2OH ⬡—OH CH_3CH_2SH $CH_3CH_2OCH_2CH_3$

an alcohol   a phenol   a thiol   an ether

4.1 Structure of Alcohols and Ethers

The characteristic structural feature of an alcohol is a **hydroxyl (—OH) group** bonded to a saturated carbon atom (Section 1.4A). The oxygen atom of an alcohol is sp^3-hybridized. Two sp^3 hybrid orbitals of oxygen form sigma bonds to atoms of carbon and hydrogen, and the remaining hybrid orbitals each hold an unshared pair of electrons. Figure 4.1 shows a Lewis structure of methanol,

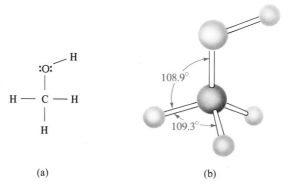

(a)   (b)

Figure 4.1 Structure of methanol, CH_3OH: (a) Lewis structure; (b) ball-and-stick model.

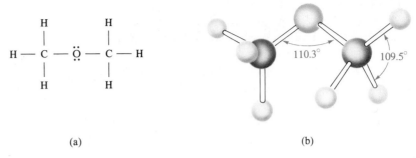

(a)   (b)

Figure 4.2 The structure of dimethyl ether, CH_3OCH_3: (a) Lewis structure; (b) ball-and-stick model.

CH_3OH, the simplest alcohol. The measured H—O—C bond angle in methanol is 108.9°, a value very close to the predicted tetrahedral angle of 109.5°. Also shown in Figure 4.1 is a ball-and-stick model of methanol.

The characteristic structural feature of an ether is an atom of oxygen bonded to two carbon chains (Section 1.4A). In an ether, sp^3 hybrid orbitals of oxygen form sigma bonds to the two hydrocarbon chains. Each of the remaining sp^3 hybrid orbitals contains an unshared pair of electrons. Figure 4.2 shows a Lewis structure and a three-dimensional representation of dimethyl ether, CH_3OCH_3, the simplest ether. The C—O—C bond angle in dimethyl ether is 110.3°, a value close to the predicted tetrahedral angle of 109.5°.

4.2 Nomenclature

A. Alcohols

In the IUPAC system, the longest chain of carbon atoms containing the —OH group is selected as the parent compound. To show that the compound is an

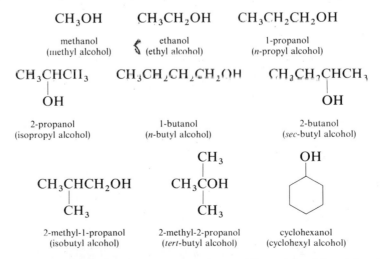

Figure 4.3 Names and structural formulas of several low-molecular-weight alcohols.

alcohol, the suffix -e is changed to -ol (Section 2.5), and a number is added to show the location of the —OH group.

Common names for **alcohols** are derived by naming the alkyl group attached to —OH and adding the word *alcohol*. Figure 4.3 (page 109) gives IUPAC and, in parentheses, common names for several low-molecular-weight alcohols.

Example 4.1

Write IUPAC names for these alcohols:

a. 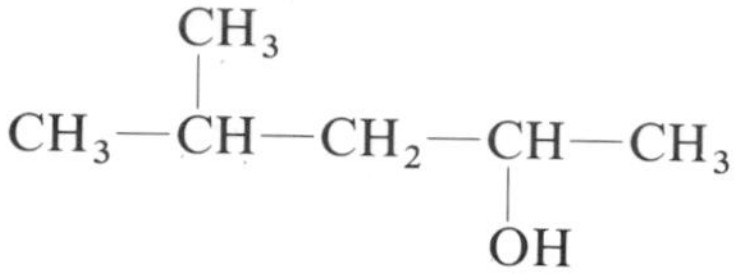 CH$_3$—CH—CH$_2$—CH—CH$_3$

with CH$_3$ on the second carbon and OH on the fourth carbon

b. 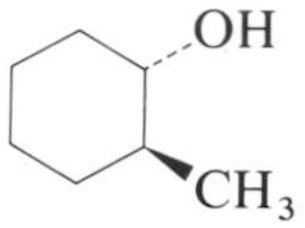 cyclohexane with OH and CH$_3$ substituents

Solution

a. The longest chain that contains the —OH group is five carbon atoms, and it must be numbered so —OH is given the lowest possible number. The IUPAC name of this alcohol is 4-methyl-2-pentanol.

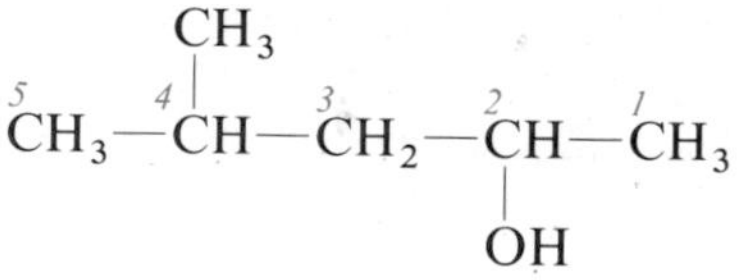

$$\overset{5}{CH_3}-\overset{4}{CH}-\overset{3}{CH_2}-\overset{2}{CH}-\overset{1}{CH_3}$$

with CH$_3$ on carbon 4 and OH on carbon 2

4-methyl-2-pentanol

b. In cyclic alcohols, the carbon atoms of the ring are numbered from the carbon bearing the —OH group. Because the —OH is automatically on carbon 1, there is no need to give a number to show its location. In this compound, the hydroxyl and methyl groups are trans to each other, as shown by the dashed line (back of the plane of the paper) for —OH, and the solid wedge (in front of the plane of the paper) for the —CH$_3$ group. The name of this alcohol is *trans*-2-methylcyclohexanol.

Problem 4.1

Give IUPAC names for these alcohols:

a. 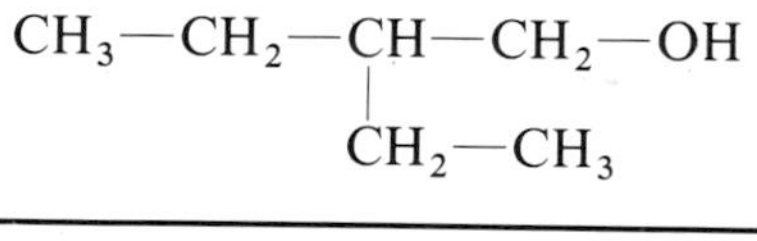 CH$_3$—CH$_2$—CH—CH$_2$—OH
with CH$_2$—CH$_3$ below the third carbon

b. 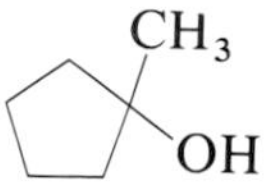 cyclopentane with CH$_3$ and OH on same carbon

---

Alcohols are classified as **primary (1°)**, **secondary (2°)**, or **tertiary (3°)**, depending on whether the —OH group is on a primary carbon, a secondary carbon, or a tertiary carbon. General formulas of 1°, 2°, and 3° alcohols are given in Figure 4.4.

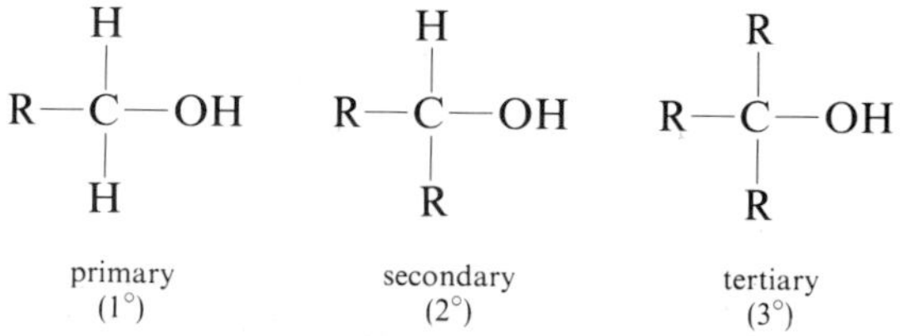

$$R-\overset{\overset{H}{|}}{\underset{\underset{H}{|}}{C}}-OH \qquad R-\overset{\overset{H}{|}}{\underset{\underset{R}{|}}{C}}-OH \qquad R-\overset{\overset{R}{|}}{\underset{\underset{R}{|}}{C}}-OH$$

primary (1°)　　　secondary (2°)　　　tertiary (3°)

Figure 4.4 Classification of alcohols: primary, secondary, and tertiary.

Classify these alcohols as primary, secondary, or tertiary:

Example 4.2

a. $CH_3-\overset{\overset{\displaystyle CH_3}{|}}{CH}-OH$

b. [cyclopentane ring]$-\overset{\overset{\displaystyle CH_3}{|}}{CH}-OH$

c. $CH_3-\overset{\overset{\displaystyle CH_3}{|}}{\underset{\underset{\displaystyle CH_3}{|}}{C}}-OH$

d. [benzene ring]$-CH_2OH$

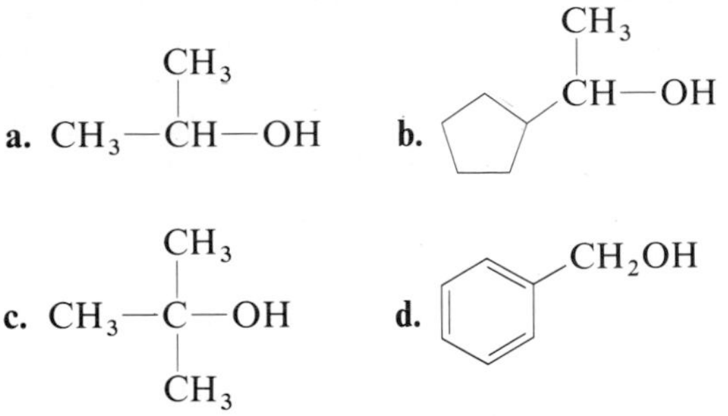

☐ *Solution*

a. Because the carbon bearing the —OH group has two attached alkyl groups, it is a secondary carbon, and the alcohol is a secondary alcohol.
b. Because the carbon bearing the —OH group has one attached cycloalkyl group and one alkyl group, it is a secondary (2°) carbon, and the alcohol is a secondary (2°) alcohol.
c. A tertiary (3°) alcohol.
d. A primary (1°) alcohol.

Classify these alcohols as primary, secondary, or tertiary:

Problem 4.2

a. $CH_3-\overset{\overset{\displaystyle CH_3}{|}}{\underset{\underset{\displaystyle CH_3}{|}}{C}}-CH_2-OH$

b. [cyclopropane ring]$-OH$

c. $CH_2{=}CH-CH_2-OH$

d. [cyclopentane ring with OH and CH_3]

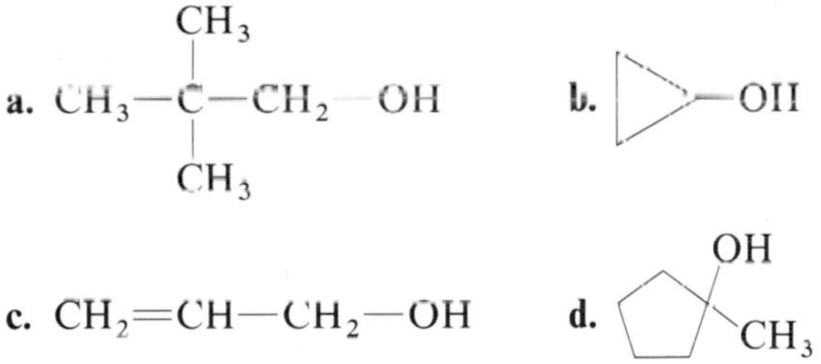

In the IUPAC system, compounds containing two hydroxyl groups are called **diols**, those containing three hydroxyl groups are called **triols**, and so on. Note that in IUPAC names for these compounds, the final -*e* (the suffix) of the parent name is retained.

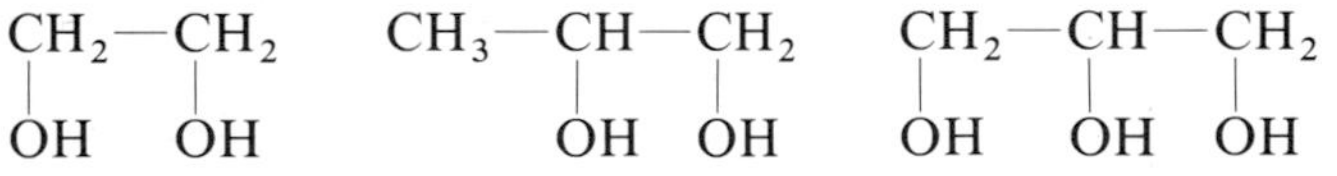

$$\underset{\underset{\displaystyle OH \quad OH}{|\qquad|}}{CH_2-CH_2} \qquad \underset{\underset{\displaystyle OH \quad OH}{\quad|\quad|}}{CH_3-CH-CH_2} \qquad \underset{\underset{\displaystyle OH \quad OH \quad OH}{|\quad\;|\quad\;|}}{CH_2-CH-CH_2}$$

| 1,2-ethanediol | 1,2-propanediol | 1,2,3-propanetriol |
| (ethylene glycol) | (propylene glycol) | (glycerol, glycerine) |

Common names for compounds containing two hydroxyl groups on adjacent carbons are often referred to as **glycols** (Section 3.7A). Ethylene glycol and propylene glycol are synthesized from ethylene and propylene respectively, hence their common names.

Compounds containing —OH and C=C groups are called unsaturated (because of the presence of the carbon-carbon double bond) alcohols. In the IUPAC system, they are named as alcohols and the parent chain is numbered to give the —OH group the lowest possible number. That the carbon chain contains a double bond is shown by changing the infix from -*an*- to -*en*-, and that it is an alcohol is indicated by changing the suffix from -*e* to -*ol* (Section 2.5). Numbers must be used to show the location of both the carbon-carbon double bond and the hydroxyl group.

Example 4.3

Write IUPAC names for these unsaturated alcohols:

a. CH_2=CH—CH_2—OH   **b.** CH_3—CH_2—CH=CH—$\overset{\displaystyle OH}{\underset{\displaystyle |}{CH}}$—$CH_3$

☐ *Solution*

a. Because the chain contains three carbons, the parent name is propane. The chain is numbered so that the —OH group is on carbon 1 and the double bond is between carbons 2 and 3. The IUPAC name of this primary, unsaturated alcohol is 2-propen-1-ol. Its common name is allyl alcohol.

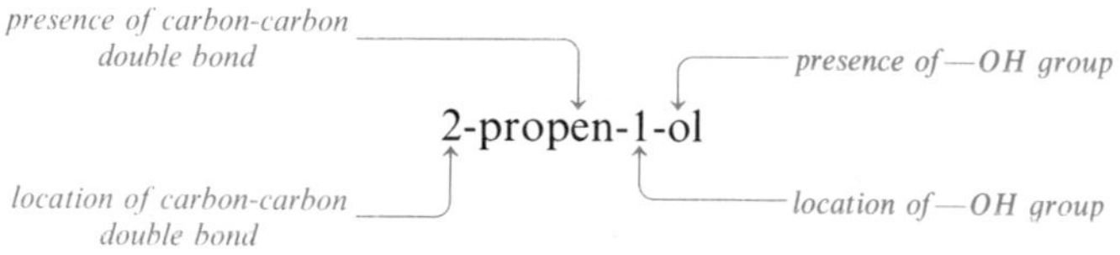

presence of carbon-carbon double bond ⎤
                              presence of —OH group ⎤

2-propen-1-ol

location of carbon-carbon double bond ⎦
                              location of —OH group ⎦

b. Because the parent chain contains six carbons, this compound is named as a derivative of hexane.

presence of carbon-carbon double bond ⎤
                              presence of —OH group ⎤

3-hexen-2-ol

location of carbon-carbon double bond ⎦
                              location of —OH group ⎦

Problem 4.3

Write IUPAC names for these unsaturated alcohols:

a. CH_3—CH=CH—CH_2—OH   **b.**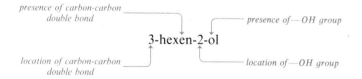

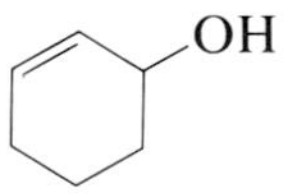

B. Ethers

In the IUPAC system, **ethers** are named by selecting the longest carbon chain as the parent compound and naming the —OR group as an alkoxy substituent. Following are IUPAC names for two low-molecular-weight ethers.

$$CH_3CH_2CH_2CH_2CHCH_2CH_2CH_3$$
$$| $$
$$OCH_3$$

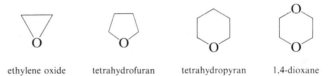

4-methoxyoctane   *trans*-2-ethoxycyclohexanol

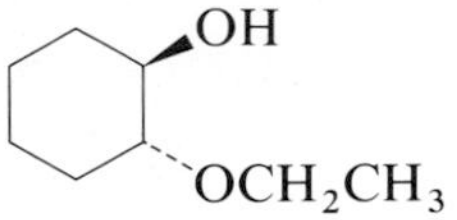

Common names are derived by specifying the alkyl groups attached to oxygen in alphabetical order and adding the word *ether*. Chemists almost invariably use common names for low-molecular-weight ethers. For example, while ethoxy-ethane is the IUPAC name for $CH_3CH_2OCH_2CH_3$, it is rarely called that but rather, diethyl ether, ethyl ether, or even more commonly, simply ether.

Heterocyclic ethers, that is, cyclic compounds in which the ether oxygen is one of the atoms in a ring, are given special names:

ethylene oxide   tetrahydrofuran   tetrahydropyran   1,4-dioxane

---

Example 4.4

Give common names for these ethers:

a. $CH_3\!-\!\overset{\displaystyle CH_3}{\underset{\displaystyle CH_3}{\overset{|}{\underset{|}{C}}}}\!-\!O\!-\!CH_2\!-\!CH_3$ b. 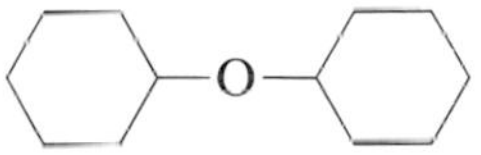 ⬡—O—⬡

c. $CH_2\!=\!CH\!-\!OCH_3$

Solution

a. *tert*-butyl ethyl ether. The IUPAC name of this ether is 2-ethoxy-2-methyl-propane.
b. Its common name is dicyclohexyl ether. Its IUPAC name is cyclohexoxy-cyclohexane.
c. Groups attached to oxygen are methyl and vinyl (Section 3.2B). Therefore, its common name is methyl vinyl ether. Its IUPAC name is methoxyethene.

Problem 4.4

Name these ethers:

a. $CH_3\!-\!\overset{\displaystyle }{\underset{\displaystyle CH_3}{\overset{}{\underset{|}{CH}}}}\!-\!CH_2\!-\!O\!-\!CH_2\!-\!CH_3$ b. 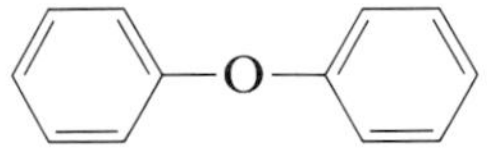 ⬡—O—⬡

c. $CH_3\!-\!O\!-\!\overset{\displaystyle CH_3}{\underset{\displaystyle CH_3}{\overset{|}{\underset{|}{C}}}}\!-\!CH_3$

4.3 Physical Properties of Alcohols and Ethers

A. Polarity of Alcohols

Because of the presence of the —OH group, alcohols are polar compounds. Oxygen is more electronegative than either carbon or hydrogen (Figure 1.1), and therefore there are partial positive charges on carbon and hydrogen and a partial negative charge on oxygen, as illustrated in Figure 4.5.

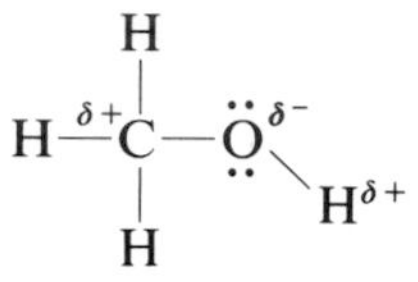

$$H\overset{\delta+}{-}C-\overset{\cdot\cdot}{\underset{\cdot\cdot}{O}}{}^{\delta-}$$

Figure 4.5 Polarity of the C—O—H bonds in alcohols.

The attraction between the positive end of one dipole and the negative end of another is called **dipole-dipole interaction**. When the positive end of one of the dipoles is a hydrogen bonded to a very electronegative atom, the interaction between dipoles is given the special name of **hydrogen bonding**. Figure 4.6 shows the association of ethanol molecules by hydrogen bonding between the partially negative oxygen atom of one alcohol and the partially positive hydrogen atom of another.

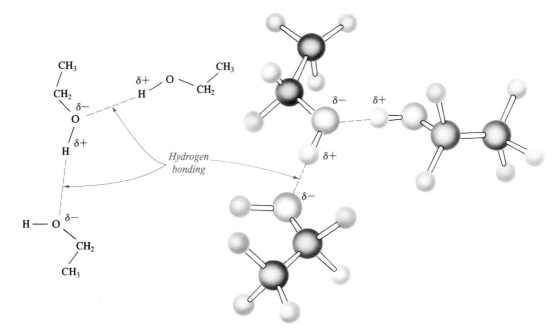

Figure 4.6 The association of ethanol in the liquid state: (a) Lewis structures; (b) ball-and-stick models. Each O—H can participate in up to three hydrogen bonds (one through hydrogen and two through oxygen). Only two of the three possible hydrogen bonds per molecule are shown in this figure.

B. Polarity of Ethers

Ethers are also polar molecules; oxygen bears a partial negative charge and each attached carbon bears a partial positive charge (Figure 4.7).

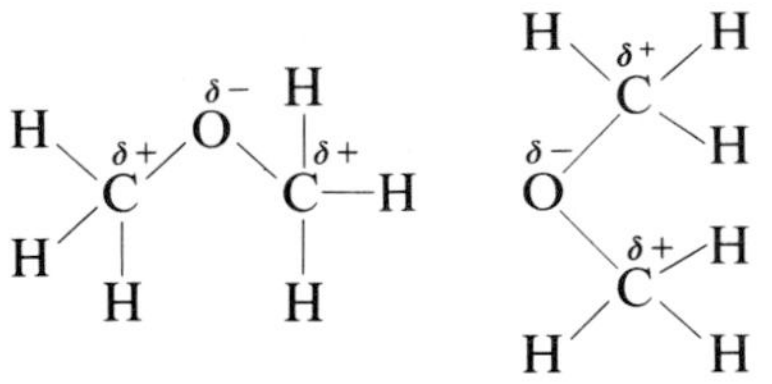

Figure 4.7 Polarity of C—O—C bonds in ethers.

Association of ether molecules by hydrogen bonding is not possible, since there is no partially positive hydrogen atom attached to oxygen to participate in hydrogen bonding. There is still the possibility, however, for ether molecules to associate by interaction between a partially positive carbon of one molecule and the partially negative oxygen of another. In fact, this type of interaction is very slight. Because each partially positive carbon is surrounded by four other atoms, it is not possible for oppositely charged dipoles to come close enough to interact. Thus, although ether molecules are polar, there is very little association between them in the pure state.

C. Relation between Structure and Physical Properties

Listed in Table 4.1 are boiling points and solubilities in water for several groups of alcohols, ethers, and hydrocarbons of similar molecular weights. Of the three classes of compounds compared in Table 4.1, alcohols have the highest boiling points because of hydrogen bonding between polar —OH groups. Ethers have boiling points close to those of nonpolar hydrocarbons of comparable molecular weight.

The effect of hydrogen bonding is illustrated dramatically by comparing the boiling points of ethanol (bp 78°C) and its structural isomer dimethyl ether (bp −24°C). The boiling points of these two compounds are different because polar O—H groups are present in the alcohol. Alcohol molecules interact by hydrogen bonding; ether molecules do not. Compare also the boiling points of 1-propanol (97°C) and ethyl methyl ether (11°C); of 1-butanol (117°C) and diethyl ether (35°C). The presence of additional hydroxyl groups in a molecule further increases the significance of hydrogen bonding, as can be seen by comparing the boiling points of hexane (bp 69°C), 1-pentanol (bp 138°C), and 1,4-butanediol (bp 230°C), all of which have approximately the same molecular weight. Because of increased dispersion forces between all the atoms, including the hydrocarbon portions, of the molecules, boiling points of alcohols and ethers increase with

Table 4.1
Boiling points and solubilities in water of several groups of alcohols, ethers, and hydrocarbons of similar molecular weight.

Structural Formula	Name	Molecular Weight	bp (°C)	Solubility in Water
CH_3OH	methanol	32	65	infinite
CH_3CH_3	ethane	30	−89	insoluble
CH_3CH_2OH	ethanol	46	78	infinite
CH_3OCH_3	dimethyl ether	46	−24	7 g/100 g
$CH_3CH_2CH_3$	propane	44	−42	insoluble
$CH_3CH_2CH_2OH$	1-propanol	60	97	infinite
$CH_3CH_2OCH_3$	ethyl methyl ether	60	11	soluble
$CH_3CH_2CH_2CH_3$	butane	58	0	insoluble
$CH_3CH_2CH_2CH_2OH$	1-butanol	74	117	8 g/100 g
$CH_3CH_2OCH_2CH_3$	diethyl ether	74	35	8 g/100 g
$CH_3CH_2CH_2CH_2CH_3$	pentane	72	36	insoluble
$CH_3CH_2CH_2CH_2CH_2OH$	1-pentanol	88	138	2.3 g/100 g
$CH_3CH_2CH_2CH_2OCH_3$	butyl methyl ether	88	71	slightly
$HOCH_2CH_2CH_2CH_2OH$	1,4-butanediol	90	230	infinite
$CH_3OCH_2CH_2OCH_3$	ethylene glycol dimethyl ether	90	84	infinite
$CH_3CH_2CH_2CH_2CH_2CH_3$	hexane	88	69	insoluble

increasing molecular weight. To see this, compare the boiling points of ethanol, 1-propanol, and 1-butanol.

Because alcohols and ethers can interact by hydrogen bonding with water, they are more soluble in water than alkanes of comparable molecular weight (Figure 4.8). Methanol, ethanol, and 1-propanol are soluble in water in all

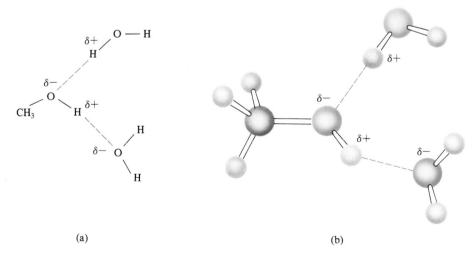

(a)   (b)

Figure 4.8 Hydrogen bonding between water and methanol: (a) Lewis structures; (b) ball-and-stick models.

proportions. As molecular weight increases, the physical properties of alcohols and ethers become more like the physical properties of hydrocarbons of comparable molecular weight. Alcohols of higher molecular weight are much less soluble in water because of the increase in size of the hydrocarbon portion of the molecule. For example, 1-decanol is insoluble in water but soluble in ethanol and in nonpolar hydrocarbon solvents, such as benzene and hexane.

Example 4.5

Arrange the following compounds in order of increasing boiling points. Explain the basis for your answer.

$$CH_3-CH_2-OH \qquad CH_3-CH_2-Cl \qquad CH_3-CH_2-CH_3$$

Solution

Propane and ethanol have similar molecular weights. Propane is a nonpolar hydrocarbon and the only interactions between molecules in the pure liquid are dispersion forces (review Section 2.8). Ethanol is a polar compound and there is extensive hydrogen bonding between ethanol molecules in the pure liquid. Therefore, ethanol has a higher boiling point than propane. Chloroethane has a higher molecular weight than ethanol and is a polar molecule. However, because it cannot associate by hydrogen bonding, chloroethane has a lower boiling point than ethanol. In order of increasing boiling point, the compounds are

$$CH_3CH_2CH_3 \qquad CH_3CH_2Cl \qquad CH_3CH_2OH$$

| bp $-42°C$ | bp $12°C$ | bp $78°C$ |
| (mw 44.1) | (mw 64.5) | (mw 46.1) |

Problem 4.5

Arrange the compounds in order of increasing boiling point. Explain the basis for your answer.

$$CH_3OCH_2CH_2OCH_3 \qquad HOCH_2CH_2OH \qquad CH_3OCH_2CH_2OH$$

Example 4.6

Arrange the following compounds in order of increasing solubility in water. Explain the basis of your answer.

$$CH_3CH_2CH_2CH_2CH_2CH_3 \qquad CH_3OCH_2CH_2OCH_3 \qquad CH_3CH_2OCH_2CH_3$$

Solution

Water is a polar solvent. Hexane, C_6H_{14}, a nonpolar hydrocarbon, has the lowest solubility in water. Both diethyl ether and ethylene glycol dimethyl ether (1,2-dimethoxyethane) are polar compounds due to the presence of C—O—C bonds, and each interacts with water molecules by hydrogen bonding. Because ethylene glycol dimethyl ether has more sites within the molecule for hydrogen bonding, it is more soluble in water than diethyl ether. The water solubilities of these substances are given in Table 4.1.

$$CH_3CH_2CH_2CH_2CH_2CH_3 \qquad CH_3CH_2OCH_2CH_3 \qquad CH_3OCH_2CH_2OCH_3$$

insoluble   8 g/100 g water   soluble in all
                                                                proportions

Problem 4.6 Arrange the following compounds in order of increasing solubility in water.

$$ClCH_2CH_2Cl \qquad CH_3CH_2CH_2OH \qquad CH_3CH_2OCH_2CH_3$$

4.4 Reactions of Alcohols

Alcohols undergo various important reactions including (1) dehydration to alkenes and (2) oxidation to aldehydes, ketones, and carboxylic acids. Therefore, alcohols are valuable starting materials for the synthesis of other classes of organic compounds.

A. Dehydration to an Alkene

An alcohol can be converted to an alkene by elimination of a molecule of water from adjacent carbon atoms. Elimination of a molecule of water is called **de-hydration**. In the laboratory, dehydration is most commonly brought about by heating the alcohol with either 85% phosphoric acid or concentrated sulfuric acid at temperatures of 100–200°C. For example, acid-catalyzed dehydration of ethanol gives ethylene:

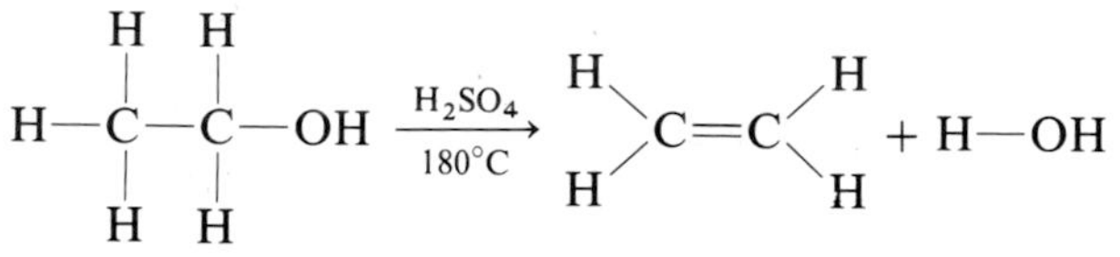

In the process, two sigma bonds (C—H and C—OH) are broken, and one pi bond (C=C) and one sigma bond (H—OH) are formed. Dehydration of cyclo-hexanol in the presence of 85% phosphoric acid yields cyclohexene:

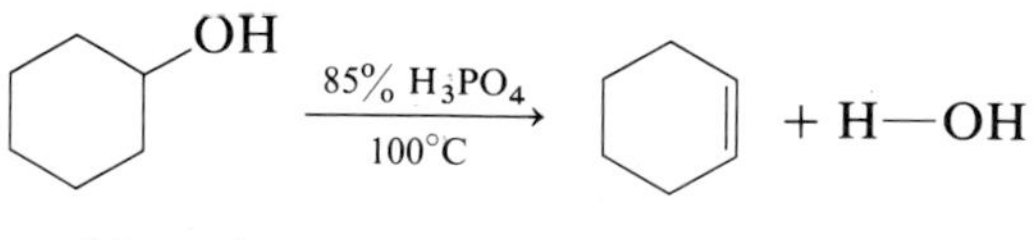

cyclohexanol   cyclohexene

Long before anyone understood the mechanism of acid-catalyzed dehydra-tion of alcohols, it was recognized that when isomeric alkenes are obtained in an elimination reaction, the alkene having the greater number of substituents on the double bond generally predominates. For example, acid-catalyzed dehydration of 2-butanol gives 80% 2-butene and 20% 1-butene:

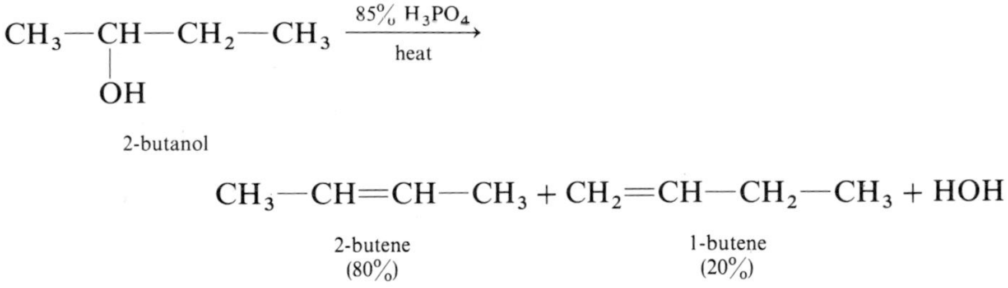

$$CH_3-CH-CH_2-CH_3 \xrightarrow[\text{heat}]{85\% \ H_3PO_4}$$
$$\qquad\quad |$$
$$\qquad\quad OH$$

2-butanol

$$CH_3-CH{=}CH-CH_3 + CH_2{=}CH-CH_2-CH_3 + HOH$$

2-butene	1-butene
(80%)	(20%)

Example 4.7

Draw structural formulas for the alkenes formed on acid-catalyzed dehydration of the following alcohols. Predict which is the major product and which the minor product.

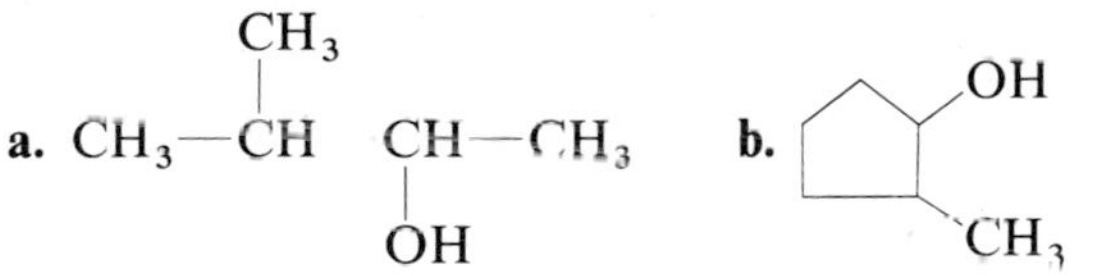

$$\qquad\qquad CH_3$$
$$\qquad\qquad |$$
a. $CH_3-CH \quad CH-CH_3$ **b.**
$$\qquad\qquad\qquad |$$
$$\qquad\qquad\qquad OH$$

☐ *Solution*

a. Dehydration of 3-methyl-2-butanol causes loss of —H and —OH from carbons 1 and 2, to produce 3-methyl-1-butene, and from carbons 2 and 3, to produce 2-methyl-2-butene.

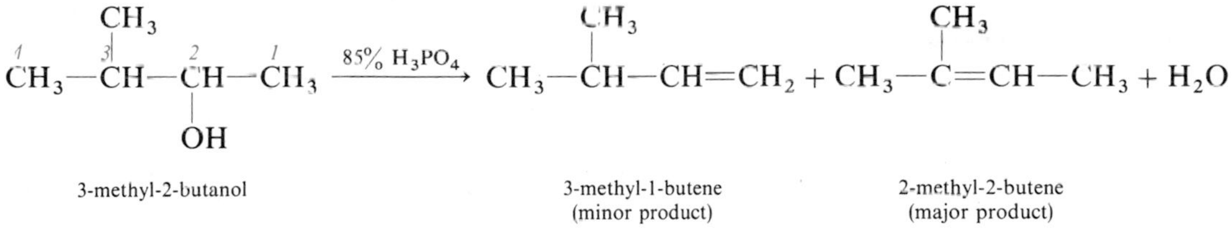

$$\overset{CH_3}{\underset{4}{CH_3}}-\overset{3}{CH}-\overset{2}{CH}-\overset{1}{CH_3} \xrightarrow{85\% \ H_3PO_4} CH_3-CH-CH{=}CH_2 + CH_3-C{=}CH-CH_3 + H_2O$$

3-methyl-2-butanol	3-methyl-1-butene (minor product)	2-methyl-2-butene (major product)

The product 3-methyl-1-butene has one alkyl substituent (an isopropyl group) on the double bond, whereas 2-methyl-2-butene has three alkyl substituents (three methyl groups) on the double bond. Therefore, the prediction is that 2-methyl-2-butene is the major product and 3-methyl-1-butene is the minor product.

b.

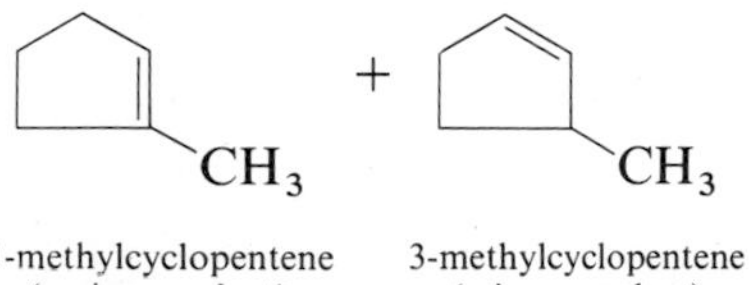

1-methylcyclopentene (major product)	3-methylcyclopentene (minor product)

The major product, 1-methylcyclopentene, has three alkyl substituents on the carbon-carbon double bond. The minor product, 3-methylcyclopentene, has only two substituents on the double bond.

■

Problem 4.7

Draw structural formulas for the alkenes formed on acid-catalyzed dehydration of the following alcohols. For each, predict which is the major product and which the minor product.

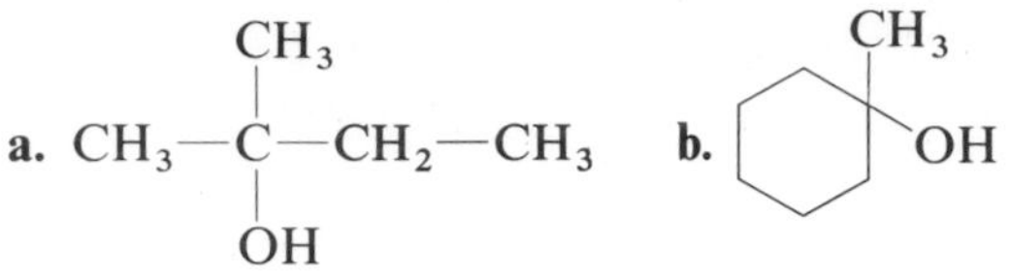

a. CH₃—C(CH₃)(OH)—CH₂—CH₃   b. cyclohexane with CH₃ and OH

In Section 3.5D we discussed a mechanism for acid-catalyzed hydration of alkenes to alcohols. This mechanism involves reaction of H⁺ and an alkene, to form a carbocation intermediate, and then reaction of the carbocation with a molecule of water, to give the alcohol. The mechanism for acid-catalyzed dehydration of an alcohol to an alkene is the reverse of this mechanism. In step 1, as shown by the curved arrow, an unshared pair of electrons on oxygen forms a new bond to H⁺, to form an oxonium ion.

Step 1

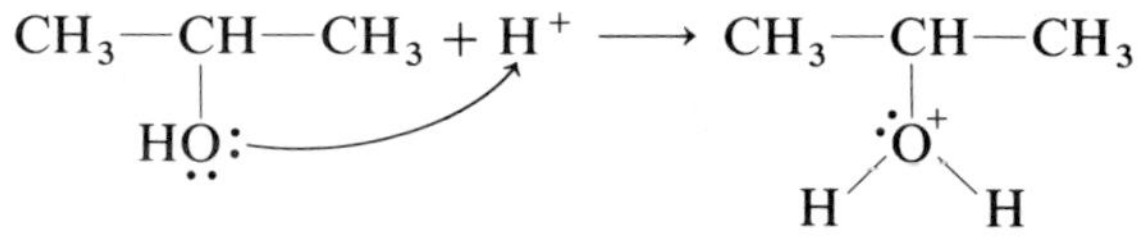

$$CH_3-CH-CH_3 + H^+ \longrightarrow CH_3-CH-CH_3$$

HO:

an oxonium ion

In step 2, the C—O bond is broken and the pair of electrons forming this bond goes to oxygen, to form a molecule of water and a secondary carbocation.

Step 2

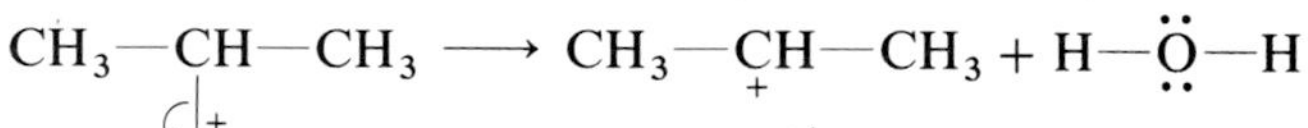

$$CH_3-CH-CH_3 \longrightarrow CH_3-\overset{+}{CH}-CH_3 + H-\overset{..}{\underset{..}{O}}-H$$

a secondary carbocation

Finally, in step 3, a C—H bond on a carbon adjacent to the carbocation breaks to form H⁺, and the pair of electrons of the C—H bond forms the pi bond of the alkene.

Step 3

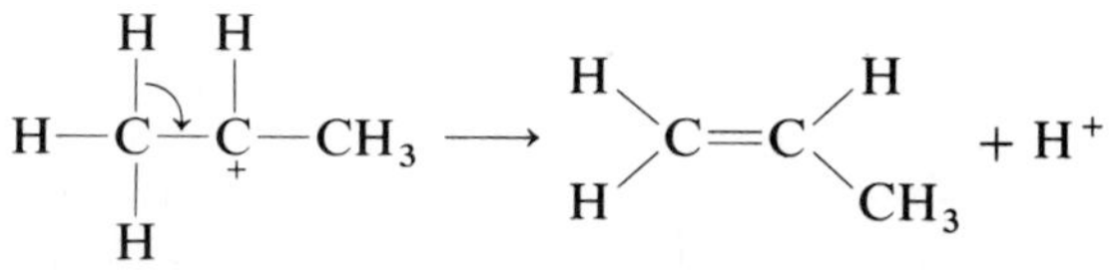

$$H-\overset{H}{\underset{H}{C}}-\overset{H}{\underset{+}{C}}-CH_3 \longrightarrow \overset{H}{\underset{H}{}}C=C\overset{H}{\underset{CH_3}{}} + H^+$$

Because the reactive intermediate in acid-catalyzed dehydration is a carbocation, the relative ease of dehydration of alcohols parallels the ease of formation of carbocations. Tertiary alcohols are dehydrated more readily than secondary alcohols because a tertiary carbocation is formed more readily than a secondary carbocation. For the same reason, secondary alcohols undergo acid-catalyzed dehydration more readily than primary alcohols. Thus, the ease of dehydration of alcohols is

$$3° \text{ alcohols} > 2° \text{ alcohols} > 1° \text{ alcohols}$$

Because hydration-dehydration reactions are reversible, alkene formation and alcohol dehydration are competing processes and the following equilibrium exists:

$$\underset{\text{alkene}}{\mathrm{>C=C<}} + H_2O \underset{}{\overset{H^+}{\rightleftharpoons}} \underset{\text{alcohol}}{-\underset{H}{\overset{|}{C}}-\underset{OH}{\overset{|}{C}}-}$$

Large amounts of water favor alcohol formation, whereas experimental conditions in which water is removed favor alkene formation. Depending on the conditions, it is possible to use the hydration-dehydration equilibrium to prepare either alcohols or alkenes, each in high yields.

B. Oxidation to Aldehydes, Ketones, or Carboxylic Acids

1. Primary Alcohols

A primary alcohol can be oxidized to either an aldehyde or a carboxylic acid depending on the experimental conditions. In oxidation of a primary alcohol to a carboxylic acid, the aldehyde is an intermediate, as shown in the following equation:

$$\underset{\substack{\text{a primary} \\ \text{alcohol}}}{H-\overset{\overset{\displaystyle H}{|}}{\underset{\underset{\displaystyle H}{|}}{C}}-\overset{\overset{\displaystyle \dot H}{|}}{\underset{\underset{\displaystyle H}{|}}{C}}-O-H} \xrightarrow{[O]} \underset{\text{an aldehyde}}{H-\overset{\overset{\displaystyle H}{|}}{\underset{\underset{\displaystyle H}{|}}{C}}-\overset{\overset{\displaystyle O}{||}}{C}-H} \xrightarrow{[O]} \underset{\substack{\text{a carboxylic} \\ \text{acid}}}{H-\overset{\overset{\displaystyle H}{|}}{\underset{\underset{\displaystyle H}{|}}{C}}-\overset{\overset{\displaystyle O}{||}}{C}-O-H}$$

Inspection of balanced half-reactions shows that each transformation in this series is a two-electron oxidation.

$$H-\overset{\overset{\displaystyle H}{|}}{\underset{\underset{\displaystyle H}{|}}{C}}-\overset{\overset{\displaystyle H}{|}}{\underset{\underset{\displaystyle H}{|}}{C}}-O-H \longrightarrow H-\overset{\overset{\displaystyle H}{|}}{\underset{\underset{\displaystyle H}{|}}{C}}-\overset{\overset{\displaystyle O}{||}}{C}-H + 2H^+ + 2e^-$$

$$H-\underset{\underset{\displaystyle H}{|}}{\overset{\overset{\displaystyle H}{|}}{C}}-\overset{\overset{\displaystyle O}{\|}}{C}-H + H_2O \longrightarrow H-\underset{\underset{\displaystyle H}{|}}{\overset{\overset{\displaystyle H}{|}}{C}}-\overset{\overset{\displaystyle O}{\|}}{C}-O-H + 2H^+ + 2e^-$$

The most common oxidizing agent in the laboratory for converting primary alcohols to aldehydes or carboxylic acids is potassium dichromate, $K_2Cr_2O_7$. It is prepared by dissolving potassium dichromate in aqueous sulfuric acid and then mixing this solution with the organic alcohol dissolved in acetone. Generally, oxidation is rapid and yields of aldehyde are good. Green chromium(III) salts precipitate and can be separated by pouring off the acetone solution.

Oxidation of a primary alcohol to an aldehyde by potassium dichromate is illustrated by oxidation of 1-propanol (*n*-propyl alcohol) to propanal:

$$CH_3-CH_2-CH_2-OH + Cr_2O_7^{2-} \xrightarrow[\text{acetone}]{H^+} CH_3-CH_2-\overset{\overset{\displaystyle O}{\|}}{C}-H + Cr^{3+}$$

$$\underset{\substack{\text{1-propanol} \\ \text{(a 1° alcohol)}}}{}\qquad\qquad \underset{\text{(orange)}}{}\qquad\qquad \underset{\substack{\text{propanal} \\ \text{(an aldehyde)}}}{}\qquad \underset{\text{(green)}}{}$$

Aldehydes are easily oxidized to carboxylic acids in a reaction that involves conversion of the aldehyde C—H bond to a C—OH bond:

$$CH_3-CH_2-\overset{\overset{\displaystyle O}{\|}}{C}-H + Cr_2O_7^{2-} \xrightarrow{H^+} CH_3-CH_2-\overset{\overset{\displaystyle O}{\|}}{C}-OH + Cr^{3+}$$

$$\underset{\substack{\text{propanal} \\ \text{(an aldehyde)}}}{}\qquad\qquad\qquad \underset{\substack{\text{propanoic acid} \\ \text{(a carboxylic acid)}}}{}$$

The product of oxidation of a primary alcohol depends on the solvent, the temperature, and other reaction conditions. We will not be concerned with these types of experimental details, noting only that either an aldehyde or a carboxylic acid can be obtained if proper conditions are chosen.

2. Secondary Alcohols

Secondary alcohols are oxidized to ketones:

menthol
(a 2° alcohol)

menthone
(a ketone)

3. *Tertiary Alcohols*

Tertiary alcohols are resistant to oxidation, because the carbon bearing the —OH has no hydrogen atom on it. It is already bonded to three other carbon atoms and therefore cannot form a carbon-oxygen double bond.

$$\text{(cyclohexane ring)}\begin{array}{c} \text{CH}_3 \\ \text{OH} \end{array} + \text{Cr}_2\text{O}_7^{2-} \xrightarrow{\text{H}^+} \text{no oxidation}$$

4.5 Alcohols of Industrial Importance

A. Methanol

By far the most important alcohol, at least in terms of bulk, is methanol. In 1985, production of methanol in the United States was just over 6 billion pounds. Virtually all methanol is manufactured by reaction of carbon monoxide and hydrogen.

$$\text{CO} + 2\text{H}_2 \xrightarrow{\text{catalyst}} \text{CH}_3\text{OH}$$
$$\text{methanol}$$

The major derivative of methanol is formaldehyde, prepared on an industrial scale by air oxidation. In a typical oxidation, air and methanol are passed over a zinc oxide–chromium oxide catalyst at elevated temperature and pressure:

$$2\text{CH}_3\text{OH} + \text{O}_2 \xrightarrow{\text{catalyst}} \overset{\displaystyle \text{O}}{\underset{}{2\text{H}-\overset{\|}{\text{C}}-\text{H}}} + 2\text{H}_2\text{O}$$

$$\text{methanol} \qquad\qquad \begin{array}{c}\text{methanal}\\ \text{(formaldehyde)}\end{array}$$

The most important end use of formaldehyde is in the manufacture of adhesives for plywood and particle board, and of a variety of polymers.

B. Ethylene Glycol

Over 4.7 billion pounds of ethylene glycol was manufactured in the United States in 1985. Starting material for its synthesis is ethylene, itself derived from refining either natural gas or petroleum. In the manufacture of ethylene glycol, a mixture of ethylene and air is passed over a silver catalyst. Reaction of ethylene oxide with water gives ethylene glycol:

$$\text{CH}_2=\text{CH}_2 \xrightarrow[\text{heat}]{\text{O}_2,\ \text{Ag}} \underset{\text{O}}{\text{CH}_2-\text{CH}_2} \xrightarrow{\text{H}_2\text{O}} \underset{\overset{|}{\text{OH}}\quad \overset{|}{\text{OH}}}{\text{CH}_2-\text{CH}_2}$$

$$\qquad\qquad\qquad\qquad \text{ethylene oxide} \qquad\qquad \text{ethylene glycol}$$

Ethylene glycol has two important uses: as an antifreeze in automobile radiators and as the monomer used along with terephthalic acid for the synthesis of polyester fibers (Dacron) and films (Mylar).

C. Ethanol

Ethanol, or simply "alcohol" in nonscientific language, has been prepared since antiquity by fermentation of sugars and starches, catalyzed by yeast:

$$C_6H_{12}O_6 \xrightarrow[\text{fermentation}]{\text{alcoholic}} 2CH_3CH_2OH + 2CO_2$$

glucose   ethanol

The sugars for this fermentation come from a variety of sources, including grains (hence the name *grain alcohol*), grape juice, various vegetables, and agricultural wastes, including corn stalks. The immediate product of fermentation is a water solution containing up to 15% alcohol. This alcohol can be concentrated by distillation. Beverage alcohol may contain traces of flavor derived from the source (grapes in brandy, grains in whiskeys) or may be essentially flavorless (like vodka).

Synthetic ethanol (1.1 billion pounds in 1985) is made by acid-catalyzed hydration of ethylene over a phosphoric acid catalyst:

$$CH_2{=}CH_2 + H_2O \xrightarrow{\text{catalyst}} CH_3{-}CH_2{-}OH$$

ethanol
(ethyl alcohol)

Ordinary commercial alcohol is a mixture of 95% alcohol and 5% water. **Absolute ethanol**, or 100% ethanol, is prepared from 95% ethanol by techniques that remove water from the mixture.

D. 2-Propanol

After methanol and ethylene glycol, the largest-volume alcohol produced by the US chemical industry is 2-propanol (isopropyl alcohol), manufactured by acid-catalyzed hydration of propene:

$$CH_3{-}CH{=}CH_2 + H_2O \xrightarrow{H_2SO_4} CH_3{-}\underset{\underset{OH}{|}}{CH}{-}CH_3$$

propene   2-propanol
(propylene)   (isopropyl alcohol)

Some 2-propanol is used in 70% aqueous solution as rubbing alcohol. However, the bulk of it is oxidized to acetone, which finds wide use as a solvent.

$$2CH_3 - \overset{\overset{\displaystyle OH}{|}}{CH} - CH_3 + O_2 \xrightarrow{\text{catalyst}} 2CH_3 - \overset{\overset{\displaystyle O}{||}}{C} - CH_3 + 2H_2O$$

2-propanone
(acetone)

4.6 Ethers

A. Reactions of Ethers

Ethers resemble hydrocarbons in their resistance to chemical reactions. They do not react with oxidizing agents such as potassium permanganate or potassium dichromate. They are not affected by most strong acids or bases at moderate temperatures. It is precisely this resistance to chemical reaction and good solvent properties that make ethers such good solvents.

B. Ethers and Anesthesia

Before the middle 1800s, surgery was performed only when absolutely necessary, because there was no truly effective general anesthetic. More often than not, patients were drugged, hypnotized, or simply tied down. In 1772, Joseph Priestley isolated nitrous oxide, N_2O, a colorless gas, and in 1799 Sir Humphrey Davy demonstrated its anesthetic effect, naming it laughing gas. In 1844, an American dentist, Horace Wells, introduced nitrous oxide into general dental practice. However, one patient awakened prematurely, screaming with pain, and another died. Wells was forced to withdraw from practice, became embittered and depressed, and committed suicide at age 33. In the same period, a Boston chemist, Charles Jackson, anesthetized himself with diethyl ether and persuaded a dentist, William Morton, to use it. Subsequently, they persuaded a surgeon, John Warren, to give a public demonstration of surgery under anesthesia. The operation was completed successfully, and soon general anesthesia by diethyl ether became routine for general surgery.

Diethyl ether is easy to use and causes excellent muscle relaxation. Blood pressure, pulse rate, and respiration are usually only slightly affected. Its chief drawbacks are its irritating effect on the respiratory passages and its aftereffect of nausea. Further, when mixed with air in the right proportions, it is explosive! Modern operating theaters are often a maze of sophisticated electrical equipment, including cautery devices, which are frequently used throughout most surgical procedures. During prolonged ether anesthesia, concentrations of the gas in a patient's fatty tissue, abdominal cavity, and bladder can reach explosive levels. An electrical spark could quite literally produce an explosion. Precautionary measures to guard against such accidents became so cumbersome that the incentive was strong to develop alternative, nonflammable, and nonexplosive anesthetics.

One such alternative is the halogenated hydrocarbon 1-bromo-1-chloro-2,2,2-trifluoroethane, or as it is more commonly known, halothane:

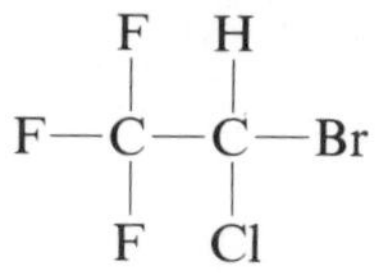

$$\begin{array}{ccc} & F & H \\ & | & | \\ F- & C- & C-Br \\ & | & | \\ & F & Cl \end{array}$$

1-bromo-1-chloro-2,2,2-trifluoroethane
(halothane)

Halothane has distinct advantages over diethyl ether in that it is nonflammable and nonexplosive, and causes minimum discomfort to the patient. Although some cases of liver damage have been reported, halothane's record as a safe anesthetic is impressive.

Both diethyl ether and now halothane have been replaced by an even newer type of inhalation anesthetic, namely halogenated ethers. The most widely used of these are marketed under the trade names Enflurane and Isoflurane:

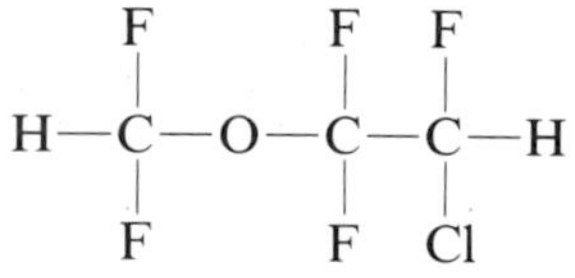

$$\begin{array}{ccccc} F & & F & F \\ | & & | & | \\ H-C-O & -C-C-H \\ | & & | & | \\ F & & F & Cl \end{array} \qquad \begin{array}{ccccc} F & & H & F \\ | & & | & | \\ H-C-O & -C-C-F \\ | & & | & | \\ F & & Cl & F \end{array}$$

Enflurane 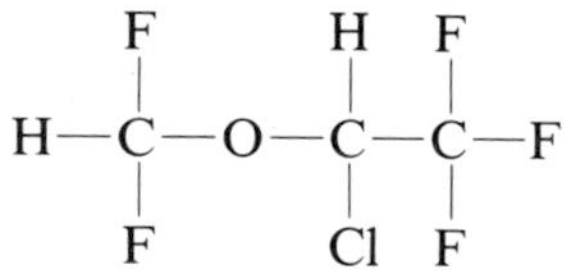   Isoflurane

4.7 Phenols

A. Structure and Nomenclature

The characteristic structural feature of a **phenol** is the presence of a hydroxyl group bonded directly to a benzene or other aromatic ring. Following is the structural formula of phenol, the simplest member of this class of compounds:

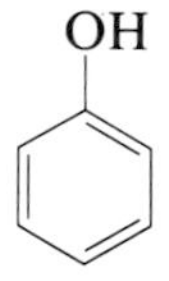

OH

phenol

Other phenols are named either as derivatives of the parent hydrocarbon or by common names.

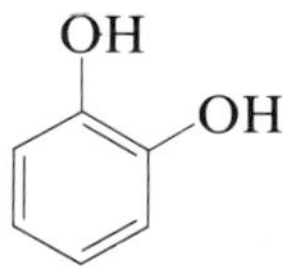

1,2-dihydroxy-
benzene
(catechol)

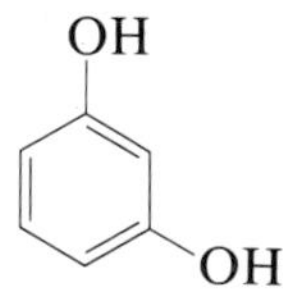

1,3-dihydroxy-
benzene
(resorcinol)

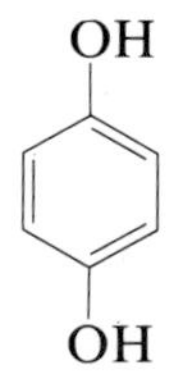

1,4-dihydroxy-
benzene
(hydroquinone)

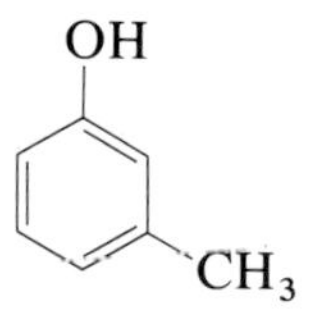

3-methylphenol
(*m*-cresol)

Phenols are widely distributed in nature. Phenol itself and the isomeric cresols (*ortho*-, *meta*-, and *para*-cresol) are found in coal tar and petroleum. Thymol and vanillin are important constituents of thyme and vanilla beans.

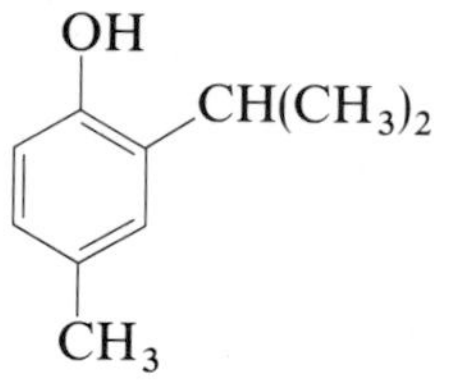

thymol 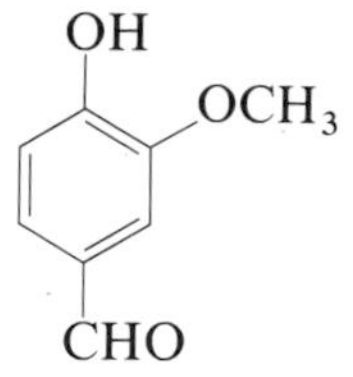   vanillin

Phenol, or carbolic acid, as it was once called, is a low-melting solid that is only slightly soluble in water. In sufficiently high concentrations, it is corrosive to all kinds of cells. In dilute solutions, it has some antiseptic properties, and was used for the first time in the nineteenth century by Joseph Lister for antiseptic surgery. Its medical use is now limited. It has been replaced by antiseptics that are more powerful and have fewer undesirable side effects. Among these is *n*-hexylresorcinol, a substance widely used in household preparations (Sucrets and mouthwashes) as a mild antiseptic and disinfectant:

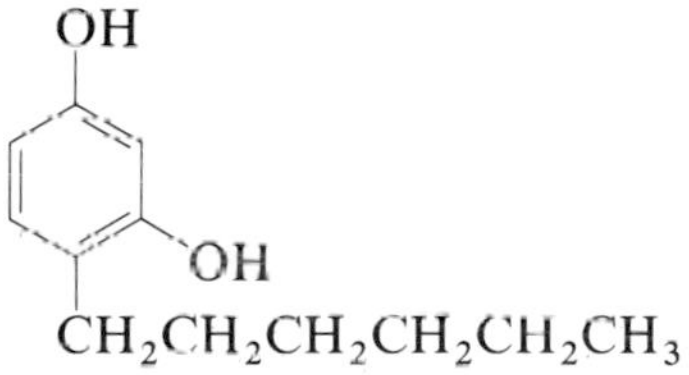

n-hexylresorcinol

B. Acidity of Phenols

Phenols and alcohols both contain a hydroxyl group, —OH. However, phenols are grouped as a separate class of compounds because their chemical properties are different from those of alcohols. One of the most important of these differences is that phenols are significantly more acidic than alcohols. The acid dissociation constant is approximately 10^6 times larger for phenol than for ethanol.

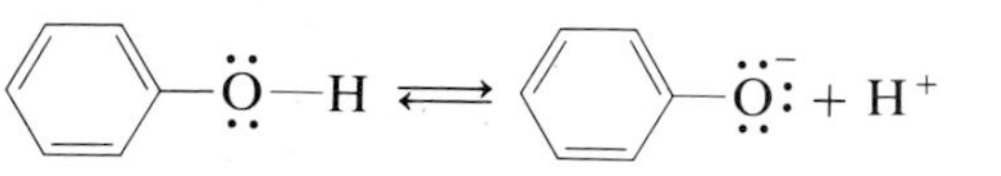

$K_a = 1.3 \times 10^{-10}$ $pK_a = 9.89$

$K_a = 1.0 \times 10^{-16}$ $pK_a = 16.0$

Another way to compare the relative acid strengths of ethanol and phenol is to look at the hydrogen ion concentration and pH of a 0.1M aqueous solution of each (Table 4.2). For comparison, the hydrogen ion concentration and the pH of 0.1M HCl are also included.

Table 4.2
Relative acidities of 0.1M ethanol, phenol, and HCl.

Dissociation Equation	[H$^+$]	pH
$CH_3CH_2OH \rightleftharpoons CH_3CH_2O^- + H^+$	10^{-7}	7.0
$C_6H_5OH \rightleftharpoons C_6H_5O^- + H^+$	3.3×10^{-6}	5.4
$HCl \longrightarrow Cl^- + H^+$	0.1	1.0

In aqueous solution, alcohols are neutral substances and the hydrogen ion concentration of 0.1M ethanol is the same as that of pure water. A 0.1M solution of phenol is slightly acidic and has a pH of 5.4. By contrast, 0.1M HCl, a strong acid (completely ionized in aqueous solution), has a pH of 1.0.

Because phenols are weak acids, they react with strong bases such as NaOH, to form salts. However, phenols do not react with weaker bases, such as sodium bicarbonate.

| phenol (stronger acid) | sodium hydroxide (stronger base) | sodium phenoxide (weaker base) | water (weaker acid) |

That phenols are weakly acidic whereas alcohols are neutral provides a very convenient way to separate phenols from water-insoluble alcohols. Suppose that we want to separate phenol from cyclohexanol. Each is only slightly soluble in water, and therefore they can not be separated on the basis of their water solubility. However, they can be separated on the basis of their differences in acidity. First, the mixture of the two is dissolved in diethyl ether or some other water-immiscible solvent. Next, the ether solution is placed in a separatory funnel and shaken with dilute NaOH. Under these conditions, phenol reacts with NaOH and is converted to sodium phenoxide, a water-soluble salt. The upper layer in the separatory funnel is now diethyl ether, containing only dissolved cyclohexanol. The lower aqueous layer contains dissolved sodium phenoxide. The layers are separated and distillation of the ether (bp 35°C) leaves pure cyclohexanol (bp 161°C). Acidification of the aqueous phase with 0.1M HCl or other strong acid converts sodium phenoxide to phenol, which is water-insoluble and can be separated and recovered in pure form. These experimental steps are summarized in the accompanying flowchart.

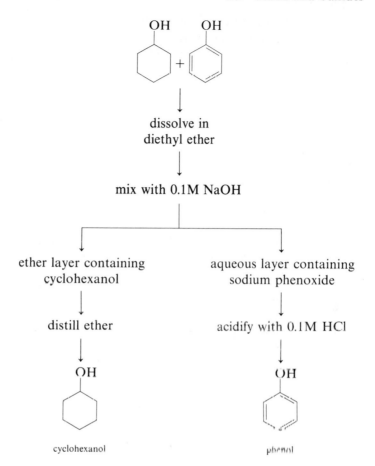

4.8 Thiols and Sulfides

A. Structure

Sulfur analogs of alcohols are called **thiols** (*thi-* from the Greek *theion*, sulfur), or in the older literature, mercaptans. The characteristic structural feature of a thiol is a **sulfhydryl (—SH) group** bonded to a carbon chain. Figure 4.9 shows a Lewis structure and a ball-and-stick representation of methanethiol, CH_3SH, the simplest thiol. The C—S—H bond angle in methanethiol is 100.3°.

B. Nomenclature

In the IUPAC system, thiols are named by selecting as the parent compound the longest chain of carbon atoms that contains the —SH group. To show that the compound is a thiol, the final -*e* in the name of the parent chain is retained and the suffix -*thiol* is added. A number must be used to locate the —SH group on the parent chain.

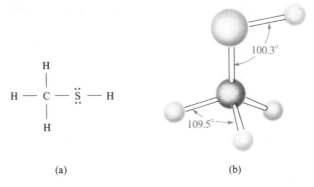

$$H - \overset{\overset{\displaystyle H}{|}}{\underset{\underset{\displaystyle H}{|}}{C}} - \overset{..}{\underset{..}{S}} - H$$

(a)   (b)

Figure 4.9 Structure of methanethiol, CH_3SH: (a) Lewis structure; (b) ball-and-stick model.

In the common system of nomenclature, thiols are known as **mercaptans**. Common names for simple thiols are derived by naming the alkyl groups attached to —SH and then adding the word *mercaptan*. Listed in Figure 4.10 are IUPAC names, and in parentheses, common names, for several low-molecular-weight thiols.

$$CH_3-CH_2-SH \qquad CH_3-CH_2-CH_2-CH_2-SH \qquad CH_3-\overset{\overset{\displaystyle |}{}}{\underset{\underset{\displaystyle CH_3}{|}}{CH}}-CH_2-SH$$

ethanethiol   1-butanethiol   2-methyl-1-propanethiol
(ethyl mercaptan)   (*n*-butyl mercaptan)   (isobutyl mercaptan)

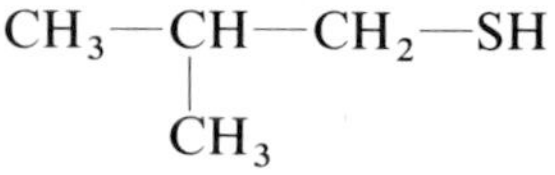

$$CH_3-CH_2-\overset{\overset{\displaystyle |}{}}{\underset{\underset{\displaystyle SH}{|}}{CH}}-CH_3 \qquad HS-CH_2-CH_2-SH$$

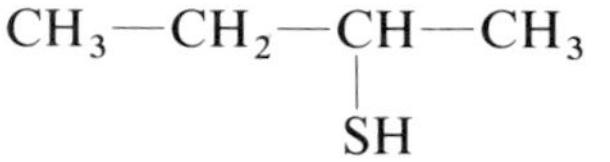

2-butanethiol   1,2-ethanedithiol
(*sec*-butyl mercaptan)   (ethylene mercaptan)

Figure 4.10 Names and structural formulas for several low-molecular-weight thiols.

In compounds containing other functional groups of higher priority, the presence of an —SH group is indicated by the prefix *mercapto-*.

$$HS-CH_2-CH_2-OH$$

2-mercaptoethanol

Sulfur analogs of ethers are named by using the word **sulfide** to show the presence of the —S— group. Following are common names of two thioethers:

$$CH_3-S-CH_3 \qquad CH_3-CH_2-S-\overset{\overset{\displaystyle |}{}}{\underset{\underset{\displaystyle CH_3}{|}}{CH}}-CH_3$$

dimethyl sulfide   ethyl isopropyl sulfide

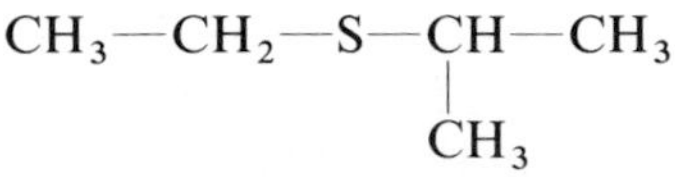

The characteristic structural feature of a **disulfide** is the presence of an —S—S— group. Common names of disulfides are derived by listing the names of the alkyl or aryl groups attached to sulfur and adding the word *disulfide*.

$$CH_3—S—S—CH_3$$

dimethyl disulfide

C. Physical Properties of Thiols

The physical properties of thiols are different from those of alcohols primarily because of the large difference in polarity of the O—H bond compared with the S—H bond. The electronegativities of sulfur and hydrogen are almost identical, and the S—H bond is nonpolar covalent. In comparison, the electronegativity difference between oxygen and hydrogen is 0.9 unit (3.0 − 2.1), and the O—H bond is polar covalent.

$$CH_3—S{\overset{\nearrow}{}}H \qquad CH_3—\overset{\delta-}{O}{\overset{\nearrow}{}}\overset{\delta+}{H}$$

a nonpolar covalent   a polar covalent
bond   bond

Because of the very low polarity of the S—H bond, thiols show little association by hydrogen bonding. Consequently, they have lower boiling points and are less soluble in water and in other polar solvents than alcohols of comparable molecular weights. Table 4.3 gives names and boiling points for several low-molecular-weight thiols. Shown for comparison are boiling points of alcohols of the same number of carbon atoms.

Table 4.3
Boiling points of several thiols and alcohols of the same number of carbon atoms.

Thiol	bp (°C)	Alcohol	bp (°C)
methanethiol	6	methanol	65
ethanethiol	35	ethanol	78
1-butanethiol	98	1-butanol	117

In Section 4.3, we illustrated the importance of hydrogen bonding in alcohols by comparing the boiling points of ethanol (bp 78°C) and its structural isomer dimethyl ether (bp −24°C). By comparison, the boiling point of ethanethiol is 35°C, and the boiling point of its structural isomer dimethyl sulfide is 37°C.

$$CH_3—CH_2—SH \qquad CH_3—S—CH_3$$

ethanethiol   dimethyl sulfide
(bp 35°C)   (bp 37°C)

That the boiling points of these isomers are almost identical indicates little or no association by hydrogen bonding between thiols in the pure liquid.

The most outstanding physical characteristic of low-molecular-weight thiols is their stench. The scent of skunks is due primarily to two thiols, 3-methyl-1-butanethiol and 2-butene-1-thiol:

$$CH_3CH-CH_2CH_2SH \qquad CH_3CH=CHCH_2SH$$

with CH_3 attached above to the second carbon.

3-methyl-1-butanethiol   2-butene-1-thiol

Further, traces of low-molecular-weight thiols are added to natural gas (which has no detectable odor) so that gas leaks can be detected by the smell of the thiol.

D. Acidity of Thiols

Hydrogen sulfide is a stronger acid than water.

$$H_2O \rightleftharpoons HO^- + H^+ \qquad pK_a = 15.7$$

$$H_2S \rightleftharpoons HS^- + H^+ \qquad pK_a = 7.04$$

Similarly, thiols are stronger acids than alcohols are. Compare, for example, the acid dissociation constants of ethanol and ethanethiol in dilute aqueous solution.

$$CH_3-CH_2-OH \rightleftharpoons CH_3-CH_2-O^- + H^+ \qquad pK_a = 15.9$$

$$CH_3-CH_2-SH \rightleftharpoons CH_3-CH_2-S^- + H^+ \qquad pK_a = 8.5$$

Thiols are sufficiently strong acids that when they are dissolved in aqueous sodium hydroxide, they are converted completely to alkylsulfide salts:

$$CH_3-CH_2-SH + Na^+OH^- \longrightarrow CH_3-CH_2-S^-Na^+ + H-OH$$

(stronger acid)   (stronger base)   sodium ethylsulfide (weaker base)   (weaker acid)

To name salts of thiols, give the name of the cation first and then the name of the alkyl group to which is attached the suffix -*sulfide*. For example, the sodium salt derived from ethanethiol is named sodium ethylsulfide.

   Like hydrogen sulfide and hydrosulfide salts, thiols form water-insoluble salts with most heavy metals:

$$Hg^{2+} + 2RSH \longrightarrow Hg(SR)_2 + 2H^+$$

In fact, the common name of this class of sulfur-containing compounds is derived from the Latin *mercurium captans*, which means mercury-capturing. Reaction with Pb(II) is often used as a qualitative test for the presence of a sulfhydryl group. Treatment of a thiol with a saturated solution of lead(II) acetate, $Pb(OAc)_2$, usually gives a yellow solid and is a positive test for the presence of a thiol.

$$R-SH + Pb^{2+} \longrightarrow Pb(S-R)_2 + 2H^+$$

(yellow precipitate)

E. Oxidation of Thiols and Sulfides

Many of the chemical properties of thiols and sulfides are related to the fact that a divalent sulfur atom is easily oxidized to several higher oxidation states. Thiols are oxidized by mild oxidizing agents such as iodine, I_2, to **disulfides**:

$$R—S—H + H—S—R + I_2 \longrightarrow R—S—S—R + 2H^+ + 2I^-$$

thiols   a disulfide

They are also oxidized to disulfides by molecular oxygen. In fact, thiols are so susceptible to oxidation that they must be protected from contact with air during storage.

$$R—S—H + H—S—R + \tfrac{1}{2}O_2 \longrightarrow R—S—S—R + H_2O$$

Disulfide bonds are an important structural feature of many biomolecules, including proteins (Chapter 13).

Oxidation of sulfides of the type R—S—R gives molecules called sulfoxides, probably the most familiar of which is dimethyl sulfoxide, abbreviated DMSO. This is a colorless, odorless liquid with a slightly bitter taste and a sweet after-taste. It solidifies at 18.5°C and boils at 189°C. Because DMSO is a byproduct of wood-pulp processing, its principal suppliers are companies associated with the paper industry. Dimethyl sulfoxide is also manufactured by air oxidation of dimethyl sulfide in the presence of oxides of nitrogen:

$$2CH_3—S—CH_3 + O_2 \xrightarrow{\text{oxides of nitrogen}} 2CH_3-\overset{\overset{\displaystyle O}{\|}}{S}—CH_3$$

dimethyl   dimethyl
sulfide   sulfoxide

It is a polar liquid, and an excellent solvent for polar organic molecules.

Key Terms and Concepts

absolute ethanol (4.5C)

acidity of phenols (4.7B)

acidity of thiols (4.8D)

alcohol (introduction and 4.2)

dehydration of alcohols (4.4A)

diol (4.2A)

dipole-dipole interaction (4.3A)

disulfide (4.8B)

ether (4.2B and 4.6)

ethers and anesthesia (4.6B)

glycol (4.2A)

hydrogen bonding (4.3A)

hydroxyl group (4.1)

mercaptan (4.8B)

nomenclature of alcohols (4.2A)

nomenclature of ethers (4.2B)

nomenclature of sulfides (4.8B)

nomenclature of thiols (4.8B)

oxidation of alcohols (4.4B)

oxidation of thiols (4.8E)

phenols (introduction and 4.7)   sulfide (4.8B)

primary alcohol (4.2A)   tertiary alcohol (4.2A)

secondary alcohol (4.2A)   thiol (4.8A)

sulfhydryl group (4.8A)   triol (4.2A)

Key Reactions

1. Acid-catalyzed dehydration of alcohols to alkenes (Section 4.4A).
2. Oxidation of primary alcohols to aldehydes (Section 4.4B).
3. Oxidation of primary alcohols to carboxylic acids (Section 4.4B).
4. Oxidation of secondary alcohols to ketones (Section 4.4B).
5. Phenols are weak acids and ionize in aqueous solution (Section 4.7B).
6. Reaction of phenols with NaOH and other strong bases to form water-soluble salts (Section 4.7B).
7. Thiols are weak acids and ionize in aqueous solution (Section 4.8D).
8. Reaction of thiols with NaOH and other strong bases to form water-soluble salts (Section 4.8D).
9. Reaction of thiols with Hg(II) and other heavy metal ions to form water-insoluble salts (Section 4.8D).
10. Oxidation of thiols by O_2 or I_2 to disulfides (Section 4.8E).

Problems

Structure and nomenclature of alcohols and ethers (Sections 4.1 and 4.2)

4.8   Name the following compounds.

a. $CH_3CH_2CH_2OH$

b. $HOCH_2CH_2CH_2CH_2OH$

c. 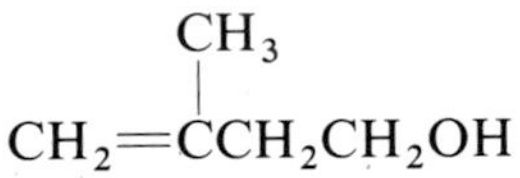 $CH_2{=}\overset{\overset{\displaystyle CH_3}{|}}{C}CH_2CH_2OH$

d. 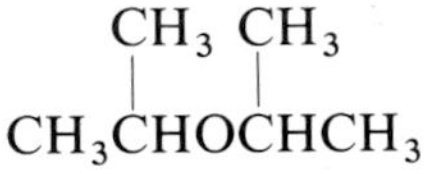 $CH_3\overset{\overset{\displaystyle CH_3}{|}}{C}H O\overset{\overset{\displaystyle CH_3}{|}}{C}HCH_3$

e. $CH_3OCH_2CH_2OH$

f. $CH_3(CH_2)_8CH_2OH$

g.

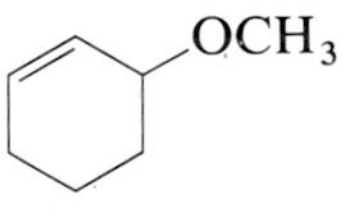

h.

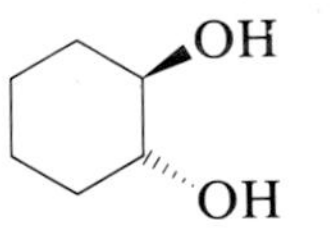

i.

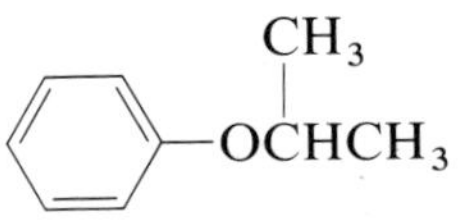

j.

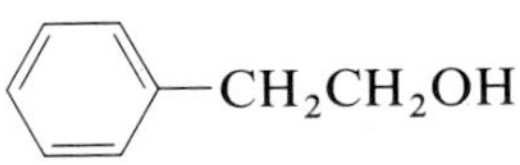

k. 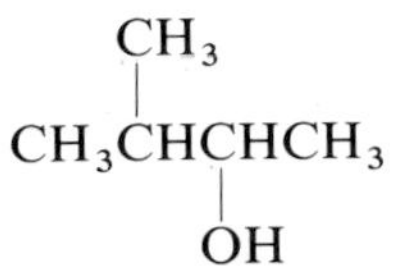 $CH_3\overset{\overset{\displaystyle CH_3}{|}}{C}H\underset{\underset{\displaystyle OH}{|}}{C}HCH_3$

l. 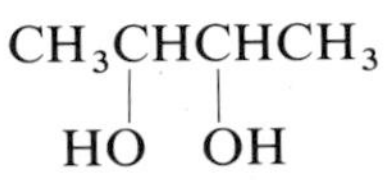 $CH_3\underset{\underset{\displaystyle HO}{|}}{C}H\underset{\underset{\displaystyle OH}{|}}{C}HCH_3$

$$\underset{\substack{|\\ OH}}{m.\ CH_3CHCH_2Cl}$$

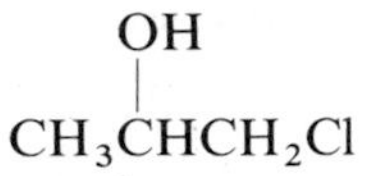

$$\underset{\substack{|\\ CH_3CH_2}}{\overset{\substack{CH_3\\ |}}{n.\ CH_3CH_2CCH_2OH}}$$

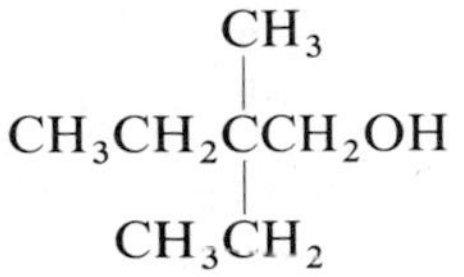

4.9   Write structural formulas for these compounds:

a. isopropyl methyl ether
b. propylene glycol
c. 2-methyl-2-propylpropane-1,3-diol
d. 1-chloro-2-hexanol
e. 5-methyl-2-hexanol
f. 2,5-dimethylcyclohexanol
g. 2,2-dimethyl-1-propanol
h. *tert*-butyl alcohol
i. methyl cyclopropyl ether
j. ethylene glycol
k. methyl phenyl ether
l. *trans*-2-ethylcyclohexanol

4.10   Name and draw structural formulas for the eight isomeric alcohols of molecular formula $C_5H_{12}O$. Classify each as primary, secondary, or tertiary.

4.11   Name and draw structural formulas for the six isomeric ethers of molecular formula $C_5H_{12}O$.

Physical properties of alcohols and ethers (Section 4.3)

4.12   In the following compounds: (1) Circle each hydrogen atom that is capable of hydrogen bonding. (2) Put a square around each atom capable of hydrogen bonding to a partially positive hydrogen atom.

a. 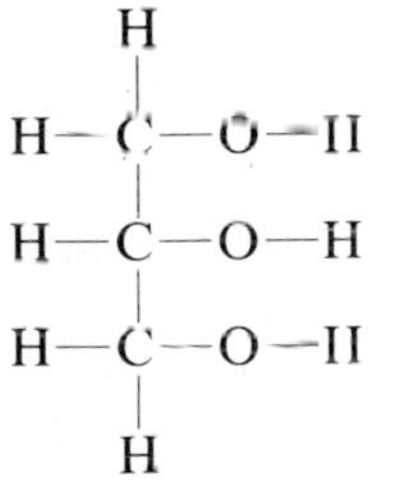
$$\begin{array}{c} H\\ |\\ H-C-O-H\\ |\\ H-C-O-H\\ |\\ H-C-O-H\\ |\\ H \end{array}$$

b.
$$\begin{array}{c} H\ \ H\\ |\ \ \ |\\ H-O-C-C-S-H\\ |\ \ \ |\\ H\ \ H \end{array}$$

c.
$$\begin{array}{c} H\\ |\\ H-C-\\ |\\ H \end{array} \bigcirc -O-H$$

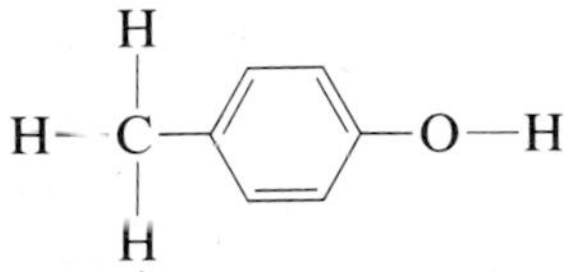

d. 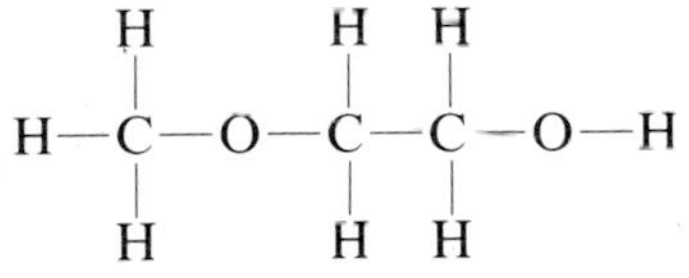
$$\begin{array}{c} H\ \ \ \ \ \ H\ \ H\\ |\ \ \ \ \ \ |\ \ \ |\\ H-C-O-C-C-O-H\\ |\ \ \ \ \ \ |\ \ \ |\\ H\ \ \ \ \ \ H\ \ H \end{array}$$

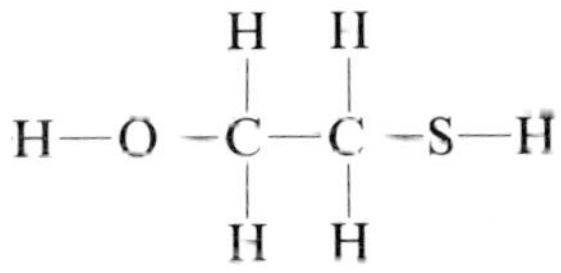

4.13   Arrange the compounds of each of the sets in order of decreasing boiling points. Explain your reasoning.

a. $CH_3CH_2CH_3$ $CH_3CH_2CH_2CH_2CH_2CH_2CH_3$
$CH_3CH_2CH_2CH_2CH_3$
b. N_2H_4 H_2O_2 CH_3CH_3
c. CH_3CO_2H CH_3CH_2OH CH_3OCH_3

d. 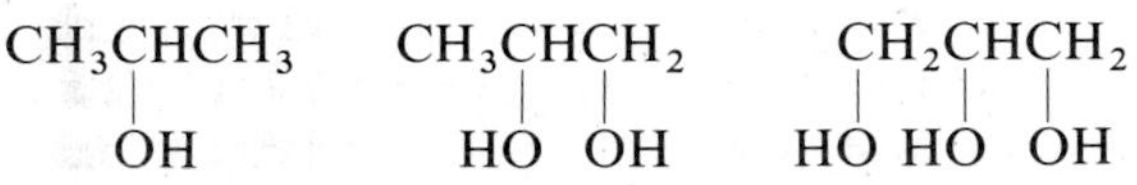
$$\underset{\substack{|\\ OH}}{CH_3CHCH_3} \quad \underset{\substack{|\ \ |\\ HO\ OH}}{CH_3CHCH_2} \quad \underset{\substack{|\ \ |\ \ |\\ HO\ HO\ OH}}{CH_2CHCH_2}$$

4.14   Arrange the compounds in each set in order of decreasing solubility in water. Explain your reasoning.

      a. ethanol, butane, diethyl ether
      b. 1-hexanol, 1,2-hexanediol, hexane

4.15   Account for the fact that the boiling point of 1-butanol is higher than that of its structural isomer, diethyl ether.

$$CH_3CH_2CH_2CH_2OH \qquad CH_3CH_2OCH_2CH_3$$

1-butanol   diethyl ether
(bp 117°C)   (bp 35°C)

4.16   Why does ethanol (mw 46, bp 78°C) have a boiling point over 43° **higher** than that of ethanethiol (mw 62, bp 35°C)?

$$CH_3CH_2OH \qquad CH_3CH_2SH$$

ethanol   ethanethiol
(bp 78°C)   (bp 35°C)

4.17   Propanoic acid and methyl acetate are structural isomers. Both are liquid at room temperature. The boiling point of one of these liquids is 57°C; of the other, 141°C.

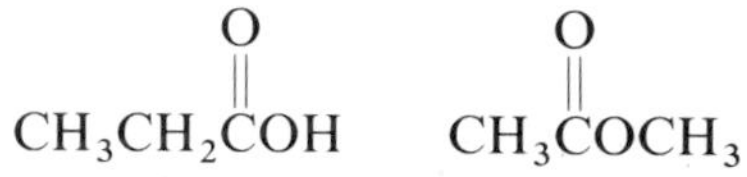

$$\overset{O}{\overset{\|}{CH_3CH_2COH}} \qquad \overset{O}{\overset{\|}{CH_3COCH_3}}$$

propanoic acid   methyl acetate

      a. Which compound has the boiling point of 141°C? of 57°C? Explain your reasoning.
      b. Which compound is more soluble in water? Explain your reasoning.

4.18   Compounds that contain NH bonds show association by hydrogen bonding. Do you expect this association to be stronger or weaker than that in compounds containing OH bonds? (*Hint*: Remember the table of relative electronegativities, Figure 1.1.)

Preparation of alcohols (Review Section 4.5C)

4.19   Write structural formulas for the alkenes of the given molecular formula that undergo acid-catalyzed hydration to give the alcohol shown as the major product.

a. 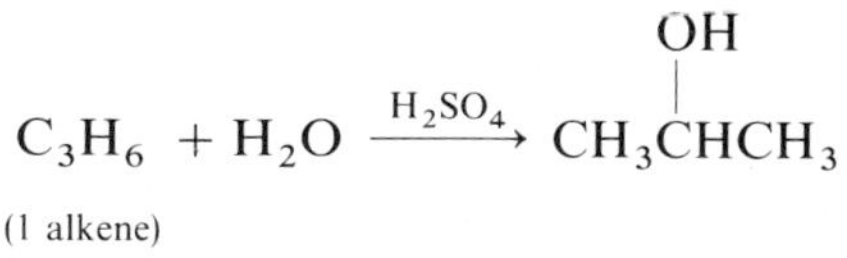 $C_3H_6 + H_2O \xrightarrow{H_2SO_4} CH_3\overset{\overset{\displaystyle OH}{|}}{C}HCH_3$

    (1 alkene)

b. 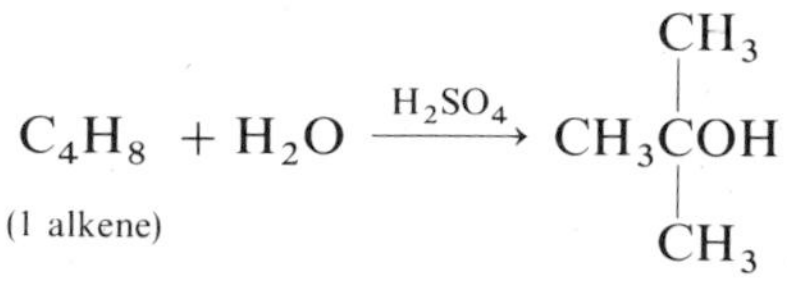 $C_4H_8 + H_2O \xrightarrow{H_2SO_4} CH_3\overset{\overset{\displaystyle CH_3}{|}}{\underset{\underset{\displaystyle CH_3}{|}}{C}}OH$

    (1 alkene)

$$\text{c.} \quad C_4H_8 + H_2O \xrightarrow{H_2SO_4} CH_3\overset{\overset{\displaystyle OH}{|}}{C}HCH_2CH_3$$

(3 alkenes)

$$\text{d.} \quad C_7H_{12} + H_2O \xrightarrow{H_2SO_4}$$

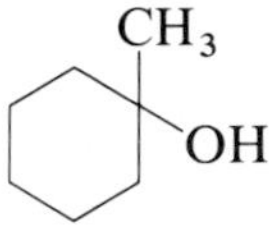

(2 alkenes)

4.20 Propose a mechanism for the acid-catalyzed hydration

$$CH_3\overset{\overset{\displaystyle CH_3}{|}}{C}=CHCH_3 + H_2O \xrightarrow{H_2SO_4} CH_3\overset{\overset{\displaystyle CH_3}{|}}{\underset{\underset{\displaystyle OH}{|}}{C}}CH_2CH_3$$

Reactions of alcohols (Section 4.4)

4.21 Draw structural formulas for the alkene or alkenes formed from acid-catalyzed dehydration of the following alcohols. Where two or more alkenes are formed, predict which is the major product and which the minor product.

a. $CH_3CH_2CH_2OH$

b. $CH_3CH_2\overset{\overset{\displaystyle }{|}}{C}HCH_3$ with OH

c. (cyclopentane with CH_3 and OH)

d. (benzene ring)$-CH_2\overset{\overset{\displaystyle OH}{|}}{C}HCH_3$

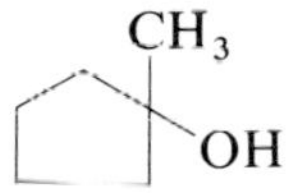

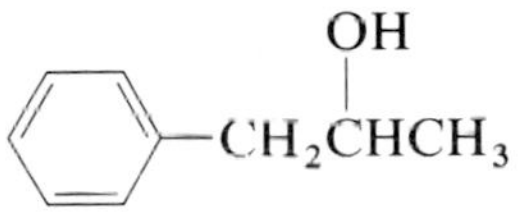

e. $HOCH_2CH_2CH_2CH_2OH$

f. $CH_3\overset{\overset{\displaystyle CH_3}{|}}{\underset{\underset{\displaystyle OH}{|}}{C}}CH_2CH_3$

4.22 Propose a mechanism for the acid-catalyzed dehydration of cyclohexanol to give cyclohexene and water.

4.23 Predict the relative ease with which the following alcohols undergo acid-catalyzed dehydration. Draw a structural formula for the major products of each dehydration.

a. $CH_3\overset{\overset{\displaystyle OH}{|}}{C}H\overset{\overset{\displaystyle }{|}}{C}HCH_2CH_3$ with CH_3

b. $CH_3CH_2\overset{\overset{\displaystyle OH}{|}}{C}CH_2CH_3$ with CH_3

c. $HOCH_2CH_2\overset{\overset{\displaystyle }{|}}{C}HCH_2CH_3$ with CH_3

4.24 One of the reactions in the metabolism of glucose is the isomerization of citric acid to isocitric acid. The isomerization is catalyzed by the enzyme aconitase.

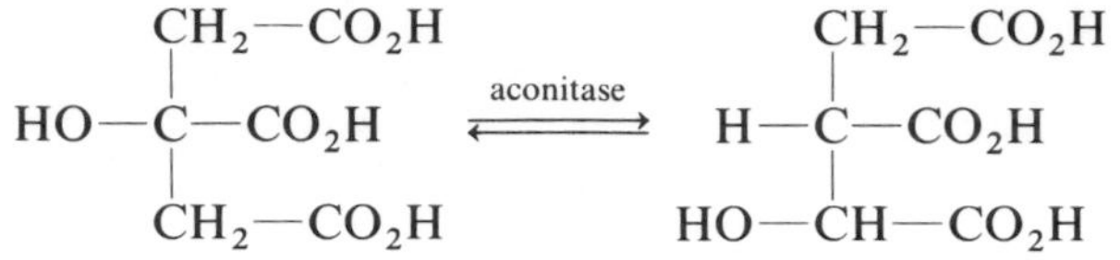

$$
\begin{array}{ccc}
\text{CH}_2\text{—CO}_2\text{H} & & \text{CH}_2\text{—CO}_2\text{H} \\
| & \xrightleftharpoons{\text{aconitase}} & | \\
\text{HO—C—CO}_2\text{H} & & \text{H—C—CO}_2\text{H} \\
| & & | \\
\text{CH}_2\text{—CO}_2\text{H} & & \text{HO—CH—CO}_2\text{H} \\
\text{citric acid} & & \text{isocitric acid}
\end{array}
$$

Propose a reasonable mechanism to account for this isomerization. (*Hint*: Within its structure, aconitase has groups that can function as acids.)

4.25 Write structural formulas for the major organic product or products of the following oxidations.

a. 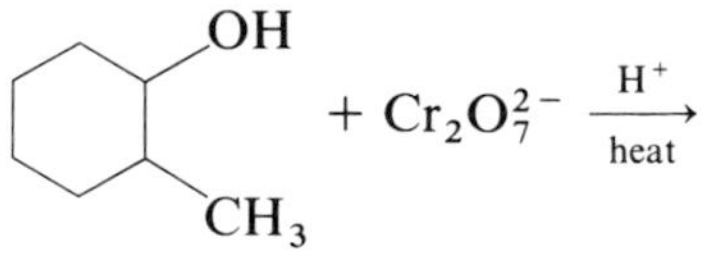 [cyclohexane ring with OH at top and CH$_3$ below] $+ \text{Cr}_2\text{O}_7^{2-} \xrightarrow[\text{heat}]{\text{H}^+}$

b. $\text{HOCH}_2\text{CH}_2\text{CH}_2\text{CH}_2\text{CH}_2\text{CH}_2\text{OH} + \text{Cr}_2\text{O}_7^{2-} \xrightarrow[\text{heat}]{\text{H}^+}$

c. $\text{CH}_3\text{CH}_2\text{OH} + \text{O}_2$ (excess) $\xrightarrow{\text{(combustion)}}$

d. 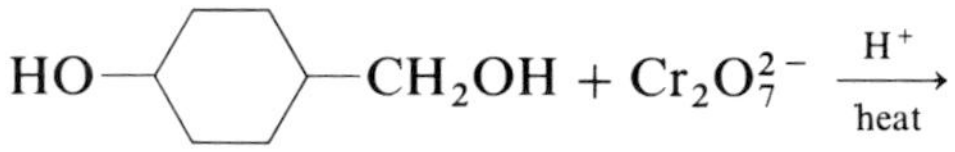 HO—[cyclohexane ring]—$\text{CH}_2\text{OH} + \text{Cr}_2\text{O}_7^{2-} \xrightarrow[\text{heat}]{\text{H}^+}$

e. 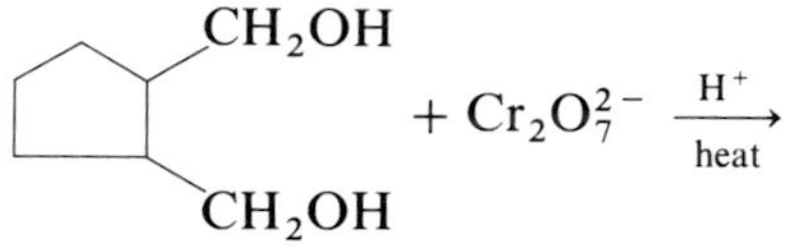 [cyclopentane ring with CH$_2$OH and CH$_2$OH] $+ \text{Cr}_2\text{O}_7^{2-} \xrightarrow[\text{heat}]{\text{H}^+}$

4.26 The following reactions are important in the metabolism of either fats or carbohydrates. State which are oxidations, which reductions, and which neither oxidation nor reduction.

a. $\text{CH}_3(\text{CH}_2)_{12}\text{CH}_2\text{CH}_2\text{CO}_2\text{H} \longrightarrow \text{CH}_3(\text{CH}_2)_{12}\text{CH}=\text{CHCO}_2\text{H}$

b. $\text{CH}_3(\text{CH}_2)_{12}\text{CH}=\text{CHCO}_2\text{H} \longrightarrow \text{CH}_3(\text{CH}_2)_{12}\overset{\overset{\displaystyle\text{OH}}{|}}{\text{C}}\text{HCH}_2\text{CO}_2\text{H}$

c. 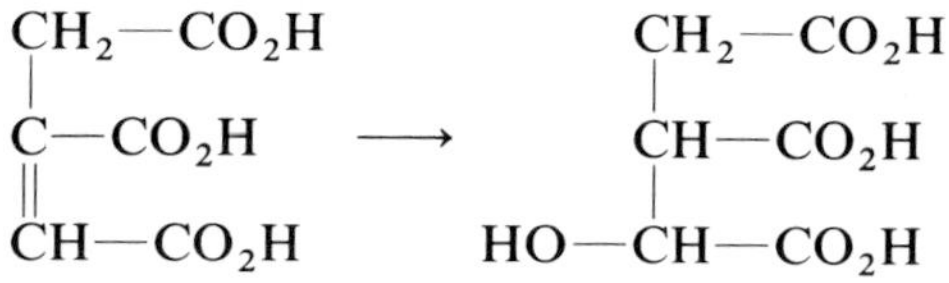
$$
\begin{array}{ccc}
\text{CH}_2\text{—CO}_2\text{H} & & \text{CH}_2\text{—CO}_2\text{H} \\
| & & | \\
\text{C—CO}_2\text{H} & \longrightarrow & \text{CH—CO}_2\text{H} \\
\| & & | \\
\text{CH—CO}_2\text{H} & & \text{HO—CH—CO}_2\text{H} \\
\text{aconitic acid} & & \text{isocitric acid}
\end{array}
$$

$$\begin{array}{ll}
\text{d.} & \begin{array}{l} CH_2\!-\!CO_2H \\ | \\ CH\!-\!CO_2H \\ | \\ HO\!-\!CH\!-\!CO_2H \end{array} \longrightarrow \begin{array}{l} CH_2\!-\!CO_2H \\ | \\ CH\!-\!CO_2H \\ | \\ O\!=\!C\!-\!CO_2H \end{array} \\
& \quad\text{isocitric acid} \qquad\qquad \text{oxalosuccinic acid}
\end{array}$$

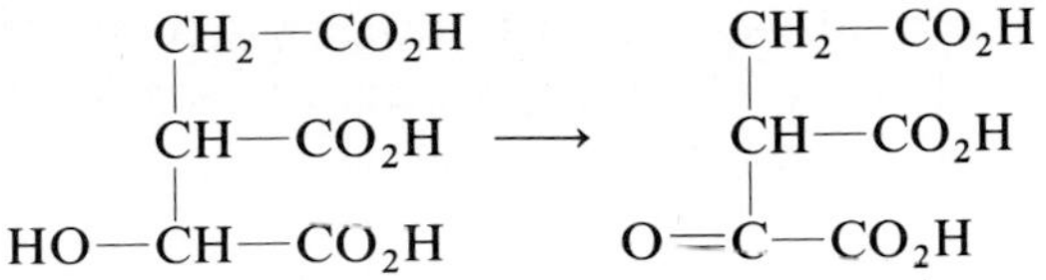

e. 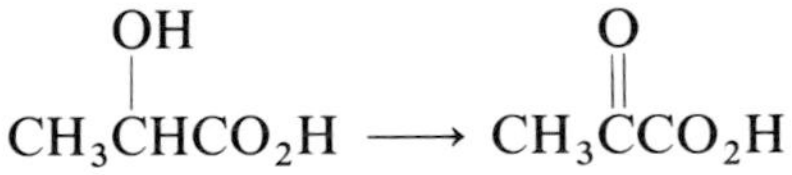 $\underset{\text{lactic acid}}{CH_3\overset{\displaystyle OH}{\underset{|}{C}}HCO_2H} \longrightarrow \underset{\text{pyruvic acid}}{CH_3\overset{\displaystyle O}{\overset{\|}{C}}CO_2H}$

4.27 The following conversions can be carried out in one step. Show the reagent you would use to bring about each conversion.

a.

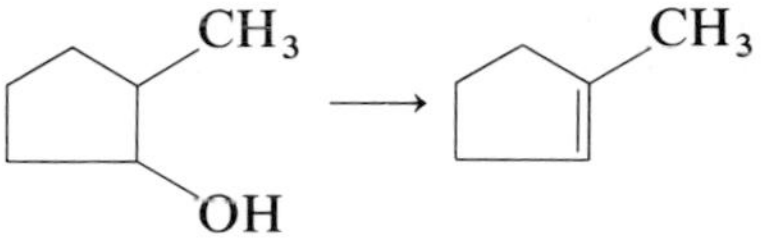

b.

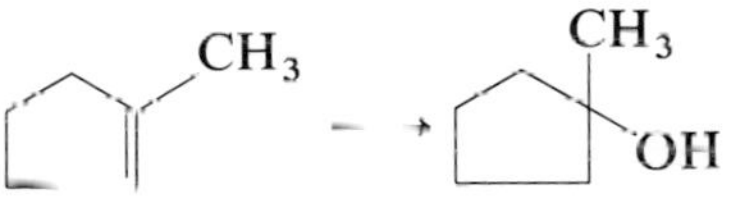

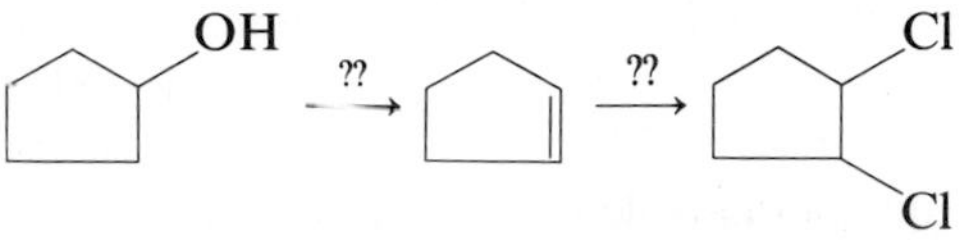

c. 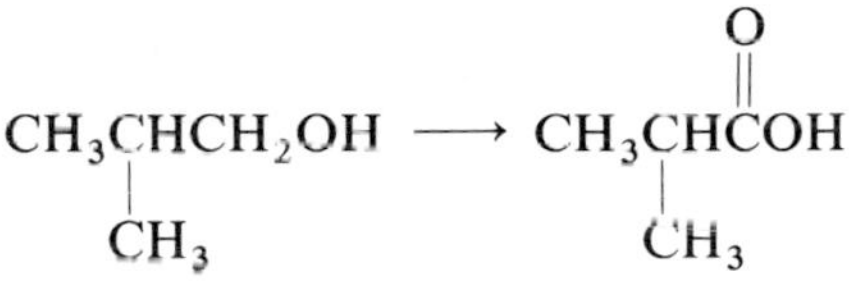 $CH_3\underset{\underset{\displaystyle CH_3}{|}}{C}HCH_2OH \longrightarrow CH_3\underset{\underset{\displaystyle CH_3}{|}}{C}H\overset{\displaystyle O}{\overset{\|}{C}}OH$

d.

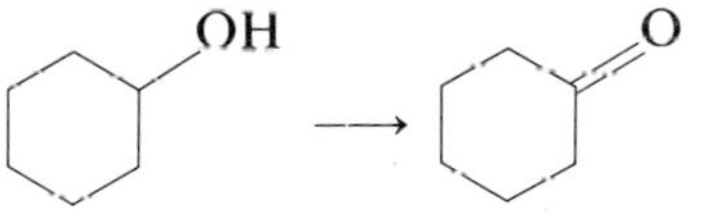

e.

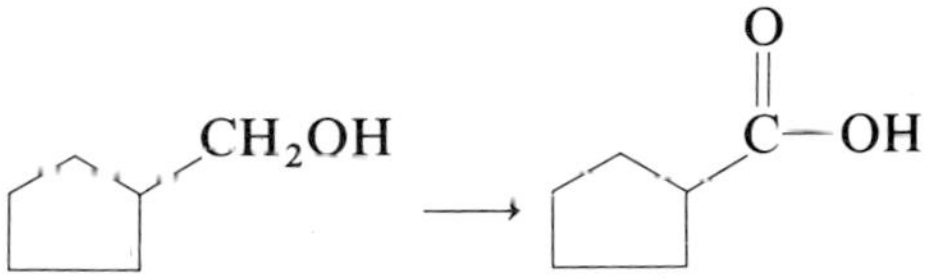

4.28 Following are a series of conversions in which a starting material is converted to the indicated product in two steps. State the reagent or reagents you would use to bring about each conversion.

a. 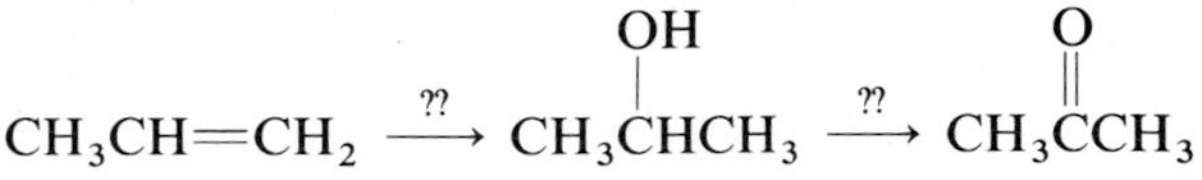 $CH_3CH\!=\!CH_2 \xrightarrow{\;??\;} CH_3\overset{\displaystyle OH}{\underset{|}{C}}HCH_3 \xrightarrow{\;??\;} CH_3\overset{\displaystyle O}{\overset{\|}{C}}CH_3$

b.

c.

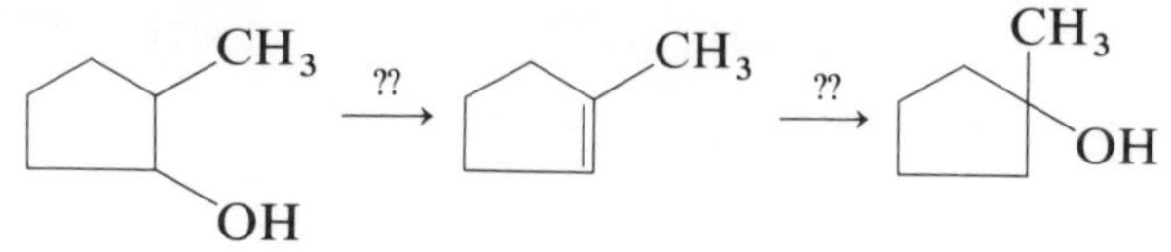

d.

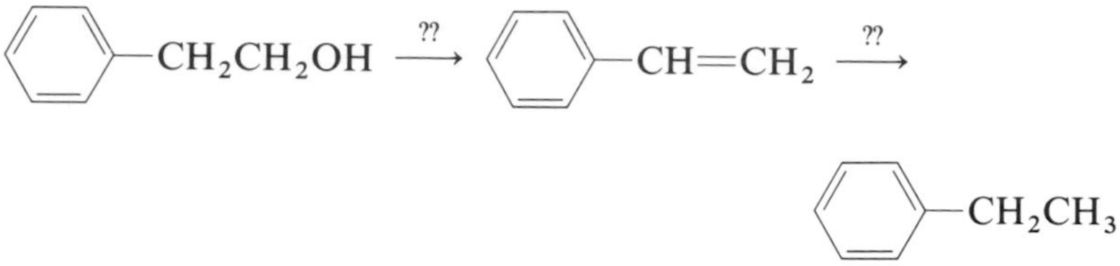

4.29 The following conversions can be carried out in two steps. Show reagents you would use and the structural formula of the intermediate formed in each conversion.

a.

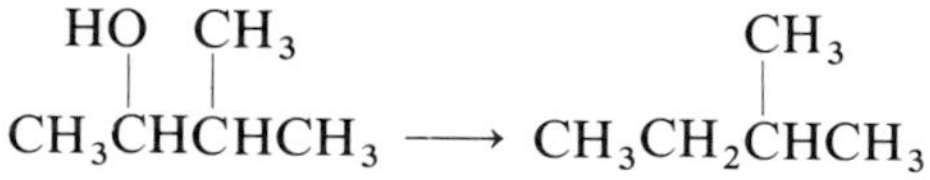

$$\underset{\substack{| \\ OH}}{HO} \underset{\substack{| \\ }}{CH_3}$$
a. $CH_3\overset{\displaystyle OH}{\underset{\displaystyle |}{C}}H\overset{\displaystyle CH_3}{\underset{\displaystyle |}{C}}HCH_3 \longrightarrow CH_3CH_2\overset{\displaystyle CH_3}{\underset{\displaystyle |}{C}}HCH_3$

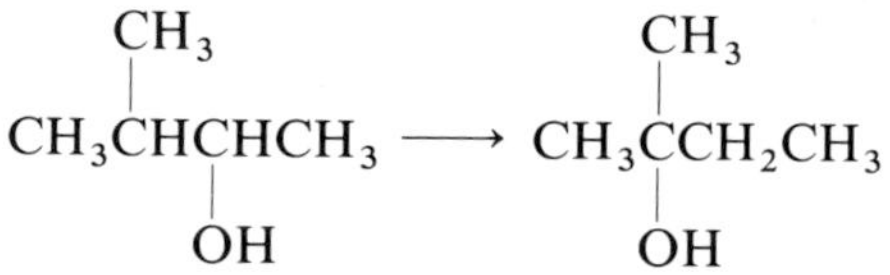

b. $CH_3\overset{\displaystyle CH_3}{\underset{\displaystyle |}{C}}H\overset{\displaystyle }{\underset{\displaystyle OH}{C}}HCH_3 \longrightarrow CH_3\overset{\displaystyle CH_3}{\underset{\displaystyle |}{\underset{\displaystyle |}{C}}}CH_2CH_3$

c. $CH_3CH_2CH_2OH \longrightarrow CH_3\overset{\displaystyle }{\underset{\displaystyle Cl}{C}}H\overset{\displaystyle }{\underset{\displaystyle Cl}{C}}H_2$

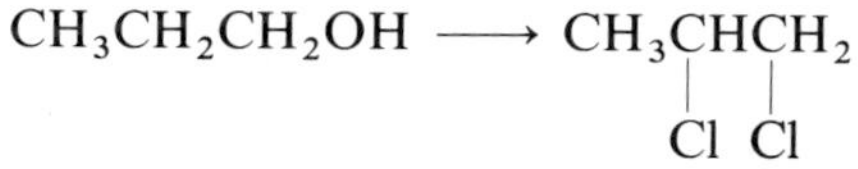

d.

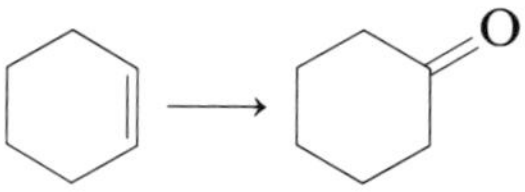

Phenols
(Section 4.7)

4.30 Name the following compounds.

a. 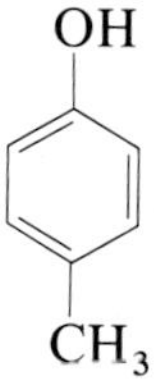   **b.** 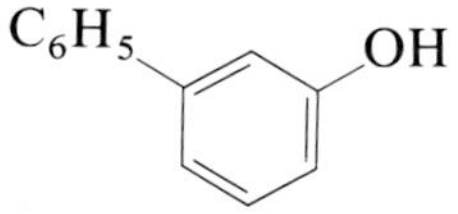   **c.**

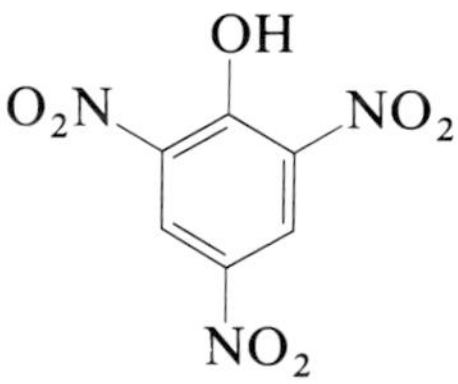

4.31 Write structural formulas for these compounds:
a. 2,4-dimethoxyphenol   **b.** sodium phenoxide
c. 2-isopropyl-4-methylphenol   **d.** *m*-cresol

4.32 Following is the structural formula of a compound known by the common name butylated hydroxytoluene (BHT). This substance is used as a preservative in food and animal feed and also as an antioxidant in soaps and petroleum products.

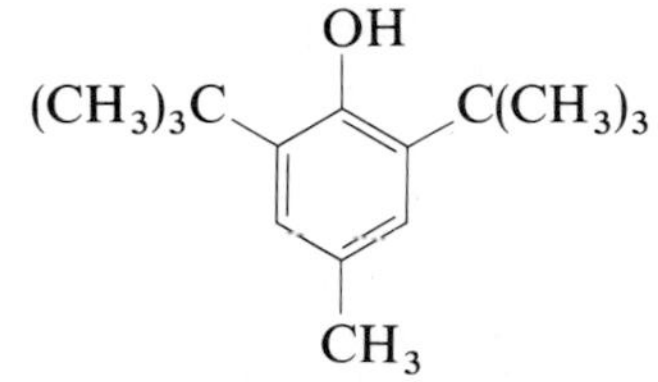

OH

$(CH_3)_3C$ $C(CH_3)_3$

CH_3

butylated hydroxytoluene
(BHT)

a. Write the IUPAC name of BHT. (*Hint*: Name it as a trisubstituted phenol.)

b. It is common to add 0.0001% BHT to anhydrous diethyl ether as an antioxidant. Calculate the number of milligrams of BHT in 1 L of diethyl ether.

c. The molecular weight of BHT is 220.3 g/mol. Calculate the number of moles of BHT in 1 pt of stabilized diethyl ether.

d. Another common antoxidant is butylated hydroxyanisole (BHA). Knowing what you do about the structure of BHT, propose a structural formula for BHA.

4.33   Identify all functional groups in the following compounds:

CH_2OH

$C=O$

H_3C $-OH$

H_3C

a.

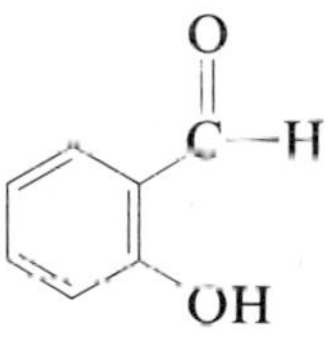

salicylaldehyde

b.

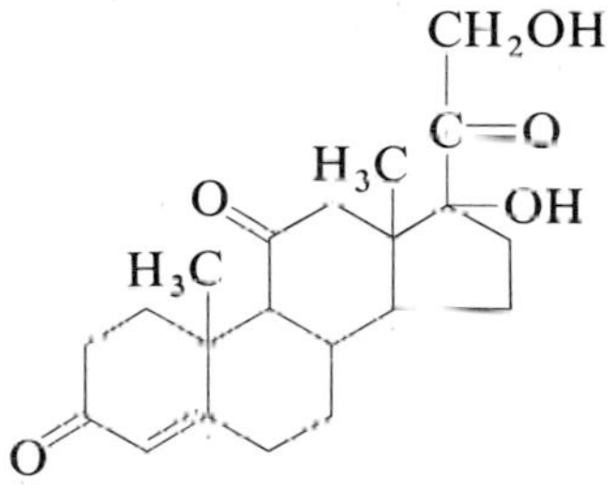

cortisone

c.

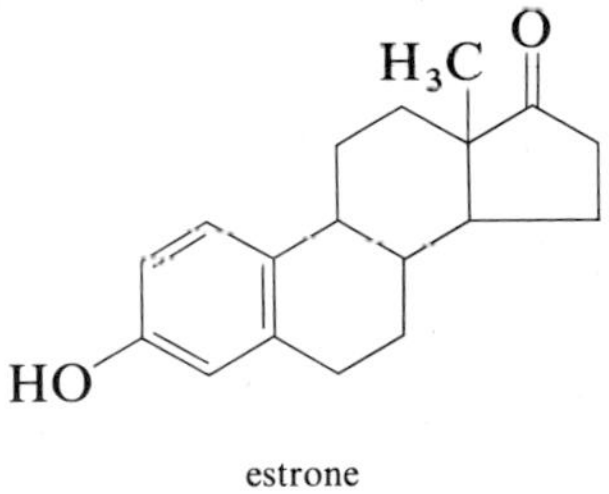

HO

estrone
(a female sex hormone)

H_3C O

d.

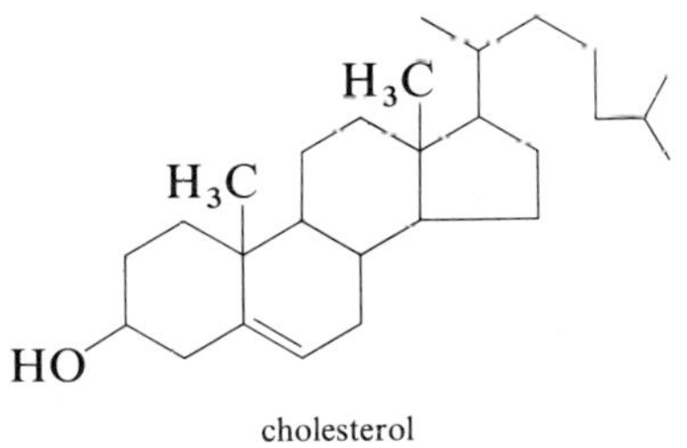

HO

cholesterol

H_3C

H_3C

O

$C-H$

e.

OCH_3

OH

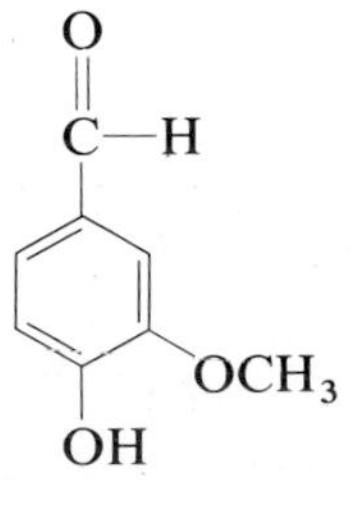

vanillin

4.34 Complete the reactions. Where you predict no reaction, write N.R.

a.  CH_3—⟨⟩—OH + NaOH ⟶

b. 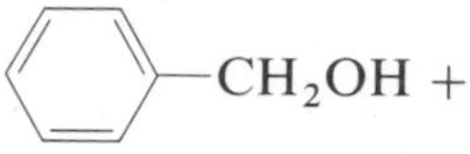 ⟨⟩—CH_2OH + NaOH ⟶

4.35 Show how you could distinguish between the following pairs of compounds by a simple chemical test. In each case, tell what test you would perform and what you would expect to observe, and write an equation for each positive test.

a. 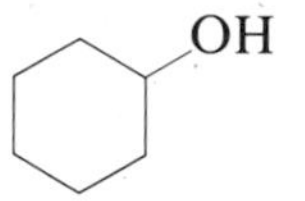 [cyclohexane with OH]   and   [benzene with OH]

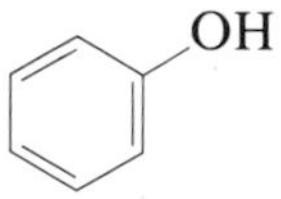

b. 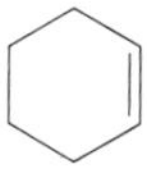 [cyclohexene]   and   [cyclohexane with OH]

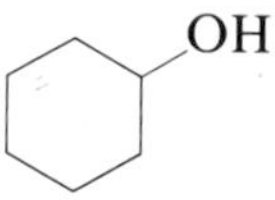

c.  [benzene]   and   [cyclohexene]

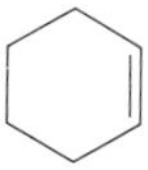

d. 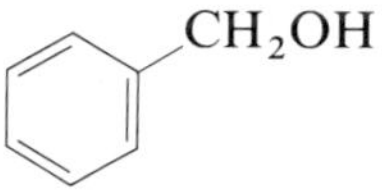 [benzene with CH_2OH]   and   [benzene with CH_3 and OH]

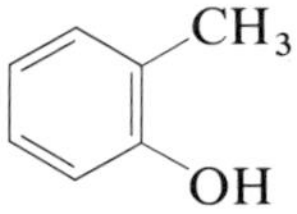

Thiols, sulfides, and disulfides (Section 4.8)

4.36 Name the compounds:

a. CH_3CH_2SH

b. 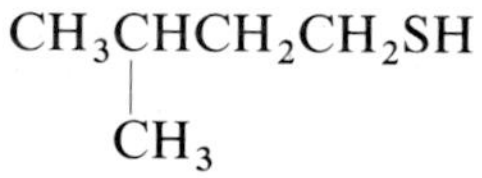 $CH_3CHCH_2CH_2SH$
        |
        CH_3

c. 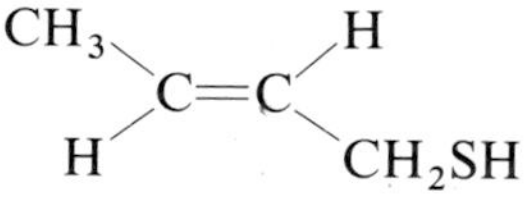 $\underset{H}{\overset{CH_3}{>}}C=C\underset{CH_2SH}{\overset{H}{<}}$

d. $HOCH_2CH_2SH$

4.37 Draw structural formulas for these compounds:
a. 2-pentanethiol   b. cyclopentanethiol
c. 1,2-ethanedithiol   d. diisobutyl sulfide
e. diisobutyl disulfide   f. 2,3-dimercapto-1-propanol

4.38 Write a balanced half-reaction for the conversion of two molecules of a thiol to a disulfide, and show that this conversion is a two-electron oxidation.

4.39 Draw structural formulas for the major organic products of the following reactions:

a. $CH_3(CH_2)_6CH_2SH$ + NaOH ⟶

b. $HSCH_2CH_2CH_2SH + \frac{1}{2}O_2 \longrightarrow$

c. $CH_3CH{=}CHCH_2CH_2SH + H_2 \xrightarrow{Pt}$

4.40 Penicillamine can be used to treat lead poisoning. Write an equation for the reaction of penicillamine with Pb^{2+} and explain how penicillamine might be used to counteract lead poisoning.

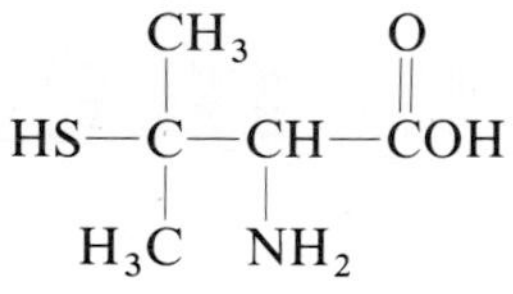

$$HS{-}\underset{\underset{H_3C}{|}}{\overset{\overset{CH_3}{|}}{C}}{-}\underset{\underset{NH_2}{|}}{CH}{-}\overset{\overset{O}{\|}}{C}OH$$

penicillamine

4.41 One treatment for mercury poisoning uses 2,3-dimercapto-1-propanol, a substance that forms a water-soluble complex with Hg(II) ion, which is then excreted in the urine. Draw a structural formula for this water-soluble complex.

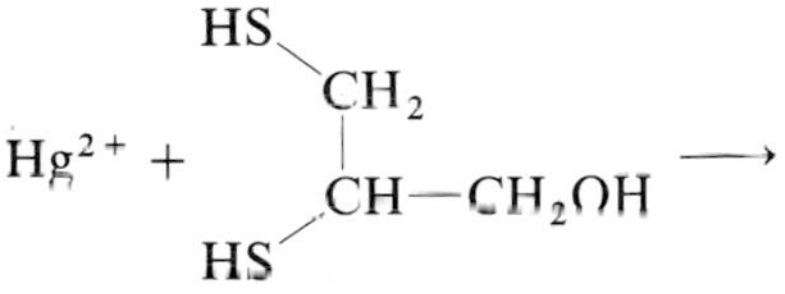

$$Hg^{2+} + \begin{array}{c} HS \\ \quad CH_2 \\ \quad | \\ \quad CH{-}CH_2OH \\ HS \end{array} \longrightarrow$$

2,3-dimercapto-1-propanol

5 Amines

5.1 Structure, Classification, and Nomenclature

Amines are derivatives of ammonia in which one or more hydrogens are replaced by alkyl or aromatic groups. The most important chemical property of amines is their basicity. As we have seen, carbon, hydrogen, and oxygen are the three most common elements in organic compounds. Because of the wide distribution of amines in the biological world, nitrogen is the fourth most common component of organic materials.

A. Classification

Amines are classified as primary, secondary, or tertiary, depending on the number of carbon atoms bonded to nitrogen. In a **primary amine**, one hydrogen of ammonia is replaced by carbon. In a **secondary amine**, two hydrogens are replaced by carbons; and in a **tertiary amine**, three hydrogens are replaced by carbons. Following are structural formulas for a primary, a secondary, and a tertiary amine:

$$H-\overset{\overset{\displaystyle H}{|}}{\underset{\underset{\displaystyle H}{|}}{N}}:\qquad CH_3-\overset{\overset{\displaystyle H}{|}}{\underset{\underset{\displaystyle H}{|}}{N}}:\qquad CH_3-\overset{\overset{\displaystyle H}{|}}{\underset{\underset{\displaystyle CH_3}{|}}{N}}:\qquad CH_3-\overset{\overset{\displaystyle CH_3}{|}}{\underset{\underset{\displaystyle CH_3}{|}}{N}}:$$

ammonia   methylamine   dimethylamine   trimethylamine
(a 1° amine)   (a 2° amine)   (a 3° amine)

Alcohols are also classified as primary, secondary, or tertiary (Section 4.2A) but the basis for classification is different from that for amines. Classification of alcohols depends on the number of carbon atoms attached to the carbon bearing the —OH group.

$$CH_3-\underset{\underset{CH_3}{|}}{\overset{\overset{CH_3}{|}}{C}}-OH$$

This is a tertiary carbon

a tertiary alcohol

$$CH_3-\underset{\underset{CH_3}{|}}{\overset{\overset{CH_3}{|}}{C}}-NH_2$$

Only one carbon attached directly to nitrogen

a primary amine

Amines are further divided into aliphatic and aromatic amines. In aliphatic amines, all the carbons attached directly to nitrogen are derived from alkyl groups; in aromatic amines, one or more of the groups attached to nitrogen are aromatic rings.

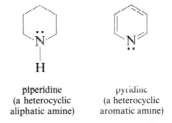

aniline
(a 1° aromatic amine)

N-methylaniline
(a 2° aromatic amine)

benzyldimethylamine
(a 3° aliphatic amine)

Amines in which the nitrogen atom is part of a ring are classified as heterocyclic amines. When a nitrogen atom replaces an atom of carbon in an aromatic ring, the amine is classified as a heterocyclic aromatic amine. Following are structural formulas for two cyclic amines, one classified as a heterocyclic aliphatic (3) amine and the other as a heterocyclic aromatic amine:

piperidine
(a heterocyclic aliphatic amine)

pyridine
(a heterocyclic aromatic amine)

B. Nomenclature

1. IUPAC System

In IUPAC nomenclature, the longest chain of carbon atoms that contains the **amino group** is taken as the parent and $-NH_2$ is considered a substituent, like $-Cl$, $-NO_2$, and so on. Its presence is indicated by the prefix *amino-*, and a number is used to show its location. If more than one alkyl group is attached to nitrogen, the one containing the longest chain of carbon atoms is taken as the parent; other substituents on nitrogen are named as alkyl groups and are preceded by *N-* to show that they are bonded to the nitrogen atom of the amine.

In these and the following examples, IUPAC names are given first. Instead of showing the same structural formulas again under "Common Names," we have listed common names, where they exist, here in parentheses under IUPAC names. As you will discover in your reading, common names for most amines are much more widely used than IUPAC names.

$$CH_3CH_2{-}NH_2$$

aminoethane
(ethylamine)

$$\underset{\underset{CH_3}{|}}{CH_3CHCH_2CH_2}{-}NH_2$$

1-amino-3-methylbutane
(isopentylamine)

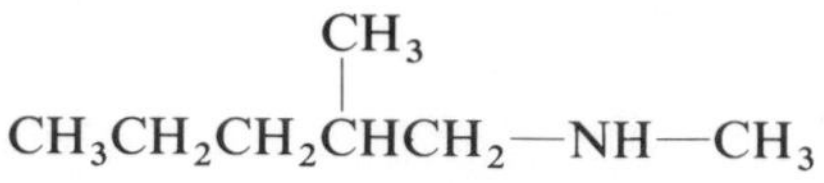

$$\underset{\underset{CH_3}{|}}{CH_3CH_2CH_2CHCH_2}{-}NH{-}CH_3$$

N-methyl-1-amino-2-methylpentane

Compounds containing two or more amino groups are named by prefixes *di-*, *tri-*, and so on, to show multiple substitution and by numbers to show the location of each substituent. Following are structural formulas for three diamines. The first, 1,6-diaminohexane is one of two raw materials for the synthesis of Nylon 66. The second two amines are products of the decomposition of animal matter, as their alternative, common names surely suggest.

$$H_2NCH_2CH_2CH_2CH_2CH_2CH_2NH_2$$

1,6-diaminohexane
(hexamethylenediamine)

$$H_2NCH_2CH_2CH_2CH_2CH_2NH_2$$

1,5-diaminopentane
(pentamethylenediamine;
cadaverine)

$$H_2NCH_2CH_2CH_2CH_2NH_2$$

1,4-diaminobutane
(tetramethylenediamine;
putrescine)

If the —NH$_2$ group is one of two or more substituents, it is shown by the prefix *amino-*, as in the following examples:

$$H_2NCH_2CH_2OH$$

2-aminoethanol
(ethanolamine)

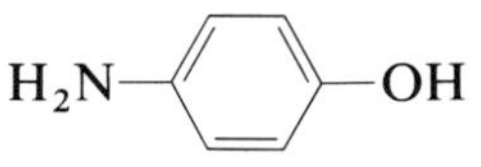

$$H_2N{-}\bigcirc{-}OH$$

4-aminophenol
(*p*-aminophenol)

The compound C$_6$H$_5$NH$_2$ is named aniline, which becomes the parent name for its derivatives. Several simple derivatives of aniline are known almost exclusively by their common names, for example, toluidine (the ortho isomer is shown below).

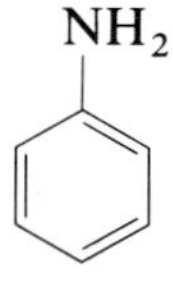

aniline

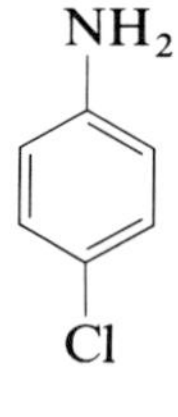

4-chloroaniline
(*p*-chloroaniline)

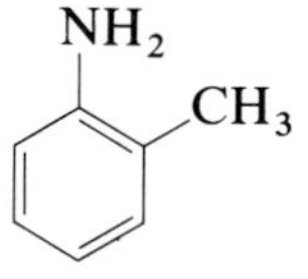

2-methylaniline
(*o*-methylaniline;
o-toluidine)

When a nitrogen atom has four organic groups attached to it (any combination of aliphatic or aromatic), the compound is named as an ammonium salt, as in the following examples:

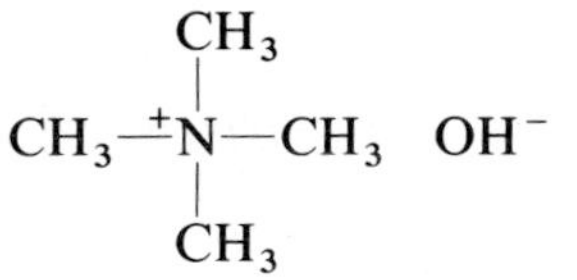

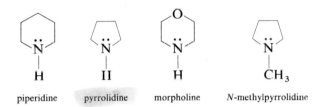

tetramethyl ammonium hydroxide 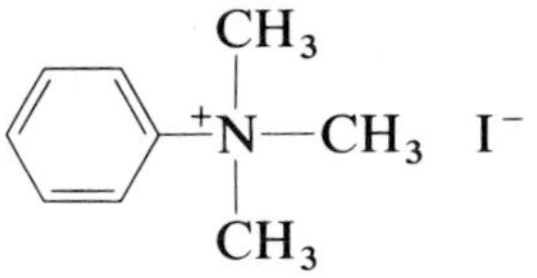   phenyl trimethyl ammonium iodide

Heterocyclic amines all have common names that the IUPAC has accepted. Structural formulas for the three most common **heterocyclic aliphatic amines** are shown below. Note that when there is an additional substituent on the nitrogen atom of the parent molecule, its location is indicated by *N*-.

piperidine   pyrrolidine   morpholine   *N*-methylpyrrolidine

Finally there are groups of **heterocyclic aromatic amines** whose common names have been retained by the IUPAC. Structural formulas, names, and numbering systems for the most common of these are shown:

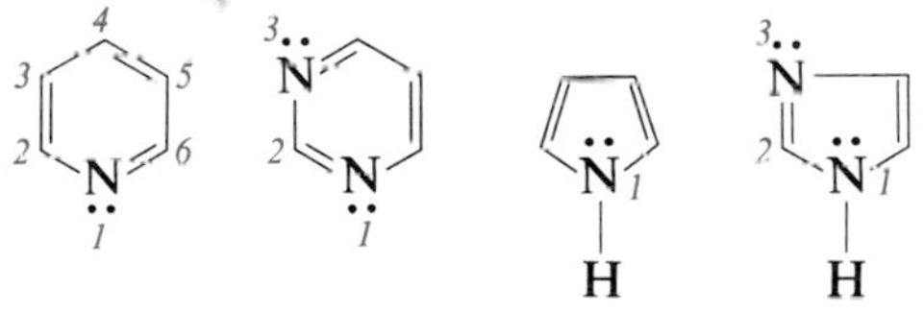

pyridine   pyrimidine   pyrrole   imidazole

2. Common Names

Common names for simple aliphatic amines are derived by listing the alkyl group or groups attached to nitrogen in alphabetical order, continuously in one long word, ending with the suffix -*amine*:

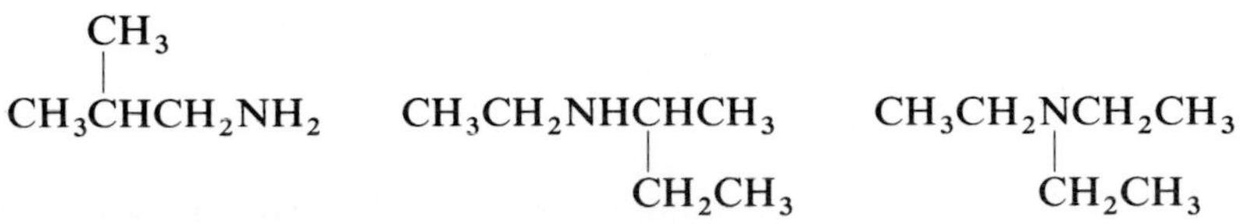

$$CH_3CHCH_2NH_2 \qquad CH_3CH_2NHCHCH_3 \qquad CH_3CH_2NCH_2CH_3$$

isobutylamine   *sec*-butylethylamine   triethylamine

Example 5.1

Give each compound an IUPAC name and where possible, a common name also.

a. $CH_3CH_2NHCH_3$

b. $CH_3CHCH_2CH_2CH_2CH_3$
$\qquad\quad |$
$\qquad\quad N(CH_3)_2$

c.

(structure: pyridine ring with NH_2 substituent)

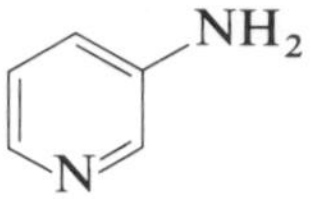

d. $CH_3CHCHCH_3$
$\qquad\ \ |\ \ |$
$\qquad\ HO\ NH_2$

☐ *Solution*

a. IUPAC: *N*-methyl-1-aminoethane. Common: ethylmethylamine.
b. IUPAC: *N,N*-dimethyl-2-aminohexane. Common: none.
c. IUPAC: 3-aminopyridine. Common: *β*-pyridylamine. In this common name, the pyridine ring is shown as a substituent on the amine group. We have little occasion to use this type of common name.
d. IUPAC: 3-amino-2-butanol. Common: none.

Problem 5.1

Give IUPAC and, where possible, common names for the following:

a. 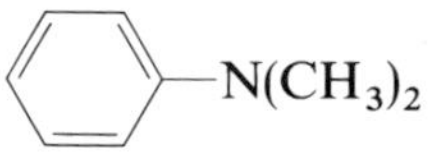 (benzene ring)$-N(CH_3)_2$

b. $CH_3CH_2CHCH_3$
$\qquad\qquad |$
$\qquad\qquad NH_2$

c. $H_2NCH_2CH_2CH_2OH$

5.2 Physical Properties of Amines

Amines are polar compounds and both primary and secondary amines can form intermolecular **hydrogen bonds.**

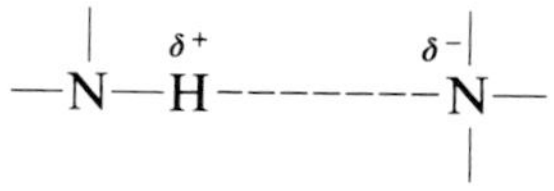

$$-\overset{|}{N}-\overset{\delta^+}{H}--------\overset{\delta^-}{\underset{|}{N}}-$$

However, because the difference in electronegativity is not so great between nitrogen and hydrogen $(3.0 - 2.1 = 0.9)$ as between oxygen and hydrogen $(3.5 - 2.1 = 1.4)$, an N—H······N hydrogen bond is not nearly so strong as an O—H······O hydrogen bond. The boiling points of ethane, methylamine, and methanol, all compounds of comparable molecular weight, are

	Mol. Wt.	bp (°C)
CH_3CH_3	30	-88
CH_3NH_2	31	-7
CH_3OH	32	65

Ethane is a nonpolar hydrocarbon, and the only interactions between molecules in the pure liquid are very weak dispersion forces. Therefore, it has the

Table 5.1 Physical properties of selected amines.

Name	Structure	mp (°C)	bp (°C)	Solubility (g/100 g H_2O)	Density of Liquid at 20°C (g/mL)
ammonia	NH_3	−78	−33	90	(a gas)
methylamine	CH_3NH_2	−95	−6	very	(a gas)
ethylamine	$CH_3CH_2NH_2$	−81	17	infinite	(a gas)
propylamine	$CH_3(CH_2)_2NH_2$	−83	48	soluble	0.717
isopropylamine	$(CH_3)_2CHNH_2$	−95	32	infinite	0.889
n-butylamine	$CH_3(CH_2)_3NH_2$	−49	78	infinite	0.741
dimethylamine	$(CH_3)_2NH$	−93	7	very	(a gas)
diethylamine	$(CH_3CH_2)_2NH$	−48	56	very	0.706
trimethylamine	$(CH_3)_3N$	−117	3	very	(a gas)
triethylamine	$(CH_3CH_2)_3N$	−114	89	soluble	0.727
cyclohexylamine	$C_6H_{11}NH_2$	−17	145	soluble	0.819
aniline	$C_6H_5NH_2$	−6	184	3.7	1.02
benzylamine	$C_6H_5CH_2NH_2$	—	185	infinite	0.981
pyridine	C_5H_5N	−42	116	infinite	0.982

lowest boiling point of the three. Both methylamine and methanol are polar molecules and interact in the pure liquid by hydrogen bonding. Hydrogen bonding is weaker in methylamine than in methanol, and therefore methylamine has a lower boiling point than methanol.

All classes of amines form hydrogen bonds with water and therefore are more soluble in water than hydrocarbons of comparable molecular weight are. Most low-molecular-weight amines are completely soluble in water. The higher-molecular-weight amines are only moderately soluble. Boiling points and solubilities in water for several amines are listed in Table 5.1.

5.3 Reactions of Amines

In this chapter, we will discuss only two reactions characteristic of amines, namely their basicity and their reaction with acids. Amines also react with other functional groups including aldehydes, ketones, carboxylic acids, and esters. We will discuss reactions of amines with these functional groups in following chapters.

A. Basicity of Amines

Like ammonia, all primary, secondary, and tertiary amines are weak bases, and aqueous solutions of amines are basic. The following reactions are written using curved arrows to emphasize that when an amine acts as a base, the unshared

pair of electrons on nitrogen forms a new bond to hydrogen, and an H—O
bond of water breaks to form hydroxide ion.

$$H-\overset{\overset{\displaystyle H}{|}}{\underset{\underset{\displaystyle H}{|}}{N}}: + H-\overset{\cdot\cdot}{\underset{\cdot\cdot}{O}}-H \rightleftharpoons H-\overset{\overset{\displaystyle H}{|}}{\underset{\underset{\displaystyle H}{|}}{\overset{+}{N}}}-H + ^{-}\!\!:\!\overset{\cdot\cdot}{\underset{\cdot\cdot}{O}}-H$$

ammonia   ammonium hydroxide

$$CH_3-\overset{\overset{\displaystyle H}{|}}{\underset{\underset{\displaystyle H}{|}}{N}}: + H-\overset{\cdot\cdot}{\underset{\cdot\cdot}{O}}-H \rightleftharpoons CH_3-\overset{\overset{\displaystyle H}{|}}{\underset{\underset{\displaystyle H}{|}}{\overset{+}{N}}}-H + ^{-}\!\!:\!\overset{\cdot\cdot}{\underset{\cdot\cdot}{O}}-H$$

methylamine   methylammonium hydroxide

The equilibrium constant for the reaction of methylamine with water to give
methylammonium hydroxide is called a **base dissociation constant** and is given
the symbol K_b.

$$K_b = \frac{[CH_3NH_3^+][OH^-]}{[CH_3NH_2]}$$

Values of K_b for some primary, secondary, and tertiary aliphatic amines and
for some aromatic amines are given in Table 5.2. All aliphatic amines, whether
primary, secondary, or tertiary have about the same base strength as ammonia.
Aromatic amines such as aniline are significantly less basic than aliphatic amines.
The K_b of aniline is less than the K_b of cyclohexylamine by a factor of one
million (10^6).

Table 5.2
Basicity of several
aliphatic and
aromatic amines.

Name	Structural Formula	K_b	pK_b	pK_a
ammonia	NH_3	1.8×10^{-5}	4.74	9.26
methylamine	CH_3NH_2	4.4×10^{-4}	3.36	10.64
ethylamine	$CH_3CH_2NH_2$	6.3×10^{-4}	3.20	10.80
diethylamine	$(CH_3CH_2)_2NH$	3.1×10^{-4}	3.51	10.49
triethylamine	$(CH_3CH_2)_3N$	1.0×10^{-4}	4.00	10.00
cyclohexylamine	⬡—NH_2	5.5×10^{-4}	3.26	10.74
aniline	⬡—NH_2	4.2×10^{-10}	9.37	4.63
pyridine	⬡N	1.8×10^{-9}	8.74	5.26

cyclohexylamine   cyclohexylammonium hydroxide

$K_b = 5.5 \times 10^{-4}$

aniline   anilinium hydroxide

$K_b = 4.2 \times 10^{-10}$

Salts of aliphatic amines are named by changing the suffix-*amine* to -*ammonium* and adding the name of the anion in the salt. Salts of aromatic amines are named by dropping the terminal -*e* and adding -*ium* followed by the name of the anion in the salt.

Until very recently, it was customary to list only K_b or pK_b for amines. Now it is becoming more and more common, particularly in biochemistry, to list only K_a and pK_a values for amines. The K_a and pK_a values are used almost exclusively in discussing the acid-base properties of amino acids and proteins (Chapter 13). For this reason, Table 5.2 also gives pK_a values for amines. Values for pK_a and pK_b are related by the following equation:

$$pK_a + pK_b = 14$$

We can illustrate the differences between pK_b and pK_a by the following equations:

$$CH_3NH_2 + H_2O \rightleftharpoons CH_3NH_3^+ + OH^- \quad K_b = 4.4 \times 10^{-4} \quad pK_b = 3.36$$
$$CH_3NH_3^+ \rightleftharpoons CH_3NH_2 + H^+ \quad K_a = 2.3 \times 10^{-11} \quad pK_a = 10.64$$

Whereas pK_b measures directly the strength of CH_3NH_2 as a base, pK_a measures the strength of $CH_3NH_3^+$ as an acid. For perspective you might compare the pK_a values for acetic acid and the methylammonium ion:

$$CH_3CO_2H \rightleftharpoons CH_3CO_2^- + H^+ \quad K_a = 1.8 \times 10^{-5} \quad pK_a = 4.74$$
$$CH_3NH_3^+ \rightleftharpoons CH_3NH_2 + H^+ \quad K_a = 2.3 \times 10^{-11} \quad pK_a = 10.64$$

By using K_a values for a carboxylic acid and an amine, we can compare their acidities directly, because in each case we are looking at the dissociation constant of an acid to form a base and a proton. It is obvious from the above data that acetic acid is a much stronger acid than the methylammonium ion.

B. Reaction of Amines with Acids

All amines, whether soluble or insoluble in water, react quantitatively with acids to form salts. In the following amine examples, curved arrows are used to show the flow of electrons from the base to the acid.

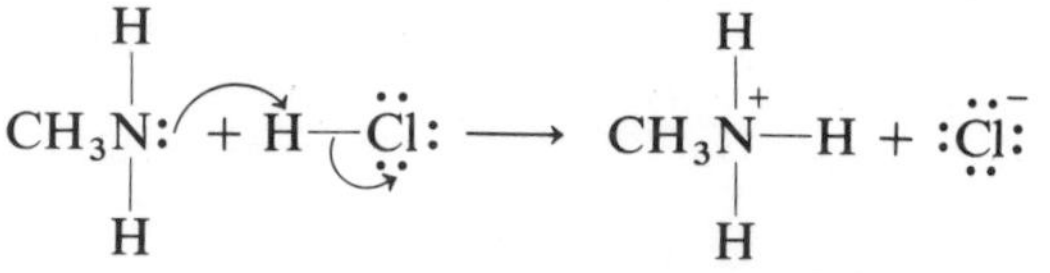

methylamine

methylammonium chloride
(methylamine hydrochloride)

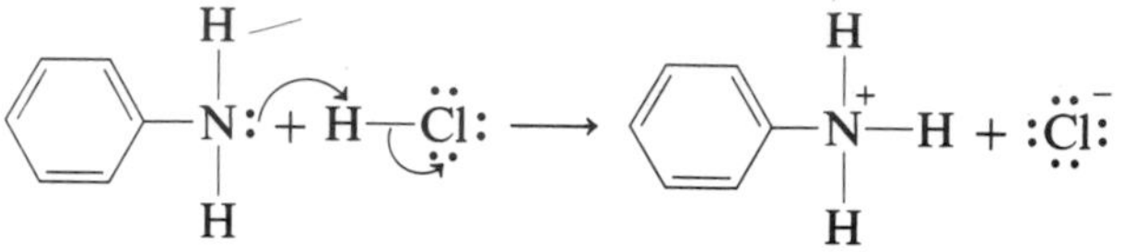

aniline

anilinium chloride
(aniline hydrochloride)

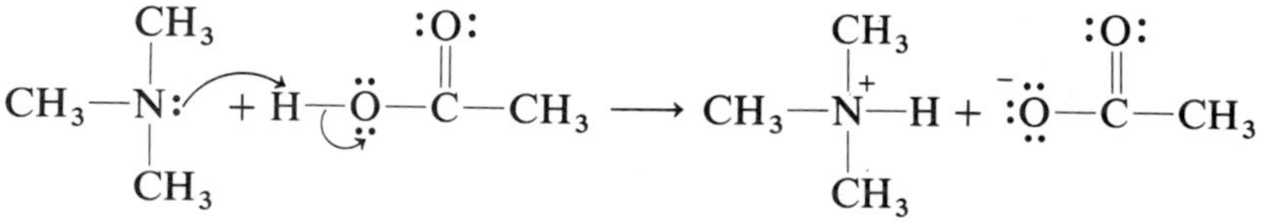

trimethylamine   acetic acid   trimethylammonium acetate

Example 5.2

Complete the acid-base reactions and name the salt formed.

a. $(CH_3CH_2)_2NH + HCl \longrightarrow$

b.  $+ CH_3CO_2H \longrightarrow$

☐ *Solution*

a. Diethylamine is a secondary aliphatic amine and reacts with HCl to form the salt diethylammonium chloride:

$$(CH_3CH_2)_2NH + HCl \longrightarrow (CH_3CH_2)_2NH_2^+ \ Cl^-$$

diethylamine   diethylammonium chloride

b. Pyridine is a heterocyclic aromatic amine and reacts with acetic acid to form the salt pyridinium acetate:

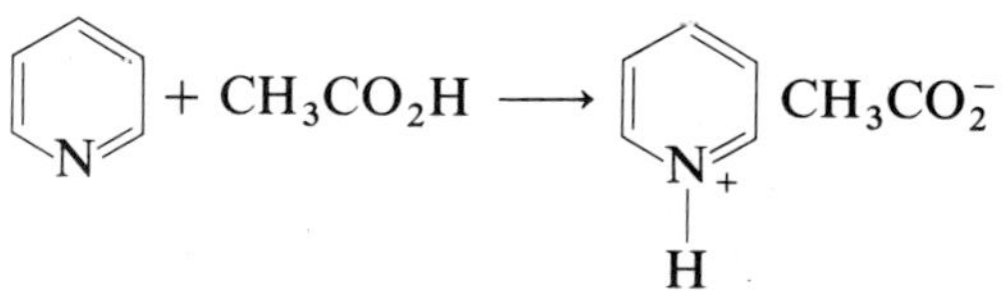

pyridine   acetic acid   pyridinium acetate

Problem 5.2

Complete the acid-base reactions and name each salt formed.

a. $(CH_3CH_2)_3N + HCl \longrightarrow$

b. 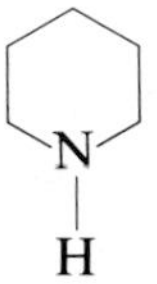 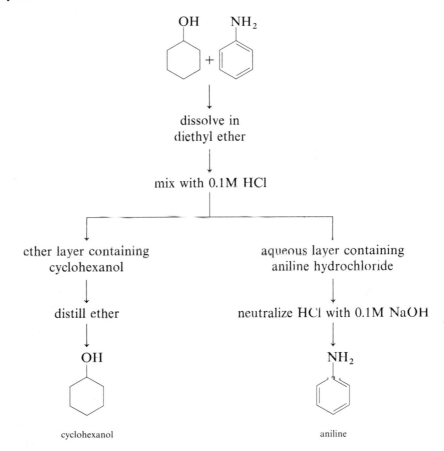 + CH₃CO₂H ⟶

The basicity of amines and the solubility of amine salts in water can be used to distinguish between amines and nonbasic, water-insoluble compounds and also to separate them. The flowchart shows the separation of aniline from cyclohexanol.

OH   NH₂

+

dissolve in
diethyl ether

mix with 0.1M HCl

ether layer containing   aqueous layer containing
cyclohexanol   aniline hydrochloride

distill ether   neutralize HCl with 0.1M NaOH

OH   NH₂

cyclohexanol   aniline

Aniline and cyclohexanol are only slightly soluble in water and therefore cannot be separated on the basis of their water solubilities. However, both dissolve in diethyl ether. When an ether solution of the two compounds is shaken with 0.1M HCl, aniline reacts to form a water-soluble salt, aniline hydrochloride. Cyclohexanol remains in the ether layer. Separation of the ether layer and distillation of the ether gives cyclohexanol. Neutralization of the HCl in the aqueous layer with 0.1M NaOH converts aniline hydrochloride to free aniline, which then separates as a water-insoluble layer and can be recovered.

5.4 Some Natural and Synthetic Amines

Structural formulas for various natural amines of both plant and animal origin are shown in Figure 5.1. These molecules are chosen to illustrate something of the structural diversity and range of physiological activity of amines, their value as drugs, and their importance in nutrition.

Coniine from the water hemlock is highly toxic. It can cause weakness, labored respiration, paralysis, and eventually death. This is the toxic substance in "poison hemlock," used by Socrates to commit suicide. Nicotine is one of the principal heterocyclic amines of the tobacco plant. In small doses, nicotine is a stimulant. However, in larger doses it causes depression, nausea, and vomiting. In still larger doses, nicotine is a poison. Solutions of nicotine in water are often used as insecticides. Note that nicotinic acid, an oxidation product of nicotine, is one of the water-soluble vitamins humans need for proper nutrition. Ingested or inhaled nicotine does not give rise to nicotinic acid in the body, because humans have no enzyme systems capable of catalyzing this conversion. Smoking will not supply any vitamins! Quinine, isolated from the bark of the cinchona tree in South America, has long been used to treat malaria.

Histamine, formed by decarboxylation of the amino acid histidine (Table 13.1), is a toxic substance present in all tissues of the body, combined in some manner with proteins. Histamine is produced extensively during hypersensitive, allergic reactions, and the symptoms of this release are unfortunately familiar to most of us, particularly those who suffer from hay fever or other seasonal allergies. The search for antihistamines—drugs that inhibit the effects of histamine—has led to the synthesis of several drugs whose trade names are well known. Structural formulas for three of the most widely used antihistamines are shown in Figure 5.2. Note the structural similarity in these three drugs; each has two aromatic rings and a dimethylaminoethyl group, $-CH_2CH_2N(CH_3)_2$. Dexbrompheniramine is the most potent.

Serotonin and acetylcholine are both neurotransmitters important in human physiology—serotonin in parts of the central nervous system mediating affective behavior, acetylcholine in certain motor neurons responsible for causing contraction of voluntary muscles. Acetylcholine is stored in synaptic vesicles and released in response to electric activity in the neuron. It diffuses across the synapse and interacts with postsynaptic receptor sites on a neighboring neuron, to cause membrane depolarization and transmission of a nerve impulse. After interaction with a postsynaptic receptor site, acetylcholine is deactivated through hydrolysis to choline and acetate ion, an action catalyzed by the enzyme acetylcholinesterase. Choline itself has no activity as a neurotransmitter.

$$CH_3{}^+\!\!\underset{\underset{\displaystyle CH_3}{|}}{\overset{\overset{\displaystyle CH_3}{|}}{N}}CH_2CH_2O\overset{\overset{\displaystyle O}{\|}}{C}CH_3 + H_2O \xrightarrow{\underset{\text{esterase}}{\text{acetylcholin-}}} CH_3{}^+\!\!\underset{\underset{\displaystyle CH_3}{|}}{\overset{\overset{\displaystyle CH_3}{|}}{N}}CH_2CH_2OH + {}^-O\overset{\overset{\displaystyle O}{\|}}{C}CH_3$$

acetylcholine   choline   acetate

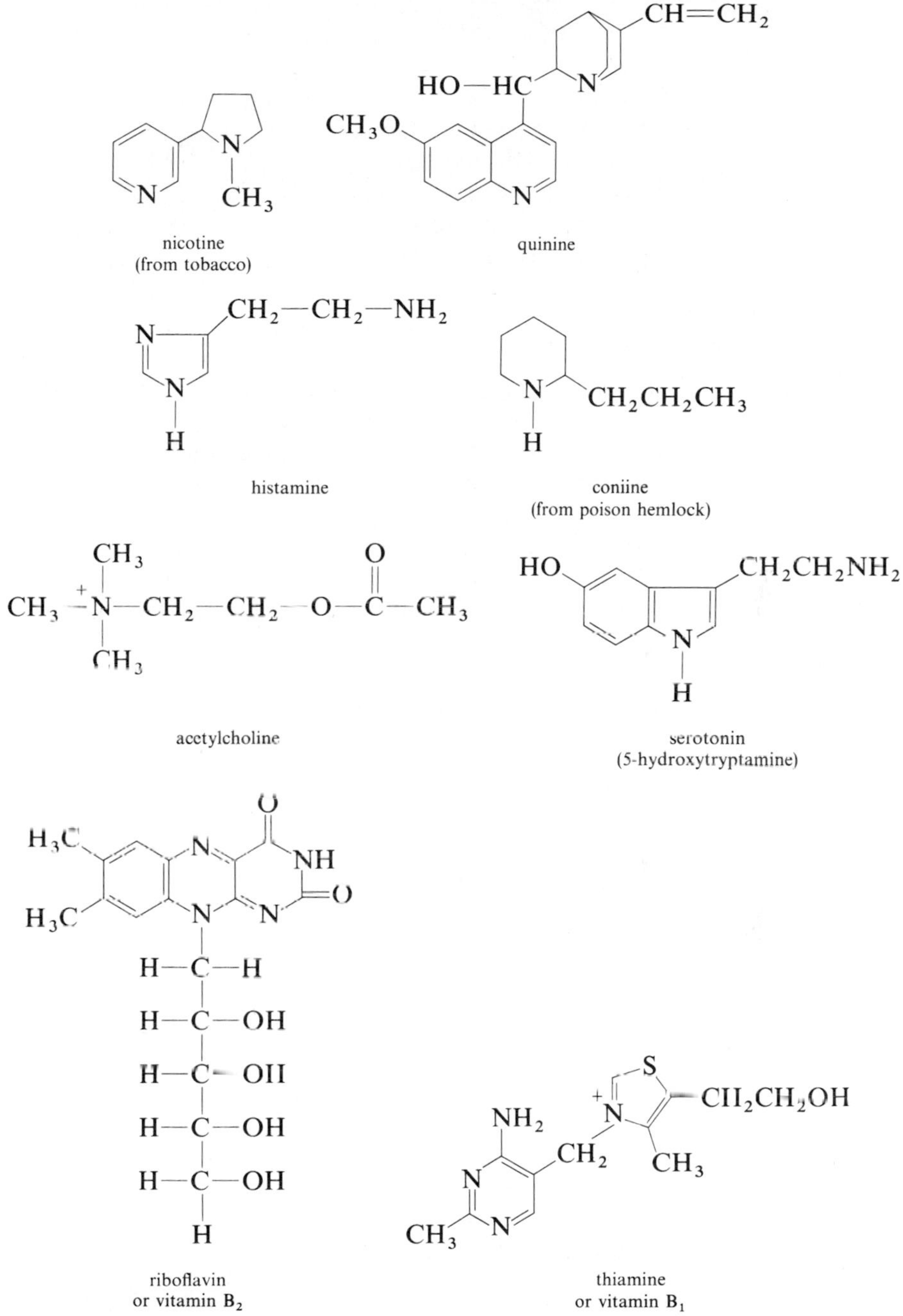

Figure 5.1 Several amines of plant and animal origin.

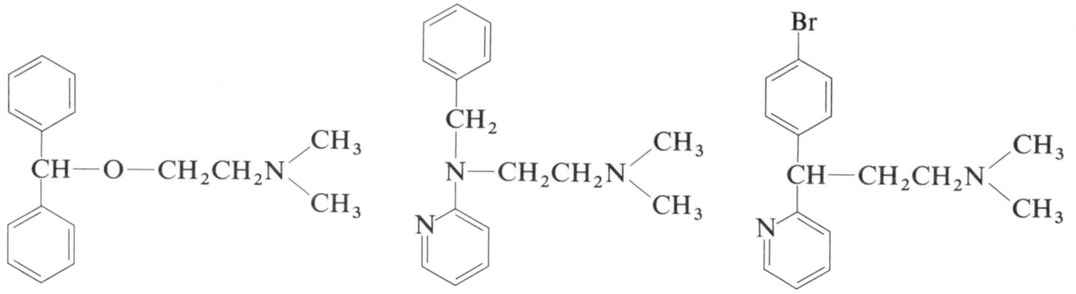

diphenylhydramine
(Benadryl)

tripelennamine
(Pyribenzamine)

dexbrompheniramine
(Disomer)

Figure 5.2 Three synthetic antihistamines.

Several other classes of synthetic compounds affect acetylcholine-mediated nerve transmission. Among the most widely known of these are the so-called nerve gases and related compounds that are now or have been used as insecticides. The nerve gases diisopropyl fluorophosphate (DFP) and Tabun are both potent inhibitors of acetylcholinesterase, and a few milligrams of either can kill a person in a few minutes through paralysis and respiratory failure.

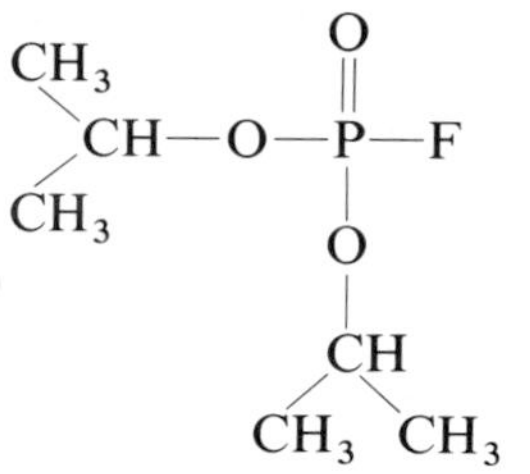

diisopropyl fluorophosphate
(DFP)

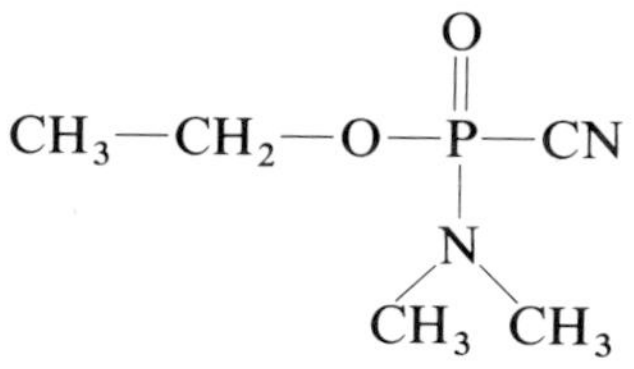

Tabun

Several water-soluble vitamins contain cyclic amines. Riboflavin (Figure 5.1) contains a fused three-ring amine called flavin. One of the nitrogen atoms of this ring system contains a five-carbon chain derived from the sugar D-ribose, hence the name riboflavin. Thiamine (vitamin B_1) contains a substituted pyrimidine ring and also a five-membered ring containing one atom each of nitrogen and sulfur. Pyridoxine, vitamin B_6, contains a substituted pyridine ring.

Key Terms and Concepts

amino group (5.1B)

base dissociation constant (5.3A)

basicity of amines (5.3A)

heterocyclic aliphatic amine (5.1B1)

heterocyclic aromatic amine (5.1B1)

hydrogen bonding between amines (5.2)

nomenclature of amines (5.1B)

primary amine (5.1A)

secondary amine (5.1A)

tertiary amine (5.1A)

Key Reactions

1. Aliphatic and aromatic amines are weak bases and ionize in aqueous solution (Section 5.3A).
2. Aliphatic and aromatic amines react with HCl and other strong acids to give water-soluble salts (Section 5.3B).

Problems

Structure, nomenclature, and classification of amines (Section 5.1)

5.3 Write structural formulas for the following compounds. In addition, classify each as primary amine, secondary amine, tertiary amine, aromatic amine, or ammonium salt.

a. diethylamine
b. aniline
c. cyclohexylamine
d. pyrrole
e. pyridine
f. tetramethylammonium iodide
g. 2-aminoethanol (ethanolamine)
h. 2-aminopropanoic acid (alanine)
i. pyrimidine
j. trimethylammonium chloride
k. p-methoxyaniline
l. triethylamine
m. N-methylaniline
n. p-chloroaniline
o. 1,4-diaminobenzene
p. p-aminophenol
q. pyridine 3-carboxylic acid (nicotine)

5.4 Give an acceptable name for these compounds:

a. 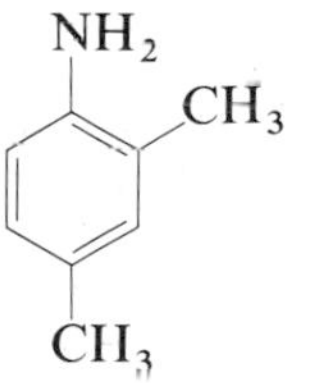 [benzene ring with NH$_2$, CH$_3$, and CH$_3$ substituents]

b. 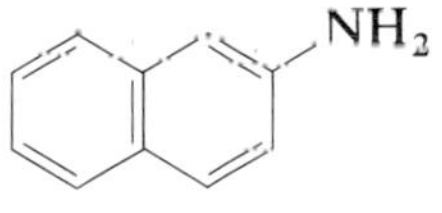 [naphthalene ring with NH$_2$ substituent]

c. $CH_3CHCHCH_2CH_3$ with HO and NH$_2$ substituents

d. 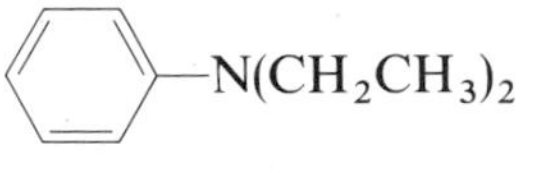 [cyclohexane ring]—$N(CH_2CH_3)_2$

e. $CH_3CH_2CH_2CH_2NH_2$

f. $(CH_3CH_2)_2NCH_3$

5.5 Draw structural formulas for the eight isomeric amines of molecular formula $C_4H_{11}N$. Name each and label as primary, secondary, or tertiary.

Physical properties of amines (Section 5.2)

5.6 Draw structural formulas to illustrate hydrogen bonding between the circled atoms.

a. CH_3—Ⓝ—H and CH_3—N—Ⓗ
with H below each N

b. 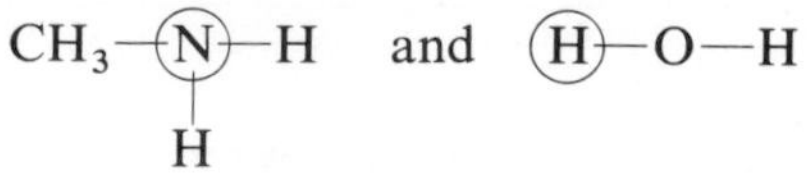 CH$_3$—Ⓝ—H and Ⓗ—O—H
       |
       H

c. 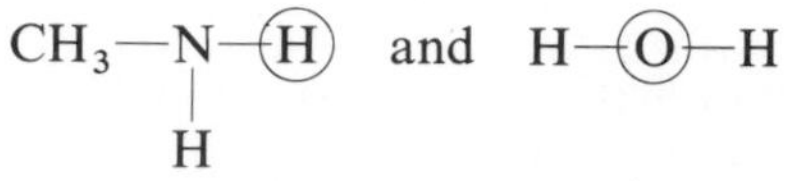 CH$_3$—N—Ⓗ and H—Ⓞ—H
      |
      H

5.7 Both 1-aminobutane and 1-butanol are liquids at room temperature. One of these compounds has a boiling point of 117°C, the other a boiling point of 78°C. Which compound has which boiling point? Explain your reasoning.

5.8 Arrange the following compounds in order of increasing boiling point:

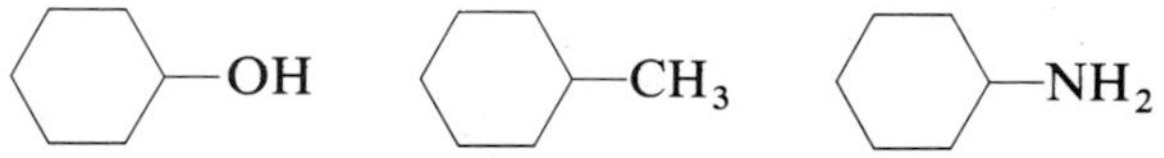

    ⬡—OH   ⬡—CH$_3$ ⬡—NH$_2$

    cyclohexanol   methylcyclohexane   cyclohexylamine

Reactions of amines (Section 5.3)

5.9 Name and write the structural formula for the salts formed by reaction of the following with HCl.

a. CH$_3$CH$_2$NH$_2$ **b.** 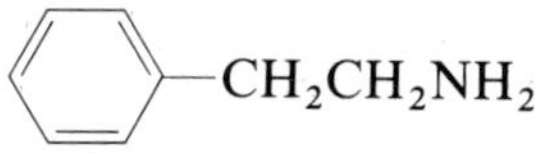 ⬡—CH$_2$CH$_2$NH$_2$

c. 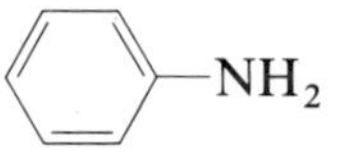 ⬡—NH$_2$ **d.** 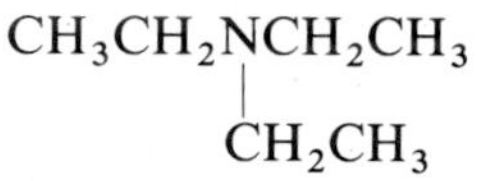 CH$_3$CH$_2$NCH$_2$CH$_3$
                         |
                         CH$_2$CH$_3$

5.10 Select the stronger base of each pair.

a. CH$_3$CH$_2$NH$_2$ or ⬡—NH$_2$

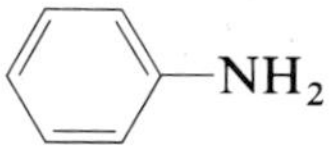

b. 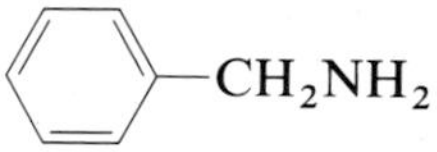 ⬡—CH$_2$NH$_2$ or ⬡—NH$_2$

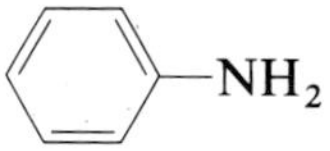

c. 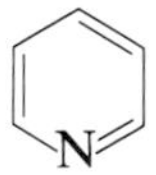 ⬡N or ⬡—NH$_2$

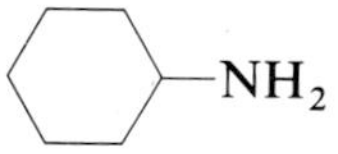

d. 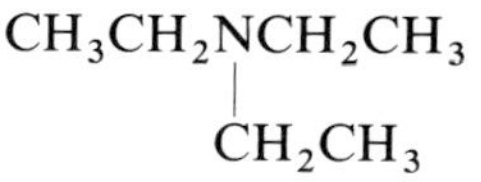 CH$_3$CH$_2$NCH$_2$CH$_3$ or ⬡—N⟨CH$_2$CH$_3$ / CH$_2$CH$_3$
           |
           CH$_2$CH$_3$

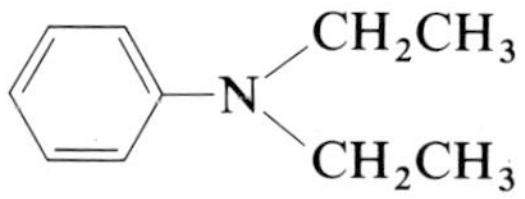

5.11 Arrange the compounds of each set in order of increasing basicity.

a. 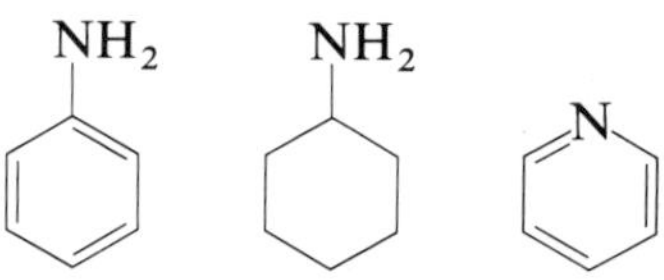 NH$_2$⬡   NH$_2$⬡   ⬡N

    (i)   (ii)   (iii)

b.

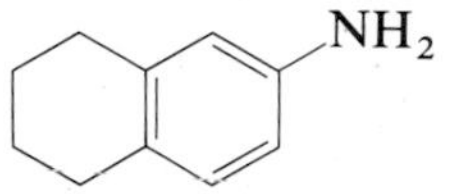

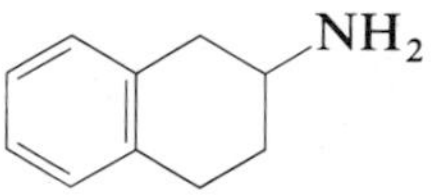

(i)   (ii)

5.12 Suppose you are given a mixture of the following three compounds. Describe a procedure you could use to separate and isolate each in a pure form.

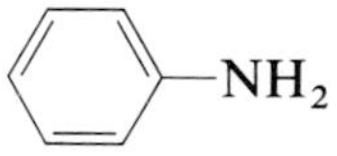

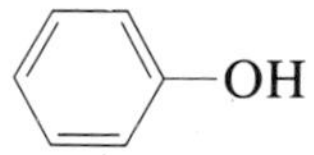

aniline   phenol   1-hexanol

5.13 Alanine (2-aminopropanoic acid) is one of the important amino acids in proteins. Is the structural formula of alanine better represented by (i) or (ii)? Explain.

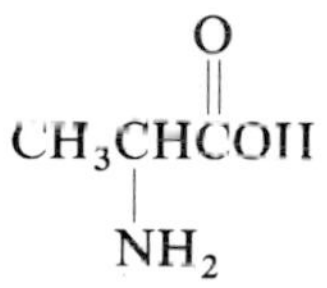

$$\text{CH}_3\text{CHCOH} \quad \text{or} \quad \text{CH}_3\text{CHCO}^-$$

          NH$_2$ NH$_3^+$

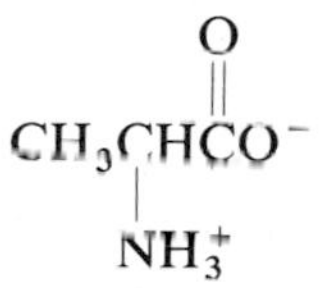

(i)   (ii)

5.14 Describe a simple chemical test by which you could distinguish between the compounds of each of the following pairs. In each case, state what test you would perform and what you would expect to observe, and write a balanced equation for each positive test.
a. phenol and aniline
b. cyclohexylamine and cyclohexanol
c. trimethylacetic acid and 2,2-dimethyl-1-aminopropane

Alkaloids

Alkaloids are nitrogen-containing compounds of plant origin, which are physiologically active when administered to humans. Examples of alkaloids are morphine from opium poppies, quinine from the bark of cinchona trees, cocaine from coca leaves, and nicotine from tobacco plants. It is estimated that alkaloids are present in 10–20% of all vascular plants.

The field of alkaloid chemistry is both vast and complex, and we shall examine only a few members of this class. In so doing, we shall try to portray the ranges of structural diversity and physiological activity that characterize alkaloids and to indicate how important these natural products are in chemotherapy and psychopharmacology.

Atropine and Cocaine

One subgroup of alkaloids consists of the tropane alkaloids, the characteristic structural feature of which is tropane, a six-membered piperidine ring with a two-carbon bridge stretching between carbons 2 and 6. Atropine from deadly nightshade (*Atropa belladonna*) is used in dilute solutions to dilate the pupil of the eye before ophthalamic examination and eye surgery.

atropine

Cocaine is a tropane alkaloid isolated from the leaves of the South American coca plant (*Erythroxylon coca*). It was first isolated in pure form in 1880 and soon thereafter its property as a local anesthetic was noted. It was introduced into medicine in 1884 by two young Viennese physicians, Sigmund Freud and Karl Koller. Unfortunately, the use of cocaine can create a dependence, as Freud himself observed when he used it to wean a colleague from morphine and thereby produced one of the first known cocaine addicts.

After determining cocaine's structure, chemists could ask the tantalizing questions: How is the structure of cocaine related to its physiological activity? Is it possible to separate the anesthetic effects from the habituation? If these questions could be answered, then it should be possible to prepare synthetic alkaloidlike drugs that incorporate only those structural elements essential for anesthetic function, simultaneously eliminating the undesirable effects. With cocaine, chemists have duplicated the essential structural features: its benzoate ester, its basic nitrogen, and something of its carbon arrangement. This search resulted in the synthesis of procaine

cocaine

(Novocaine) in 1905; it almost immediately replaced cocaine in dentistry and surgery. Lidocaine (Xylocaine) was introduced in 1948 and today is one of the most widely used local anesthetics. More recently, mepivacaine (Carbocaine) was also introduced.

procaine

lidocaine

mepivacaine

Thus, seizing on clues provided by nature, chemists have been able to synthesize drugs far more suitable for a specific function than anything known to be produced by nature itself.

Morphine

Of all the alkaloids, probably the most widely known and completely studied are the so-called morphine alkaloids. Opium, the source of morphine alkaloids, is obtained from the opium poppy (*Papaver somniferum*) by cutting the unripe seed capsule. The milky juice that exudes is dried and powdered to make the opium of commerce, which contains well over twenty alkaloids including about 15% morphine and 0.5% codeine.

R = H in morphine
R = CH₃ in codeine

The two —OH groups of morphine are particularly important because many semisynthetic derivatives can be made by modifications of either or both of these groups. For example, codeine is methylmorphine, the methyl substituent being on the phenolic —OH group. Diacetylmorphine, or heroin, is made from morphine by acetylation of both —OH groups.

For the relief of severe pain of virtually every kind, morphine and its synthetic analogs remain the most potent drugs known. In 1680, Thomas Sydenham, an English physician, wrote: "Among the remedies which it has pleased Almighty God to give man to relieve his suffering, none is so universal and so efficacious as opium." Even today, morphine, the alkaloid that gives opium its analgesic action, remains the standard against which newer analgesics are measured.

We can ask the same type of question about morphine as we did of cocaine, namely, Is it possible, using the structural features of morphine as a guide, to design drugs that possess the desirable analgesic effects of morphine and yet are free from the undesirable side effects of respiratory depression, hallucinogenic activity, and addiction? With this in mind, chemists have created numerous synthetic analgesics related in structure to morphine, the oldest and perhaps best known of which is meperidine (Demerol, Figure III-1). Meperidine was at first thought to be free of many of the morphinelike undesirable side effects. It is now clear, however, that meperidine is definitely addictive. In spite of much determined research, there are as yet no agents as effective as morphine against severe pain that are also entirely free from the risks of

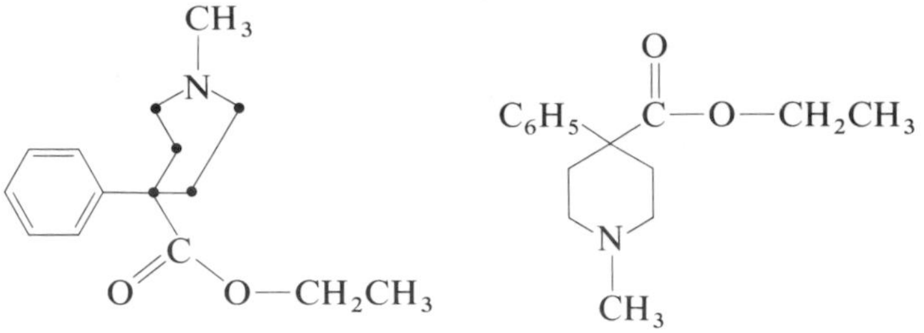

Figure III-1 Meperidine (Demerol) drawn in two different representations, the first of which suggests a structural similarity to morphine.

addiction. Still, the hope remains that some day the ideal analgesic will be prepared.

How and in what regions of the brain does morphine act? In 1973 scientists demonstrated that there are specific receptors for morphine and other opiates; these sites are clustered mainly in the brain's limbic system, an area involved with emotion and pain perception. It was further shown that these opioid receptors do not interact with any known neurotransmitters. Why then, scientists asked, does the human brain have receptors specific for morphine? Could it be that the brain produces its own opiates? In 1974, scientists discovered that opiatelike compounds are indeed present in the brain, and in 1975 Solomon Snyder at Johns Hopkins University isolated the first known brain opioids, and named them enkephalins, meaning "in the brain." Other brain opiates have since been isolated and named endorphins. The human endorphin anodynin is about as potent an analgesic as morphine. Scientists have yet to understand the role of these natural brain opioids. Perhaps when we do understand their biochemistry and function, we will discover clues that lead us toward the synthesis of more potent but less addictive analgesics.

Indole Alkaloids

During the last four decades, several indole alkaloids have received much attention, not only from chemists but from pharmacologists, psychologists, and physicians. This interest has evolved mainly from the discovery of the remarkable physiological properties of reserpine. The story of reserpine begins long ago. For at least 3000 years, the people of India have used the root of the climbing shrub *Rauwolfia serpentina* as a folk remedy to treat various afflictions. It is called *sarpagandha* in Sanskrit, referring to its use as an antidote for snakebite; it is called *chandra*, meaning "moon" in Hindi, referring to its calming effect on certain forms of insanity.

The modern story of rauwolfia began in the late 1920s, when Indian scientists undertook to study it and other botanical preparations used by native practitioners. By 1931, chemists had isolated a crystalline powder from dry rauwolfia root, and physicians reported that use of this powder brought relief in cases of acute insomnia accompanied by fits of insanity. According to one report: "Symptoms such as headache, a sense of heat, and insomnia disappear quickly and blood pressure can be reduced in a matter of weeks . . . ." Further clinical testing verified the potency of the ancient snakeroot remedy, and soon rauwolfia became an important drug in India for treating high blood pressure and certain forms of mental illness.

In 1949, Dr. Rustrom Jal Vakil, a physician at the King Edward Memorial Hospital in Bombay, reported in the *British Medical Journal* on his research with rauwolfia therapy for patients with high blood pressure. These studies were read by Dr. Robert Wilkins, director of the Hypertension Clinic at Massachusetts Memorial Hospital, who decided to try rauwolfia to see whether it would help certain of his patients who

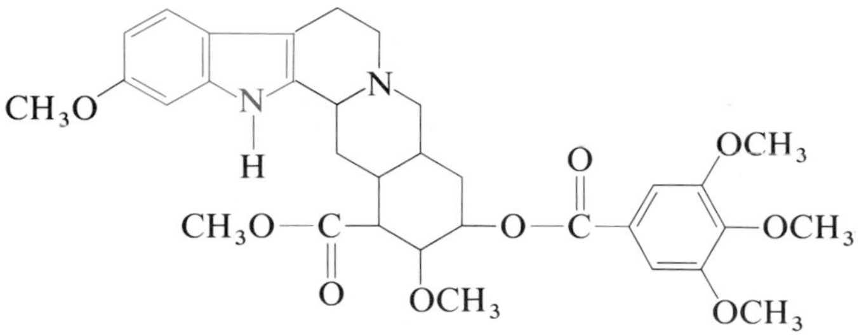

Figure III-2 Reserpine, a rauwolfia alkaloid. The indole ring is shown in color.

were not responding to other medication. In 1952 he confirmed reports from India of its mildly hypotensive (blood pressure–lowering) effect. He reported further that rauwolfia has a sedative action different from that of any other drug known at the time. Unlike barbiturates and other standard sedatives, rauwolfia does not produce grogginess, stupor, or lack of coordination. Patients on rauwolfia therapy appeared to be relaxed, quiet, and tranquil. Reports such as this generated interest among psychiatrists, and soon rauwolfia was recognized as an entirely new class of drug for the treatment of mental illness. It was the first of the so-called tranquilizers. (See Mini-Essay II, "Biogenic Amines and Emotion.")

In 1952 chemists isolated the active compound from rauwolfia and named it reserpine. Reserpine, which proved to be 10,000 times more effective than the same weight of crude snakeroot extract, contains an indole nucleus as part of five fused rings (see Figure III-2).

In 1954, the first full year of reserpine therapy, two dozen companies in the United States were preparing rauwolfia products and by 1960, the cost of prescriptions for these totaled $30 million per year. Thus, in slightly more than two decades, rauwolfia had advanced from a folk remedy to a widely used and highly effective drug for the treatment of high blood pressure and certain forms of mental illness.

Along with reserpine and reserpinelike alkaloids, another group of alkaloids, either containing the indole nucleus or containing structural features suggesting the indole nucleus, has been under intense research (Figure III-3). All characteristically produce behavioral aberrations—hallucinations, delusions, disturbances in thinking, and changes in mood.

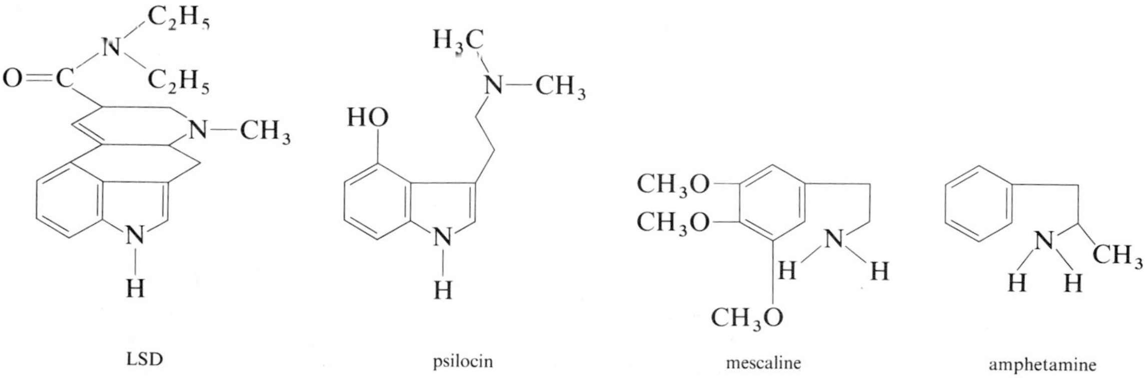

LSD   psilocin   mescaline   amphetamine

Figure III-3 Structural relations among several psychotomimetic drugs. The indole ring appears in LSD and psilocin. Mescaline and amphetamine are drawn so as to suggest an indole nucleus.

The principal source of lysergic acid from which lysergic acid diethylamide (LSD) is made is ergot, a fungus that grows parasitically on rye. Mescaline is obtained from the cactus known as peyote, or mescal (*Lophophora williamsii*) and is named for the Mescalero Apaches of the Great Plains, who developed a religious rite in which its use was common. Because mescaline does not contain the indole nucleus, it does not properly belong to the class of indole alkaloids. Yet there is a suggestive chemical similarity and certainly strong pharmacological similarity between it and the indole psychotomimetics. Psilocin is the hallucinogenic principal of *Psilocybe aztecorum*, the narcotic mushroom of the Aztecs.

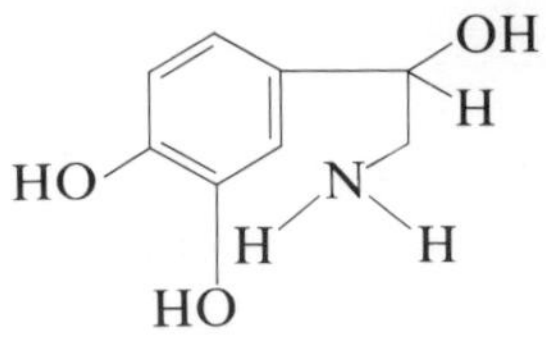

Serotonin and Norepinephrine

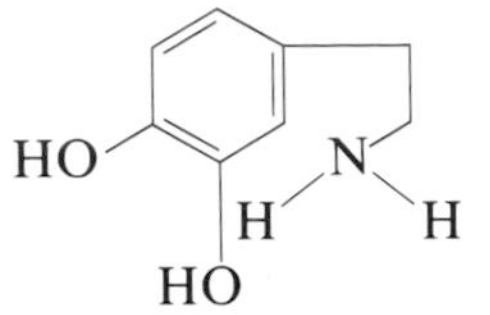

Of the various amines found in the brain, studies of serotonin, dopamine, and norepinephrine have been very important in currently evolving psychopharmacological concepts. Each compound functions in the central nervous system as a neurotransmitter.

An examination of the molecular structures

norepinephrine

dopamine

of the psychotomimetic drugs reveals certain striking similarities between them and serotonin, dopamine, and norepinephrine. One theory holds that all known hallucinogens exert the effects they do because they can assume a conformation that simulates a structural characteristic of one or more of these three neurotransmitters.

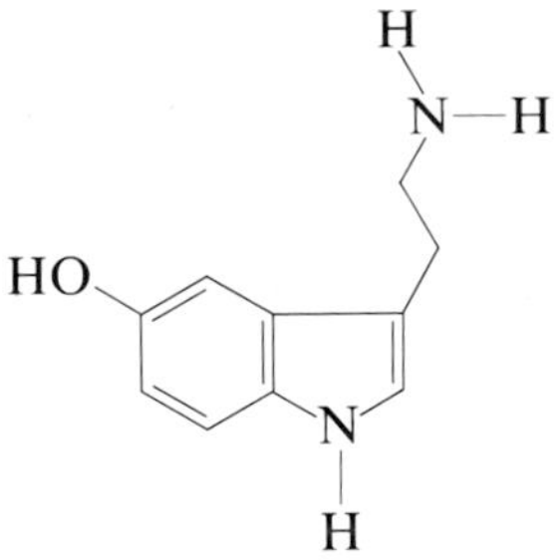

The interest aroused by the discoveries of drugs that act selectively to affect mood and behavior has focused attention on possible chemical bases for disturbances in behavior. Underlying this effort is the hope that research on brain biochemistry will benefit the understanding and treatment of mental disease.

serotonin

Sources

Goodman, L. S., and Gilman, A. 1980. *The Pharmacological Basis of Therapeutics*, 6th ed. New York: Macmillan.

Guillemin, R. 1978. Peptides in the brain: The new endocrinology of the neuron. *Science* 202:390.

Snyder, S. H. 1978. The brain's own opiates. *Chem. & Eng. News* 55:26.

Biogenic Amines and Emotion

Over the past several decades, much study has been given to biochemical correlates of emotion. Initial research in this field concentrated largely on levels of chemical modulators in blood and excretion of hormones and metabolites in the urine. The historic studies of Walter Cannon after 1900 suggested that epinephrine (adrenaline) is secreted in response to stimuli that produce fear and rage reactions in animals. In the ensuing years, studies in both animals and humans clearly indicated an increased secretion of epinephrine as well as norepinephrine in various types of stress, including parachute jumping, competitive sports, aggressive behavior, and viewing emotion-laden movies.

An area of more interest, but one far less amenable to direct experimental observation, is the pattern of biochemical changes that takes place in the brain itself in relation to emotional states. The reasons are obvious. It is far easier to examine blood levels of chemical modulators or the excretion of hormones and metabolites in the urine than to examine the biochemistry of neurons themselves. Yet it has been possible to do just this, in a limited way, in the past four decades. The catecholamines norepinephrine and dopamine (each derived from the parent molecule catechol) and the indole amine serotonin have received the most attention. We shall concentrate on norepinephrine and dopamine because more is known about them than any other class of neurotransmitters.

Biosynthesis of Norepinephrine and Dopamine

Norepinephrine and dopamine are limited to certain cells in the central nervous system where they are present in the cytoplasm, the axon, and synaptic vesicles. Synthesis of norepinephrine and dopamine begins with the amino acid tyrosine (Chapter 13). Oxidation of tyrosine, catalyzed by the enzyme tyrosine hydroxylase, forms dihydroxyphenylalanine (DOPA), which in turn undergoes loss of CO_2, catalyzed by DOPA decarboxylase, to give dihydroxyphenylethylamine (dopamine). Oxidation of dopamine, catalyzed by dopamine beta-hydroxylase, gives norepinephrine (Figure IV-1).

Norepinephrine and dopamine are protected from further reaction by being bound in synaptic vesicles. When an action potential arrives at a nerve ending, its neurotransmitters are released and they pass through the presynaptic membrane into the synaptic cleft. There they interact with specific receptor sites on the postsynaptic membrane.

Once a neurotransmitter has interacted with a receptor site on the next nerve cell, its further interaction must be prevented. Otherwise its effects would continue for too long and precise control of nerve function would be lost. Dopamine and norepinephrine are rapidly inactivated in two ways: (1) by reuptake through the presynaptic membranes into synaptic vesicles, and (2) by enzyme-catalyzed transformation in the synapse itself. A substantial fraction of each neurotransmitter is returned to synaptic vesicles for reuse. Some of each, however, undergo oxidation catalyzed by the enzyme monoamine oxidase (MAO), in which the $-CH_2NH_2$ group is converted to a $-CO_2H$ group. The resulting carboxylic acid leaves via the blood and is excreted in the urine. Norepinephrine within synapses is metabolized principally by methyla-

159F

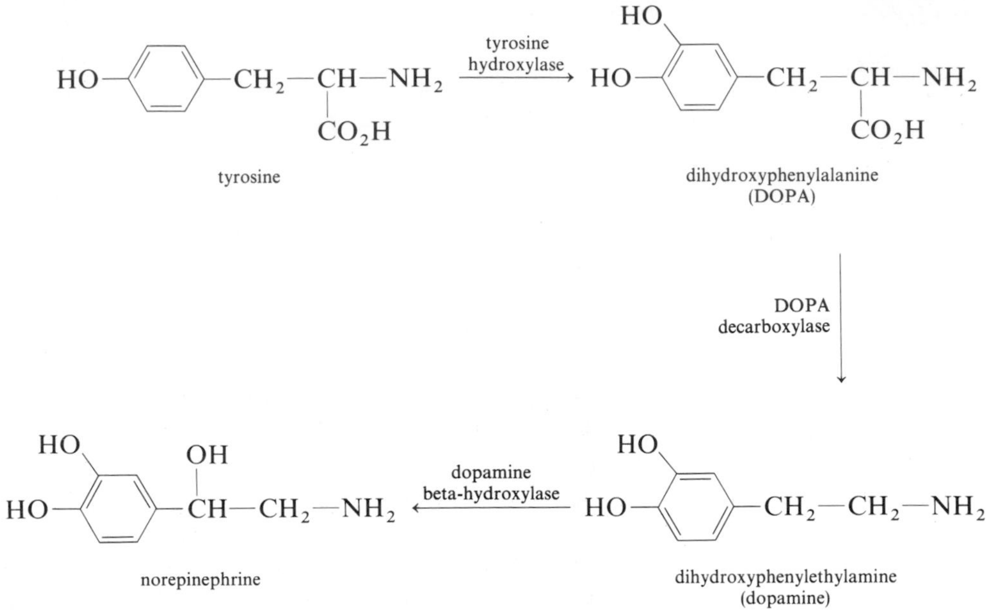

Figure IV-1 Biosynthesis of norepinephrine and dopamine from tyrosine.

tion catalyzed by the enzyme catecholamine-O-methyl transferase (COMT). The —OH on carbon 3 of the catechol ring is converted to —OCH$_3$. This product also enters the bloodstream and eventually is excreted in the urine (Figure IV-2).

Dopamine and Parkinson's Disease

An understanding of how dopamine acts as a neurotransmitter has led to an effective treatment for the crippling affliction Parkinson's disease. In 1959, the Swedish pharmacologist Arvid Carlsson discovered that when rats are given large doses of reserpine, they develop Parkinson-like tremors. He further discovered that there is an associated decrease in dopamine in the brain's caudate nucleus, a center involved with coordination and integration of fine muscle movement. Soon thereafter, George Cotzias of Brookhaven National Laboratory discovered that DOPA counteracts dopamine deficiency in

Parkinson patients and effectively relieves symptoms of the disease. Today, even more effective anti-Parkinson drugs have been developed. The story of unraveling the biochemistry of dopamine is a success story of how basic research can lead to a new treatment for a disease.

Psychopharmacology

Although interest in the psychological effects of drugs is almost as old as mankind, the use of drugs to treat psychiatric disorders has become widespread only since the mid-1950s, initiated largely by the great wave of enthusiasm for the use of reserpine for treating mania and excitement (see Mini-Essay III, "Alkaloids").

Almost simultaneously, scientists discovered the remarkable antipsychotic properties of chlorpromazine. Henri Laborit, a French physician, was looking for drugs to calm patients before surgery. He tried chlorpromazine (Thorazine), a drug synthesized by the French pharmaceutical company Specia in a search for newer

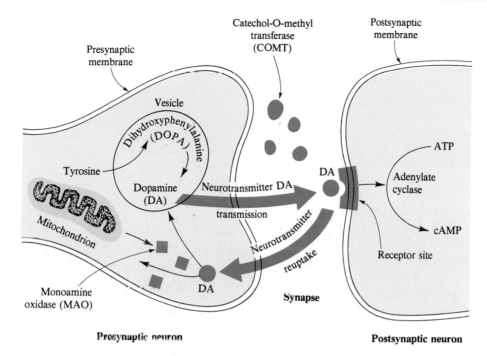

Figure IV-2 Schematic diagram of a synapse.

and more effective antihistamines. Testing of chlorpromazine as an antihistamine was not pursued because initial trials showed that it was too sedating. However, Laborit discovered that for his purposes, the drug was ideal, and he went on to suggest to his colleagues that they try it on heretofore unmanageable patients in mental institutions. Much to their surprise, psychiatrists discovered that the drug calmed schizophrenics and seemed to relieve their symptoms. Chlorpromazine drugs produce sedation, emotional

quieting, and relaxation, without clouding consciousness or intellectual functioning.

Chlorpromazine quickly usurped the position of reserpine when it became clear that chlorpromazine is easier to control and more effective than reserpine. Within a few years, chlorpromazine and its derivatives became one of the most widely used drugs in medicine. It is estimated that between 1955, when it was first introduced, and 1965 at least 50 million patients received the drug (Figure IV-3).

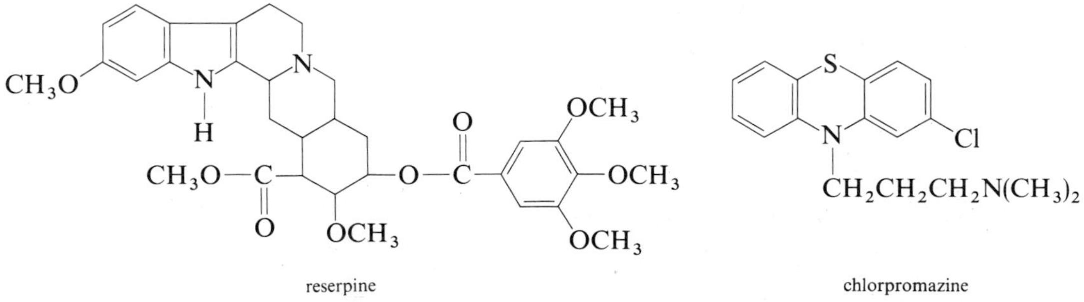

reserpine

chlorpromazine

Figure IV-3 Tranquilizers. Reserpine depletes synaptic vesicles of norepinephrine. Chlorpromazine impairs norepinephrine-receptor interaction.

The Catecholamine Hypothesis of Affective Illnesses

In attempts to correlate these observations with the underlying biochemistry, scientists have proposed the "catecholamine hypothesis" of affective illness. In outline, it has been proposed that:

1. Depression is caused by a functional decrease in either norepinephrine or dopamine; the major antidepressants act either by increasing the amount of norepinephrine or dopamine, or both, released into synapses or by enhancing their accumulation at appropriate receptor sites.
2. Mania is caused by a functional increase in either norepinephrine or dopamine; the major tranquilizing drugs act either by depleting the supply of available norepinephrine or dopamine, or both, or by reducing the effectiveness of their interaction at specific receptor sites.

What is the evidence for the hypotheses that norepinephrine or dopamine have anything to do with emotional states? Research on both animals and people suggests that reserpine, by some mechanism not yet understood, acts on synaptic vesicles and impairs their ability to store norepinephrine and dopamine. As a consequence, catecholamine neurotransmitters are released; they diffuse freely through the cytoplasm and onto mitochondrial-bound MAO, where they are oxidized to inactive compounds and eventually excreted in the urine. Thus, instead of being stored in synaptic vesicles for later release on nerve stimulation, catecholamine supplies are prematurely depleted. Even though there is continuing synthesis, the depletion may last for days or even weeks.

The weight of evidence suggests that the principal action of chlorpromazine is on norepinephrine and dopamine receptor sites rendering them incapable of responding. The result is similar to a decrease in supply of the neurotransmitters. Thus, although both reserpine and chlorpromazine are powerful tranquilizers, the mechanism of action of each is different.

Discovery of the tranquilizing effects of reserpine and chlorpromazine was followed within a few years by the discovery that iproniazid, a drug developed for the treatment of tuberculosis, had mood-elevating effects on tuberculosis patients. The norepinephrine hypothesis was reinforced with the finding that iproniazid is a powerful inhibitor of monoamine oxidase (MAO), the enzyme responsible for metabolic degradation of norepinephrine, dopamine, and other brain monoamines. Inhibition of MAO permits an increase in the concentration of catecholamine neurotransmitters at nerve endings and presumably at synapses as well. This accumulation is thought to produce the antidepressant action of the drug in the human body. Elucidation of the structure of iproniazid sparked research efforts to synthesize new compounds with even greater clinical effectiveness. Among those synthesized and marketed was isocarboxazid (Marplan).

Imipramine and closely related tricyclic compounds are the drugs most widely used today to treat depression. It is interesting that imipramine (an antidepressant) has a ring structure that differs from that of chlorpromazine (a tranquilizer) only in replacement of the atom of sulfur in the middle ring by a two-carbon ethylene bridge. The mechanism by which imipramine functions as an antidepressant is not fully understood. It is thought, however, that it favors accumulation of norepinephrine at receptor sites by inhibiting uptake of intercellular norepinephrine. Thus imipramine artificially increases the concentration of norepinephrine at receptor sites and thereby potentiates its action. Antidepressants are shown in Figure IV-4.

In summary, a significant body of experimental evidence is at least consistent with the hypothesis that the effects of the major tranquilizers and antidepressants are related to their effects on norepinephrine and dopamine and that these catecholamine neurotransmitters are important in mental and behavioral states. But by no means does this evidence prove the hypothesis. For one thing, it is estimated that to date scientists have isolated and identified only a fraction of all cen-

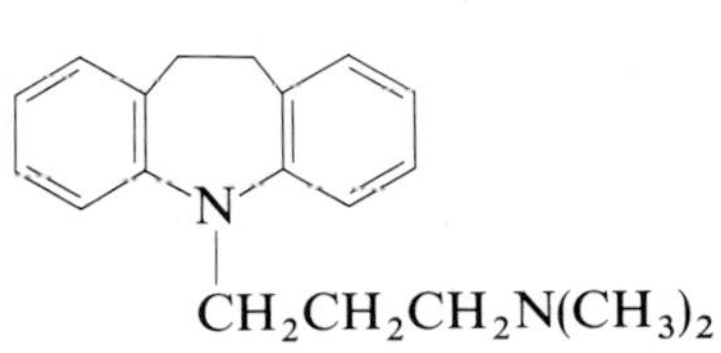

CH$_2$CH$_2$CH$_2$N(CH$_3$)$_2$

imipramine

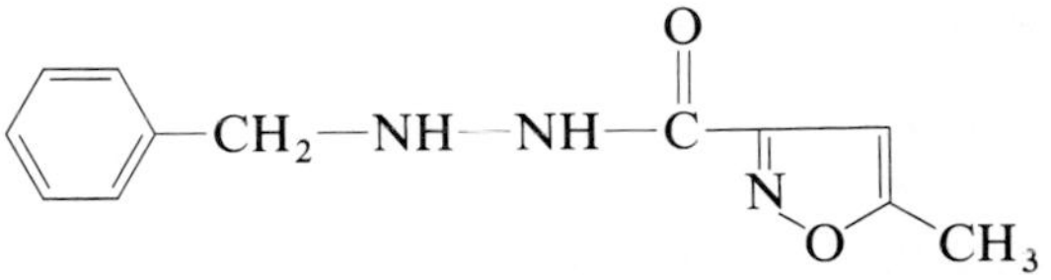

isocarboxazid

Figure IV-4 Antidepressants. Imipramine (Tofranil), an inhibitor of reuptake of norepinephrine. Isocarboxazid (Marplan), an inhibitor of monoamine oxidase.

tral-nervous-system neurotransmitters. It would be remarkable indeed if the few studied to date are the most important ones involved in mediation of emotions.

Further, it is unlikely that norepinephrine or dopamine, or for that matter any other single neurotransmitter, is entirely responsible for a specific emotional state. Rather, it is more likely that other factors are important too—the interaction of certain amines at particular sites within the central nervous system, and environmental and psychological determinants.

Finally we must be wary of oversimplification in a subject as complex and multifaceted as human behavior.

Sources

Baldessarine, R. J. 1957. The basis for amine hypotheses in affective disorders. *Arch. Gen. Psychiatry* 32:285.

Cooper, J. R., Bloom, F. E., and Roth, R. H. 1978. *The Biochemical Basis of Neuropharmacology.* New York: Oxford University Press.

Krassner, M. B. 1983. Brain chemistry. *Chem. & Eng. News* 61:22.

Schildkraut, J. J., and Kety, S. S. 1967. Biogenic amines and emotion. *Science* 156:21.

6

Aldehydes and Ketones

In Chapter 3 we studied the physical and chemical properties of compounds containing carbon-carbon double bonds. In this and the following two chapters, we will study the physical and chemical properties of compounds containing the **carbonyl group** (C=O). Because the carbonyl group is the central structural feature of aldehydes, ketones, carboxylic acids, and their functional derivatives, it is one of the most important functional groups in organic chemistry. The chemical properties of this group are straightforward, and an understanding of its few characteristic reaction themes leads very quickly to an understanding of a wide variety of reactions in organic and biochemistry.

6.1 The Structure of Aldehydes and Ketones

A. Characteristic Structural Features

The characteristic structural feature of an **aldehyde** is the presence of a carbonyl group bonded to a hydrogen atom (Section 1.4B). In methanal (formaldehyde), the simplest aldehyde, the carbonyl group is bonded to two hydrogen atoms. In other aldehydes, it is bonded to one hydrogen atom and one carbon atom. Following are Lewis structures for methanal and ethanal. Under each in parentheses is its common name.

$$\underset{\substack{\text{methanal} \\ \text{(formaldehyde)}}}{H-\overset{\displaystyle :O:}{\overset{\|}{C}}-H} \qquad \underset{\substack{\text{ethanal} \\ \text{(acetaldehyde)}}}{CH_3-\overset{\displaystyle :O:}{\overset{\|}{C}}-H}$$

The characteristic structural feature of a **ketone** is a carbonyl group bonded to two carbon atoms (Section 1.4B). Following is a Lewis structure for 2-propanone (acetone), the simplest ketone.

$$:O:$$
$$\|$$
$$C$$

$$CH_3 \qquad CH_3$$

2-propanone
(acetone)

B. Covalent Bonding in Aldehydes and Ketones

In forming bonds to three other atoms, carbon uses sp^2 hybrid orbitals. The carbon-oxygen double bond consists of one sigma bond formed by overlap of sp^2 atomic orbitals of carbon and oxygen, and one pi bond formed by the overlap of parallel $2p$ orbitals. The two nonbonding pairs of electrons on oxygen lie in the remaining sp^2 orbitals of oxygen. Figure 6.1(b) shows an orbital-overlap diagram of the covalent bonding in formaldehyde. From the sp^2 hybridization of the carbonyl carbon, bond angles about this atom are predicted to be 120°. Actual bond angles in formaldehyde are shown in Figure 6.1(a).

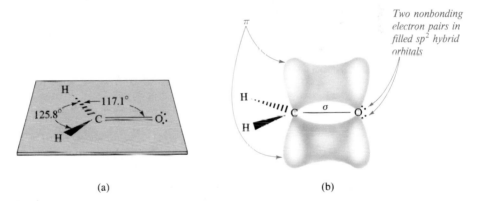

(a)   (b)

Figure 6.1 Covalent bonding in formaldehyde: (a) Lewis structure showing observed bond angles, and (b) covalent bond formation by overlap of atomic orbitals.

6.2 Nomenclature of Aldehydes and Ketones

A. IUPAC Nomenclature

The IUPAC system of nomenclature for aldehydes and ketones follows the familiar pattern of selecting as the parent compound the longest chain of carbon atoms that contains the functional group. The aldehyde group is shown by changing the suffix -*e* of the parent name to -*al* (Section 2.5). Because the aldehyde group can appear only at the end of a hydrocarbon chain and because numbering must start with it as carbon 1, its position is unambiguous and therefore there is no need to use a number to locate it. Following are IUPAC names for several low-molecular-weight aldehydes.

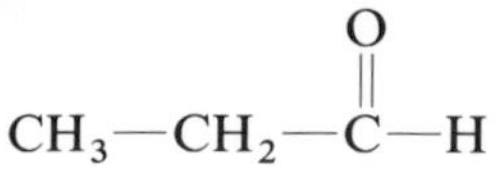

$$CH_3-CH_2-\overset{\overset{\displaystyle O}{\|}}{C}-H \qquad CH_3-\underset{\underset{\displaystyle CH_3}{|}}{CH}-CH_2-\overset{\overset{\displaystyle O}{\|}}{C}-H \qquad CH_3-CH_2-\underset{\underset{\displaystyle CH_3}{|}}{CH}-\overset{\overset{\displaystyle O}{\|}}{C}-H$$

propanal 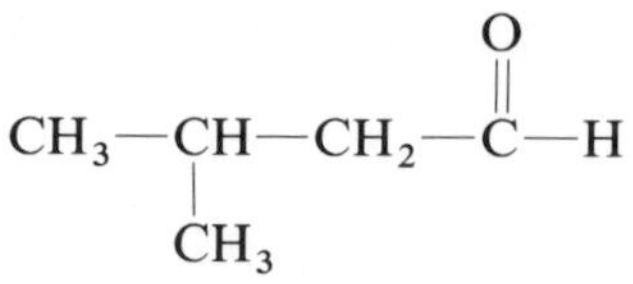   3-methylbutanal 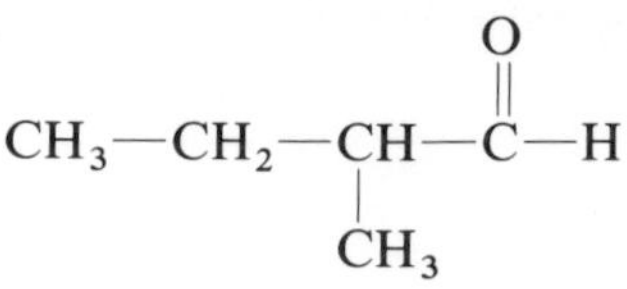   2-methylbutanal

For unsaturated aldehydes (that is, those also containing an alkene), the presence of the carbon-carbon double bond is indicated by the infix *-en-*, as shown in the following examples. As with other molecules with both an infix and a suffix, the location of the suffix determines the numbering pattern.

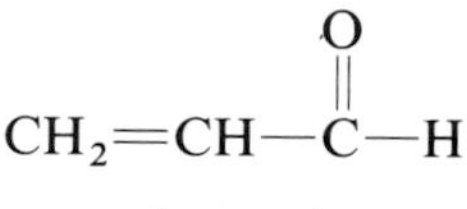

$$CH_2{=}CH-\overset{\overset{\displaystyle O}{\|}}{C}-H \qquad CH_3-\underset{\underset{\displaystyle CH_3}{|}}{C}{=}CH-CH_2-CH_2-\underset{\underset{\displaystyle CH_3}{|}}{C}{=}CH-\overset{\overset{\displaystyle O}{\|}}{C}-H$$

2-propenal 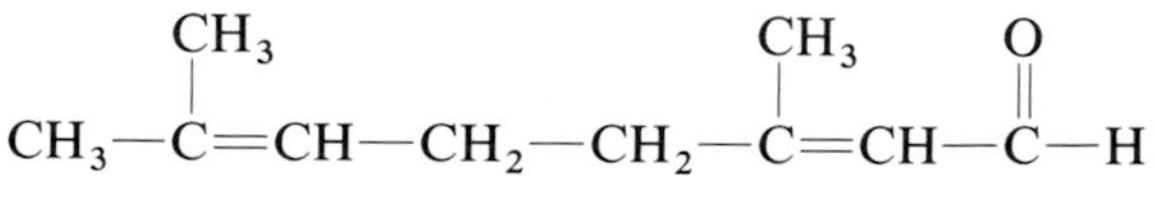   3,7-dimethyl-2,6-octadienal

Among the aldehydes for which the IUPAC system retains common names are benzaldehyde and cinnamaldehyde.

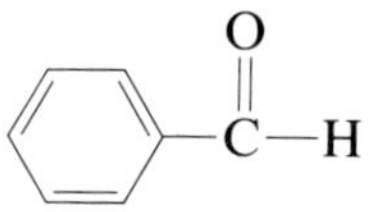

benzaldehyde 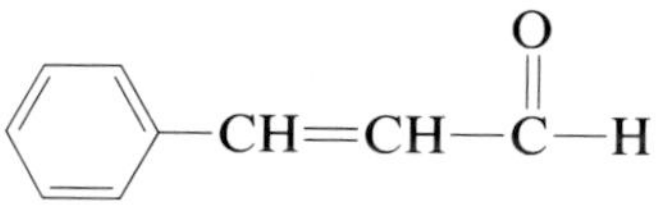   cinnamaldehyde

In the IUPAC system, ketones are named by selecting as the parent compound the longest chain that contains the carbonyl group and then indicating the presence of the carbonyl group by changing the suffix from *-e* to *-one*. The parent chain is numbered from the direction that gives the carbonyl group the lowest number. The IUPAC system retains the common names acetone and acetophenone.

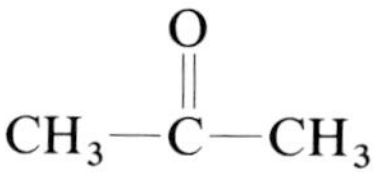

$$CH_3-\overset{\overset{\displaystyle O}{\|}}{C}-CH_3 \qquad CH_3-\overset{\overset{\displaystyle O}{\|}}{C}-CH_2-CH_3 \qquad CH_3-CH_2-\overset{\overset{\displaystyle O}{\|}}{C}-\underset{\underset{\displaystyle CH_3}{|}}{CH}-CH_2-CH_3$$

2-propanone   2-butanone 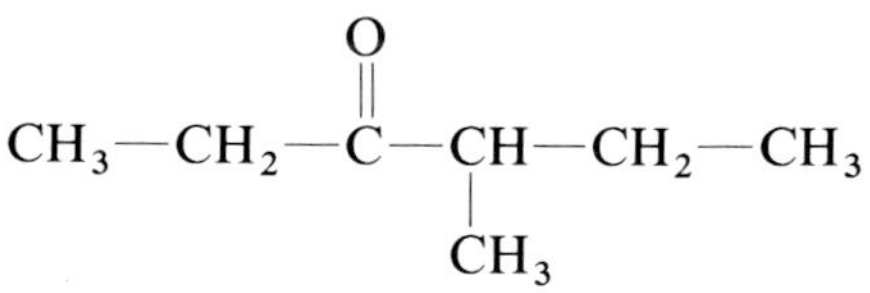   4-methyl-3-hexanone
(acetone)

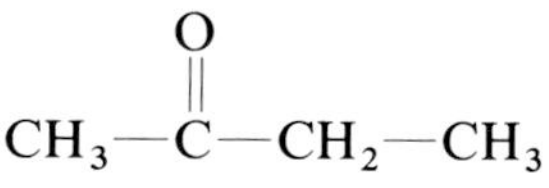

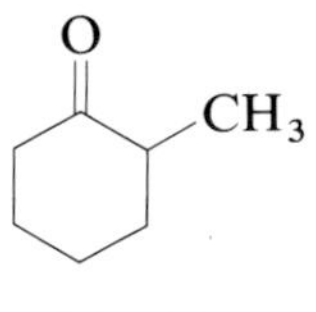

2-methylcyclohexanone 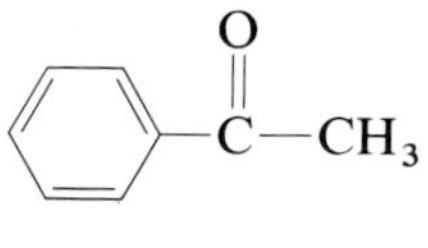   acetophenone 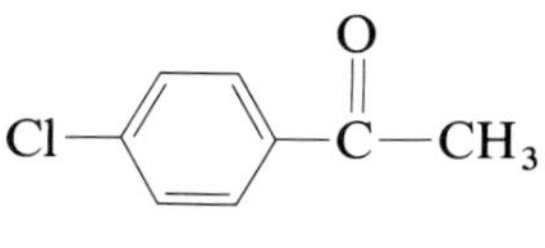   *p*-chloroacetophenone

Example 6.1

Give IUPAC names for these compounds:

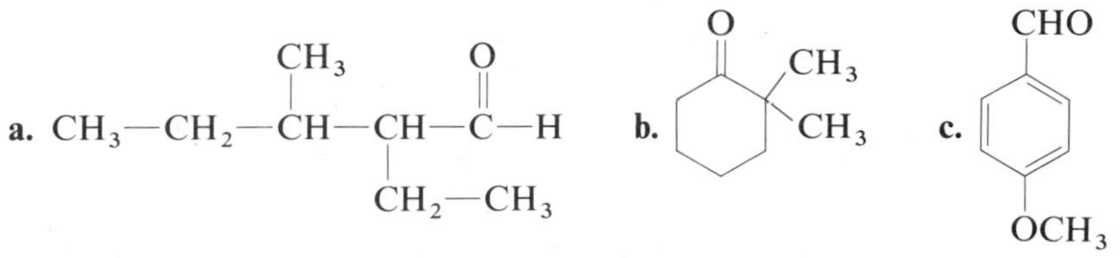

a. $CH_3-CH_2-CH-CH-C-H$ (with CH_3 on carbon 3, O on the carbonyl, and CH_2-CH_3 below)

b. (cyclohexanone with two CH_3 groups and O)

c. (benzene ring with CHO top and OCH_3 bottom)

Solution

a. In this molecule, the longest chain is six carbons, but the longest chain that contains the aldehyde is five carbons. Therefore the parent chain is pentane.

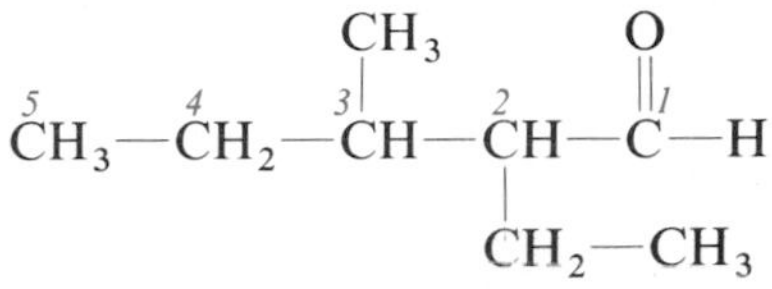

$$\overset{5}{C}H_3-\overset{4}{C}H_2-\overset{3}{C}H-\overset{2}{C}H-\overset{1}{C}-H$$

with CH_3 on C3, O on C1, and CH_2-CH_3 on C2

2-ethyl-3-methylpentanal

b. Number the six-membered ring beginning with the carbon bearing the carbonyl group. The IUPAC name is 2,2-dimethylcyclohexanone.

c. This molecule is derived from benzaldehyde. Its IUPAC name is 4-methoxybenzaldehyde. It may also be named *p*-methoxybenzaldehyde.

Problem 6.1

Give IUPAC names for the following.

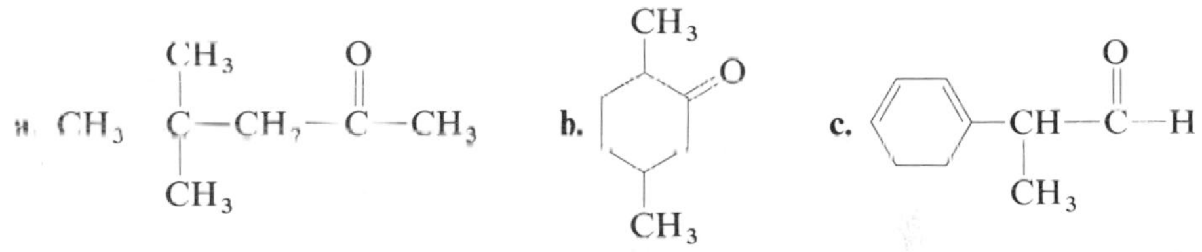

a. $CH_3-\underset{CH_3}{\overset{CH_3}{C}}-CH_2-\overset{O}{C}-CH_3$

b. (cyclohexanone with CH_3 groups and O)

c. (benzene ring)$-CH-C-H$ with CH_3 and O

B. IUPAC Names for More Complex Aldehydes and Ketones

For naming compounds that contain more than one functional group, that is, more than one that might be indicated by a suffix, the IUPAC system has established what is known as an order of precedence of functions. The order of precedence for the functional groups we will deal with most often is given in Table 6.1. There you will find how to indicate a given functional group if it has the highest precedence, and also how to show its presence if it has a lower precedence.

Following are several examples illustrating precedence rules.

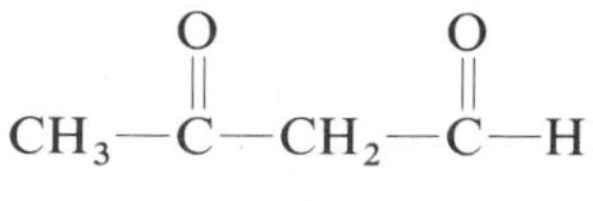

$$CH_3-\overset{O}{C}-CH_2-\overset{O}{C}-H \qquad CH_3-\overset{OH}{CH}-CH_2-CH_2-\overset{O}{C}-H$$

3-oxobutanal 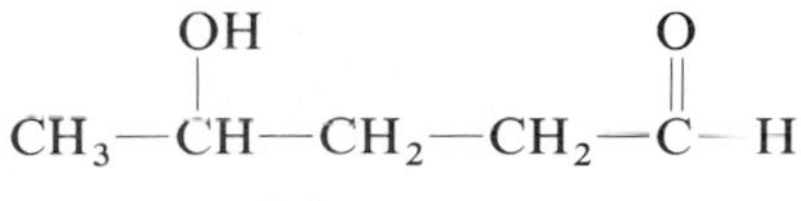   4-hydroxypentanal

Table 6.1
Order of
precedence of
several functional
groups.

Functional Group	Suffix If Highest Precedence	Prefix If Lower Precedence
$-\overset{\overset{O}{\|\|}}{C}-OH$	oic acid	—
$-\overset{\overset{O}{\|\|}}{C}-H$	al	oxo
$-\overset{\overset{O}{\|\|}}{C}-$	one	oxo
$-OH$	ol	hydroxy
$-NH_2$	amine	amino
$-SH$	thiol	mercapto

(decreasing precedence)

$$CH_3-\overset{\overset{O}{\|\|}}{C}-CH_2-\overset{\overset{O}{\|\|}}{C}-OH$$

3-oxobutanoic acid

$$CH_3-CH_2-CH_2-CH_2-CH_2-CH_2-CH_2-\overset{\overset{OH}{\|}}{CH}-CH_2-\overset{\overset{O}{\|\|}}{C}-OH$$

3-hydroxydecanoic acid

C. Common Names

The common name for an aldehyde is derived from the common name of the corresponding carboxylic acid by changing the suffix -*ic* to -*aldehyde*. Because we have not yet studied common names for carboxylic acids, we cannot at this point give common names to aldehydes. However, we can illustrate how common names are derived by reference to a few common names you are familiar with. The common name formaldehyde is derived from formic acid, and the name acetaldehyde is derived from acetic acid:

$$H-\overset{\overset{O}{\|\|}}{C}-H \qquad H-\overset{\overset{O}{\|\|}}{C}-OH \qquad CH_3-\overset{\overset{O}{\|\|}}{C}-H \qquad CH_3-\overset{\overset{O}{\|\|}}{C}-OH$$

formaldehyde   formic acid   acetaldehyde   acetic acid

Common names for ketones are derived by naming the two alkyl or aryl groups attached to the carbonyl group, followed by the word *ketone*, as shown in the following examples. Note that each alkyl or aryl group is listed in alphabetical order as a separate word followed by the word *ketone*.

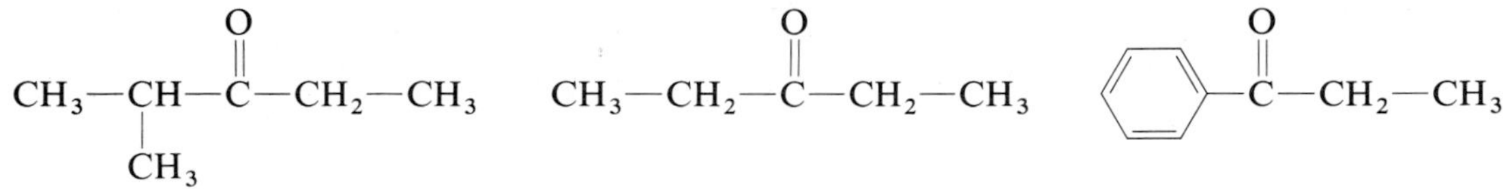

$$CH_3-CH-\overset{\overset{\displaystyle O}{\|}}{C}-CH_2-CH_3$$
$$|$$
$$CH_3$$

ethyl isopropyl ketone

$$CH_3-CH_2-\overset{\overset{\displaystyle O}{\|}}{C}-CH_2-CH_3$$

diethyl ketone

ethyl phenyl ketone

6.3 Physical Properties of Aldehydes and Ketones

Oxygen is more electronegative than carbon (3.5 compared with 2.5); therefore, the carbon-oxygen double bond is polar covalent. The oxygen atom of a carbonyl group bears a partial negative charge and the carbon atom a partial positive charge, as illustrated here for formaldehyde:

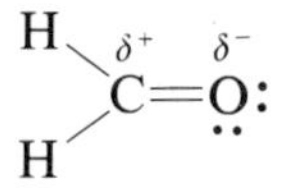

Alternatively, the **carbonyl group** may be pictured as a hybrid of two major contributing structures. Structure (b) places a negative charge on the more electronegative oxygen atom and a positive charge on the less electronegative carbon atom.

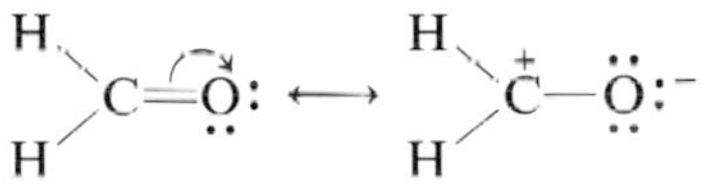

(a)   (b)

Because aldehydes and ketones are polar compounds and can interact in the pure state by dipole-dipole interaction, they have higher boiling points than nonpolar compounds of comparable molecular weight. Table 6.2 lists boiling points of six compounds of comparable molecular weight.

Pentane, a nonpolar hydrocarbon, has the lowest boiling point. Although methyl propyl ether is a polar compound, there is little association between molecules in the liquid state. Hence it has a boiling point only slightly higher than that of pentane. Both butanal and 2-butanone are polar compounds, and because of the association between a partially positive carbon of one molecule and a partially negative oxygen of another, their boiling points are higher than the boiling points of pentane and methyl propyl ether. Aldehydes and ketones have no partially positive hydrogen atom attached to oxygen and cannot associate by hydrogen bonding. Therefore, they have lower boiling points than those of alcohols and carboxylic acids, compounds that can associate by hydrogen bonding.

Aldehydes and ketones can interact with water molecules as hydrogen-bond acceptors (Figure 6.2), and therefore low-molecular-weight aldehydes and ketones are more soluble in water than are nonpolar compounds of comparable

Table 6.2
Boiling points of compounds of comparable molecular weight.

Compound	Structural Formula	Mol. Wt.	bp (°C)
pentane	$CH_3CH_2CH_2CH_2CH_3$	72	36
methyl propyl ether	$CH_3CH_2CH_2OCH_3$	74	39
butanal	$CH_3CH_2CH_2\overset{\overset{\displaystyle O}{\|\|}}{C}H$	72	76
2-butanone	$CH_3CH_2\overset{\overset{\displaystyle O}{\|\|}}{C}CH_3$	72	80
1-butanol	$CH_3CH_2CH_2CH_2OH$	74	117
propanoic acid	$CH_3CH_2\overset{\overset{\displaystyle O}{\|\|}}{C}OH$	74	151

molecular weight. Listed in Table 6.3 are values for boiling point and solubility in water for several low-molecular-weight aldehydes and ketones.

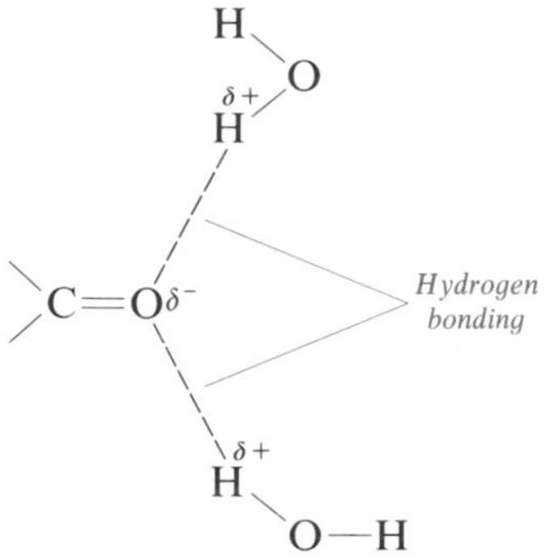

Figure 6.2 Hydrogen bonding between a carbonyl oxygen and water molecules.

Table 6.3
Physical properties of selected aldehydes and ketones.

Name	Structural Formula	bp (°C)	Solubility (g/100 g water)
formaldehyde	HCHO	−21	infinite
acetaldehyde	CH_3CHO	20	infinite
propanal	CH_3CH_2CHO	49	16
butanal	$CH_3CH_2CH_2CHO$	76	7
hexanal	$CH_3(CH_2)_4CHO$	129	slight
acetone	CH_3COCH_3	56	infinite
2-butanone	$CH_3CH_2COCH_3$	80	26
2-pentanone	$CH_3CH_2CH_2COCH_3$	101	5

6.4 Reactions of Aldehydes and Ketones

A carbonyl group, because of its polarity, reacts with both electrophiles and nucleophiles:

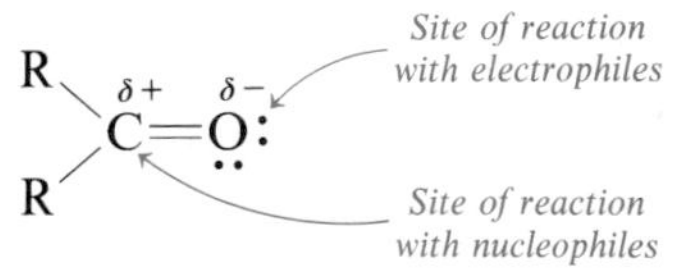

Site of reaction with electrophiles

Site of reaction with nucleophiles

A **nucleophile** is any atom or group of atoms with an unshared pair of electrons that can be shared with another atom or group of atoms to form a new covalent bond. One of the most common reaction themes of the carbonyl group is **nucleophilic addition**, a reaction in which a nucleophile adds to the carbonyl carbon, to form a tetrahedral carbonyl addition compound. In the following general reaction, the nucleophilic reagent is written as H—Nu: to emphasize the presence of the unshared pair of electrons in the nucleophile.

Nucleophilic Addition

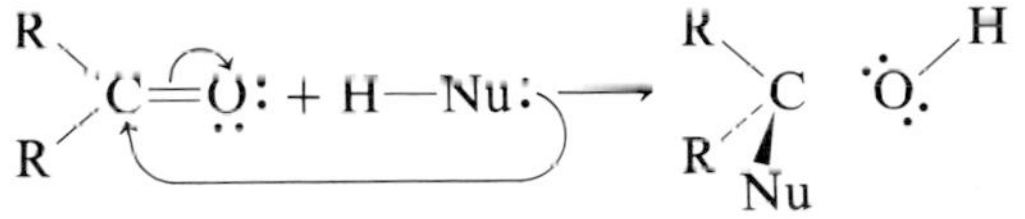

tetrahedral carbonyl
addition compound

The most common electrophilic reagent in carbonyl addition reactions is the proton, H^+. Reaction of a carbonyl group with a proton gives a resonance-stabilized cation:

Electrophilic Addition

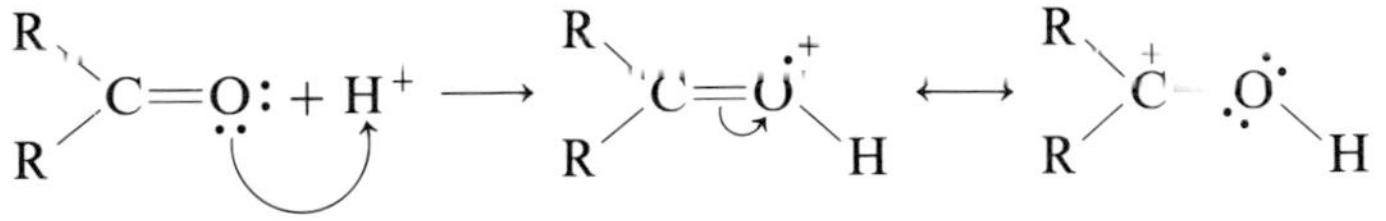

resonance-stabilized cation

Protonation increases the electron deficiency of the carbonyl carbon and makes it even more reactive toward nucleophiles.

A. Addition of Water

Addition of water to the carbonyl group of an aldehyde or ketone forms a 1,1-diol. Note that in this designation, the numbers 1,1- do not refer to an IUPAC numbering system, but indicate that the two hydroxyl groups are on the same carbon. They are also called hydrates:

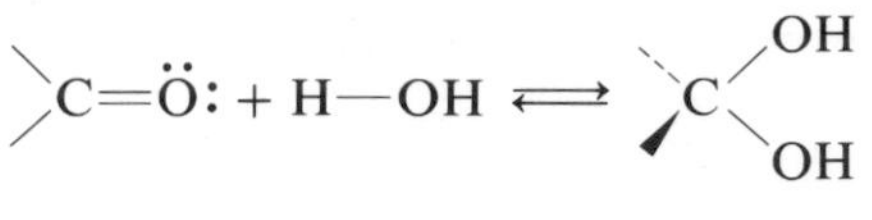

$$\begin{array}{c}
\diagdown \\
\diagup
\end{array} C = \ddot{O}: + H-OH \rightleftharpoons \begin{array}{c}
\diagdown \\
\diagup
\end{array} C \begin{array}{c}
\diagup OH \\
\diagdown OH
\end{array}$$

a 1,1-diol
(a hydrate)

When formaldehyde is dissolved in water at 20°C, it is more than 99% hydrated. The equilibrium constant for this reaction is approximately 10^3.

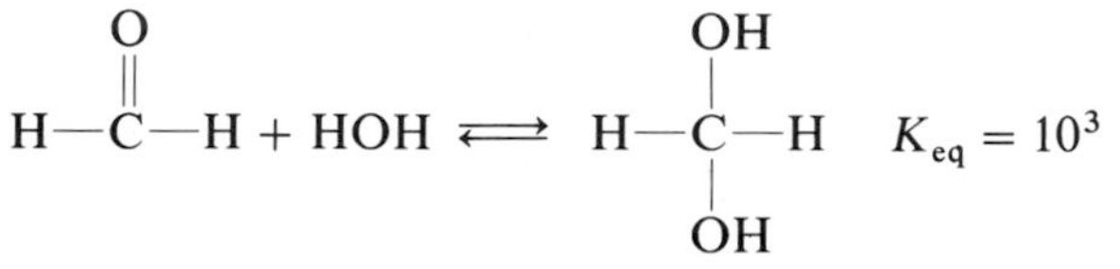

$$\underset{\displaystyle \text{O}}{\overset{\displaystyle \parallel}{H-C-H}} + HOH \rightleftharpoons \underset{\displaystyle \text{OH}}{\overset{\displaystyle \text{OH}}{H-C-H}} \quad K_{eq} = 10^3$$

greater than 99%

For acetaldehyde and most other aldehydes, equilibrium constants for hydration are approximately 1. For example, under experimental conditions comparable with those described for formaldehyde, acetaldehyde is approximately 58% hydrated.

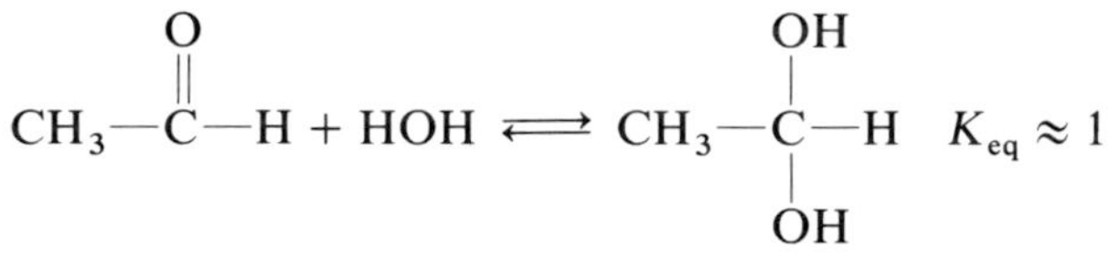

$$\underset{\displaystyle \text{O}}{\overset{\displaystyle \parallel}{CH_3-C-H}} + HOH \rightleftharpoons \underset{\displaystyle \text{OH}}{\overset{\displaystyle \text{OH}}{CH_3-C-H}} \quad K_{eq} \approx 1$$

58%

For most simple ketones, equilibrium constants for hydration are 10^{-3} or less. Thus, in aqueous solutions, ketones are almost entirely in the keto form instead of the hydrate.

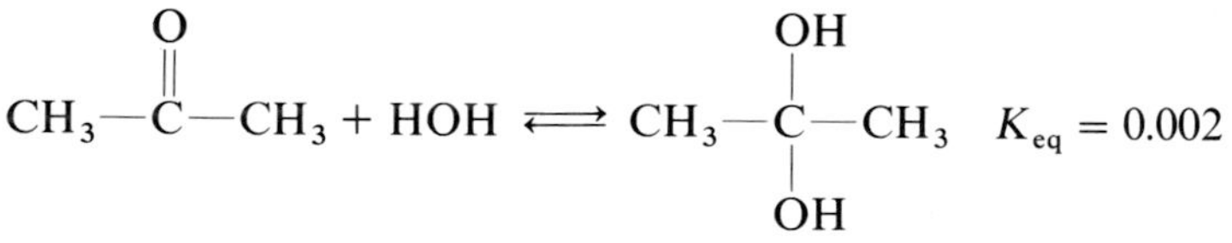

$$\underset{\displaystyle \text{O}}{\overset{\displaystyle \parallel}{CH_3-C-CH_3}} + HOH \rightleftharpoons \underset{\displaystyle \text{OH}}{\overset{\displaystyle \text{OH}}{CH_3-C-CH_3}} \quad K_{eq} = 0.002$$

very low

We can interpret the position of equilibria for carbonyl hydration by the relative sizes of atoms and groups of atoms bonded to the carbonyl carbon. Bond angles about a carbonyl carbon are approximately 120°; those for the same carbon in a hydrate are approximately 109.5°. Thus in hydration, there is a crowding as the groups attached to the carbonyl carbon are brought closer together in space. In formaldehyde, the two groups brought closer are hydrogen atoms; in acetaldehyde they are methyl and hydrogen; and in acetone they are two methyl groups. Because the increase in crowding is greatest for ketones, compared with aldehydes, equilibrium constants for their hydration are much smaller than for aldehydes.

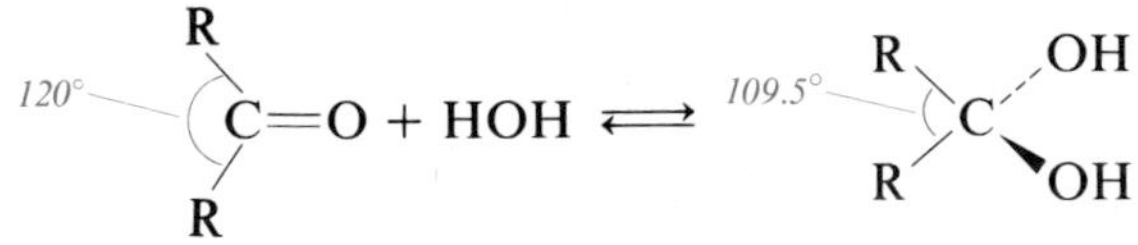

$$120° \quad \underset{R}{\overset{R}{\diagdown}} C{=}O + HOH \rightleftharpoons 109.5° \quad \underset{R}{\overset{R}{\diagdown}} C \overset{OH}{\underset{OH}{\diagup}}$$

B. Addition of Alcohols: Formation of Acetals and Ketals

Alcohols add to aldehydes and ketones in the manner described for addition of water. Addition of one molecule of alcohol to an aldehyde forms a **hemiacetal** (a half-acetal). The comparable reaction with a ketone forms a **hemiketal** (a half-ketal).

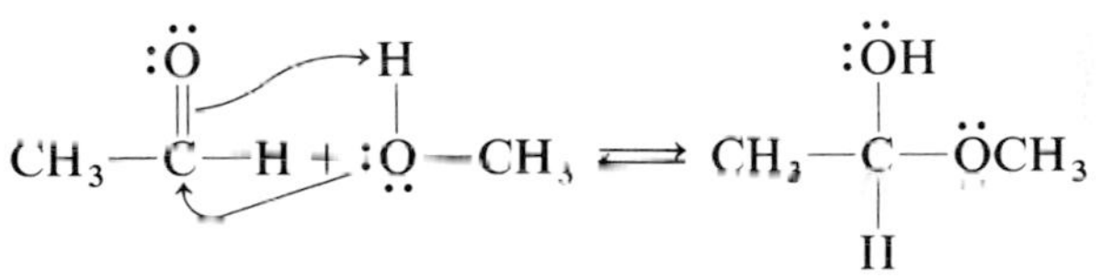

$$CH_3{-}\overset{\overset{\ddot{O}}{\parallel}}{C}{-}H + :\overset{..}{\underset{..}{O}}{-}CH_3 \rightleftharpoons CH_3{-}\overset{:\overset{..}{O}H}{\underset{\underset{H}{|}}{C}}{-}\overset{..}{\underset{..}{O}}CH_3$$

(a hemiacetal)

$$CH_3{-}\overset{\overset{\ddot{O}}{\parallel}}{C}{-}CH_3 + :\overset{..}{\underset{..}{O}}{-}CH_2CH_3 \rightleftharpoons CH_3{-}\overset{:\overset{..}{O}H}{\underset{\underset{CH_3}{|}}{C}}{-}\overset{..}{\underset{..}{O}}CH_2CH_3$$

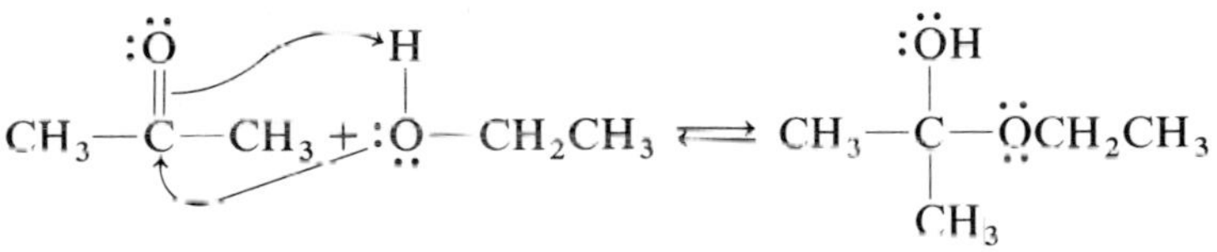

(a hemiketal)

Following are the characteristic structural features of a hemiacetal and a hemiketal. In each instance, the R group may be alkyl or aryl.

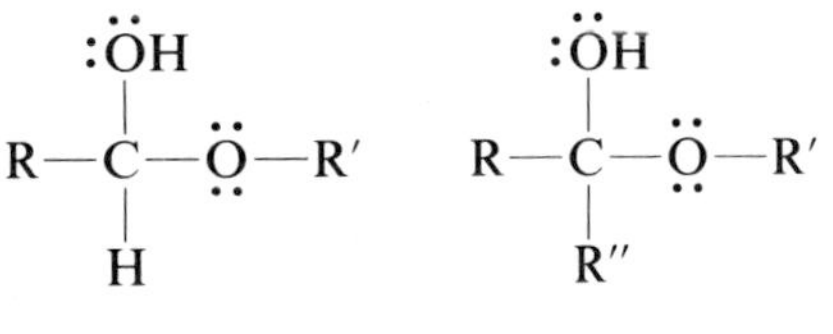

$$R{-}\overset{:\overset{..}{O}H}{\underset{\underset{H}{|}}{C}}{-}\overset{..}{\underset{..}{O}}{-}R' \qquad R{-}\overset{:\overset{..}{O}H}{\underset{\underset{R''}{|}}{C}}{-}\overset{..}{\underset{..}{O}}{-}R'$$

a hemiacetal   a hemiketal
(characteristic   (characteristic
structural feature)   structural feature)

Hemiacetals and hemiketals are only minor components of an equilibrium mixture except in one very important type of molecule. When a hydroxyl group is part of the same molecule that contains the carbonyl group, and a five- or six-membered ring can form, the compound exists almost entirely in the cyclic hemiacetal or cyclic hemiketal form.

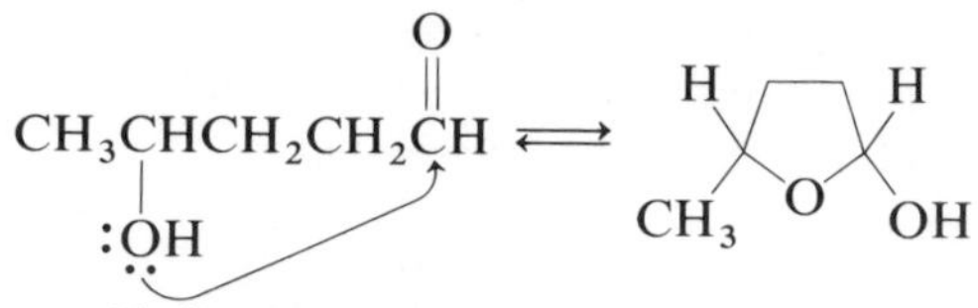

$$CH_3CHCH_2CH_2\overset{\overset{\displaystyle O}{\|}}{C}H \rightleftharpoons$$
$$:\overset{\cdot\cdot}{O}H$$

4-hydroxypentanal
(minor)

a cyclic hemiacetal
(major form present
at equilibrium)

Hemiacetals and hemiketals react further with alcohols to form **acetals** and **ketals** plus a molecule of water. Each of these reactions is acid-catalyzed.

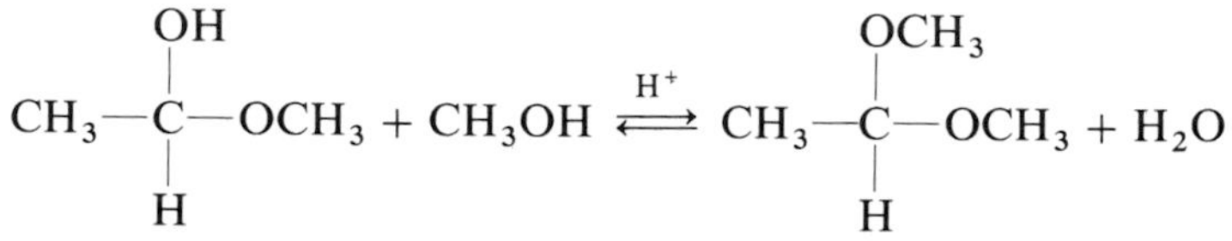

$$CH_3\underset{\underset{\displaystyle H}{|}}{\overset{\overset{\displaystyle OH}{|}}{C}}OCH_3 + CH_3OH \overset{H^+}{\rightleftharpoons} CH_3\underset{\underset{\displaystyle H}{|}}{\overset{\overset{\displaystyle OCH_3}{|}}{C}}OCH_3 + H_2O$$

(a hemiacetal)

1,1-dimethoxyethane
(a dimethyl acetal)

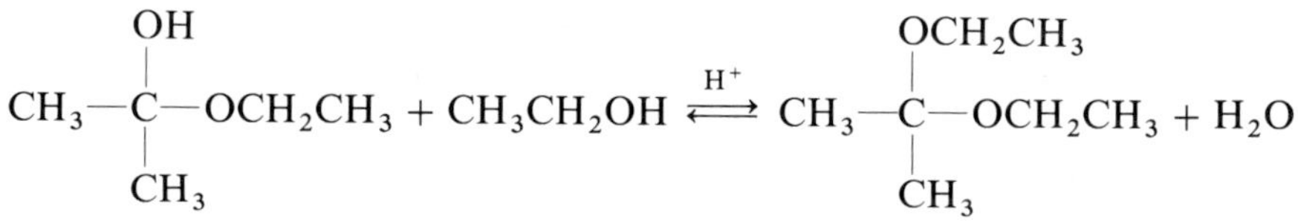

$$CH_3\underset{\underset{\displaystyle CH_3}{|}}{\overset{\overset{\displaystyle OH}{|}}{C}}OCH_2CH_3 + CH_3CH_2OH \overset{H^+}{\rightleftharpoons} CH_3\underset{\underset{\displaystyle CH_3}{|}}{\overset{\overset{\displaystyle OCH_2CH_3}{|}}{C}}OCH_2CH_3 + H_2O$$

(a hemiketal)

2,2-diethoxypropane
(a diethyl ketal)

Following are general formulas showing the characteristic structural feature of an acetal and a ketal:

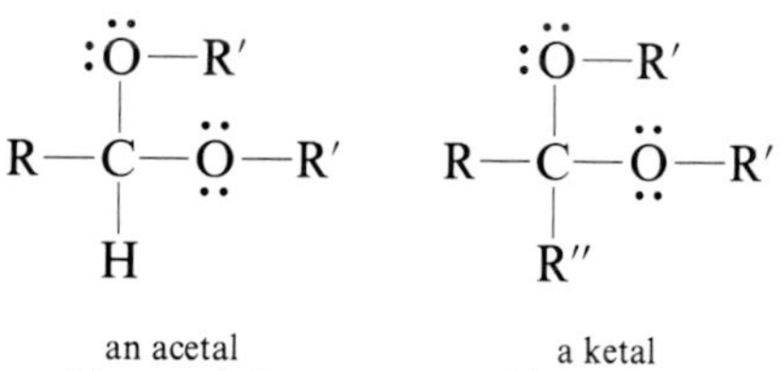

$$R\underset{\underset{\displaystyle H}{|}}{\overset{\overset{\displaystyle :\overset{\cdot\cdot}{O}-R'}{|}}{C}}\overset{\cdot\cdot}{\underset{\cdot\cdot}{O}}-R'$$

$$R\underset{\underset{\displaystyle R''}{|}}{\overset{\overset{\displaystyle :\overset{\cdot\cdot}{O}-R'}{|}}{C}}\overset{\cdot\cdot}{\underset{\cdot\cdot}{O}}-R'$$

an acetal
(characteristic
structural feature)

a ketal
(characteristic
structural feature)

Acetal and ketal formation are equilibrium reactions, and to obtain high yields, it is necessary to remove water from the reaction mixture so as to favor product formation.

Example 6.2

For the following, show the reaction of each carbonyl compound with one molecule of alcohol to form a hemiacetal or hemiketal, and then with a second molecule of alcohol to form an acetal or ketal. Note that in part (b), ethylene glycol is a diol and one molecule provides both —OH groups.

a. 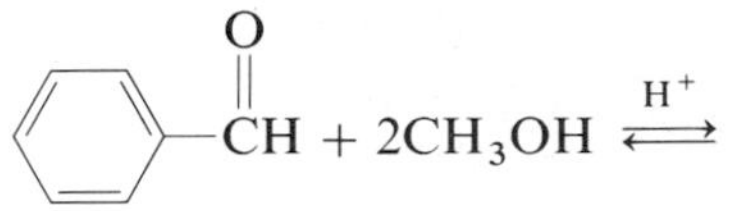 (phenyl)—$\overset{\overset{\text{O}}{\|}}{\text{CH}}$ + 2CH$_3$OH $\overset{H^+}{\rightleftharpoons}$

b.  (cyclohexyl)=O + HOCH$_2$CH$_2$OH $\overset{H^+}{\rightleftharpoons}$

☐ *Solution*

a. 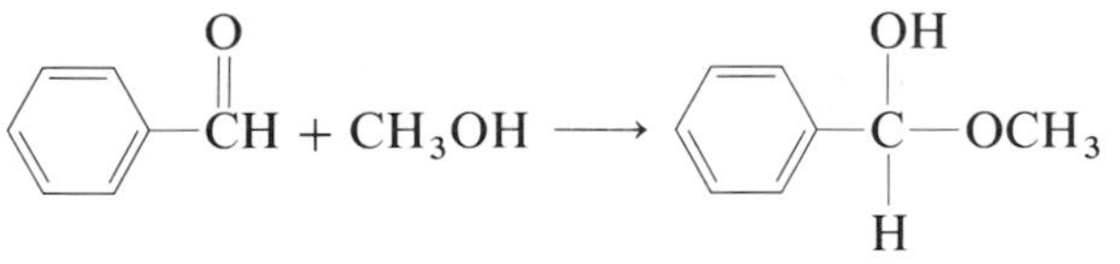 (phenyl)—$\overset{\overset{\text{O}}{\|}}{\text{CH}}$ + CH$_3$OH $\longrightarrow$ (phenyl)—$\overset{\overset{\text{OH}}{|}}{\underset{\text{H}}{\text{C}}}$—OCH$_3$

a hemiacetal

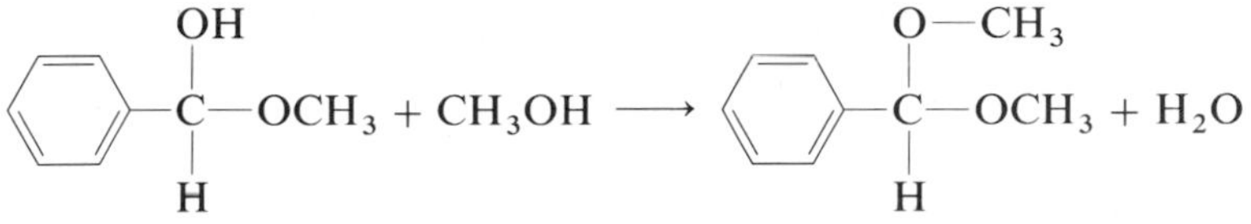

(phenyl)—$\overset{\overset{\text{OH}}{|}}{\underset{\text{H}}{\text{C}}}$—OCH$_3$ + CH$_3$OH $\longrightarrow$ (phenyl)—$\overset{\overset{\text{O—CH}_3}{|}}{\underset{\text{H}}{\text{C}}}$—OCH$_3$ + H$_2$O

an acetal

b. 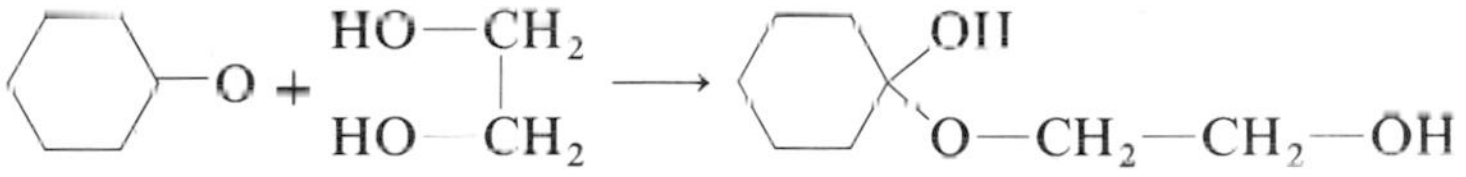 (cyclohexyl)—O + $\overset{\text{HO—CH}_2}{\underset{\text{HO—CH}_2}{|}}$ $\longrightarrow$ (cyclohexyl)$\overset{\text{OH}}{\underset{\text{O—CH}_2—\text{CH}_2—OH}{}}$

a hemiketal

(cyclohexyl)$\overset{\text{OH}}{\underset{\text{O—CH}_2—\text{CH}_2 \quad OH}{}}$ $\longrightarrow$ (cyclohexyl)$\overset{\text{O—CH}_2}{\underset{\text{O—CH}_2}{|}}$ + H$_2$O

a cyclic ketal

■ **Problem 6.2** The reaction of an acetal or ketal with water to form an aldehyde or a ketone and two molecules of alcohol is called hydrolysis. Following are structural formulas for one ketal and one acetal. Draw the structural formulas for the products of hydrolysis of each.

a. 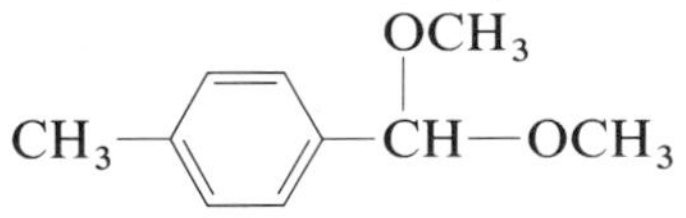 CH$_3$—(phenyl)—$\overset{\overset{\text{OCH}_3}{|}}{\text{CH}}$—OCH$_3$ **b.** 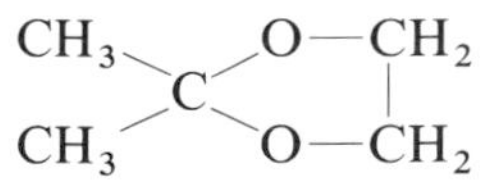 $\overset{\text{CH}_3}{\underset{\text{CH}_3}{}}C\overset{\text{O—CH}_2}{\underset{\text{O—CH}_2}{}}$ ■

As noted earlier, formation of acetals and ketals is catalyzed by acid. Their hydrolysis in water is also catalyzed by acid. Both acetals and ketals, however, are stable and unreactive in aqueous base.

In practice, formation of acetals and ketals is often carried out using the alcohol as the solvent and anhydrous acid, often dry HCl (hydrogen chloride gas), dissolved in the alcohol. Because the alcohol is both a reactant and solvent, it is present in large molar excess, which forces the position of equilibrium to the right and favors acetal/ketal formation.

C. Addition of Ammonia and Its Derivatives

Ammonia and amines of the type $R—NH_2$ react with the carbonyl group of aldehydes and ketones in the presence of an acid catalyst to give products that contain a carbon-nitrogen double bond, as shown below. A molecule containing a carbon-nitrogen double bond is called an **imine**, or alternatively, a **Schiff base**. The characteristic structural feature of an imine is a C=N group. Following are examples of conversion of an aldehyde and a ketone to imines.

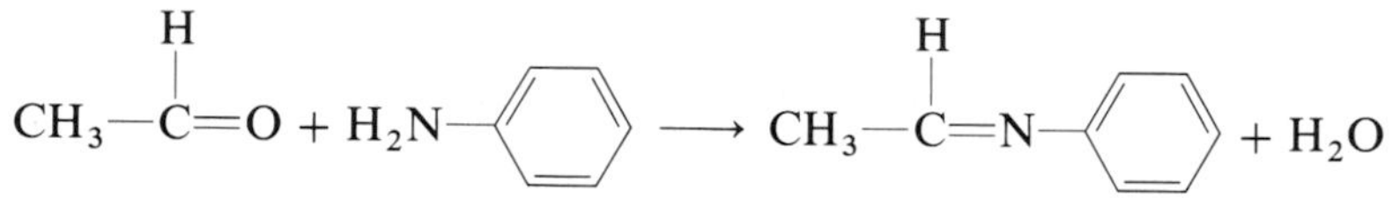

$$CH_3—\overset{\overset{\displaystyle H}{|}}{C}=O + H_2N—\bigcirc \longrightarrow CH_3—\overset{\overset{\displaystyle H}{|}}{C}=N—\bigcirc + H_2O$$

an imine
(a Schiff base)

$$\bigcirc=O + H_2N—CH_3 \longrightarrow \bigcirc=N—CH_3 + H_2O$$

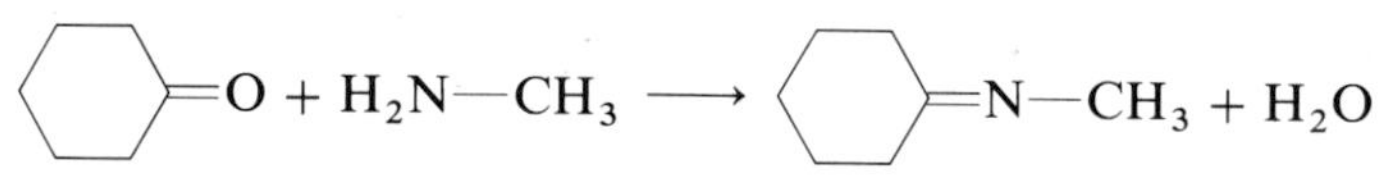

an imine
(a Schiff base)

The mechanism of imine formation can be divided into two steps. In step 1, the nitrogen atom of ammonia or the amine adds to the carbonyl carbon to form a tetrahedral carbonyl addition compound. In step 2, loss of water gives the imine.

Step 1: Formation of a tetrahedral carbonyl addition compound

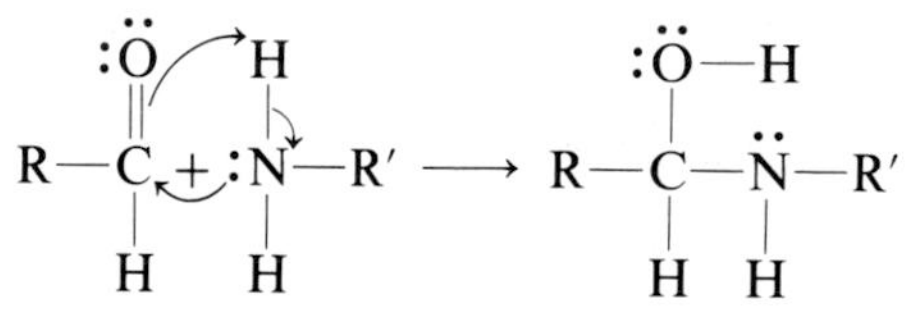

$$R—\overset{\overset{\displaystyle :\ddot{O}}{\|}}{\underset{\underset{\displaystyle H}{|}}{C}}+:\overset{\overset{\displaystyle H}{|}}{\underset{\underset{\displaystyle H}{|}}{N}}—R' \longrightarrow R—\overset{\overset{\displaystyle :\ddot{O}—H}{|}}{\underset{\underset{\displaystyle H}{|}}{C}}—\overset{\overset{\displaystyle ..}{}}{\underset{\underset{\displaystyle H}{|}}{N}}—R'$$

tetrahedral carbonyl
addition intermediate

Step 2: Loss of water

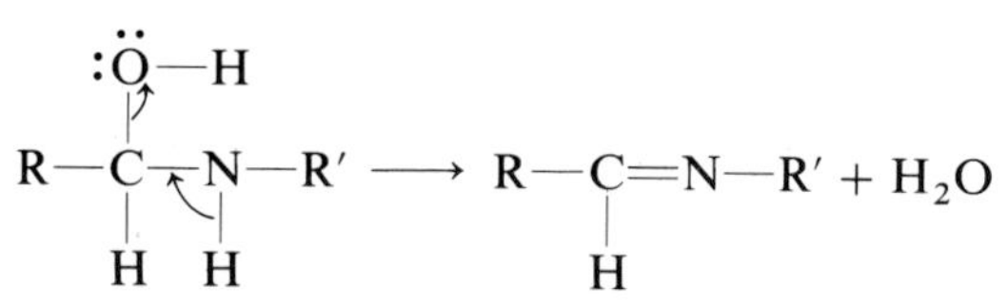

$$R—\overset{\overset{\displaystyle :\ddot{O}—H}{}}{\underset{\underset{\displaystyle H}{|}}{C}}—\overset{}{\underset{\underset{\displaystyle H}{|}}{N}}—R' \longrightarrow R—\overset{}{\underset{\underset{\displaystyle H}{|}}{C}}=N—R' + H_2O$$

an imine
(a Schiff base)

As one example of the importance of imines in biological systems, the active form of vitamin A aldehyde (retinal) is bound to the protein opsin in the form of an imine. The $-NH_2$ group for imine formation is provided by the side chain of the amino acid lysine (Section 13.1). The imine formed by combination of vitamin A aldehyde and opsin is called rhodopsin, or alternatively, visual purple.

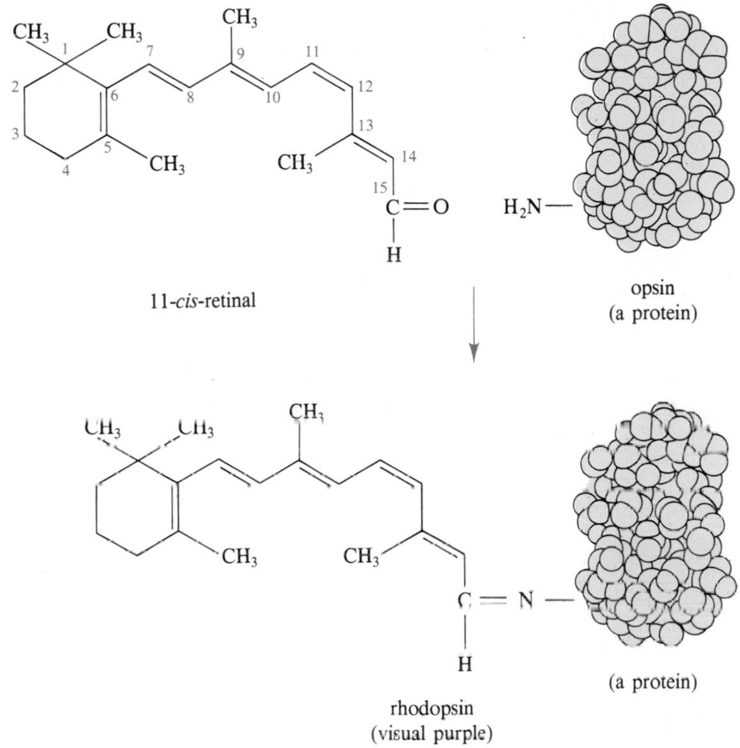

11-*cis*-retinal

opsin
(a protein)

rhodopsin
(visual purple)

(a protein)

Example 6.3

For the following, write structural formulas for the tetrahedral carbonyl addition compound and the imine (Schiff base) formed from it by loss of water.

a.

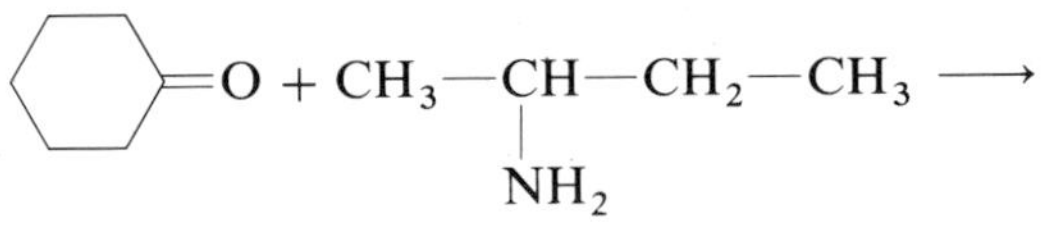

b.

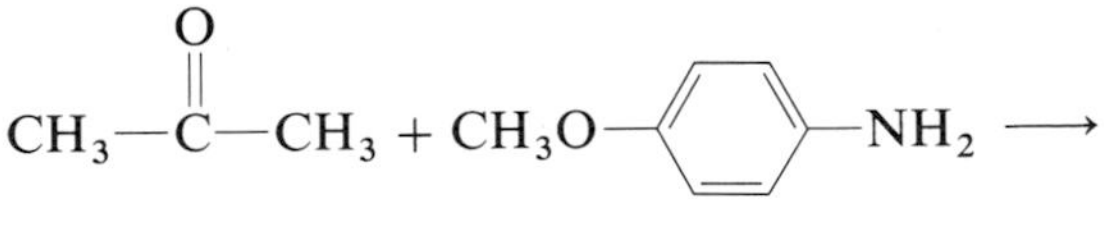

Solution

Following are structural formulas for the tetrahedral carbonyl addition compound and the imine.

a.

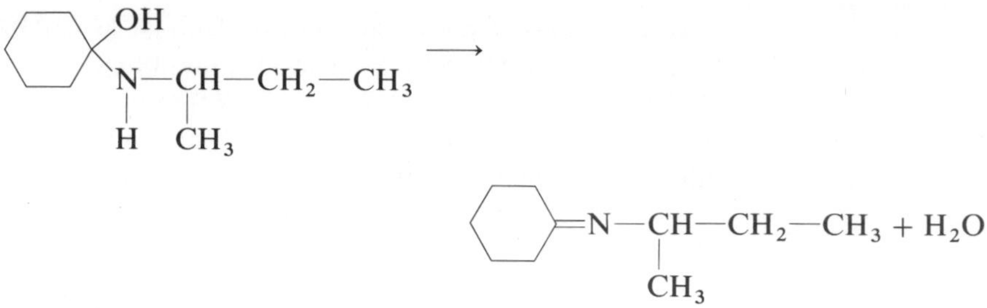

$$\text{[cyclohexane ring]}\overset{\text{OH}}{\underset{\overset{|}{\text{H}}}{\text{N}}}-\overset{}{\underset{\overset{|}{\text{CH}_3}}{\text{CH}}}-\text{CH}_2-\text{CH}_3 \longrightarrow$$

$$\text{[cyclohexane ring]}=\text{N}-\underset{\overset{|}{\text{CH}_3}}{\text{CH}}-\text{CH}_2-\text{CH}_3 + \text{H}_2\text{O}$$

b.

$$\underset{\text{CH}_3}{\overset{\text{CH}_3}{\diagdown}}\overset{\text{OH}}{\underset{\underset{\text{H}}{\overset{|}{\text{N}}}}{\overset{|}{\text{C}}}}-\text{[benzene ring]}-\text{OCH}_3 \longrightarrow$$

$$\underset{\text{CH}_3}{\overset{\text{CH}_3}{\diagdown}}\text{C}=\text{N}-\text{[benzene ring]}-\text{OCH}_3 + \text{H}_2\text{O}$$

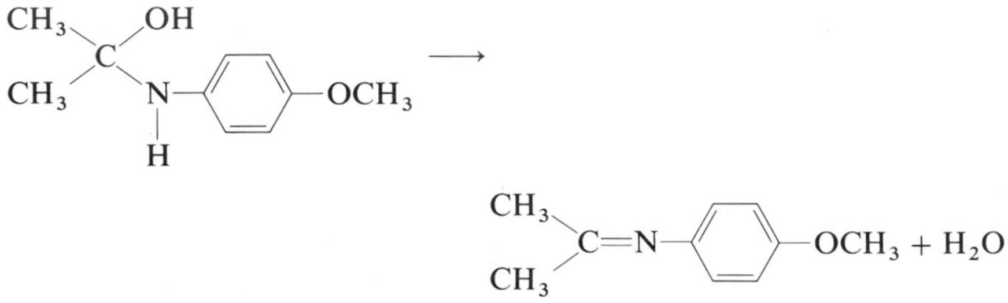

Problem 6.3   The reaction of an imine (a Schiff base) with water to form an amine and an aldehyde or ketone is called hydrolysis. Write structural formulas for the products of hydrolysis of the following imines.

a.

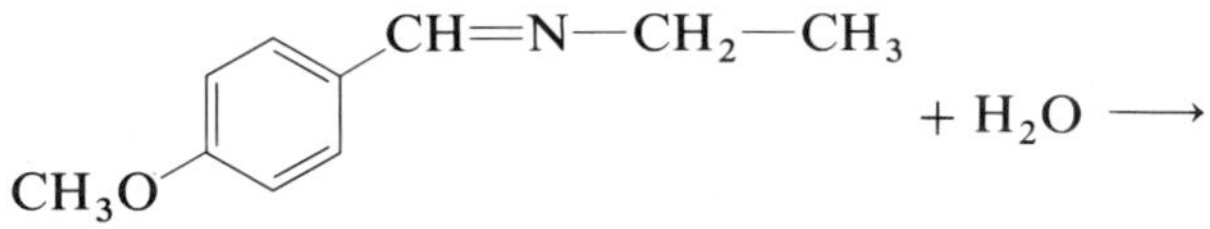

$$\text{CH}_3\text{O}-\text{[benzene ring]}-\text{CH}=\text{N}-\text{CH}_2-\text{CH}_3 \quad + \text{H}_2\text{O} \longrightarrow$$

b.

$$\text{[benzene ring]}-\text{CH}_2-\text{N}=\text{[cyclohexane ring]} + \text{H}_2\text{O} \longrightarrow$$

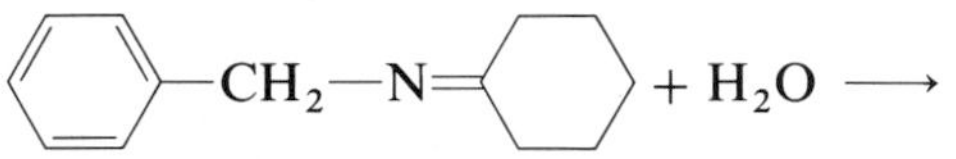

6.5 Oxidation of Aldehydes and Ketones

A. Oxidation of Aldehydes

Aldehydes are oxidized to carboxylic acids by a wide variety of oxidizing agents. In fact, they are one of the most easily oxidized of all functional groups. **Oxidation of an aldehyde** to a carboxylic acid is a two-electron oxidation, as shown by the balanced half-reaction

$$\overset{\overset{\text{O}}{\|}}{\text{R}-\text{C}-\text{H}} + \text{H}_2\text{O} \longrightarrow \overset{\overset{\text{O}}{\|}}{\text{R}-\text{C}-\text{O}-\text{H}} + 2\text{H}^+ + 2\text{e}^-$$

Oxidizing agents commonly used to oxidize aldehydes are potassium permanganate and chromic acid.

Aldehydes are also oxidized to carboxylic acids by silver(I) dissolved in ammonium hydroxide (**Tollens reagent**). The reagent is prepared by dissolving silver nitrate in water, adding sodium hydroxide to precipitate silver(I) as Ag_2O, and then adding ammonium hydroxide to redissolve silver(I) as the silver–ammonia complex ion. When a few drops of this test solution is added to an aldehyde, the aldehyde is oxidized to a carboxylate anion and silver(I) is reduced to metallic silver. If the Tollens test is done properly, silver precipitates as a smooth, mirrorlike deposit, hence the name silver-mirror test.

$$\underset{\text{O}}{\overset{\text{O}}{R-C-H}} + Ag^+ \xrightarrow{\text{NH}_4\text{OH}} \underset{\text{O}}{\overset{\text{O}}{R-C-O^-}} + Ag$$

precipitates
as silver mirror

Silver(I) is rarely used now because silver is expensive and because other, more convenient methods exist for oxidizing aldehydes. This oxidation, however, is still used to silver mirrors.

Copper(II) is also used for the oxidation of aldehydes to carboxylic acids. One copper-containing oxidizing agent is known as Benedict's solution, another as Fehling's solution. We will discuss oxidation by copper(II) in Section 11.2D within the context of oxidation of monosaccharides.

B. Oxidation of Ketones

Ketones are not normally oxidized by either potassium dichromate or potassium permanganate. At higher temperatures, however, they can be oxidized further with cleavage of one of the bonds to the carbonyl group. The most useful synthetic application of ketone oxidation is that of symmetrical cyclic ketones. For example, cyclohexanone is oxidized to hexanedioic acid (adipic acid), one of the two monomers required for the synthesis of the polymer Nylon 66. In the industrial process, the oxidizing agent is nitric acid.

$$\text{(cyclohexanone)} + HNO_3 \longrightarrow \underset{\text{O}}{\overset{\text{O}}{HOCCH_2CH_2CH_2CH_2COH}} + \text{oxides of nitrogen}$$

hexanedioic acid
(adipic acid)

6.6 Reduction of Aldehydes and Ketones

A. Catalytic Reduction (Catalytic Hydrogenation)

Aldehydes are reduced to primary alcohols and ketones to secondary alcohols by hydrogen in the presence of a heavy metal catalyst, most often finely divided palladium, platinum, nickel, ruthenium, or a copper-chromium complex.

Reductions are generally possible at temperatures of 25–100°C and at pressures of hydrogen of 1–5 atm. Sometimes, however, it is necessary to use much higher pressures of hydrogen. Under suitable conditions, cyclohexanone is reduced to cyclohexanol, and 3-hydroxybutanal is reduced to 1,3-butanediol:

$$\text{cyclohexanone} + H_2 \xrightarrow[\text{25°C, 2 atm}]{\text{Pt}} \text{cyclohexanol}$$

cyclohexanone   cyclohexanol

$$CH_3-\overset{OH}{\underset{|}{CH}}-CH_2-\overset{O}{\overset{\|}{C}}-H + H_2 \xrightarrow{\text{Ni}} CH_3-\overset{OH}{\underset{|}{CH}}-CH_2-CH_2OH$$

3-hydroxybutanal   1,3-butanediol

The advantage of catalytic reduction (hydrogenation) of aldehydes and ketones is that it is simple to carry out, yields are generally very high, and isolation of the final product is very easy. The disadvantage is that some other functional groups are also reduced under these conditions, for example, carbon-carbon double and triple bonds; this can be seen in the commercial synthesis of 1-butanol from 2-butenal:

$$CH_3-CH=CH-\overset{O}{\overset{\|}{C}}-H + 2H_2 \xrightarrow{\text{Ni}} CH_3-CH_2-CH_2-CH_2OH$$

2-butenal   1-butanol
(crotonaldehyde)   (n-butyl alcohol)

B. Metal Hydride Reductions

By far the most common laboratory reagents for reducing aldehydes and ketones to alcohols are sodium borohydride, lithium aluminum hydride (LAH), and their derivatives. These compounds behave as sources of hydride ion.

$$Na^+ \quad H-\overset{H}{\underset{H}{\overset{|}{\underset{|}{B}}}}-H \qquad Li^+ \quad H-\overset{H}{\underset{H}{\overset{|}{\underset{|}{\overset{-}{Al}}}}}-H \qquad H:^-$$

sodium   lithium aluminum   hydride
borohydride   hydride (LAH)   ion

Lithium aluminum hydride is a very powerful reducing agent; it reduces not only aldehydes and ketones rapidly but also other functional groups such as carboxylic acids (Section 7.5C) and their functional derivatives (Section 8.6A). Sodium borohydride is a much more selective reducing agent; it reduces only aldehydes and ketones rapidly.

Reductions with sodium borohydride are usually effected in aqueous methanol, diethyl ether, or tetrahydrofuran. The initial product of reduction is a metal alkoxide, which on reaction with water is converted to an alcohol and sodium and borate salts. One mole of sodium borohydride reduces four moles of aldehyde.

$$4CH_3\overset{\overset{\displaystyle O}{\|}}{C}H + NaBH_4 \longrightarrow (CH_3CH_2O)_4B^-Na^+ \xrightarrow{H_2O} 4CH_3CH_2OH$$

metal alkoxide

Unlike sodium borohydride, lithium aluminum hydride reacts violently with water, methanol, and other solvents containing —OH groups, to liberate hydrogen gas and form metal hydroxides. Therefore, reductions of aldehydes and ketones with this reagent must be carried out in nonhydroxylic solvents, usually diethyl ether or tetrahydrofuran. The stoichiometry for lithium aluminum hydride reductions is the same as for sodium borohydride reductions, namely 1 mol of lithium aluminum hydride per 4 mol of aldehyde/ketone:

$$4R\overset{\overset{\displaystyle O}{\|}}{\underset{}{C}}R + LiAlH_4 \xrightarrow{ether} (R_2CHO)_4Al^-Na^+ \xrightarrow{H_2O} 4R\overset{\overset{\displaystyle OH}{|}}{\underset{}{C}}H\!-\!R$$

The key step in metal hydride reductions of aldehydes or ketones is transfer of a hydride ion from the reducing agent to the carbonyl carbon, to form a tetrahedral carbonyl addition compound.

$$Na^+ \quad H\!-\!\underset{\underset{\displaystyle H}{|}}{\overset{\overset{\displaystyle H}{|}}{B}}\!-\!H + R\!-\!\overset{\overset{\displaystyle \ddot{O}:}{\|}}{C}\!-\!R \longrightarrow R\!-\!\overset{\overset{\displaystyle :\ddot{O}\!-\!BH_3^-\ Na^+}{|}}{\underset{\underset{\displaystyle H}{|}}{C}}\!-\!R$$

a metal alkoxide

Hydride transfer from boron or aluminum is repeated three times until all reducing equivalents have been used. In reduction of an aldehyde or ketone to an alcohol, only the hydrogen atom attached to carbon comes from the hydride reducing agent; the hydrogen atom attached to oxygen comes from water during hydrolysis of the metal alkoxide salt.

$$R\!-\!\overset{\overset{\displaystyle O\!-\!H}{|}}{\underset{\underset{\displaystyle H}{|}}{C}}\!-\!R$$

This hydrogen comes from water during hydrolysis

This hydrogen comes from the hydride reducing agent

Lithium aluminum hydride and sodium borohydride are reagents used in the laboratory to reduce aldehydes and ketones to alcohols. As we shall see in

Section 17.4, nicotinamide adenine dinucleotide, a biological reducing agent, also reduces aldehydes and ketones by transfer of a hydride ion. Thus, hydride ion reductions are common in both the laboratory and the biological world.

6.7 Properties of Alpha-Hydrogens

A. Acidity of Alpha-Hydrogens

A carbon atom adjacent to a carbonyl group is called an α-carbon (alpha-carbon), and hydrogen atoms attached it are called α-hydrogens (alpha-hydrogens).

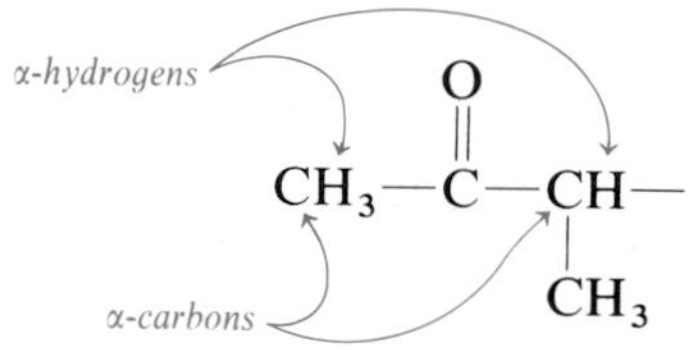

Because carbon and hydrogen have comparable electronegativities, there is normally no appreciable polarity to a C—H bond and no tendency for it to ionize; that is, a hydrogen atom attached to carbon shows no acidity. However, the situation is different for α-hydrogens. In the presence of a very strong base, an α-hydrogen can be removed to form an anion, as shown in the following equation:

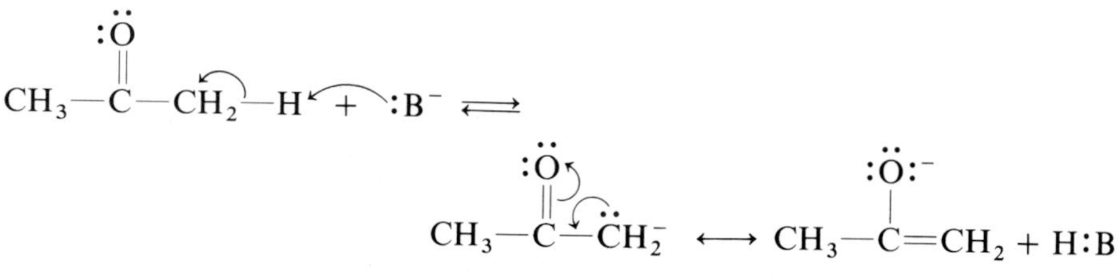

(a resonance-stabilized anion)

Two factors contribute to the increased acidity of α-hydrogens relative to other C—H bonds. First, the presence of the adjacent polar covalent C=O bond polarizes the electron pair of the C—H bond, so the hydrogen can be removed as a proton by a strong base. Second, and more important, is the resonance stabilization of the resulting anion, which can be written as a hybrid of two major contributing structures.

When the resonance-stabilized anion reacts with a proton, it may do so either on oxygen or on the α-carbon. Protonation on carbon gives the original molecule in what is called the keto form. Protonation on oxygen gives an **enol** (*en-* to show that it is an alkene and *-ol* to show it is an alcohol). The resonance-stabilized anion is called an enolate anion. Keto and enol forms have the same molecular formula but different structural formulas; therefore they are structural isomers.

$$CH_3-\overset{\displaystyle :\overset{..}{O}:}{\overset{\|}{C}}-CH_2-H \;\underset{}{\overset{H-B}{\rightleftharpoons}}\; CH_3-\overset{\displaystyle :\overset{..}{O}:}{\overset{\|}{C}}-CH_2^{-} \;\longleftrightarrow\; CH_3-\overset{\displaystyle :\overset{..}{O}:^{-}}{C}=CH_2 \;\overset{H-B}{\longrightarrow}$$

keto form   (an enolate anion)

$$CH_3-\overset{\displaystyle :\overset{..}{O}H}{C}=CH_2$$

enol form

B. Keto-Enol Tautomerism

Under ordinary conditions, all aldehydes and ketones are in equilibrium with the corresponding enol forms. Interconversion of these isomers is catalyzed by acids and bases, but even the surface of ordinary laboratory glassware is acidic enough to catalyze this interconversion. Interconversion of keto and enol forms is an example of **tautomerism**, the rearrangement of a proton and a double bond. Keto-enol interconversion is the most common form of tautomerism.

For most simple aldehydes and ketones, the position of equilibrium in keto-enol tautomerism lies on the side of the keto form simply because a carbon-oxygen double bond is stronger than a carbon-carbon double bond. Thus, for acetaldehyde and acetone, the keto form predominates by more than 99% at equilibrium.

$$CH_3-\overset{\displaystyle O}{\overset{\|}{C}}-H \;\rightleftharpoons\; CH_2=\overset{\displaystyle HO}{\overset{|}{C}}-H \qquad CH_3-\overset{\displaystyle O}{\overset{\|}{C}}-CH_3 \;\rightleftharpoons\; CH_3-\overset{\displaystyle OH}{\overset{|}{C}}=CH_2$$

(> 99%)   (> 99%)

For certain types of molecules, the enol form is the major and in some cases the only form present at equilibrium. For example, in molecules where the α-carbon is flanked by two carbonyl groups, as in 2,4-pentanedione, the position of equilibrium shifts in favor of the enol form. The reason for this shift in position of equilibrium is that the enol is stabilized by hydrogen bonding between O—H and the other carbonyl oxygen.

(20%)   (80%)

Phenols (Section 4.7) may be looked on as highly stable enols. The enol form in this equilibrium is, of course, favored by the gain in resonance stabilization of the aromatic ring.

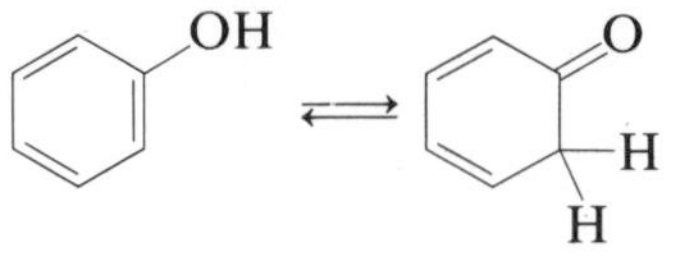

enol form   keto form

Example 6.4

Write two enol forms for these compounds:

a. $CH_3-CH_2-\overset{\overset{\displaystyle O}{\|}}{C}-CH_3$ b.

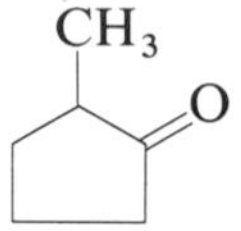

☐ *Solution*

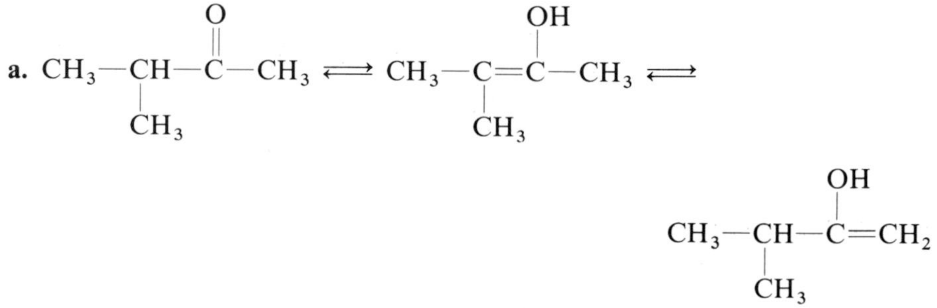

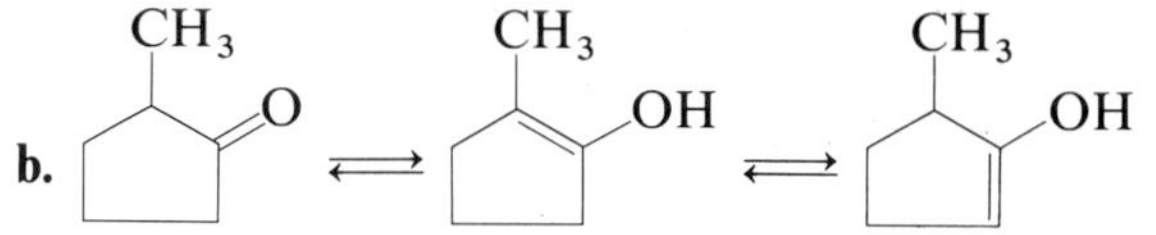

a. $CH_3-\underset{\underset{\displaystyle CH_3}{|}}{CH}-\overset{\overset{\displaystyle O}{\|}}{C}-CH_3 \rightleftharpoons CH_3-\overset{\overset{\displaystyle OH}{|}}{C}=\underset{\underset{\displaystyle CH_3}{|}}{C}-CH_3 \rightleftharpoons$

$CH_3-\underset{\underset{\displaystyle CH_3}{|}}{CH}-\overset{\overset{\displaystyle OH}{|}}{C}=CH_2$

b.

Problem 6.4

Following are enol forms. Draw the structural formula for the corresponding keto form.

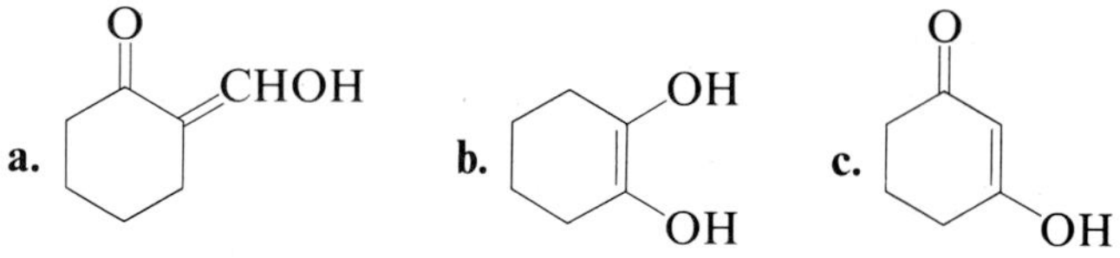

a.   b.   c.

6.8 The Aldol Condensation

Unquestionably the most important reaction of an anion derived from the α-carbon of an aldehyde or ketone is nucleophilic addition to the carbonyl group of another carbonyl-containing compound, as illustrated by the following reactions. While such reactions may be catalyzed by either acid or base, base catalysis is more common.

$$CH_3\overset{\overset{O}{\|}}{-C}-H + CH_3\overset{\overset{O}{\|}}{-C}-H \xrightarrow{NaOH} CH_3\overset{\overset{OH}{\overset{\beta|}{}}}{-CH}\overset{\alpha}{-CH_2}\overset{\overset{O}{\|}}{-C}-H$$

3-hydroxybutanal
(aldol)
(a β-hydroxyaldehyde)

$$CH_3\overset{\overset{O}{\|}}{-C}-CH_3 + CH_3\overset{\overset{O}{\|}}{-C}-CH_3 \xrightarrow{NaOH} CH_3\overset{\overset{OH}{\overset{\beta|}{}}}{\underset{\underset{CH_3}{|}}{-C}}\overset{\alpha}{-CH_2}\overset{\overset{O}{\|}}{-C}-CH_3$$

4-hydroxy-4-methyl-2-pentanone
(a β-hydroxyketone)

These reactions are called **aldol condensations**. The name aldol shows that the product contains both an aldehyde (*ald-*) and an alcohol (*-ol*). The product of the aldol condensation of acetaldehyde itself is called aldol. The name *condensation* tells that these reactions join, or condense, two molecules. The characteristic structural feature of a product of an aldol condensation is the presence of a β-hydroxyaldehyde or a β-hydroxyketone.

Chemists have proposed a three-step mechanism for base-catalyzed aldol condensations, the key step in which is *nucleophilic addition* of an α-carbanion from one carbonyl-containing compound to the carbonyl group of another, to form a **tetrahedral carbonyl addition compound**. This mechanism is illustrated by the aldol condensation between two molecules of acetaldehyde.

Step 1: Formation of anion at an α-carbon

$$H-\overset{..}{\underset{..}{O}}:^- + H-CH_2\overset{\overset{\overset{..}{O}:}{\|}}{-C}-H \longrightarrow$$

$$H_2O + {}^-\overset{..}{C}H_2\overset{\overset{\overset{..}{O}:}{\|}}{-C}-H \longleftrightarrow CH_2=\overset{\overset{:\overset{..}{O}:^-}{|}}{C}-H$$

resonance-stabilized
enolate anion

Step 2: Nucleophilic addition of the anion from one aldehyde or ketone to the carbonyl group of another, to form a tetrahedral carbonyl addition compound

$$CH_3\overset{\overset{:\overset{..}{O}}{\|}}{-C}-H + {}^-\overset{..}{C}H_2\overset{\overset{\overset{..}{O}:}{\|}}{-C}-H \longrightarrow CH_3\overset{\overset{:\overset{..}{O}:^-}{|}}{-CH}-CH_2\overset{\overset{\overset{..}{O}:}{\|}}{-C}-H$$

Step 3: Reaction of the oxygen anion with a proton donor

$$CH_3\overset{\overset{:\overset{..}{O}:^-}{|}}{-CH}-CH_2\overset{\overset{\overset{..}{O}:}{\|}}{-C}-H + H-OH \longrightarrow CH_3\overset{\overset{:\overset{..}{O}H}{|}}{-CH}-CH_2\overset{\overset{\overset{..}{O}:}{\|}}{-C}-H + OH^-$$

The ingredients in the key step of an aldol condensation are an anion and a carbonyl acceptor. In self-condensation, both roles are played by one kind of molecule. Mixed aldol condensations are also possible, like the mixed aldol condensation between acetone and formaldehyde. Formaldehyde cannot function as an anion because it has no α-hydrogen, but can function as a particularly good anion acceptor because of the unhindered nature of its carbonyl group. Acetone forms an anion, but its carbonyl group, bonded to two alkyl groups, is a poorer anion acceptor than the carbonyl group of formaldehyde. Consequently, mixed aldol condensation between acetone and formaldehyde gives 4-hydroxy-2-butanone:

$$CH_3-\overset{\overset{\displaystyle O}{\|}}{C}-CH_3 + H-\overset{\overset{\displaystyle O}{\|}}{C}-H \xrightarrow{\text{NaOH}} CH_3-\overset{\overset{\displaystyle O}{\|}}{C}-CH_2-\overset{\overset{\displaystyle OH}{|}}{CH_2}$$

4-hydroxy-2-butanone

In mixed aldol condensations, where the difference in reactivity between the two carbonyl-containing compounds is not appreciable, mixtures of products result. For example, in the condensation between equimolar concentrations of propanal and butanal, both α-carbons are similar, and so are the carbonyls. Consequently, aldol condensation between these two aldehydes gives a mixture of all four possible aldol condensation products:

$$CH_3CH_2\overset{\overset{\displaystyle O}{\|}}{CH} \quad + \quad CH_3CH_2CH_2\overset{\overset{\displaystyle O}{\|}}{CH}$$

self-condensation   mixed condensation   self-condensation

$$CH_3CH_2\underset{\underset{\displaystyle CH_3}{|}}{\overset{\overset{\displaystyle OH}{|}}{CH}}CH\overset{\overset{\displaystyle O}{\|}}{CH} + CH_3CH_2CH_2\underset{\underset{\displaystyle CH_3}{|}}{\overset{\overset{\displaystyle OH}{|}}{CH}}CH\overset{\overset{\displaystyle O}{\|}}{CH} + CH_3CH_2\underset{\underset{\displaystyle CH_2CH_3}{|}}{\overset{\overset{\displaystyle OH}{|}}{CH}}CH\overset{\overset{\displaystyle O}{\|}}{CH} + CH_3CH_2CH_2\underset{\underset{\displaystyle CH_2CH_3}{|}}{\overset{\overset{\displaystyle OH}{|}}{CH}}CH\overset{\overset{\displaystyle O}{\|}}{CH}$$

β-hydroxyaldehydes and β-hydroxyketones are very easily dehydrated, and often the conditions necessary for aldol condensation (acid or base catalysis) are enough to cause dehydration. Alternatively, warming the aldol product in dilute mineral acid leads to dehydration. The major product from dehydration of an aldol condensation product is one in which the double bond is in conjugation with the carbonyl group; that is, the product is an α,β-unsaturated aldehyde or ketone.

$$CH_3-\overset{\overset{\displaystyle OH}{|}}{CH}-CH_2-\overset{\overset{\displaystyle O}{\|}}{C}-H \xrightarrow{\text{HCl}} CH_3-\overset{\beta}{CH}=\overset{\alpha}{CH}-\overset{\overset{\displaystyle O}{\|}}{C}-H + H_2O$$

3-hydroxybutanal   2-butenal
(aldol)   (crotonaldehyde)

Example 6.5

Name and draw structural formulas for the products of the following aldol condensation products and for the unsaturated compounds produced by dehydration.

a. $CH_3-\overset{\overset{O}{\|}}{C}-CH_3 \xrightarrow{\text{base}}$

b. 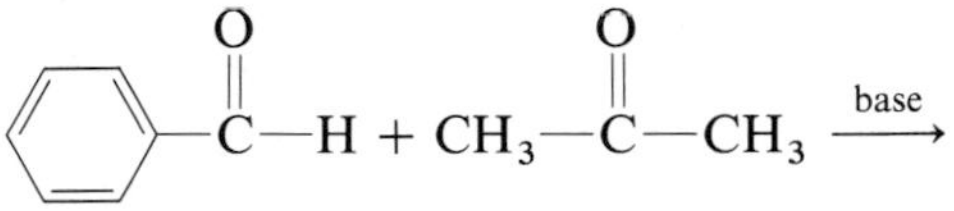 $C_6H_5-\overset{\overset{O}{\|}}{C}-H + CH_3-\overset{\overset{O}{\|}}{C}-CH_3 \xrightarrow{\text{base}}$

Solution

a. $CH_3-\overset{\overset{OH}{|}}{\underset{\overset{|}{CH_3}}{C}}-CH_2-\overset{\overset{O}{\|}}{C}-CH_3 \longrightarrow CH_3-\overset{\underset{\overset{|}{CH_3}}{C}=CH}{-}\overset{\overset{O}{\|}}{C}-CH_3 + H_2O$

4-hydroxy-4-methyl-2-pentanone   4-methyl-3-penten-2-one

b. 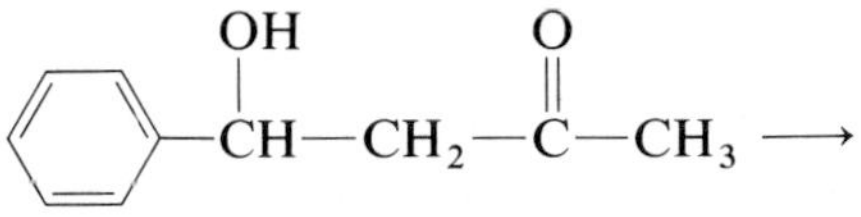 $C_6H_5-\overset{\overset{OH}{|}}{CH}-CH_2-\overset{\overset{O}{\|}}{C}-CH_3 \longrightarrow$

4-hydroxy-4-phenyl-2-butanone

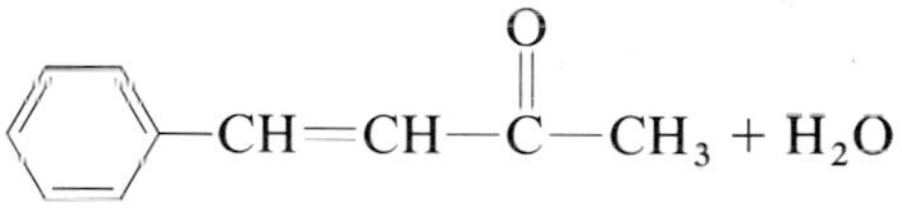

$C_6H_5-CH=CH-\overset{\overset{O}{\|}}{C}-CH_3 + H_2O$

4-phenyl-3-buten-2-one

Problem 6.5

Draw structural formulas for the two carbonyl-containing compounds that on aldol condensation give the following:

a. 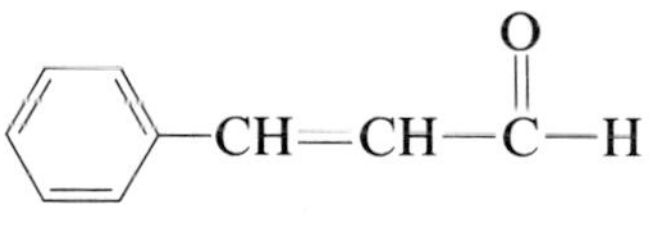 $C_6H_5-CH=CH-\overset{\overset{O}{\|}}{C}-H$

cinnamaldehyde

b. $CH_3(CH_2)_6CH=\overset{\underset{\overset{|}{CH_2(CH_2)_4CH_3}}}{C}\overset{\overset{O}{\|}}{C}H$

The double bonds of alkenes, aldehydes, and ketones can be reduced to single bonds. Hence aldol condensation is often used for preparing saturated alcohols. For example, acetaldehyde is converted to 1-butanol by the following series of steps: aldol condensation, dehydration, and catalytic reduction of both the carbon-carbon and the carbon-oxygen double bonds:

$2CH_3\overset{\overset{O}{\|}}{C}H \xrightarrow{\text{NaOH}} CH_3\overset{\overset{OH}{|}}{C}HCH_2\overset{\overset{O}{\|}}{C}H \xrightarrow{-H_2O}$

3-hydroxybutanal

$CH_3CH=CH\overset{\overset{O}{\|}}{C}H \xrightarrow{2H_2/Ni} CH_3CH_2CH_2CH_2OH$

2-butenal   1-butanol

Alternatively, if the β-hydroxyaldehyde is isolated, the aldehyde may be reduced by hydrogen in the presence of a metal catalyst, LiAlH$_4$, or NaBH$_4$ to give 1,3-butanediol, or the aldehyde may be oxidized, to give 3-hydroxybutanoic acid.

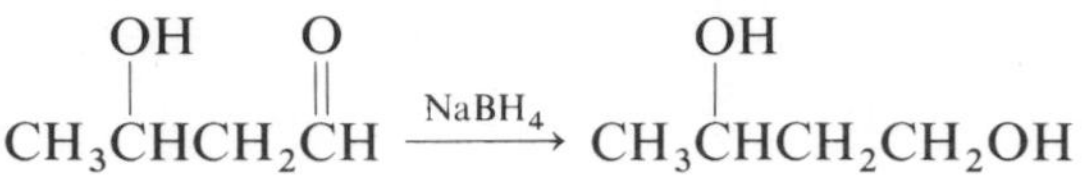

$$
\underset{\substack{|\\ OH}}{CH_3CHCH_2} \overset{\substack{O\\ ||}}{CH} \xrightarrow{NaBH_4} \underset{\substack{|\\ OH}}{CH_3CHCH_2CH_2OH}
$$

1,3-butanediol

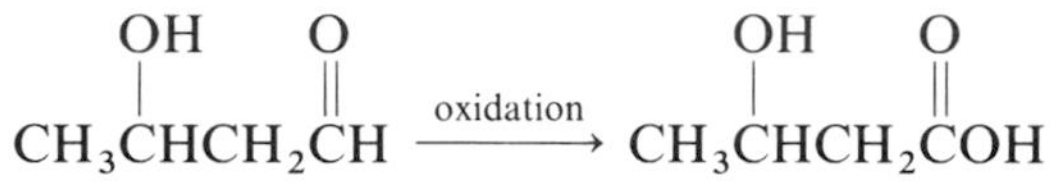

$$
\underset{\substack{|\\ OH}}{CH_3CHCH_2} \overset{\substack{O\\ ||}}{CH} \xrightarrow{oxidation} \underset{\substack{|\\ OH}}{CH_3CHCH_2} \overset{\substack{O\\ ||}}{COH}
$$

3-hydroxybutanoic acid

Example 6.6

Show reagents and conditions to illustrate how the following products can be synthesized from the given starting materials. Use an aldol condensation at some step in each synthesis.

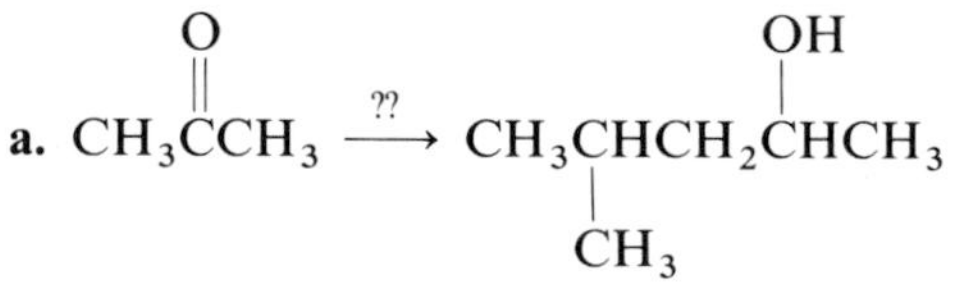

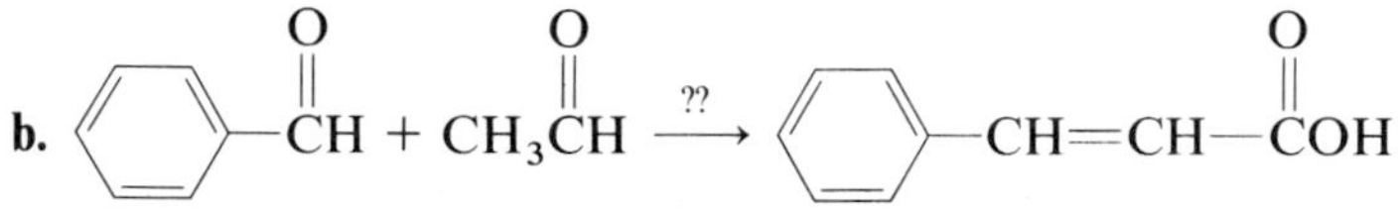

a. $CH_3\overset{\substack{O\\ ||}}{C}CH_3 \xrightarrow{??} CH_3\underset{\substack{|\\ OH}}{CH}CH_2\underset{\substack{|\\ CH_3}}{CH}CH_3$

b. [benzene ring]—$\overset{\substack{O\\||}}{CH}$ + $CH_3\overset{\substack{O\\||}}{CH}$ $\xrightarrow{??}$ [benzene ring]—$CH=CH-\overset{\substack{O\\||}}{COH}$

Solution

a. Aldol condensation of acetone in the presence of NaOH or other strong base gives 4-hydroxy-4-methyl-2-pentanone. Warming this β-hydroxyketone in acid leads to dehydration, to form 4-methyl-3-penten-2-one. Catalytic reduction of this α,β-unsaturated ketone by hydrogen in the presence of a heavy metal catalyst reduces both double bonds and gives the desired product.

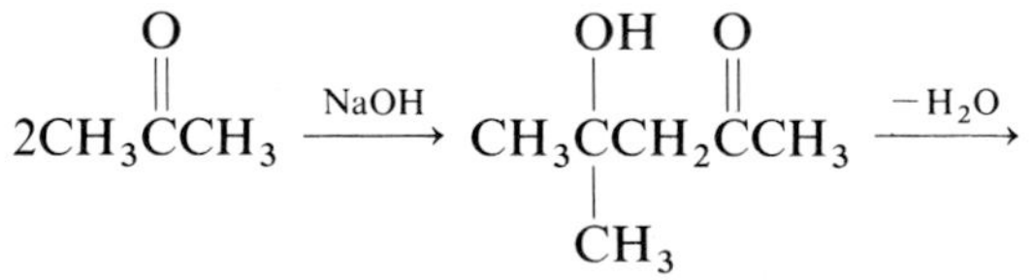

$$
2CH_3\overset{\substack{O\\||}}{C}CH_3 \xrightarrow{NaOH} CH_3\underset{\substack{|\\CH_3}}{\overset{\substack{OH\\|}}{C}}CH_2\overset{\substack{O\\||}}{C}CH_3 \xrightarrow{-H_2O}
$$

acetone

4-hydroxy-4-methyl-2-pentanone

$$
CH_3\underset{\substack{|\\CH_3}}{C}=CH\overset{\substack{O\\||}}{C}CH_3 \xrightarrow{2H_2/Pt} CH_3\underset{\substack{|\\OH}}{CH}CH_2\underset{\substack{|\\CH_3}}{CH}CH_3
$$

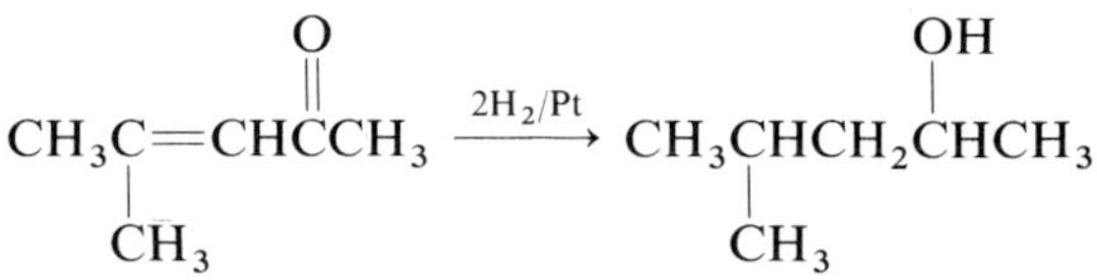

4-methyl-3-penten-2-one   4-methyl-2-pentanol

b. Mixed aldol condensation between benzaldehyde and acetaldehyde gives 3-hydroxy-3-phenylpropanal. Dehydration of this α,β-unsaturated aldehyde gives cinnamaldehyde. Finally, oxidation of the aldehyde with silver nitrate in ammonium hydroxide (Tollens reagent) gives cinnamic acid:

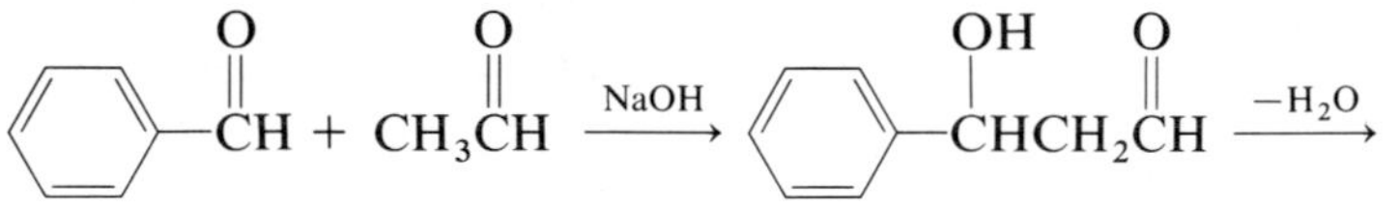

benzaldehyde   acetaldehyde   3-hydroxy-3-phenylpropanal

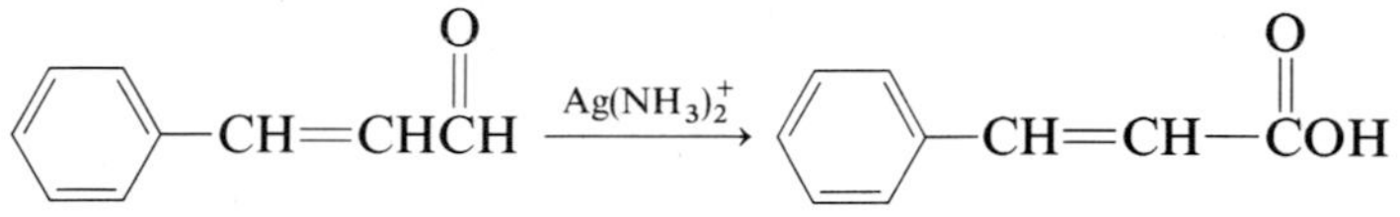

cinnamaldehyde   cinnamic acid

Problem 6.6

Show reagents and conditions to illustrate how the following products can be obtained from the given starting materials. Use an aldol condensation at some stage in each synthesis.

a. $CH_3CH_2\overset{\displaystyle O}{\overset{\|}{C}}H \overset{?}{\longrightarrow} CH_3CH_2CH_2CHCH_2OH$
                                                                                $\underset{\displaystyle CH_3}{|}$

b. $CH_3CH_2CH_2\overset{\displaystyle O}{\overset{\|}{C}}H \longrightarrow CH_3CH_2CH_2CH{=}CCOH$
                                                                                    $\underset{\displaystyle CH_2CH_3}{|}$

Key Terms and Concepts

acetal (6.4B)

aldehyde (6.1A)

aldol condensation (6.8)

carbonyl group (introduction)

enol (6.7A)

hemiacetal (6.4B)

hemiketal (6.4B)

imine (6.4C)

ketal (6.4B)

ketone (6.1A)

nucleophile (6.4)

nucleophilic addition to a carbonyl group (6.4)

oxidation of aldehydes (6.5A)

reduction of aldehydes and ketones (6.6)

Schiff base (6.4C)

tautomerism (6.7B)

tetrahedral carbonyl addition intermediate (6.4)

Tollens test (6.5A)

Key Reactions

1. Addition of water: hydration (Section 6.4A).
2. Addition of alcohols: formation of acetals and ketals (Section 6.4B).
3. Addition of ammonia and amines: formation of Schiff bases (Section 6.4C).
4. Oxidation of aldehydes to carboxylic acids (Section 6.5A).
5. Oxidation and cleavage of ketones to two carboxylic acids (Section 6.5B).
6. Catalytic reduction of aldehydes and ketones to alcohols (Section 6.6A).
7. Metal hydride reduction of aldehydes and ketones to alcohols (Section 6.6B).
8. Keto-enol tautomerism (Section 6.7B).
9. Aldol condensation of aldehydes and ketones (Section 6.8).
10. Dehydration of the products of aldol condensations (Section 6.8).

Problems

Structure and nomenclature of aldehydes and ketones (Sections 6.1 and 6.2)

6.7   Name the following compounds:

a. 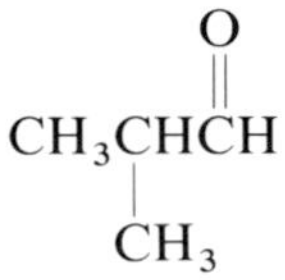
$$CH_3CHCH$$
with =O above the CH and CH_3 below

b. 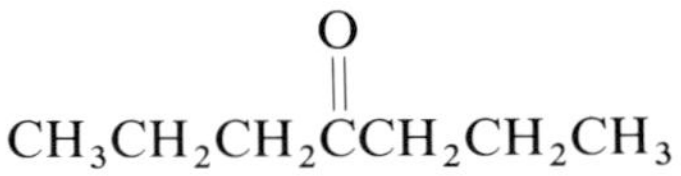
$$CH_3CH_2CH_2CCH_2CH_2CH_3$$
with =O above the C

c.
cyclopentanone with CH_3 substituent

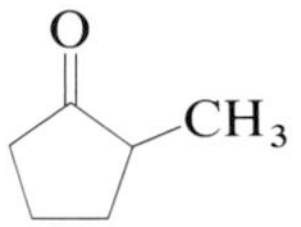

d. 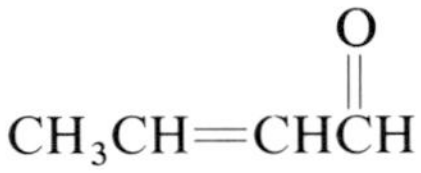
$$CH_3CH=CHCH$$
with =O above the last CH

e. 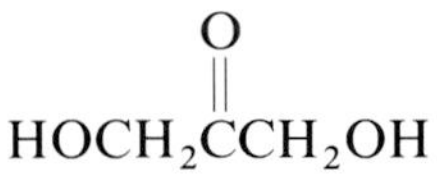
$$HOCH_2CCH_2OH$$
with =O above the central C

f. 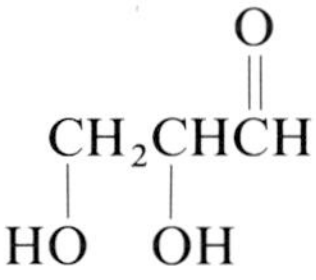
$$CH_2CHCH$$
with =O above the last CH, HO under the first CH$_2$, OH under the middle CH

g. 
$$CH_3OCH_2CH_2CH$$
with =O above the last CH

h. 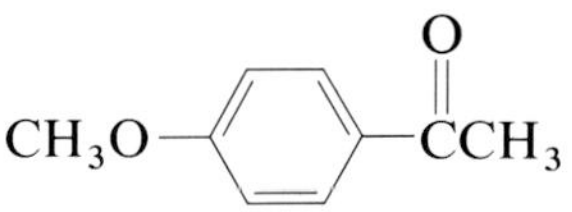
$$CH_3O-\text{(benzene ring)}-CCH_3$$
with =O above the C

i. 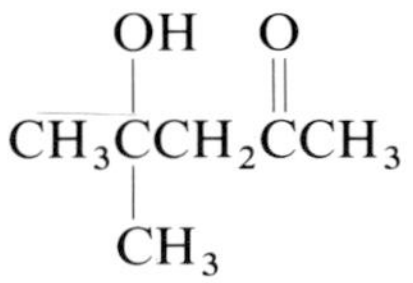
$$CH_3CCH_2CCH_3$$
with OH above the first C, O above the second C, CH_3 below the first C

j. 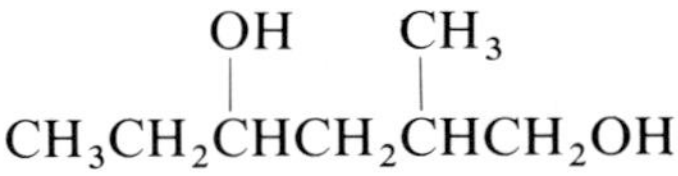
$$CH_3CH_2CHCH_2CHCH_2OH$$
with OH above the third C, CH_3 above the fifth C

k. 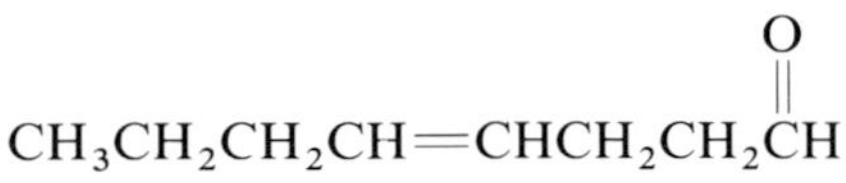
$$CH_3CH_2CH_2CH=CHCH_2CH_2CH$$
with =O above the last CH

l. 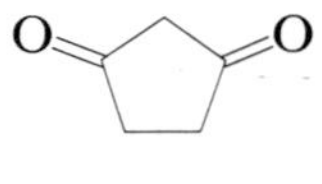
cyclopentane-1,3-dione structure with two O= groups

6.8 Write structural formulas for these compounds:
  a. cycloheptanone   **b.** benzaldehyde
  c. 3,3-dimethyl-2-butanone   **d.** 2,4-pentanedione
  e. hexanal   **f.** 2-decanone
  g. o-hydroxybenzaldehyde
  h. 3-methoxy-4-hydroxybenzaldehyde (vanillin from the vanilla bean)
  i. 3-phenylpropenal (from oil of cinnamon)

6.9 Name and draw structural formulas for all:
  a. aldehydes of formula C_4H_8O **b.** aldehydes of formula $C_5H_{10}O$
  c. ketones of formula C_4H_8O **d.** ketones of formula $C_5H_{10}O$

Reactions of
aldehydes
and ketones
(Sections
6.4–6.6)

6.10 Complete the following reactions. Where you predict no reaction, write NR.

a. $\overset{\text{O}}{\underset{\parallel}{CH_3CH_2CH_2CH}} + H_2 \overset{Pt}{\longrightarrow}$

b. $\underset{\underset{HO\quad OH}{\mid\quad\mid}}{CH_2CHCH} \overset{\text{O}}{\overset{\parallel}{}} + Ag^+ \overset{NH_4OH}{\longrightarrow}$

c. $\underset{\underset{OH}{\mid}}{CH_3CHCH_2}\overset{\text{O}}{\overset{\parallel}{CCH_3}} + K_2Cr_2O_7 \overset{H^+}{\longrightarrow}$

d. $\underset{\underset{OH}{\mid}}{CH_3CHCH_2}\overset{\text{O}}{\overset{\parallel}{CCH_3}} + H_2 \overset{Pt}{\longrightarrow}$

e. 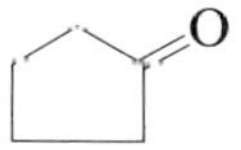 [cyclopentanone-type ketone] $+ Ag^+ \overset{NH_4OH}{\longrightarrow}$

f. 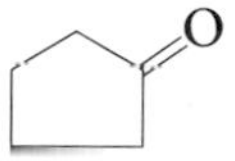 [cyclopentanone] $+ H_2 \overset{Pt}{\longrightarrow}$

g. $\overset{\text{O}}{\overset{\parallel}{CH_3CH_2CH_2CH}} + 1CH_3CH_2OH \longrightarrow$

h. $\overset{\text{O}}{\overset{\parallel}{CH_3CH_2CH_2CH}} + 2CH_3CH_2OH \overset{H^+}{\longrightarrow}$

i. 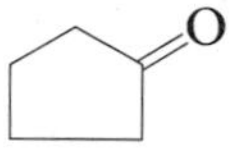 [cyclopentanone] $+ 1CH_3CH_2OH \longrightarrow$

j. 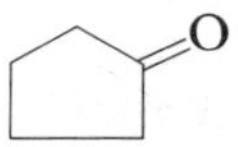 [cyclopentanone] $+ 2CH_3CH_2OH \overset{H^+}{\longrightarrow}$

k. $CH_3\overset{\overset{\displaystyle OCH_3}{|}}{\underset{\underset{\displaystyle OCH_3}{|}}{C}}CH_3 + H_2O \xrightarrow{H^+}$

l. [cyclic ketal structure] $+ H_2O \xrightarrow{H^+}$

m. [cyclopentane with OCH_3 and OCH_3] $+ H_2O \xrightarrow{H^+}$

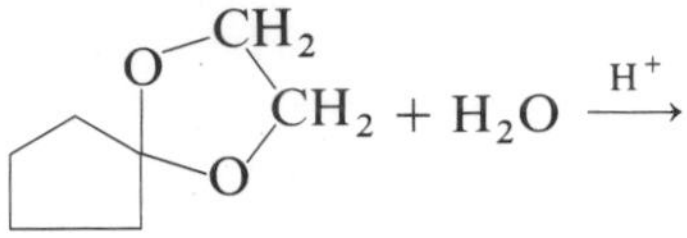

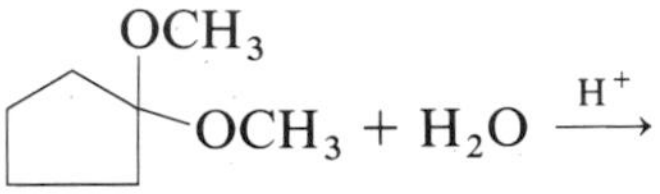

n. 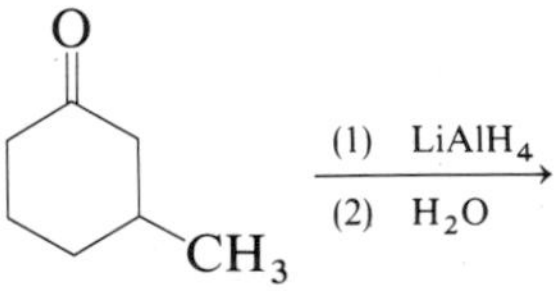 [cyclohexanone with CH₃ substituent] $\xrightarrow[\text{(2) } H_2O]{\text{(1) LiAlH}_4}$

6.11 Show how you could distinguish between the following pairs of compounds by a simple chemical test. In each case, tell what test you would perform and what you would expect to observe, and write an equation for each positive test.

a. benzaldehyde and acetophenone

b. benzaldehyde and benzyl alcohol

c.
$$\begin{array}{cc}
CH_2OH & CHO \\
| & | \\
CHOH \quad\text{and}\quad & CHOH \\
| & | \\
CH_2OH & CH_2OH \\
\text{glycerol} & \text{glyceraldehyde}
\end{array}$$

6.12 The compound 4-hydroxypentanal forms a 5-member cyclic hemiacetal (Section 6.4B) that reacts with a molecule of methanol to form a 5-member cyclic acetal. Draw structural formulas for the 5-member cyclic hemiacetal and the 5-member cyclic acetal.

6.13 5-hydroxyhexanal readily forms a six-member hemiacetal:

$$CH_3\underset{\underset{\displaystyle OH}{|}}{C}HCH_2CH_2CH_2\overset{\overset{\displaystyle O}{\|}}{C}H \longrightarrow \text{a cyclic hemiacetal}$$

5-hydroxyhexanal

a. Draw a structural formula for this cyclic hemiacetal.

b. How many cis-trans isomers are possible for this cyclic hemiacetal?

c. Draw planar hexagon representations for each cis and trans isomer.

d. Draw a chair conformation for the cyclic hemiacetal in which the hydroxyl and methyl groups are equatorial. Is this a cis or a trans isomer?

6.14 Acetaldehyde reacts with ethylene glycol in the presence of a trace of sulfuric acid, to give a cyclic acetal of formula $C_4H_8O_2$. Draw a structural formula for this acetal.

6.15 The following conversions can be carried out in either one step or two. Show reagents you would use to bring about each conversion, and draw structural formulas for the intermediate formed in any conversion that requires two steps.

a.
$$\underset{\underset{\text{OH}}{|}}{CH_3CHCO_2H} \longrightarrow \underset{\underset{\text{O}}{||}}{CH_3CCO_2H}$$

b.

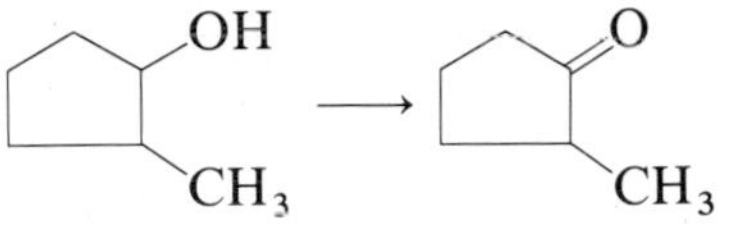

c.

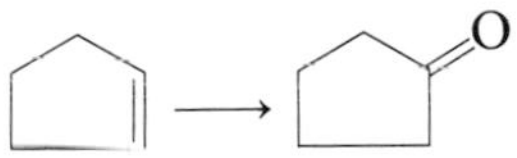

d.
$$\underset{\underset{\underset{\text{CH}_3}{|}}{\underset{\text{OH}}{|}}}{CH_3CHCHCH_2CH} \longrightarrow \underset{\underset{\underset{\text{CH}_3}{|}}{O\quad\quad O}}{CH_3CCHCH_2COH}$$

e.

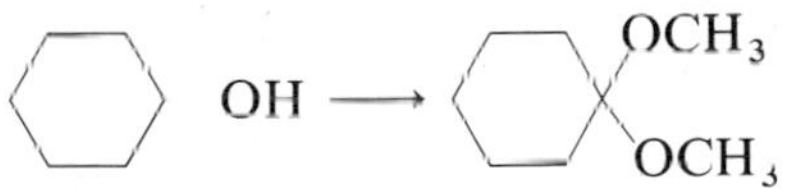

f.

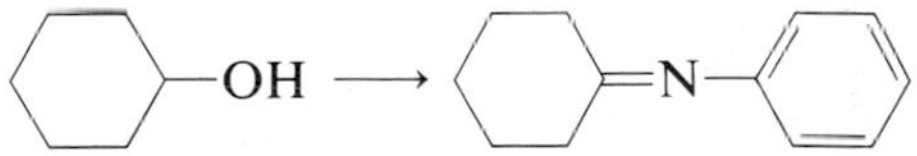

g.
$$CH_3CH{=}CH_2 \longrightarrow \underset{\underset{\text{O}}{||}}{CH_3CCH_3}$$

6.16 Pyridoxal phosphate is one of the metabolically active forms of vitamin B_6. Draw structural formulas for the Schiff bases formed by reaction of pyridoxal phosphate with the primary amines of (a) tyrosine and (b) glutamic acid.

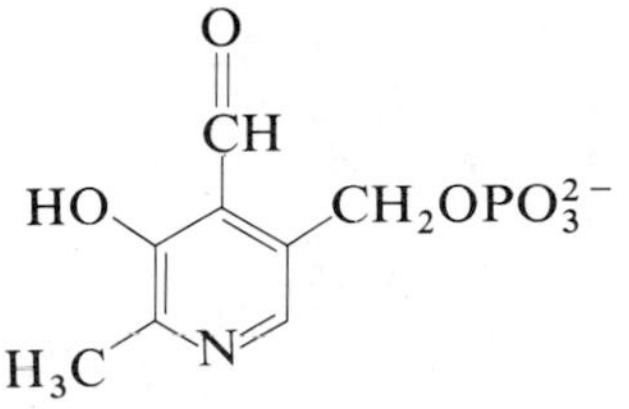

pyridoxal phosphate

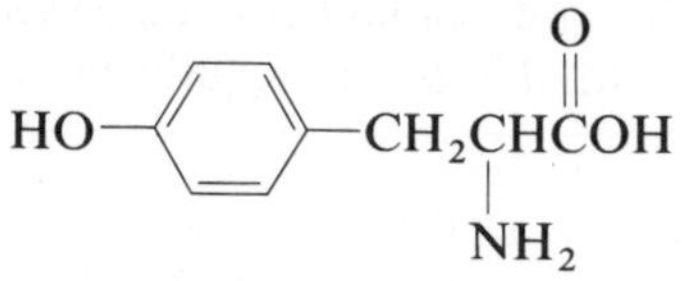

$$\text{HO}-\!\!\left\langle\;\right\rangle\!\!-\text{CH}_2\text{CHCOH} \qquad \text{HOCCH}_2\text{CH}_2\text{CHCOH}$$

tyrosine 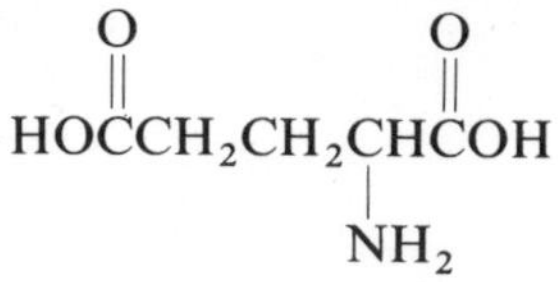   glutamic acid

6.17 Another of the metabolically active forms of vitamin B_6 is pyridoxamine phosphate. Draw structural formulas for the Schiff bases formed by re-action of pyridoxamine phosphate with the ketones of (a) pyruvic acid and (b) oxaloacetic acid.

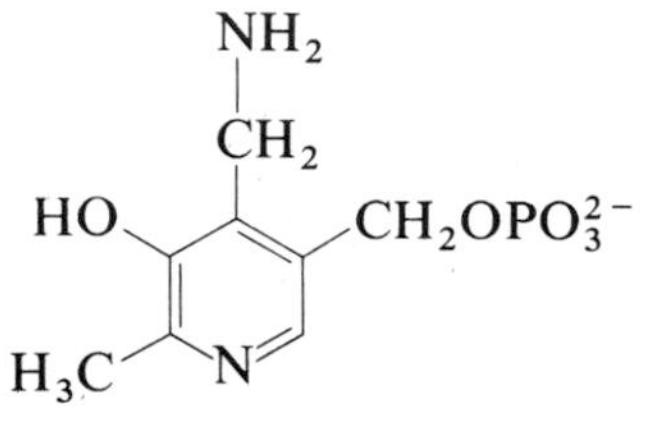

pyridoxamine   pyruvic   oxaloacetic
phosphate   acid   acid

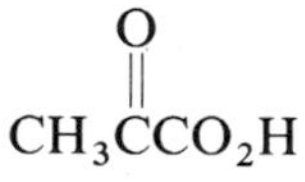

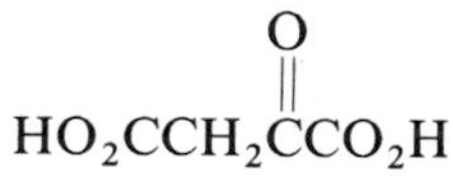

$$\text{CH}_3\text{CCO}_2\text{H} \qquad \text{HO}_2\text{CCH}_2\text{CCO}_2\text{H}$$

Reactions of aldehydes and ketones—tautomerism (Section 6.7)

6.18 What is meant by the term *keto-enol tautomerism*?

6.19 Draw the indicated number of enol structures for the following aldehydes and ketones. Draw them to show bond angles of approximately 120° about the carbon-carbon double bond of each enol form.

a. 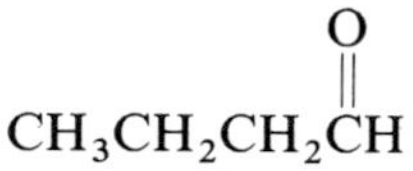 $\text{CH}_3\text{CH}_2\text{CH}_2\text{CH}$ **b.** $\text{CH}_3\text{CH}_2\text{CCH}_3$ **c.**

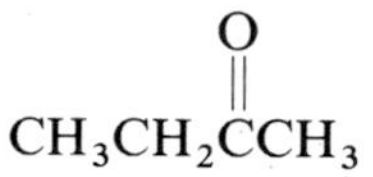

(1 enol form)   (2 enol forms)   (2 enol forms)

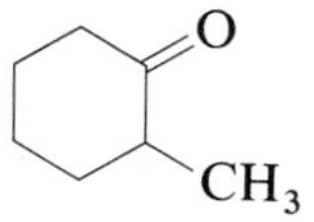

6.20 The following are enols. Draw structural formulas for the keto form of each.

a. 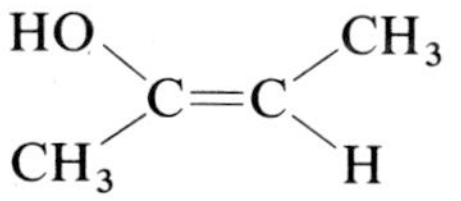   **b.** 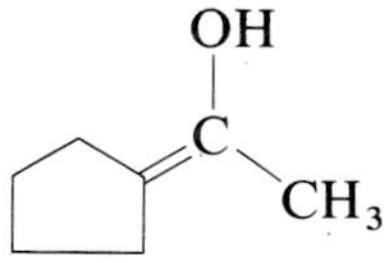   **c.** 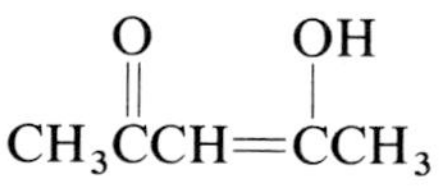 $\text{CH}_3\text{CCH}\!=\!\text{CCH}_3$

6.21 The following compound is an enediol, a compound with two hydroxyl groups on a carbon-carbon double bond. Draw structural formulas for the two carbonyl-containing compounds with which the enediol is in equilibrium.

$$\text{a hydroxyketone} \;\rightleftharpoons\; \text{CH}_3-\overset{\text{HO}}{\underset{}{\text{C}}}=\overset{\text{OH}}{\underset{}{\text{C}}}-\text{H} \;\rightleftharpoons\; \text{a hydroxyaldehyde}$$

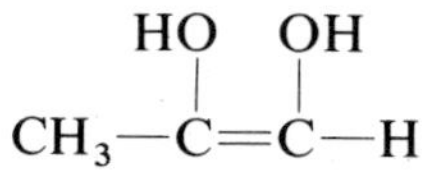

an enediol

6.22   How could you account for the conversion of glyceraldehyde, in dilute aqueous NaOH, to an equilibrium mixture of glyceraldehyde and dihydroxyacetone?

$$
\begin{array}{ccc}
\text{CHO} & \text{CHO} & \text{CH}_2\text{OH}\\
| & | & |\\
\text{CHOH} & \overset{\text{dilute base}}{\rightleftharpoons} \quad \text{CHOH} \quad + \quad \text{C}=\text{O}\\
| & | & |\\
\text{CH}_2\text{OH} & \text{CH}_2\text{OH} & \text{CH}_2\text{OH}\\
\text{glyceraldehyde} & \text{glyceraldehyde} & \text{dihydroxyacetone}
\end{array}
$$

The aldol condensation (Section 6.8)

6.23   Draw structural formulas for the products of the following aldol condensations and for the unsaturated compounds formed by loss of water from the aldol condensation product:

a. $2\text{CH}_3\text{CH}_2\overset{\overset{\text{O}}{\|}}{\text{CH}} \overset{\text{base}}{\longrightarrow}$

b. $2 \; \text{C}_6\text{H}_5{-}\overset{\overset{\text{O}}{\|}}{\text{C}}\text{CH}_2\text{CH}_3 \overset{\text{base}}{\longrightarrow}$

c. 2 (cyclohexanone) $\overset{\text{base}}{\longrightarrow}$

6.24   Draw structural formulas for the products of these mixed aldol condensations and for the unsaturated compounds formed by loss of water from the mixed aldol condensation product:

a. $\text{C}_6\text{H}_5{-}\overset{\overset{\text{O}}{\|}}{\text{CH}} + \text{CH}_3\overset{\overset{\text{O}}{\|}}{\text{C}}{-}\text{C}_6\text{H}_5 \overset{\text{base}}{\longrightarrow}$

b. (cyclohexanone) $+ \text{H}\overset{\overset{\text{O}}{\|}}{\text{CH}} \overset{\text{base}}{\longrightarrow}$

6.25   Show reagents and conditions to illustrate how the following products can be synthesized from the indicated starting materials by way of an aldol condensation:

a. $\text{CH}_3\overset{\overset{\text{O}}{\|}}{\text{CH}} \overset{??}{\longrightarrow} \text{CH}_3\text{CH}_2\text{CH}_2\text{CH}_2\text{OH}$

b. $\text{CH}_3\overset{\overset{\text{O}}{\|}}{\text{CH}} \overset{??}{\longrightarrow} \text{CH}_3\text{CH}{=}\text{CH}\overset{\overset{\text{O}}{\|}}{\text{C}}\text{OH}$

c. $\text{CH}_3\overset{\overset{\text{O}}{\|}}{\text{C}}\text{CH}_3 \overset{??}{\longrightarrow} \text{CH}_3\overset{\overset{\text{OH}}{|}}{\underset{\underset{\text{CH}_3}{|}}{\text{C}}}\text{CH}_2\overset{\overset{\text{OH}}{|}}{\text{CH}}\text{CH}_3$

Pheromones

Chemical communication abounds in nature: the clinging, penetrating odor of the skunk's defensive spray; the hound, nose to the ground, in pursuit of prey; the female dog's making known her sexual availability; the female moth's attracting males from great distances for mating. As biologists and chemists cooperate to extend our knowledge of other animals, it is becoming increasingly clear that chemical communication is the primary mode of communication in most animals.

Before 1950, the isolation of enough biologically active material to permit us to decipher any chemical communications seemed an insurmountable task. Rapid progress in instrumental techniques, however, particularly in chromatography and spectroscopy, has now made it possible to isolate and carry out structural determinations on as little as a few micrograms of material. Even with these advances, isolating and identifying the components of pheromones remains a challenge to technical and experimental expertise. For example, obtaining a mere 12 milligrams of gypsy moth sex attractant required processing 500,000 virgin female moths, each yielding only 0.02 microgram of attractant. In other insect species, it is not uncommon to process at least 20,000 insects to get enough material for chemical identification.

The term *pheromone* (from the Greek *pherin*, to carry, and *horman*, to excite) is the accepted name for chemicals secreted by an organism of one species to evoke a response in another member of the same species. We will look at insect pheromones, because these have been the most widely studied. Pheromones are generally divided into two classes: primer and releaser pheromones, depending on mode of action. Primer pheromones cause important physiological changes that affect an organism's development and later behavior. The most clearly understood primer pheromones regulate caste systems in social insects (bees, ants, and termites). A typical colony of honey bees (*Apis mellifera*) consists of one queen, several hundred drones (males), and thousands of workers (underdeveloped females). The queen bee is the only fully developed female in the colony. She secretes a "queen substance," which prevents the development of workers' ovaries and promotes the construction of royal colony cells for the rearing of new queens. One of the components of the primer pheromone in the queen substance has been identified as 9-keto-*trans*-2-decenoic acid (Figure V-1). This same substance also serves as a sex pheromone, attracting drones to the queen during her mating flight.

Releaser pheromones produce rapid, reversible changes in behavior, such as sexual attraction and stimulation, aggregation, trail marking, territorial and home-range marking, and other social behaviors. Some of the earliest observations of releaser pheromones were recorded in the alarm pheromones of the honey bee. Beekeepers, and perhaps some of the rest of us too, are well aware that the sting of one bee often causes swarms of angry workers to attack the same spot. When a worker stings an intruder, it discharges, along with venom, an alarm pheromone, which evokes the aggressive attack of other bees. One component of this alarm pheromone is isoamyl acetate, a sweet-smelling substance with an odor like that of banana oil (Figure V-2).

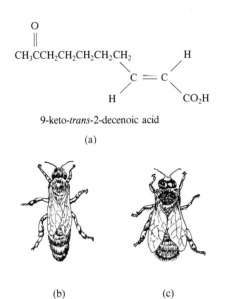

O
‖
CH₃CCH₂CH₂CH₂CH₂CH₂

$$CH_3CCH_2CH_2CH_2CH_2CH_2$$

9-keto-*trans*-2-decenoic acid

(a)

(b)   (c)

Figure V-1 (a) 9-keto-*trans*-2-decenoic acid, a component of the queen substance. (b) *Apis mellifera*, a queen bee, and (c) a drone.

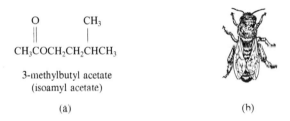

O   CH₃
‖ |
CH₃COCH₂CH₂CHCH₃

3-methylbutyl acetate
(isoamyl acetate)

(a)   (b)

Figure V-2 (a) 3-methylbutyl acetate (isoamyl acetate), a component of the alarm pheromone of the honey bee. (b) A worker (female).

Of all classes of pheromones, sex and aggregating pheromones have received the greatest attention in both the scientific community and the popular press. Larvae of certain insects that release these classes of pheromones, particularly the moths and beetles, are among the world's most serious agricultural and forestland pests.

Sex pheromones are commonly referred to as sex attractants, but this term is misleading because it implies only attraction. Actually, the behavior elicited by sex pheromones is much more complex. Low levels of sex pheromone sti-

mulation cause orientation and flight of a male toward a female (or in some species, flight of a female toward a male). If the level of stimulation is high enough, copulation follows.

One of the aggregating pheromones recently identified and studied is that of the *Ips* genus of bark beetles. *Ips paraconfusus*, an insect especially destructive to ponderosa pines in the Sierra Nevada range of California, lives in the soil during the winter. In the early spring when the temperature begins to rise, a few males emerge from the ground and seek out ponderosa pines in which to construct breeding chambers. The few males bore into trees, and during this process, a pheromone produced in their hind gut is emitted, triggering a massive secondary invasion of both males and females. As these bark beetles bore into ponderosa pine, they infect the trees with fungal spores, and it is the fungal-spore infection that actually kills the trees. After fertilization and hatching, *Ips* larvae grow and develop behind the bark. In autumn, they leave the tree and return to the soil to begin another life cycle.

Investigation of the aggregating pheromone of *Ips paraconfusus* led to isolation and identification of three components (Figure V-3), all of which are terpene alcohols. Ipsenol and ipsdienol have carbon skeletons identical to that of myrcene (see Mini-Essay II, "Terpenes") and differ only in the presence of a carbon-carbon double bond. The third component, *cis*-verbenol, has a carbon skeleton identical to that of alpha-pinene (see Mini-Essay II, "Terpenes"). It has been shown more recently that *Ips paraconfusus*

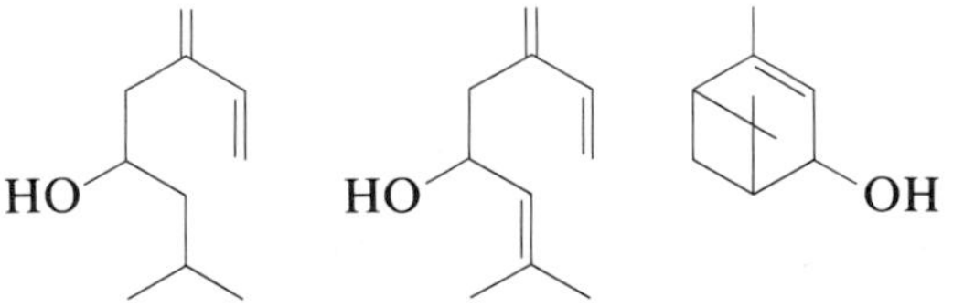

HO   HO   OH

ipsenol   ipsdienol   *cis*-verbenol

Figure V-3 Components of the aggregating pheromone of the bark beetle *Ips paraconfusus*. A mixture of all three is necessary for attraction of males and females in the field.

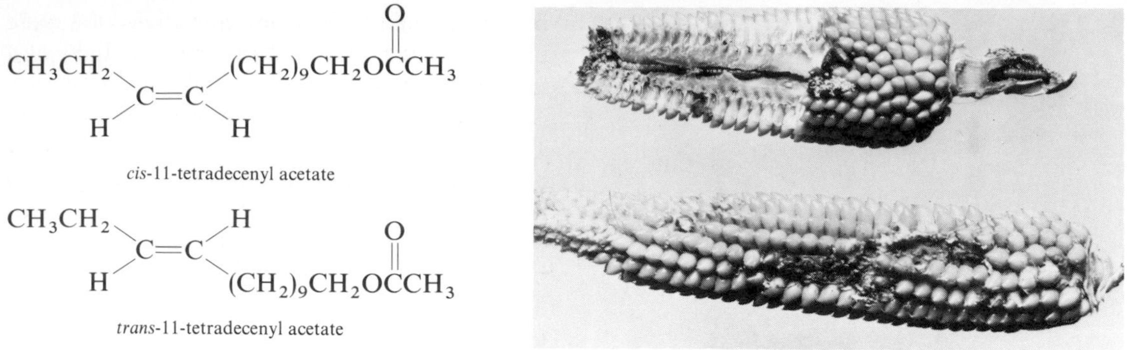

Figure V-4 (a) 11-tetradecenyl acetate, a component of the sex pheromone of the European corn borer. (b) Corn infested with European corn borers.

synthesizes verbenol from the alpha-pinene it encounters in the thick resin that flows from an injured ponderosa pine.

Several groups of scientists have studied the components of the sex pheromone of both Iowa and New York strains of the European corn borer. Females of these closely related species secrete the same sex attractant, 11-tetradecenyl acetate (Figure V-4). Males of the Iowa strain show maximum response to a mixture containing about 96% of the cis isomer and 4% of the trans isomer. When the pure cis isomer is used alone, males are only weakly attracted. Males of the New York strain show an entirely different response pattern. They respond maximally to a mixture containing 3% of the cis isomer and 97% of the trans isomer. There is evidence that optimum response to a narrow range of stereoisomers as we see here, or to a mixture of components as in *Ips paraconfusus*, is widespread in nature; also, at least some species of insects maintain species isolation, at least for mating and reproduction, by the stereochemistry of their pheromones.

Within the last decade, scientists have developed several practical applications of pheromone systems to monitor and control selected insect pests. First, pheromone-baited traps can be placed in the field to monitor populations of selected insect pests. In this way, changes in population levels can be determined and large or potentially large areas of infestation defined. The great value of this information is that large-scale spraying of conventional insecticides can be drastically reduced or even avoided in areas where populations are below threshold levels. Several companies here and abroad are marketing pheromone-baited traps to monitor population levels of such insect pests as the Japanese beetle, the gypsy moth, the boll weevil, and the Mediterranean fruit fly.

Second, pheromones can be used for mass-trapping and population suppression of particular insect pests. Probably the largest single effort to date involving trapping and population suppression was undertaken in Norway and Sweden, to prevent a potentially catastrophic infestation of *Ips typographus*, a bark beetle largely confined to the coniferous forests of Europe and Asia and particularly attracted to the commercially valuable Norway spruce. The aggregation pheromone of *Ips typographus* consists of three components (Figure V-5).

In the three years before 1979, severe drought in Norway and Sweden affected a huge number of spruce trees, and in that year bark beetles killed or severely damaged an estimated 5 million trees. The governments of Norway and Sweden initiated a large-scale program to control *Ips typographus*; in the summer of 1979 they placed about a million pheromone-baited traps in infested forests. The beetle catch that year was

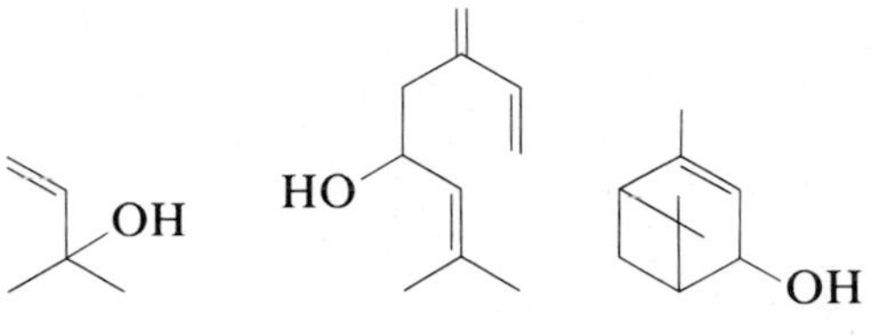

2-methyl-3-
buten-2-ol   ipsdienol   verbenol

Figure V-5 The three components of the aggregating pheromone of the bark beetle *Ips typographus*.

estimated at 2.9 million. The program was repeated in 1980, with an estimated catch in Norway alone of 4.5 billion beetles. Although tree mortality during these years remained high, the feared catastrophic infestation of *Ips typographus* did not occur.

In a third use of pheromones in the field, insects can be lured to traps and treated there with insecticides, insect juvenile hormones or juvenile hormone analogs, or species-specific pathogenic organisms, all of which are then spread throughout the local population of that particular pest when the treated insects are released.

In a fourth use, specific pheromones can be spread throughout the air to disrupt mating or aggregation. Clearly, this means of population control and suppression requires an understanding of the growth and behavior patterns of the particular pest and also good timing.

From this information on pheromones, it should be clear that they are becoming an integral part of more environmentally sound means of insect pest control.

Sources

Klun, J. A., et al. 1973. Insect sex pheromones: Minor amounts of opposite geometrical isomer critical to attraction. *Science* 181:661.

O'Sullivan, D.A. 1979. Pheromone lures help control bark beetles. *Chem. & Eng. News.*

Shorey, H. H. 1976. *Animal Communication by Pheromones.* New York: Academic Press.

Silverstein, R. M. 1981. Pheromones: Background and potential use in pest control. *Science* 213.1326.

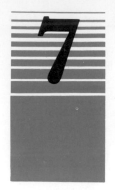

7 Carboxylic Acids

The most important chemical property of carboxylic acids, another class of organic compounds containing the carbonyl group, is their acidity. Further, carboxylic acids form numerous important derivatives, including esters, amides, and anhydrides. In this chapter we will concentrate on carboxylic acids themselves and, in the following chapter, on derivatives of carboxylic acids.

7.1 Structure of Carboxylic Acids

The characteristic structural feature of a carboxylic acid is the presence of a **carboxyl group** (Section 1.4C). Shown in Figure 7.1 is a Lewis structure for formic acid, the simplest organic compound containing a carboxyl group. Also shown is a ball-and-stick model. Predicted bond angles of the carboxyl group are 120° about the carbonyl carbon and 109.5° about the hydroxyl oxygen. Observed bond angles in formic acid are close to the predicted angles.

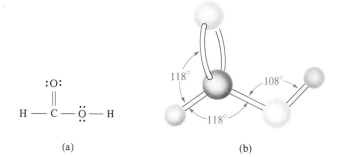

(a) (b)

Figure 7.1 The structure of formic acid. (a) Lewis structure; (b) ball-and-stick model, showing observed bond angles.

7.2 Nomenclature of Carboxylic Acids

A. IUPAC Names

The IUPAC system of nomenclature for carboxylic acids selects as the parent compound the longest chain of carbon atoms that contains the carboxyl group.

The carboxyl group is indicated by changing the suffix -*e* of the parent compound to -*oic acid* (Section 1.4). Because the carbon of the carboxyl group is always carbon 1 of the parent compound, there is no need to give it a number. Following are structural formulas and IUPAC names for several carboxylic acids. Note that the IUPAC system retains the common names formic acid and acetic acid.

$$
\overset{O}{\overset{\|}{HCOH}} \qquad \overset{O}{\overset{\|}{CH_3COH}} \qquad \overset{O}{\overset{\|}{CH_3CH_2COH}} \qquad \overset{CH_3 \quad O}{\overset{|\qquad \|}{CH_3CHCH_2COH}}
$$

methanoic acid (formic acid)   ethanoic acid (acetic acid)   propanoic acid   3-methylbutanoic acid

If the carboxylic acid contains a carbon-carbon double bond, the infix is changed from -*an*- to -*en*- to indicate a double bond. The position of the double bond is shown by a number, just as for simple alkenes. Following are structural formulas for two carboxylic acids, each also containing carbon-carbon double bonds. In parentheses is the common name of each acid.

$$
\overset{O}{\overset{\|}{CH_2=CH-COH}} \qquad\qquad \overset{O}{\overset{\|}{\langle\text{benzene}\rangle-CH=CH-COH}}
$$

2-propenoic acid (acrylic acid)   3-phenyl-2-propenoic acid (cinnamic acid)

In the IUPAC system, a carboxyl group takes precedence over most other functional groups, including the hydroxyl group and also the carbonyl group of aldehydes and ketones (Table 6.1). In a substituted carboxylic acid, the presence of an — OH group is indicated by the prefix *hydroxy-*; the presence of a carbonyl group of an aldehyde or ketone is indicated by the prefix *oxo-*, as illustrated in the following examples.

$$
\overset{OH}{\overset{|}{CH_3CHCH_2CH_2CH_2COH}} \overset{O}{\overset{\|}{}} \qquad\qquad \overset{O}{\overset{\|}{CH_3CCH_2CH_2CH_2COH}} \overset{O}{\overset{\|}{}}
$$

5-hydroxyhexanoic acid   5-oxohexanoic acid

Dicarboxylic acids are named by adding the suffix -*dioic acid* to the name of the parent compound that contains both carboxyl groups. The IUPAC system retains certain common names including oxalic, malonic, succinic, and tartaric acids. In Table 7.1 are structural formulas, IUPAC names, and common names for the most common dicarboxylic acids found in the biological world. The name oxalic acid is derived from one of its sources in the biological world, namely plants of the genus *Oxalis*, one of which is rhubarb. Oxalic acid is used as a cleansing agent for automobile radiators, as a laundry bleach, and in textile finishing and cleaning. Tartaric acid is a byproduct of fermentation of grape juice to wine. It is collected as the potassium salt and sold under the name cream of tartar. Adipic acid is one of the two monomers required for the synthesis of

Table 7.1
Dicarboxylic acids.

Structural Formula	IUPAC Name	Common Name
$\overset{O}{\overset{\|}{HOC}}-\overset{O}{\overset{\|}{COH}}$	ethanedioic acid	oxalic acid
$\overset{O}{\overset{\|}{HOCCH_2}}\overset{O}{\overset{\|}{COH}}$	propanedioic acid	malonic acid
$\overset{O}{\overset{\|}{HOCCH_2}}CH_2\overset{O}{\overset{\|}{COH}}$	butanedioic acid	succinic acid
$\overset{O}{\overset{\|}{HOC}}-\underset{HO}{\overset{}{CH}}-\underset{OH}{\overset{}{CH}}-\overset{O}{\overset{\|}{COH}}$	2,3-dihydroxy-butanedioic acid	tartaric acid
$\overset{O}{\overset{\|}{HOCCH_2}}CH_2CH_2\overset{O}{\overset{\|}{COH}}$	pentanedioic acid	glutaric acid
$\overset{O}{\overset{\|}{HOCCH_2}}CH_2CH_2CH_2\overset{O}{\overset{\|}{COH}}$	hexanedioic acid	adipic acid

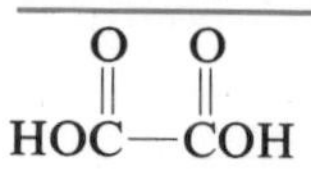

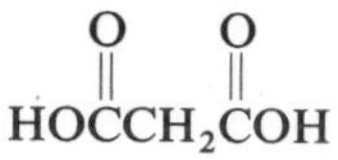

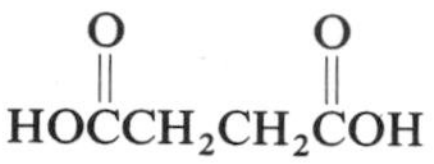

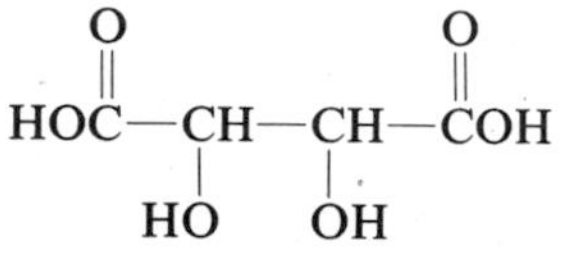

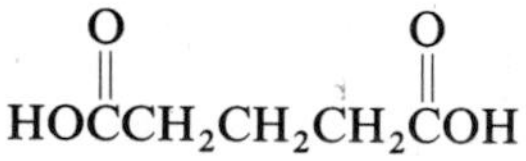

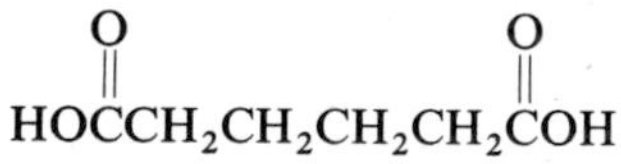

the polymer Nylon 66. In 1985, the U.S. chemical industry produced 1.63 billion pounds of adipic acid, solely for the synthesis of Nylon 66.

Tri- and higher carboxylic acids are named by using the suffixes *-tricarboxylic acid*, *-tetracarboxylic acid*, and so on. An example of a tricarboxylic acid is 2-hydroxy-1,2,3-propanetricarboxylic acid, whose common name, citric acid, is also retained by the IUPAC system. Citric acid is important in a metabolic pathway known as the tricarboxylic acid (TCA) cycle, the citric acid cycle, or the Krebs cycle (Section 17.5).

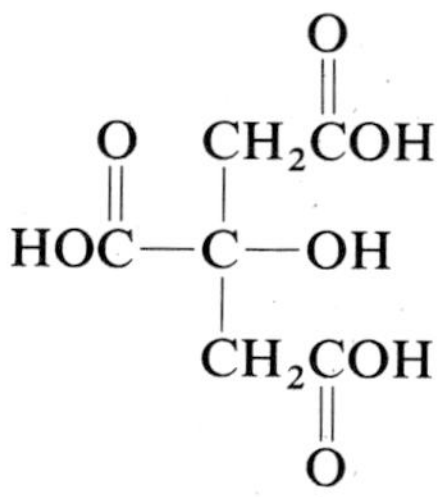

$$\underset{\substack{| \\ O}}{\overset{\substack{CH_2\overset{O}{\overset{\|}{COH}} \\ |}}{HOC-C-OH}}\quad\text{with }CH_2\overset{O}{\overset{\|}{COH}}$$

2-hydroxy-1,2,3-propane-
tricarboxylic acid
(citric acid)

The simplest aromatic carboxylic acid is benzoic acid. Derivatives are named by using numbers and prefixes to show the presence and location of substituents. Certain aromatic carboxylic acids have common names by which they are more usually known. For example, 2-hydroxybenzoic acid is more often called salicylic acid, a name derived from the fact that this carboxylic acid was first isolated from the bark of the willow, a tree of the genus *Salix*.

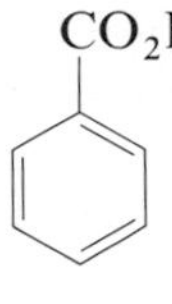

CO_2H

benzoic
acid

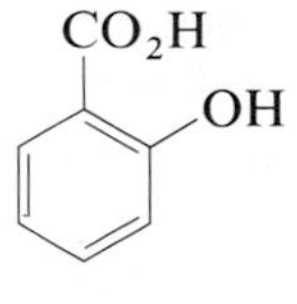

CO_2H
OH

2-hydroxybenzoic
acid
(salicylic acid)

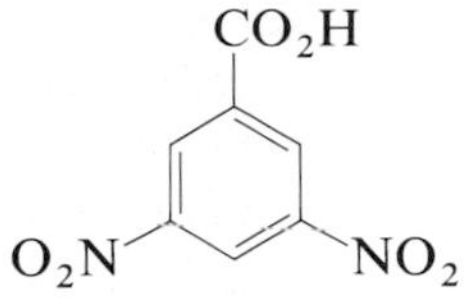

CO_2H

O_2N NO_2

3,5-dinitrobenzoic
acid

Aromatic dicarboxylic acids are named using the suffix -*dicarboxylic acid*. Following are structural formulas for 1,2-benzenedicarboxylic acid and 1,4-benzenedicarboxylic acid. Each of these carboxylic acids has a common name by which it is more usually known—phthalic acid and terephthalic acid. The mono-potassium salt of phthalic acid (potassium hydrogen phthalate, or KHP) is widely used as a standard in preparing solutions for acid-base titrations.

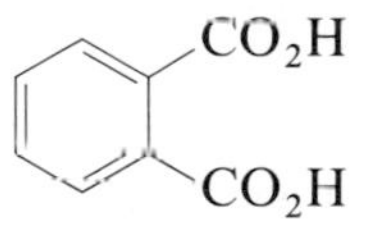

CO_2H
CO_2H

1,2-benzenedicarboxylic
acid
(phthalic acid)

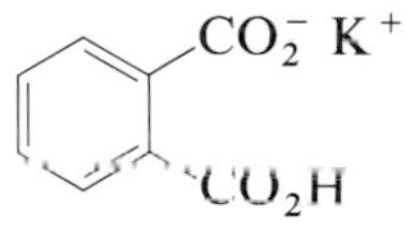

CO_2^- K^+
CO_2H

potassium hydrogen
phthalate
(KHP)

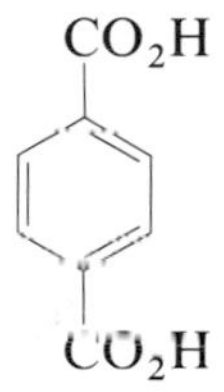

CO_2H
CO_2H

1,4-benzenedicarboxylic
acid
(terephthalic acid)

Terephthalic acid is one of the two organic components required for synthesizing the textile fiber known as Dacron polyester, or Dacron. The U.S. chemical industry produced over 6 billion pounds of terephthalic acid in 1985. The raw material for its synthesis is *p*-xylene, a compound derived exclusively from the refining of petroleum. Oxidation of *p*-xylene by nitric acid gives terephthalic acid:

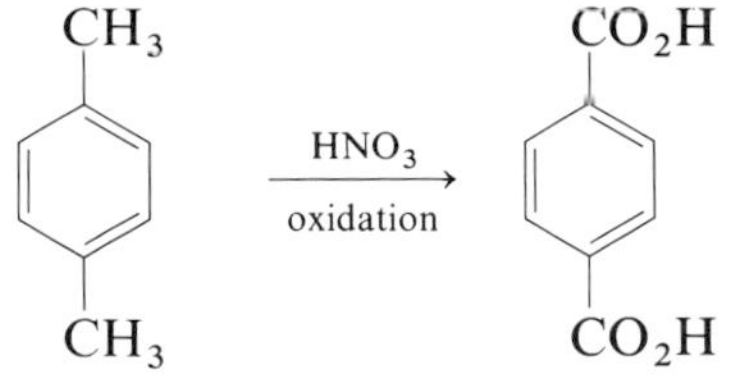

CH_3
CH_3

p-xylene
(a product of
petroleum refining)

$\xrightarrow[\text{oxidation}]{HNO_3}$

CO_2H
CO_2H

terephthalic acid

B. Common Names

Aliphatic carboxylic acids, many of which were known long before the development of structural theory and IUPAC nomenclature, were named according to their source or for some characteristic property. Formic acid was so named because it was first isolated from ants (Latin, *formica*). Acetic acid is a component of vinegar (Latin, *acetum*). Propionic acid was the first acid to be classified as a fatty acid (Greek, *pro*, first, and *pion*, fat). Butyric acid was first isolated from butter (Latin, *butyrum*). Oxalic acid was first isolated from a plant of the genus *Oxalis*. Valeric acid was first isolated from a plant of the genus *Valeriana*, native to Eurasia, widely cultivated in gardens and generally known as garden heliotrope. Several of the unbranched carboxylic acids most often found in the biological world along with their IUPAC and common names are listed in Table 7.2.

Table 7.2 Several carboxylic acids, and their common names and derivation.

Structure	IUPAC Name	Common Name	Derivation
HCO_2H	methanoic acid	formic acid	Latin: *formula*, ant
CH_3CO_2H	ethanoic acid	acetic acid	Latin: *acetum*, vinegar
$CH_3CH_2CO_2H$	propanoic acid	propionic acid	Greek: *propion*, first fatty acid
$CH_3(CH_2)_2CO_2H$	butanoic acid	butyric acid	Latin: *butyrum*, butter
$CH_3(CH_2)_3CO_2H$	pentanoic acid	valeric acid	Latin: *valeriana*, a flowering plant
$CH_3(CH_2)_4CO_2H$	hexanoic acid	caproic acid	Latin: *caper*, goat
$CH_3(CH_2)_6CO_2H$	octanoic acid	caprylic acid	Latin: *caper*, goat
$CH_3(CH_2)_8CO_2H$	decanoic acid	capric acid	Latin: *caper*, goat
$CH_3(CH_2)_{10}CO_2H$	dodecanoic acid	lauric acid	Latin: *laurus*, laurel
$CH_3(CH_2)_{12}CO_2H$	tetradecanoic acid	myristic acid	Greek: *muristikos*, fragrant
$CH_3(CH_2)_{14}CO_2H$	hexadecanoic acid	palmitic acid	Latin: *palma*, palm tree
$CH_3(CH_2)_{16}CO_2H$	octadecanoic acid	stearic acid	Greek: *stear*, solid fat
$CH_3(CH_2)_{18}CO_2H$	eicosanoic acid	arachidic acid	Greek: *arachne*, spider

When common names are used, Greek letters *alpha*, *beta*, *gamma*, and *delta* are often attached to locate substituents. The alpha position is the one next to the carboxyl group and an alpha substituent in a common name is equivalent to a 2 substituent in a IUPAC name. Following are two examples of the use of Greek letters to show the position of substituents. In each example, the IUPAC name is given first, followed by the common names. 2-aminopropanoic acid (alanine) is one of the 20 alpha–amino acids from which proteins are constructed (Chapter 13).

$$\overset{\delta}{C}-\overset{\gamma}{C}-\overset{\beta}{C}-\overset{\alpha}{C}-\overset{\overset{\textstyle O}{\|}}{C}OH$$

$$\underset{\underset{\textstyle OH}{|}}{CH_2}CH_2CH_2\overset{\overset{\textstyle O}{\|}}{C}OH$$

4-hydroxybutanoic acid
(γ-hydroxybutyric acid)

$$CH_3\underset{\underset{\textstyle NH_2}{|}}{CH}\overset{\overset{\textstyle O}{\|}}{C}OH$$

2-aminopropanoic acid
(α-aminopropionic acid; alanine)

In common names, the presence of a ketone in a substituted carboxylic acid is indicated by the prefix *keto-*, as illustrated by the common name β-ketobutyric acid.

$$CH_3\overset{\overset{\displaystyle O}{\|}}{C}CH_2\overset{\overset{\displaystyle O}{\|}}{C}OH \qquad CH_3\overset{\overset{\displaystyle O}{\|}}{C}-$$

3-oxobutanoic acid   an aceto group
(β-ketobutyric acid;
acetoacetic acid)

In practice, 3-oxobutanoic acid is known by two common names, *β-ketobutyric acid* and *acetoacetic acid*. The latter name, which is more common in the biological literature, is derived from the frequent reference to CH_3CO—as an aceto group. Thus, in the common nomenclature, 3-oxobutanoic acid is a substituted acetic acid and the name of the substituent is *aceto-*.

Example 7.1

Give IUPAC names for the following. Where possible, also give common names.

a. $CH_3\overset{\overset{\displaystyle CH_3}{|}}{C}HCO_2H$

b. $Cl-\overset{\overset{\displaystyle Cl}{|}}{\underset{\underset{\displaystyle Cl}{|}}{C}}-CO_2H$

c. $H_2N-\langle\!\!\bigcirc\!\!\rangle-CO_2H$

d. $CH_3\overset{\overset{\displaystyle HO}{|}}{C}HCH_2CO_2H$

Solution

a. The longest carbon chain is three atoms. The IUPAC name of this carboxylic acid is 2-methylpropanoic acid. The common name of the parent hydrocarbon is isobutane; therefore, the common name of this carboxylic acid is isobutyric acid.

b. Trichloroacetic acid. Acetic acid is retained by the IUPAC system. It is not necessary to use the numbers 2,2,2- to show the location of the three chlorine atoms because they can only be on the second carbon of the two-carbon chain. Trichloroacetic acid is often used in the clinical chemistry laboratory to precipitate proteins before analyzing blood or other biological fluids for other components.

c. 4-aminobenzoic acid. Its common name is *para*-aminobenzoic acid, or as it is often abbreviated, PABA. This substance is a growth factor needed by most microorganisms for the synthesis of folic acid (Section 14.6C). It is also used as a sunscreen in many tanning lotions.

d. 3-hydroxybutanoic acid. Its common name is beta-hydroxybutyric acid. This substituted carboxylic acid is one of three substances known as ketone bodies (Section 18.5).

Problem 7.1

Give IUPAC names for these compounds:

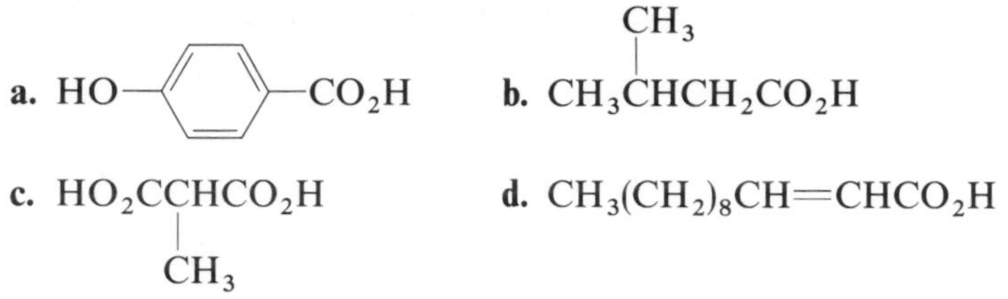

a. HO—⟨benzene ring⟩—CO$_2$H

b. $\overset{\displaystyle CH_3}{CH_3CHCH_2CO_2H}$

c. HO$_2$CCHCO$_2$H
    |
    CH$_3$

d. CH$_3$(CH$_2$)$_8$CH=CHCO$_2$H

7.3 Physical Properties of Carboxylic Acids

Because the carboxyl group contains three polar covalent bonds, carboxylic acids are polar compounds. The carbonyl oxygen and the hydroxyl oxygen bear partial negative charges, and the carbonyl carbon and hydroxyl hydrogen bear partial positive charges, as shown in Figure 7.2.

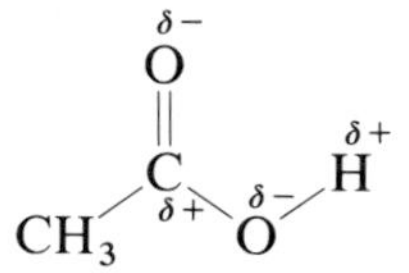

$$
\overset{\delta-}{O} \\
\parallel \\
\underset{CH_3}{\overset{}{}} \overset{\delta+}{C} \overset{\delta-}{O} \overset{\delta+}{H}
$$

Figure 7.2 Polarity of the carboxyl group.

Carboxylic acids can participate in hydrogen bonding through both C=O and —OH groups, as shown in Figure 7.3 for acetic acid.

Because carboxylic acids are even more extensively hydrogen-bonded than alcohols, their boiling points are higher relative to alcohols of comparable

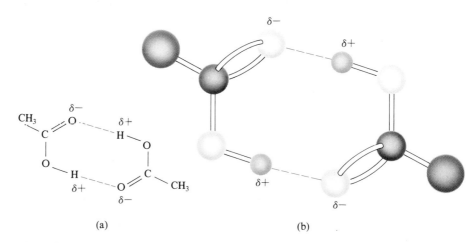

(a)   (b)

Figure 7.3 Hydrogen bonding between acetic acid molecules in pure liquid acetic acid. (a) Lewis structures; (b) ball-and-stick model.

molecular weight. For example, propanoic acid and 1-butanol have almost identical molecular weights, but because of more extensive hydrogen bonding, propanoic acid has a boiling point 24° higher than that of 1-butanol.

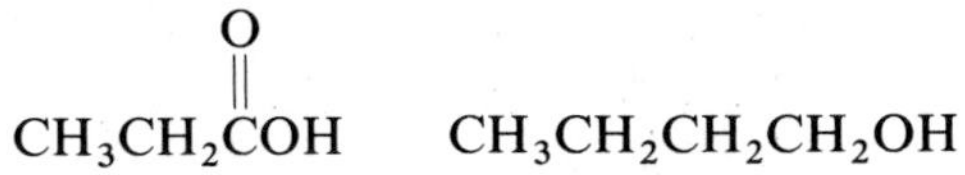

$$CH_3CH_2\overset{\displaystyle O}{\overset{\displaystyle \|}{C}}OH \qquad CH_3CH_2CH_2CH_2OH$$

propanoic acid;   1-butanol
(propionic acid;   (*n*-butyl alcohol;
mw 74, bp 141°C)   mw 74, bp 117°C)

Carboxylic acids also interact with water molecules by hydrogen bonding through both the carboxyl oxygen and the hydroxyl group:

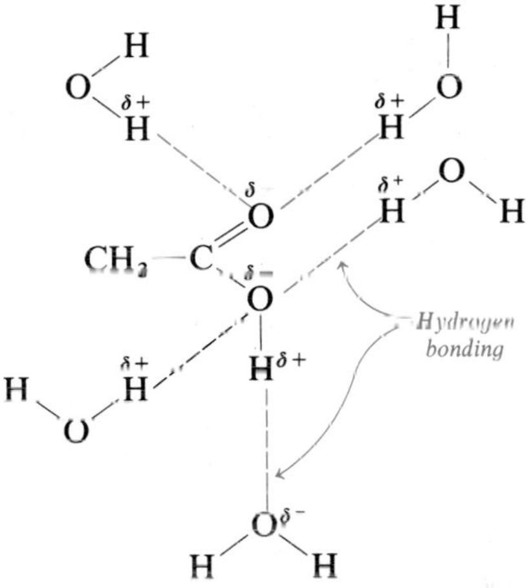

Because of these hydrogen-bonding interactions, carboxylic acids are more soluble in water than alkanes, ethers, alcohols, aldehydes, or ketones of comparable molecular weight. For example, propanoic acid (mw 74) is infinitely soluble in water, whereas the solubility of 1-butanol (mw 74) is only 8 g/100 g water.

As shown in Table 7.3, carboxylic acids with one to four carbon atoms are infinitely soluble in water. As molecular weight increases further, water solubility decreases. We can account for this trend in water solubility in the following

Table 7.3
Physical properties of some monocarboxylic acids.

Name	Structural Formula	mp (°C)	bp (°C)	Solubility in Water (g/100 g water)
formic acid	HCO_2H	8	100	infinite
acetic acid	CH_3CO_2H	16	118	infinite
propanoic acid	$CH_3CH_2CO_2H$	−22	141	infinite
butanoic acid	$CH_3(CH_2)_2CO_2H$	−6	164	infinite
hexanoic acid	$CH_3(CH_2)_4CO_2H$	−3	205	1.0
decanoic acid	$CH_3(CH_2)_8CO_2H$	32	—	insoluble

way. A carboxylic acid consists of two distinct parts, a polar hydrophilic carboxyl group, and except for formic acid, a nonpolar hydrophobic hydrocarbon chain. (**Hydrophilic** means "having an affinity for water; capable of dissolving in water." **Hydrophobic** means "tending not to combine with water; incapable of dissolving in water.") The hydrophilic carboxyl group increases water solubility; the hydrophobic hydrocarbon chain decreases water solubility.

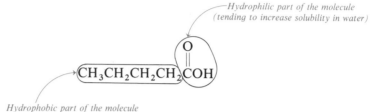

—Hydrophilic part of the molecule (tending to increase solubility in water)

Hydrophobic part of the molecule (tending to decrease solubility in water)

The first four aliphatic carboxylic acids are infinitely soluble in water because the hydrophobic effect of the hydrocarbon chain is more than counterbalanced by the hydrophilic character of the carboxyl group. As the size of the hydrophobic hydrocarbon chain increases relative to the size of the hydrophilic carboxyl group, water solubility decreases. The solubility of hexanoic acid in water is only 1.0 g per 100 g water. Decanoic acid is insoluble in water.

Example 7.2

Arrange the compounds in order of increasing boiling point:

$$\text{O}$$
a. $CH_3CH_2CH_2\overset{\|}{C}OH$ **b.** $CH_3CH_2CH_2CH_2\overset{\overset{\text{O}}{\|}}{C}H$

butanoic acid   pentanal

c. $CH_3CH_2CH_2CH_2CH_2OH$

1-pentanol

Solution

All three compounds are polar molecules of comparable molecular weight. Pentanal has no polar —OH group, cannot participate in hydrogen bonding in the pure liquid, and has the lowest boiling point. 1-Pentanol participates in hydrogen bonding through the polar —OH group and is next in boiling point. Butanoic acid participates in hydrogen bonding through both the polar C=O group and the —OH group; it has the highest boiling point.

pentanal	1-pentanol	butanoic acid
(mw 82; bp 103°C)	(mw 84; bp 137°C)	(mw 86; bp 164°C)

Problem 7.2

Arrange in order of increasing solubility in water:

$$\overset{O}{\underset{\|}{}}$$

a. $CH_3CH_2OCH_2CH_3$ **b.** $CH_3CH_2CH_2\overset{\overset{\displaystyle O}{\|}}{C}OH$ **c.** $CH_3(CH_2)_8\overset{\overset{\displaystyle O}{\|}}{C}OH$

         diethyl ether   butanoic acid   decanoic acid

7.4 Preparation of Carboxylic Acids

A. Oxidation of Primary Alcohols

Oxidation of a primary alcohol (Section 4.4B) yields a carboxylic acid. In the laboratory, the most common oxidizing agents are potassium dichromate and potassium permanganate.

$$CH_3(CH_2)_5CH_2OH + Cr_2O_7^{2-} \xrightarrow{H_3O^+} CH_3(CH_2)_5\overset{\overset{\displaystyle O}{\|}}{C}OH + Cr^{3+}$$

     1-heptanol   heptanoic acid
    (*n*-heptyl alcohol)

$$HOCH_2CH_2CH_2CH_2OH + MnO_4^- \xrightarrow{H_3O^+} HO\overset{\overset{\displaystyle O}{\|}}{C}CH_2CH_2\overset{\overset{\displaystyle O}{\|}}{C}OH + Mn^{2+}$$

    1,4-butanediol   butanedioic acid
                                     (succinic acid)

B. Oxidation of Aldehydes

Aldehydes are oxidized to carboxylic acids by potassium permanganate, potassium dichromate, and even such weak oxidizing agents as Ag(I) and Cu(II), as described in Section 6.5A. In the following oxidation, silver ion is reduced to silver metal. Note that under these conditions, neither the primary nor the secondary alcohol of the starting material is oxidized.

$$\begin{matrix} \overset{\overset{\displaystyle O}{\|}}{C}H \\ | \\ CHOH \\ | \\ CH_2OH \end{matrix} + Ag^+ \xrightarrow{NH_4OH} \begin{matrix} \overset{\overset{\displaystyle O}{\|}}{C}OH \\ | \\ CHOH \\ | \\ CH_2OH \end{matrix} + Ag$$

   2,3-dihydroxy-   2,3-dihydroxy-
      propanal   propanoic acid
   (glyceraldehyde)   (glyceric acid)

C. Oxidation of Alkenes

Oxidation of disubstituted alkenes of the type RCH=CHR by potassium dichromate or potassium permanganate causes cleavage of the carbon-carbon double bond and formation of two carboxylic acids:

$$CH_3CH_2CH_2CH{=}CHCHCH_3 + Cr_2O_7^{2-} \xrightarrow{H_3O^+} CH_3CH_2CH_2\overset{\overset{\displaystyle O}{\|}}{C}OH + HO\overset{\overset{\displaystyle O}{\|}}{C}CHCH_3 + Cr^{3+}$$

with CH_3 groups below the respective carbons.

2-methyl-3-heptene

butanoic acid
(butyric acid)

2-methyl-
propanoic acid
(isobutyric acid)

Oxidation of cyclohexene cleaves the carbon-carbon double bond and yields adipic acid, a dicarboxylic acid:

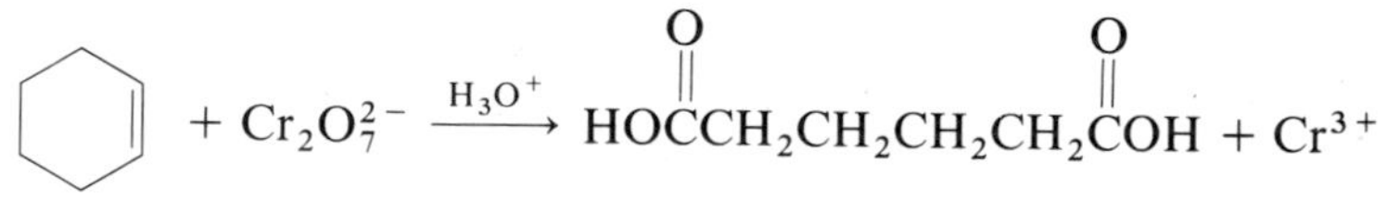

$$\text{cyclohexene} + Cr_2O_7^{2-} \xrightarrow{H_3O^+} HO\overset{\overset{\displaystyle O}{\|}}{C}CH_2CH_2CH_2CH_2\overset{\overset{\displaystyle O}{\|}}{C}OH + Cr^{3+}$$

cyclohexene

hexanedioic acid
(adipic acid)

Oxidation of a trisubstituted alkene yields a ketone and a carboxylic acid. For example, oxidation of 3-methyl-3-heptene gives

$$CH_3CH_2CH_2CH{=}CCH_2CH_3 + Cr_2O_7^{2-} \xrightarrow{H_3O^+} CH_3CH_2CH_2\overset{\overset{\displaystyle O}{\|}}{C}OH + O{=}CCH_2CH_3 + Cr^{3+}$$

with CH_3 groups below the respective carbons.

3-methyl-3-heptene

butanoic acid
(butyric acid)

2-butanone
(ethyl methyl ketone)

Example 7.3

Draw structural formulas for the products of these oxidations:

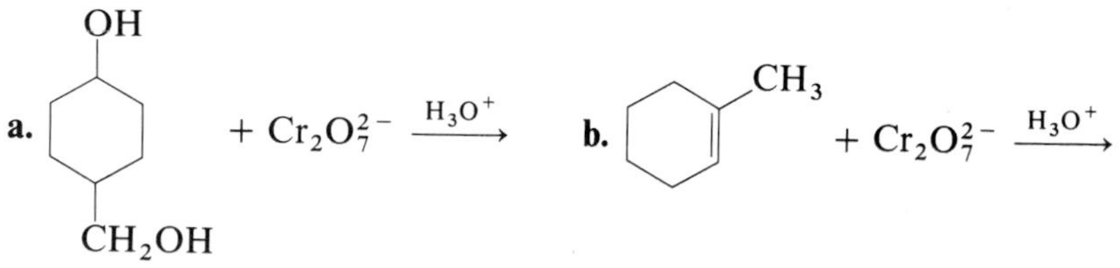

a.   [cyclohexane ring with OH at top and CH₂OH at bottom] $+ Cr_2O_7^{2-} \xrightarrow{H_3O^+}$

b.   [cyclohexene ring with CH₃] $+ Cr_2O_7^{2-} \xrightarrow{H_3O^+}$

☐ *Solution*   **a.** The starting material contains both a primary alcohol and a secondary alcohol. In the presence of potassium dichromate, the primary alcohol is oxidized to a carboxylic acid and the secondary alcohol to a ketone:

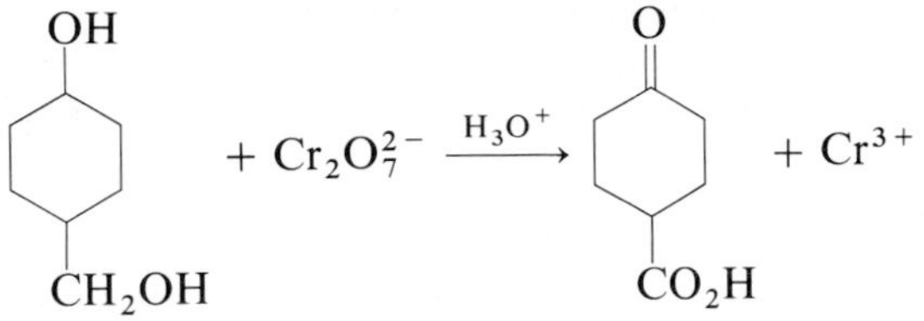

b. Oxidation of 1-methylcyclohexene yields a ketone and a carboxylic acid. The structural formula of the product is drawn below in two different ways: the first to emphasize where the ring is cleaved, the second to show the molecule as a chain of seven carbon atoms.

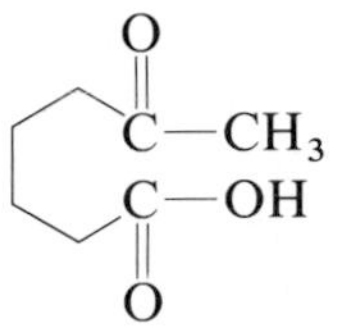

Problem 7.3

Draw structural formulas for the products of these oxidations:

a. $CH_3\underset{\underset{CH_3}{|}}{C}{=}CHCH_2CH_2\underset{\underset{CH_3}{|}}{CH}{=}CCH_3 + Cr_2O_7^{2-} \xrightarrow{H_3O^+}$

b. $HOCH_2CH_2CH_2CH_2\overset{\overset{O}{\parallel}}{CH} + Cr_2O_7^{2-} \xrightarrow{H_3O^+}$

7.5 Reactions of Carboxylic Acids

A. Acidity

Carboxylic acids ionize in water, to give acidic solutions. However, carboxylic acids are different in acidity from inorganic acids such as HCl, HBr, HNO_3, and H_2SO_4. Because these inorganic acids are 100% ionized in aqueous solution, they are classified as strong (meaning completely ionized) acids.

$$HCl \xrightarrow{H_2O} H^+ + Cl^-$$

Carboxylic acids are only slightly ionized in aqueous solution and therefore are classified as weak acids. When a carboxylic acid ionizes in water, an equilibrium is established between the carboxylic acid, the carboxylate ion, and H^+, as illustrated for the ionization of acetic acid:

$$CH_3\overset{\overset{O}{\parallel}}{C}OH \underset{H_2O}{\rightleftharpoons} CH_3\overset{\overset{O}{\parallel}}{C}O^- + H^+ \qquad K_a = \frac{[H^+][CH_3CO_2^-]}{[CH_3CO_2H]} = 1.8 \times 10^{-5}$$

The equilibrium constant for this ionization K_a is called an acid dissociation constant. The value of the acid dissociation constant for acetic acid is 1.8×10^{-5}, a typical value for simple carboxylic acids.

The ionization illustrated above is for acetic acid in water. A practical example of this equilibrium is vinegar, a 5% solution of acetic acid in water. Expressed in other units, the concentration of acetic acid in vinegar is approximately 50 g/L, or 0.83 mol/L. The hydrogen ion concentration in this solution is 3.8×10^{-3} mol/L, and its pH is approximately 2.4. By comparison, the concentration of hydrogen ion 0.83M HCl is 0.83M, and the pH of this solution is 0.081.

The other class of organic acids we have studied so far are the phenols (Section 4.7B). The difference in acidity between HCl and weak organic acids such as acetic acid and phenol can be seen by comparing the hydrogen ion concentrations and pH of a 0.1M solution of each of these acids in water, as shown in Table 7.4. Hydrochloric acid is a strong acid and 100% ionized in aqueous solution. By comparison, acetic acid and phenol are weak acids, and only slightly ionized in water. Carboxylic acids, however, are much stronger acids than phenols. It is carboxylic acids that give most common biological materials their acid pH. One exception is human gastric (stomach) juice, whose acidity is due to hydrochloric acid. Shown in Table 7.5 are approximate pH values for some biological materials and foods and beverages.

Table 7.4
Relative acid strengths of 0.1M HCl, acetic acid, and phenol.

Acid	K_a	Ionization in Water	$[H^+]$	pH
HCl	very large	100%	0.1M	1.0
CH_3CO_2H	1.8×10^{-5}	1.3%	0.0013M	2.9
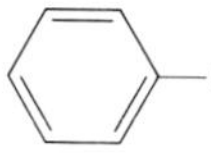 ⬡—OH	3.3×10^{-10}	0.0033%	3.3×10^{-6}	5.5

Table 7.5
Approximate pH values for some foods, beverages, and biological materials.

Substance	pH	Substance	pH
blood plasma (human)	7.3–7.5	grapefruit	3.0–3.3
saliva (human)	6.5–7.5	lemons	2.2–2.4
stomach fluids (human)	1.0–3.0	oranges	3.0–4.0
milk (human)	6.6–7.6	potatoes	5.6–6.0
apples	2.9–3.3	skin (human)	4.5–5.5
bananas	4.5–4.7	soft drinks	2.4–4.0
beans	5.0–6.0	tomatoes	4.0–4.4
beer	4.0–5.0	vinegar	2.4–3.4
cheese	4.8–6.4	water, drinking	6.5–8.0
cider	2.9–3.3	wines	2.8–3.8
eggs	7.6–8.0		

B. Reaction with Bases

All carboxylic acids, whether soluble or insoluble in water, react quantitatively with NaOH, KOH, and other strong bases, to form salts.

$$
\text{(benzoic acid)}\!-\!\overset{\overset{\displaystyle O}{\|}}{C}\!OH + NaOH \longrightarrow \text{(sodium benzoate)}\!-\!\overset{\overset{\displaystyle O}{\|}}{C}\!O^- \, Na^+ + H_2O
$$

benzoic acid (only slightly soluble in water)	sodium benzoate (very soluble in water)

Carboxylic acids also react with ammonia, and with primary, secondary, and tertiary amines to form salts.

$$
\text{(benzoic acid)}\!-\!\overset{\overset{\displaystyle O}{\|}}{C}\!OH + NH_4OH \longrightarrow \text{(ammonium benzoate)}\!-\!\overset{\overset{\displaystyle O}{\|}}{C}\!O^- NH_4^+ + H_2O
$$

benzoic acid (only slightly soluble in water)	ammonium benzoate (very soluble in water)

Sodium, potassium, and ammonium salts of carboxylic acids are ionic compounds and are much more soluble in water than the carboxylic acids from which they are derived. Benzoic acid, for example, is only very slightly soluble in water at room temperature. By contrast, the solubility of ammonium benzoate is 20 g/100 g water and of sodium benzoate is 66 g/100 g water.

Salts of carboxylic acids are named like the salts of inorganic acids; the cation is named first and then the anion. The name of the anion is derived from the carboxylic acid by dropping the suffic -ic acid and adding the suffix -ate, as illustrated by these examples:

Carboxylic Acid	Anion	Salt
$CH_3CH_2CO_2H$	$CH_3CH_2CO_2^-$	$(CH_3CH_2CO_2^-)_2Ca^{2+}$
propanoic acid (propionic acid)	propanoate (propionate)	calcium propanoate (calcium propionate)
$CH_3(CH_2)_{14}CO_4H$	$CH_3(CH_2)_{14}CO_2^-$	$CH_3(CH_2)_{14}CO_2^- \, Na^+$
hexadecanoic acid (palmitic acid)	hexadecanoate (palmitate)	sodium hexadecanoate (sodium palmitate)

Calcium propanoate is often added to bread and other baked goods "to retard spoilage." Hexadecanoic acid (palmitic acid) is one of the most abundant long-chain carboxylic acids in the biological world. Its concentration in palm oil is particularly high, hence its common name, palmitic acid. Because sodium hexadecanoate (sodium palmitate) is relatively abundant in animal fats, from which natural soaps are made, it is an important component of natural soaps (Section 12.2).

Example 7.4

Name these salts:

a. $CH_3CO_2^-(CH_3)_3NH^+$ b. 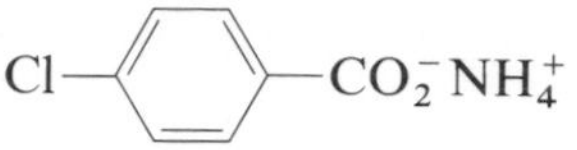 $Cl-\langle\bigcirc\rangle-CO_2^-NH_4^+$

Solution

a. The carboxylic acid is acetic acid and its anion is acetate ion. The cation is derived from trimethylamine and is named trimethylammonium ion. The name of the salt is trimethylammonium acetate.
b. The acid is 4-chlorobenzoic acid or *p*-chlorobenzoic acid. The name of this salt is ammonium 4-chlorobenzoate, or ammonium *p*-chlorobenzoate.

Problem 7.4

Draw structural formulas for these salts:

a. potassium 2-methylpropanoate (potassium isobutyrate)
b. calcium hexadecanoate (calcium palmitate)

Carboxylic acids also react with weaker bases such as sodium bicarbonate and sodium carbonate, to form salts. In these reactions, the carboxylic acid is converted to a sodium salt; bicarbonate and carbonate ions are converted to carbonic acid, H_2CO_3, which spontaneously breaks down to form carbon dioxide and water:

$$2CH_3\overset{O}{\overset{\|}{C}}OH + Na_2CO_3 \longrightarrow 2CH_3\overset{O}{\overset{\|}{C}}O^-Na^+ + CO_2 + H_2O$$

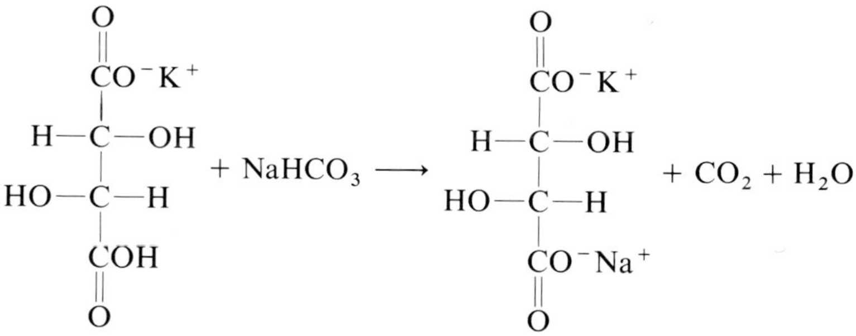

acetic acid   sodium carbonate   sodium acetate

potassium hydrogen tartrate (cream of tartar)   potassium sodium tartrate

By this last reaction, you can see why sodium carbonate or sodium bicarbonate or both are used in baking. Baking powder is a combination of sodium carbonate or sodium bicarbonate and potassium hydrogen tartrate. When water is added, the acid and base in baking powder react, to liberate carbon dioxide, which forms bubbles in the batter or dough and causes it to "rise." Baking soda (sodium bicarbonate) can also be mixed with vinegar (a solution of acetic acid in water) or with lemon juice or orange juice (a solution containing citric acid) to produce carbon dioxide. Other recipes call for mixing baking soda with sour cream, which contains 2-hydroxypropanoic acid, known almost exclusively by its common name, lactic acid. Lactic acid and sodium bicarbonate react, to produce a salt plus carbon dioxide and water:

$$\underset{\substack{\text{2-hydroxypropanoic} \\ \text{acid} \\ \text{(lactic acid)}}}{CH_3\overset{\overset{\displaystyle OH}{|}}{C}HCO_2H} + NaHCO_3 \longrightarrow \underset{\substack{\text{sodium 2-hydroxy-} \\ \text{propanoate} \\ \text{(sodium lactate)}}}{CH_3\overset{\overset{\displaystyle OH}{|}}{C}HCO_2^- Na^+} + CO_2 + H_2O$$

Example 7.5

Complete and balance the acid-base equations:

a. $CH_3\overset{\overset{\displaystyle HO}{|}}{C}H\overset{\overset{\displaystyle O}{||}}{C}OH + NaOH \longrightarrow$

b. $CH_3\overset{\overset{\displaystyle O}{||}}{C}CH_2CH_2CH_2\overset{\overset{\displaystyle O}{||}}{C}OH + NaHCO_3 \longrightarrow$

c. $HO\overset{\overset{\displaystyle O}{||}}{C}CH_2CH_2\overset{\overset{\displaystyle O}{||}}{C}OH + Na_2CO_3 \longrightarrow$

☐ *Solution*

a. 2-hydroxypropanoic acid (lactic acid) is a monocarboxylic acid and reacts in a 1:1 molar ratio with sodium hydroxide, to form sodium lactate and water:

$$\underset{\substack{\text{2-hydroxy-} \\ \text{propanoic acid} \\ \text{(lactic acid)}}}{CH_3\overset{\overset{\displaystyle HO}{|}}{C}H\overset{\overset{\displaystyle O}{||}}{C}OH} + NaOH \longrightarrow \underset{\substack{\text{sodium 2-hydroxy-} \\ \text{propanoate} \\ \text{(sodium lactate)}}}{CH_3\overset{\overset{\displaystyle HO}{|}}{C}H\overset{\overset{\displaystyle O}{||}}{C}O^- Na^+} + H_2O$$

b. 5-oxohexanoic acid is also a monocarboxylic acid and reacts in a 1:1 molar ratio with sodium bicarbonate, to form a sodium salt, carbon dioxide, and water:

$$\underset{\text{5-oxohexanoic acid}}{CH_3\overset{\overset{\displaystyle O}{||}}{C}CH_2CH_2CH_2\overset{\overset{\displaystyle O}{||}}{C}OH} + NaHCO_3 \longrightarrow \underset{\text{sodium 5-oxohexanoate}}{CH_3\overset{\overset{\displaystyle O}{||}}{C}CH_2CH_2CH_2\overset{\overset{\displaystyle O}{||}}{C}O^- Na^+} + CO_2 + H_2O$$

c. Butanedioic acid (succinic acid) is a dicarboxylic acid and reacts with sodium carbonate in a 1:1 molar ratio:

$$\underset{\substack{\text{butanedioic acid} \\ \text{(succinic acid)}}}{HO\overset{O}{\overset{\|}{C}}CH_2CH_2\overset{O}{\overset{\|}{C}}OH} + Na_2CO_3 \longrightarrow \underset{\substack{\text{sodium butanedioate} \\ \text{(sodium succinate)}}}{Na^+ {}^-O\overset{O}{\overset{\|}{C}}CH_2CH_2\overset{O}{\overset{\|}{C}}O^- Na^+} + CO_2 + H_2O$$

Problem 7.5

Complete and balance these acid-base reactions.

a. $CH_3CH_2CH{=}CHCO_2H + NaOH \longrightarrow$

b. $2\langle\!\!\!\bigcirc\!\!\!\rangle{-}CO_2H + Na_2CO_3 \longrightarrow$

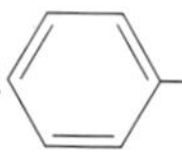

c. $HO{-}\underset{\underset{\displaystyle CH_2{-}CO_2H}{|}}{\overset{\overset{\displaystyle CH_2{-}CO_2H}{|}}{C}}{-}CO_2H \ \ + 3NaHCO_3 \longrightarrow$

citric acid

C. Reduction

Carboxyl groups are among the most difficult organic functional groups to reduce. They are not affected by catalytic hydrogenation under conditions that easily reduce aldehydes and ketones to alcohols, say, temperatures of 25–100°C and pressures of hydrogen of 1–5 atm (Section 6.6A). Thus it is possible to reduce aldehydes or ketones to alcohols in the presence of carboxylic acids:

$$\underset{\text{5-oxohexanoic acid}}{CH_3\overset{O}{\overset{\|}{C}}CH_2CH_2CH_2\overset{O}{\overset{\|}{C}}OH} + H_2 \xrightarrow[\text{25°C,2 atm}]{Pt} \underset{\text{5-hydroxyhexanoic acid}}{CH_3\overset{OH}{\overset{|}{C}}HCH_2CH_2CH_2\overset{O}{\overset{\|}{C}}OH}$$

It is, however, possible to reduce carboxylic acids to primary alcohols, using the metal hydride reducing agent lithium aluminum hydride, $LiAlH_4$ (Section 6.6B). With this reagent, terephthalic acid is reduced to a diol:

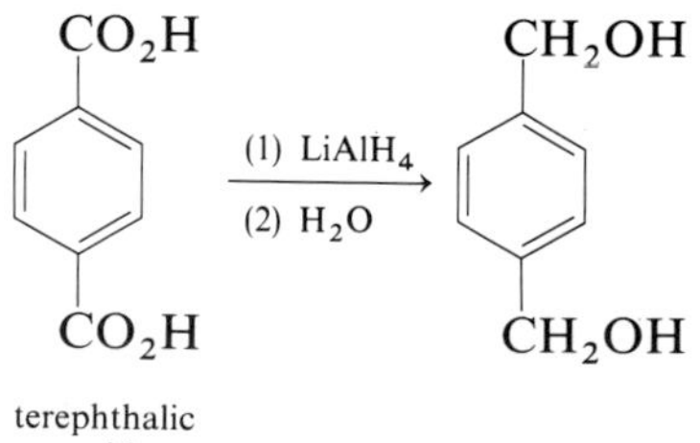

$$\underset{\substack{\text{terephthalic} \\ \text{acid}}}{\underset{CO_2H}{\overset{CO_2H}{\bigcirc}}} \xrightarrow[\text{(2) H}_2\text{O}]{\text{(1) LiAlH}_4} \underset{CH_2OH}{\overset{CH_2OH}{\bigcirc}}$$

Recall (Section 6.6B) that $LiAlH_4$ also reduces aldehydes and ketones to alcohols but does not reduce alkenes or alkynes.

Example 7.6

☐ *Solution*

Draw structural formulas for the products formed by reduction of the following with lithium aluminum hydride.

a. $\overset{\displaystyle CH_3}{\underset{\displaystyle |}{CH_3C}}=CHCH_2CH_2\overset{\displaystyle O}{\overset{\displaystyle \|}{C}}OH$ b. $H\overset{\displaystyle O}{\overset{\displaystyle \|}{C}}CH_2CH_2CH_2\overset{\displaystyle O}{\overset{\displaystyle \|}{C}}OH$

a. $\overset{\displaystyle CH_3}{\underset{\displaystyle |}{CH_3C}}=CHCH_2CH_2CH_2OH$ b. $HOCH_2CH_2CH_2CH_2CH_2OH$

Problem 7.6

Draw the structural formula for a compound of given molecular formula that after reduction by lithium aluminum hydride gives the following molecules:

a. $C_6H_8O_2 \xrightarrow[\text{(2) H}_2\text{O}]{\text{(1) LiAlH}_4} HO-\langle\bigcirc\rangle-OH$

b. $C_6H_{10}O_4 \xrightarrow[\text{(2) H}_2\text{O}]{\text{(1) LiAlH}_4} HOCH_2CH_2CH_2CH_2CH_2CH_2OH$

D. Decarboxylation of Beta-Ketoacids and Beta-Dicarboxylic Acids

Carboxylic acids that have a carbonyl group on the carbon atom beta to the carboxyl group lose CO_2 on heating. This reaction causes loss of the carboxyl group and is called decarboxylation. For example, when 3-oxobutanoic acid is heated, it decarboxylates, to give acetone and carbon dioxide:

This carbonyl group is beta to the carboxyl group

$$CH_3-\overset{\overset{\displaystyle O}{\displaystyle \|}}{\underset{\displaystyle \beta}{C}}-\overset{\displaystyle \alpha}{C}H_2-\overset{\overset{\displaystyle O}{\displaystyle \|}}{C}-OH \xrightarrow{\text{heat}} CH_3-\overset{\overset{\displaystyle O}{\displaystyle \|}}{C}-CH_3 + CO_2$$

3-oxobutanoic acid
(β-ketobutyric acid;
acetoacetic acid)   acetone

3-Oxobutanoic, or acetoacetic acid as it is more commonly named in biochemistry, and its reduction product, 3-hydroxybutanoic acid (β-hydroxybutyric acid) are synthesized in the liver by partial oxidation of fatty acids; they are known collectively as ketone bodies. (We shall discuss the synthesis and metabolism of ketone bodies in Section 18.5.) The concentration of ketone bodies in the blood of healthy, well-fed humans is approximately 0.01 mM/L. However, in starvation or diabetes mellitus, the concentration of ketone bodies may increase to as much as 500 times normal. Under these conditions, the concentration of 3-oxobutanoic acid (acetoacetic acid) increases to the point where it undergoes spontaneous decarboxylation, to form acetone and carbon dioxide. Acetone is not metabolized by humans and is excreted through the kidneys and lungs. The odor of acetone is responsible for the characteristic "sweet smell" on the breath of severely diabetic patients.

Decarboxylation on heating is a unique property of 3-oxocarboxylic acids (β-ketocarboxylic acids), and is not observed with other classes of ketoacids.

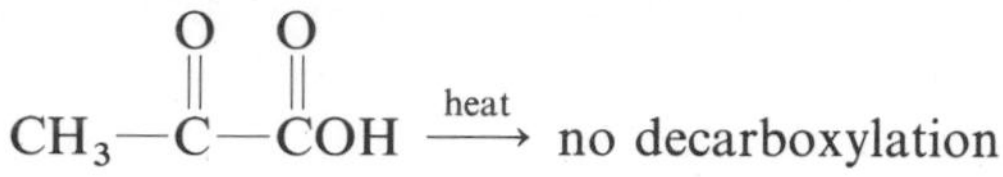

$$CH_3-\overset{O}{\underset{\|}{C}}-\overset{O}{\underset{\|}{C}}OH \xrightarrow{\text{heat}} \text{no decarboxylation}$$

2-oxopropanoic acid
(pyruvic acid)

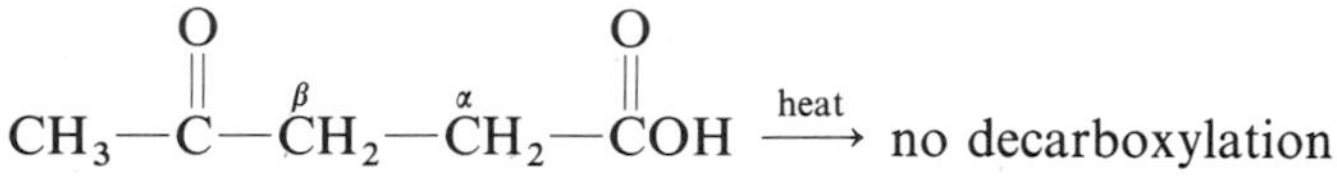

$$CH_3-\overset{O}{\underset{\|}{C}}-\overset{\beta}{C}H_2-\overset{\alpha}{C}H_2-\overset{O}{\underset{\|}{C}}OH \xrightarrow{\text{heat}} \text{no decarboxylation}$$

4-oxopentanoic acid
(γ-ketovaleric acid)

The mechanism for decarboxylation of β-ketoacids is illustrated by the decarboxylation of 3-oxobutanoic acid. The reaction is thought to involve a cyclic six-membered transition state, which by rearrangement of electron pairs, leads to the enol form of acetone and carbon dioxide. The enol form of acetone is in equilibrium with the keto form.

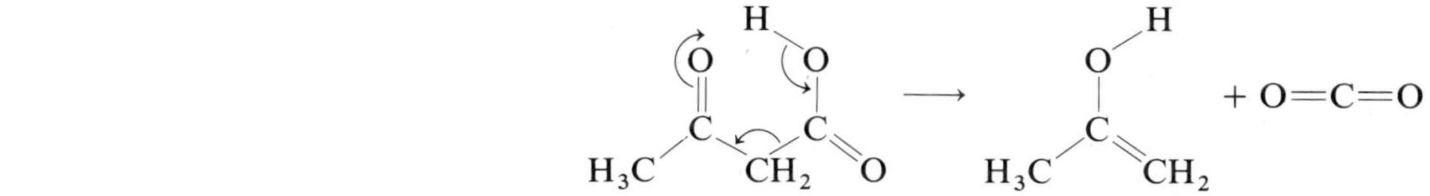

3-ketobutanoic acid   enol form
of acetone

Another important decarboxylation of a β-ketoacid in the biological world occurs during the oxidation of foodstuffs in the tricarboxylic acid cycle (Section 17.5). One of the intermediates in this cycle is 1-oxo-1,2,3-propanetricarboxylic acid, known almost exclusively by its common name, oxalosuccinic acid. Oxalosuccinic acid undergoes spontaneous decarboxylation to produce 2-oxopentanedioic acid (α-ketoglutaric acid). Only one of the three carboxyl groups of oxalosuccinic acid has a carbonyl group in the β-position to it. It is this carboxyl group that is lost as CO_2.

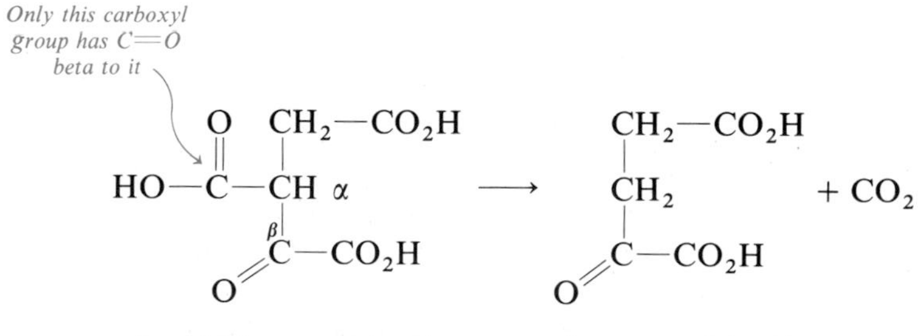

Only this carboxyl group has C=O beta to it

1-oxo-1,2,3-propanetrioic acid   2-oxopentanedioic acid
(oxalosuccinic acid)   (α-ketoglutaric acid)

Key Terms and Concepts

carboxyl group (7.1)   hydrophobic (7.3)

hydrophilic (7.3)   beta-ketoacid (7.5D)

Key Reactions

1. Carboxylic acids are weak acids and ionize in water to give acidic solutions (Section 7.5).

2. Reaction with NaOH and other strong bases to give water-soluble salts salts (Section 7.5B).

3. Metal hydride reduction to primary alcohols (Section 7.5C).

4. Decarboxylation of β-ketoacids (Section 7.5D).

Problems

Structure and nomenclature of carboxylic acids (Sections 7.1 and 7.2)

7.7   Name the compounds:

a. $CH_3CHCH_2CH_2CO_2H$
         |
         OH

b. $C_6H_5CH_2CH_2CH_2CO_2H$

c. $ClCH_2CO_2H$

d. CH_3CHCO_2H
        |
        CH_3

e. 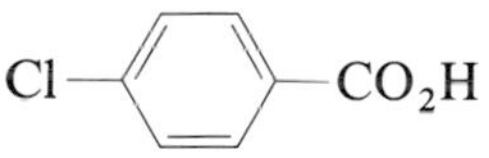 Cl—⟨ ⟩—CO_2H

f. 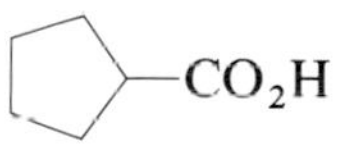 ⟨ ⟩—CO_2H

g. $C_6H_5CO_2^-\ Na^+$

h. $CH_3CH_2CH_2CH_2CO_2^-\ NH_4^+$

i. $HO_2CCH_2CH_2CH_2CH_2CO_2H$

j. CF_3CO_2H

k. $(CH_3CH_2CO_2^-)_2\ Ca^{2+}$

l. $CH_3(CH_2)_7CH{=}CH(CH_2)_7CO_2H$

7.8   Draw structural formulas for these compounds:

a. 3-hydroxybutanoic acid   **b.** sodium oxalate

c. trichloroacetic acid   **d.** 4-aminobutanoic acid

e. sodium hexadecanoate   **f.** calcium octanoate

g. potassium phenylacetate   **h.** octanoic acid

i. 2-aminopropanoic acid (alanine)

j. 2-hydroxypropanoic acid (lactic acid)

k. *p*-methoxybenzoic acid

l. potassium 2,4-hexadienoate (the food preservative, potassium sorbate)

*Physical
properties of
carboxylic
acids
(Section 7.3)*

7.9   Draw structural formulas to illustrate hydrogen bonding between the circled atoms.

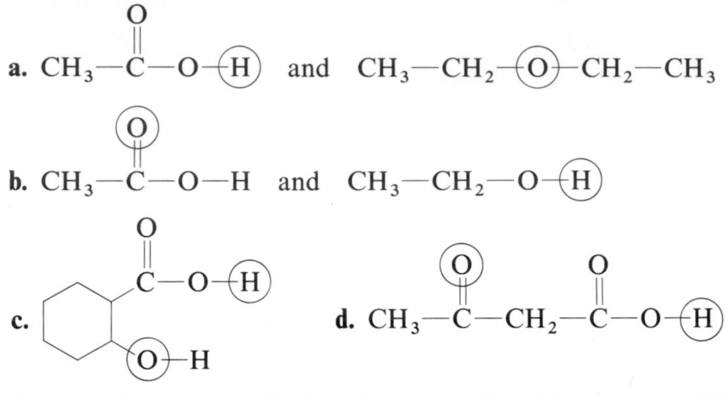

a. $CH_3-\overset{\overset{\displaystyle O}{\|}}{C}-O-\boxed{H}$ and $CH_3-CH_2-\boxed{O}-CH_2-CH_3$

b. $CH_3-\overset{\overset{\displaystyle \boxed{O}}{\|}}{C}-O-H$ and $CH_3-CH_2-O-\boxed{H}$

c.

d. $CH_3-\overset{\overset{\displaystyle \boxed{O}}{\|}}{C}-CH_2-\overset{\overset{\displaystyle O}{\|}}{C}-O-\boxed{H}$

7.10   Arrange the compounds of each set in order of increasing boiling points.

a. $CH_3CH_2\overset{\overset{\displaystyle O}{\|}}{C}OH$ $CH_3CH_2CH_2CH_2OH$ $CH_3CH_2OCH_2CH_3$

b. $CH_3(CH_2)_8CO_2H$ —OH $CH_3(CH_2)_4CO_2H$

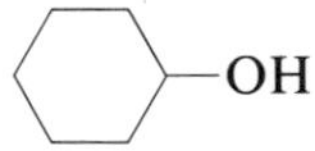

*Preparation of
carboxylic
acids
(Section 7.4)*

7.11   Complete the following reactions:

a. $CH_3(CH_2)_4CH_2OH + Cr_2O_7^{2-} \xrightarrow[\text{heat}]{H_2O,\ H^+}$

b. 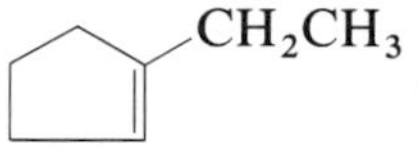 $+ Cr_2O_7^{2-} \xrightarrow[\text{heat}]{H_2O,\ H^+}$

c. 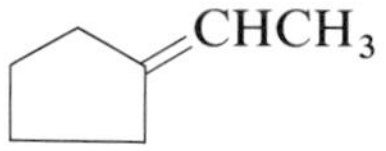 $+ Cr_2O_7^{2-} \xrightarrow[\text{heat}]{H_2O,\ H^+}$

d. $CH_3(CH_2)_7CH=CH(CH_2)_7\overset{\overset{\displaystyle O}{\|}}{C}OH + Cr_2O_7^{2-} \xrightarrow[\text{heat}]{H_2O,\ H^+}$

e. 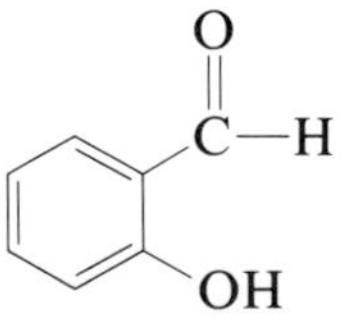 $+ Ag^+ \xrightarrow{NH_4OH}$

salicylaldehyde

f. 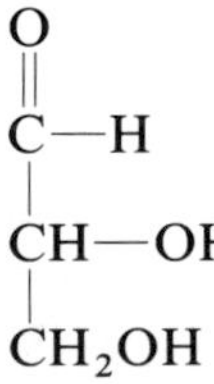 $\overset{\overset{\displaystyle O}{\|}}{\underset{\underset{\displaystyle CH_2OH}{\displaystyle CH-OH}}{C}}-H + Ag^+ \xrightarrow{NH_4OH}$

glyceraldehyde

7.12   Draw the structural formula of a compound of the given molecular formula that on oxidation gives the carboxylic acid or dicarboxylic acid shown.

a. $C_6H_{14}O$ $\xrightarrow{\text{oxidation}}$ $CH_3(CH_2)_4\overset{\overset{\displaystyle O}{\|}}{C}OH$

b. $C_6H_{12}O$ $\xrightarrow{\text{oxidation}}$ $CH_3(CH_2)_4\overset{\overset{\displaystyle O}{\|}}{C}OH$

c. $C_6H_{14}O_2$ $\xrightarrow{\text{oxidation}}$ $HO\overset{\overset{\displaystyle O}{\|}}{C}CH_2CH_2CH_2CH_2\overset{\overset{\displaystyle O}{\|}}{C}OH$

d. C_6H_{10} $\xrightarrow{\text{oxidation}}$ $HO\overset{\overset{\displaystyle O}{\|}}{C}CH_2CH_2CH_2CH_2\overset{\overset{\displaystyle O}{\|}}{C}OH$

Reactions of carboxylic acids (Section 7.5)

7.13 Arrange the compounds in order of increasing acidity:

a. ⬡—OH   b. ⬡—CO$_2$H   c. ⬡—OH

7.14 Complete these reactions. Where there is no reaction, write N.R.

a. $CH_3CO_2H + NaOH \longrightarrow$

b. ⬡—$CO_2H + NaOH \longrightarrow$

c. $CH_3(CH_2)_{14}CO_2Na + H_2SO_4 \longrightarrow$

d. $CH_3CH_2CH_2CO_2H + NaHCO_3 \longrightarrow$

e. ⬡ with CH_2OH and OH + NaOH $\longrightarrow$

f. HO—⬡—$CO_2H + NaHCO_3 \longrightarrow$

g. $CH_3CH_2\overset{\overset{\displaystyle O}{\|}}{C}\underset{\underset{\displaystyle CH_3}{|}}{CH}\overset{\overset{\displaystyle O}{\|}}{C}OH \xrightarrow{\text{heat}}$ h. $HO\overset{\overset{\displaystyle O}{\|}}{C}CH_2\overset{\overset{\displaystyle OO}{\|\|}}{C}OH \xrightarrow{\text{heat}}$

7.15 Show how you could distinguish between the following pairs of compounds by a simple chemical test. In each case, tell what test you would perform and what you would expect to observe, and write an equation for each positive test.
a. acetic acid and acetaldehyde
b. hexanoic acid and 1-hexanol
c. benzoic acid and phenol
d. sodium salicylate and salicylic acid

    e. oleic acid and stearic acid (see Table 12.1 for structural formulas of these fatty acids)

    f. phenylacetic acid and acetophenone (methyl phenyl ketone)

7.16 Decarboxylation is a general reaction for any molecule that has a carbonyl group on the carbon atom beta to a carboxylic acid. One such example is malonic acid, which on heating loses carbon dioxide:

$$\underset{\text{malonic acid}}{HOCCH_2COH} \xrightarrow{\text{heat}} \text{a carboxylic acid} + CO_2$$

    a. Draw a structural formula for the carboxylic acid formed on decarboxylation of malonic acid.

    b. Do you expect that succinic acid would undergo the same type of decarboxylation?

7.17 The following conversions can be carried out in either one step or two. Show the reactants you would use and draw structural formulas for the intermediate formed in any conversion that requires two steps.

a. $CH_3(CH_2)_6CH \xrightarrow{} CH_3(CH_2)_6COH$ (each with $=O$)

b.  (cyclopentane) $\xrightarrow{}$ $HOCCH_2CH_2CH_2COH$

    1,5-pentanedicarboxylic acid
    (glutaric acid)

c. 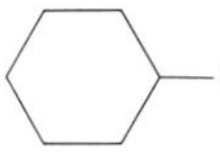 (cyclohexane)$-OH \xrightarrow{}$ $HOCCH_2CH_2CH_2CH_2COH$

    1,6-hexanedicarboxylic acid
    (adipic acid)

d. $HOCCH_2CH_2COH \xrightarrow{} Na^+\ {}^-OCCH_2CH_2CO^-\ Na^+$

    1,4-butanedioic acid   sodium 1,4-butanedioate
    (succinic acid)   (sodium succinate)

e. $HO_2CCH_2CH_2CO_2H \xrightarrow{} HOCH_2CH_2CH_2CH_2OH$

f.
$$\begin{array}{l} CHO \\ | \\ CHOH \\ | \\ CHOH \\ | \\ CH_2OH \end{array} \xrightarrow{} \begin{array}{l} CO_2H \\ | \\ CHOH \\ | \\ CHOH \\ | \\ CH_2OH \end{array}$$

g. $CH_3(CH_2)_6CH_2OH \xrightarrow{} CH_3(CH_2)_6CO_2^-\ NH_4^+$

Prostaglandins

The prostaglandins are a group of natural substances, all having the 20-carbon skeleton of prostanoic acid:

prostanoic acid

Prostaglandins and prostaglandin-derived materials have been found in virtually all human tissues examined thus far, and they are intimately involved in a host of bodily processes. For example, they are involved in both the induction of the inflammatory response and in its relief. The medical significance of these facts becomes obvious when we realize that more than 5 million Americans suffer from rheumatoid arthritis, an inflammatory disease. Prostaglandins are also involved in almost every phase of reproductive physiology.

The discovery of prostaglandins and determination of their structure began in 1930, when Raphael Kurzrok and Charles Lieb, gynecologists practicing in New York, observed that human seminal fluid stimulates contraction of isolated human uterine muscle. A few years later in Sweden, Ulf von Euler confirmed this report and noted that human seminal fluid also produces contraction of intestinal smooth muscle and lowers blood pressure when injected into the bloodstream. Von Euler proposed the name prostaglandin for the mysterious substance or substances responsible for such diverse effects, because at the time it was believed that they originated in the prostate gland. We now know

that prostaglandin production is by no means limited to the prostate gland. However, the name has stuck. By 1960, several prostaglandins had been isolated in pure crystalline form and their structural formulas had been determined. Structural formulas for three common prostaglandins are given in Figure VI-1.

Prostaglandins are abbreviated PG, with an additional letter and numerical subscript to indicate the type and series. The various types differ in the functional groups present in the five membered ring. Those of the A type are alpha,beta-unsaturated ketones; those of the E type are beta-hydroxyketones; and those of the F type are 1,3-diols. The subscript alpha in the F type indicates that the hydroxyl group at carbon 9 is below the plane of the five-membered ring and on the same side as the hydroxyl at carbon 11. The various series of prostaglandins differ in the number of double bonds on the two side chains. Those of the 1 series have only one double bond; those of the 2 series have two double bonds; and those of the 3 series have three double bonds.

Concurrently with investigations of the chemical structure of prostaglandins, clinical scientists began to study the biochemistry of these remarkable compounds and their potential as drugs. Initially, research was hampered by the high cost and great difficulty of isolating and purifying them. If they could not be isolated easily, could they be synthesized instead? The first totally synthetic prostaglandins became available in 1968, when both Dr. John Pike of the Upjohn Company and Professor E. J. Corey of Harvard University announced laboratory syntheses of several prostaglandins and pro-

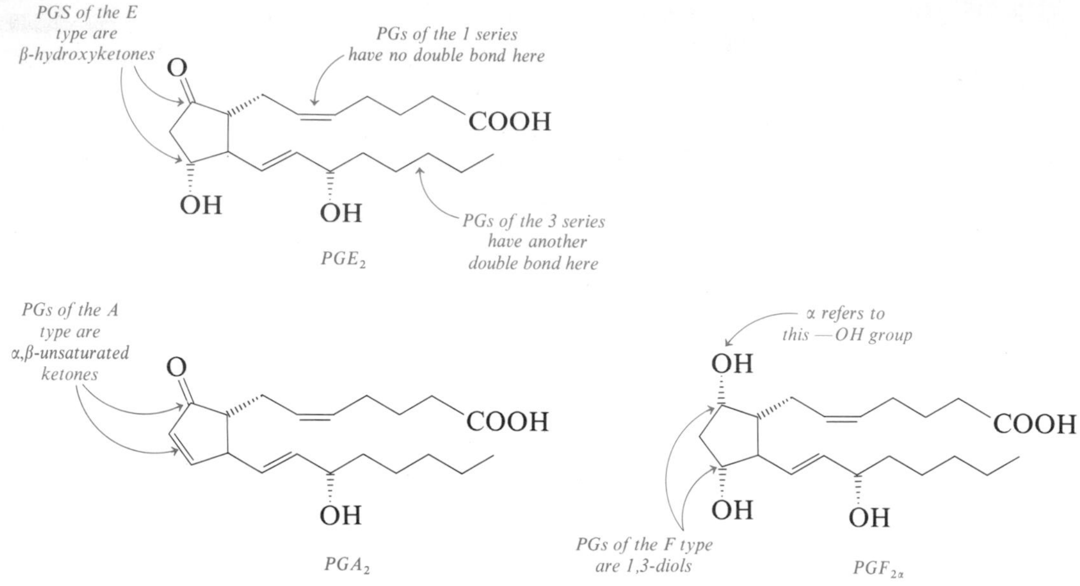

PGS of the E
type are
β-hydroxyketones

PGs of the 1 series
have no double bond here

PGs of the 3 series
have another
double bond here

PGE₂

PGs of the A
type are
α,β-unsaturated
ketones

PGA₂

α refers to
this —OH group

PGs of the F type
are 1,3-diols

PGF₂ₐ

Figure VI-1 Prostaglandins PGA₁, PGE₂, and PGF₂ₐ.

staglandin analogs. However, costs were still high. Then, in 1969, the price of prostaglandins dropped dramatically with the discovery that the gorgonian sea whip or sea fan, *Plexaura homomalla*, a coral that grows on reefs off the coast of Florida and in the Caribbean, is a rich source of prostaglandinlike materials (Figure VI-2). The concentration of PG-like substances in this marine organism is about 100 times the normal concentration found in most mammalian sources. In the laboratory, the PG-like compounds were extracted and then transformed to prostaglandins and prostaglandin analogs. Now, however, there is no need to depend on this natural source, because chemists have developed highly effective and stereospecific laboratory schemes for synthesizing almost any prostaglandin or prostaglandinlike substance.

Prostaglandins are not stored as such in tissues; instead, they are synthesized in response to specific environmental or physiological triggers. Starting materials for prostaglandin synthesis are unsaturated fatty acids of twenty carbon atoms. Those of the 2 series are derived from

Figure VI-2 Found in the Caribbean Sea, *Plexaura homomalla*, known as the sea whip or gorgonian, contains the highest concentration of prostaglandinlike compounds so far found in nature. Before economical laboratory syntheses of prostaglandins became possible, Upjohn extracted the rare substances from this coral.

arachidonic acid (5,8,11,14-eicosatetraenoic acid), an unsaturated fatty acid containing four carbon-carbon double bonds. Steps in the biochemical pathways by which arachidonic acid is converted to several key prostaglandins are summarized in Figure VI-3. Arachidonic acid is drawn in this figure to show the relation between its structural formula and that of the prostaglandins derived from it.

A key step in the biosynthesis of prostaglandins of the 2 series is reaction of arachidonic acid with two molecules of oxygen, O_2, to form PGG_2.

Enzyme-catalyzed reduction of PGG_2 gives PGH_2, a key intermediate from which all other prostaglandins of the 2 series are synthesized. Within minutes, it is converted to other prostaglandin and prostaglandin-derived compounds. Shown in Figure VI-3 is the biosynthesis of types A, E, F, G, and H. There are other types, also derived from the key intermediate PGH_2. One of these is thromboxane A_2. Precisely which prostaglandins are produced depends on the enzymes present in the particular tissue.

Now that we have seen how the body synthesizes prostaglandins, let us look at several functions of these substances in the body. First is participation of prostaglandins in blood clotting. There are three distinct phases to the physiological mechanisms that operate within the body to stop bleeding from a ruptured blood vessel. The first phase, called platelet aggregation, is initiated at the site of the injury by agents such as thrombin. During platelet aggregation, blood platelets become sticky and form a platelet plug at the site of the injury. If damage is minor and the blood vessel is small, this platelet plug may be sufficient to stop loss of blood from the vessel. If it is not sufficient, the platelets are stimulated to release a group of substances (the platelet-release reaction), which in turn promote a second wave of platelet aggregation and constriction of the injured vessel. The third phase is the triggering of the actual blood coagulation.

We have learned within the past few years that among the substances released in platelet release reactions is thromboxane A_2. This pro-

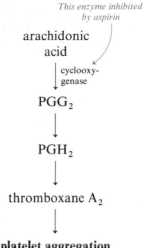

This enzyme inhibited by aspirin

arachidonic acid

↓ cyclooxygenase

PGG_2

↓

PGH_2

↓

thromboxane A_2

↓

platelet aggregation

staglandin-derived molecule is a very potent vasoconstrictor and the key substance that triggers platelet aggregation. Just as we have known for some time that thrombin stimulates the second, irreversible phase of platelet aggregation, we have also known that aspirin and aspirinlike drugs, such as indomethacin, inhibit this second phase. How these drugs are able to do this remained a mystery until it was discovered that aspirin inhibits cyclooxygenase, the enzyme that initiates the synthesis of thromboxane A_2.

There is now good evidence that the ability of aspirin to reduce inflammation is also related to its ability to inhibit prostaglandin synthesis. Further research on the prostaglandins may help us understand even more about inflammatory diseases such as rheumatoid arthritis, asthma, and other allergic responses.

The first recorded observations on the biological activity of prostaglandins were those of gynecologists Kurzrok and Lieb. The first widespread clinical application of these substances was also by gynecologists and obstetricians. The observation that prostaglandins stimulate contraction of uterine smooth muscle led to the suggestion that these substances could be used for termination of second-trimester pregnancy. One problem with the use of natural prostaglandins for this purpose is that they are very rapidly degraded within the body. Therefore, their use

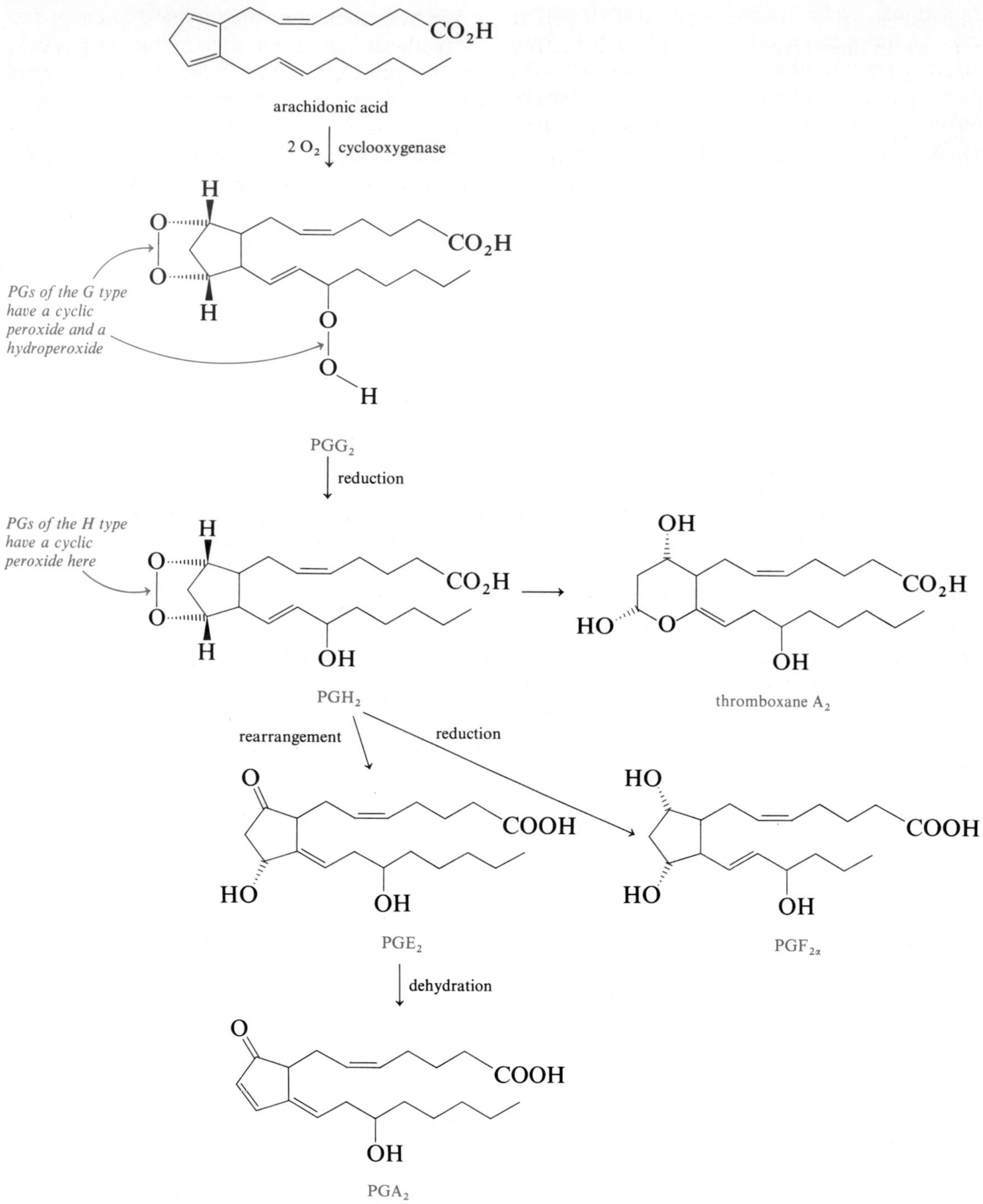

Figure VI-3 Biosynthesis of several prostaglandins from arachidonic acid.

required repeated administration over a period of hours. In the search for less rapidly degrading prostaglandins, a number of semisynthetic prostaglandin analogs were prepared. One of the most effective was 15-methyl prostaglandin $F_{2\alpha}$, which is longer-acting and has 10 to 20 times the potency of $PGF_{2\alpha}$.

An extra methyl group at carbon 15

15-methyl prostaglandin $F_{2\alpha}$

The potential clinical use of prostaglandins and prostaglandin analogs for termination of second-trimester pregnancy was explored in a study designed and conducted by the World Health Organization Task Force on the Use of Prostaglandins for the Regulation of Fertility. This multicenter, multinational study, entitled "Prostaglandins and Abortion," is described in the *American Journal of Obstetrics and Gynecology* (1977). It concludes that a single intra-amniotic injection of 15-methyl $PGF_{2\alpha}$ is a safe and effective way of terminating second-trimester pregnancy.

We have looked at only a few aspects of the biosynthesis and importance in human physiology of prostaglandins. From even this brief study, it should be clear that we are only beginning to understand the chemistry and biochemistry of this group of compounds. And it should also be clear that the enormous prostaglandin research effort now under way offers great promise for even deeper insight into human physiology and for the development of new and highly effective drugs for clinical use.

Sources

Bergstrom, S. 1967. Prostaglandins: Members of a new hormonal system. *Science* 157:382.

Bergstrom, S., Carlson, L. A., and Weeks, J. R. 1968. The prostaglandins: A family of biologically active lipids. *Pharm. Rev.* 20:1.

Goodman, L. S., and Gilman A. 1980. *The Pharmacological Basis of Therapeutics.* 6th ed. New York: Macmillan.

Kuehl, F. A., and Egan, R. W. 1980. Prostaglandins, arachidonic acid, and inflammation. *Science* 210: 978–984.

Needleman, P., et al. 1976. Prostaglandins and abortion. *Nature* 261:558–560.

Ramwell, P. W., ed. 1973. *The Prostaglandins.* Vol. 1. New York: Plenum Press.

Samuelsson, B., et al. 1975. Prostaglandins. *Ann. Rev. Biochem.* 44:669–695.

World Health Organization. 1977. Prostaglandins and abortion. *Am. J. Gyn.* 129:593–606.

8

Derivatives of Carboxylic Acids

Anhydrides, **esters**, and **amides** are all functional derivatives of carboxylic acids. A general formula for the characteristic structural feature of each of these derivatives is

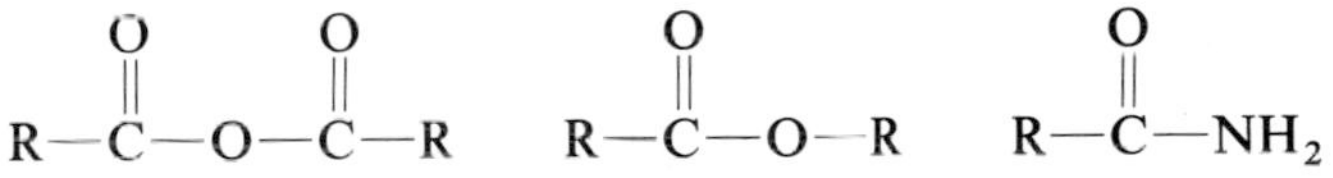

$$R-\overset{\overset{\displaystyle O}{\|}}{C}-O-\overset{\overset{\displaystyle O}{\|}}{C}-R \qquad R-\overset{\overset{\displaystyle O}{\|}}{C}-O-R \qquad R-\overset{\overset{\displaystyle O}{\|}}{C}-NH_2$$

an anhydride   an ester   an amide

In each of these functional groups, the —OH of the carboxyl group has been replaced by another group of atoms. One way to relate the structural formulas of these functional groups to the structural formula of a carboxylic acid is to imagine a reaction in which the —OH from the carboxyl and —H from an acid, an alcohol, or an amine is removed as water and the remaining atoms are joined in the following ways:

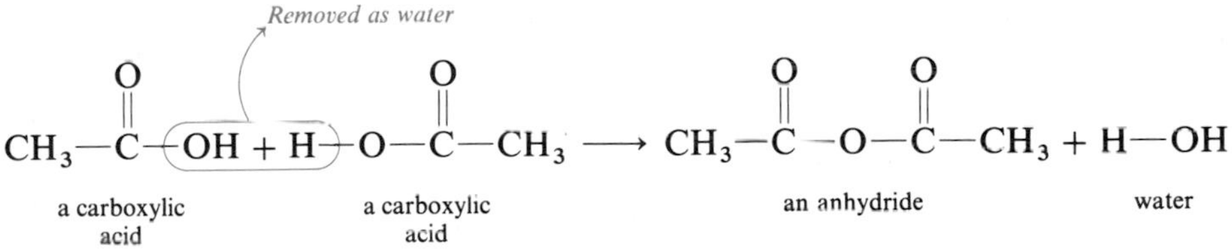

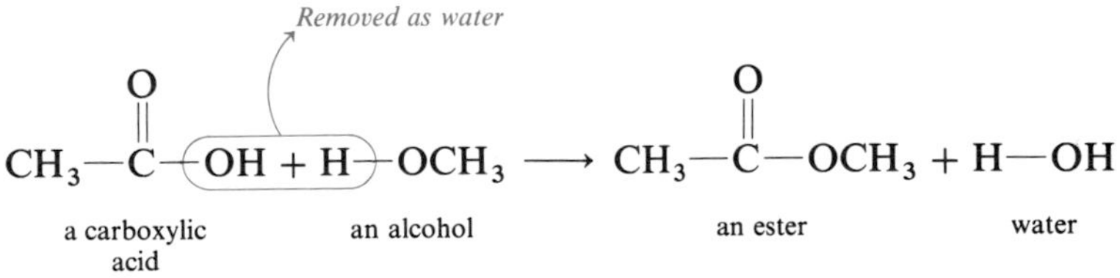

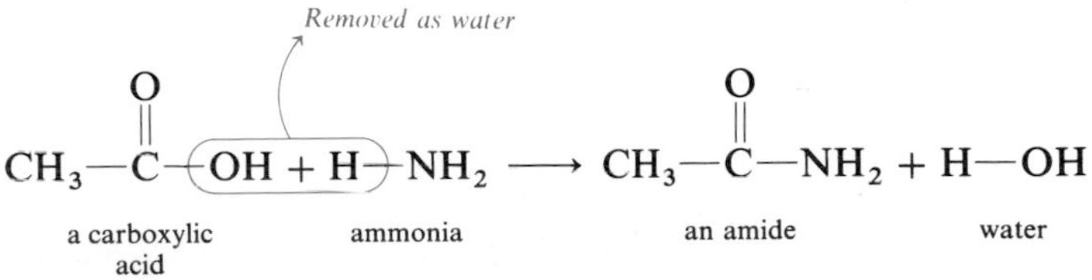

Removed as water

$$CH_3-\overset{\overset{\displaystyle O}{\|}}{C}-(OH + H)-O-\overset{\overset{\displaystyle O}{\|}}{C}-CH_3 \longrightarrow CH_3-\overset{\overset{\displaystyle O}{\|}}{C}-O-\overset{\overset{\displaystyle O}{\|}}{C}-CH_3 + H-OH$$

a carboxylic   a carboxylic   an anhydride   water
acid   acid

Removed as water

$$CH_3-\overset{\overset{\displaystyle O}{\|}}{C}-(OH + H)-OCH_3 \longrightarrow CH_3-\overset{\overset{\displaystyle O}{\|}}{C}-OCH_3 + H-OH$$

a carboxylic   an alcohol   an ester   water
acid

Removed as water

$$CH_3-\overset{\overset{\displaystyle O}{\|}}{C}-(OH + H)-NH_2 \longrightarrow CH_3-\overset{\overset{\displaystyle O}{\|}}{C}-NH_2 + H-OH$$

a carboxylic   ammonia   an amide   water
acid

215

The equations are shown only to illustrate the relation between a carboxylic acid and the three functional groups we shall study. They are not meant to illustrate methods for the synthesis of these functional groups.

8.1 Nomenclature

A. Anhydrides

The characteristic structural feature of an organic anhydride is the presence of a —CO—O—CO— group. Two examples of symmetrical anhydrides, that is, anhydrides derived from just one carboxylic acid, are

$$
\underset{\substack{\text{characteristic}\\\text{structural feature}}}{-\overset{\overset{\text{O}}{\|}}{\text{C}}-\text{O}-\overset{\overset{\text{O}}{\|}}{\text{C}}-} \qquad
\underset{\text{acetic anhydride}}{\text{CH}_3-\overset{\overset{\text{O}}{\|}}{\text{C}}-\text{O}-\overset{\overset{\text{O}}{\|}}{\text{C}}-\text{CH}_3} \qquad
\underset{\text{benzoic anhydride}}{\bigcirc-\overset{\overset{\text{O}}{\|}}{\text{C}}-\text{O}-\overset{\overset{\text{O}}{\|}}{\text{C}}-\bigcirc}
$$

In the IUPAC system, anhydrides are named by adding the word *anhydride* to the name of the parent acid. The anhydride derived from two molecules of acetic acid is named acetic anhydride; that derived from two molecules of benzoic acid is named benzoic anhydride.

Phosphoric acid also forms anhydrides with the following characteristic structural feature:

$$
\underset{\substack{\text{characteristic}\\\text{structural feature}\\\text{of a phosphate}\\\text{anhydride}}}{-\overset{\overset{\text{O}}{\|}}{\underset{\underset{\text{HO}}{|}}{\text{P}}}-\text{O}-\overset{\overset{\text{O}}{\|}}{\underset{\underset{\text{OH}}{|}}{\text{P}}}-} \qquad
\underset{\substack{\text{phosphoric anhydride}\\\text{(pyrophosphoric acid)}}}{\text{HO}-\overset{\overset{\text{O}}{\|}}{\underset{\underset{\text{HO}}{|}}{\text{P}}}-\text{O}-\overset{\overset{\text{O}}{\|}}{\underset{\underset{\text{OH}}{|}}{\text{P}}}-\text{OH}} \qquad
\underset{\text{pyrophosphate}}{{}^-\text{O}-\overset{\overset{\text{O}}{\|}}{\underset{\underset{{}^-\text{O}}{|}}{\text{P}}}-\text{O}-\overset{\overset{\text{O}}{\|}}{\underset{\underset{\text{O}^-}{|}}{\text{P}}}-\text{O}^-}
$$

The anhydride derived from two molecules of phosphoric acid is named phosphoric anhydride, or more commonly, pyrophosphoric acid. Phosphoric anhydride, like phosphoric acid, is a strong acid. At pH 7, it is completely ionized and has a net charge of -4. The anion is named pyrophosphate. Pyrophosphate and other anhydrides of phosphoric acid are especially important in the biological world.

B. Esters

In the IUPAC system, an ester is named as a derivative of the carboxylic acid from which it is derived. The alkyl or aryl group attached to oxygen is named first. Then the acid is named by dropping the suffix *-ic acid* and adding the

Table 8.1
Example showing
the derivation of
ester names.

Structural Formula	Alkyl Group Attached to Oxygen	Name of Carboxylic Acid	Name of Ester
$\overset{O}{\overset{\|\|}{CH_3C}}OCH_2CH_3$	ethyl	acetic acid	ethyl acetate
$C_6H_5{-}\overset{O}{\overset{\|\|}{C}}OCHCH_3$ with $\|$ CH_3	isopropyl	benzoic acid	isopropyl benzoate
$\overset{O}{\overset{\|\|}{CH_2C}}OCH_2CH_3$ $\overset{\|}{CH_2\overset{O}{\underset{\|\|}{C}}OCH_2CH_3}$	diethyl	butanedioic acid (succinic acid)	diethyl butanedioate (diethyl succinate)

suffix -*ate*. In Table 8.1, names of esters are derived in a stepwise manner. First is given the name of the alkyl or aryl group attached to oxygen, then the IUPAC and common names of the acid from which the ester is derived, and finally the name of the ester.

Example 8.1

Name the following esters.

a. $CH_3\overset{CH_3}{\overset{\|}{C}}HCH_2\overset{O}{\overset{\|\|}{C}}OCH_3$

b. $H\overset{O}{\overset{\|\|}{C}}OCHCH_2CH_3$ with $\overset{\|}{CH_3}$

c. $CH_3\overset{O}{\overset{\|\|}{C}}CH_2\overset{O}{\overset{\|\|}{C}}OCH_2CH_3$

Solution

Each name is derived in a stepwise manner. First the name of the alkyl group attached to oxygen, then the IUPAC name (and where appropriate, common name as well) of the acid, and finally the name of the ester.

Alkyl Group Attached to Oxygen	Name of Carboxylic Acid	Name of Ester
a. methyl	3-methylbutanoic acid (isovaleric acid)	methyl 3-methylbutanoate (methyl isovalerate)
b. *sec*-butyl	formic acid	*sec*-butyl formate
c. ethyl	3-oxobutanoic acid (β-ketobutyric acid; acetoacetic acid)	ethyl 3-oxobutanoate (ethyl β-ketobutyrate; ethyl acetoacetate)

Name the following esters.

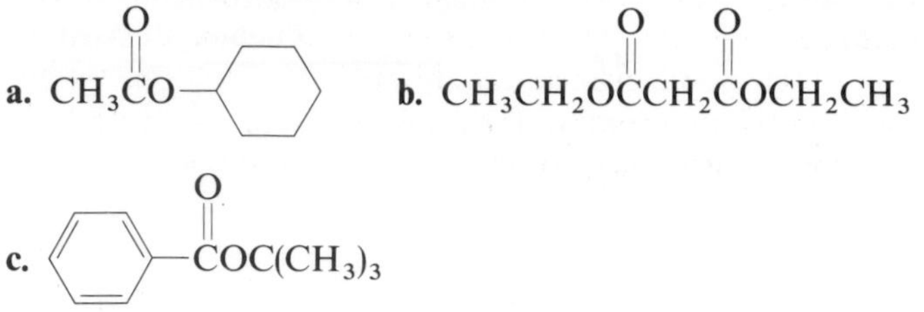

a. $CH_3\overset{\displaystyle O}{\overset{\|}{C}}O$—⟨cyclohexyl⟩

b. $CH_3CH_2O\overset{\displaystyle O}{\overset{\|}{C}}CH_2\overset{\displaystyle O}{\overset{\|}{C}}OCH_2CH_3$

c. ⟨phenyl⟩—$\overset{\displaystyle O}{\overset{\|}{C}}OC(CH_3)_3$

Esters of phosphoric acid are especially important in the biological world. Phosphoric acid has three —OH groups and can form mono-, di-, and triesters. These are named by giving the name or names of the groups attached to oxygen followed by the word *phosphate*. Examples are:

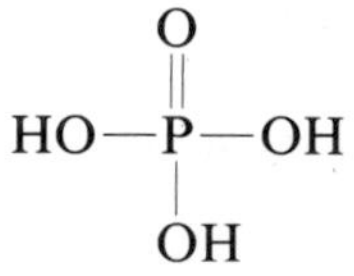

$$HO\overset{\displaystyle O}{\underset{\displaystyle OH}{\overset{\|}{-P-}}}OH \qquad CH_3O\overset{\displaystyle O}{\underset{\displaystyle OH}{\overset{\|}{-P-}}}OH \qquad CH_3O\overset{\displaystyle O}{\underset{\displaystyle OH}{\overset{\|}{-P-}}}OCH_3 \qquad CH_3O\overset{\displaystyle O}{\underset{\displaystyle OCH_3}{\overset{\|}{-P-}}}O-CH_3$$

phosphoric acid	methyl phosphate (a monoester)	dimethyl phosphate (a diester)	trimethyl phosphate (a triester)

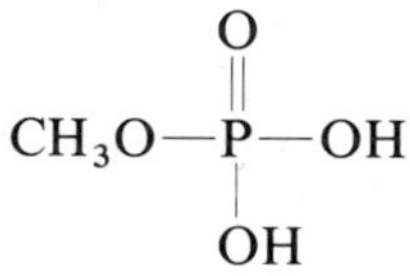

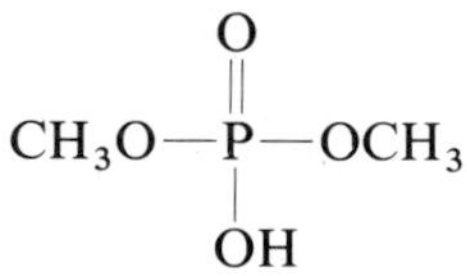

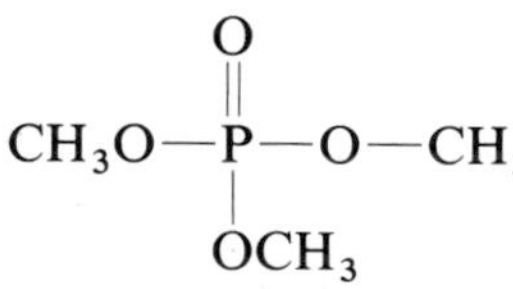

In more complex phosphate esters, it is often easier to name the organic molecule itself and indicate the presence of a phosphate ester by adding the word *phosphate* as shown in the following examples. Glyceraldehyde 3-phosphate is an intermediate in glycolysis. Pyridoxal, more commonly called vitamin B_6, is metabolically active only after it is converted to its phosphate ester, pyridoxal phosphate.

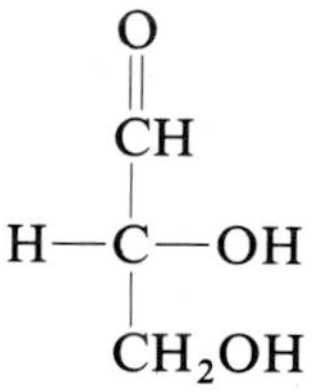

$$\begin{array}{c} \overset{\displaystyle O}{\overset{\|}{CH}} \\ H-\overset{\displaystyle }{\underset{\displaystyle }{C}}-OH \\ CH_2OH \end{array} \qquad \begin{array}{c} \overset{\displaystyle O}{\overset{\|}{CH}} \\ H-\overset{\displaystyle }{\underset{\displaystyle }{C}}-OH\;\;O \\ CH_2O-\overset{\displaystyle \|}{\underset{\displaystyle O^-}{P}}-O^- \end{array}$$

glyceraldehyde 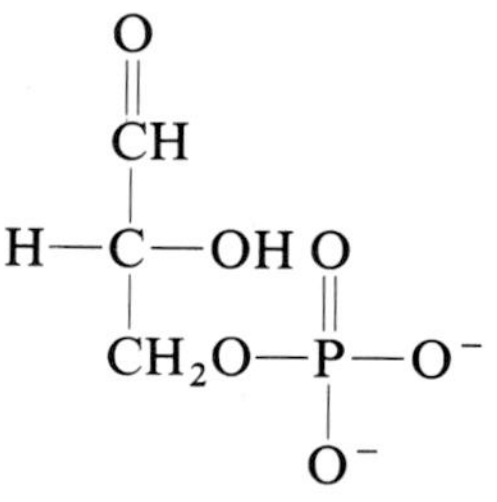   glyceraldehyde 3-phosphate

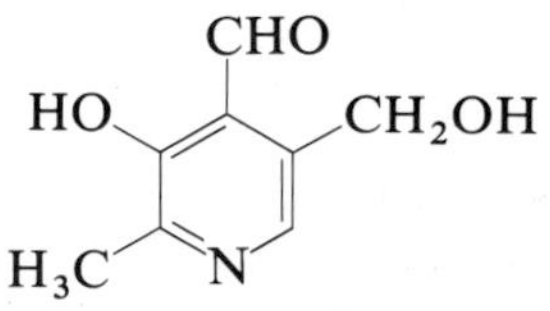

pyridoxal (vitamin B_6) 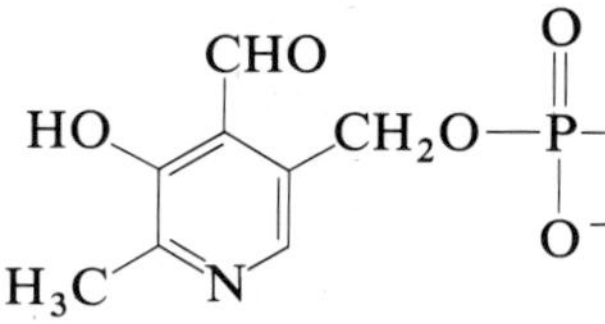   pyridoxal phosphate

In drawing structural formulas for phosphate anhydrides and esters found in biological systems, we will show the state of ionization at pH 7.4, the pH of normal blood plasma. At this pH, the two remaining hydrogen atoms of each phosphate group in the above esters are completely ionized, each having, therefore, a net charge of -2.

C. Amides

Amides are named as derivatives of their specific carboxylic acid by dropping the suffix *-oic acid* from the IUPAC name or *-ic acid* from the common name and adding *-amide*. In the examples, the IUPAC name is given first, and then in parentheses where appropriate, the common name.

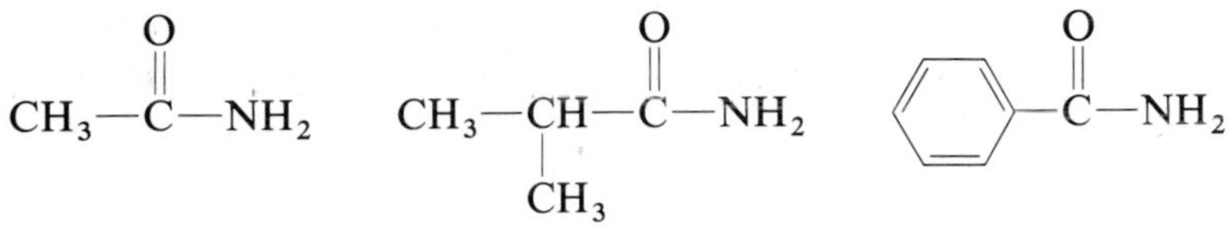

$$CH_3-\overset{\overset{\displaystyle O}{\|}}{C}-NH_2 \qquad CH_3-\underset{\underset{\displaystyle CH_3}{|}}{CH}-\overset{\overset{\displaystyle O}{\|}}{C}-NH_2 \qquad \underset{}{\bigcirc}-\overset{\overset{\displaystyle O}{\|}}{C}-NH_2$$

acetamide   2-methylpropanamide   benzamide
                      (isobutyramide)

If a hydrogen atom of an amide is replaced by an alkyl or an aryl group, the substituent is named and its location on nitrogen is indicated by *N*-. Two substituents on an amide nitrogen are named and indicated by *N,N*-.

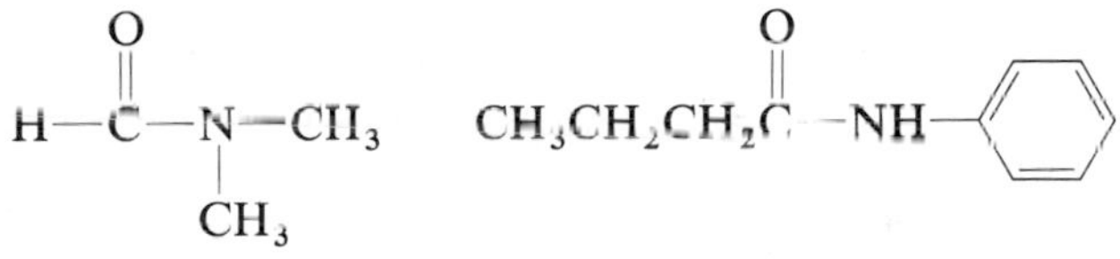

$$H-\overset{\overset{\displaystyle O}{\|}}{C}-\underset{\underset{\displaystyle CH_3}{|}}{N}-CH_3 \qquad CH_3CH_2CH_2\overset{\overset{\displaystyle O}{\|}}{C}-NH-\bigcirc$$

N,N-dimethylformamide   *N*-phenylbutanamide

Example 8.2

Name these amides:

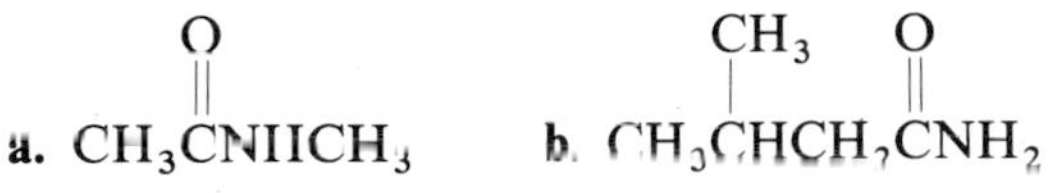

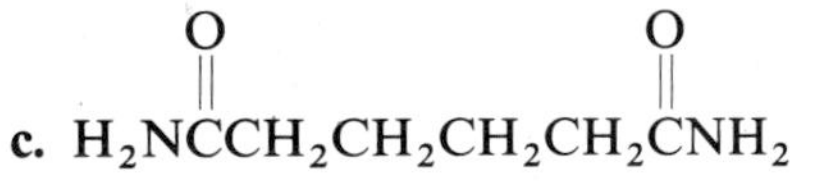

a. $CH_3\overset{\overset{\displaystyle O}{\|}}{C}NHCH_3$ **b.** $CH_3\underset{\underset{\displaystyle}{|}}{\overset{\overset{\displaystyle CH_3}{|}}{CH}}CH_2\overset{\overset{\displaystyle O}{\|}}{C}NH_2$

c. $H_2N\overset{\overset{\displaystyle O}{\|}}{C}CH_2CH_2CH_2CH_2\overset{\overset{\displaystyle O}{\|}}{C}NH_2$

Solution

Each is named below in a stepwise manner, just as we did for esters in Example 8.1.

Group Attached to Nitrogen	Name of the Carboxylic Acid	Name of the Amide
a. *N*-methyl	acetic acid	*N*-methylacetamide
b. _____	3-methylbutanoic acid (isovaleric acid)	3-methylbutanamide (isovaleramide)
c. _____	hexanedioic acid (adipic acid)	hexanediamide (adipamide)

■
Problem 8.2

Name the following amides:

a. $CH_3\overset{O}{\overset{\|}{C}}NH\!-\!\!\bigcirc$

b. $CH_3(CH_2)_6\overset{O}{\overset{\|}{C}}NH_2$

c. (structure: benzene ring with $\overset{O}{\overset{\|}{C}}NH_2$ group and OH group)

8.2 Esters of Nitrous and Nitric Acids

Several esters of nitric acid and nitrous acid have been used as drugs for more than a hundred years. Esters of nitric acid are named nitrates. Glyceryl nitrate, usually known as nitroglycerine, is a triester of glycerol.

$$HO\!-\!NO_2$$

$$\begin{array}{l} CH_2\!-\!OH \\ | \\ CH\!-\!OH \\ | \\ CH_2\!-\!OH \end{array}$$

$$\begin{array}{l} CH_2\!-\!O\!-\!NO_2 \\ | \\ CH\!-\!O\!-\!NO_2 \\ | \\ CH_2\!-\!O\!-\!NO_2 \end{array}$$

nitric acid   1,2,3-propanetriol   glyceryl nitrate
          (glycerol or glycerine)   (nitroglycerine)

Esters of nitrous acid are named nitrites. The structural formula of 3-methylbutyl nitrite, more often known as isopentyl nitrite or isoamyl nitrite, is

$$HO\!-\!NO$$

$$\overset{\displaystyle CH_3}{\underset{\displaystyle CH_3CHCH_2CH_2\!-\!OH}{|}}$$

$$\overset{\displaystyle CH_3}{\underset{\displaystyle CH_3CHCH_2CH_2\!-\!O\!-\!NO}{|}}$$

nitrous acid   3-methyl-1-butanol   3-methyl-1-butyl nitrite
     (isopentyl alcohol; isoamyl alcohol)   (isopentyl nitrite; isoamyl nitrite)

The most important medical use of nitroglycerine and isoamyl nitrite is relaxation of the smooth muscle of blood vessels and dilation of all large and small arteries of the heart. Both esters are used in treating angina pectoris, a heart disease characterized by spasms of the coronary artery and agonizing chest pains.

8.3 Physical Properties of Esters and Amides

Esters are polar compounds and are attracted to each other in the pure state by a combination of dipole-dipole interactions between polar —COO— groups and dispersion forces between nonpolar hydrocarbon chains. Most esters are insoluble in water because of the more dominant, hydrophobic character of the hydrocarbon portions of the molecule. Esters are soluble in polar organic solvents such as diethyl ether and acetone.

The low-molecular-weight esters have pleasant odors. The characteristic fragrances of many flowers and fruits is due to the presence of esters, either singly or more often, in mixtures. Fragrances associated with several esters are:

Ester	Fragrance
ethyl formate	artificial rum flavor
methyl butanoate	apples
ethyl butanoate	pineapples
octyl acetate	oranges

Amides are also polar molecules, and because of the polar character of the $C=O$ and $N-H$ bonds, hydrogen bonding is possible between a partially positive hydrogen of one amide group and a partially negative oxygen of another.

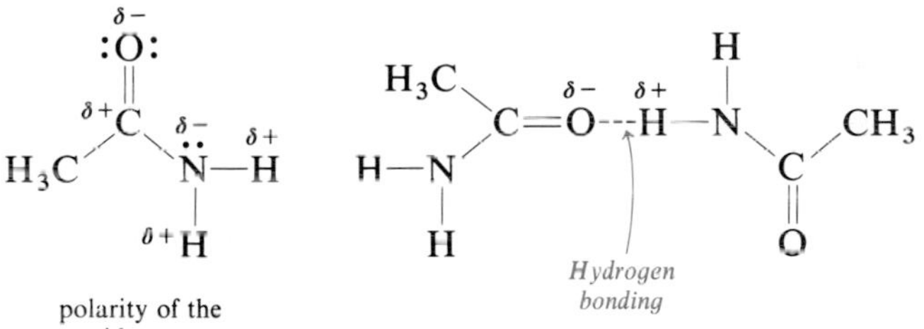

polarity of the
amide group

Hydrogen bonding

Because of this polarity and association by hydrogen bonding, amides have higher boiling points and are more soluble in water compared with esters of similar molecular weight. Virtually all amides are solids at room temperature.

8.4 Reactions of Anhydrides, Esters, and Amides

The basic reaction theme common to the carbonyl group of aldehydes, ketones, carboxylic acids, anhydrides, esters, and amides is nucleophilic addition to the carbonyl group. In aldehydes and ketones, the carbonyl addition product is often isolated as such. For example, in aldol condensations (Section 6.8), the carbonyl group undergoing reaction is transformed to an alcohol, and this nucleophilic addition product is the final product of the reaction:

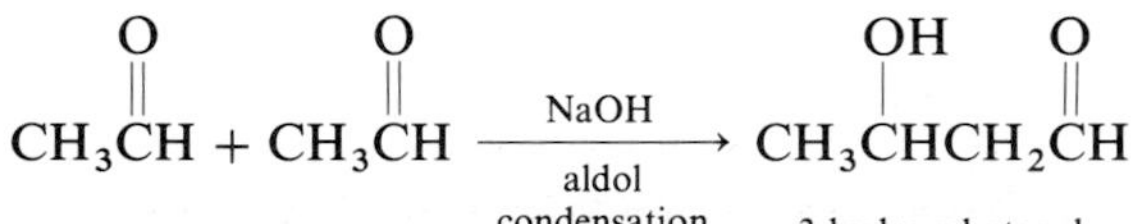

$$CH_3\overset{O}{\overset{\|}{C}}H + CH_3\overset{O}{\overset{\|}{C}}H \xrightarrow[\text{condensation}]{\text{NaOH aldol}} CH_3\overset{OH}{\overset{|}{C}}HCH_2\overset{O}{\overset{\|}{C}}H$$

3-hydroxybutanal
(aldol)

In other reactions of aldehydes and ketones, a carbonyl addition product is formed but then undergoes loss of H_2O, to yield a new functional group. For example, reaction of an aldehyde or a ketone with a primary amine forms a carbonyl addition product that then loses a molecule of water, forming an imine (a Schiff base); in this reaction, a C=O group is transformed to a C=N— group (Section 6.4C).

$$CH_3\overset{\overset{\displaystyle O}{\|}}{C}H + H_2N\!-\!\!\langle \rangle \longrightarrow \left[CH_3\overset{\overset{\displaystyle HO}{|}}{\underset{\underset{\displaystyle H}{|}}{C}}\!-\!\overset{\overset{\displaystyle H}{|}}{N}\!-\!\!\langle \rangle \right] \longrightarrow CH_3\overset{}{C}\!=\!\overset{\overset{\displaystyle }{}}{\underset{\underset{\displaystyle H}{|}}{N}}\!-\!\!\langle \rangle + H_2O$$

a tetrahedral carbonyl   an imine
addition intermediate   (a Schiff base)

With the new functional groups to be studied in this chapter, the carbonyl addition product collapses to regenerate the carbonyl group.

$$R\!-\!\overset{\overset{\displaystyle O}{\|}}{C}\!-\!X + H\!-\!Nu: \longrightarrow \left[R\!-\!\overset{\overset{\displaystyle OH}{|}}{\underset{\underset{\displaystyle X}{|}}{C}}\!-\!Nu \right] \longrightarrow R\!-\!\overset{\overset{\displaystyle O}{\|}}{C}\!-\!Nu + H\!-\!X$$

tetrahedral carbonyl
addition intermediate

The effect of this reaction is for a new atom or group of atoms to be substituted for one already attached to the carbonyl group. That is why we characterize these reactions as **nucleophilic substitution at a carbonyl carbon.**

8.5 Preparation of Esters: Fischer Esterification

A carboxylic acid can be converted to an ester by being heated with an alcohol in the presence of an acid catalyst, usually concentrated sulfuric acid, dry hydrogen chloride, or an ion-exchange resin in the acid form. Conversion of a carboxylic acid and an alcohol to an ester in the presence of an acid catalyst is called **Fischer esterification.** As an example, reaction of acetic acid and ethanol in the presence of concentrated sulfuric acid gives ethyl acetate and water:

$$CH_3\overset{\overset{\displaystyle O}{\|}}{C}OH + HOCH_2CH_3 \underset{}{\overset{H^+}{\rightleftharpoons}} CH_3\overset{\overset{\displaystyle O}{\|}}{C}OCH_2CH_3 + HOH$$

acetic acid   ethanol   ethyl acetate

Acid-catalyzed esterification is reversible, and generally at equilibrium the quantities of both ester and alcohol are appreciable. If 60.1 g (1.0 mol) of acetic acid and 60.1 g (1.0 mol) of 1-propanol are heated under reflux in the presence of a few drops of concentrated sulfuric acid until equilibrium is reached, the reaction mixture will contain about 0.67 mol each of propyl acetate and water, and about

0.33 mol each of acetic acid and 1-propanol. At equilibrium, about 67% of the acid and alcohol has converted to the desired ester.

$$\underset{\text{acetic acid}}{CH_3\overset{O}{\overset{\|}{C}}OH} + \underset{\text{1-propanol}}{HOCH_2CH_2CH_3} \overset{H^+}{\rightleftharpoons} \underset{\text{propyl acetate}}{CH_3\overset{O}{\overset{\|}{C}}OCH_2CH_2CH_3} + \underset{\text{water}}{H_2O}$$

	acetic acid	1-propanol	propyl acetate	water
initial:	1.00 mol	1.00 mol	0.00 mol	0.00 mol
equilibrium:	0.33 mol	0.33 mol	0.67 mol	0.67 mol

By careful control of reaction conditions, direct esterification can be used to prepare esters in high yields. For example, if the alcohol is inexpensive compared with the acid, a large excess of it can be used to drive the reaction to the right and achieve a high conversion of carboxylic acid to its ester. Or it may be possible to take advantage of a situation in which the boiling points of the reactants and ester are higher than the boiling point of water. In this case, heating the reaction mixture somewhat above 100°C removes water as it is formed and shifts the position of equilibrium toward the production of a higher yield of ester.

Fischer esterification is but one of the general methods for preparing esters. Another is reaction of an alcohol or phenol with an anhydride (Section 8.7).

Example 8.3

Name and draw structural formulas for the esters produced in these reactions:

a. $\overset{CO_2H}{\underset{OH}{}}$ $+ CH_3OH \overset{H^+}{\rightleftharpoons}$

b. $HO\overset{O}{\overset{\|}{C}}CH_2\overset{O}{\overset{\|}{C}}OH + 2CH_3CH_2OH \overset{H^+}{\rightleftharpoons}$

☐ *Solution*

a. $\overset{\overset{O}{\|}}{\underset{OH}{C}OH}$ $+ CH_3OH \overset{H^+}{\rightleftharpoons}$ $\overset{\overset{O}{\|}}{\underset{OH}{C}OCH_3}$ $+ H_2O$

2-hydroxybenzoic acid   methanol   methyl 2-hydroxybenzoate
(salicylic acid)   (methyl salicylate; oil of wintergreen)

b. $\underset{\substack{\text{propanedioic acid}\\\text{(malonic acid)}}}{HO\overset{O}{\overset{\|}{C}}CH_2\overset{O}{\overset{\|}{C}}OH} + \underset{\text{ethanol}}{2CH_3CH_2OH} \overset{H^+}{\rightleftharpoons} \underset{\substack{\text{diethyl propanedioate}\\\text{(diethyl malonate)}}}{CH_3CH_2O\overset{O}{\overset{\|}{C}}CH_2\overset{O}{\overset{\|}{C}}OCH_2CH_3} + H_2O$

Problem 8.3

Name and draw structural formulas for the esters formed in these reactions:

a.

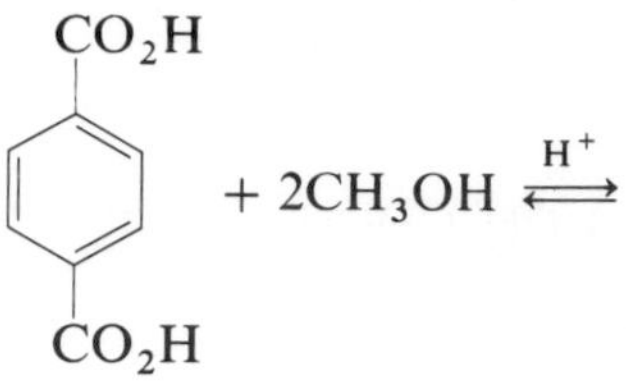

$$CO_2H$$

(benzene ring with CO_2H groups) $+ 2CH_3OH \xrightarrow{H^+}$

$$CO_2H$$

b. $CH_3CO_2H + CH_3CH_2CH_2CH_2CH_2OH \xrightarrow{H^+}$

Following is a mechanism for acid-catalyzed esterification. Protonation of the carbonyl oxygen in step 1 gives a carbocation with a positive charge on the carbonyl carbon. Reaction of this ion in step 2 with a pair of electrons on the oxygen atom of the alcohol gives an oxonium ion, which loses H^+ in step 3, to give a tetrahedral carbonyl addition intermediate. Reaction of an —OH of this intermediate with H^+ in step 4, followed by loss of a molecule of water in step 5, gives a new carbocation, which then loses H^+ in step 6, to give the ester.

Step 1: Reaction of the carbonyl oxygen with H^+ to give a carbocation

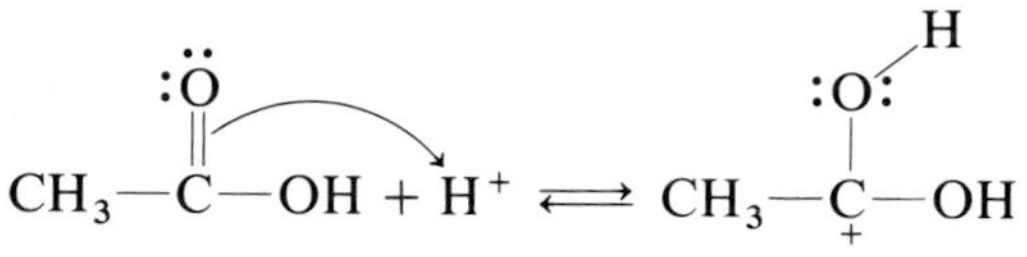

$$CH_3-\overset{\overset{\displaystyle :\ddot{O}:}{\|}}{C}-OH + H^+ \rightleftharpoons CH_3-\overset{\overset{\displaystyle :\overset{H}{\diagup}O:}{|}}{\underset{+}{C}}-OH$$

a carbocation

Step 2: Reaction of an unshared pair of electrons from oxygen to form an oxonium ion

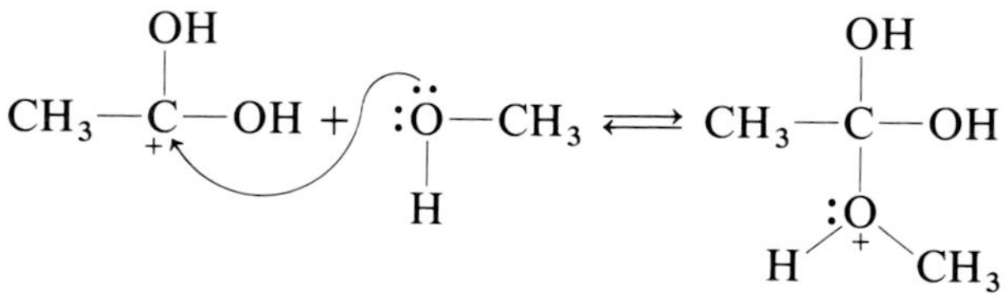

$$CH_3-\overset{\overset{\displaystyle OH}{|}}{\underset{+}{C}}-OH + \overset{\displaystyle :\ddot{O}-CH_3}{\underset{\displaystyle H}{}} \rightleftharpoons CH_3-\overset{\overset{\displaystyle OH}{|}}{\underset{\overset{\displaystyle :O}{\underset{\displaystyle H \overset{+}{\diagup}\diagdown CH_3}{}}}{C}}-OH$$

an oxonium ion

Step 3: Loss of H^+ to form a tetrahedral carbonyl addition intermediate

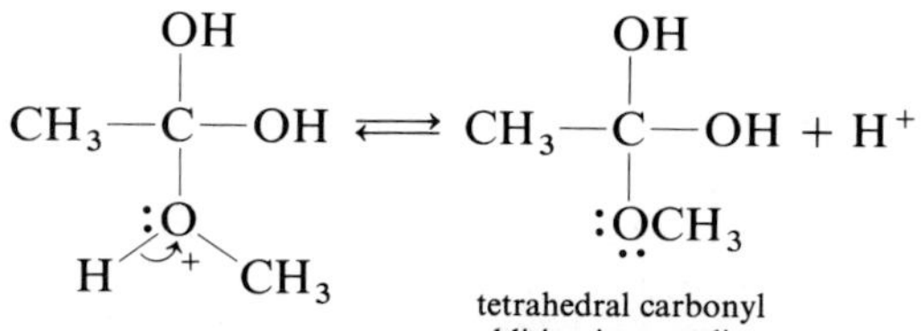

$$CH_3-\overset{\overset{\displaystyle OH}{|}}{\underset{\overset{\displaystyle :O}{\underset{\displaystyle H \overset{+}{\diagup}\diagdown CH_3}{}}}{C}}-OH \rightleftharpoons CH_3-\overset{\overset{\displaystyle OH}{|}}{\underset{\overset{\displaystyle |}{:\ddot{O}CH_3}}{C}}-OH + H^+$$

tetrahedral carbonyl
addition intermediate

Step 4: Reaction of an —OH of the tetrahedral carbonyl addition intermediate with H^+ to give a new oxonium ion

$$CH_3-\overset{\overset{\displaystyle OH}{|}}{\underset{\underset{\displaystyle OCH_3}{|}}{C}}-\ddot{O}H + H^+ \rightleftharpoons CH_3-\overset{\overset{\displaystyle OH}{|}}{\underset{\underset{\displaystyle OCH_3}{|}}{C}}-\overset{+}{\ddot{O}}\overset{\diagup H}{\diagdown H}$$

a new oxonium ion

Step 5: Loss of H_2O from this oxonium ion to form a carbocation

$$CH_3-\overset{\overset{\displaystyle OH}{|}}{\underset{\underset{\displaystyle OCH_3}{|}}{C}}-\overset{+}{\ddot{O}}\overset{\diagup H}{\diagdown CH_3} \rightleftharpoons CH_3-\overset{\overset{\displaystyle OH}{|}}{\underset{\underset{\displaystyle OCH_3}{|}}{\overset{+}{C}}} + H_2\ddot{O}:$$

a carbocation

Step 6: Loss of H^+ from the carbocation to give the ester

$$CH_3-\overset{\overset{\displaystyle :\ddot{O}\diagup H}{|}}{\underset{\underset{\displaystyle OCH_3}{|}}{\overset{+}{C}}} \rightleftharpoons CH_3-\overset{\overset{\displaystyle :\ddot{O}}{||}}{\underset{\underset{\displaystyle OCH_3}{|}}{C}} + H^+$$

an ester

This mechanism is a specific example of the more general mechanism we proposed in Section 8.4 for nucleophilic substitution at a carbonyl carbon. A key step in Fischer esterification is formation of a tetrahedral carbonyl addition intermediate:

$$CH_3-\overset{\overset{\displaystyle O}{||}}{C}-OH + \overset{\overset{\displaystyle H}{|}}{O}-CH_3 \rightleftharpoons \left[CH_3-\overset{\overset{\displaystyle O\diagup H}{|}}{\underset{\underset{\displaystyle OCH_3}{|}}{C}}-OH \right] \rightleftharpoons CH_3-\overset{\overset{\displaystyle O}{||}}{C}-OCH_3 + H_2O$$

tetrahedral carbonyl
addition intermediate

The six-step mechanism we have proposed for Fischer esterification predicts that the oxygen atom of the alcohol is incorporated into the ester and that the oxygen atom appearing in the water molecule is derived from one of the two oxygen atoms of the carboxyl group. This prediction has been tested in the following way. Oxygen in nature is a mixture of three isotopes.

Isotope	Natural Abundance
oxygen-16	99.76%
oxygen-17	0.037%
oxygen-18	0.204%

Through the use of modern techniques for separating isotopes, it is possible to prepare compounds significantly enriched in oxygen-18. Such compounds are said to be isotopically enriched, or isotopically labeled. One easily prepared compound now commercially available is isotopically labeled methanol. When methanol enriched with oxygen-18 is caused to react with benzoic acid containing only naturally occurring amounts of oxygen-18, all the isotope enrichment (the isotope label) is found in the ester.

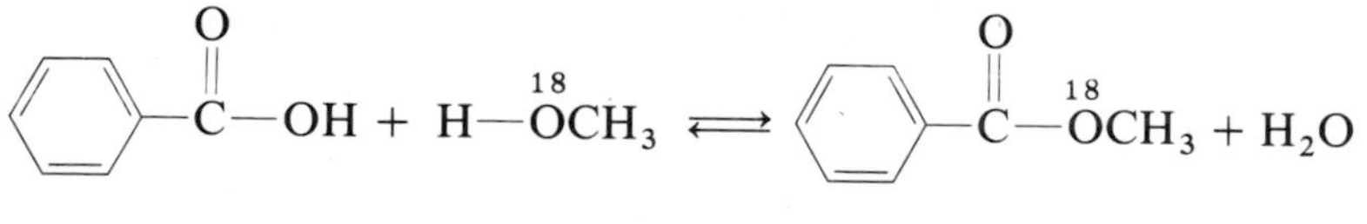

$$\text{C}_6\text{H}_5\text{C(=O)}-\text{OH} + \text{H}-\overset{18}{\text{O}}\text{CH}_3 \rightleftharpoons \text{C}_6\text{H}_5\text{C(=O)}-\overset{18}{\text{O}}\text{CH}_3 + \text{H}_2\text{O}$$

labeled methanol   labeled ester

Example 8.4

Draw structural formulas for the tetrahedral carbonyl addition intermediates formed in these acid-catalyzed esterifications:

a. $\text{CH}_3\text{CH}_2\text{C(=O)OH} + \text{HOCH}_2\text{CH}_3 \xrightarrow{\text{H}^+} \text{CH}_3\text{CH}_2\text{C(=O)OCH}_2\text{CH}_3 + \text{H}_2\text{O}$

b. 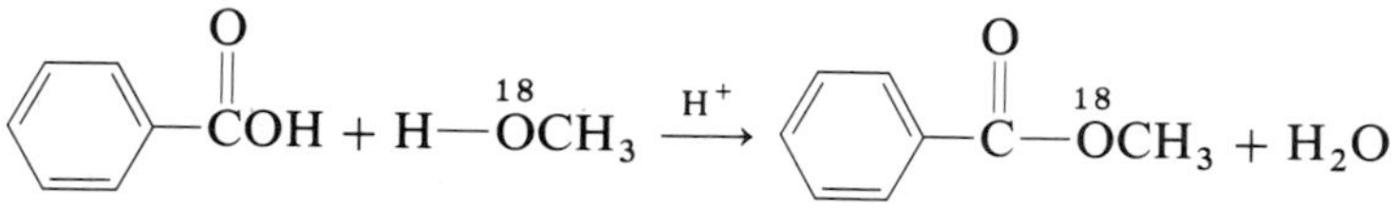 $\text{C}_6\text{H}_5\text{C(=O)OH} + \text{H}-\overset{18}{\text{O}}\text{CH}_3 \xrightarrow{\text{H}^+} \text{C}_6\text{H}_5\text{C(=O)}-\overset{18}{\text{O}}\text{CH}_3 + \text{H}_2\text{O}$

☐ *Solution*

a. $\text{CH}_3\text{CH}_2\text{C(=O)OH} + \text{HOCH}_2\text{CH}_3 \xrightarrow{\text{H}^+} \left[\text{CH}_3\text{CH}_2\overset{\text{OH}}{\underset{\text{OH}}{\text{C}}}-\text{OCH}_2\text{CH}_3 \right]$

b. 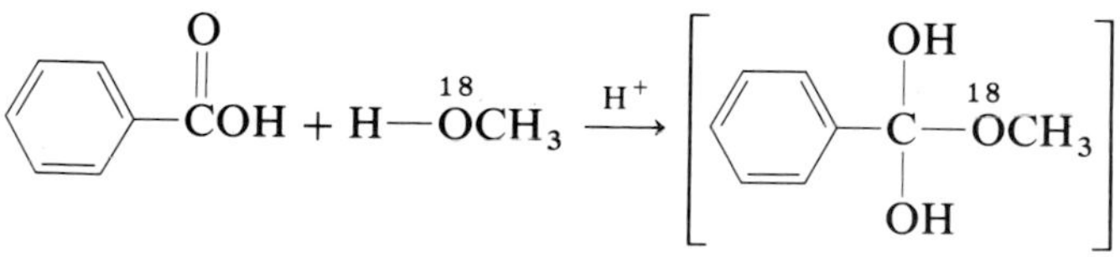 $\text{C}_6\text{H}_5\text{C(=O)OH} + \text{H}-\overset{18}{\text{O}}\text{CH}_3 \xrightarrow{\text{H}^+} \left[\text{C}_6\text{H}_5\overset{\text{OH}}{\underset{\text{OH}}{\text{C}}}-\overset{18}{\text{O}}\text{CH}_3 \right]$

Problem 8.4

Draw structural formulas for the tetrahedral carbonyl addition intermediates formed in the following acid-catalyzed esterifications:

a. $\text{HC(=O)OH} + \text{HOCH(CH}_3)\text{CH}_3 \xrightarrow{\text{H}^+} \text{HC(=O)OCH(CH}_3)\text{CH}_3 + \text{H}_2\text{O}$

b. 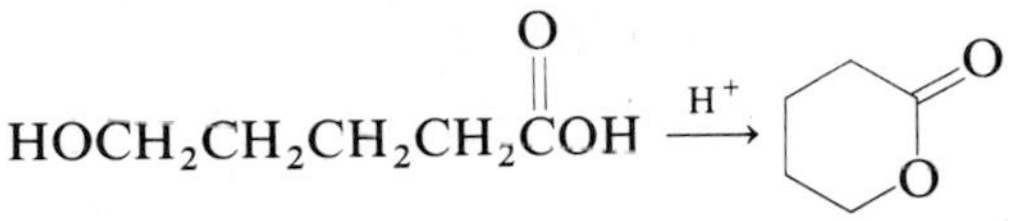 $HOCH_2CH_2CH_2CH_2COH \xrightarrow{H^+}$

8.6 Reactions of Esters

Of the three classes of carboxylic acid derivatives so far discussed, esters are intermediate in reactivity. Under most conditions, they are less reactive than anhydrides, but more reactive than amides.

A. Reduction

Esters are reduced by hydrogen in the presence of a heavy metal catalyst to two alcohols, one derived from the carboxyl portion of the ester, the other from the alkyl or aryl group. Such reductions often require high pressures of hydrogen gas and high temperatures. Esters are also reduced smoothly and easily by lithium aluminum hydride at room temperature:

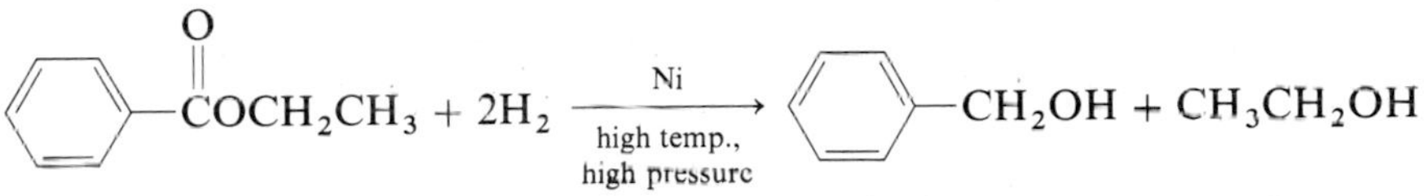

$$\text{ethyl benzoate} + 2H_2 \xrightarrow[\text{high pressure}]{\text{Ni}}{\atop\text{high temp.,}} \text{benzyl alcohol} + CH_3CH_2OH$$

ethyl benzoate   benzyl alcohol   ethanol

$$CH_3OC(CH_2)_4COCH_3 \xrightarrow[\text{(2) } H_2O]{\text{(1) LiAlH}_4} HOCH_2(CH_2)_4CH_2OH + 2CH_3OH$$

dimethyl hexanedioate   1,6-hexanediol   methanol
(dimethyl adipate)

B. Hydrolysis

Esters are normally unreactive with water. However, in the presence of either aqueous acid (most commonly HCl or H_2SO_4) or aqueous base (most commonly NaOH or KOH), they are split into a carboxylic acid and an alcohol.

1. Hydrolysis in Aqueous Acid

Following is an equation for acid-catalyzed hydrolysis of ethyl acetate to give acetic acid and ethanol:

$$CH_3COCH_2CH_3 + H_2O \underset{}{\overset{H^+}{\rightleftharpoons}} CH_3COH + CH_3CH_2OH$$

Because the mechanism we proposed in Section 8.5 for acid-catalyzed esterification is reversible, formation of the same tetrahedral carbonyl addition intermediate accounts equally well for acid-catalyzed hydrolysis.

$$\underset{\text{O}}{\overset{\text{O}}{\underset{\|}{CH_3C}}}-OCH_2CH_3 + H-OH \rightleftharpoons \left[\underset{\text{OH}}{\overset{\text{OH}}{CH_3\overset{|}{\underset{|}{C}}-OCH_2CH_3}} \right] \rightleftharpoons CH_3\overset{\text{O}}{\overset{\|}{C}}OH + CH_3CH_2OH$$

$$\xleftarrow{\hspace{1cm}} \text{hydrolysis} \xrightarrow{\hspace{1cm}}$$
$$\xleftarrow{\hspace{1cm}} \text{esterification} \xrightarrow{\hspace{1cm}}$$

If acid-catalyzed hydrolysis is carried out in a large excess of water, the position of equilibrium is shifted in favor of formation of a carboxylic acid and an alcohol.

2. Hydrolysis in Aqueous Base

Esters are hydrolyzed in aqueous base to a carboxylate anion and an alcohol. Alkaline hydrolysis of esters is often referred to as **saponification**, a name derived from the Latin root *sapon-*, soap (Section 12.2). For saponification, one mole of base is required for each mole of ester, as illustrated in the equation,

$$CH_3\overset{\text{O}}{\overset{\|}{C}}OCH_2CH_3 + NaOH \xrightarrow[\text{(saponification)}]{H_2O} CH_3\overset{\text{O}}{\overset{\|}{C}}O^-Na^+ + CH_3CH_2OH$$

For all practical purposes, hydrolysis of an ester in aqueous base is irreversible, because the carboxylate anion, once formed, has no tendency to react with an alcohol.

Example 8.5

Complete and balance the following equations. Be certain to show each product as it would be ionized under the conditions specified in the problem.

a. $CH_3\overset{\text{O}}{\overset{\|}{C}}O-$⬡ $\xrightarrow{\substack{\text{hydrolysis in} \\ \text{aqueous acid}}}$

b. $CH_3\overset{\text{O}}{\overset{\|}{C}}OCH_2CH_2O\overset{\text{O}}{\overset{\|}{C}}CH_3 \xrightarrow[\text{(saponification)}]{\substack{\text{hydrolysis in} \\ \text{aqueous NaOH}}}$

Solution

a. The products of hydrolysis of an ester in aqueous acid are a carboxylic acid and an alcohol, in this case acetic acid and cyclohexanol:

$$CH_3\overset{\text{O}}{\overset{\|}{C}}O-⬡ + H_2O \xrightarrow{H^+} CH_3\overset{\text{O}}{\overset{\|}{C}}OH + HO-⬡$$

cyclohexyl acetate   acetic acid   cyclohexanol

b. This compound is a diester. Hydrolysis in aqueous NaOH requires two moles of NaOH per mole of ester and gives two moles of sodium acetate and one mole of ethylene glycol:

$$CH_3\overset{\overset{\displaystyle O}{\|}}{C}OCH_2CH_2O\overset{\overset{\displaystyle O}{\|}}{C}CH_3 + 2NaOH \longrightarrow 2CH_3\overset{\overset{\displaystyle O}{\|}}{C}O^-Na^+ + HOCH_2CH_2OH$$

sodium acetate   1,2-ethanediol
(ethylene glycol)

Problem 8.5

Complete and balance the following equations. Be certain to show each product as it would be ionized under the conditions specified in the problem.

a.

hydrolysis in aqueous acid $\longrightarrow$

b. $CH_3\overset{\overset{\displaystyle O}{\|}}{C}CH_2CH_2CH_2\overset{\overset{\displaystyle O}{\|}}{C}OCH_2CH_3$ $\xrightarrow[\text{(saponification)}]{\substack{\text{hydrolysis in} \\ \text{aqueous NaOH}}}$

C. Reaction with Ammonia and Amines: Formation of Amides

Reaction with ammonia converts an ester to an amide. This reaction is similar to hydrolysis (splitting apart by water). Because the "splitting" agent in this case is ammonia, the reaction is called **ammonolysis**. Ammonia is a strong nucleophile and adds directly to the carbonyl carbon to form a tetrahedral carbonyl addition intermediate, which in turn collapses to lose a molecule of alcohol and form an amide:

$$CH_3\overset{\overset{\displaystyle O}{\|}}{C}OCH_2CH_3 + NH_3 \longrightarrow \left[CH_3\overset{\overset{\displaystyle OH}{|}}{\underset{\underset{\displaystyle NH_2}{|}}{C}}-OCH_2CH_3 \right] \longrightarrow CH_3\overset{\overset{\displaystyle O}{\|}}{C}NH_2 + CH_3CH_2OH$$

ethyl acetate   tetrahedral carbonyl   acetamide
addition intermediate

Although ammonolysis is an equilibrium reaction, the position of the equilibrium lies very far to the right; the concentration of ester present at equilibrium is so small that it may be regarded as zero. Thus, it is possible to prepare an amide from an ester; but it is not possible to prepare an ester from an amide.

Another example of ammonolysis of an ester is the laboratory synthesis of barbituric acid and barbiturates. Heating urea (the diamide of carbonic acid) and diethyl malonate at 110°C in the presence of sodium ethoxide (the sodium salt of ethanol) gives barbituric acid:

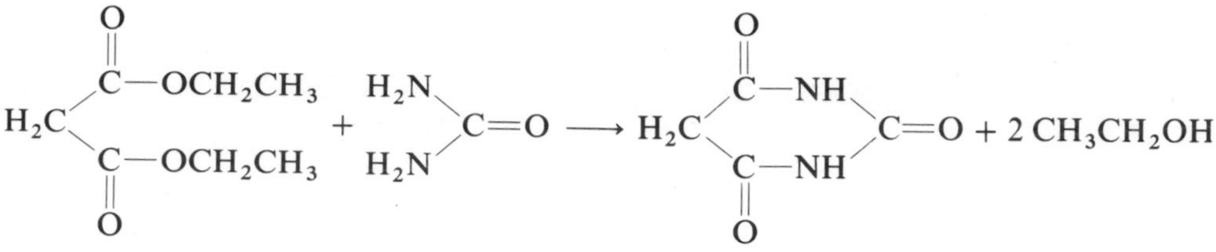

diethyl malonate   urea   barbituric acid

Mono- and disubstituted malonic esters yield substituted barbituric acids known as barbiturates.

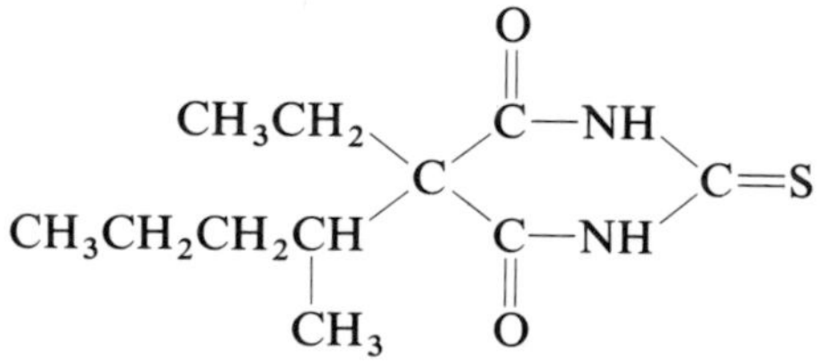

thiopental
(Penthothal)

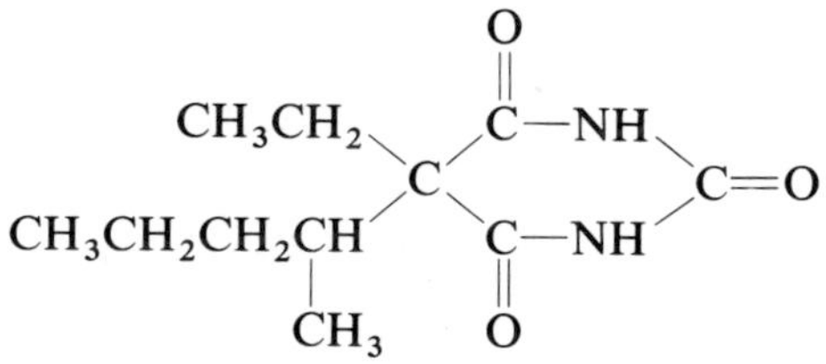

pentobarbital
(Nembutal)

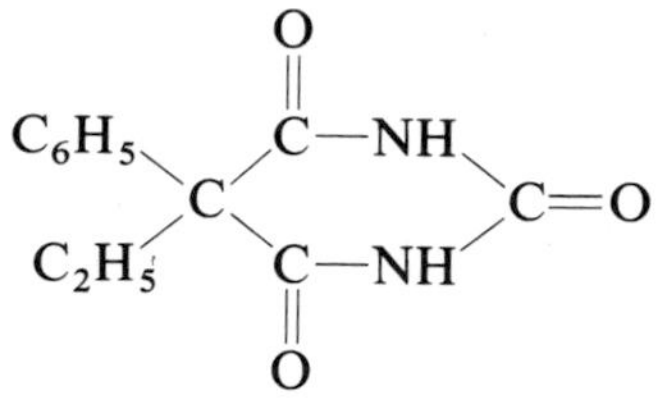

phenobarbital
(Luminal)

Barbiturates produce effects ranging from mild sedation to deep anesthesia, and even death, depending on the dose and the particular barbiturate. Sedation, long- or short-acting, depends on the structure of the barbiturate. Phenobarbital is long-acting, whereas pentobarbital acts for only about three hours. Thiopental is very fast-acting and is used as an anesthetic for producing deep sedation quickly.

Example 8.6

Complete equations for these ammonolysis reactions. The stoichiometry of each is given in the problem.

a. $\overset{O}{\overset{\|}{H}}COCH_2CH_3 + NH_3 \longrightarrow$

b. $CH_3CH_2O\overset{O}{\overset{\|}{C}}OCH_2CH_3 + 2NH_3 \longrightarrow$

Solution

a. The starting material is ethyl formate. Reaction with a mole of ammonia gives formamide:

$$\underset{\text{ethyl formate}}{HCOCH_2CH_3} + NH_3 \longrightarrow \underset{\text{formamide}}{HCNH_2} + HOCH_2CH_3$$

b. Reaction of one mole of diethyl carbonate with two moles of ammonia gives urea:

$$\underset{\text{diethyl carbonate}}{CH_3CH_2OCOCH_2CH_3} + 2NH_3 \longrightarrow \underset{\text{urea}}{H_2NCNH_2} + 2CH_3CH_2OH$$

Problem 8.6

Write equations for these ammonolysis reactions:

a. CH_3OC—⬡—$COCH_3 + 2NH_3 \longrightarrow$

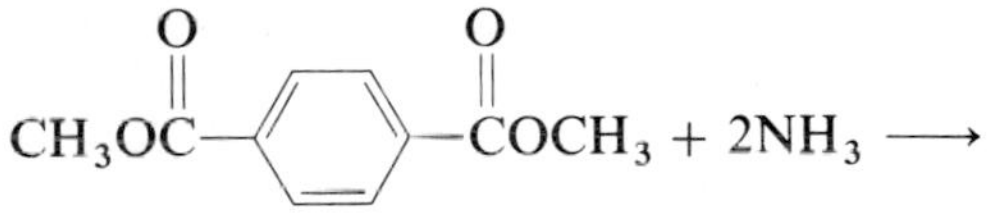

diethyl terephthalate

b. [pyridine ring]—$COCH_2CH_3$ $+ NH_3 \longrightarrow$

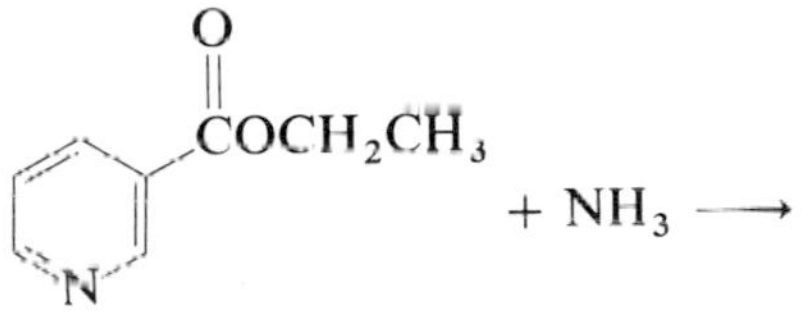

ethyl nicotinate

8.7 Reactions of Anhydrides

Of the three classes of functional derivatives of carboxylic acid we have studied, acid anhydrides are by far the most reactive. Anhydrides react with water to form carboxylic acids, with alcohols to form esters, and with ammonia and primary and secondary amines to form amides. Thus, acid anhydrides are valuable starting materials for preparing these other functional groups.

A. Hydrolysis

In hydrolysis, an anhydride is cleaved, forming two molecules of carboxylic acid, as illustrated by the hydrolysis of acetic anhydride:

$$\underset{\text{acetic anhydride}}{CH_3-C-O-C-CH_3} + H_2O \longrightarrow \underset{\text{acetic acid}}{CH_3-C-OH} + \underset{\text{acetic acid}}{HO-C-CH_3}$$

Acetic anhydride and other low-molecular-weight anhydrides react so readily with water that they must be protected from moisture during storage.

B. Reaction with Alcohols: Formation of Esters

Anhydrides react with alcohols to give one molecule of ester and one molecule of a carboxylic acid. Thus, reaction of an alcohol with an anhydride is a useful method for the synthesis of esters.

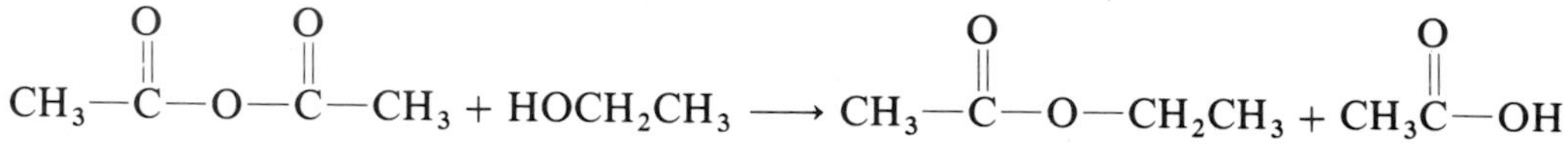

$$CH_3-\overset{\displaystyle O}{\overset{\|}{C}}-O-\overset{\displaystyle O}{\overset{\|}{C}}-CH_3 + HOCH_2CH_3 \longrightarrow CH_3-\overset{\displaystyle O}{\overset{\|}{C}}-O-CH_2CH_3 + CH_3\overset{\displaystyle O}{\overset{\|}{C}}-OH$$

acetic anhydride   ethyl acetate

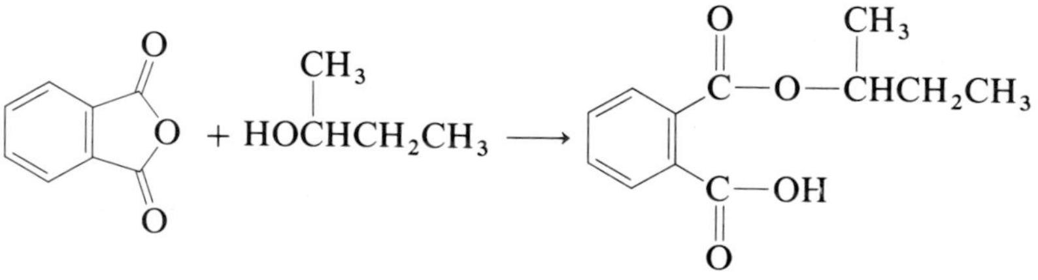

phthalic anhydride   sec-butyl hydrogen phthalate

Aspirin is prepared by reaction of acetic anhydride with the phenolic —OH group of salicylic acid. The CH_3CO— group is commonly called an acetyl group, a name derived from acetic acid by dropping the -ic from the name of the acid and adding -yl. Therefore, the chemical name of the product formed by reaction of acetic anhydride and salicylic acid is acetyl salicylic acid.

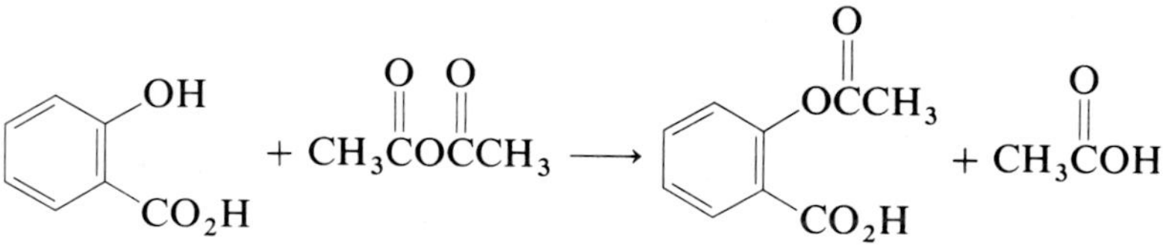

salicylic acid   acetyl salicylic acid
                                     (aspirin)

Aspirin is one of the few drugs produced on an industrial scale. In 1977, the United States produced 35 million pounds of it. Aspirin has been used since the turn of the century for relief of minor pain and headaches and for the reduction of fever. Compared with other drugs, aspirin is safe and well tolerated. However, it does have side effects. Because of its relative insolubility and acidity, it can irritate the stomach wall. These effects can be partially overcome by using its more soluble sodium salt instead. Because of these side effects, there has been increasing use of newer nonprescription analgesics, such as acetaminophen and Ibuprofen.

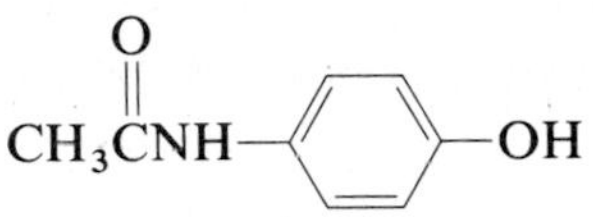

$$CH_3\overset{\displaystyle O}{\overset{\displaystyle \|}{C}}NH-\!\!\!\left\langle\!\!\!\bigcirc\!\!\!\right\rangle\!\!\!-OH \qquad CH_3\overset{\displaystyle CH_3}{\overset{\displaystyle |}{CH}}CH_2-\!\!\!\left\langle\!\!\!\bigcirc\!\!\!\right\rangle\!\!\!-\overset{\displaystyle CH_3}{\overset{\displaystyle |}{CH}}CO_2H$$

N-acetyl-4-aminophenol
(acetaminophen)

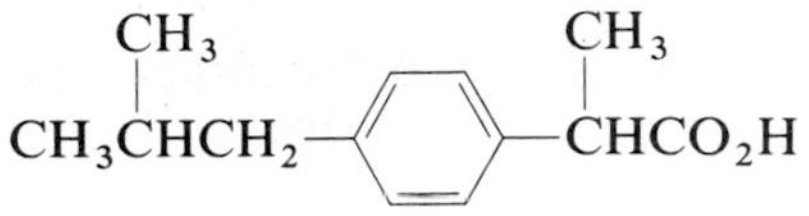

2-(4-isobutylphenyl)-propanoic acid
(Ibuprofen)

Ibuprofen was introduced in the United Kingdom in 1969 by the Boots Company as an anti-inflammatory agent for the treatment of rheumatoid arthritis and allied conditions. The Upjohn Company introduced it in the United States in 1974, and it is now marketed in over 120 countries. As an analgesic, it is approximately 28 times more potent than aspirin, and it is approximately 20 times more potent as a fever-reducing agent.

C. Reaction with Ammonia and Amines: Formation of Amides

Acid anhydrides react with ammonia, as well as with primary and secondary amines, to form amides. For complete conversion of an acid anhydride to an amide, two moles of amine are required, the first to form the amide and the second to neutralize the carboxylic acid byproduct. Reaction of an acid anhydride with an amine is one of the most common laboratory methods for synthesizing amides.

$$CH_3-\overset{\displaystyle O}{\overset{\displaystyle \|}{C}}-O-\overset{\displaystyle O}{\overset{\displaystyle \|}{C}}-CH_3 + 2NH_3 \longrightarrow CH_3-\overset{\displaystyle O}{\overset{\displaystyle \|}{C}}-NH_2 + CH_3\overset{\displaystyle O}{\overset{\displaystyle \|}{-C}}-O^-\ NH_4^+$$

acetic anhydride   acetamide   ammonium acetate

$$CH_3\overset{\displaystyle O}{\overset{\displaystyle \|}{C}}O\overset{\displaystyle O}{\overset{\displaystyle \|}{C}}CH_3 + HN(CH_3)_2 \longrightarrow CH_3\overset{\displaystyle O}{\overset{\displaystyle \|}{C}}N(CH_3)_2 + CH_3\overset{\displaystyle O}{\overset{\displaystyle \|}{C}}O^-\ H_2\overset{+}{N}(CH_3)_2$$

acetic
anhydride

dimethylamine

N,N-dimethyl-
acetamide

dimethylammonium
acetate

Example 8.7

Complete the following reactions. The stoichiometry of each is shown in the problem.

a. $2CH_3\overset{\displaystyle O}{\overset{\displaystyle \|}{C}}O\overset{\displaystyle O}{\overset{\displaystyle \|}{C}}CH_3 + HOCH_2CH_2OH \longrightarrow$

b. 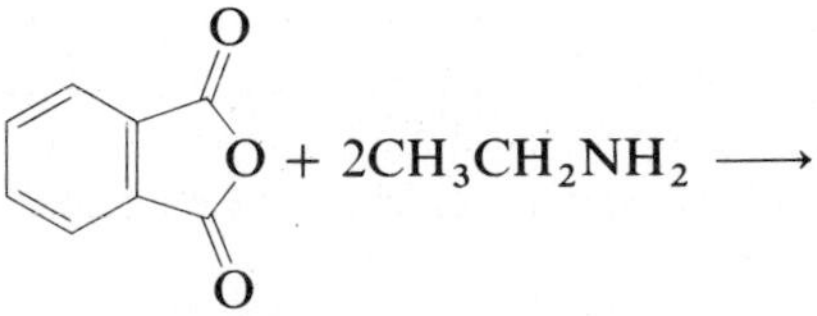 $\quad\quad \text{(phthalic anhydride)} + 2CH_3CH_2NH_2 \longrightarrow$

Solution

a. Ethylene glycol is a diol and reacts with two moles of acetic anhydride to produce a diester:

$$2CH_3\overset{O}{\overset{\|}{C}}O\overset{O}{\overset{\|}{C}}CH_3 + HOCH_2CH_2OH \longrightarrow CH_3\overset{O}{\overset{\|}{C}}OCH_2CH_2O\overset{O}{\overset{\|}{C}}CH_3 + 2CH_3\overset{O}{\overset{\|}{C}}OH$$

acetic anhydride   ethylene glycol   (a diester)

b. Phthalic anhydride reacts with two moles of ethylamine, the first to form the amide bond of the product and the second to form the ethylammonium salt of the remaining carboxyl group.

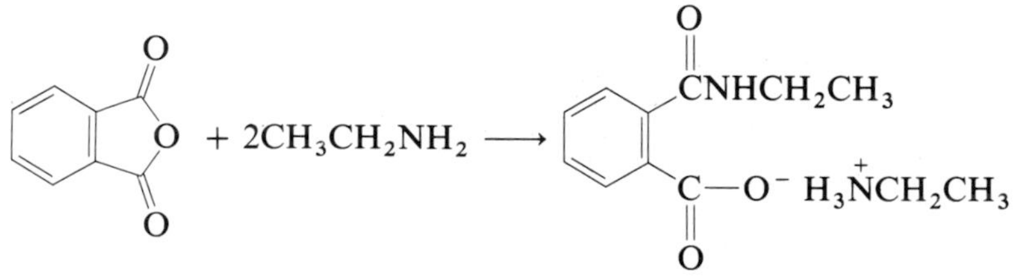

$$\text{phthalic anhydride} + 2CH_3CH_2NH_2 \longrightarrow$$

phthalic anhydride   ethylammonium *N*-ethylphthalamide

Problem 8.7

Write equations to show how you could prepare the following compounds by reaction of an anhydride with an alcohol or a phenol.

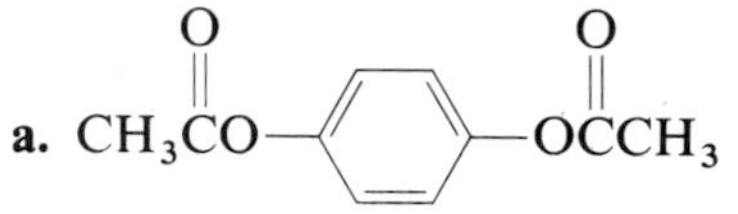

a. $CH_3\overset{O}{\overset{\|}{C}}O$—⟨benzene⟩—$O\overset{O}{\overset{\|}{C}}CH_3$ **b.** $CH_3\overset{CH_3}{\overset{|}{C}}{=}CHCH_2O\overset{O}{\overset{\|}{C}}CH_3$

8.8 Reactions of Amides

Amides are by far the least reactive of the functional groups we are considering. Their only important reaction for our purposes is hydrolysis brought about by heating the amide under reflux with concentrated aqueous acid (most commonly HCl) or aqueous base (most commonly NaOH or KOH). Because amides are so resistant to hydrolysis, it is often necessary to treat them under these conditions for several hours to bring about reaction.

Amides are hydrolyzed in aqueous acid to a carboxylic acid and an ammonium ion or an amine salt:

$$H{-}\overset{O}{\overset{\|}{C}}{-}NH_2 + H_2O + HCl \longrightarrow H{-}\overset{O}{\overset{\|}{C}}{-}O{-}H + NH_4^+ \; Cl^-$$

formamide   formic acid   ammonium chloride

$$\underset{\substack{\text{N,N-dimethyl-}\\\text{formamide}}}{\overset{\displaystyle O}{\overset{\|}{H-C-N-CH_3}}}\;\underset{CH_3}{} + H_2O + HCl \longrightarrow \underset{\text{formic acid}}{\overset{\displaystyle O}{\overset{\|}{H-C-O-H}}} + \underset{\substack{\text{dimethylammonium}\\\text{chloride}}}{\overset{\displaystyle H}{\overset{|+}{CH_3-N-H}}\;Cl^-}$$

In aqueous base, the products of amide hydrolysis are ammonia or free amine and a carboxylate salt:

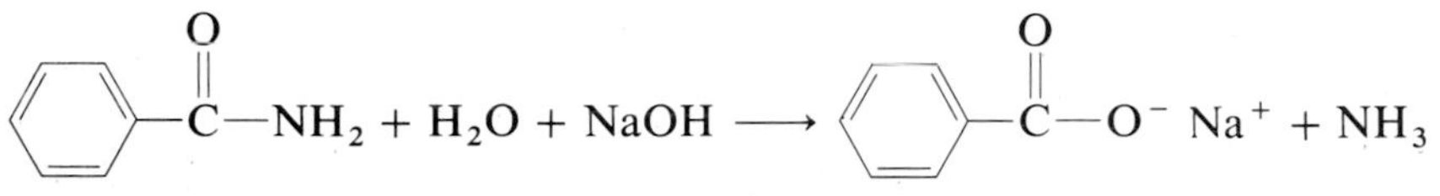

$$\underset{\text{benzamide}}{\overset{\displaystyle O}{\overset{\|}{\bigcirc\!\!-C-NH_2}}} + H_2O + NaOH \longrightarrow \underset{\text{sodium benzoate}}{\overset{\displaystyle O}{\overset{\|}{\bigcirc\!\!-C-O^-\;Na^+}}} + NH_3$$

$$\underset{\substack{\text{N-methyloctadecanamide}\\\text{(N-methylpalmitamide)}}}{\overset{\displaystyle O}{\overset{\|}{CH_3(CH_2)_{16}C-NHCH_3}}} + H_2O + NaOH \longrightarrow \underset{\substack{\text{sodium octadecanoate}\\\text{(sodium palmitate;}\\\text{a natural soap)}}}{\overset{\displaystyle O}{\overset{\|}{CH_3(CH_2)_{16}C-O^-\;Na^+}}} + CH_3NH_2$$

Thus, in hydrolysis of amides, one mole of either acid or base is required for each mole of amide hydrolyzed.

■
Example 8.8

Complete equations for hydrolysis of the following amides in concentrated aqueous HCl. Show all products as they would exist under these conditions and indicate in your equation the number of moles of HCl required for hydrolysis of each amide.

a. $\overset{\displaystyle O}{\overset{\|}{CH_3CNH_2}}$ b. 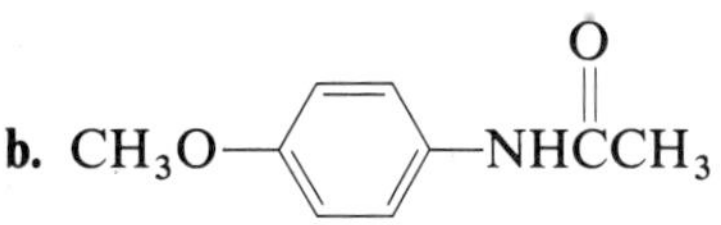 $CH_3O-\bigcirc\!\!-NH\overset{\displaystyle O}{\overset{\|}{C}}CH_3$

☐ *Solution*

a. Hydrolysis of acetamide gives acetic acid and ammonia. Ammonia is a base and in aqueous HCl is protonated to form an ammonium ion. The product is shown here as ammonium chloride:

$$\overset{\displaystyle O}{\overset{\|}{CH_3CNH_2}} + H_2O + HCl \longrightarrow \overset{\displaystyle O}{\overset{\|}{CH_3COH}} + NH_4^+\;Cl^-$$

b. 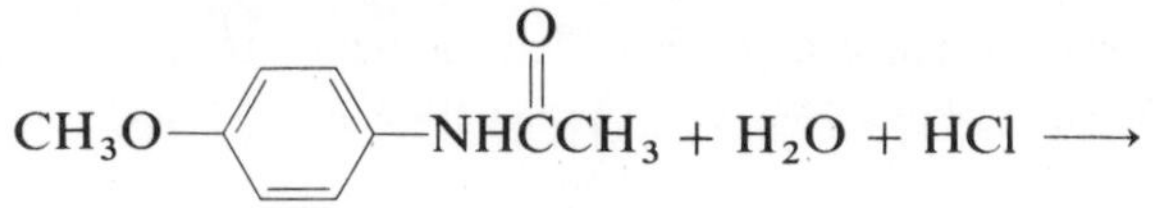 $CH_3O-\underset{}{\bigcirc}-\overset{\overset{O}{\parallel}}{N}HCCH_3 + H_2O + HCl \longrightarrow$

N-acetyl-4-methoxyaniline
(N-acetyl-p-methoxyaniline)

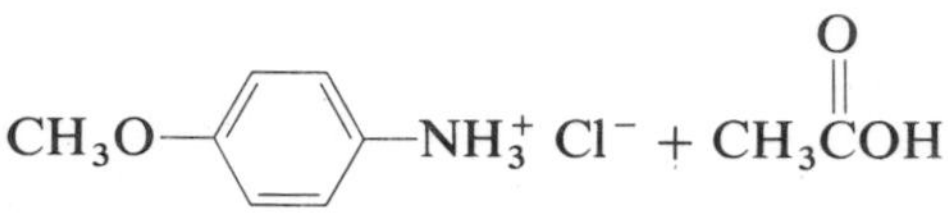

$CH_3O-\underset{}{\bigcirc}-NH_3^+ \ Cl^- + CH_3\overset{\overset{O}{\parallel}}{C}OH$

4-methoxyanilinium chloride
(p-methoxyanilinium chloride)

Problem 8.8

Complete the equations for hydrolysis of the following amides in concentrated aqueous NaOH. Show all products as they would exist under these conditions and indicate in your equation the number of moles of NaOH required for hydrolysis of each amide.

a. $CH_3\overset{\overset{O}{\parallel}}{C}NH_2$ **b.** 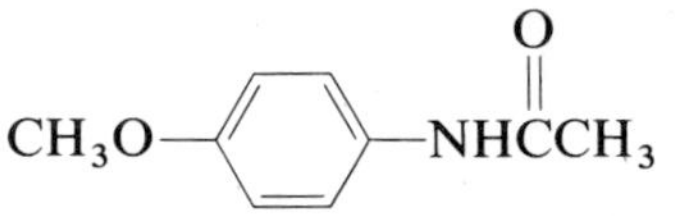 $CH_3O-\underset{}{\bigcirc}-NH\overset{\overset{O}{\parallel}}{C}CH_3$

8.9 The Claisen Condensation: Synthesis of Beta-Ketoesters

The **Claisen condensation** involves condensation of the α-carbon of one molecule of ester with the carbonyl carbon of a second molecule of ester. For example, when ethyl acetate is heated with sodium ethoxide in ethanol, the α-carbon of one molecule of ethyl acetate forms a new carbon-carbon bond with the carbonyl group of a second molecule of ethyl acetate, and $-OCH_2CH_3$ is displaced.

New carbon-carbon bond formed in the Claisen condensation

$CH_3\overset{\overset{O}{\parallel}}{C}OCH_2CH_3 + CH_3\overset{\overset{O}{\parallel}}{C}OCH_2CH_3 \xrightarrow[CH_3CH_2OH]{CH_3CH_2O^-Na^+} CH_3\overset{\overset{O}{\parallel}}{C}-CH_2\overset{\overset{O}{\parallel}}{C}OCH_2CH_3 + CH_3CH_2OH$

Carbonyl group to which α-carbon adds   *α-carbon*

ethyl 3-oxobutanoate
(ethyl acetoacetate)

Thus, the Claisen condensation is another example of the general mechanism of nucleophilic substitution at a carbonyl carbon (Section 8.4).

The characteristic structural feature of the product of a Claisen condensation is a ketone on carbon 3 of an ester chain. In common nomenclature, carbon 3 of the carboxylic acid chain is called the beta-carbon (β-carbon). Thus, products of Claisen condensations are often called **β-ketoesters**.

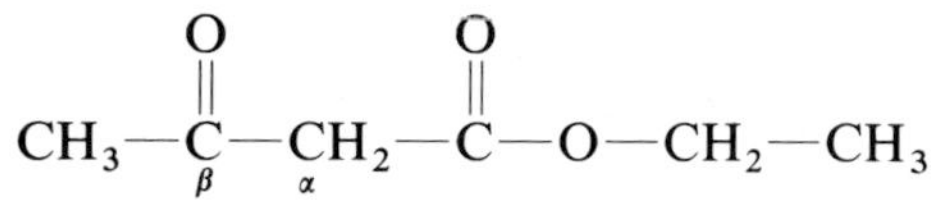

$$CH_3-\overset{\overset{\displaystyle O}{\|}}{\underset{\beta}{C}}-\underset{\alpha}{CH_2}-\overset{\overset{\displaystyle O}{\|}}{C}-O-CH_2-CH_3$$

a beta-ketoester

Claisen condensation of two molecules of ethyl propanoate gives the following beta-ketoester. In this reaction, the structural formulas of the two ester molecules are written to emphasize that the step forming the new carbon-carbon bond involves the α-carbon of one ester and the carbonyl carbon of the other.

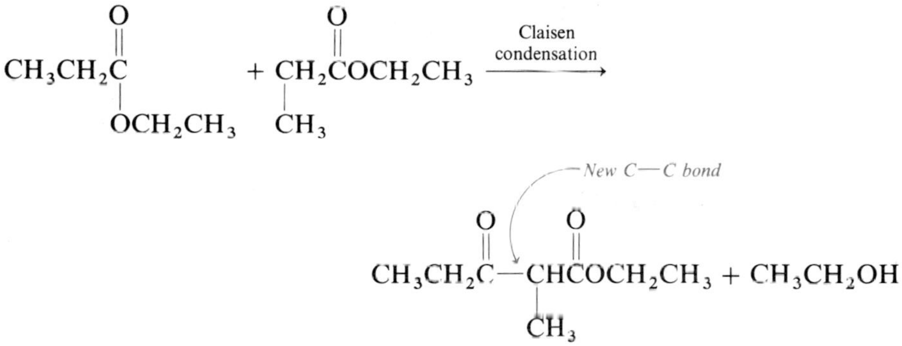

$$CH_3CH_2\overset{\overset{\displaystyle O}{\|}}{\underset{\underset{\displaystyle OCH_2CH_3}{|}}{C}} \quad + \quad CH_2\overset{\overset{\displaystyle O}{\|}}{\underset{\underset{\displaystyle CH_3}{|}}{C}}OCH_2CH_3 \quad \xrightarrow[\text{condensation}]{\text{Claisen}}$$

— New C—C bond

$$CH_3CH_2\overset{\overset{\displaystyle O}{\|}}{C}-\overset{\overset{\displaystyle O}{\|}}{\underset{\underset{\displaystyle CH_3}{|}}{CH}}COCH_2CH_3 + CH_3CH_2OH$$

ethyl 2-methyl-3-oxopentanoate
(a beta-ketoester)

The mechanism for the Claisen condensation is similar to the three-step mechanism we proposed for the aldol condensation (Section 6.8). Both begin in step 1 by formation of an anion on an α-carbon. This is followed in step 2 of the Claisen condensation by addition of this anion to the carbonyl carbon of another ester, to form a tetrahedral carbonyl addition intermediate. Collapse of this intermediate in step 3 gives the beta-ketoester.

Step 1: Reaction of an α-hydrogen with base to form an anion

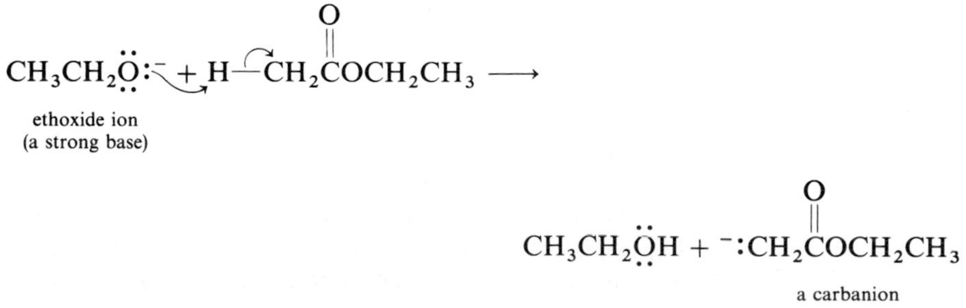

$$CH_3CH_2\ddot{\overset{..}{O}}: + H-CH_2\overset{\overset{\displaystyle O}{\|}}{C}OCH_2CH_3 \longrightarrow$$

ethoxide ion
(a strong base)

$$CH_3CH_2\overset{..}{O}H + {}^-:CH_2\overset{\overset{\displaystyle O}{\|}}{C}OCH_2CH_3$$

a carbanion

Step 2: Addition of the carbanion to a carbonyl carbon to form a tetrahedral carbonyl addition intermediate

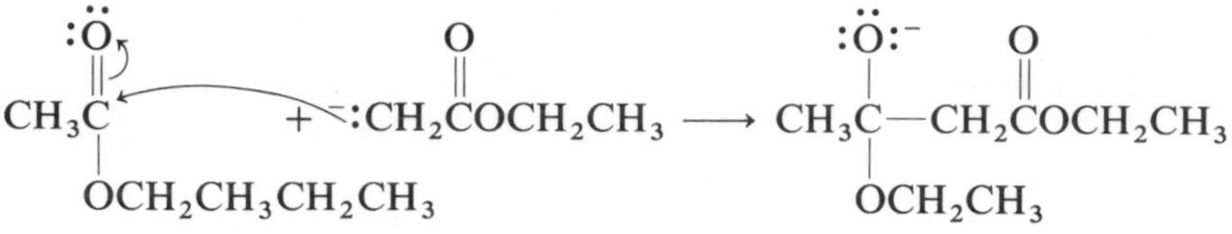

$$CH_3C\overset{:\overset{..}{O}:}{\underset{OCH_2CH_3CH_2CH_3}{\|}} + \ :CH_2COCH_2CH_3 \longrightarrow CH_3\overset{:\overset{..}{O}:^-}{\underset{OCH_2CH_3}{C}}-CH_2COCH_2CH_3$$

tetrahedral carbonyl
addition intermediate

Step 3: Collapse of the tetrahedral carbonyl addition intermediate to give the beta-ketoester and regenerate ethoxide ion

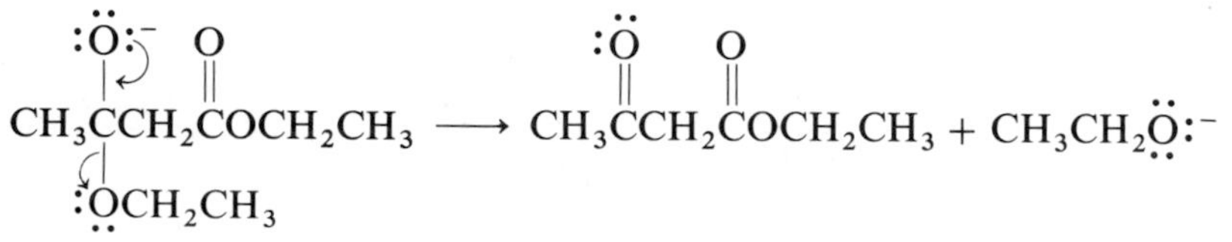

$$CH_3\overset{:\overset{..}{O}:^-}{\underset{:OCH_2CH_3}{C}}CH_2COCH_2CH_3 \longrightarrow CH_3\overset{:\overset{..}{O}}{C}CH_2\overset{O}{C}OCH_2CH_3 + CH_3CH_2\overset{..}{\underset{..}{O}}:^-$$

Note that ethoxide ion is a catalyst in this reaction; it is used as a reactant in step 1 but regenerated as a product in step 3.

In the case of mixed Claisen condensations—that is, a condensation between two different esters—a mixture of four possible products is possible unless there is an appreciable difference in reactivity between one ester and the other. One such difference is if one of the esters has no α-hydrogens and therefore cannot serve as an anion. Examples of esters without α-hydrogens are ethyl formate, ethyl benzoate, and diethyl carbonate:

$$\overset{O}{HCOCH_2CH_3} \qquad \overset{O}{\underset{}{\bigcirc}-COCH_2CH_3} \qquad \overset{O}{CH_3CH_2OCOCH_2CH_3}$$

ethyl formate   ethyl benzoate   diethyl carbonate

Example 8.9

Draw structural formulas for the products of the following Claisen condensations.

a. 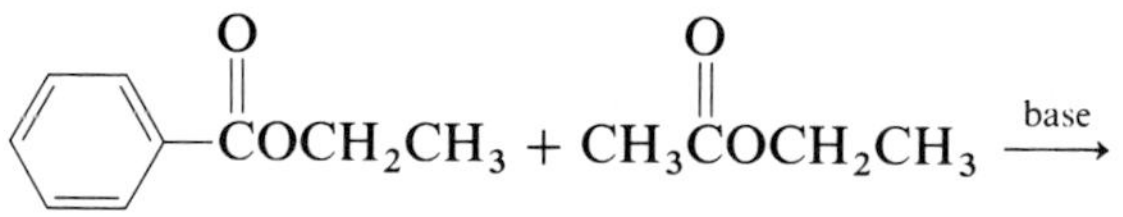 $\bigcirc-\overset{O}{COCH_2CH_3} + CH_3\overset{O}{COCH_2CH_3} \xrightarrow{\text{base}}$

b. 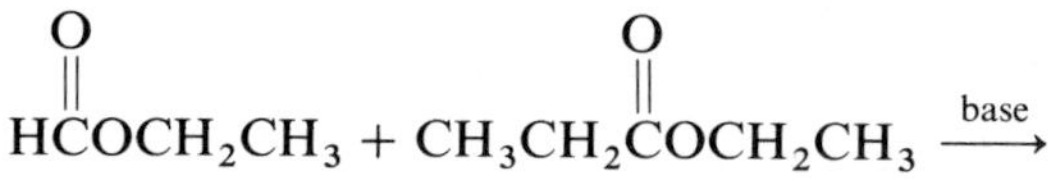 $\overset{O}{HCOCH_2CH_3} + CH_3CH_2\overset{O}{COCH_2CH_3} \xrightarrow{\text{base}}$

☐ *Solution*

a. Ethyl benzoate has no α-hydrogens and therefore can function only as an anion acceptor in a Claisen condensation. Ethyl acetate has three α-hydrogens and forms an anion, which then completes the reaction.

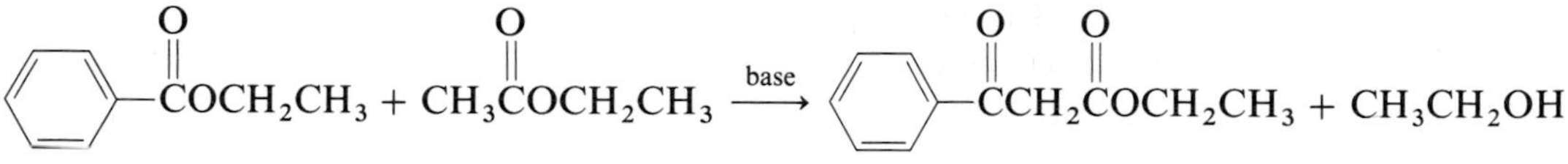

ethyl benzoate   ethyl acetate   ethyl 3-oxo-3-phenyl-propanoate (a beta-ketoester)

b. Ethyl formate has no α-hydrogens and can function only as a carbanion acceptor for the anion formed on the α-carbon of ethyl propanoate.

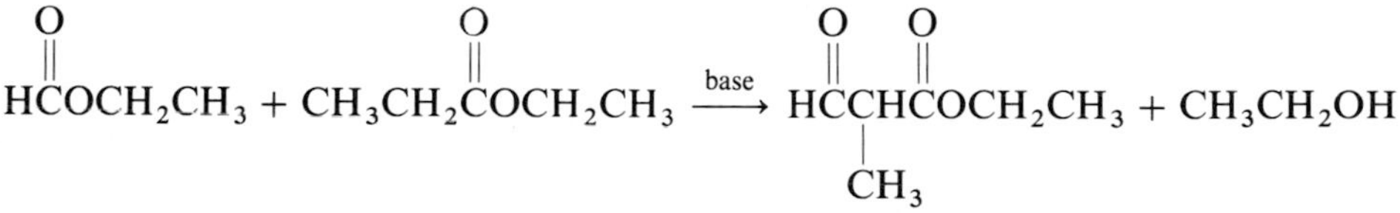

cthyl formate   ethyl propanoate   (a beta-ketoester)

Problem 8.9

Following are structural formulas for two beta-ketoesters. Show how each could be formed by a Claisen condensation.

a. 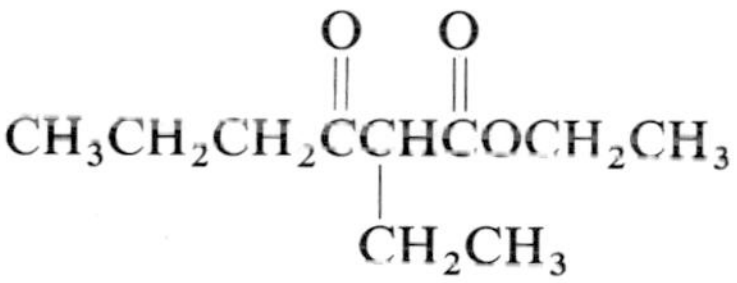 $CH_3CH_2CH_2CCHCOCH_2CH_3$ **b.**

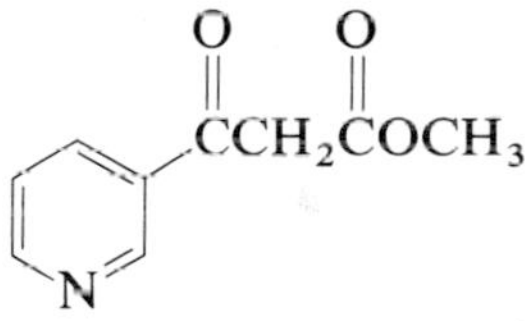

Key Terms and Concepts

amide (introduction)	Fischer esterification (8.5)
ammonolysis (8.6C)	β-ketoester (8.9)
anhydride (introduction)	nucleophilic substitution at a
Claisen condensation (8.9)	carbonyl carbon (8.4)
ester (introduction)	saponification (8.6B)

Key Reactions

1. Acid-catalyzed esterification: Fischer esterification (Section 8.5).
2. Catalytic reduction of esters to two alcohols (Section 8.6A).
3. Metal hydride reduction of esters to two alcohols (Section 8.6A).
4. Hydrolysis of esters in aqueous acid to a carboxylic acid and an alcohol (Section 8.6B).
5. Hydrolysis of esters in aqueous base to a carboxylate salt and an alcohol (Section 8.6B).
6. Reaction of esters with ammonia and amines: formation of amides (Section 8.6C).

7. Hydrolysis of anhydrides (Section 8.7A).

8. Reaction of anhydrides with alcohols: formation of **esters** (Section 8.7B).

9. Reaction of anhydrides with ammonia and amines: formation of **amides** (Section 8.7C).

10. Hydrolysis of amides in aqueous acid to a carboxylic acid and an ammonium salt (Section 8.8).

11. Hydrolysis of amides in aqueous base to a carboxylate salt and ammonia or an amine (Section 8.8).

12. Claisen condensation: formation of β-ketoesters (Section 8.9).

Problems

Structure and nomenclature of esters, amides, and anhydrides (Section 8.1)

8.10   Name the compounds:

a. $CH_3CH_2\overset{O}{\overset{\|}{C}}OCHCH_3$
    $\quad\quad\quad\quad\quad\underset{CH_3}{|}$

b. $CH_3\overset{O}{\overset{\|}{C}}NH_2$

c. 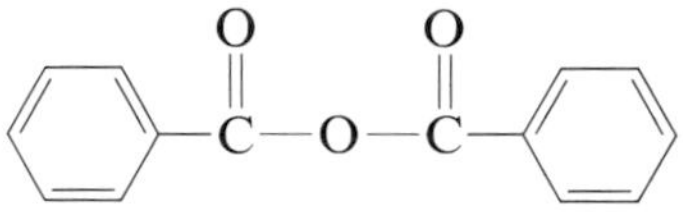 $\text{C}_6\text{H}_5-\overset{O}{\overset{\|}{C}}-O-\overset{O}{\overset{\|}{C}}-\text{C}_6\text{H}_5$

d. $CH_2{=}CH\overset{O}{\overset{\|}{C}}OCH_3$

e. $CH_3CH_2CH_2CH_2\overset{O}{\overset{\|}{C}}NHCH_3$

f. $H_2N\overset{O}{\overset{\|}{C}}NH_2$

g. $CH_3CH_2\overset{O}{\overset{\|}{C}}OC_6H_5$

h. $CH_3O\overset{O}{\overset{\|}{C}}CH_2CH_2\overset{O}{\overset{\|}{C}}OCH_3$

i. 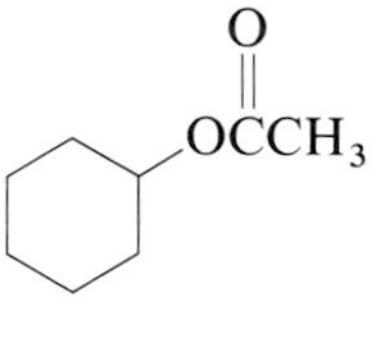 (cyclohexyl)$\text{O}\overset{O}{\overset{\|}{C}}CH_3$

j. 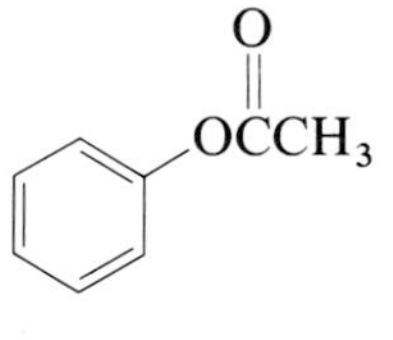 (phenyl)$\text{O}\overset{O}{\overset{\|}{C}}CH_3$

8.11   Draw structural formulas for the following compounds:

a. phenyl benzoate
b. diethyl carbonate
c. benzamide
d. cyclobutyl butanoate
e. methyl 3-methylbutanoate
f. isopropyl 3-methylhexanoate
g. diethyl oxalate
h. ethyl *cis*-2-pentenoate
i. *N*-phenylacetamide
j. *N,N*-dimethylacetamide
k. acetic anhydride
l. *N*-phenylbutanamide
m. diethyl malonate
n. formamide
o. ethyl 3-hydroxybutanoate
p. methyl formate
q. trimethyl citrate
r. *p*-nitrophenyl acetate
s. ethyl *p*-hydroxybenzoate
t. ethyl *p*-aminobenzoate

8.12 Draw structural formulas for the nine isomeric esters of molecular formula $C_5H_{10}O_2$. Give each an IUPAC name.

Physical properties of esters, amides, and anhydrides (Section 8.3)

8.13 Acetic acid and methyl formate are structural isomers. Both are liquids at room temperature. One has a boiling point of 32°C, the other a boiling point of 118°C. Which compound has which boiling point? Explain your reasoning.

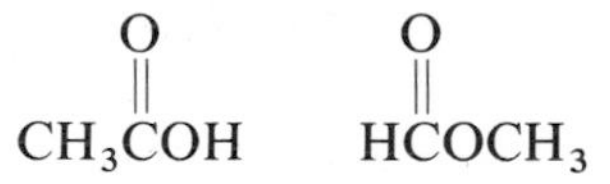

$$\underset{\text{acetic acid}}{CH_3\overset{\overset{\displaystyle O}{\|}}{C}OH} \qquad \underset{\text{methyl formate}}{H\overset{\overset{\displaystyle O}{\|}}{C}OCH_3}$$

8.14 Draw structural formulas to show hydrogen bonding between the circled atoms.

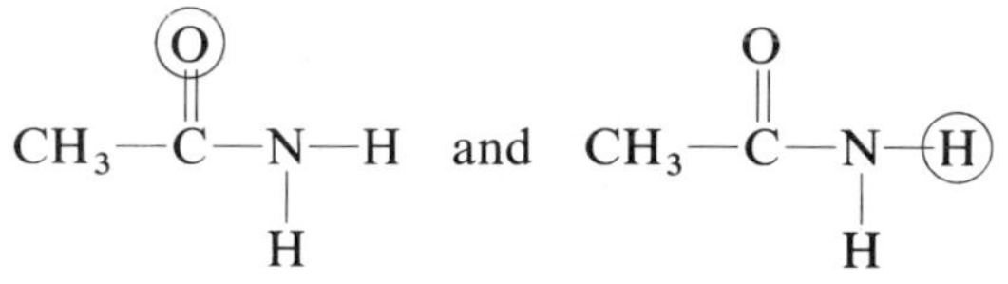

$$CH_3-\overset{\overset{\displaystyle \text{(O)}}{\|}}{C}-\underset{\underset{\displaystyle H}{|}}{N}-H \quad \text{and} \quad CH_3-\overset{\overset{\displaystyle O}{\|}}{C}-\underset{\underset{\displaystyle H}{|}}{N}-\text{(H)}$$

8.15 Following are melting and boiling points for acetamide and ethyl acetate.

$$\underset{\substack{\text{acetamide}\\\text{(mp 82.3°C; bp 221°C)}}}{CH_3\overset{\overset{\displaystyle O}{\|}}{C}NH_2} \qquad \underset{\substack{\text{ethyl acetate}\\\text{(mp   83.6°C; bp 77°C)}}}{CH_3\overset{\overset{\displaystyle O}{\|}}{C}OCH_2CH_3}$$

a. What is the physical state (solid, liquid, or gas) of each compound at room temperature?

b. How could you account for the considerably higher boiling point of acetamide relative to ethyl acetate?

Preparation of esters (Section 8.5)

8.16 a. Write an equation for the equilibrium established when acetic acid and 1-propanol are heated under reflux in the presence of a few drops of concentrated sulfuric acid.

b. Using the data in Section 8.5, calculate the equilibrium constant for this reaction.

8.17 If 15 g of salicylic acid is caused to react with excess methanol, how many grams of methyl salicylate (oil of wintergreen) could be formed?

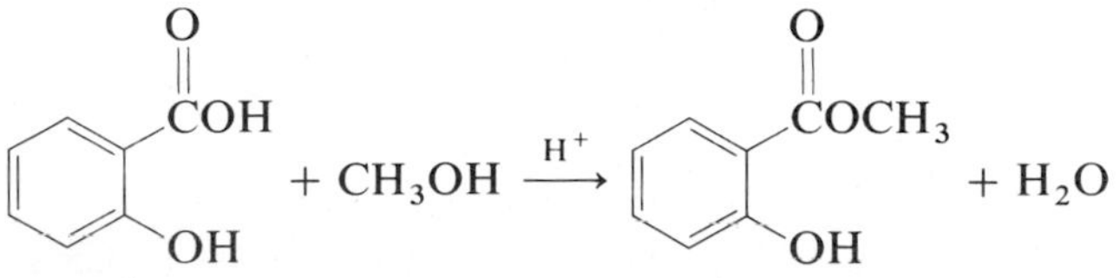

salicylic acid   methyl salicylate

8.18 When a carboxylic acid contains more than one —COOH group and the alcohol contains more than one —OH group, then under appropriate experimental conditions, hundreds of molecules can be linked to give a polyester. Dacron is a polyester of terephthalic acid and ethylene glycol:

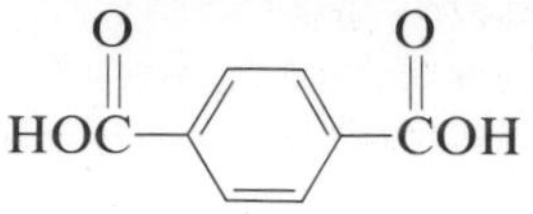

$$HOC\!\!-\!\!\bigcirc\!\!-\!\!COH \qquad HOCH_2CH_2OH$$

terephthalic acid   ethylene glycol

a. Formulate a structure of Dacron polyester. Be certain to show in principle how several hundred molecules can be hooked together to form the polyester.

b. Write an equation for the chemistry involved when a drop of concentrated hydrochloric acid makes a hole in a Dacron polyester shirt or blouse.

c. From what starting materials do you think the condensation fiber Kodel polyester is made?

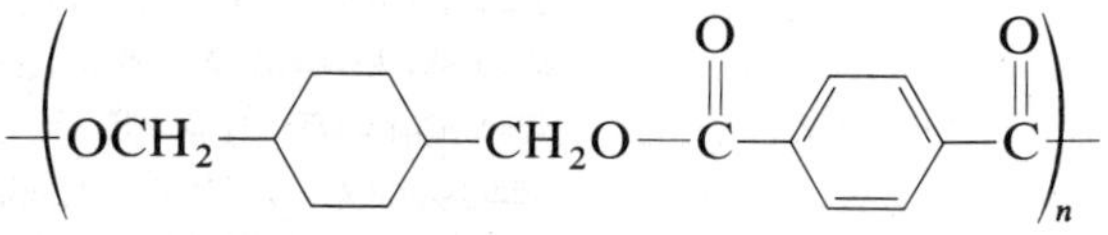

$$\left(\!\!-OCH_2\!\!-\!\!\bigcirc\!\!-\!\!CH_2O\!\!-\!\!\overset{O}{\underset{\|}{C}}\!\!-\!\!\bigcirc\!\!-\!\!\overset{O}{\underset{\|}{C}}\!\!-\!\!\right)_{\!n}$$

Kodel polyester

Reactions of esters, amides, and anhydrides (Sections 8.6–8.8)

8.19 Complete the equations for the hydrolysis of the following esters, amides, and anhydrides.

$$\text{a. } CH_3\overset{O}{\underset{\|}{C}}OCH_2CH_3 + H_2O \longrightarrow$$

$$\text{b. } CH_3\overset{O}{\underset{\|}{C}}O\underset{\underset{CH_2OCCH_3}{\overset{\|}{O}}}{\overset{\overset{\|}{O}}{\overset{CH_2OCCH_3}{C}}}H \qquad + 3H_2O \longrightarrow$$

$$\text{c. } CH_3CH_2\overset{O}{\underset{\|}{C}}OCH_2CH_2O\overset{O}{\underset{\|}{C}}CH_2CH_3 + 2H_2O \longrightarrow$$

$$\text{d. } CH_3CH_2O\overset{O}{\underset{\|}{C}}CH_2CH_2\overset{O}{\underset{\|}{C}}OCH_2CH_3 + 2H_2O \longrightarrow$$

$$\text{g. } 2CH_3\overset{\overset{\displaystyle O}{\|}}{C}O\overset{\overset{\displaystyle O}{\|}}{C}CH_3 + HOCH_2CH_2OH \longrightarrow$$

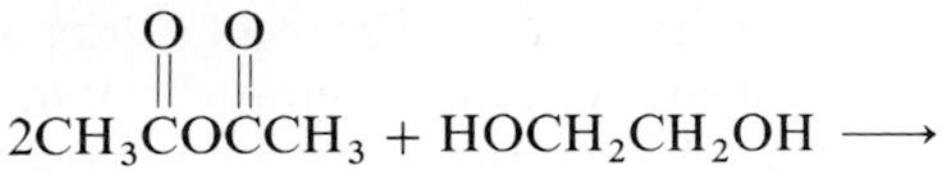

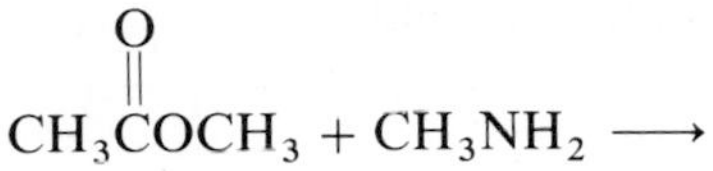

$$\text{h. } CH_3\overset{\overset{\displaystyle O}{\|}}{C}OCH_3 + CH_3NH_2 \longrightarrow$$

$$\text{i. } CH_3\overset{\overset{\displaystyle O}{\|}}{C}OCH_3 + (CH_3)_2NH \longrightarrow$$

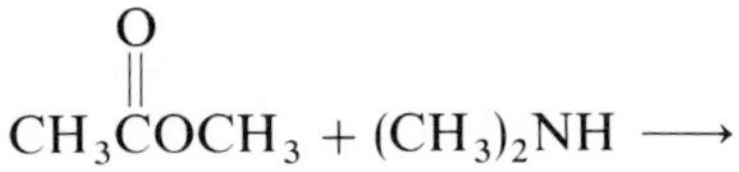

8.24 Draw structural formulas for the products of complete hydrolysis of the following phosphate esters, amides, and anhydrides. In each case, name the type of functional group undergoing hydrolysis.

a. 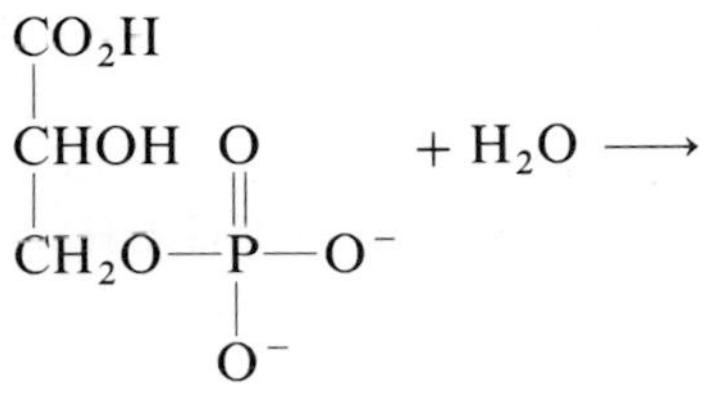
$$\begin{array}{c} CO_2H \\ | \\ CHOH \quad O \\ | \qquad\quad \| \\ CH_2O-P-O^- \\ | \\ O^- \end{array} \quad + H_2O \longrightarrow$$

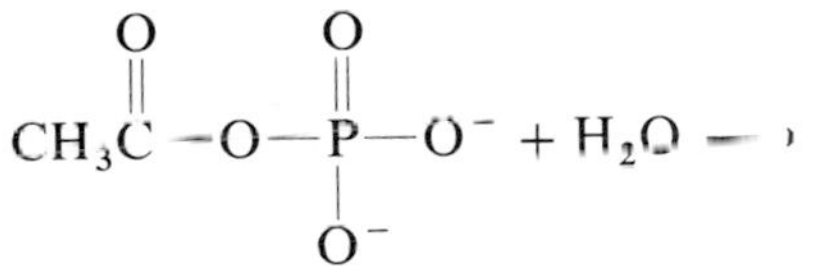

b.
$$CH_3\overset{\overset{\displaystyle O}{\|}}{C}-O-\overset{\overset{\displaystyle O}{\|}}{\underset{\underset{\displaystyle O^-}{|}}{P}}-O^- + H_2O \longrightarrow$$

c.
$$^-O-\overset{\overset{\displaystyle O}{\|}}{\underset{\underset{\displaystyle O^-}{|}}{P}}-O-\overset{\overset{\displaystyle O}{\|}}{\underset{\underset{\displaystyle O^-}{|}}{P}}-O^- + H_2O \longrightarrow$$

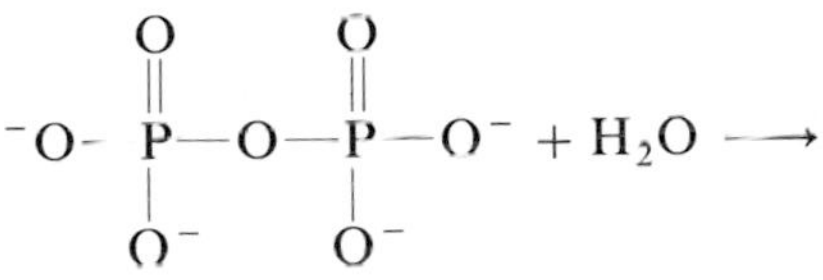

d. 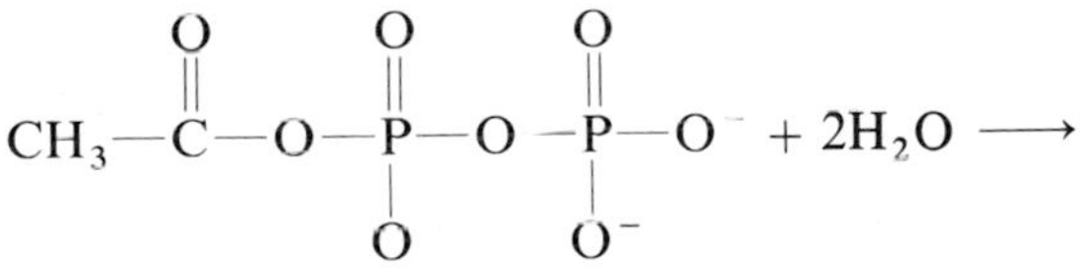
$$CH_3-\overset{\overset{\displaystyle O}{\|}}{C}-O-\overset{\overset{\displaystyle O}{\|}}{\underset{\underset{\displaystyle O}{\|}}{P}}-O-\overset{\overset{\displaystyle O}{\|}}{\underset{\underset{\displaystyle O^-}{|}}{P}}-O^- + 2H_2O \longrightarrow$$

e. 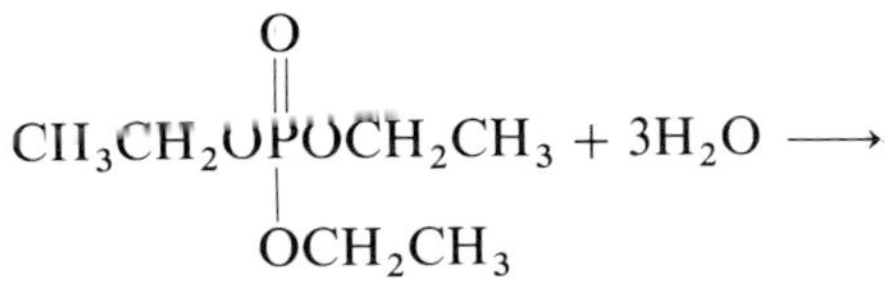
$$CH_3CH_2O\overset{\overset{\displaystyle O}{\|}}{\underset{\underset{\displaystyle OCH_2CH_3}{|}}{P}}OCH_2CH_3 + 3H_2O \longrightarrow$$

8.25 Show how you could distinguish between the following pairs of compounds by a simple chemical test. In each case, tell what test you would perform and what you would expect to observe, and write an equation for each positive test.
a. isopropyl formate and 2-methylpropanoic acid (isobutyric acid)
b. butanoic acid and butanamide
c. methyl hexanoate and hexanal

d. 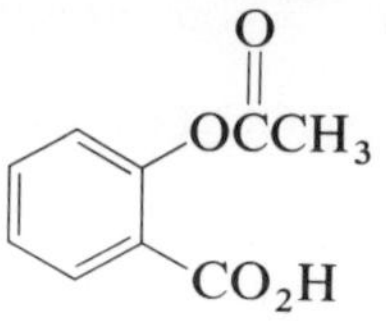   and

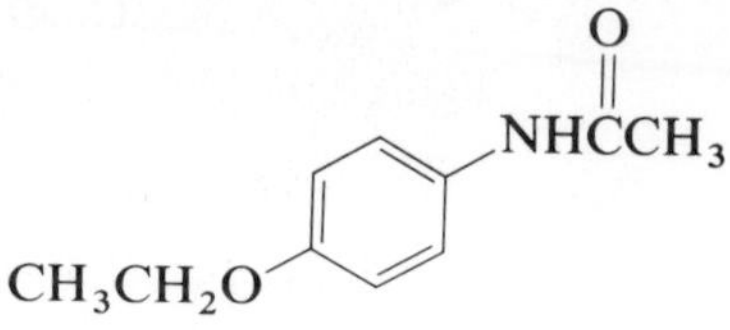

aspirin   phenacetin

8.26 The following conversions can be done in either one step or two. Show the reagents you would use and draw structural formulas for any intermediate involved in a two-step reaction. Use any necessary inorganic and organic compounds in addition to the indicated starting material.

a. $CH_3CH_2CH_2CH_2CH_2OH \longrightarrow CH_3CH_2CH_2CH_2\overset{\displaystyle O}{\overset{\|}{C}}OH$

b. $CH_3CH_2CH_2CH_2OH \longrightarrow CH_3CH_2CH_2\overset{\displaystyle O}{\overset{\|}{C}}OCH_2CH_2CH_2CH_3$

c. $CH_3CH_2CH_2CH_2OH \longrightarrow CH_3CH_2CH{=}CH_2$

d. $CH_3CH_2CH_2CH_2OH \longrightarrow CH_3CH_2\underset{\displaystyle OH}{CH}CH_3$

e.

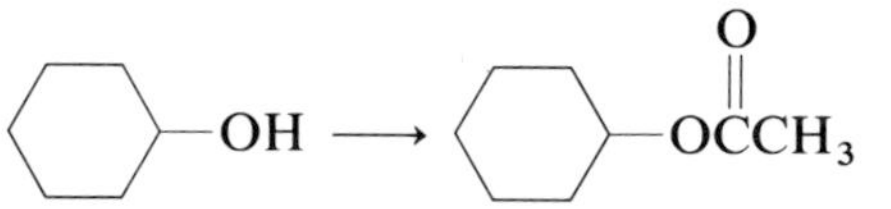

f. $HO\overset{\displaystyle O}{\overset{\|}{C}}CH_2CH_2CH_2CH_2\overset{\displaystyle O}{\overset{\|}{C}}OH \longrightarrow CH_3O\overset{\displaystyle O}{\overset{\|}{C}}CH_2CH_2CH_2CH_2\overset{\displaystyle O}{\overset{\|}{C}}OCH_3$

g. $2CH_3\overset{\displaystyle O}{\overset{\|}{C}}CH_3 \longrightarrow CH_3\underset{\displaystyle CH_3}{\overset{\displaystyle OH}{C}}CH_2\overset{\displaystyle O}{\overset{\|}{C}}CH_3$

h.  $\longrightarrow CH_3CH_2O\overset{\displaystyle O}{\overset{\|}{C}}CH_2CH_2CH_2CH_2\overset{\displaystyle O}{\overset{\|}{C}}OCH_2CH_3$

i.

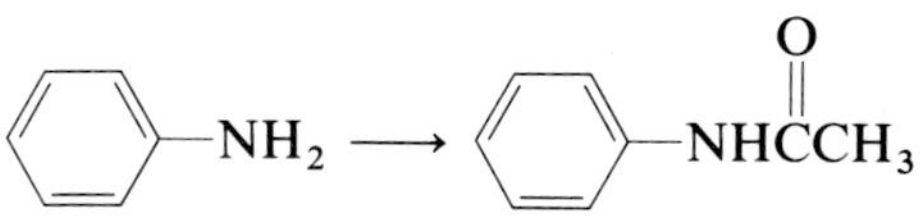

j. $CH_3CH_2CH_2OH \longrightarrow (CH_3CH_2CO_2^-)_2\,Ca^{2+}$

The Claisen condensation (Section 8.9)

8.27 What is the characteristic structural feature of the product of a Claisen condensation?

8.28 What is the characteristic structural feature of the product of an aldol condensation?

8.29   Draw structural formulas for the products of the following Claisen condensations.

a. $CH_3CH_2\overset{\overset{\displaystyle O}{\|}}{C}OCH_2CH_3$ $\xrightarrow[\text{CH}_3\text{CH}_2\text{OH}]{\text{CH}_3\text{CH}_2\text{O}^-\ \text{Na}^+}$

b. 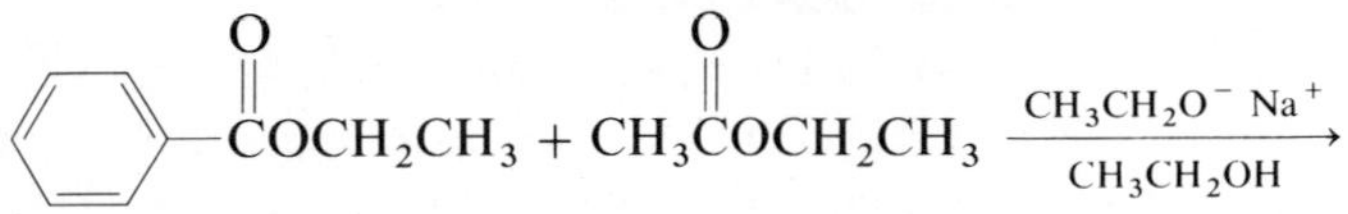   (benzene ring)$-\overset{\overset{\displaystyle O}{\|}}{C}OCH_2CH_3 + CH_3\overset{\overset{\displaystyle O}{\|}}{C}OCH_2CH_3$ $\xrightarrow[\text{CH}_3\text{CH}_2\text{OH}]{\text{CH}_3\text{CH}_2\text{O}^-\ \text{Na}^+}$

8.30   Write equations to show how ethyl propanoate could be converted to these compounds:

a. $CH_3CH_2\overset{\overset{\displaystyle O}{\|}}{C}\underset{\underset{\displaystyle CH_3}{|}}{C}H\overset{\overset{\displaystyle O}{\|}}{C}OCH_2CH_3$

b. $CH_3CH_2\overset{\overset{\displaystyle O}{\|}}{C}\underset{\underset{\displaystyle CH_3}{|}}{C}H\overset{\overset{\displaystyle O}{\|}}{C}OH$

c. $CH_3CH_2\overset{\overset{\displaystyle O}{\|}}{C}CH_2CH_3$

Nylon
and Dacron

After World War I, many chemists recognized the need for developing a basic knowledge of polymer chemistry. One of the most creative of these pioneers was Wallace M. Carothers. In the early 1930s, Carothers and his associates at E. I. du Pont de Nemours & Co., Inc., began fundamental research into the reactions of aliphatic dicarboxylic acids and dialcohols. From adipic acid and ethylene glycol, they obtained a polyester of high molecular weight that could be drawn into fibers. However, melting points of these first polyester fibers were too low for them to be used as textile fibers, and they were not investigated further. Carothers then turned his attention to the reactions of dicarboxylic acids and diamines, and in 1934 synthesized Nylon 66,

to-one salt, called nylon salt. Nylon salt is then heated in an autoclave to 250°C. As the temperature increases in the closed system, the internal pressure rises to about 15 atm. Under these conditions, $-CO_2^-$ groups from adipic acid and $-NH_3^+$ groups from hexamethylenediamine react to form amides. Water is a byproduct. As polymerization proceeds and more water is formed, steam and alcohol vapors are continuously bled from the autoclave to maintain a constant internal pressure. The temperature is gradually raised to about 275°C, and when all water vapor is removed, the internal pressure of the reaction vessel falls to 1 atm. Nylon 66 formed under these conditions melts at 250–260°C and has a molecular weight range of 10,000–20,000.

$$\underset{\text{adipic acid}}{HOC(CH_2)_4COH} + \underset{\substack{\text{hexamethylene-}\\\text{diamine}\\\text{(HMDA)}}}{H_2N(CH_2)_6NH_2} \longrightarrow$$

$$\underset{\text{nylon salt}}{^-OC(CH_2)_4CO^- \, H_3\overset{+}{N}(CH_2)_6NH_3^+} \xrightarrow{\text{heat}} \underset{\text{Nylon 66}}{\left[C(CH_2)_4CNH(CH_2)_6NH \right]_n} + H_2O$$

the first purely synthetic fiber. Nylon 66 is so named because it is synthesized from two different organic starting materials, each of six carbon atoms.

In the synthesis of Nylon 66, hexanedioic acid (adipic acid) and 1,6-diaminohexane (hexamethylene diamine, HMDA) are dissolved in aqueous ethanol, and they react to form a one-

In the first stage of fiber production, crude Nylon 66 is melted, spun into fibers, and cooled to room temperature. Next melt-spun fibers are drawn at room temperature (cold-drawn) to about four times their original length. As the fibers are drawn, crystalline regions are oriented in the direction of the fiber axis and hydrogen bonds are formed between carbonyl oxygens of

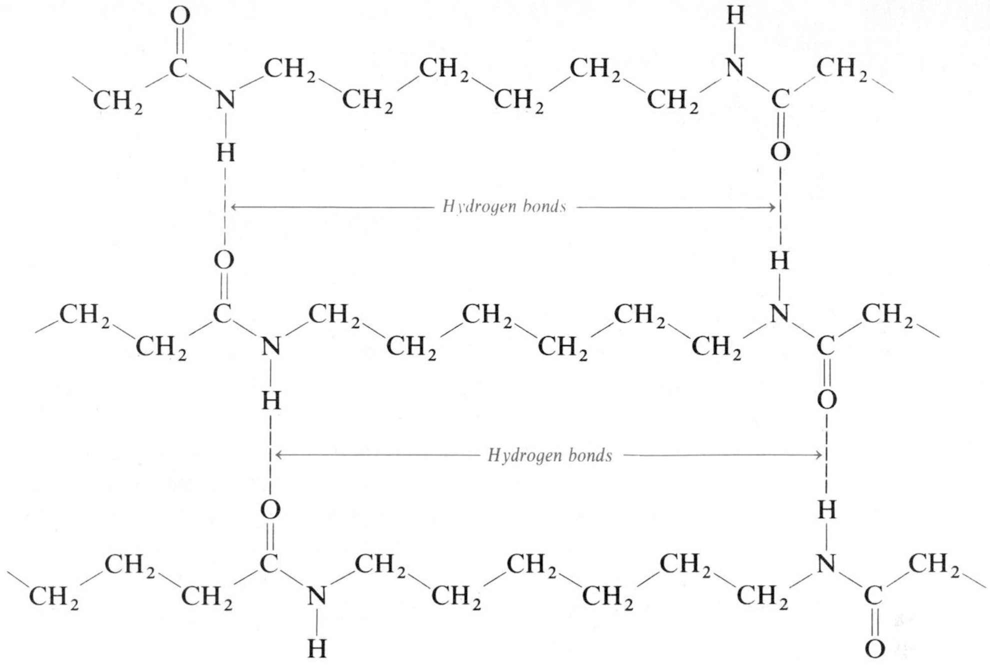

Figure VII-1 The structure of cold-drawn Nylon 66. Hydrogen bonds between adjacent chains hold molecules together.

one chain and amide hydrogens of another chain (Figure VII-1). The effects of orientation of polyamide molecules on the physical properties of the fiber are dramatic; both tensile strength and stiffness are increased markedly. Cold-drawing is an important step in the production of all synthetic fibers.

When Du Pont's management decided to begin production of Nylon 66, only adipic acid was commercially available. The raw material base for the production of adipic acid was benzene. Catalytic reduction of benzene to cyclohexane followed by air oxidation yields a mixture of cyclohexanol and cyclohexanone. Oxidation of this mixture by nitric acid gives adipic acid.

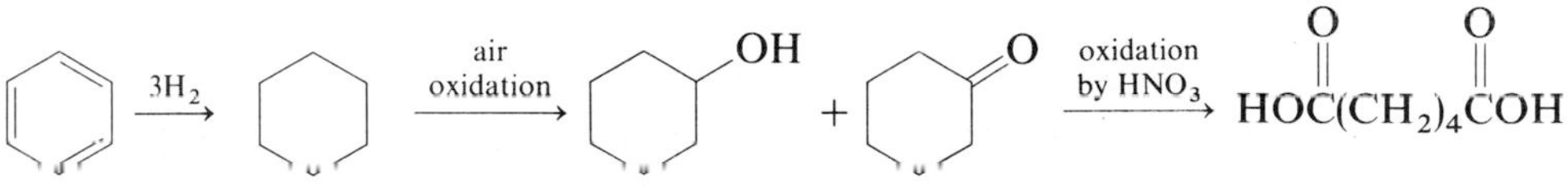

benzene cyclohexane cyclohexanol cyclohexanone adipic acid

Adipic acid was in turn the starting material for the synthesis of hexamethylenediamine. Reaction of adipic acid with ammonia gives a diammonium salt, which when heated gives adipamide. Catalytic reduction of adipamide gives hexamethylenediamine.

$$\underset{\text{adipic acid}}{\overset{\text{O}\quad\quad\text{O}}{\text{HOC(CH}_2\text{)}_4\text{COH}}} \xrightarrow{\text{2NH}_3} \underset{\text{diammonium adipate}}{\overset{\text{O}\quad\quad\quad\text{O}}{\text{NH}_4^+ \; {}^-\text{OC(CH}_2\text{)}_4\text{CO}^- \; \text{NH}_4^+}} \xrightarrow{\text{heat}}$$

$$\underset{\text{adipamide}}{\overset{\text{O}\quad\quad\text{O}}{\text{H}_2\text{NC(CH}_2\text{)}_4\text{CNH}_2}} \xrightarrow{\text{4H}_2} \underset{\text{HMDA}}{\text{H}_2\text{N(CH}_2\text{)}_6\text{NH}_2}$$

At the time, benzene was derived largely from coal, and both nitric acid and ammonia were derived from nitrogen present in air (air is approximately 80% N_2). Thus, Du Pont could rightly claim that Nylon 66 was derived from coal, air, and water.

two moles of sodium cyanide, to form 1,4-dicyano-2-butene. Reaction of this intermediate with hydrogen in the presence of a catalyst reduces both the carbon-carbon double bond and the carbon-nitrogen triple bonds and gives hexamethylenediamine.

$$\underset{\text{butadiene}}{\text{CH}_2\text{=CHCH=CH}_2} \xrightarrow{\text{Cl}_2} \underset{\substack{\text{1,4-dichloro-2-}\\\text{butene}}}{\text{ClCH}_2\text{CH=CHCH}_2\text{Cl}} \xrightarrow{\text{NaCN}}$$

$$\underset{\substack{\text{1,4-dicyano-2-}\\\text{butene}}}{\text{N}\equiv\text{CCH}_2\text{CH=CHCH}_2\text{C}\equiv\text{N}} \xrightarrow{\text{5H}_2} \underset{\text{hexamethylenediamine}}{\text{H}_2\text{N(CH}_2\text{)}_6\text{NH}_2}$$

Following World War II, Du Pont embarked on a great expansion of its Nylon 66 capacity, a move that demanded a similar expansion in facilities for producing hexamethylenediamine. The problem facing management was whether to continue to make this key starting material from benzene or to search for a more economical raw-material base and a more economical synthesis. Within a few years, Du Pont developed a new synthesis of hexamethylenediamine, using butadiene as a starting material. Recall that butadiene is a coproduct of thermal cracking of ethane and other light hydrocarbons extracted from natural gas. It is also obtained from catalytic cracking and re-forming of naphtha and other petroleum fractions.

Reaction of butadiene with chlorine under carefully controlled conditions forms 1,4-dichloro-2-butene, which is, in turn, reacted with

By the early 1950s, almost all hexamethylenediamine was being made from butadiene derived from petroleum and natural gas. The starting material for the production of adipic acid was benzene and still is. Today, however, benzene is derived almost entirely from catalytic cracking and re-forming of petroleum. Thus, the raw-material base for the synthesis of Nylon 66 has shifted from coal, air, and water to petroleum and natural gas.

Nylon 66 has been the primary nylon fiber synthesized in the United States and Canada, while that synthesized in many other parts of the world, particularly Germany, Italy and Japan is Nylon 6. The manufacture of Nylon 6 uses only one starting material, caprolactam, a six-carbon cyclic amide. The raw material base for the synthesis of caprolactam is benzene. During the synthesis of Nylon 6, caprolactam is partially hydrolyzed and heated to 250°C to drive off water and bring about polymerization.

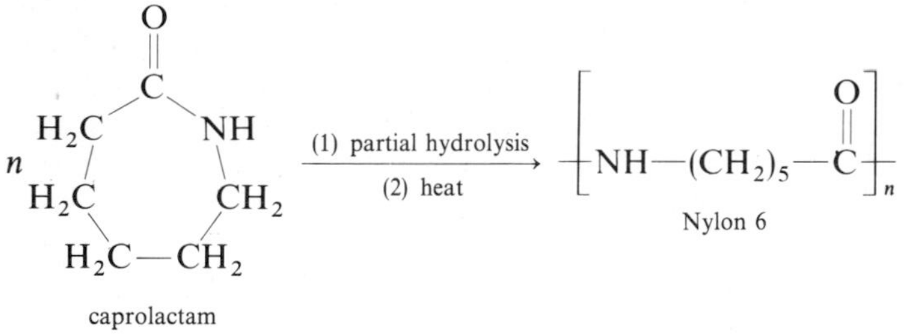

caprolactam

(1) partial hydrolysis
(2) heat

Nylon 6

Why is Nylon 66 the primary nylon fiber produced in the United States and Canada and Nylon 6 the primary nylon fiber produced in Germany, Italy, and Japan? The answer lies chiefly in the availability of raw materials. In the United States and Canada, butadiene is readily available from thermal cracking of light hydrocarbons extracted from natural gas, itself a vast natural resource. Because natural gas is not so plentiful in Europe and Japan, these countries are forced to depend on petrochemicals as a raw material base for their synthesis of nylon. It is more economical for them to synthesize Nylon 6 from caprolactam than to synthesize Nylon 66 from adipic acid and hexamethylenediamine.

By the 1940s, scientists were beginning to understand some of the relationships between molecular structure and bulk physical proper-ties, and polyester condensations were reexamined. Recall that Carothers and his associates had already concluded that the polyester fibers from aliphatic dicarboxylic acids and ethylene glycol were not suitable for textile use because they were too low-melting. Winfield and Dickson at the Calico Printers Association in England reasoned, correctly as it turned out, that a greater resistance to rotation in the polymer backbone would stiffen the polymer, raise its melting point, and thereby lead to a more acceptable polyester fiber. To create stiffness in the polymer chain, they used terephthalic acid, an aromatic dicarboxylic acid (Figure VII-2).

The crude polyester is first spun into fibers and then cold-drawn to form a textile fiber with the trade name Dacron polyester. The outstanding feature of Dacron polyester is its stiffness (about four times that of Nylon 66), very high strength, and a remarkable resistance to creasing

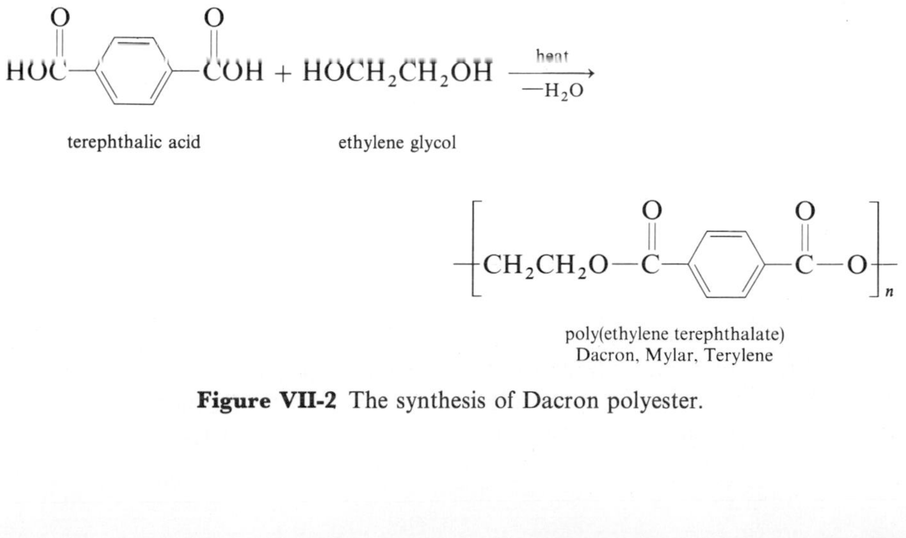

terephthalic acid   ethylene glycol

poly(ethylene terephthalate)
Dacron, Mylar, Terylene

Figure VII-2 The synthesis of Dacron polyester.

and wrinkling. Because Dacron polyesters are harsh to the touch (because of their stiffness), they are usually blended with cotton or wool to make acceptable textile fibers. Crude polyester is also fabricated into films and marketed under the trade name Mylar.

Ethylene glycol is prepared by air oxidation of ethylene followed by hydrolysis. Terephthalic acid is obtained by oxidation of p-xylene, an aromatic hydrocarbon obtained along with benzene and toluene from catalytic cracking and re-forming of naphtha and other petroleum fractions (Figure VII-3).

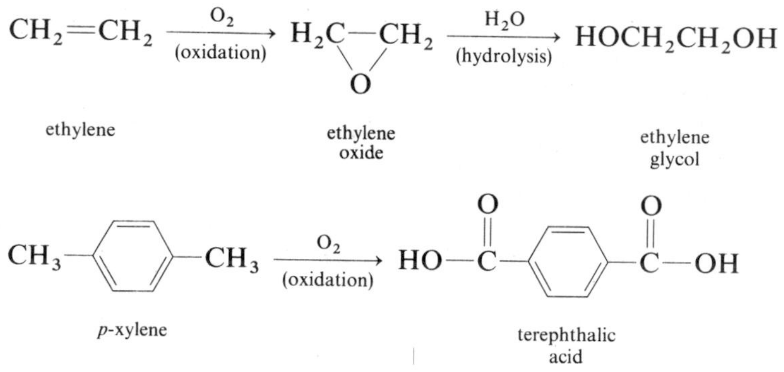

synthesized from terephthalic acid and 1,4-diaminobenzene; this was marketed under the trade name Kevlar (Figure VII-4). One of the remarkable features of Kevlar is its extremely light weight compared with other materials of similar strength. For example, a 3-in. cable woven of Kevlar has a strength equal to that of a similarly woven 3-in. steel cable. But whereas the steel cable weighs about 20 lb/ft, a comparable Kevlar cable weighs only 4 lb/ft. Kevlar now finds use in such things as anchor cables for offshore drilling rigs and reinforcement fibers for automobile tires. Kevlar can also be woven

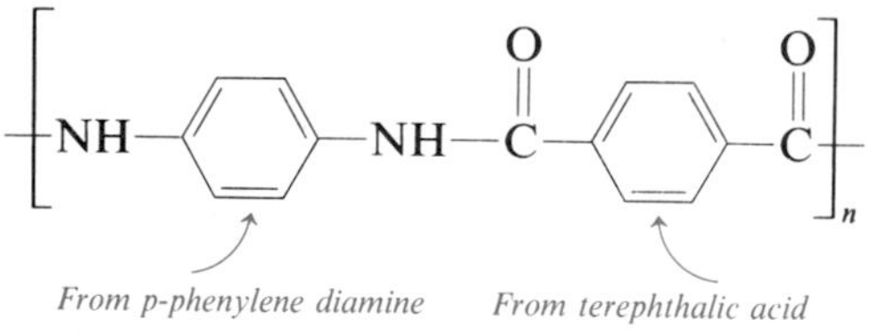

Figure VII-3 The starting materials for the synthesis of Dacron polyester and Mylar are derived from petroleum and natural gas.

In 1981 production of man-made fibers in the United States exceeded 10 billion pounds. Heading the list were polyester fibers (4.8 billion pounds) and polyamide fibers (2.7 billion pounds).

The fruits of the research into the relations between molecular structure and physical properties of polymers, and of advances in fabrication techniques are nowhere better illustrated than by the polyaromatic amides (aramids) introduced by Du Pont in the early 1960s. Researchers reasoned that a polyamide composed of aromatic rings would be stiffer and stronger than a polyamide such as Nylon 66 or Nylon 6. In early 1960, Du Pont introduced a polyaromatic amide fiber

Figure VII-4 Kevlar, an aramid polymer.

into a fabric that stretches almost like a trampoline, when struck by a bullet, absorbing the impact. Today there is a rapidly growing market among VIP's for Kevlar-lined vests, jackets, and raincoats (Figure VII-5).

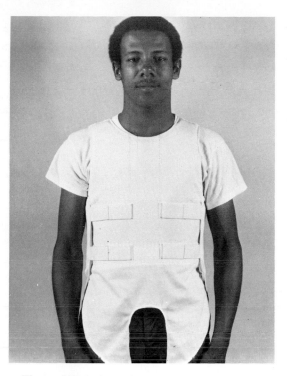

Figure VII-5 A Kevlar-lined bulletproof vest.

Sources

Anderson, B. C., Barton, L. R., and Collette, J. W. 1980. Trends in polymer development. *Science* 208:807–812.

Deanin, R. D. 1979. *New Industrial Polymers.* Washington, D.C.: American Chemical Society.

Encyclopedia of Polymer Science and Technology. 1976. New York: Wiley.

Witcoff, H. A., and Reuben, B. G. 1980. *Industrial Organic Chemicals in Perspective.* New York: Wiley.

9 pH and Acid-Base Buffers

One characteristic of living systems is their ability to maintain an internal environment of relatively constant hydrogen ion concentration. In a healthy human, for example, blood plasma is maintained within the pH range 7.3–7.5. Increases or decreases in pH beyond this range become threatening to health and to life itself. The ability of blood and other biological fluids to remain at a relatively constant pH depends on the presence of buffers. Two of the most important biological buffers are the bicarbonate buffer and the phosphate buffer.

9.1 Ionization of Weak Acids

An acid is a compound that dissociates in aqueous solution to produce a proton, H^+. Acids that are completely dissociated in aqueous solution are classified as **strong acids.** Examples of strong acids are HCl, HBr, HI, HNO_3, and H_2SO_4. Acids that are only partially dissociated in aqueous solution are classified as **weak acids.** Carboxylic acids, phenols, thiols, the ammonium ion, and amine salts are all weak acids. In aqueous solution, an equilibrium is established between the weak acid, its **conjugate base**, and H^+.

$$HA \rightleftharpoons H^+ + A^-$$

weak   proton   conjugate
acid   base

The equilibrium constant for this dissociation is called an **acid dissociation constant,** K_a, and is given by the expression

$$K_a = \frac{[H^+][A^-]}{[HA]} = \frac{[H^+][\text{conjugate base}]}{[\text{weak acid}]}$$

The relative strengths of weak acids can be compared by looking at the values of their acid dissociation constants. The acid with the largest K_a is the most dissociated in aqueous solution and therefore the strongest acid. Following are acid dissociation equations and values of K_a for three weak acids.

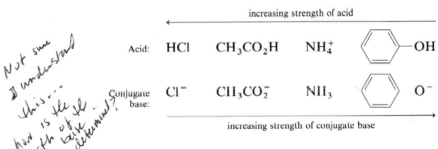

Of these three weak acids, acetic acid has the largest value of K_a and therefore is the strongest acid. Phenol has the smallest value of K_a and is the weakest acid.

Comparison of the relative strengths of HCl (a strong acid), acetic acid, ammonium ion, and phenol illustrates a general rule of acid-base behavior. The stronger the acid, the weaker its conjugate base; in contrast, the weaker the acid, the stronger its conjugate base.

increasing strength of acid

| Acid: | HCl | CH_3CO_2H | NH_4^+ | ⬡—OH |
| Conjugate base: | Cl^- | $CH_3CO_2^-$ | NH_3 | ⬡—O^- |

increasing strength of conjugate base

Of these four compounds, HCl is the strongest acid and Cl^- is the weakest base; phenol is the weakest acid and phenoxide ion is the strongest base.

9.2 Ionization of Water

Pure water ionizes, to a small but very important extent, to give H^+ and OH^-.

$$H_2O \rightleftharpoons H^+ + OH^-$$

The equilibrium constant for this reaction at 25°C is

$$K_{eq} = \frac{[H^+][OH^-]}{[H_2O]} = 1.8 \times 10^{-16}$$

In any aqueous solution, the concentration of water is very high (55.5 mol/L), and the number of ionized water molecules is so low that the concentration of the water molecules remains constant at 55.5 mol/L. This value and the value of K_{eq} are multiplied to give K_w, the ion product for water:

$$K_w = K_{eq} \times [55.5] = [H^+][OH^-] = 1.00 \times 10^{-14}$$

At 25°C, the value for the ion product of water is 1.00×10^{-14}. In any aqueous solution, concentration of H^+ times the concentration of OH^- must be 1.00×10^{-14}. Thus, if we know the value of $[H^+]$, we can calculate the value of $[OH^-]$, and vice versa.

Example 9.1

Calculate the hydroxide ion concentration in the following solutions.

a. $[H^+] = 1.0 \times 10^{-3}$ **b.** $[H^+] = 2.5 \times 10^{-8}$

☐ *Solution*

The ion product of water can be rewritten in the following form, and then used to calculate the hydroxide ion concentration of each solution.

$$[OH^-] = \frac{1.0 \times 10^{-14}}{[H^+]}$$

a. $[OH^-] = \dfrac{1.0 \times 10^{-14}}{1.0 \times 10^{-3}} = 1.0 \times 10^{-11}$

b. $[OH^-] = \dfrac{1.0 \times 10^{-14}}{2.5 \times 10^{-8}} = 4.0 \times 10^{-7}$

Problem 9.1

Calculate the hydrogen ion concentration in the following solutions:

a. $[OH^-] = 5.2 \times 10^{-4}$ **b.** $[OH^-] = 7.1 \times 10^{-8}$

9.3 pH

The hydrogen ion concentration in solutions of weak acids is almost always very low, and therefore, when expressed in exponential notation, contains a negative exponent. Similarly, values of K_a and K_b of weak acids and bases, when expressed in exponential notation, also contain negative exponents. Many people find numbers with negative exponents confusing, and answer with hesitation such questions as, Is 1.8×10^{-4} larger or smaller than 3.6×10^{-5}? To make it easier to deal with numbers containing negative exponents, scientists have adopted a system for writing them, not in exponential notation, but as negative logarithms of the numbers. The letter p stands for "negative logarithm of." For example, the symbol pH stands for the negative logarithm of the hydrogen ion concentration:

$$pH = -\log[H^+]$$

Following are definitions of **pOH**, **pK_a**, and **pK_b**:

$$pOH = -\log[OH^-]$$
$$pK_a = -\log[K_a]$$
$$pK_b = -\log[K_b]$$

Example 9.2

The hydrogen ion concentration in a 1.0M solution of acetic acid is 4.2×10^{-3}. Calculate the pH of this solution.

Solution

We define pH as the negative logarithm of $[H^+]$.

$$[H^+] = 4.2 \times 10^{-3}$$
$$\log[H^+] = \log 4.2 + \log 10^{-3}$$
$$= 0.62 + (-3) = -2.38$$
$$-\log[H^+] = +2.38$$
$$pH = 2.38$$

Problem 9.2

Complete the table:

$[H^+]$	$[OH^-]$	pH	pOH
1.0×10^{-2}	———	———	———
———	1.0×10^{-4}	———	———
———	———	2.3	———
7.3×10^{-11}	———	———	———
———	———	———	5.6

9.4 Calculation of pH for Solutions of Weak Acids

If we are given the value of K_a for a weak acid, and the concentration of the weak acid in solution, we can calculate the hydrogen ion concentration in the solution. Consider, as an example, a 1.00M solution of acetic acid in water. Such a solution can be prepared by dissolving 60.0 g (one mole) of acetic acid in enough water to make one liter of solution. The first step in calculating the hydrogen ion concentration of this solution is to write an equation for the dissociation of acetic acid and its corresponding acid dissociation constant.

$$CH_3CO_2H \rightleftharpoons CH_3CO_2^- + H^+$$

$$K_a = \frac{[H^+][CH_3CO_2^-]}{[CH_3CO_2H]} = 1.76 \times 10^{-5}$$

Inspection of the balanced equation shows that dissociation of one molecule of acetic acid gives one hydrogen ion and one acetate ion. Therefore, in a solution prepared by dissolving acetic acid in water, $[H^+]$ and $[CH_3CO_2^-]$ are equal. Initially, the concentration of acetic acid is 1.00M, and the concentrations of H^+ and $CH_3CO_2^-$ are zero. Let the concentration of H^+ at equilibrium be x. The concentration of $CH_3CO_2^-$ is also x, and the concentration of un-ionized acetic acid is $1.00 - x$.

$$CH_3CO_2H \rightleftharpoons CH_3CO_2^- + H^+$$

Initial:	1.00	0	0
At equilibrium:	$1.00 - x$	x	x

Substituting these values in the acid dissociation constant gives

$$\frac{[x][x]}{[1.00 - x]} = 1.76 \times 10^{-5}$$

The final step is to solve this equation for x. An exact solution for x requires solving a quadratic equation. We can solve this equation more easily, however, by making a straightforward approximation. Acetic acid is a weak acid, and therefore only slightly dissociated in aqueous solution. Because acetic acid is only slightly dissociated, the concentration of hydrogen ions and acetate ions is small compared with the concentration of un-ionized acetic acid molecules; therefore, $1.00 - x$ is approximately equal to 1.00.

$$\text{Approximation:} \quad 1.00 - x = 1.00$$

With this approximation, the original quadratic equation can be simplified to

$$x^2 = 1.76 \times 10^{-5}$$

Solving this equation gives

$$x = \sqrt{1.76 \times 10^{-5}} = 4.20 \times 10^{-3}$$

Therefore

$$[H^+] = [CH_3CO_2^-] = 4.20 \times 10^{-3}$$

We can now calculate the percentage of ionization of acetic acid in a 1.00M solution:

$$\text{Percentage ionization} = \frac{4.20 \times 10^{-3}}{1.00} \times 100 = 0.42\%$$

The percentage of ionization is only 0.42%, and thus our approximation that $1.00 - x$ equals 1.00 is valid.

■ **Example 9.3**   Calculate $[H^+]$, $[OH^-]$, and pH in solutions of the following weak acids.

a. 0.10M CH_3CO_2H $K_a = 1.76 \times 10^{-5}$
b. 6.0×10^{-2}M NH_4Cl $K_a = 5.5 \times 10^{-10}$

Solution

a. First write an equation for the dissociation of acetic acid, a weak acid, and let x equal the concentration of H^+ at equilibrium.

$$CH_3CO_2H \rightleftharpoons CH_3CO_2^- + H^+$$

Initial:	0.10	0	0
At equilibrium:	$0.10 - x$	x	x

Because acetic acid is a weak acid and only slightly dissociated in aqueous solution, we can make the approximation that $0.10 - x$ equals 0.10. Therefore, the equation becomes

$$\frac{x^2}{0.10} = 1.76 \times 10^{-5}$$

Solving gives

$$x^2 = 1.76 \times 10^{-6}$$
$$x = 1.33 \times 10^{-3}$$

Therefore

$$[H^+] = 1.33 \times 10^{-3}$$
$$pH = -\log 1.33 \times 10^{-3}$$
$$= -(0.124 - 3) = -(-2.88)$$
$$= 2.88$$

The hydroxide ion concentration in this solution is

$$[OH^-] = \frac{1 \times 10^{-14}}{1.33 \times 10^{-3}} = 7.52 \times 10^{-12}$$

b. The ammonium ion is a weak acid and dissociates according to the following equation:

$$NH_4^+ \rightleftharpoons NH_3 + H^+$$

Initial:	6.0×10^{-2}	0	0
At equilibrium:	$6.0 \times 10^{-2} - x$	x	x
Approximation:	6.0×10^{-2}	x	x

$$K_a = \frac{[x][x]}{[6.0 \times 10^{-2}]} = 5.5 \times 10^{-10}$$

Solving gives

$$[H^+] = 5.7 \times 10^{-6}$$
$$pH = -\log 5.7 - \log(10^{-6}) = 5.24$$

$$[OH^-] = \frac{1.0 \times 10^{-14}}{5.7 \times 10^{-6}} = 1.7 \times 10^{-9}$$

Problem 9.3

Calculate $[H^+]$, $[OH^-]$, and pH for solutions of the following weak acids.

a. 0.80M CH_3CO_2H $K_a = 1.76 \times 10^{-5}$
b. 0.10M phenol $K_a = 1.28 \times 10^{-10}$
c. 0.50M CH_3NH_3Cl $K_a = 2.7 \times 10^{-11}$

9.5 Titration Curves

The curve obtained when 50 mL of 0.10M acetic acid is titrated with 0.10M sodium hydroxide is shown in Figure 9.1. This curve is obtained by measuring the pH of 0.10M acetic acid before and after the addition of successive amounts of 0.10M sodium hydroxide. Let us examine this curve in some detail, especially the region around point B, the midpoint of the titration.

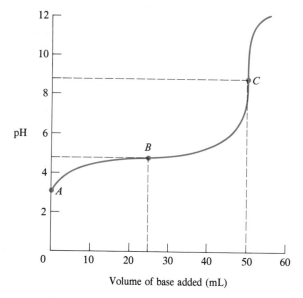

Figure 9.1 Titration of 50 mL of 0.10M acetic acid with 0.10M NaOH.

Point A of the curve represents the pH of 0.10M acetic acid, and as we calculated in Example 9.3(a), it falls at pH 2.88. Point C represents the endpoint of the titration. At this point, exactly enough 0.10M NaOH has been added to react with, or neutralize, the 50 mL of 0.10M acetic acid. The pH at the endpoint is 8.72.

Point B represents the midpoint of the titration; exactly enough 0.10M NaOH has been added to neutralize one-half of the acetic acid. At the midpoint, the concentrations of CH_3CO_2H and $CH_3CO_2^-$ are equal, and the observed pH of the solution is 4.74. Notice that around point B, the change in pH per milliliter of NaOH added is relatively small. In contrast, the change in pH per milliliter of NaOH added around point C, the endpoint of the titration, is very large.

9.6 Buffer Solutions

A buffer is anything that resists change. In chemistry, a pH buffer, or acid-base buffer, is a solution that resists changes in pH when either H^+ or OH^- is added. It must be emphasized that a buffered solution does change pH when either acid or base is added, but the change is very small compared with that of an unbuffered solution.

An acid-base buffer contains two compounds: a weak acid and its conjugate base. An acetate buffer, for example, contains acetic acid and acetate ion and can be prepared by dissolving acetic acid and sodium acetate in water. To see that an acetate buffer resists changes in pH when OH^- is added, look at the titration curve of acetic acid with sodium hydroxide (Figure 9.1). Point B represents the point of maximum buffering action. The titration curve is almost flat on either side of this point, which means that there is only a small change in pH even though a large volume of NaOH has been added. As an example, the pH of the solution when 20 mL of NaOH has been added is 4.56, and when 30 mL of NaOH has been added, it is 4.92. Even though 10 mL of 0.10M NaOH has been added, the change in pH is only 0.36. By comparison, the change in pH between the addition of 40 mL and 50 mL of 0.10M NaOH is 3.38 (from 5.34 to 8.72).

How do we account for the solution's action as a buffer at the midpoint of its acid-base titration? If hydrogen ions, H^+, are added to an acetate buffer, they react with acetate ion to form acetic acid:

$$CH_3CO_2^- + H^+ \longrightarrow CH_3CO_2H$$

Thus, added hydrogen ions are removed from solution by forming an equivalent amount of acetic acid.

If hydroxide ions, OH^-, are added to an acetate buffer, they react with acetic acid to form acetate ions and water:

$$CH_3CO_2H + OH^- \longrightarrow CH_3CO_2^- + HOH$$

Added hydroxide ions are removed from solution by formation of an equivalent amount of acetate ion. Thus, addition of the ions H^+ or OH^- to a buffered solution causes a change in the relative concentrations of weak acid and its conjugate base, but only small changes in the concentrations of H^+ and OH^-.

The pH of a buffered solution can be calculated using the **Henderson-Hasselbalch equation**. This useful equation is derived from the equation for the dissociation of a weak acid in the following way:

$$\text{Weak acid} \rightleftharpoons H^+ + \text{conjugate base} \qquad K_a = \frac{[H^+]\,[\text{conjugate base}]}{[\text{weak acid}]}$$

Taking the logarithm of both sides of this equation gives

$$\log K_a = \log [H^+] + \log \frac{[\text{conjugate base}]}{[\text{weak acid}]}$$

Rearranging this expression gives

$$-\log [H^+] = -\log K_a + \log \frac{[\text{conjugate base}]}{[\text{weak acid}]}$$

By definition, $-\log [H^+]$ equals pH and $-\log K_a$ equal pK_a. Substituting these values gives the Henderson-Hasselbalch equation:

$$pH = pK_a + \log \frac{[\text{conjugate base}]}{[\text{weak acid}]}$$

This equation can be used for several types of calculations. If both the pK_a for a weak acid and the pH of a buffered solution are known, the Henderson-Hasselbalch equation can be used to calculate the ratio of the concentrations of the conjugate base to its weak acid. Conversely, if the concentrations of conjugate base and weak acid and the pK_a of the weak acid are given, the Henderson-Hasselbalch equation can be used to calculate the pH of the solution.

Example 9.4

Calculate the pH of a solution prepared by dissolving 0.10 mol acetic acid and 0.010 mol sodium acetate in enough water to make 1.0 L solution.

☐ *Solution*

In this solution, the concentration of acetic acid is 0.10M and of the acetate ion is 0.010M. The K_a of acetic acid is 1.76×10^{-5}, and its pK_a is 4.74.

$$pH = pK_a + \log \frac{[\text{acetate ion}]}{[\text{acetic acid}]}$$

$$= 4.74 + \log \frac{0.010}{0.10}$$

$$= 4.74 + \log 0.10$$

$$= 4.74 + (-1.00)$$

$$= 3.74$$

Problem 9.4

Calculate the ratio of $[CH_3NH_2]$ to $[CH_3NH_3^+]$ in a solution of pH 10.5. The equation for the dissociation of methylammonium ion is

$$CH_3NH_3^+ \rightleftharpoons CH_3NH_2 + H^+ \qquad K_a = 2.7 \times 10^{-11}$$

(weak acid)   (conjugate base)

Two factors determine the capacity or effectiveness of a buffer. First, **buffer capacity** is directly proportional to the molar concentration of the buffer; the greater the concentration of the buffer, the greater its buffering capacity. By convention, buffer concentration is the sum of the concentrations of the weak acid and its conjugate base. A 0.100M acetate buffer, for example, may be made up of 0.025 mol acetic and 0.075 mol sodium acetate in 1 L of solution, or any other concentrations of the two that total 0.100 mol/L. The second factor affecting the capacity of a buffer is the ratio of the concentrations of the conjugate base to the weak acid; the closer the ratio to 1, the greater the buffering capacity of the solution.

For practical purposes, a solution acts as an effective buffer when its pH is within the range ± 1 of the pK_a of the weak acid. We can see how this range is determined by using the Henderson-Hasselbalch equation. When the concentration of the weak acid, HA, is 10 times the concentration of its conjugate base, A^-, then

$$pH = pK_a + \log \frac{[A^-]}{[HA]}$$

$$= pK_a + \log \frac{1}{10}$$

$$= pK_a - 1$$

In other words, when HA has a concentration 10 times that of A^-, the pH of the buffer solution is one unit less than the pK_a of the weak acid.

When the conjugate base, A^-, has a concentration 10 times that of the weak acid, HA, then

$$pH = pK_a + \log \frac{[A^-]}{[HA]}$$

$$= pK_a + \log \frac{10}{1}$$

$$= pK_a + 1$$

Thus, when the concentration of A^- is 10 times the concentration of HA, the pH of the buffer solution is one unit greater than the pK_a of the weak acid.

The pK_a of acetic acid is 4.74. Therefore, a solution of acetic acid and sodium acetate functions as a buffer within the pH range 3.74–5.74. The most effective acetate buffer, however, is one with equal concentrations of acetic acid and acetate ion.

The metabolism of carbohydrates, fats, proteins, and other biomolecules produces a variety of inorganic and organic acids and bases (Chapters 17–19). For example, metabolism of glucose under anaerobic conditions produces lactic acid, which dissociates, to produce lactate and hydrogen ion:

$$C_6H_{12}O_6 \longrightarrow 2CH_3\overset{\displaystyle OH}{\underset{\displaystyle |}{C}}HCO_2H \rightleftharpoons 2CH_3\overset{\displaystyle OH}{\underset{\displaystyle |}{C}}HCOCO_2^- + 2H^+$$

glucose   lactic acid   lactate

The ability of blood to transport lactic acid and other acids and bases as well without appreciable changes in pH depends on the presence of effective buffers in intracellular and extracellular fluids of the body. The principal buffers responsible for the constant pH of most biological fluids are the bicarbonate buffer and the phosphate buffer.

9.7 The Bicarbonate Buffer

In a healthy person, the pH of blood plasma remains within the remarkably constant range 7.35–7.45. The bicarbonate buffer is the principal buffer of blood plasma. Carbonic acid is a diprotic acid, and dissociates in two steps:

$$H_2CO_3 \rightleftharpoons H^+ + HCO_3^- \qquad K_1 = \frac{[H^+][HCO_3^-]}{[H_2CO_3]}$$

$$= 8.0 \times 10^{-7} \qquad pK_1 = 6.1$$

$$HCO_3^- \rightleftharpoons H^+ + CO_3^{2-} \qquad K_2 = \frac{[H^+][CO_3^{2-}]}{[HCO_3^-]}$$

$$= 5.6 \times 10^{-11} \qquad pK_2 = 10.3$$

The pK_1 of carbonic acid in blood depends on the constant input of CO_2 from respiration, as shown in the following equation:

$$\underset{\substack{\text{(from} \\ \text{respiration)}}}{CO_2} + H_2O \rightleftharpoons H_2CO_3 \rightleftharpoons H^+ + HCO_3^-$$

In accord with the above equation, the first acid dissociation constant for carbonic acid can also be written

$$K_1 = \frac{[H^+][HCO_3^-]}{[CO_2]} = 8.0 \times 10^{-7} \qquad pK_1 = 6.1$$

Because carbonic acid is a diprotic acid, it can buffer in the ranges pH = 6.1 ± 1, and pH = 10.3 ± 1. It is the first pH range that is important in buffering blood. The components of the bicarbonate buffer in blood are carbonic acid and the bicarbonate ion.

Under normal conditions, the pH of blood plasma is 7.4. By using the Henderson-Hasselbalch equation, we can calculate the ratio of bicarbonate ion to carbonic acid at this pH:

$$pH = pK_1 + \log \frac{[HCO_3^-]}{[H_2CO_3]}$$

$$7.4 = 6.1 + \log \frac{[HCO_3^-]}{[H_2CO_3]}$$

$$1.3 = \log \frac{[HCO_3^-]}{[H_2CO_3]}$$

$$\frac{20}{1} = \frac{[HCO_3^-]}{[H_2CO_3]}$$

Thus, in blood plasma at pH 7.4, the ratio of bicarbonate ion to carbonic acid is 20 to 1. Under normal conditions, the concentration of bicarbonate in plasma is about 0.0245 mol/L, and of carbonic acid is about 0.0012 mol/L. Therefore, the concentration of the bicarbonate buffer is approximately 0.026M, mainly as bicarbonate.

Recall from our discussion in Section 9.6 that a buffer consisting of a weak acid and its conjugate base is most effective in the concentration range 10% A^-/90% HA to 90% A^-/10% HA, that is, in the region pH = pK_a ± 1. At pH 7.4, the bicarbonate buffer is approximately 95% bicarbonate and 5% carbonic acid. We should expect this buffer system to be at the outer limits of its useful range and therefore not very effective. Yet the bicarbonate buffer in blood plasma is very effective, primarily because it is coordinated with the respiratory system, which provides a means for rapidly adjusting the concentration of carbon dioxide in the blood.

$$H_2CO_3 \rightleftharpoons CO_2 + H_2O$$

In addition, the kidneys provide a means for long-term adjustments in the concentration of bicarbonate. Through the cooperative interaction of these two systems, the bicarbonate-to-carbonic-acid ratio is kept very close to 20:1.

Clinically, the two main disturbances in the acid-base balance of blood plasma are **acidosis** and **alkalosis**. In acidosis, the pH of the blood plasma falls below 7.30. Acidosis is brought about by any abnormal condition that causes acid to accumulate in the body or excessive loss of alkali. Alkalosis is brought about by any abnormal condition that leads to loss of acid or accumulation of alkali. In alkalosis, the pH of blood rises above 7.50. Disturbances in blood pH due to abnormal breathing patterns are called respiratory acidosis and respiratory alkalosis. Disturbances in blood pH due to abnormal metabolism are called **metabolic acidosis** and **metabolic alkalosis**.

The immediate effect of a respiratory problem is a change (either increase or decrease) in the concentration of H_2CO_3. Hyperventilation due to abnormally rapid breathing brings about a loss, or "blowoff," of CO_2 and consequently a decrease in the concentration of H_2CO_3:

$$\frac{[HCO_3^-]}{[H_2CO_3]}$$ *In hyperventilation, this ratio increases because the concentration of carbonic acid decreases.*

As the concentration of H_2CO_3 decreases, the pH of the blood increases (becomes more basic); this condition is known as respiratory alkalosis. Hypoventilation due to abnormally shallow breathing, such as might be caused by lung disease, increases the concentration of CO_2 and therefore of H_2CO_3 in the blood:

$$\frac{[HCO_3^-]}{[H_2CO_3]}$$ *In hypoventilation, this ratio decreases because the concentration of carbonic acid increases.*

As the concentration of H_2CO_3 increases, the pH of the blood decreases (becomes more acidic); this condition is called respiratory acidosis. In response to more long-term changes in the concentration of H_2CO_3, the body makes compensating changes by controlling the excretion or retention of HCO_3^- through the kidneys. For example, to compensate for respiratory alkalosis and the increase in blood pH, the kidneys increase the excretion of HCO_3^- and the retention of H^+. To compensate for respiratory acidosis, the kidneys respond by reabsorbing HCO_3^- and excretion of H^+ in the urine.

The pH of blood can also be altered by metabolic disorders. Production of abnormally large amounts of acidic substances, like lactic acid, produces excessive amounts of H^+ and thereby decreases the concentration of HCO_3^-. As $[HCO_3^-]$ decreases, $[H_2CO_3]$ increases according to the reaction

$$H^+ + HCO_3^- \longrightarrow H_2CO_3 \qquad \text{(in metabolic acidosis)}$$

Metabolic acidosis results from several types of disorders, for example, production of ketone bodies in diabetes (Chapter 17 and 18), ingestion of excessive amounts of acidic substances, kidney failure and reduced excretion of H^+, and diarrhea. In diarrhea, bicarbonate is lost as a result of the loss of pancreatic secretions. Hyperventilation, one way by which the body can compensate for the loss of HCO_3^- and the increase in H_2CO_3 due to metabolic acidosis, "blows off" CO_2, thereby decreasing the concentration of H_2CO_3.

Metabolic alkalosis leads to an increase in the concentration of bicarbonate ion. Metabolic alkalosis can result from vomiting (loss of stomach acid) or from ingesting excessive amounts of conjugate bases of weak acids (for example, $NaHCO_3$).

Example 9.5

You are given a blood sample of pH 7.6.

a. Calculate the ratio of $[HCO_3^-]$ to $[H_2CO_3]$ at this pH.
b. The concentration of bicarbonate in this sample is 24.5mM. Calculate the concentration of H_2CO_3.
c. Is this alkalosis more likely to be a respiratory disorder or a metabolic disorder? (*Hint:* Compare $[H_2CO_3]$ in this sample and in a sample of normal pH.)

☐ *Solution*

a. $\text{pH} = \text{p}K_a + \log \dfrac{[HCO_3^-]}{[H_2CO_3]}$

$7.6 = 6.1 + \log \dfrac{[HCO_3^-]}{[H_2CO_3]}$

$1.5 = \log \dfrac{[HCO_3^-]}{[H_2CO_3]}$

$\dfrac{32}{1} = \dfrac{[HCO_3^-]}{[H_2CO_3]}$

b. If the ratio of $[HCO_3^-]$ to $[H_2CO_3]$ is 32:1 and the concentration of HCO_3^- is 24.5mM, then the concentration of H_2CO_3 is

$$[H_2CO_3] = \frac{24.5\text{mM}}{32} = 0.77\text{mM}$$

c. The concentration of H_2CO_3 in normal plasma is 1.2mM. In this sample $[H_2CO_3]$ is only 0.77mM, or 36% lower than normal. The concentration of HCO_3^- in this sample is normal. Because $[H_2CO_3]$ is directly related to $[CO_2]$, this alkalosis is most likely due to a respiratory problem.

Problem 9.5

You are given a blood sample of pH 6.8.

a. Calculate what the ratio of $[HCO_3^-]$ to $[H_2CO_3]$ is at this pH.
b. Calculate what the concentration of HCO_3^- in this blood sample is if the concentration of H_2CO_3 is 2.5mM.
c. Is the cause of this acidosis most likely due to a respiratory disorder or to a metabolic disorder?

9.8   The Phosphate Buffer

The most important buffer of intracellular fluids is the phosphate buffer. Phosphoric acid is a triprotic acid and dissociates in three successive steps:

$$H_3PO_4 \rightleftharpoons H_2PO_4^- + H^+ \qquad K_1 = 7.5 \times 10^{-3} \qquad pK_1 = 2.12$$
$$H_2PO_4^- \rightleftharpoons HPO_4^{2-} + H^+ \qquad K_2 = 2.0 \times 10^{-7} \qquad pK_2 = 6.70$$
$$HPO_4^{2-} \rightleftharpoons PO_4^{3-} + H^+ \qquad K_3 = 2.2 \times 10^{-13} \qquad pK_3 = 12.66$$

Because phosphoric acid is a triprotic acid, it can function as a buffer at three different ranges: pH = 2.12 ± 1, pH = 6.70 ± 1, and pH = 12.66 ± 1. Although it is difficult to measure the pH within cells with any certainty, the value is estimated to be approximately 6.7. Therefore, for buffering action within cells, the second pH range of phosphoric acid is the important one. The main components of the **phosphate buffer** in this pH range are $H_2PO_4^-$ and HPO_4^{2-}. The ratio of $H_2PO_4^-$ to HPO_4^{2-} at 6.7 can be calculated using the Henderson-Hasselbalch equation:

$$pH = pK_2 + \log \frac{[HPO_4^{2-}]}{[H_2PO_4^-]}$$
$$6.7 = 6.7 + \log \frac{[HPO_4^{2-}]}{[H_2PO_4^-]}$$
$$0.0 = \log \frac{[HPO_4^{2-}]}{[H_2PO_4^-]}$$
$$\frac{1.0}{1.0} = \frac{[HPO_4^{2-}]}{[H_2PO_4^-]}$$

Therefore at pH 6.7, the phosphate buffer is 50% in the acid form ($H_2PO_4^-$) and 50% in the conjugate base form (HPO_4^{2-}).

Example 9.6

Calculate the pH of a solution containing $20.0 \times 10^{-3}M$ Na_2HPO_4 and $26.6 \times 10^{-3}M$ NaH_2PO_4.

☐ *Solution*

$$pH = pK_2 + \log \frac{[HPO_4^{2-}]}{[H_2PO_4^-]}$$

$$= 6.7 + \log \frac{[20.0 \times 10^{-3}]}{[26.6 \times 10^{-3}]}$$

$$= 6.7 + \log 0.75$$

$$= 6.6$$

Problem 9.6

Calculate the ratio of Na_2HPO_4 to NaH_2PO_4 required to prepare a buffer of pH 7.0.

The phosphate buffer also is important in the kidneys and urine for controlling acid-base balance, and for regulating the concentration of Na^+ and H^+. Both Na_2HPO_4 and NaH_2PO_4 are present in blood plasma, at which pH the ratio of HPO_4^{2-} to $H_2PO_4^-$ is 5 to 1.

$$pH = pK_2 + \log \frac{[HPO_4^{2-}]}{[H_2PO_4^-]}$$

$$7.4 = 6.7 + \log \frac{[HPO_4^{2-}]}{[H_2PO_4^-]}$$

$$0.7 = \log \frac{[HPO_4^{2-}]}{[H_2PO_4^-]}$$

$$\frac{5}{1} = \frac{[HPO_4^{2-}]}{[H_2PO_4^-]}$$

The pH of urine is normally 5.6, and at this pH, the ratio of HPO_4^{2-} to $H_2PO_4^-$ is 0.079:1.0 (or 1:12):

$$pH = pK_2 + \log \frac{[HPO_4^{2-}]}{[H_2PO_4^-]}$$

$$5.6 = 6.7 + \log \frac{[HPO_4^{2-}]}{[H_2PO_4^-]}$$

$$-1.1 = \log \frac{[HPO_4^{2-}]}{[H_2PO_4^-]}$$

$$\frac{0.079}{1.0} = \frac{1}{12} = \frac{[HPO_4^{2-}]}{[H_2PO_4^-]}$$

Thus, in urine at pH 5.6, the concentration of $H_2PO_4^-$ is approximately 12 times the concentration of HPO_4^{2-}. The change in the relative concentrations of HPO_4^{2-} and $H_2PO_4^-$ between blood and urine occurs because of the following reaction, in which a sodium ion of Na_2HPO_4 is replaced by a hydrogen ion:

$$Na_2HPO_4 + H^+ \longrightarrow NaH_2PO_4 + Na^+$$

The ability of the kidneys to replace sodium ions by hydrogen ions serves two functions: first, sodium ions are retained, and second, acid in the form of $H_2PO_4^-$ is excreted. Excretion of acidic urine (pH 5.6 or less) by the kidneys also enables the body to neutralize the effects of metabolically produced acids.

Example 9.7

Under normal conditions, the pH of blood is 7.4 and of urine is 5.6. Following are the concentrations of Na_2HPO_4 and NaH_2PO_4 at these pH values.

	$[NaH_2PO_4]$	$[Na_2HPO_4]$
Urine, pH 5.6	43.2×10^{-3}	3.4×10^{-3}
Blood, pH 7.4	7.65×10^{-3}	39.0×10^{-3}

How many moles per liter of sodium ions are conserved by the kidneys by this change in pH between blood and urine?

Solution

The moles of sodium ion used as a phosphate counterion per liter of blood at pH 7.4 are

$$7.65 \times 10^{-3} + 2(39.0 \times 10^{-3}) = 85.7 \times 10^{-3} \frac{\text{mol Na}^+}{\text{L blood}}$$

The moles of sodium ion used as a phosphate counterion per liter of urine at pH 5.6 are

$$43.2 \times 10^{-3} + 2(3.4 \times 10^{-3}) = 50.0 \times 10^{-3} \frac{\text{mol Na}^+}{\text{L urine}}$$

The moles of sodium ion conserved by this change in pH between blood and urine are

$$85.7 \times 10^{-3} - 50.0 \times 10^{-3} = 35.7 \times 10^{-3} \text{ mol Na}^+$$

Problem 9.7

a. Calculate the pH of urine in which the concentration of NaH_2PO_4 is 32.72×10^{-3}M and the concentration of Na_2HPO_4 is 13.88×10^{-3}M.
b. Calculate the moles of sodium ion conserved per liter of urine excreted at this pH

Key Terms and Concepts

acid dissociation constant (9.1)
acidosis (9.7)
alkalosis (9.7)
bicarbonate buffer (9.7)
buffer (9.6)
buffer capacity (9.6)

conjugate base (9.1)
Henderson-Hasselbalch equation (9.6)
ionization of water (9.2)
metabolic acidosis (9.7)
metabolic alkalosis (9.7)

phosphate buffer (9.8)

pH (9.3)

pK_a (9.3)

pK_b (9.3)

pOH (9.3)

respiratory acidosis (9.7)

respiratory alkalosis (9.7)

strong acid (9.1)

weak acid (9.1)

Problems

Weak acids and pH (Sections 9.1–9.5)

9.8 Calculate $[H^+]$, $[OH^-]$, and pH for the following solutions:

a. 0.20M HCl

b. 0.20M CH_3CO_2H

c. 0.20M  —OH

d. 0.20M CH_3NH_3Cl

e. 0.20M CH_3SH ($K_a = 2 \times 10^{-8}$)

f. 0.20M NaOH

g. 0.20M $Ca(OH)_2$

h. 0.20M NaCl

9.9 Following are approximate pH ranges for a several human biological fluids. Calculate the corresponding hydrogen ion concentration ranges for each:

a. spinal fluid   7.3–7.5   **b.** milk   6.6–7.6

c. gastric contents   1.0–3.0   **d.** urine   4.8–8.4

e. saliva   6.5–7.5   **f.** bile   6.8–7.0

9.10 Following are approximate pH ranges for some foods. Calculate the corresponding hydrogen ion concentration ranges for each.

a. apples   2.9–3.3   **b.** asparagus   5.4–5.8

c. beer   4.0–5.0   **d.** cheese   4.8–6.4

e. eggs   7.6–8.0   **f.** grapefruit   2.8–3.0

g. limes   1.8–2.0   **h.** oysters   6.1–6.6

i. soft drinks   2.0–4.0   **j.** vinegar   2.4–3.4

9.11 Following are some weak organic acids and their K_a's. Calculate the pK_a for each.

a. HCO_2H $K_a = 1.77 \times 10^{-4}$

formic acid

$$\overset{\displaystyle OH}{\underset{\displaystyle |}{}}$$

b. CH_3CHCO_2H $K_a = 8.4 \times 10^{-4}$

lactic acid

$$\begin{array}{l} CH_2CO_2H \\ | \\ \text{c. } HO-C-CO_2H \\ | \\ CH_2CO_2H \end{array}$$

$K_1 = 7.10 \times 10^{-4}$

$K_2 = 1.68 \times 10^{-5}$

$K_3 = 6.40 \times 10^{-6}$

citric acid

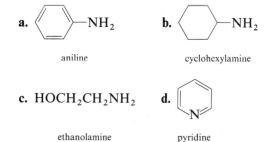

d. $K_a = 6.3 \times 10^{-11}$

o-cresol

9.12   For these weak organic bases, write the structural formula of the corresponding acid form, and then write an equation for the ionization of this weak acid to give a proton and a conjugate base.

a. ⬡—NH₂

aniline

b. ⬡—NH₂

cyclohexylamine

c. HOCH₂CH₂NH₂

ethanolamine

d. ⬡N

pyridine

9.13   Following are acid dissociation constants for the organic bases given in Problem 9.12. Calculate pK_a's for each.
a. aniline, $K_a = 2.34 \times 10^{-5}$
b. cyclohexylamine, $K_a = 2.19 \times 10^{-11}$
c. ethanolamine, $K_a = 3.16 \times 10^{-10}$
d. pyridine, $K_a = 5.62 \times 10^{-6}$

**Buffers
(Sections
9.6–9.8)**

9.14   Write equations to show how these buffers remain at a relatively constant pH when small amounts of HCl are added; small amounts of NaOH.
a. a bicarbonate buffer composed of HCO_3^- and H_2CO_3.
b. a phosphate buffer composed of HPO_4^{2-} and $H_2PO_4^-$.

9.15   Calculate the ratio of conjugate base to weak acid for the intracellular phosphate buffer at pH 6.9.

9.16   Calculate the pH of a phosphate buffer prepared by dissolving 4.662 g of KH_2PO_4 and 16.595 g of K_2HPO_4 in enough water to make 2 L of solution.

9.17   A bicarbonate buffer is prepared by adding 6.36 g of sodium carbonate, Na_2CO_3, and 3.36 g of sodium bicarbonate, $NaHCO_3$, to enough water to make 1 L of solution. Calculate the pH of this buffer.

9.18   Using the Henderson-Hasselbalch equation, calculate the pH of 1 L of these buffer solutions:
a. 0.50M CH_3CO_2H and 0.10M CH_3CO_2Na
b. 0.50M CH_3CO_2H and 0.50M CH_3CO_2Na
c. 0.25M $H_2PO_4^-$ and 0.40M HPO_4^{2-}
d. 0.10M $HOCH_2CH_2NH_2$ and 0.20M $HOCH_2CH_2NH_3^+$ Cl^-
(for $HOCH_2CH_2NH_2$, $K_a = 3.16 \times 10^{-10}$.)

9.19   Under normal conditions, the concentration of H_2CO_3 in blood is $1.2 \times 10^{-3}M$, and of HCO_3^- is $24.5 \times 10^{-3}M$, and the pH of blood is 7.4. Predict, by looking at the following numbers, whether blood pH will be greater (more basic) or less (more acidic) than 7.4 under the following conditions. Further, indicate whether any change in pH is more likely due to a respiratory disorder or to a metabolic disorder.

	$[H_2CO_3]$	$[HCO_3^-]$
a.	2.0×10^{-3}M	24.5×10^{-3}M
b.	1.2×10^{-3}M	38.0×10^{-3}M
c.	0.66×10^{-3}M	11.0×10^{-3}M

9.20 How might blood pH be returned to normal in the case of the acid-base imbalances shown in parts (a) and (b) of Problem 9.19?

9.21 Under normal conditions, urine starts out at pH 7.4 and is eventually excreted by the kidneys at pH 5.6. As a result of this pH change, 85.7×10^{-3} mol/L of Na^+ is reabsorbed by the kidneys. Will a higher concentration of Na^+ or a lower concentration be reabsorbed by the kidneys if urine is excreted instead at a pH of (a) 7.9? (b) 6.2? (c) 4.3?

9.22 Narcotic overdose and resulting shallow breathing can lead to respiratory acidosis. Do you expect that in this type of acidosis, $[H_2CO_3]$ would be elevated, depressed, or unchanged? Would $[HCO_3^-]$ be elevated, depressed, or unchanged?

9.23 A patient with a severe metabolic alkalosis has a blood pH of 8.2.
a. Calculate what the ratio of $[HCO_3^-]$ to $[H_2CO_3]$ is at this pH.
b. Calculate $[HCO_3^-]$, assuming that $[H_2CO_3]$ is 1.2×10^{-3} (a normal value).
c. Do you expect that the patient would hyperventilate or hypoventilate in an attempt to compensate for this metabolic problem? (Actually, compensation would begin well before blood pH reached 8.2!)

10

Optical Isomerism

10.1 Review of Isomerism

Thus far, we have encountered three types of isomerism.

1. Structural isomers have the same molecular formula but a different order of attachment of their atoms (Section 2.2). Examples of structural isomers are pentane and 2-methylbutane; cyclohexane and 1-hexene; and ethanol and dimethyl ether.

$$CH_3CH_2CH_2CH_2CH_3 \quad \text{and} \quad CH_3\overset{\overset{\displaystyle CH_3}{|}}{C}HCH_2CH_3$$

pentane   2-methylbutane

cyclohexane   and $CH_2{=}CHCH_2CH_2CH_2CH_3$

cyclohexane   1-hexene

$$HOCH_2CH_3 \qquad \text{and} \qquad CH_3OCH_3$$

ethanol   dimethyl ether

2. Conformational isomers have the same molecular formula and the same order of attachment of atoms, but a different arrangement of their atoms in space. Further, they are readily interconvertible by rotation about carbon-carbon single bonds (Section 2.6). Examples of conformational isomers are the eclipsed and staggered forms of butane.

eclipsed butane   and   staggered butane

3. Cis-trans isomers, like conformational isomers, have the same molecular formula and the same order of attachment of atoms, but different orientation of atoms in space. Unlike conformational isomers, cis-trans isomers cannot be interconverted by rotation about carbon-carbon single bonds. We have encountered cis-trans isomerism in two different types of molecules—cycloalkanes and alkenes. All cycloalkanes with different substituents on two or more carbon atoms of the ring show cis-trans isomerism (Section 2.7). Examples are the cis and trans isomers of 1,4-dimethylcyclohexane.

cis-1,4-dimethylcyclohexane   trans-1,4-dimethylcyclohexane

All alkenes in which each carbon of the double bond has two different substituents show cis-trans isomerism (Section 3.3). Examples are the cis and trans isomers of 2-butene.

cis-2-butene   trans-2-butene

A fourth type of isomerism is known as **optical isomerism**. It is called optical isomerism because of the effects these isomers have on the plane of polarized light (Section 10.4). The structural feature responsible for optical isomerism is chirality.

10.2 Molecules with One Chiral Center

A. Chiral Molecules and Enantiomers

Every object in nature has a **mirror image**. Shown in Figure 10.1 are stereorepresentations of a lactic acid molecule and its mirror image. All bond angles about the central carbon are approximately 109.5° and the four bonds from this carbon create the shape of a regular tetrahedron (Section 1.3). The question is, What is the relation between these two structural formulas? Do they represent the same compound or different compounds? The answer is that they represent different compounds. The lactic acid molecule drawn in Figure 10.1(a) can be turned any direction in space, but so long as no bonds are broken and rearranged, only two of the four groups attached to the central carbon can be made to coincide with those in 10.1(b), as shown in Figure 10.2.

The mirror images of lactic acid differ from each other just as a right hand differs from a left hand. They are related by reflection, but not superposable on

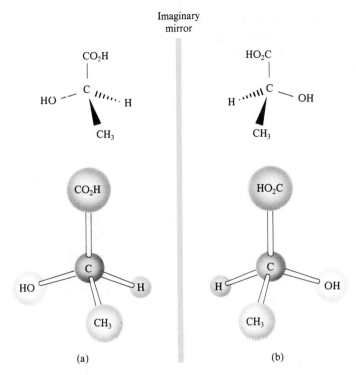

Figure 10.1 Stereorepresentations and ball-and-stick models of (a) a lactic acid molecule and (b) its mirror image. In these drawings, — represents a bond in the plane of the page, ➤ represents a bond projecting in front of the plane of the page, and ⸬ represents a bond projecting behind the plane of the page.

each other. (Superposable objects can be placed one over the other so that all like parts coincide.)

Objects that are not identical with their mirror images are said to be **chiral** (pronounced ki-ral, to rhyme with spiral; from the Greek *cheir*, hand). All chiral molecules show optical isomerism. In the molecules we will deal with in this chapter, chirality and hence optical isomerism arise because there is at least one carbon atom in the molecule that has four different groups attached to it. A carbon atom that has four different groups attached to it is called a chiral carbon, or alternatively, an **asymmetric carbon**.

**Example
10.1**

Which of these molecules contain chiral (asymmetric) carbons?

a. $CH_3CH_2CH_2CH_2OH$

b. $CH_3CH_2\overset{\displaystyle OH}{\underset{\displaystyle |}{C}}HCH_3$

c.

d. $CH_3CH=CHCH_2CH_3$

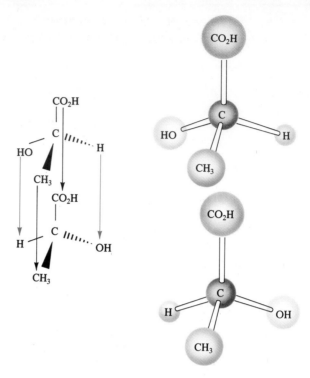

Figure 10.2 A molecule of lactic acid and its mirror image are not superposable on each other. No matter how the mirror image is turned in space, only two of the four groups attached to the central carbon can be made to coincide.

$$
\text{e. } CH_3\overset{\displaystyle O}{\overset{\displaystyle \|}{C}}CH_2CH_2CH_2CH_3
\qquad
\text{f. } CH_3\overset{\displaystyle O}{\overset{\displaystyle \|}{C}}CH_2\underset{\displaystyle CH_3}{CH}CH_2CH_3
$$

□ *Solution*

To be chiral (asymmetric), a carbon atom must have four different groups attached to it.

a. 1-Butanol. There is no chiral (asymmetric) carbon atom in this molecule.

b. 2-Butanol. Carbon 2 has four different groups attached to it and is therefore a chiral (asymmetric) carbon atom.

$$
CH_3-CH_2-\underset{\displaystyle H}{\overset{\displaystyle OH}{C}}-CH_3 \quad \textit{A chiral carbon atom}
$$

c. Methylcyclohexane. No chiral carbon atoms.

d. 2-Pentene. No chiral carbon atoms.

e. 2-Hexanone. No chiral carbon atoms.

f. 4-Methyl-2-hexanone contains one chiral carbon atom.

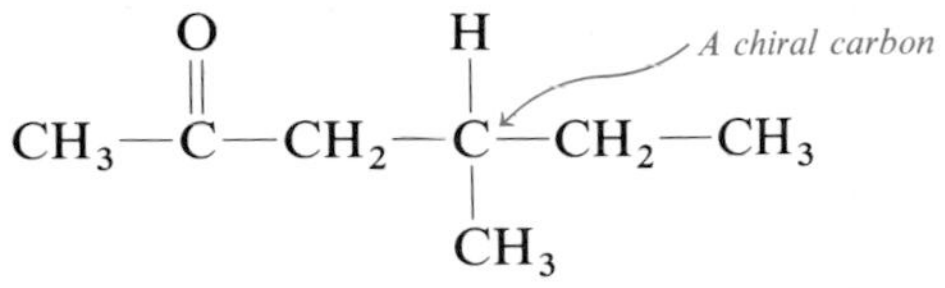

$$CH_3-\overset{\displaystyle O}{\overset{\displaystyle \|}{C}}-CH_2-\overset{\displaystyle H}{\underset{\displaystyle CH_3}{\overset{\displaystyle |}{C}}}-CH_2-CH_3 \quad \textit{A chiral carbon}$$

Problem 10.1

Which molecules contain chiral (asymmetric) carbons?

a. 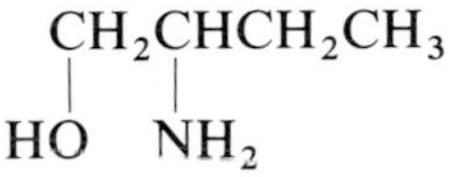 $CH_2CHCH_2CH_3$
  $\overset{}{\underset{\displaystyle HO}{|}}\ \overset{}{\underset{\displaystyle NH_2}{|}}$

b. 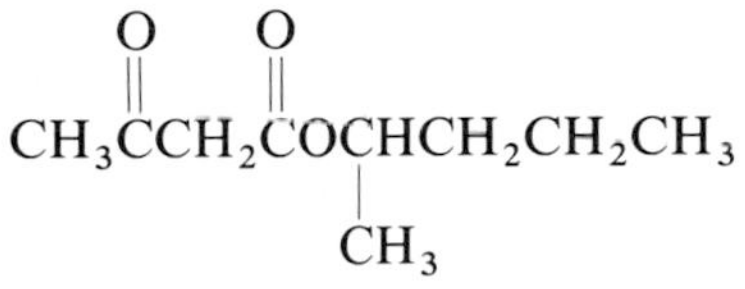 $CH_3\overset{\displaystyle O}{\overset{\displaystyle \|}{C}}CH_2\overset{\displaystyle O}{\overset{\displaystyle \|}{C}}OCHCH_2CH_2CH_3$
  $\overset{}{\underset{\displaystyle CH_3}{|}}$

c. $CH_3CH_2\overset{\displaystyle OH}{\overset{\displaystyle |}{C}}HCH=CH_2$

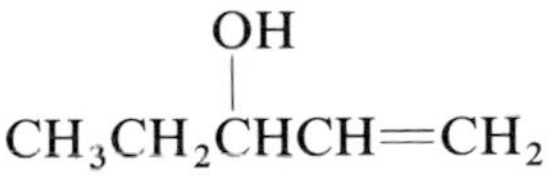

d.

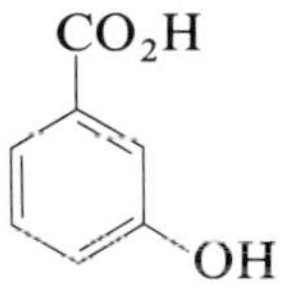

We use the term **enantiomer** (Greek, *enantios*, opposite, and *meros*, parts) to refer to molecules that are not superposable on their mirror images. Enantiomers always come in pairs; the molecule and its nonsuperposable mirror image. Many of the properties of enantiomers are identical; they have the same melting points, the same boiling points, the same solubility in solvents. Yet they are isomers, and we must expect them to show some differences in their properties. One important difference is their effect on the plane of polarized light.

Example 10.2

Draw stereorepresentations for the enantiomers of:

a. 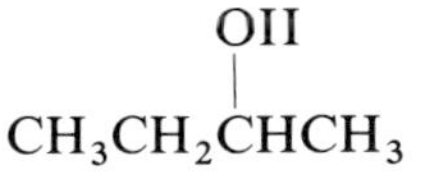 $CH_3CH_2\overset{\displaystyle OII}{\overset{\displaystyle |}{C}}HCH_3$ **b.** 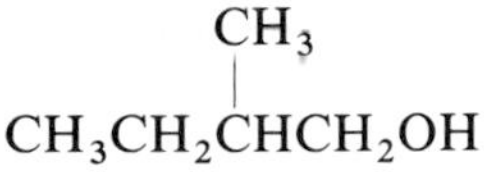 $CH_3CH_2\overset{\displaystyle CH_3}{\overset{\displaystyle |}{C}}HCH_2OH$

☐ *Solution*

First locate and draw the chiral (asymmetric) carbon atom and the four bonds from it arranged to show the tetrahedral geometry. Next, draw the four groups attached to the chiral carbon. Finally, draw the nonsuperposable mirror image.

a. 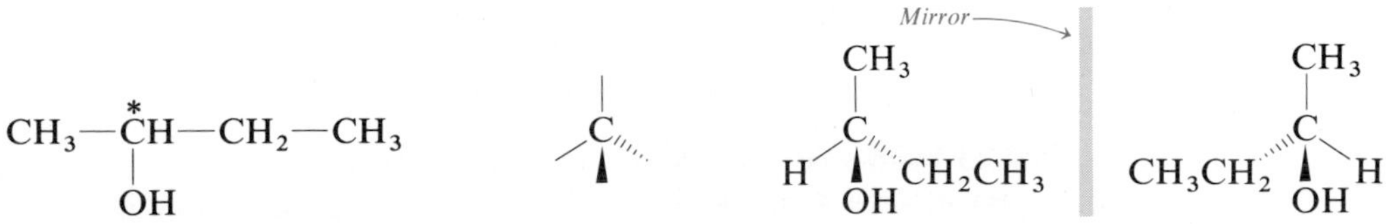 $CH_3-\overset{*}{\overset{\displaystyle |}{C}}H-CH_2-CH_3$
  $\overset{}{\underset{\displaystyle OH}{|}}$

b. 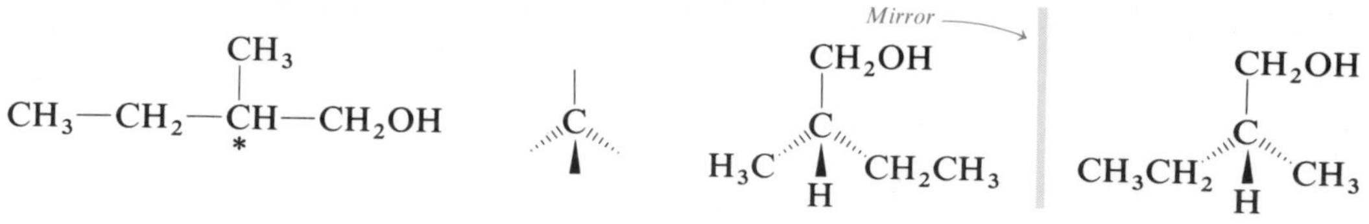 $CH_3—CH_2—\overset{\underset{*}{|}}{C}H—CH_2OH$

(the chiral carbon atom marked by an asterisk)

tetrahedral geometry of the chiral carbon

Mirror

a pair of enantiomers (nonsuperposable mirror images)

Problem 10.2

Draw stereorepresentations for the enantiomers of:

a. 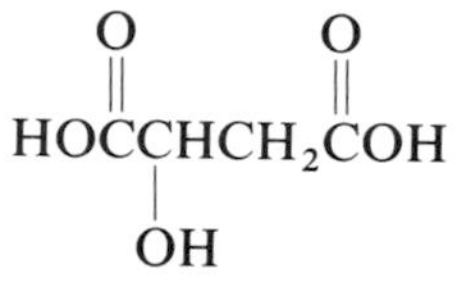 $HO\overset{\overset{O}{\|}}{C}CHCH_2\overset{\overset{O}{\|}}{C}OH$
     $\underset{OH}{|}$

malic acid

b. 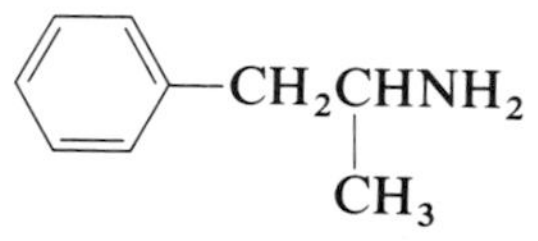 ⬡—$CH_2\overset{\underset{|}{}}{C}HNH_2$
          $\underset{CH_3}{|}$

amphetamine

B. Achiral Molecules

For all molecules that contain a single chiral carbon atom, the mirror images are nonsuperposable and the compounds show enantiomerism. For achiral (without chirality) compounds, however, the molecule and its mirror image are superposable. Consider, for example, the amino acid glycine. Figure 10.3 shows stereorepresentations of (a) glycine and (b) its mirror image. The question is, What is the relation between the two? Are they superposable or nonsuperposable? The answer is that they are superposable. One way to see this is to turn (b) by 180° about the C—CO_2H bond. When this is done, it is possible to place (b) on top of (a) in such a way that all like groups coincide. Molecules that are superposable on their mirror images are identical and therefore do not show chirality; they are achiral (without handedness); achiral molecules do not show optical isomerism.

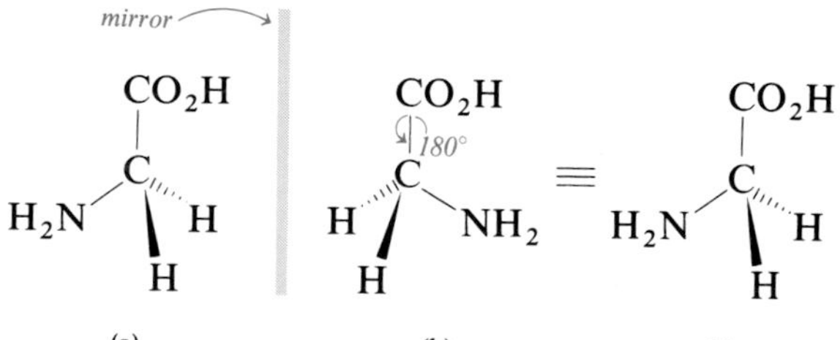

mirror

(a)   (b)   (c)

Figure 10.3 Stereorepresentations of (a) glycine and (b) its mirror image. If (b) is turned by 180° about the C—CO_2H bond (c), it becomes superposable on and hence identical to (a). Glycine does not show chirality.

C. Plane of Symmetry

A **plane of symmetry** is defined as a plane (often visualized as a mirror) cleaving an object in such a way that one side of the object is the mirror image of the other. Illustrated in Figure 10.4 are planes of symmetry in a chair and a cup. The

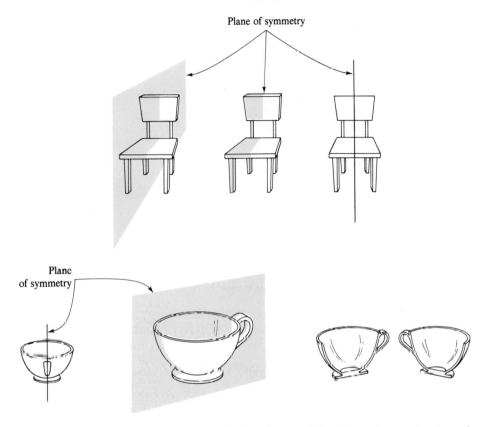

Figure 10.4 Planes of symmetry in a chair and a cup. The plane of symmetry in each object divides it so that one side is the mirror image of the other side.

plane of symmetry in the chair passes vertically through it and divides it so that one half is the mirror reflection of the other half. In the cup, the plane of symmetry passes through the handle.

Shown in Figure 10.5 is a stereorepresentation of glycine. The plane of symmetry in this molecule runs through the axis of the C—C—N bonds.

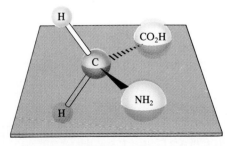

Figure 10.5 A stereorepresentation of glycine. The plane of symmetry in this molecule runs through the axis of the C—C—N bonds.

10.3 How to Predict Optical Isomerism

There are three methods you can use to determine whether a molecule shows optical isomerism.

1. The first and most direct test is to build a model of the molecule and one of its mirror image. If the two are superposable and therefore identical, then the molecule does not show optical isomerism; it has no enantiomer. If the two are nonsuperposable, the molecule shows optical isomerism; it can exist as a pair of enantiomers.

2. A second test is to look for a plane of symmetry. If the molecule has a plane of symmetry, it and its mirror image are superposable. It has no enantiomer and does not show optical isomerism. If there is no plane of symmetry, the molecule has a nonsuperposable mirror image; it can exist as a pair of enantiomers and shows optical isomerism.

3. A third and perhaps the simplest test is to look for a chiral (asymmetric) carbon atom. If the molecule has a chiral carbon atom, it has a nonsuperposable mirror image and shows optical isomerism.

You should arrive at the same answer from whichever test you apply. For example, if you conclude from test 3 that a molecule is chiral and shows optical isomerism, then you should arrive at the same conclusion from both tests 1 and 2 and vice versa.

Example 10.3

Which of the following molecules show optical isomerism? For each that does, label the chiral carbon and draw stereorepresentations for the pair of enantiomers. For each that does not, draw a stereorepresentation to show the plane of symmetry.

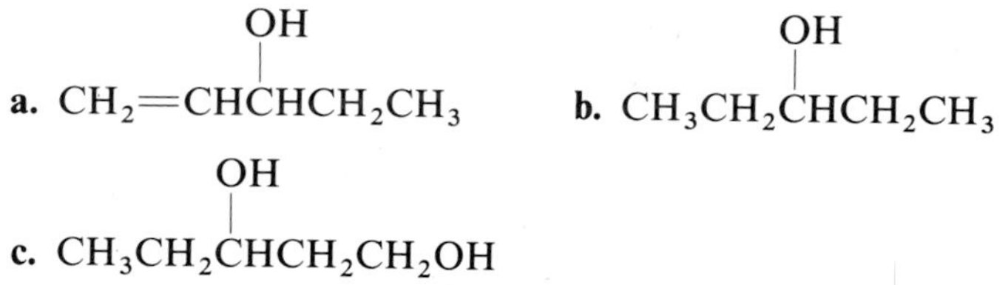

a. $CH_2{=}CHCHCH_2CH_3$ (with OH on the third carbon)   b. $CH_3CH_2CHCH_2CH_3$ (with OH on the third carbon)

c. $CH_3CH_2CHCH_2CH_2OH$ (with OH on the third carbon)

☐ *Solution*

Both (a) and (c) have chiral carbon atoms and show optical isomerism. Compound (b) has no chiral carbon and does not show optical isomerism.

Mirror

a. $CH_2{=}CH$ — C(H)(OH) — CH_2CH_3 CH_3CH_2 — C(H)(OH) — $CH{=}CH_2$

Chiral carbon atom

b. CH_3CH_2 — C(H)(OH) — CH_2CH_3

The plane of symmetry bisects the molecule through the O—C—H bonds

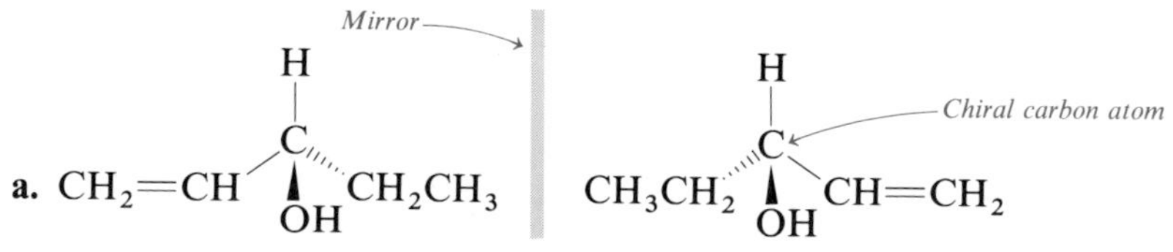

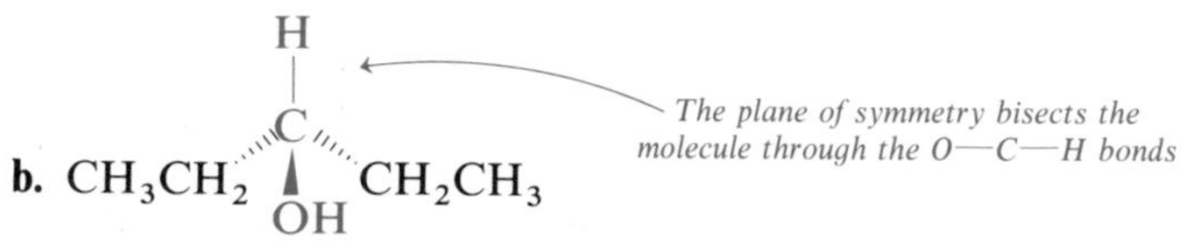

$$\text{c.}\quad \underset{HO}{\overset{CH_2CH_3}{\underset{H}{\overset{|}{C}}}}\quad CH_2CH_2OH \qquad HOCH_2CH_2\underset{H}{\overset{CH_2CH_3}{\overset{|}{\underset{|}{C}}}}\!\!\!\!\!\!OH$$

Chiral carbon atom

Problem 10.3

Which of the following molecules show optical isomerism? For each that does, label the chiral carbon and draw stereorepresentations for the pair of enantiomers. For each that does not, draw a stereorepresentation to show the plane of symmetry.

a. $CH_3\underset{NH_2}{\overset{|}{C}}HCO_2H$ b. (cyclopentane) $CHCH_3$ with OH   c. (cyclopentane) CH_3, OH

10.4 How Optical Isomerism Is Detected in the Laboratory

A. Plane-Polarized Light

Ordinary light consists of waves vibrating in all planes perpendicular to its path (Figure 10.6). Certain materials, such as Polaroid sheet, a plastic film containing properly oriented crystals of an organic substance embedded in it, selectively transmit light waves vibrating in one specific plane. Light that vibrates in only one specific plane is said to be **plane-polarized**.

If you place a polarizing disc in the path of plane-polarized light so that its polarizing axis is parallel to the axis of the plane-polarized light, the maximum

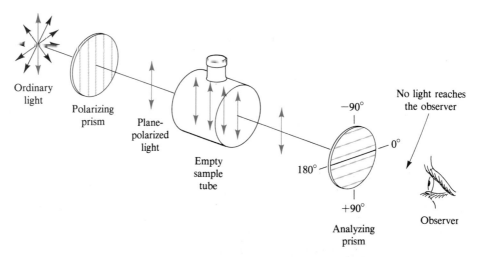

Figure 10.6 Schematic diagram of a polarimeter with the sample tube empty.

intensity of light passes through to you. If, however, the axis of the polarizing lens is perpendicular (that is, at an angle of 90°) to the axis of the polarized light, the minimum of polarized light passes through to you.

No doubt you are familiar with the effect of polarizing sheets on light from your experiences with sunglasses or camera filters made of this material. Polaroid sunglasses are particularly effective in blocking glare from water, windows, and other reflective surfaces. Glare is largely light polarized by the surface from which it is reflected, and if you rotate your head so that the polarizing axis of your Polaroid sunglasses is perpendicular to the axis of the polarized light, the glare will not pass through the lens.

B. The Polarimeter

A **polarimeter** consists of a light source, a polarizing prism, a sample tube, and an analyzing prism (Figure 10.6). If the sample tube is empty, the intensity of light reaching an observer is the maximum when the polarizing axes of the two prisms are parallel. If the analyzing prism is turned either clockwise or counterclockwise, less light is transmitted. When the axis of the analyzing prism is at right angles to the axis of the polarizing prism, the field of view is dark. This position is taken as 0° on the optical scale.

A polarimeter is used in the laboratory to detect and measure optical activity. One of the effects of an optically active compound is that it has the ability to rotate the plane of plane-polarized light, as can be observed in the following way. First, an empty sample tube is placed in the polarimeter and the analyzing prism is adjusted so that no light passes through to the observer, that is, it is set to 0°. When a solution of optically active compound is placed in the sample tube, a certain amount of light is observed now to pass through the analyzing

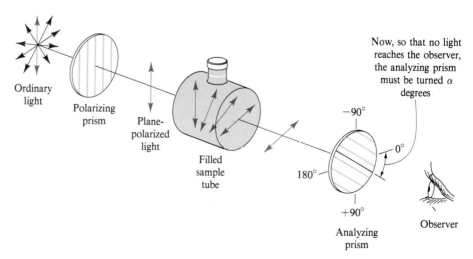

Figure 10.7 Schematic diagram of a polarimeter with the sample tube containing a solution of an optically active compound. The analyzing prism has been turned clockwise by α degrees to restore the dark field.

prism; the optically active compound has rotated the plane of light from the polarizing prism, so it is now no longer at an angle of 90° to the analyzing prism. Turning the analyzing prism a few degrees either clockwise or counterclockwise restores darkness to the field of view (Figure 10.7). The number of degrees α through which the analyzing prism must be turned to restore darkness to the field of view is called the **observed rotation**. If the analyzing prism must be turned to the right (clockwise) to restore darkness, then we say that the compound is **dextrorotatory** (Latin, *dexter*, on the right side). If the analyzing prism must be turned to the left (counterclockwise), then the compound, we say, is **levorotatory** (Latin, *laevus*, on the left side). In either case, the compound is optically active. Because of this effect of chiral compounds on the plane of polarized light, they are said to be **optically active** and to show **optical isomerism**.

C. Specific Rotation

Specific rotation $[\alpha]$ is defined as the rotation caused by a compound at a concentration of 1 g/mL in a sample tube 10 cm long. Specific rotation depends on the temperature and the wavelength of the light source, and these values must be reported as a part of the measurement. The most common light source is the D line of sodium, the same line that is responsible for the yellow color of excited sodium vapor lamps. It is common in reporting either observed or specific rotation to indicate a dextrorotatory compound by a positive sign $(+)$ and a levorotatory compound by a negative sign $(-)$. By these conventions, the specific rotation of sucrose (table sugar) dissolved in water at 25°C, with the D line of sodium as the light source, is reported in the following way:

$$[\alpha]_D^{25°C} = +66.5° \,(H_2O)$$

Enantiomers are different compounds and have different properties; and one important difference is their optical activity. One member of a pair of enantiomers is dextrorotatory and the other is levorotatory. For each, the number of degrees of the specific rotation is the same, but the sign is different, as illustrated by the dextro- and levorotatory isomers of 2-butanol.

$$
\begin{array}{c}
\text{OH} \\
| \\
\text{CH}_3-\text{C}-\text{CH}_2-\text{CH}_3 \\
| \\
\text{H}
\end{array}
$$

$(+)$-2-butanol $[\alpha]_D^{20°C} = +13.9°$
$(-)$-2-butanol $[\alpha]_D^{20°C} = -13.9°$

10.5 Racemic Mixtures

Lactic acid (Figure 10.1) was one of the first optically active compounds detected. It was originally isolated in 1780 from sour milk and found to be optically inactive; its specific rotation was 0°. In 1807, lactic acid was isolated from muscle tissue. It was found to be dextrorotatory and was designated $(+)$-lactic acid. We have already demonstrated that lactic acid shows optical isomerism. How can it be, then, that lactic acid from fermentation of milk is optically inactive while the lactic acid from muscle is optically active and dextrorotatory? The answer is that optically inactive lactic acid from fermentation of milk contains equal numbers of dextrorotatory and levorotatory molecules, and therefore has a rotation of 0°. A mixture containing equal amounts of a pair of enantiomers is called a **racemic mixture**.

10.6 Molecules with Two or More Chiral Centers

A. Enantiomers and Diastereomers

Compounds that contain two or more chiral (asymmetric) carbons can exist in more than two optical isomers. The maximum number of optical isomers is 2^n, where n is the number of chiral carbons. Consider, as an example, 2,3,4-trihydroxybutanal, a molecule that contains two chiral carbons, marked by asterisks:

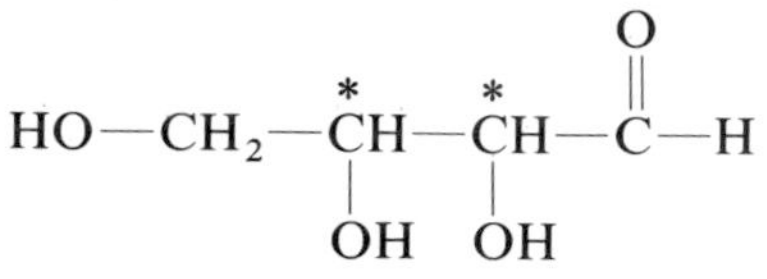

$$\overset{\text{O}}{\underset{\substack{|\\ \text{OH}}}{\text{HO}-\text{CH}_2-\overset{*}{\text{CH}}}-\overset{*}{\underset{\substack{|\\ \text{OH}}}{\text{CH}}}-\overset{\|}{\text{C}}-\text{H}$$

2,3,4-trihydroxybutanal

There are $2^2 = 4$ optical isomers of this compound, each of which is drawn in Figure 10.8. Formulas A and B are nonsuperposable mirror images and represent one pair of enantiomers. Formulas C and D are also nonsuperposable mirror images and represent a second pair of enantiomers. We can describe the optical isomers of this compound by saying that they consist of two pairs of enantiomers.

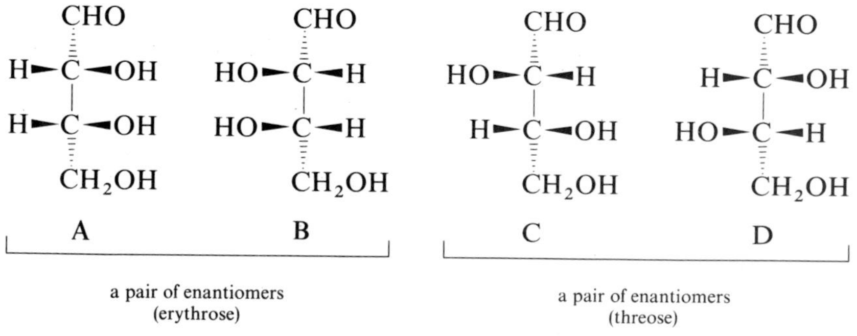

Figure 10.8 The four optical isomers of 2,3,4-trihydroxybutanal, a compound with two chiral carbons.

We know what the relation is between A and B, and between C and D; they are enantiomers. But how do we describe the relation between A and C and between A and D? The answer is that they are diastereomers. **Diastereomers** are optical isomers that are not mirror images of each other. Thus, for 2,3,4-trihydroxybutanal, A is the diastereomer of C and D; C is the diastereomer of A and B; and so on. Diastereomers have different physical and chemical properties and are sometimes given different names. The diastereomers of 2,3,4-trihydroxybutanal are given the common names erythrose and threose.

B. Meso Compounds

Certain molecules have special symmetry properties that reduce the number of optical isomers to fewer than what is predicted by the 2^n rule. One such example is tartaric acid. Both carbons 2 and 3 of this molecule are chiral (each has four different groups attached to it), and the 2^n rule predicts four optical isomers. In fact, only three optical isomers of this formula are known (Figure 10.9). Formulas E and F are nonsuperposable mirror images and represent a pair of enantiomers. Formulas G and H are mirror images of each other but are superposable. One way to see the superposability of G and H is to turn H by 180° in the plane of the page. When this is done, H can be made to lie on top of G, and all like groups coincide. Therefore, G and H are the same molecule and do not represent a pair of enantiomers.

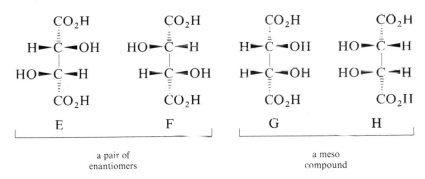

Figure 10.9 The three optical isomers of tartaric acid

A structure that contains two or more chiral carbons but is superposable on its mirror image is called a **meso compound**. A meso compound is said to be an optical isomer even though it has no effect on the plane of polarized light. The physical properties of the three optical isomers of tartaric acid are given in Table 10.1.

Table 10.1
Physical properties of the optical isomers of tartaric acid.

Acid	mp (°C)	$[\alpha]_D^{25°C}$
(+)-tartaric acid	170	+12°
(−)-tartaric acid	170	−12°
meso-tartaric acid	140	0°

10.7 Significance of Chirality in the Biological World

Except for inorganic salts and a relatively few low-molecular-weight organic molecules, most molecules in living organisms, both plant and animal, are chiral. Although these molecules can exist as a mixture of optical isomers, almost invariably only one optical isomer is found in nature. There are, of course, instances where more than one optical isomer is found, but these rarely exist together in the same biological system.

We can generalize further that only one enantiomer can be metabolized by an organism. Louis Pasteur discovered in 1858–1860, as one example of this phenomenon, that when *Penicillium glaucum*, a green mold found in aging cheese and rotting fruit, is grown in a solution containing racemic tartaric acid, the solution slowly becomes levorotatory. Pasteur concluded that the microorganism preferentially metabolizes (+)-tartaric acid. If the process is interrupted at the right time, (−)-tartaric acid can be crystallized from solution in pure form. If the process is allowed to continue, however, the microorganism eventually metabolizes (−)-tartaric acid as well. Thus while both enantiomers are metabolized by *Penicillium glaucum*, the (+)-enantiomer is metabolized much more rapidly.

A. Chirality in the Biological World

The generalization that only one enantiomer is found in a given biological system and one enantiomer is metabolized in preference to the other should be enough to convince us that we live in a chiral world. At least it is chiral at the molecular level. Essentially all chemical reactions in the biological world take place in a chiral environment. Perhaps the most conspicuous examples of chirality among biological molecules are the enzymes, all of which have multiple centers of chirality. An illustration is chymotrypsin, an enzyme that functions very efficiently in the intestine of animals at pH 7–8 in catalyzing the digestion of proteins. Chymotrypsin contains 251 separate chiral carbon atoms. The number of possible optical isomers is 2^{251}, a number large beyond comprehension. Fortunately nature does not squander its precious energies and resources unnecessarily; only one of these optical isomers is produced and used by any given organism.

B. How an Enzyme Distinguishes between a Molecule and Its Enantiomer

Enzymes catalyze biological reactions by first adsorbing on their surfaces the molecule or molecules about to undergo reaction. These molecules may be held on the enzyme surface by a combination of hydrogen bonds, ionic bonds, dispersion forces, or even covalent bonds. Thus, whether the molecules about to undergo reaction are chiral or not, they are held in a chiral enivronment.

It is generally agreed that an enzyme with specific binding site for three of the four substituents on a chiral carbon can distinguish between a molecule and its enantiomer or diastereomer. Assume for example that an enzyme involved in catalyzing a reaction of glyceraldehyde has three binding sites, one specific for —H, another specific for —OH, and the third specific for —CHO. Assume further that the three sites are arranged on the enzyme surface as shown in Figure 10.10. The enzyme can "recognize" (+)-glyceraldehyde (the natural or biologically active form) and distinguish it from (−)-glyceraldehyde, because the correct enantiomer can be adsorbed with three groups attached to their appropriate binding sites; the other enentiomer can, at best, bind to only two of these sites.

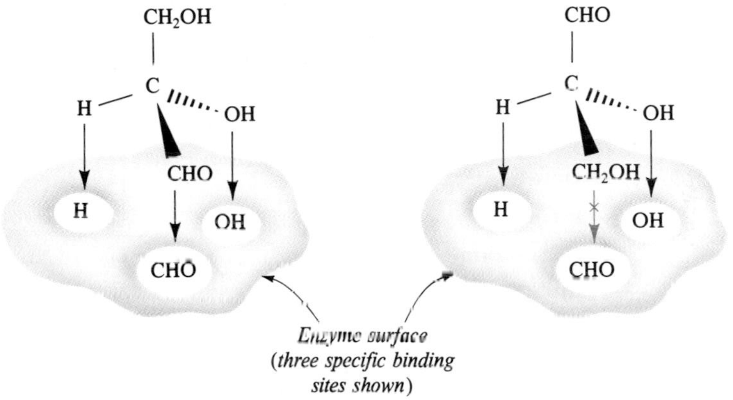

Enzyme surface
(*three specific binding
sites shown*)

(+)-glyceraldehyde   ( ) glyceraldehyde

Figure 10.10 A schematic diagram of an enzyme surface capable of interacting with (+)-glyceraldehyde at three specific binding sites, but with (−)-glyceraldehyde at only two of these sites.

Since interactions between molecules in living systems take place in a chiral environment, it should be no surprise that a molecule and its enantiomer have different physiological properties. It should be no surprise that *Penicillium glaucum* selectively metabolizes (+)-tartaric acid compared with (−)-tartaric acid; or that (+)-leucine tastes sweet, and its enantiomer (−)-leucine has a bitter taste.

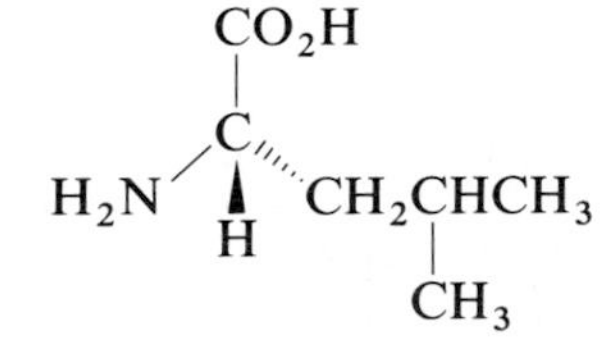

(+)-leucine
$[\alpha]_D^{25°C} = +10.42°$

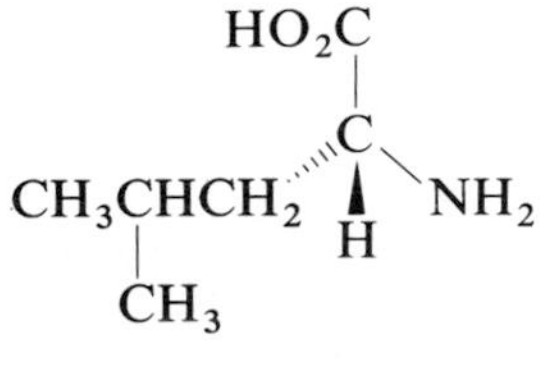

(−)-leucine
$[\alpha]_D^{25°C} = -10.42°$

That the interactions between molecules in the biological world are very specific in stereochemistry is not surprising, but just how these interactions take place at the molecular level with such precision and efficiency is one of the great puzzles that modern science has only recently begun to unravel.

Key Terms and Concepts

asymmetric carbon (10.2A)

chiral (10.2A)

cis-trans isomerism (introduction)

conformational isomerism
  (introduction)

dextrorotatory (10.4B)

diastereomer (10.6A)

enantiomer (10.2A)

levorotatory (10.4B)

meso compound (10.6B)

mirror image (10.2A)

nonsuperposable mirror image (10.2A)

observed rotation (10.4B)

optical activity (10.4B)

optical isomerism (introduction
  and 10.4B)

plane of symmetry (10.2C)

plane-polarized light (10.4A)

polarimeter (10.4B)

racemic mixture (10.5)

specific rotation (10.4C)

structural isomerism (introduction)

Problems

Molecules with one chiral center (Sections 10.2 and 10.3)

10.4   Draw mirror images for the following:

a. 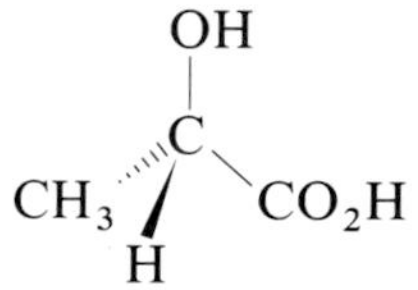
OH — C (CH$_3$, CO$_2$H, H)

b. 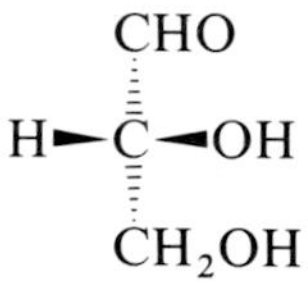 H—C—OH (CHO, CH$_2$OH)

c. H$_2$N—C—H (CO$_2$H, CH$_3$)

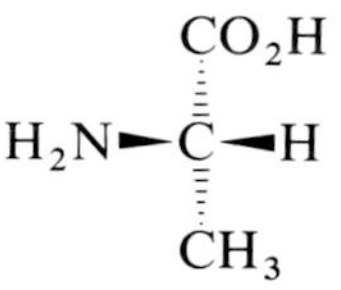

d.

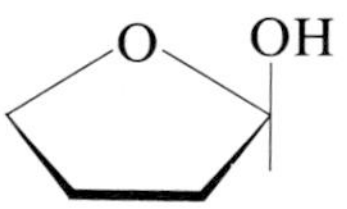

e.

f. 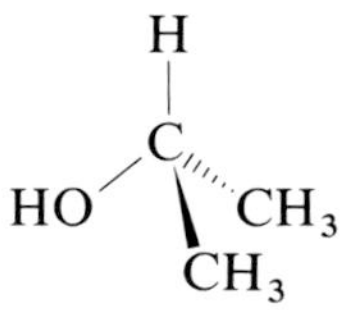 HO—C (H, CH$_3$, CH$_3$)

10.5   Following are several stereorepresentations of lactic acid. Take (a) as a reference structure. Which of the stereorepresentations are identical to (a) and which are mirror images of (a)?

a. 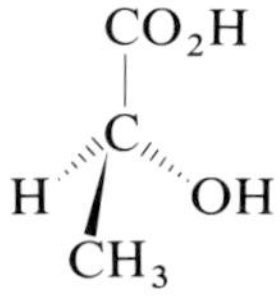 H—C—OH (CO$_2$H, CH$_3$)

b. HO—C—H (CH$_3$, CO$_2$H)

c. HO—C—CH$_3$ (CO$_2$H, H)

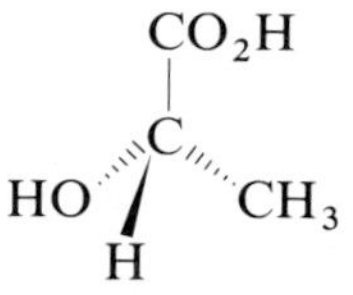

d. 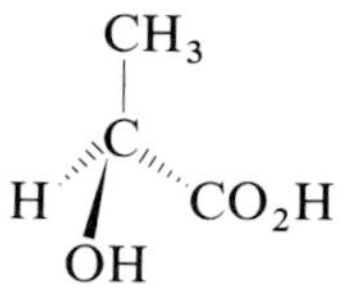 H—C—CO$_2$H (CH$_3$, OH)

e. 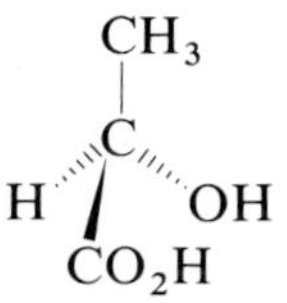 H—C—OH (CH$_3$, CO$_2$H)

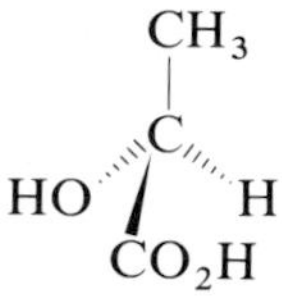

10.6   Which of the following molecules contain a chiral carbon? For each that does, draw stereorepresentations for both enantiomers.

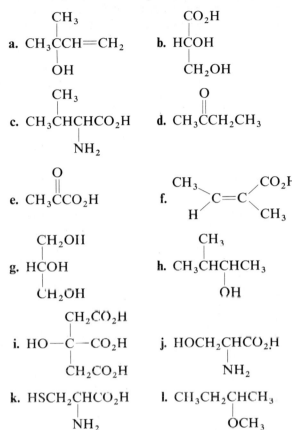

a. $CH_3\overset{\underset{\displaystyle CH_3}{|}}{C}CH=CH_2$
   $\underset{\displaystyle OH}{|}$

b. $HC\overset{\underset{\displaystyle OH}{|}}{\overset{\displaystyle CO_2H}{|}}$
   $\underset{\displaystyle CH_2OH}{|}$

c. $CH_3\overset{\underset{\displaystyle NH_2}{|}}{CH}\overset{\overset{\displaystyle CH_3}{|}}{CH}CO_2H$

d. $CH_3\overset{\displaystyle O}{\overset{||}{C}}CH_2CH_3$

e. $CH_3\overset{\displaystyle O}{\overset{||}{C}}CO_2H$

f. $\underset{\displaystyle H}{\overset{\displaystyle CH_3}{}}C=C\underset{\displaystyle CH_3}{\overset{\displaystyle CO_2H}{}}$

g. $HC\overset{\underset{\displaystyle CH_2OH}{|}}{\overset{\displaystyle CH_2OH}{|}}OH$

h. $CH_3\overset{\overset{\displaystyle CH_3}{|}}{CH}\overset{\underset{\displaystyle OH}{|}}{CH}CH_3$

i. $HO-\overset{\overset{\displaystyle CH_2CO_2H}{|}}{\underset{\displaystyle CH_2CO_2H}{|}}C-CO_2H$

j. $HOCH_2\overset{\underset{\displaystyle NH_2}{|}}{CH}CO_2H$

k. $HSCH_2\overset{\underset{\displaystyle NH_2}{|}}{CH}CO_2H$

l. $CH_3CH_2\overset{\underset{\displaystyle OCH_3}{|}}{CH}CH_3$

10.7   Draw the structural formula of at least one alkene of molecular formula C_5H_9Br that shows:
a. neither cis-trans isomerism nor optical isomerism.
b. cis-trans isomerism but not optical isomerism.
c. optical isomerism but not cis-trans isomerism.
d. both cis-trans isomerism and optical isomerism.

Molecules with two or more chiral carbons (Section 10.6)

10.8   Mark each chiral carbon in the molecules with an asterisk:

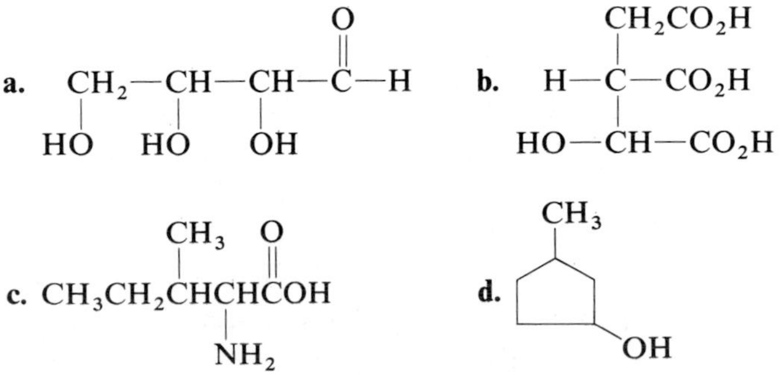

a. $\underset{\displaystyle HO}{\overset{\displaystyle}{CH_2}}-\underset{\displaystyle HO}{\overset{\displaystyle}{CH}}-\underset{\displaystyle OH}{\overset{\displaystyle}{CH}}-\overset{\displaystyle O}{\overset{||}{C}}-H$

b. $H-\overset{\overset{\displaystyle CH_2CO_2H}{|}}{C}-CO_2H$
      $HO-\overset{\displaystyle}{CH}-CO_2H$

c. $CH_3CH_2\overset{\overset{\displaystyle CH_3}{|}}{CH}\overset{\displaystyle O}{\overset{||}{\underset{\underset{\displaystyle NH_2}{|}}{C}}}H$ ... $CH_3CH_2\overset{\overset{\displaystyle CH_3}{|}}{CH}\overset{\underset{\displaystyle NH_2}{|}}{CH}\overset{\displaystyle O}{\overset{||}{C}}OH$

d.   (cyclopentane ring with CH_3 and OH substituents)

e.

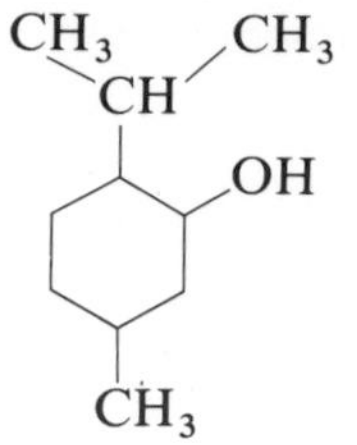

f.

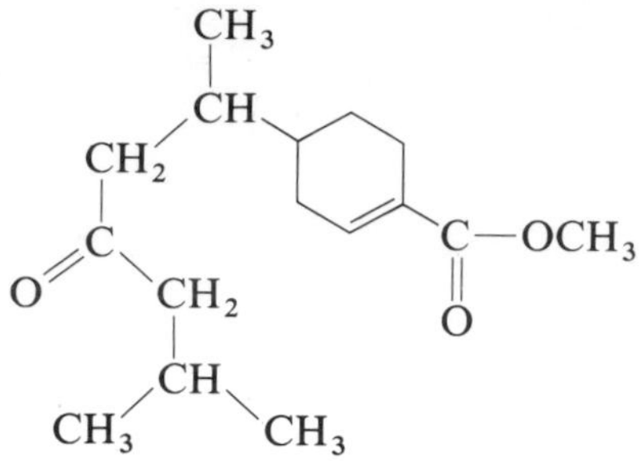

g.

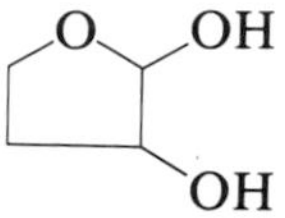

h.

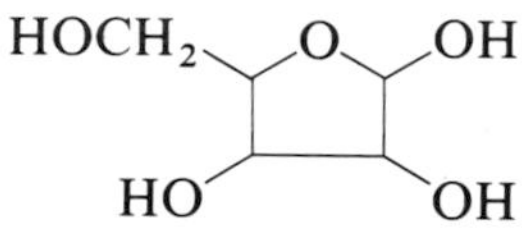

i.

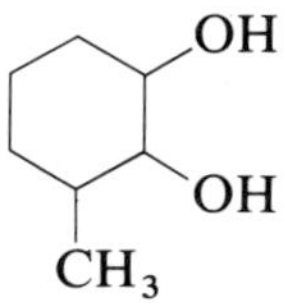

$$\begin{array}{c} O \\ \parallel \\ CH \\ \mid \\ CHOH \\ \mid \\ \text{j. } CHOH \\ \mid \\ CHOH \\ \mid \\ CH_2OH \end{array}$$

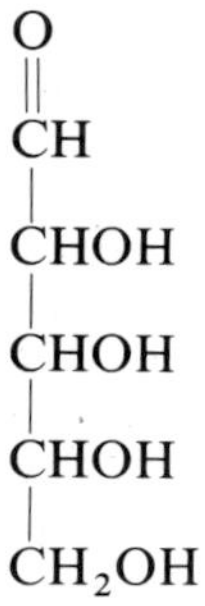

$$\begin{array}{c} CH_2OH \\ \mid \\ C{=}O \\ \mid \\ \text{k. } CHOH \\ \mid \\ CHOH \\ \mid \\ CH_2OH \end{array}$$

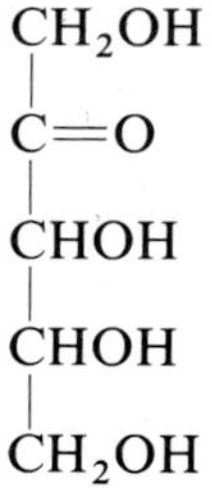

10.9 **a.** How many optical isomers are possible for each molecule in Problem 10.8?

b. How many pairs of enantiomers are possible for each molecule in Problem 10.8?

10.10 4-hydroxypentanal forms a five-member cyclic hemiacetal (Section 6.4B):

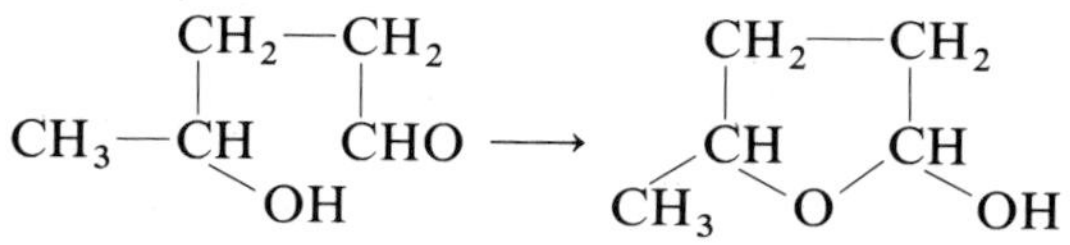

4-hydroxypentanal   a cyclic hemiacetal

How many stereoisomers are possible for this cyclic hemiacetal? Draw stereorepresentations of each.

10.11   5-hydroxyhexanal readily forms a six-member cyclic hemiacetal:

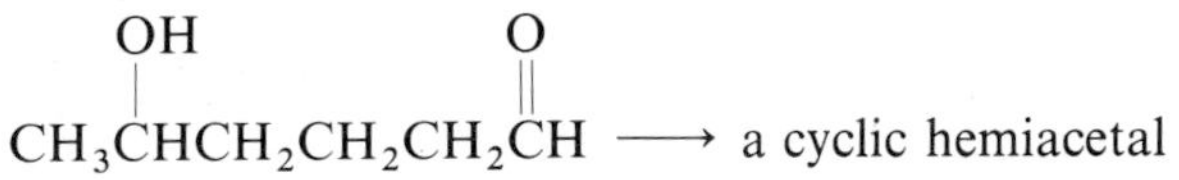

$$\underset{\text{5-hydroxyhexanal}}{CH_3\overset{\displaystyle OH}{\overset{|}{C}}HCH_2CH_2CH_2\overset{\displaystyle O}{\overset{\|}{C}}H} \longrightarrow \text{ a cyclic hemiacetal}$$

    a. Draw a structural formula for the cyclic hemiacetal.
    b. How many stereoisomers are possible for 5-hydroxyhexanal?
    c. How many stereoisomers are possible for the cyclic hemiacetal?
    d. Draw planar hexagon representations for each stereoisomer of the cyclic hemiacetal.

10.12   Glucose, a polyhydroxyaldehyde, forms a six-member cyclic hemiacetal in which the oxygen on carbon 5 of the chain reacts with the aldehyde on carbon 1.

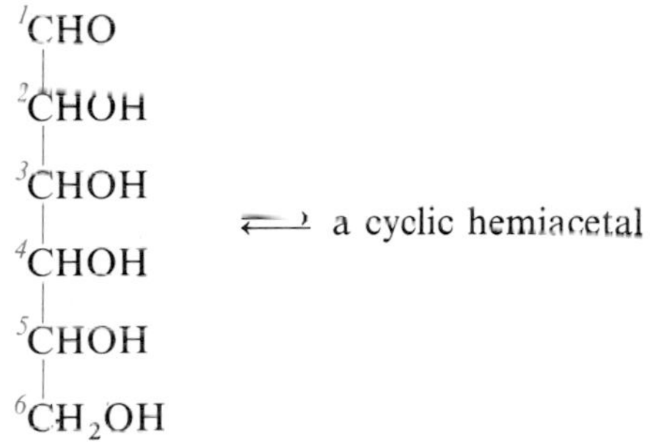

$$\begin{array}{l} ^1CHO \\ ^2CHOH \\ ^3CHOH \\ ^4CHOH \\ ^5CHOH \\ ^6CH_2OH \end{array} \rightleftharpoons \text{ a cyclic hemiacetal}$$

glucose
(open-chain form)

    a. How many chiral carbon atoms are there in the open-chain form of glucose? How many stereoisomers are possible for a molecule of this structure?
    b. Draw a structural formula for the cyclic hemiacetal of glucose (do not worry about showing stereochemistry).
    c. How many chiral carbon atoms are there in the cyclic hemiacetal formed by glucose? How many stereoisomers are possible for the cyclic hemiacetal?

10.13   Explain the difference in molecular structure between *meso*-tartaric acid and racemic tartaric acid.

10.14   Which of the following are meso compounds?

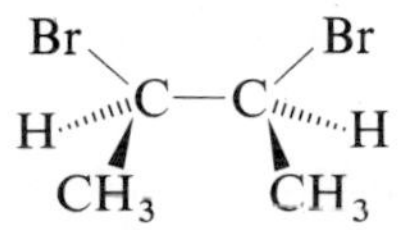

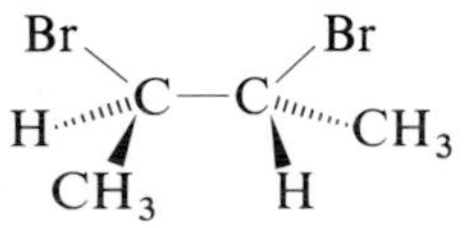

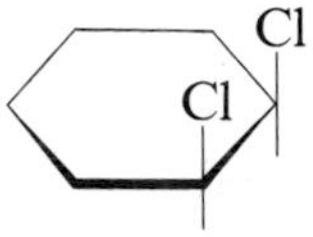

d.

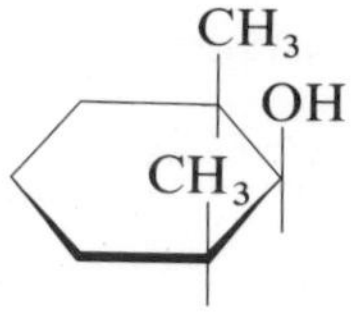

CH_3
OH
CH_3

e.

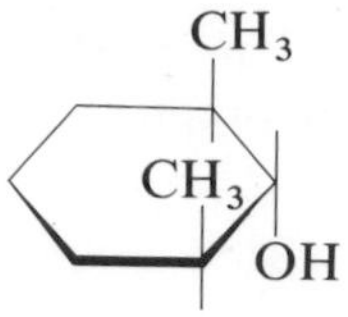

CH_3
CH_3
OH

f.

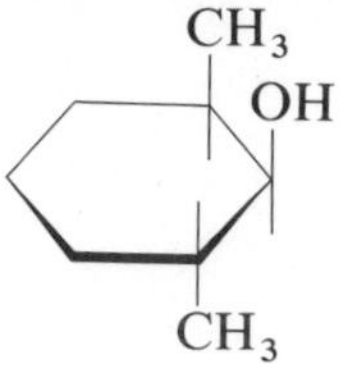

CH_3
OH
CH_3

g.

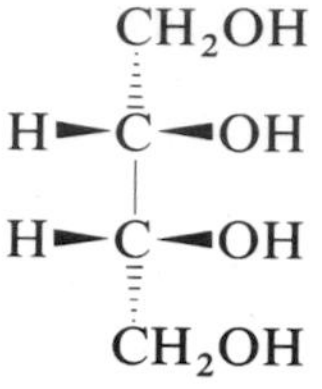

CH_2OH
H—C—OH
H—C—OH
CH_2OH

h.

CH_2OH
HO—C—H
H—C—OH
CH_2OH

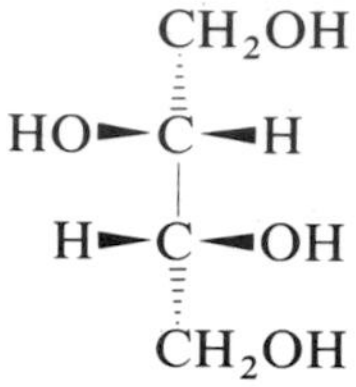

i.

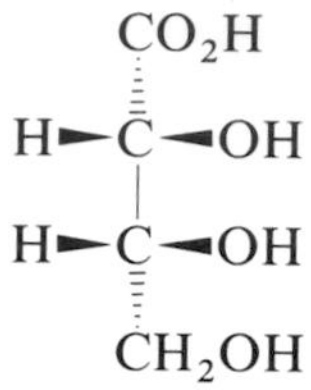

CO_2H
H—C—OH
H—C—OH
CH_2OH

j.

CO_2H
H—C—OH
HO—C—H
CH_2OH

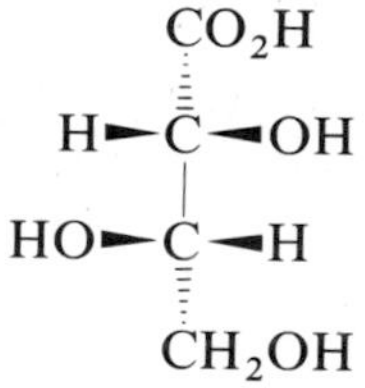

10.15 Inositol is a growth factor for animals and microorganisms. It is used medically for treating cirrhosis of the liver, hepatitis, and fatty infiltration of the liver. The most prevalent natural form is *cis*-1,2,3,5-*trans*-4,6-cyclohexanehexol. In this cis-trans designation, the —OH groups on carbons 2, 3, and 5 are cis to the —OH on carbon 1; the —OH groups on carbons 4 and 6 are trans to the —OH on carbon 1. Draw a stereorepresentation of the natural isomer (show the cyclohexane ring as a planar hexagon) and determine whether it shows enantiomerism or is a meso compound.

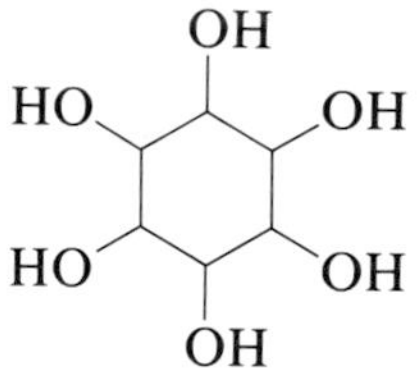

OH
HO OH
HO OH
OH

inositol

10.16 Draw all stereoisomers for the following compounds. Classify them into pairs of enantiomers or meso compounds.

a.  $CH_3CHCHCH_3$
  H_2N OH

b. 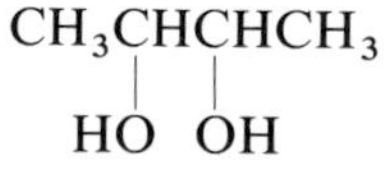 $CH_3CHCHCH_3$
  HO OH

c. 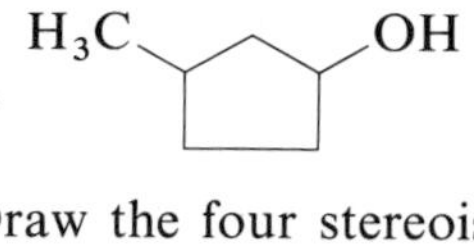 H_3C OH

d. 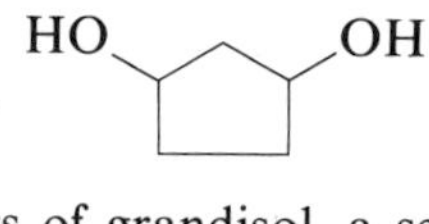 HO OH

10.17 Draw the four stereoisomers of grandisol, a sex hormone secreted by the hind gut of the male boll weevil (*Anthonomus grandis*).

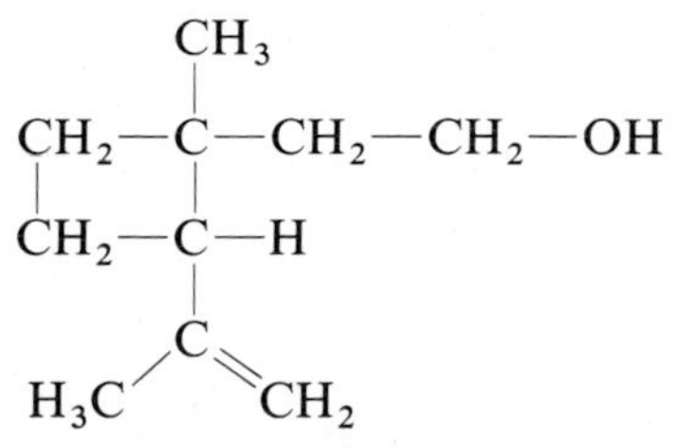

$$CH_3$$
$$CH_2-C-CH_2-CH_2-OH$$
$$CH_2-C-H$$
$$H_3C \diagdown \diagup CH_2$$

grandisol

Significance of chirality in the biological world (Section 10.7)

10.18 How can you explain the following observations:

a. An enzyme is able to distinguish between a pair of enantiomers, and catalyze a biochemical reaction of one enantiomer but not of its mirror image.

b. The microorganism *Penicillium glaucum* metabolizes (+)-tartaric acid preferentially over (−)-tartaric acid.

Carbohydrates

Carbohydrates are the most abundant organic molecules in plants and animals. They perform many vital functions—as storehouses of chemical energy (glucose, starch, glycogen); as components of supportive structures in plants (cellulose) and bacterial cell walls (mucopolysaccharides); and as essential components of nucleic acids (D-ribose and 2-deoxy-D-ribose), thereby affecting mechanisms for the genetically controlled development and growth of living cells. Further, plasma membranes of animal cells have bound to them numerous relatively small carbohydrates that mediate interactions between cells. For example, A, B, and O blood types are determined by specific membrane-bound carbohydrates.

Carbohydrates are often referred to as **saccharides**, because of the sweet taste of the simpler members of the family, the sugars (Latin, *saccharum*, sugar). The name carbohydrate is derived from the fact that many members of this class have the formula $C_n(H_2O)_m$; hence, their name "hydrates of carbon." Two examples of carbohydrates whose molecular formulas can be written alternatively as hydrates of carbon are:

Carbohydrate	Molecular Formula	Molecular Formula as Hydrate of Carbon
glucose (blood sugar)	$C_6H_{12}O_6$	$C_6(H_2O)_6$
sucrose (table sugar)	$C_{12}H_{22}O_{11}$	$C_{12}(H_2O)_{11}$

Not all carbohydrates have this general formula, however. Some also contain nitrogen. But the term carbohydrate has become firmly rooted in chemical nomenclature, and although it is not completely accurate, it persists as the name of this class of compounds.

At the molecular level, carbohydrates are polyhydroxyaldehydes, polyhydroxyketones, or compounds that yield either polyhydroxyaldehydes or polyhydroxyketones after hydrolysis. Complex carbohydrates are polymers of monosaccharides joined by acetal bonds (Section 6.4B). Therefore, the chemistry of carbohydrates is essentially the chemistry of two functional groups—

hydroxyl groups and carbonyl groups—and of acetal bonds formed between these two functional groups.

However, that carbohydrates have only two types of functional groups belies the complexity of their chemistry. All but the most simple carbohydrates contain multiple centers of chirality. For example, glucose, the most common carbohydrate in the biological world, contains one aldehyde, four secondary alcohols, one primary alcohol, and five centers of chirality. Dealing with molecules of this complexity presents enormous challenges to organic chemists and biochemists alike.

11.1 Monosaccharides

A. Structure

Monosaccharides are the monomers from which all more complex carbohydrates are constructed. They have the general formula $C_nH_{2n}O_n$, where n varies from 3 to 8. The suffix *-ose* indicates that a molecule is a carbohydrate, and the prefixes *tri-*, *tetra-*, *penta-* and so on, indicate the number of carbon atoms in the chain.

triose: $C_3H_6O_3$ hexose: $C_6H_{12}O_6$
tetrose: $C_4H_8O_4$ heptose: $C_7H_{14}O_7$
pentose: $C_5H_{10}O_5$ octose: $C_8H_{16}O_8$

There are only two trioses, glyceraldehyde and dihydroxyacetone. Glyceraldehyde contains two —OH groups and an aldehyde; dihydroxyacetone contains two —OH groups and a ketone. An aldehyde is shown to be present in a monosaccharide by the prefix *aldo-*, and a ketone by *keto-*. Thus glyceraldehyde belongs to the class of monosaccharides known as aldotrioses and dihydroxyacetone to the class known as ketotrioses.

$$
\begin{array}{cc}
\text{CHO} & \text{CH}_2\text{OH} \\
| & | \\
\text{CHOH} & \text{C}=\text{O} \\
| & | \\
\text{CH}_2\text{OH} & \text{CH}_2\text{OH} \\
\end{array}
$$

glyceraldehyde   dihydroxyacetone
(an aldotriose)   (a ketotriose)

Often the designations *aldo-* and *keto-* are omitted, and these molecules are referred to simply as trioses. While these designations do not tell the nature of the carbonyl group, at least they indicate that the monosaccharide contains three carbon atoms.

B. Nomenclature

Glyceraldehyde is a common name; the IUPAC name for this monosaccharide is 2,3-dihydroxypropanal. Similarly, dihydroxyacetone is a common name; its IUPAC name is 1,3-dihydroxypropanone. However, the common names for these and other monosaccharides are so firmly rooted in the literature of organic chemistry and biochemistry, that they are used almost exclusively whenever these compounds are referred to. Therefore, throughout our discussions of the chemistry of carbohydrates, we will use the names most common to chemistry and biochemistry.

C. Chirality of Monosaccharides

Dihydroxyacetone has no chiral carbon and does not show optical isomerism. Glyceraldehyde, however, contains one chiral carbon and can exist as a pair of enantiomers (Figure 11.1). The configuration IA is named D-glyceraldehyde and its enantiomer, IB, is named L-glyceraldehyde.

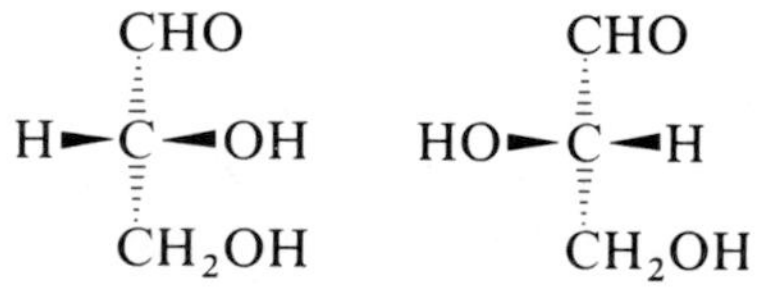

$$
\begin{array}{cc}
\text{CHO} & \text{CHO} \\
\text{H} \blacktriangleright \text{C} \blacktriangleleft \text{OH} & \text{HO} \blacktriangleright \text{C} \blacktriangleleft \text{H} \\
\text{CH}_2\text{OH} & \text{CH}_2\text{OH}
\end{array}
$$

D-glyceraldehyde   L-glyceraldehyde
$[\alpha]_D^{25°C} = +13.5°$ $[\alpha]_D^{25°C} = -13.5°$
IA   IB

Figure 11.1 The enantiomers of glyceraldehyde.

In the three-dimensional formulas in Figure 11.1, the configuration of each atom or group of atoms bonded to the chiral carbon is indicated by a combination of dashed and solid wedges. Structural formulas for monosaccharides can also be drawn as **Fischer projections**. According to this convention, the carbon chain is written vertically with the most highly oxidized carbon atom at the "top." Horizontal lines show groups projecting above the plane of the page; vertical lines show groups projecting behind the plane of the page. Applying these rules gives the following Fischer projections for the enantiomers of glyceraldehyde:

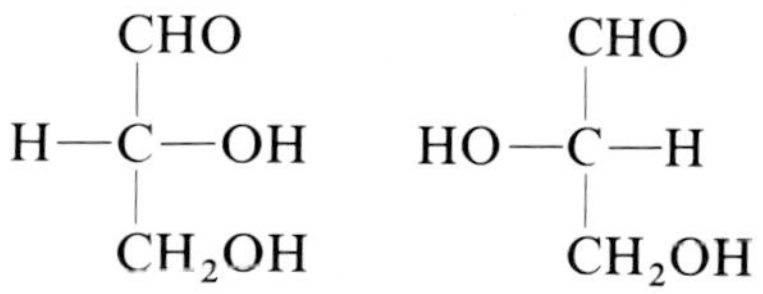

$$
\begin{array}{cc}
\text{CHO} & \text{CHO} \\
\text{H}-\text{C}-\text{OH} & \text{HO}-\text{C}-\text{H} \\
\text{CH}_2\text{OH} & \text{CH}_2\text{OH}
\end{array}
$$

D-glyceraldehyde   L-glyceraldehyde

At times there is confusion about whether a structural formula is or is not a Fischer projection. The problem arises because Fischer projections can be mistaken for Lewis structures, and whereas Fischer projection formulas do show stereochemistry about chiral carbons, Lewis structures do not. Therefore, the original Fischer convention has been modified to avoid this potential problem. In this modification, all chiral carbon atoms are represented by crossing points of bonds; chiral carbons are not shown. Following are representations of D- and

L-glyceraldehyde according to this convention. We shall use this convention throughout.

$$
\begin{array}{cc}
\text{CHO} & \text{CHO} \\
\text{H}-\!\!-\text{OH} & \text{HO}-\!\!-\text{H} \\
\text{CH}_2\text{OH} & \text{CH}_2\text{OH} \\
\text{D-glyceraldehyde} & \text{L-glyceraldehyde}
\end{array}
$$

(drawn according to the modified Fischer convention)

D-glyceraldehyde and L-glyceraldehyde serve as reference points for the assignment of configuration to all other aldoses and ketoses. Those that have the same configuration as D-glyceraldehyde about the chiral carbon farthest from the aldehyde or ketone are called **D-monosaccharides**; those that have the same configuration as L-glyceraldehyde about the chiral carbon farthest from the aldehyde or ketone are called **L-monosaccharides**.

In 1891, when Emil Fischer proposed the use of D and L to specify absolute configurations in monosaccharides, it was known that glyceraldehyde exists as a pair of enantiomers and that one of them has a specific rotation of $+13.5°$; the other, a specific rotation of $-13.5°$. The question Fischer and others faced was, Which enantiomer has which specific rotation? Is the specific rotation of D-glyceraldehyde $+13.5°$ or $-13.5°$? Because there was no experimental way to answer the question at that time, Fischer did the only thing he could—he guessed. He guessed that IA is the absolute configuration of the dextrorotatory enantiomer and named it D-glyceraldehyde (D for dextrorotatory). He named the other enantiomer L-glyceraldehyde (L for levorotatory). Fischer could have been wrong, but by a stroke of good fortune, he wasn't. In 1952, his assignment of absolute configuration was proved correct by a special application of X-ray crystallography. Thus, Fischer projections of monosaccharides give absolute configurations of all chiral centers.

The following aldohexose contains 4 chiral carbon atoms, marked by asterisks, and according to the 2^n rule, it can exist as 16 optical isomers (8 pairs of enantiomers):

$$
\begin{array}{c}
\text{CHO} \\
\text{*CHOH} \\
\text{*CHOH} \\
\text{*CHOH} \\
\text{*CHOH} \\
\text{CH}_2\text{OH}
\end{array}
$$

an aldohexose
[16 optical isomers (8 pairs of enantiomers) are possible]

Table 11.1
Configurational relationships between the isomeric D-aldotetroses, D-aldopentoses, and D-aldohexoses derived from D-glyceraldehyde.

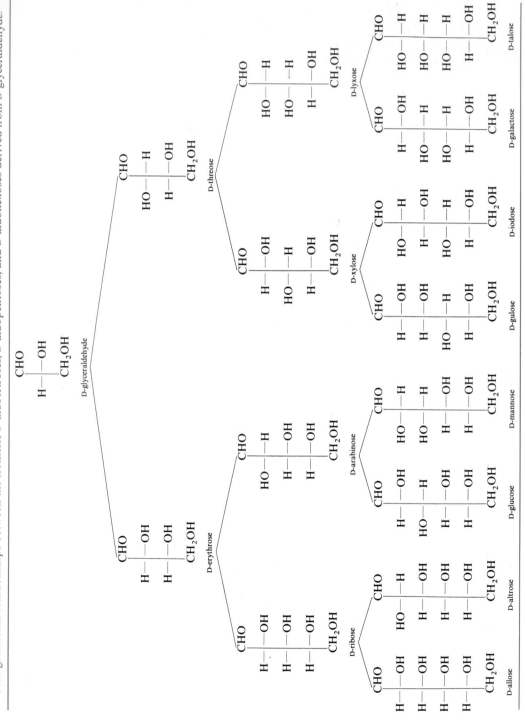

In Table 11.1 you will find Fischer projection formulas for eight of the optical isomers of this aldohexose. The other eight optical isomers are the enantiomers of those shown, that is, they are L-allose, L-altrose, L-glucose, and so on.

Shown in Tables 11.1 and 11.2 are names and Fischer projection formulas for all D-tetroses, D-pentoses and D-hexoses derived from D-glyceraldehyde. Each name consists of three parts. The letter D specifies the absolute configuration at the penultimate carbon; prefixes such as *rib-*, *arabin-*, and *gluc-* specify the configuration of all other chiral centers relative to that of the penultimate carbon; and *-ose* shows that the molecule belongs to the class called carbohydrates.

Table 11.2
Configurational relationships between the D-ketopentoses and D-ketohexoses derived from dihydroxyacetone and D-erythrulose.

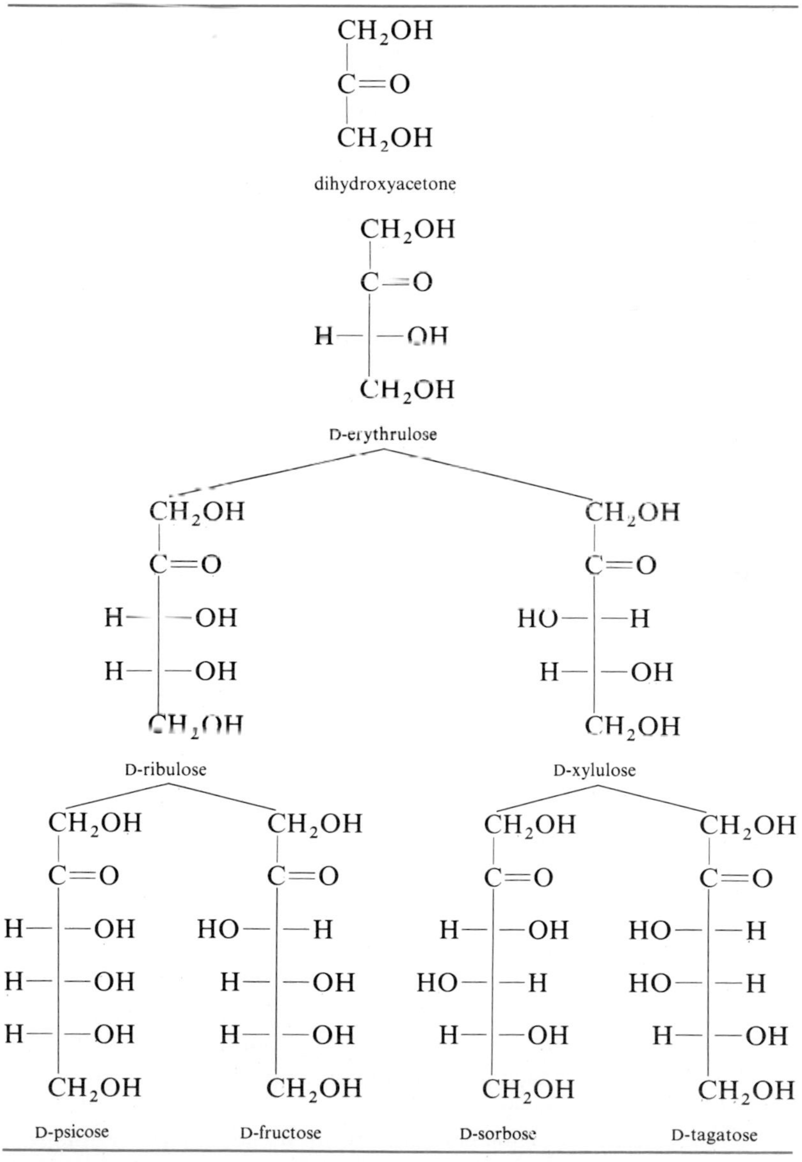

D-tetroses, D-pentoses and D-hexoses tree:

dihydroxyacetone

D-erythrulose

D-ribulose — D-xylulose

D-psicose — D-fructose — D-sorbose — D-tagatose

D-ribose and 2-deoxy-D-ribose (below), the most abundant pentoses in the biological world, are essential building blocks of nucleic acids: D-ribose in ribonucleic acids (RNAs) and 2-deoxy-D-ribose in deoxyribonucleic acids (DNAs).

$$
\begin{array}{cc}
\text{CHO} & \text{CHO} \\
\text{H}-\!\!-\text{OH} & \text{H}-\!\!-\text{H} \\
\text{H}-\!\!-\text{OH} & \text{H}-\!\!-\text{OH} \\
\text{H}-\!\!-\text{OH} & \text{H}-\!\!-\text{OH} \\
\text{CH}_2\text{OH} & \text{CH}_2\text{OH} \\
\text{D-ribose} & \text{2-deoxy-D-ribose}
\end{array}
$$

D-ribose but without oxygen at carbon 2

The three most abundant hexoses in the biological world are D-glucose, D-galactose, and D-fructose. The first two are D-aldohexoses; the third is a D-ketohexose. Note that D-galactose differs from D-glucose only in the configuration at carbon 4. Compounds that have multiple chiral centers but differ from each other in the configuration at only one of these chiral centers are called epimers. Thus, D-glucose and D-galactose are epimers.

$$
\begin{array}{ccc}
\overset{1}{\text{CHO}} & \text{CHO} & \text{CH}_2\text{OH} \\
\text{H}\overset{2}{-}\text{OH} & \text{H}-\text{OH} & \text{C}=\text{O} \\
\text{HO}\overset{3}{-}\text{H} & \text{HO}-\text{H} & \text{HO}-\text{H} \\
\text{H}\overset{4}{-}\text{OH} & \text{HO}-\text{H} & \text{H}-\text{OH} \\
\text{H}\overset{5}{-}\text{OH} & \text{H}-\text{OH} & \text{H}-\text{OH} \\
\overset{6}{\text{CH}_2\text{OH}} & \text{CH}_2\text{OH} & \text{CH}_2\text{OH} \\
\text{D-glucose} & \text{D-galactose} & \text{D-fructose} \\
 & \text{(C-4 isomer of} & \\
 & \text{of D-glucose)} &
\end{array}
$$

Glucose, by far the most common hexose, is also known as dextrose because it is dextrorotatory. Other names for this monosaccharide are grape sugar, blood sugar, and corn sugar, names that indicate its sources in an uncombined state. Human blood normally contains 65–110 mg of glucose per 100 mL.

Fructose is found combined with glucose in the disaccharide sucrose (table sugar, Section 11.4C). D-galactose is found combined with glucose in the disaccharide lactose (milk sugar). Lactose is found only in milk.

Example 11.1

a. Draw Fischer projections for all aldoses of four carbon atoms.
b. Show which are D-monosaccharides, which are L-monosaccharides, and which are enantiomers.
c. Refer to Table 11.1 and write names for the aldotetroses you have drawn.

Solution
Following are Fischer projections for the four possible aldotetroses. The designations D and L refer to the arrangement of groups attached to the penultimate carbon, which in aldotetroses is carbon 3. In the Fischer projection of a D-aldotetrose, the —OH on carbon 3 is on the right; for an L-aldotetrose, it is on the left.

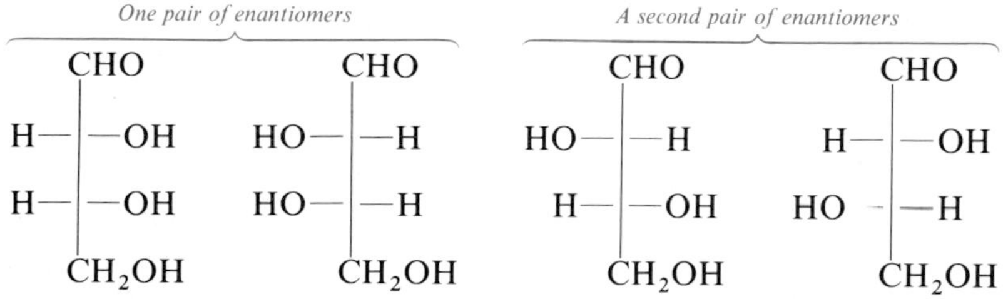

One pair of enantiomers   A second pair of enantiomers

```
   CHO              CHO              CHO              CHO
H──┼──OH      HO──┼──H        HO──┼──H         H──┼──OH
H──┼──OH      HO──┼──H         H──┼──OH       HO ──┼──H
   CH₂OH           CH₂OH            CH₂OH            CH₂OH
```

a D-aldose   an L-aldose   a D-aldose   an L-aldose
(D-erythose)   (L-erythrose)   (D-threose)   (L-threose)

Problem 11.1

a. Draw Fischer projections for all 2-ketoses with five carbon atoms.
b. Show which are D-ketopentoses, which are L-ketopentoses, and which are enantiomers.
c. Refer to Table 11.2 and write names for the ketopentoses you have drawn.

D. Amino Sugars

Amino sugars contain an —NH₂ group in place of an —OH group. Only three amino sugars are common in nature: D-glucosamine, D-mannosamine, and D-galactosamine.

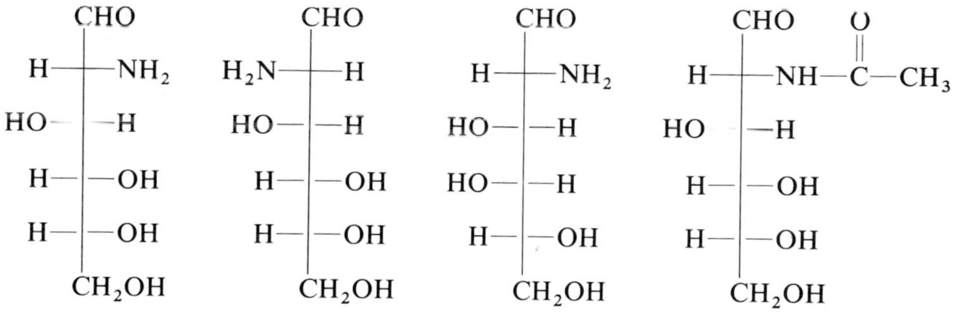

```
    CHO              CHO              CHO              CHO    O
                                                             ‖
 H──┼──NH₂      H₂N──┼──H        H──┼──NH₂       H──┼──NH──C──CH₃
HO──┼──H        HO──┼──H        HO──┼──H        HO──┼──H
 H──┼──OH        H──┼──OH       HO──┼──H         H──┼──OH
 H──┼──OH        H──┼──OH        H──┼──OH        H──┼──OH
    CH₂OH           CH₂OH            CH₂OH            CH₂OH
```

D-glucosamine   D-mannosamine   D-galactosamine   N-acetyl-D-
                   (C-2 isomer of   (C-4 isomer of   glucosamine
                   D-glucosamine)   D-glucosamine)

N-Acetyl-D-glucosamine, a derivative of D-glucosamine, is a component of many polysaccharides, including chitin, the hard shell-like exoskeleton of lobsters, crabs, shrimp, and other crustaceans.

E. Cyclic Structure of Monosaccharides

Aldehydes and ketones react with alcohols to form hemiacetals and hemiketals. Cyclic hemiacetals and hemiketals form very readily when hydroxyl and carbonyl groups are part of the same molecule. For example, 4-hydroxypentanal forms a five-membered cyclic hemiacetal. Note that 4-hydroxypentanal contains one chiral center and that in hemiacetal formation, a second chiral center is generated.

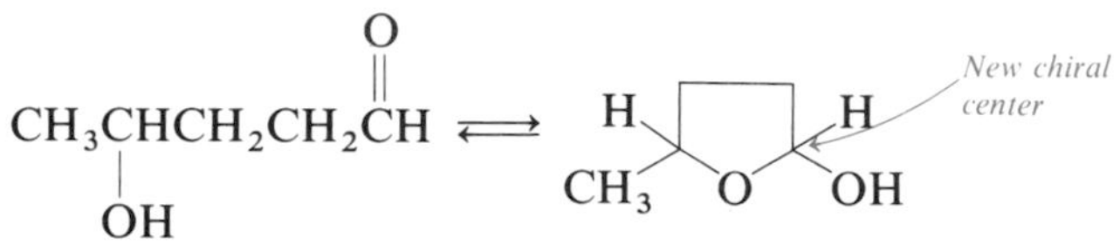

4-hydroxypentanal (minor form present at equilibrium)	a cyclic hemiacetal (major form present at equilibrium)

Monosaccharides have hydroxyl and carbonyl groups in the same molecule and they too form cyclic hemiacetals and hemiketals.

D-glucose can be isolated in two crystalline forms called α and β, which have different chemical and physical properties. One of the most easily measured physical properties of these isomers is their optical rotation. One form has a specific rotation of $+112°$, the other a specific rotation of $+19°$. These α and β isomers are formed by reaction between the —OH on carbon 5 and the carbonyl group on carbon 1, to give a pair of six-member cyclic hemiacetals. The carbonyl carbon at which cyclic hemiacetal formation takes place is given the special name **anomeric carbon**. The diastereomers thus formed are given the special name anomers.

In the terminology of carbohydrate chemistry, α indicates that the —OH on the newly created anomeric carbon is trans to the terminal —CH_2OH (carbon 6 in glucose). The term beta (β) indicates that the —OH on the anomeric carbon is cis to the terminal —CH_2OH. Structural formulas for α-D-glucose and β-D-glucose are drawn in Figure 11.2 as strain-free chair conformations. Also drawn is the open-chain, or free aldehyde, form with which the cyclic hemiacetals are

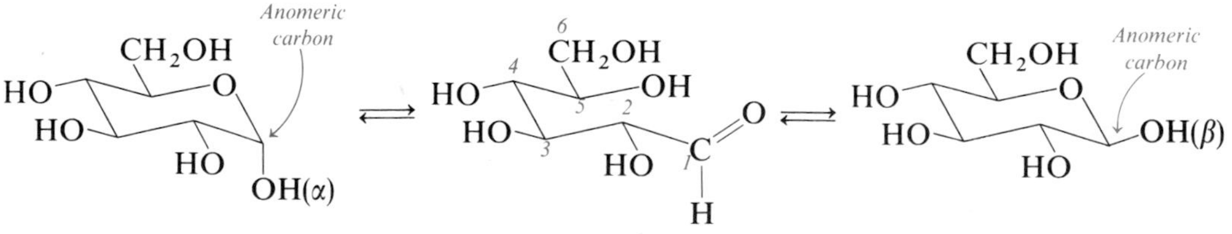

α-D-glucopyranose (α-D-glucose: mp 146°C $[\alpha] = +112°$)	open-chain, or free aldehyde, form of D-glucose	β-D-glucopyranose (β-D-glucose; mp 190°C $[\alpha] = +19°$)

Figure 11.2 The cyclic hemiacetal forms of D-glucose.

in equilibrium in aqueous solution. The equilibrium constant for hemiacetal formation is greater than 190, which means that at equilibrium, little free aldehyde is present. Note that in the chair conformations of α- and β-D-glucose, substituents on carbons 2, 3, and 4 of each ring are equatorial (Section 2.6B). The —OH on the anomeric carbon of α-D-glucose is axial; the —OH on the anomeric carbon of β-D-glucose is equatorial.

The size of a monosaccharide hemiacetal or hemiketal ring is shown by reference to the molecules pyran and furan.

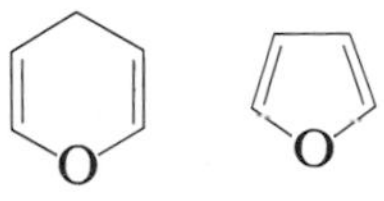

pyran   furan

Six-member hemiacetal or hemiketal rings are shown by the infix -*pyran*-, and five-member hemiacetal or hemiketal rings are shown by the infix -*furan*-. Thus, the alpha and beta anomers of D-glucose are properly named α-D-glucopyranose and β-D-glucopyranose. However, for convenience they are often named simply α-D-glucose and β-D-glucose.

Structural formulas showing these cyclic hemiacetals can also be drawn as planar hexagons (Figure 11.3). Such representations are called **Haworth structures**, after English chemist Walter N. Haworth (Nobel laureate, 1937).

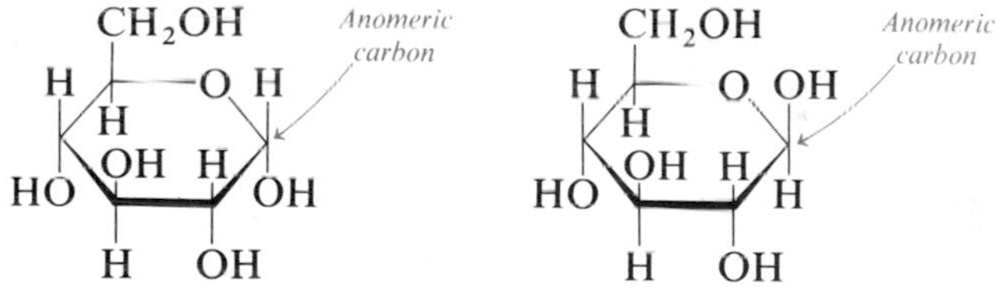

α-D-glucopyranose
(α-D-glucose)

β-D-glucopyranose
(β-D-glucose)

Figure 11.3 Haworth structures for the cyclic hemiacetal forms of D-glucose.

In Haworth structures of aldohexopyranoses, the —OH on the anomeric carbon is below the plane of the ring (trans to the terminal —CH$_2$OH) in the α-anomer and above the plane of the ring (cis to the terminal —CH$_2$OH) in the β-anomer.

Other monosaccharides also form cyclic hemiacetals and hemiketals. Following are structural formulas for those formed by D-fructose. Cyclization (in this case hemiketal formation) between the carbonyl group on carbon 2 and the hydroxyl on carbon 5 gives a pair of anomers called α-D-fructofuranose and β-D-fructofuranose. Fructose also forms a pair of **pyranoses** by cyclization between the carbonyl group and the hydroxyl on carbon 6. Structural formulas of these compounds along with approximate percentages of each at equilibrium in aqueous solution are shown in Figure 11.4. In the **furanose** forms, the most common in the biological world, the —OH on the anomeric carbon is below the plane of the ring in the α-anomer and above the plane of the ring in the β-anomer.

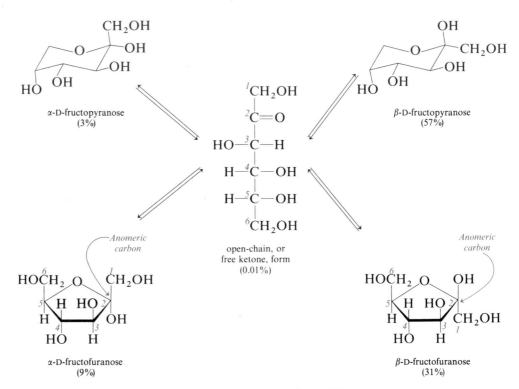

Figure 11.4 The principal forms of D-fructose in equilibrium in aqueous solution.

Example 11.2

D-galactose forms a cyclic hemiacetal containing a six-member ring. Draw chair and Haworth structures for α-D-galactopyranose and β-D-galactopyranose. Label the anomeric carbon in each cyclic hemiacetal.

☐ *Solution*

One way of drawing structures for the six-member cyclic hemiacetals of D-galactose is to use the α and β forms of D-glucopyranose as reference, and to remember, or discover by looking at Table 11.1, that D-galactose differs from D-glucose only in the configuration at carbon 4; that is, it is the epimer of glucose at carbon 4. Another way of arriving at the correct pyranose form is to build a molecular model of D-galactose, twist it to create the required hemiacetals, and examine the models to discover how each ring substituent is oriented in space. Chair structures for the alpha and beta anomers of D-galactose are drawn below, along with the open-chain form with which they are in equilibrium in aqueous solution.

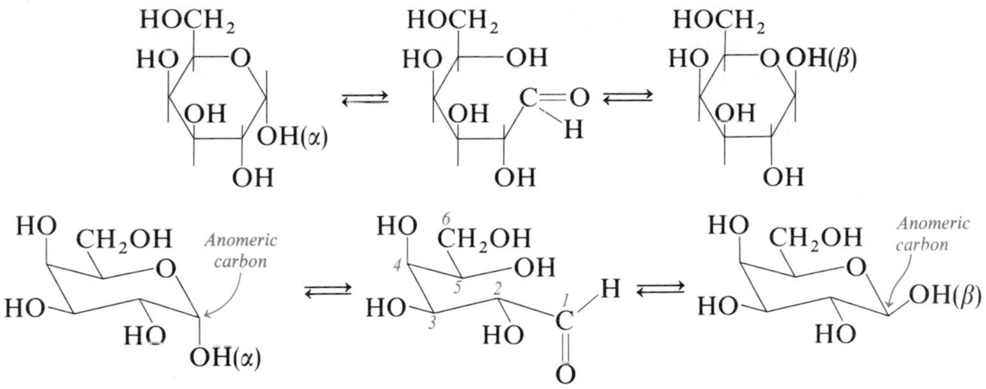

α-D-galactopyranose
([α] = +191°)

open-chain form

β-D-galactopyranose
([α] = −53°)

Problem 11.2

D-mannose forms a cyclic hemiacetal containing a six-member ring. Draw chair and Haworth structures for α-D-mannopyranose and β-D-mannopyranose. Label the anomeric carbon atom in each.

The most prevalent forms of D-ribose and other pentoses in the biological world are furanoses. Shown in Figure 11.5 are structural formulas for α-D-ribofuranose (α-D-ribose) and β-2-deoxy-D-ribofuranose (β-2-deoxy-D-ribose). Units of D-ribose and 2-deoxy-D-ribose in nucleic acids and most other biological molecules are found almost exclusively in the β configuration.

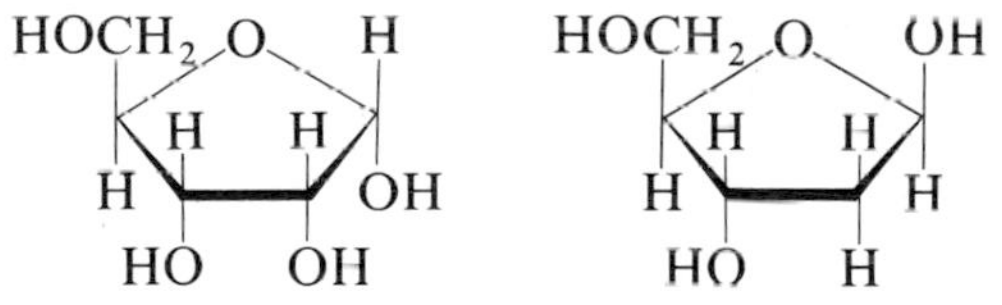

α-D-ribofuranose

β-D-2-deoxy-D-ribofuranose

Figure 11.5 Furanose forms of D-ribose and 2-deoxy-D-ribose.

F. Mutarotation

The α- and β-anomers of monosaccharides are interconvertible in aqueous solution and the change in specific rotation that accompanies this interconversion is known as **mutarotation**. As an example, a freshly prepared solution of α-D-glucose shows an initial rotation of +112°, which gradually decreases to +52° as α-D-glucose reaches an equilibrium with β-D-glucose. The equilibrium mixture consists of 64% β-D-glucose and 36% α-D-glucose. A solution of β-D-glucose also undergoes mutarotation, during which the specific rotation changes from an initial value of +19° to the same equilibrium value of +52°.

Mutarotation is common to all carbohydrates that exist in α and β forms. Shown in Table 11.3 are specific rotations for the α and β forms of D-galactose and D-mannose, along with equilibrium values for the specific rotation of each after mutarotation.

Table 11.3
Specific rotations of the α and β forms of D-galactose and D-mannose before and after mutarotation.

Monosaccharide	Specific Rotation		Specific Rotation after Mutarotation
α-D-galactose	+190.7°	⟶	+80.2°
β-D-galactose	+52.8°	⟶	+80.2°
α-D-mannose	+29.3°	⟶	+14.5°
β-D-mannose	−16.3°	⟶	+14.5°

G. Physical Properties

Monosaccharides are colorless, crystalline solids, and because hydrogen bonding is possible between polar —OH groups and water, all monosaccharides are very soluble in water. They are only slightly soluble in alcohol and insoluble in nonhydroxylic solvents, such as ether, chloroform, and benzene.

Although all monosaccharides are sweet to the taste, some are sweeter than others (Table 11.4). Of the monosaccharides, D-fructose tastes the sweetest, even more than sucrose (table sugar). The sweet taste of honey is due largely to D-fructose, and of corn syrup, to D-glucose. Molasses is a byproduct of table-sugar manufacture. In the production of table sugar, sugar cane or sugar beet is boiled with water and then cooled. As the mixture cools, sucrose crystals separate and are collected. Subsequent boilings and coolings yield a dark, thick syrup known as molasses.

Table 11.4
Relative sweetness. Sucrose is taken as a standard and assigned the value 100.

Monosaccharides		Disaccharides		Other Carbohydrate Sweetening Agents	
D-fructose	174	sucrose (table sugar)	100	honey	97
D-glucose	74	lactose (milk sugar)	0.16	molasses	74
D-xylose	0.40			corn syrup	74
D-galactose	0.22				

11.2 Reactions of Monosaccharides

A. Formation of Glycosides (Acetals)

Reaction of a monosaccharide hemiacetal or hemiketal (Section 11.1E) with an alcohol forms an acetal or a ketal, as illustrated by the reaction of β-D-gluco-pyranose (β-D-glucose) with methanol:

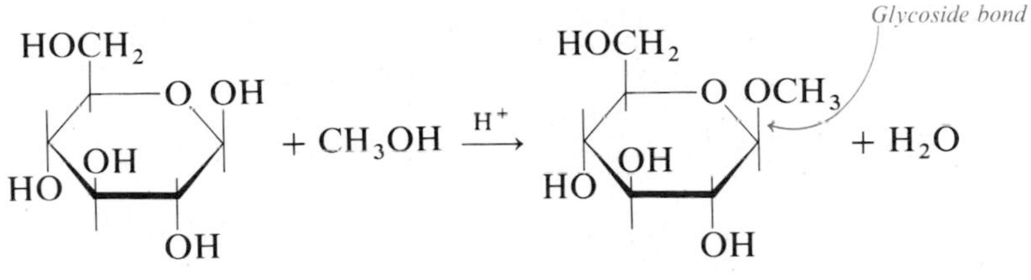

Glycoside bond

β-D-glucopyranose
(β-D-glucose)

methyl β-D-glucopyranoside
(methyl β-D-glycoside)

A cyclic acetal or ketal derived from a monosaccharide is called a **glycoside**, and the bond from the anomeric carbon to the —OR group is called a **glycoside bond**.

Glycosides are named by listing the alkyl or aryl group attached to oxygen, and following this by the name of the carbohydrate involved. The name of a particular glycoside is derived by dropping the terminal -*e* from the name of the monosaccharide and adding -*ide*. For example, glycosides derived from D-glucose are named D-glucosides; those derived from D-ribose are named D-ribosides. In Haworth structures, the —OR on an anomeric carbon of a glycoside is below the plane of the ring in an α-anomer and above the plane of the ring in a β-anomer. In chair conformations, the —OR on an anomeric carbon is axial in an α-anomer and equatorial in a β-anomer.

Example 11.3

Draw structural formulas for these glycosides. In each, label the anomeric carbon and the glycoside bond.

a. methyl β-D-ribofuranoside (methyl β-D-riboside)
b. methyl α-D-galactopyranoside (methyl α-D-galactoside)

□ *Solution*

a. D-ribose forms a five member cyclic hemiacetal that reacts with methanol to form a cyclic acetal. The —OCH₃ group on the anomeric carbon is above the plane of the ring (on the same side as the terminal —CH₂OH group) in a β-ribofuranoside.

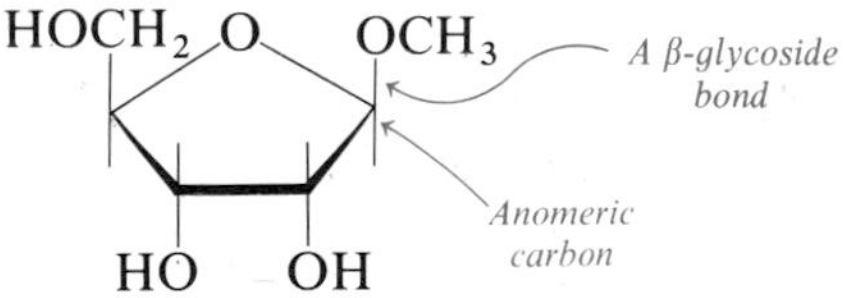

A β-glycoside bond

Anomeric carbon

methyl β-D-ribofuranoside
(methyl β-D-riboside)

b. D-galactose forms a six-member cyclic hemiacetal that reacts with methanol to form a cyclic acetal. The —OCH₃ group is below the plane of the ring (trans to the terminal —CH₂OH group) in an α-D-galactopyranoside. Following are Haworth and chair structures for methyl α-D-galactopyranoside:

methy α-D-galactopyranoside
(methyl α-D-galactoside)

**Problem
11.3**

Draw structural formulas for these glycosides. In each, label the anomeric carbon and the glycoside bond.

a. methyl β-D-fructofuranoside (methyl β-D-fructoside)
b. methyl α-D-mannopyranoside (methyl α-D-mannoside)

Glycosides are stable in water and in aqueous base, but like other acetals and ketals (Section 6.4B), they are hydrolyzed in aqueous acid to an alcohol and a monosaccharide.

Just as the anomeric carbon of a cyclic hemiacetal or hemiketal reacts with R—OH to form a glycoside, it also reacts with an N—H group to form an N-glycoside. Especially important in the biological world are the **N-glycosides** formed between D-ribose and 2-deoxy-D-ribose, each as furanoses, and the heterocyclic aromatic amines uracil, cytosine, thymine, adenine, and guanine (Figure 11.6). N-glycosides of these purine and pyrimidine bases are structural units of nucleic acids (Chapter 15).

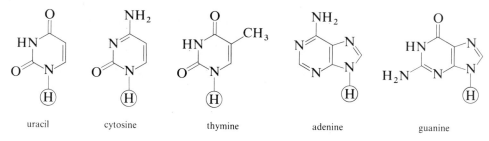

| uracil | cytosine | thymine | adenine | guanine |

Figure 11.6 Structural formulas of the most important purine and pyrimidine bases found in DNA and RNA. The circled hydrogen atom is lost in forming an N-glycoside.

**Example
11.4**

Draw structural formulas for the N-glycosides formed between the compounds in each of the sets. In each structure, label the anomeric carbon and the N-glycoside bond.

a. β-D-ribofuranoside (β-D-ribose) and cytosine
b. β-2-deoxy-D-ribofuranoside (β-2-deoxy-D-ribose) and adenine

Solution

a.

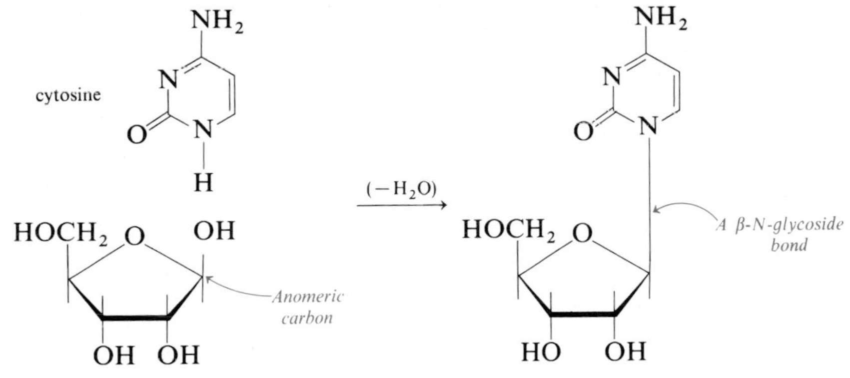

cytosine

$(-H_2O)$

A β-N-glycoside bond

HOCH₂

OH

Anomeric carbon

β-D-ribose

HOCH₂

HO OH

b.

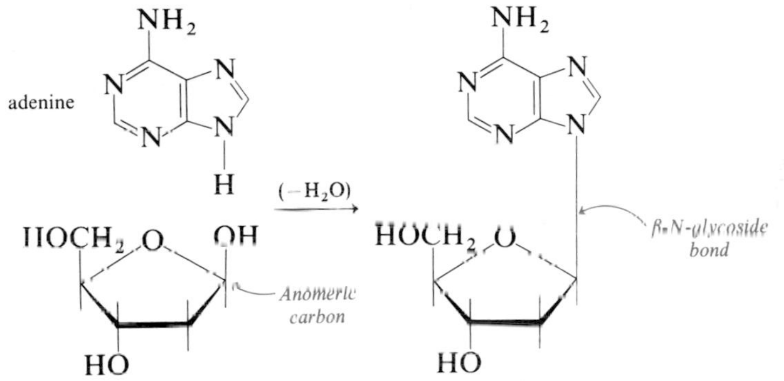

adenine

$(-H_2O)$

β-N-glycoside bond

IIOCH₂ OH

Anomeric carbon

HO

β-2-deoxy-D-ribose

HOCH₂

HO

Problem 11.4

Draw structural formulas for the *N*-glycosides formed between the following compounds. In each, label the anomeric carbon and the *N*-glycoside bond.

a. β-2-deoxy-D-ribofuranoside (β-2-deoxy-D-ribose) and uracil
b. β-D-ribofuranoside (β-D-ribose) and guanine

B. Formation of Phosphate Esters

Phosphate esters of monosaccharides are especially important in the biological world, particularly in the metabolism of monosaccharides. Following are structural formulas for D-glyceraldehyde 3-phosphate and dihydroxyacetone phosphate, two triose phosphate esters involved in glycolysis, the metabolic pathway that converts glucose to pyruvate (Section 17.3). Each is shown as it would be ionized at pH 7.0

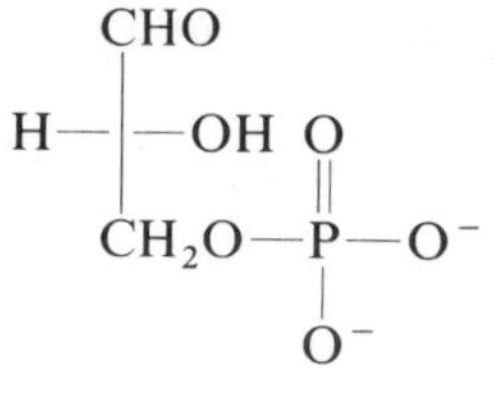

D-glyceraldehyde
3-phosphate
(net charge −2)

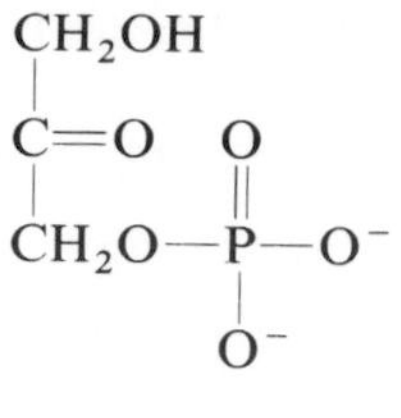

dihydroxyacetone
phosphate
(net charge −2)

Phosphate esters are also formed by monosaccharide cyclic hemiacetals and hemiketals. Following is a structural formula for β-D-glucose 1-phosphate. Glucose is in the form of a six-membered cyclic hemiacetal with the —OH on the anomeric carbon β (above the plane of the ring). The phosphate ester is formed by the —OH on carbon 1. Also shown is a structural formula for β-D-ribose 5-phosphate. Ribose is in the form of a five-membered cyclic hemiacetal with the —OH on the anomeric carbon above the plane of the ring. The phosphate ester is formed by the —OH on carbon 5 of ribose.

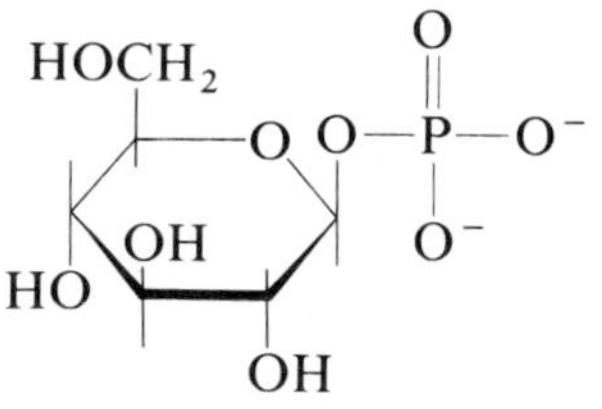

β-D-glucose 1-phosphate
(net charge −2)

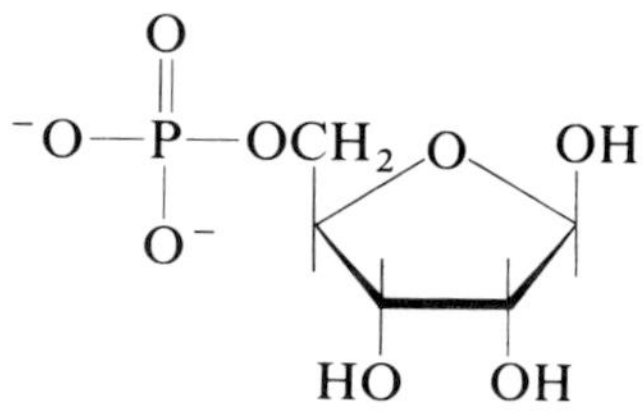

β-D-ribose 5-phosphate
(net charge −2)

Example 11.5

Draw structural formulas for the following monosaccharide phosphate esters and state the net charge on each at pH 7.

a. α-D-glucose 1-phosphate
b. β-D-fructose 1,6-diphosphate. Show fructose first as a five-membered cyclic hemiketal, then in a second structural formula as an open-chain, or free ketone, form (Figure 11.4)

Solution

a. Draw glucose as a six-membered cyclic hemiacetal with the —OH on the anomeric carbon below the plane of the ring. Show the phosphate ester bond between the oxygen atom of carbon 1 and phosphoric acid. At pH 7, the net charge on this ester is −2.

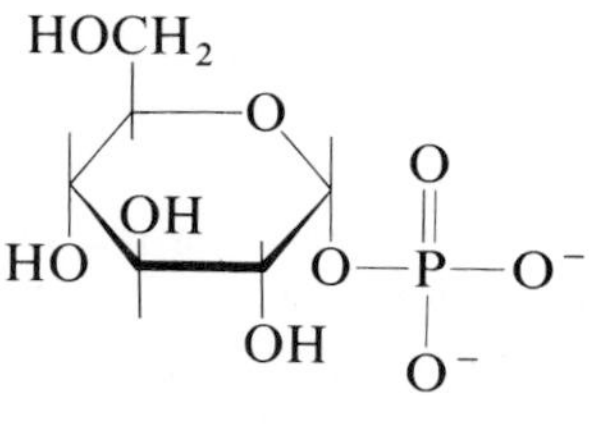

b. At pH 7, the net charge on this diester is −4.

Problem 11.5

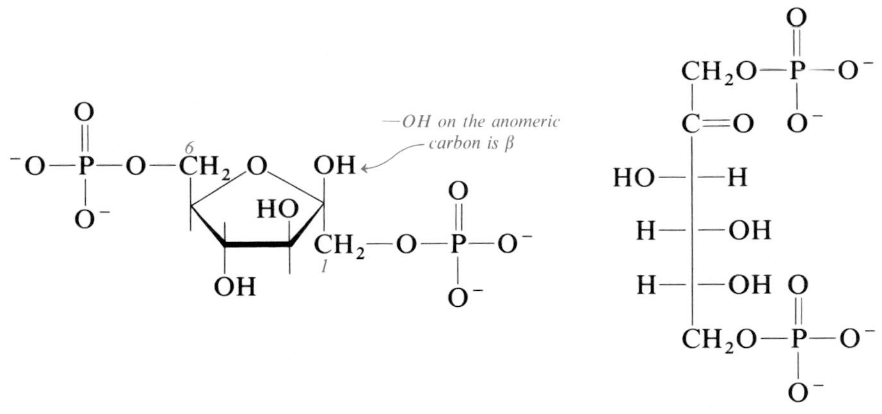

Draw structural formulas for these esters:

a. β-D-glucose 6-phosphate
b. β-D-ribose 1,5 diphosphate

C. Reduction

The carbonyl group of a monosaccharide can be reduced to an alcohol by various reducing agents, including $NaBH_4$ and hydrogen in the presence of a metal catalyst. Reduction products are known as alditols. Reduction of D-glucose gives D-glucitol, more commonly known as sorbitol.

D-glucose D-glucitol (sorbitol)

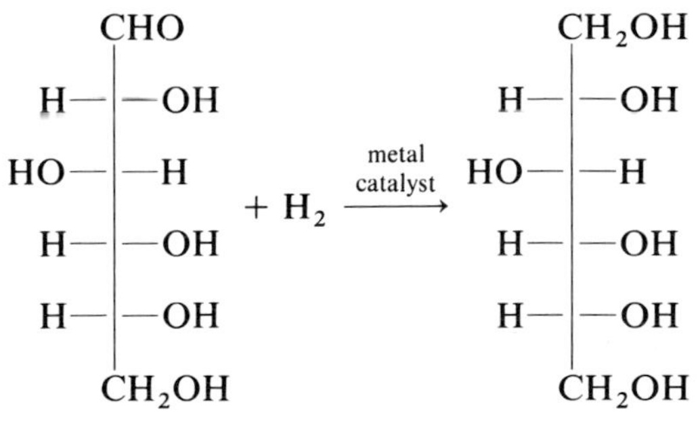

Sorbitol is found throughout the plant world. It is often used as a non-glucose-containing sweetener for foods and candies for diabetics. Also common in the biological world are erythritol and D-mannitol.

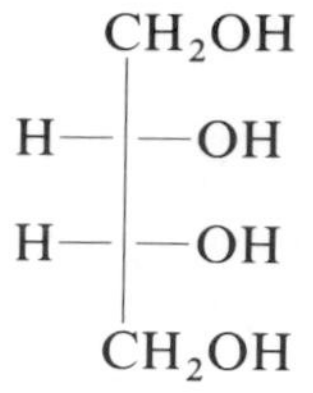

$$
\begin{array}{c}
CH_2OH \\
HO \!-\!\!|\!-\! H \\
HO \!-\!\!|\!-\! H \\
H \!-\!\!|\!-\! OH \\
H \!-\!\!|\!-\! OH \\
CH_2OH
\end{array}
$$

$$
\begin{array}{c}
CH_2OH \\
H \!-\!\!|\!-\! OH \\
H \!-\!\!|\!-\! OH \\
CH_2OH
\end{array}
$$

erythritol 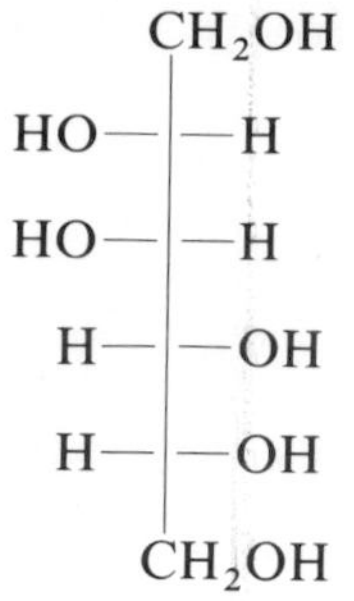   D-mannitol

At one time, xylitol was used as a sweetening agent in "sugarless" gum, candy, and sweet cereals. It has been removed from the market, however, because tests showed it to be potentially carcinogenic.

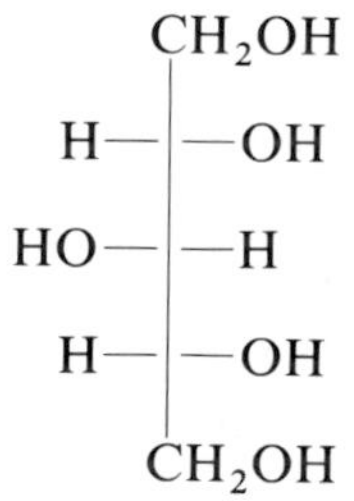

$$
\begin{array}{c}
CH_2OH \\
H \!-\!\!|\!-\! OH \\
HO \!-\!\!|\!-\! H \\
H \!-\!\!|\!-\! OH \\
CH_2OH
\end{array}
$$

xylitol

D. Oxidation of Monosaccharides

Monosaccharides (and carbohydrates in general) are classified as reducing or nonreducing according to their behavior toward Cu(II) (Benedict's solution) or toward Ag(I) in ammonium hydroxide (Tollens solution). Tollens solution is prepared by dissolving silver nitrate in ammonium hydroxide (Section 6.5A). A positive Tollens test is indicated by precipitation of metallic silver in the form of a silver mirror:

$$
\underset{}{R-\overset{\displaystyle O}{\overset{\|}{C}}-H} + Ag^+ \xrightarrow{NH_4OH} R-\overset{\displaystyle O}{\overset{\|}{C}}-O^- + \quad Ag
$$

precipitates as
a silver mirror

Benedict's solution is prepared by adding copper(II) sulfate to a solution of sodium carbonate and sodium citrate. The function of citrate is to buffer the pH of the solution and to form a complex ion with copper(II). A positive test is indicated by formation of copper(I) oxide, which precipitates as a brick-red solid.

$$R-\overset{\overset{\textstyle O}{\|}}{C}-H + 2Cu^{2+} + 5OH^- \xrightarrow{\text{citrate}} R-\overset{\overset{\textstyle O}{\|}}{C}-O^- + 3H_2O + \quad Cu_2O$$

precipitates as a
brick-red solid

Carbohydrates that reduce copper(II) ion to Cu_2O or silver(I) to metallic silver are classified as **reducing carbohydrates**. Those that do not reduce these reagents are classified as **nonreducing carbohydrates**. The chemical basis for this classification depends on two things. First, all monosaccharides contain either an aldehyde or an α-hydroxyketone; second, in dilute base, the conditions of these tests, ketoses are in equilibrium with aldoses via enediol intermediates (Problems 6.21 and 6.22).

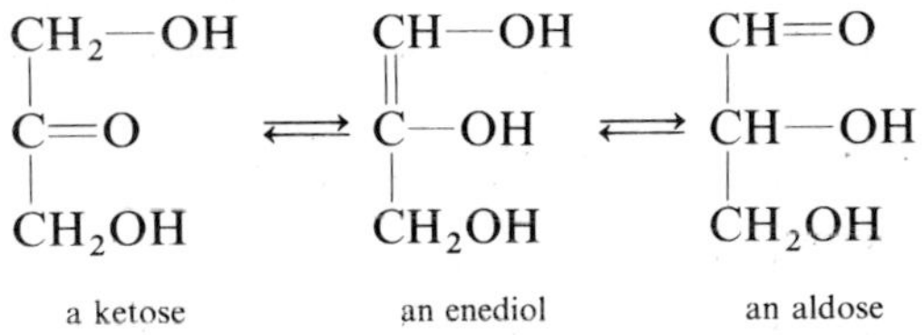

$$\begin{array}{ccccc}
CH_2-OH & & CH-OH & & CH=O \\
| & & \| & & | \\
C=O & \rightleftharpoons & C-OH & \rightleftharpoons & CH-OH \\
| & & | & & | \\
CH_2OH & & CH_2OH & & CH_2OH \\
\text{a ketose} & & \text{an enediol} & & \text{an aldose}
\end{array}$$

Oxidation of an aldose or 2-ketose by Benedict's or Tollens solutions yields a monocarboxylic acid known as an aldonic acid. For example, D-glucose is oxidized to D-gluconic acid:

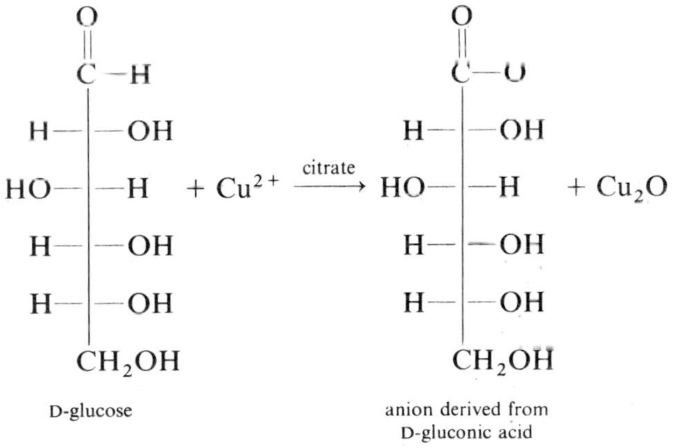

$$\begin{array}{ccc}
\overset{\overset{\textstyle O}{\|}}{C}-H & & \overset{\overset{\textstyle O}{\|}}{C}-O \\
H-\!\!-OH & & H-\!\!-OH \\
HO-\!\!-H \quad + Cu^{2+} & \xrightarrow{\text{citrate}} & HO-\!\!-H \quad + Cu_2O \\
H-\!\!-OH & & H-\!\!-OH \\
H-\!\!-OH & & H-\!\!-OH \\
CH_2OH & & CH_2OH \\
\text{D-glucose} & & \text{anion derived from} \\
& & \text{D-gluconic acid}
\end{array}$$

11.3 L-Ascorbic Acid (Vitamin C)

L-ascorbic acid (vitamin C) resembles a monosaccharide in its structural formula. In fact, it is synthesized both biochemically and industrially from D-glucose. Let us examine the individual steps in the biosynthesis of this vitamin to see how (1) D-glucose, a D-monosaccharide, is converted to L-ascorbic acid and (2) to see the types of reactions involved.

All steps in the biochemical synthesis of L-ascorbic acid are enzyme-catalyzed. As we look at them, we will be concerned only with recognizing types of reactions (oxidation, reduction, and the like), not with the particular enzymes or oxidizing and reducing agents involved in each reaction.

Oxidation of carbon 6 of D-glucose followed by reduction of the aldehyde at carbon 1 gives L-gulonic acid. This acid is of the L-series, not because of inversion of configuration at the penultimate carbon of D-glucose but because of the rules for writing Fischer projections. Carbon 1 of what was D-glucose is now —CH$_2$OH and carbon 6 is now —CO$_2$H. According to the Fischer convention, the carbon chain must be turned in the plane of the page and renumbered so that the carbonyl group (the most highly oxidized carbon) is uppermost and appears as carbon 1. When this is done, the —OH group on the penultimate carbon is now on the left and therefore the resulting monosaccharide belongs to the L series. If you compare the orientations of the —OH groups on carbons 2, 3, and 4 with that of the —OH group on carbon 5, and then refer to Table 11.1, you will discover that the monosaccharide you are now dealing with has the name gulose. The acid is therefore called gulonic acid since it is a carboxylic acid.

In the following step, L-gulonic acid is converted to a cyclic ester by reaction of the —OH on carbon 4 with the —CO$_2$H group. Cyclic esters are given the special name lactone. Therefore, the product formed in this reaction is called L-gulonolactone. Because humans, other primates, and guinea pigs lack the enzyme system necessary to convert L-gulonic acid to L-gulonolactone, they cannot synthesize ascorbic acid. Oxidation of the secondary alcohol at carbon 3, followed by enolization, gives L-ascorbic acid (vitamin C).

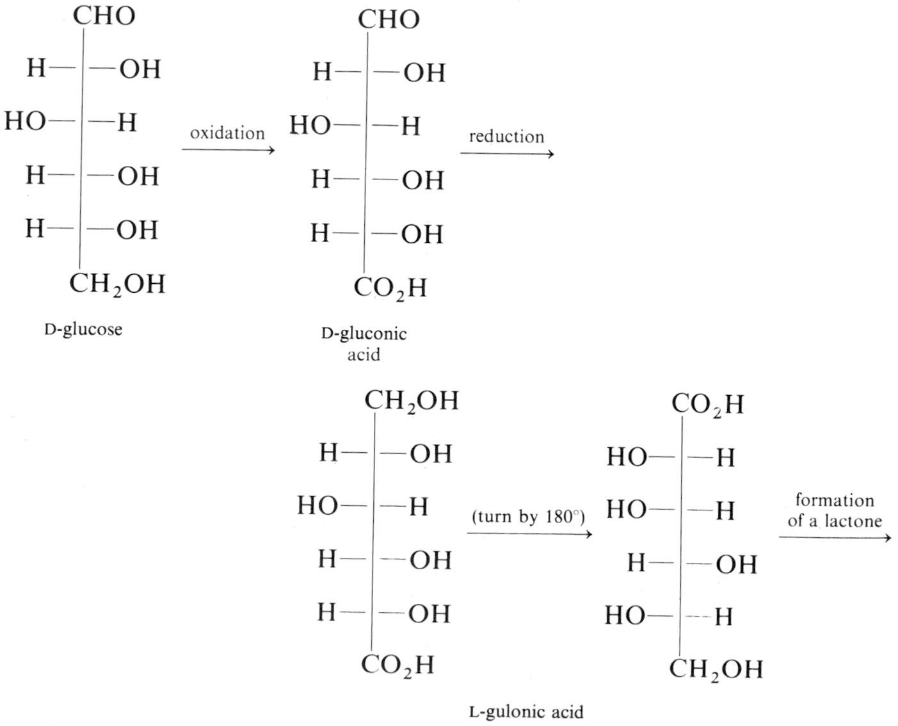

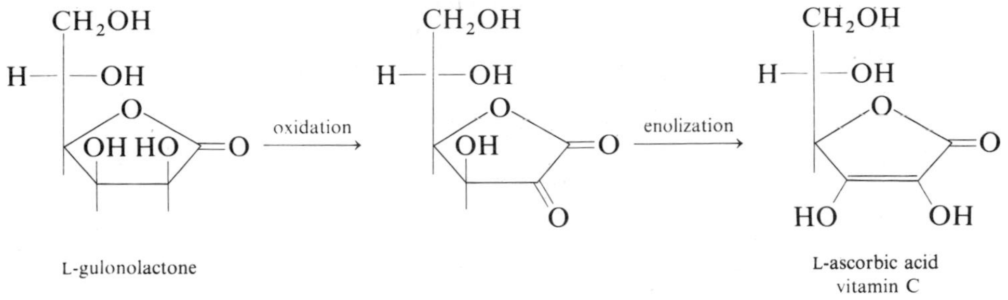

L-gulonolactone   oxidation →   enolization →   L-ascorbic acid / vitamin C

L-ascorbic acid is very easily oxidized to L-dehydroascorbic acid, a diketone. Both L-ascorbic acid and L-dehydroascorbic acid are physiologically active and are found together in most body fluids. Activity is lost if the lactone is hydrolyzed to a carboxylic acid.

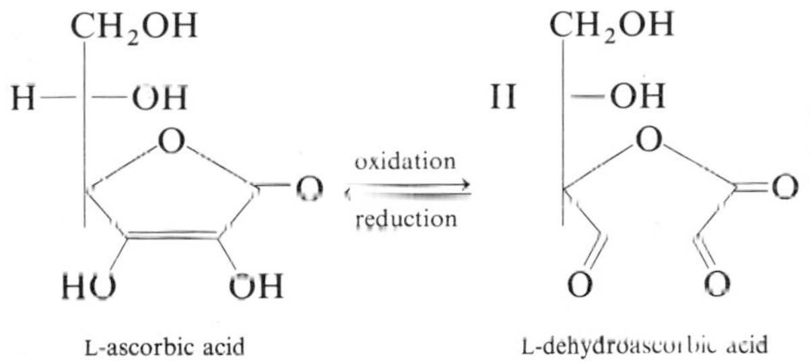

L-ascorbic acid   oxidation / reduction ⇌   L-dehydroascorbic acid

Approximately 30 million pounds of vitamin C is synthesized per year in the United States, starting from D-glucose. Within the drug industry, only aspirin exceeds vitamin C in quantity produced.

11.4 Disaccharides and Oligosaccharides

Most carbohydrates in nature contain more than one monosaccharide unit. Those that contain two units are called **disaccharides**, those that contain three units are called **trisaccharides**, and so on. The more general term **oligosaccharide** is often used for carbohydrates that contain from 4 to 10 monosaccharides. Carbohydrates containing larger numbers of monosaccharides are called **polysaccharides**.

In a disaccharide, two monosaccharide units are joined by a glycoside bond between the anomeric carbon of one unit and an —OH of the other. Three important disaccharides are maltose, lactose, and sucrose.

A. Maltose

Maltose derives its name from its presence in malt liquors, the juice from sprouted barley, and other cereal grains. Maltose consists of two molecules of D-glucose joined by a glycoside bond between carbon 1 (the anomeric carbon) of one glucose and carbon 4 of the second glucose. Because the oxygen atom on the anomeric carbon of the first glucose unit is alpha, the bond joining the two glucose units is called an α-1,4-glycoside bond. Following are Haworth and chair formulas for β-maltose, so named because the —OH on the anomeric carbon of the rightmost glucose unit is beta.

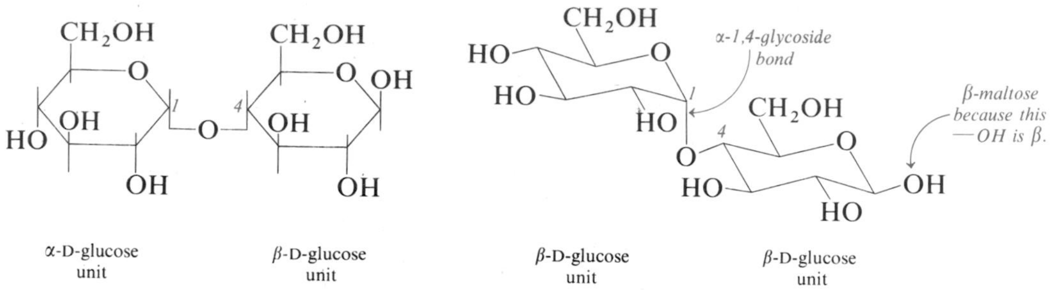

α-D-glucose unit	β-D-glucose unit

β-maltose (from the hydrolysis of starch)

Maltose is a reducing sugar, because the anomeric carbon on the right unit of D-glucose is in equilibrium with the free aldehyde and can be oxidized to a carboxylic acid.

B. Lactose

Lactose is the principal sugar present in milk. It makes up about 5–8% of human milk and 4–6% of cow's milk. Hydrolysis of lactose yields D-glucose and D-galactose. In lactose, a unit of D-galactopyranose is joined by a β-glycoside bond to carbon 4 of D-glucopyranose. Lactose is a reducing sugar.

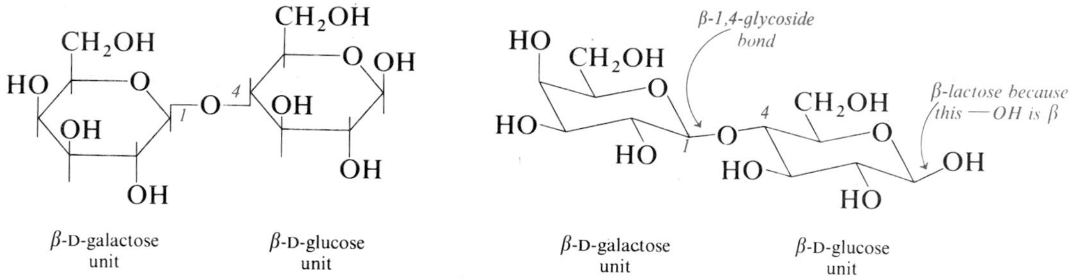

β-lactose (from the milk of mammals)

C. Sucrose

Sucrose (table sugar) is the most abundant disaccharide in the biological world. It is obtained principally from the juice of sugar cane and sugar beets. In sucrose, carbon 1 of D-glucose is joined to carbon 2 of D-fructose by an α-1,2-glycoside bond. Glucose is in a six-member (pyranose) ring form and fructose is in a five-member (furanose) ring form. Because the anomeric hemiacetal carbons of both glucose and fructose are involved in formation of the glycoside bond, sucrose is a nonreducing sugar.

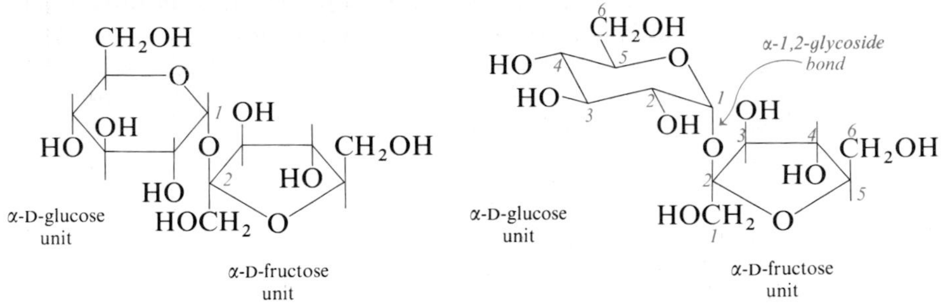

sucrose (cane or beet sugar)

Example 11.6

Draw Haworth and chair formulas for the alpha anomer of a disaccharide in which two units of D-glucopyranose are joined by an α-1,6-glycoside bond.

☐ *Solution*

First draw the structural formula of α-D-glucopyranose. Then connect the anomeric carbon of this monosaccharide to carbon 6 of a second D-glucopyranose glucose unit by an α-glycoside bond. The resulting molecule is either alpha or beta, depending on the orientation of the —OH group on the reducing end of the disaccharide.

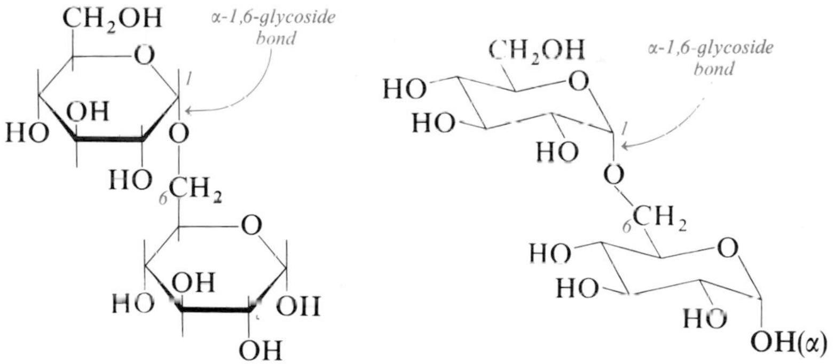

Problem 11.6

Draw Haworth and chair formulas for the beta anomer of a disaccharide in which two units of D-glucopyranose are joined by a β-1,6-glycoside bond.

D. Blood Group Substances

Plasma membranes of animal cells have large numbers of relatively small carbohydrates bound to them. In fact, it appears that the outsides of most plasma membranes are literally "sugar-coated." These membrane-bound carbohydrates are a part of the mechanism by which cells types recognize each other, and in effect act as biochemical markers (antigenic determinants). Typically, these membrane-bound carbohydrates contain from 4 to 20 monosaccharides, among which only a few kinds predominate. These include D-galactose (Gal), D-mannose (Man), L-fucose (Fuc), N-acetyl-D-glucosamine (NAGlu) and N-acetyl-D-galactosamine (NAGal). L-Fucose is a 6-deoxymonosaccharide.

L-fucose

Among the first discovered and best understood of these membrane-bound carbohydrates are the so-called blood group substances. Although blood group substances are found chiefly on the surface of erythrocytes, they are also found on proteins and lipids in other parts of the body. In the ABO system, first described in 1900, individuals are classified according to four blood types: A, B, AB, and O. Blood from individuals of the same type can be mixed without clumping (agglutination) of erythrocytes. However, if serum of type A blood is mixed with type B blood, or vice versa, the erythrocytes will clump. Serum from a type O individual causes clumping of both type A and type B blood. At the cellular level, the chemical basis for this classification is a relatively small, membrane-bound carbohydrate. Following is the composition of the tetrasaccharide found on erythrocytes of individuals with type A blood. The configurations of the glycoside bonds between the monosaccharides and of the bond between the tetrasaccharide and an —OH on the erythrocyte surface are shown in parentheses.

$$N\text{-acetyl-D-galactosamine} \xrightarrow{(\alpha\text{-}1,4)} \text{D-galactose} \xrightarrow{(\beta\text{-}1,3)} N\text{-acetyl-D-glucosamine} \xrightarrow{(\beta\text{-}1,\ldots)} \text{cell wall}$$

(NAGal) (Gal) (NAGlu)

$\uparrow (\alpha\text{-}1,2)$

L-fucose

(Fuc)

This tetrasaccharide has several distinctive features. First, it contains a monosaccharide of the "unnatural," or L series, namely L-fucose. Second, it contains D-galactose to which two other monosaccharides are bonded, one by an α-1,2-glycoside bond, the other by an α-1,4-glycoside bond. This last monosaccharide is what determines the ABO classification. In blood of type A, the chain terminates in N-acetyl-D-galactosamine (NAGal); in type B blood, it terminates instead in D-galactose (Gal); and in type O blood, it is missing completely. The saccharides of type AB blood contain both tetrasaccharides.

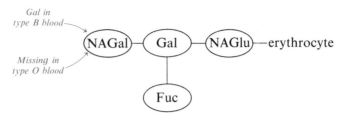

the biochemical marker (antigenic determinant) for type A blood

11.5 Polysaccharides

A. Starch: Amylose and Amylopectin

Starch is the reserve carbohydrate for plants. It is found in all plant seeds and tubers and is the form in which glucose is stored for later use by plants. Starch can be separated into two principal polysaccharides, amylose and amylopectin. While the starch from each plant is unique, most starches contain 19–25% amylose and 75–80% amylopectin. Complete hydrolysis of both amylose and amylopectin yields only D-glucose. X-Ray diffraction studies show that amylose is composed of continuous, unbranched chains of up to 4000 D-glucose monomers joined by α-1,4-glycoside bonds (Figure 11.7).

Amylopectin has a highly branched structure and contains two types of glycoside bonds. It contains the same type of chains of D-glucose joined by α-1,4-glycoside bonds as amylose does, but chain lengths vary from only 24 to 30 units. (See Figure 11.8.) In addition, there is considerable branching from this linear network. At branch points, new chains are started by α-1,6-glycoside bonds between carbon 1 of one glucose unit and carbon 6 of another glucose unit. In fact, amylopectin has such a branched structure that it is hardly possible to distinguish between main chains and branch chains.

Why are carbohydrates stored in plants as polysaccharides rather than monosaccharides, a more directly usable form of energy? The answer has to do with osmotic pressure which is proportional to molar concentration, not the molecular weight of a solute. If we assume that 1000 molecules of glucose are assembled in one starch macromolecule, then we can predict that a solution containing 1 g of starch per 10 mL will have only $\frac{1}{1000}$ the osmotic pressure of a solution of 1 g of glucose in the same volume of solution. This feat of packaging is of tremendous advantage because it reduces the strain on various membranes enclosing such macromolecules.

B. Glycogen

Glycogen is the reserve carbohydrate for animals. Like amylopectin, glycogen is a nonlinear polymer of D-glucose joined by α-1,4- and α-1,6-glycoside bonds,

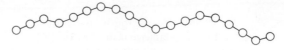

amylose
(long unbranched chains of glucose units
joined by α-1,4-glycoside bonds)

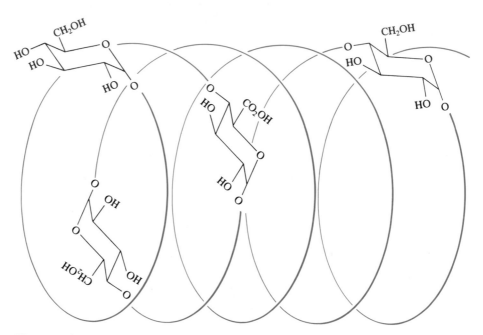

Figure 11.7 Amylose is a linear polymer of up to 4000 monomers of D-glucose joined by α-1,4,-glycoside bonds.

amylopectin

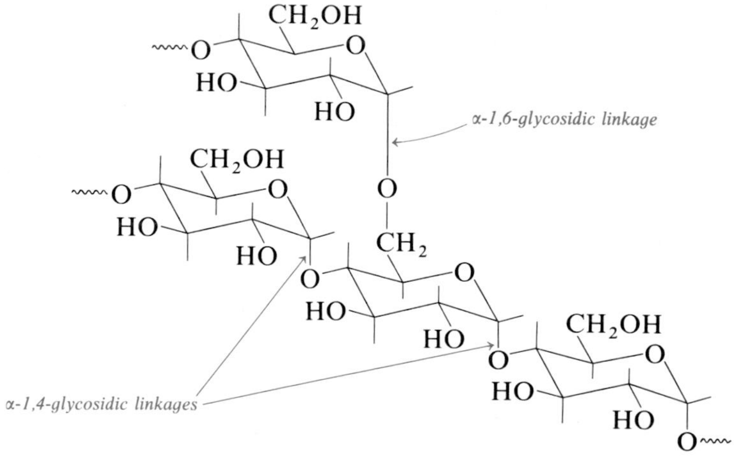

Figure 11.8 Amylopectin is a highly branched polymer with chains of 24–30 units of D-glucose joined by α-1,4-glycoside bonds and branch points created by α-1,6-glycoside bonds.

but it has a lower molecular weight and an even more highly branched structure (Figure 11.9). The total amount of glycogen in the body of a well-nourished adult is about 350 g, divided almost equally between liver and muscle.

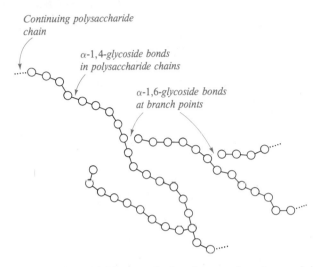

Figure 11.9 Glycogen is a highly branched polymer of D-glucose joined by α-1,4-glycoside bonds. Branch points created by α-1,6-glucoside bonds contain 12–18 units of glucose.

C. Cellulose

Cellulose, the most widely distributed skeletal polysaccharide, constitutes almost half the cell wall material of wood. Cotton is almost pure cellulose. Cellulose is a linear polymer of D-glucose monomers joined by β-1,4-glycoside bonds (Figure 11.10). It has an average molecular weight of 400,000, corresponding to approximately 2800 glucose units. Cellulose fibers consist of bundles of parallel polysaccharide chains held together by hydrogen bonding between hydroxyl groups on adjacent chains. This arrangement of parallel chains in bundles and the resulting hydrogen bonding gives cellulose fibers their high mechanical strength.

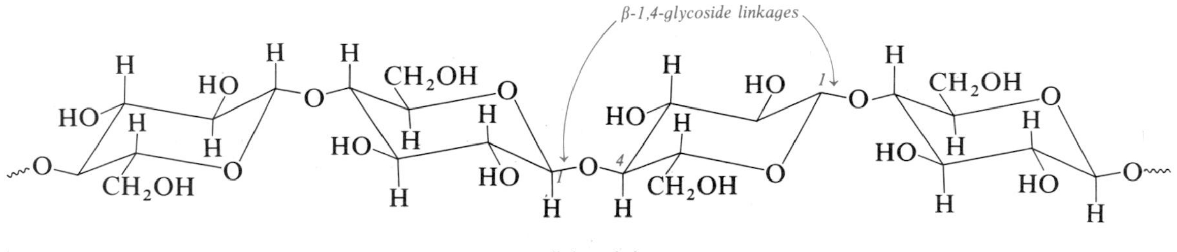

cellulose chain

Figure 11.10 Cellulose is a linear polymer of up to 3000 units of D-glucose joined by β-1,4-glycoside bonds.

Humans and other animals cannot use cellulose as a food. The reason is that our digestive systems do not contain β-glycosidases, enzymes that catalyze hydrolysis of β-glycoside bonds. They contain only α-glycosidases; hence, the polysaccharides we use as sources of glucose are starch and glycogen. On the other hand, many bacteria and microorganisms do contain β-glycosidases and can digest cellulose. Termites are fortunate in having such bacteria in their intestine and can use wood as their principal food. Ruminants (cud-chewing animals) can also digest grasses and wood because β-glycosidase-containing microorganisms are present in their alimentary systems.

D. Textile Fibers from Cellulose

Cotton, an important textile fiber, is almost pure cellulose. However, cotton represents only an insignificant fraction of the cellulose in the biological world. Both rayon and acetate rayon are made from chemically modified and regenerated cellulose and were the first man-made textile fibers to become commercially important.

In the production of **rayon**, cellulose is converted to a water-soluble derivative by reaction with carbon disulfide to form an alkali-soluble xanthate ester. A solution of cellulose xanthate is then extruded through a spinnerette, a metal disc with many tiny holes, into dilute sulfuric acid to hydrolyze the xanthate esters and precipitate free, or "regenerated," cellulose. Regenerated cellulose extruded as a filament is called viscose rayon thread; extruded as a sheet, it is called cellophane.

carbon
disulfide

sodium salt of
a xanthate ester

In the industrial synthesis of acetate rayon, cellulose is acetylated with acetic anhydride. Acetylated cellulose is then dissolved in a suitable solvent, precipitated, and drawn into fibers known as acetate rayon. Cellulose acetate, acetylated to the extent of about 80%, became commercial in Europe about 1920 and in the United States a few years later. Cellulose triacetate, which has about 97% of the hydroxyls converted to acetate esters, became commercial in the United States in 1954. Today, acetate fibers rank fourth in production in the United States, surpassed only by Dacron polyester, Nylon, and rayon fibers.

one glucose unit of a
fully acetylated cellulose

Key Terms and Concepts

amino sugar (11.1D)

anomeric carbon (11.1E)

Benedict's solution (11.2D)

blood group substances (11.4D)

carbohydrate (introduction)

disaccharide (11.4)

Fischer projection formula (11.1C)

furanose (11.1E)

glycoside (11.2A)

glycoside bond (11.2A)

N-glycoside (11.2A)

Haworth structure (11.1E)

hexose (11.1A)

monosaccharide (11.1A)

D-monosaccharide (11.1C)

L-monosaccharide (11.1C)

mutarotation (11.1F)

nonreducing carbohydrate (11.2D)

oligosaccharide (11.4)

pentose (11.1A)

polysaccharide (11.4 and 11.5)

pyranose (11.1E)

rayon (11.5D)

reducing carbohydrate (11.2D)

saccharide (introduction)

tetrose (11.1A)

trisaccharide (11.4)

Key Reactions

1. Formation of cyclic hemiacetals and hemiacetals (Section 11.1E).
2. Reaction of monosaccharides with alcohols: formation of glycosides (Section 11.2A).
3. Reaction of monosaccharides with heterocyclic aromatic amines to form N-glycosides (Section 11.2A).
4. Reduction of monosaccharides to alditols (Section 11.2C).
5. Oxidation of monosaccharides to aldonic acids: reducing sugars (Section 11.2D).

Problems

Structure of monosaccharides (Section 11.1)

11.7 The term carbohydrate is derived from "hydrates of carbon." Show the origin of this term by reference to the molecular formulas of D-ribose, D-fructose, and lactose.

11.8 Explain the meaning of the designations D and L as used to specify the stereochemistry (configuration) of monosaccharides.

11.9 List the rules for drawing Fischer projection formulas.

11.10 Table 11.1 shows a Fischer projection formula of D-arabinose. Draw a Fischer projection formula of L-arabinose, a natural aldopentose of the "unnatural" L-configuration.

11.11 **a.** Build a molecular model of D-glucose in the open-chain form.
   b. Using this molecular model, show the reaction of the —OH on carbon 5 with the aldehyde of carbon 1 to form a cyclic hemiacetal. Show that either alpha-D-glucose or beta-D-glucose can be formed,

depending on the direction from which the —OH group interacts with the aldehyde group.

11.12   Explain the conventions alpha and beta as used to designate the stereochemistry of cyclic forms of monosaccharides.

11.13   Explain the phenomenon of mutarotation with reference to carbohydrates. How is mutarotation detected?

11.14   A solution of alpha-D-glucose has a specific rotation of $+112°$; a beta-D-glucose solution has a specific rotation of $+19°$. On mutarotation, the specific rotation of each solution changes to an equilibrium value of $+52°$. Calculate the percentage of beta-D-glucose in the equilibrium mixture.

Reactions of monosaccharides (Section 11.2)

11.15   There are four isomeric D-aldopentoses (Table 11.1). Suppose the aldehyde in each is reduced to a primary alcohol. Which yield optically inactive alditols? Which yield optically active alditols?

11.16   An important technique for establishing relative configurations among isomeric aldoses is to convert both terminal carbon atoms to the same functional group. This can be done by either selective oxidation or selective reduction. As a specific example, nitric acid oxidation of D-erythrose gives *meso*-tartaric acid. Oxidation of D-threose under similar conditions gives D-tartaric acid.

$$\text{D-threose} \xrightarrow[\text{oxidation}]{\text{HNO}_3} \text{D-tartaric acid}$$

$$\text{D-erythrose} \xrightarrow[\text{oxidation}]{\text{HNO}_3} meso\text{-tartaric acid}$$

Using this information, show which of the following structural formulas is D-erythrose and which is D-threose. Check your answer by referring to Table 11.1.

```
        CHO                CHO
 a. H——|——OH     b. HO——|——H
        |                  |
    H——|——OH         H——|——OH
        |                  |
      CH₂OH              CH₂OH
```

11.17   Classify the following as reducing or nonreducing sugars.
a. alpha-D-glucose   **b.** beta-D-ribose
c. 2-deoxy-D-ribose   **d.** alpha-methyl-D-glucoside

11.18   Treatment of D-glucose in dilute aqueous base at room temperature yields an equilibrium mixture of D-glucose, D-mannose, and D-fructose. How might you account for this conversion? (*Hint:* Review Section 6.7B and your answers to Problems 6.21 and 6.22.)

11.19   Ketones are not oxidized by mild oxidizing agents. However, both dihydroxyacetone and fructose give a positive Benedict's test and are classified as reducing sugars. How could you explain that these ketoses are reducing sugars? (*Hint:* These tests are done in dilute aqueous base.)

11.20   L-fucose is one of several monosaccharides commonly found in the surface polysaccharides of animal cells. This 6-deoxyaldohexose is synthesized from D-mannose in a series of eight steps, shown below.

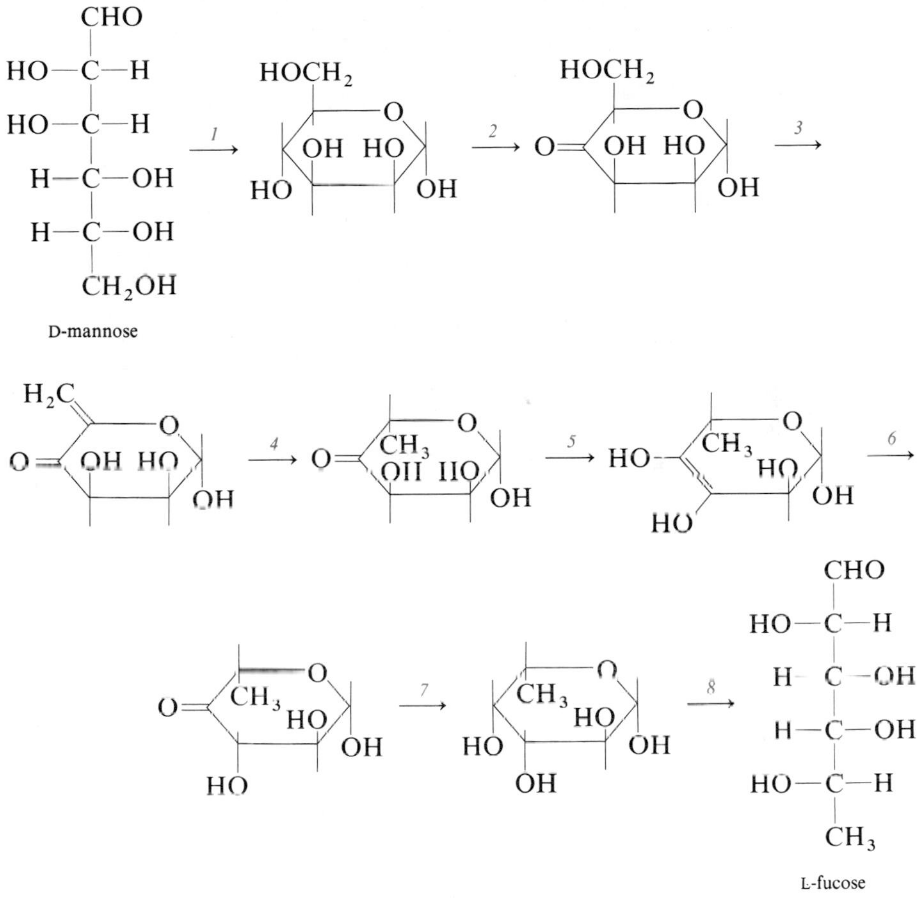

CHO
HO—C—H
HO—C—H
H—C—OH
H—C—OH
CH₂OH

D-mannose

$\xrightarrow{1}$

HOCH₂, O, OH HO, HO, OH

$\xrightarrow{2}$

HOCH₂, O, O= OH HO, OH

$\xrightarrow{3}$

H₂C, O, O= QH HO, OH

$\xrightarrow{4}$

O= CH₃, OH HO, OH

$\xrightarrow{5}$

HO— CH₃, HO, OH, HO

$\xrightarrow{6}$

O= CH₃, HO, OH, HO

$\xrightarrow{7}$

CH₃, HO, HO, OH, OH

$\xrightarrow{8}$

CHO
HO—C—H
H—C—OH
H—C—OH
HO—C—H
CH₃

L-fucose

a. Describe the type of reaction (oxidation, reduction, hydration, and so on) involved in each step.

b. Explain why this monosaccharide belongs to the L series even though it is derived biochemically from a D sugar.

11.21   Draw structural formulas for the following phosphate esters.
a. beta-D-galactose 6-phosphate
b. beta-D-ribose 3-phosphate
c. beta-2-deoxy-D-ribose 5-phosphate
d. beta-D-ribose 1,3-diphosphate
e. glycerol 1-phosphate

11.22   The backbone of ribonucleic acid (RNA) consists of units of beta-D-ribose joined by phosphate ester bonds between the hydroxyl on carbon

3 of one ribose and the hydroxyl on carbon 5 of another ribose. Draw the structural formula of two units of beta-D-ribose joined in this manner.

(beta-D-ribose)—(phosphate)—(beta-D-ribose)

Disaccharides and oligosac- charides (Section 11.4)

11.23 Classify the following as reducing or nonreducing sugars.
a. sucrose   b. lactose
c. alpha-methyl-lactoside   d. maltose
e. beta-methyl-maltoside

11.24 Trehalose, a disaccharide consisting of two glucose units joined by an alpha-1,1-glycoside bond, is found in young mushrooms and is the chief carbohydrate in the blood of certain insects.

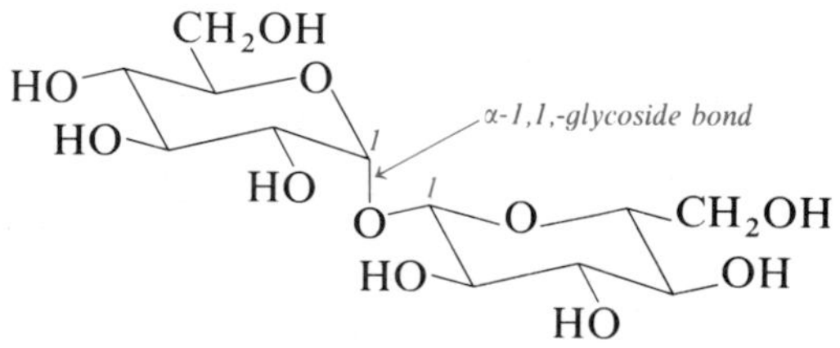

From its structural formula, would you expect trehalose (a) to be a reducing sugar? (b) to undergo mutarotation?

11.25 Raffinose is the most abundant trisaccharide in nature.
a. Name the three monosaccharide units in raffinose.
b. There are two glycoside bonds in raffinose. Describe each as you have already done for other disaccharides.
c. Would you expect raffinose to be a reducing sugar?
d. Would you expect raffinose to undergo mutarotation?

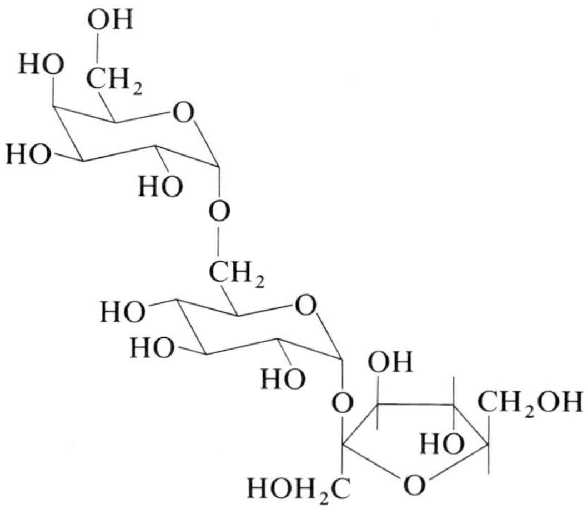

raffinose

11.26 Following is the Fischer projection formula for *N*-acetyl-D-glucosamine. This substance forms a six-member cyclic hemiacetal.

$$\begin{array}{c} \text{CHO} \\ | \\ \text{H}\!-\!|\!-\!\text{NH}\!-\!\overset{\displaystyle \text{O}}{\overset{\displaystyle \|}{\text{C}}}\text{CH}_3 \\ | \\ \text{HO}\!-\!|\!-\!\text{H} \\ | \\ \text{H}\!-\!|\!-\!\text{OH} \\ | \\ \text{H}\!-\!|\!-\!\text{OH} \\ | \\ \text{CH}_2\text{OH} \end{array}$$

N-acetyl-D-glucosamine

a. Draw Haworth and chair structures for the alpha and beta forms of this monosaccharide.

b. Draw Haworth and chair structures for the disaccharide formed by joining two units of *N*-acetyl-D-glucosamine by a beta-1,4-glycoside bond. (If you have done this correctly, you have drawn the structural formula of the repeating dimer of chitin, the polysaccharide component of the shells of lobster and other crustaceans.)

Polysac- charides (Section 11.5)

11.27 What is the main difference in structure between cellulose and starch? Why are humans unable to digest cellulose?

11.28 Propose a likely structure for the following polysaccharides.

a. Alginic acid, isolated from seaweed, is used as a thickening agent in ice cream and other foods. Alginic acid is a polymer of D-mannuronic acid units joined by beta-1,4-glycoside bonds.

b. Pectic acid is the main constituent of pectin, which is responsible for the formation of jellies from fruits and berries. Pectic acid is a polymer of D-galacturonic acid units joined by α-1,4-glycoside bonds.

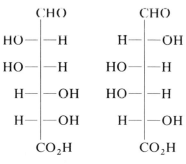

D-mannuronic acid   D-galacturonic acid

Clinical Chemistry— The Search for Specificity

The analytical procedure most often performed in the clinical chemistry laboratory is the determination of glucose in blood, urine, or other biological fluid. The need for a rapid and reliable test for blood glucose stems from the high incidence of the disease diabetes mellitus. There are approximately 2 million known diabetics in the United States, and it is estimated that another 2 million more are undiagnosed.

Diabetes mellitus is characterized by insufficient blood levels of the polypeptide hormone insulin (Section 13.6F). If insulin is deficient, glucose cannot enter muscle and liver cells, which in turn leads to increased levels of blood glucose (hyperglucosemia), impaired metabolism of fats and proteins, ketosis, and possibly diabetic coma. Thus, it is critical for the early diagnosis and effective management of this disease to have

a rapid and reliable procedure for determining blood glucose.

Over the past 70 years, many such tests have been developed. We will discuss four of these, each chosen to illustrate something of the problems involved in developing suitable clinical laboratory tests and how these problems can be solved. Furthermore, these tests will illustrate the use of both chemical and enzymatic techniques in the modern clinical chemistry laboratory.

The first widely used glucose test was based on the activity of glucose as a reducing sugar. Specifically, the aldehyde group of glucose is oxidized by ferricyanide ion to a carboxyl group. In the process Fe(III) in ferricyanide ion is reduced to Fe(II) in ferrocyanide ion (Figure VIII-1). The reaction is carried out in the pres-

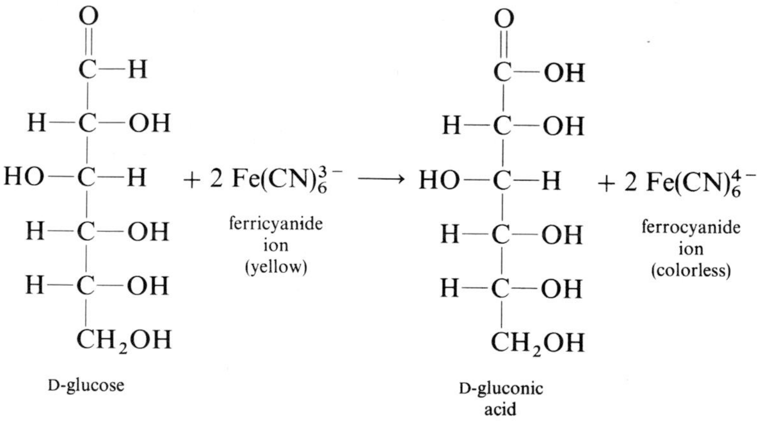

$$3 Fe(CN)_6^{4-} + 4 Fe^{3+} \longrightarrow Fe_4[Fe(CN)_6]_3$$

ferrocyanide ion   Prussian blue

Figure VIII-1 The first widely used glucose test was based on the ability of glucose to reduce Fe(III) in ferricyanide ion to Fe(II) in ferrocyanide ion. This in turn reacts with excess Fe(III) to form Prussian blue.

ence of excess Fe(III). Under these conditions, the ferrocyanide ion reacts further with Fe(III) to form ferric ferrocyanide, more familiar as Prussian blue. The concentration of glucose in the test sample is measured spectrophotometrically. In this test, the absorbance of Prussian blue is directly proportional to the concentration of glucose in the test sample.

Although this method can be used to measure glucose concentration, it has the disadvantage that ferricyanide also oxidizes several other reducing substances found in blood, including ascorbic acid, uric acid, certain amino acids, and phenols. In addition, any other aldoses present in blood also reduce ferricyanide. All these substances are said to give false positive results. The ferricyanide and other oxidative tests first developed often gave values as much as 30% or higher over the so-called true glucose value.

A more satisfactory approach in the search for specificity lay in attacking the problem in a completely different way, namely, by taking advantage of a chemical reactivity of glucose other than its property as a reducing sugar. One of the most successful and widely used of these nonoxidative methods involves reaction of glucose with o-toluidine to form a blue green Schiff base. The absorbance of this Schiff base can be measured spectrophotometrically at 625 nm and is directly proportional to glucose concentration (Figure VIII-2).

The o-toluidine method can be applied directly to serum, plasma, cerebrospinal fluid, and urine, and to samples as small as 20 microliters (20×10^{-6} L). In addition, it does not give false positive results with other reducing substances, since the procedure itself does not involve oxidation. However, galactose and mannose, and to a lesser extent lactose and xylose, are potential sources of false positive results, because they also react with o-toluidine to give colored Schiff bases. This, however, is generally not a problem, since these mono- and disaccharides are normally present in serum and plasma only in very low concentrations.

In recent years the search for even greater specificity in glucose determinations has led to the introduction of enzyme-based glucose assay procedures. What was needed was an enzyme that catalyzes a specific reaction of glucose but not comparable reactions of any other substance normally present in biological fluids. The enzyme glucose oxidase meets these requirements. It catalyzes the oxidation of beta-D-glucose to D-gluconic acid (Figure VIII-3). Glucose oxidase is specific for beta-D-glucose. Therefore, complete oxidation of any sample containing both beta-D-glucose and alpha-D-glucose requires conversion of the alpha form to the beta form. Fortunately, this interconversion is rapid and complete in the short time required for the test. Molecular oxygen, O_2, is the oxidizing agent in

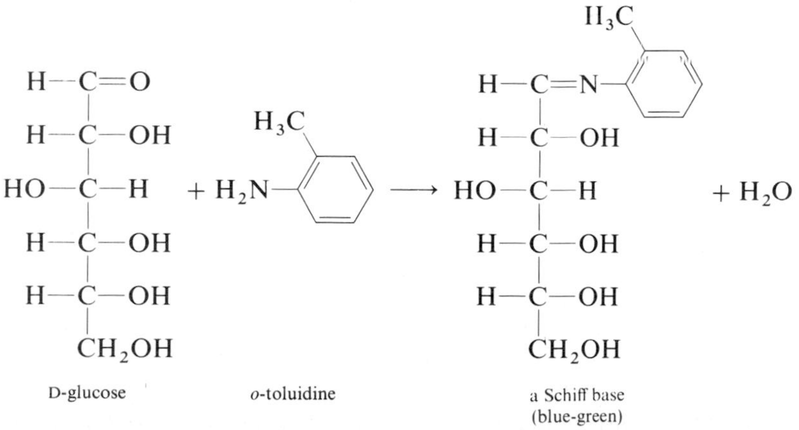

Figure VIII-2 For many years, the o-toluidine test was the standard clinical chemistry laboratory test for glucose.

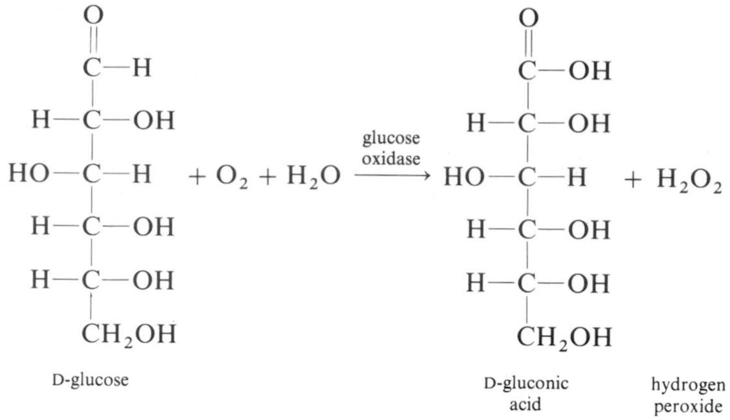

Figure VIII-3 The glucose oxidase method is the most specific test yet developed for measuring glucose concentration in biological fluids.

this reaction; it is reduced to hydrogen peroxide, H_2O_2, which in turn is used to oxidize another substance whose concentration can be determined spectrophotometrically. In one procedure, hydrogen peroxide is caused to react with iodide ion, to form molecular iodine, I_2:

$$2I^- + H_2O_2 + 2H^+ \longrightarrow I_2 + 2H_2O$$

The absorbance at 420 nm is used to calculate iodine concentration and then glucose concentration. In another procedure, hydrogen peroxide is used to oxidize o-toluidine to a colored product in a reaction catalyzed by the enzyme peroxidase. The concentration of the colored oxidation product is determined spectrophotometrically.

Recall that the first assay method we looked at in this mini-essay was also an oxidative procedure. However, it gave positive errors in the presence of other reducing substances. In a sense, the search for specificity has now come full circle, because now the most highly specific and accurate assay, and the one which is said to give "true" glucose values, is also an oxidative method. Unlike the earlier, ferricyanide method, however, this newer, oxidative method is highly specific because it is catalyzed by beta-D-glucose oxidase.

Any determination of glucose in blood or a 24-hr urine sample reflects glucose levels during the sample period only. There is now a simple, convenient laboratory method that can be used

$$o\text{-toluidine} + H_2O_2 \xrightarrow{\text{peroxidase}} \text{(colored products)} + H_2O$$

Several commercially available test kits use the glucose oxidase reaction for qualitative determination of glucose in urine. One of these, Clinistix (produced by the Ames Co., Elkhart, Ind.), consists of a filter-paper strip impregnated with glucose oxidase, peroxidase, and o-toluidine. The test end of the paper is dipped in urine, removed, and examined after 10 sec. A blue color develops if the concentration of glucose in the urine exceeds about 1 mg/mL.

to monitor long-term glucose levels. This method depends on the measurement of the relative amounts of hemoglobin and certain hemoglobin derivatives normally present in blood. Hemoglobin A (HbA) is the main type of hemoglobin present in normal red blood cells. In addition, there are several lesser components, including glycosylated hemoglobins (HbA$_1$). Glycosylated hemoglobins (see Figure VIII-4) are synthesized within red blood cells in two steps. In step 1, the

$$
\begin{array}{c}
\begin{array}{c}
\text{H}-\overset{\displaystyle \text{O}}{\overset{\|}{\text{C}}} \\
\text{H}-\text{C}-\text{OH} \\
\text{HO}-\text{C}-\text{H} \\
\text{H}-\text{C}-\text{OH} \\
\text{H}-\text{C}-\text{OH} \\
\text{CH}_2\text{OH}
\end{array}
\quad + \text{H}_2\text{N}-\text{Hb}(\beta) \rightleftharpoons
\begin{array}{c}
\text{H}-\text{C}=\text{N}-\text{Hb}(\beta) \\
\text{H}-\text{C}-\text{OH} \\
\text{HO}-\text{C}-\text{H} \\
\text{H}-\text{C}-\text{OH} \\
\text{H}-\text{C}-\text{OH} \\
\text{CH}_2\text{OH}
\end{array}
\quad \longrightarrow
\begin{array}{c}
\text{CH}_2-\text{NH}-\text{Hb}(\beta) \\
\text{C}=\text{O} \\
\text{HO}-\text{C}-\text{H} \\
\text{H}-\text{C}-\text{OH} \\
\text{H}-\text{C}-\text{OH} \\
\text{CH}_2\text{OH}
\end{array}
\end{array}
$$

D-glucose | the terminal —NH$_2$ group of a beta chain of normal hemoglobin | a Schiff base (unstable) | glycosylated hemoglobin (stable)

Figure VIII-4 Formation of glycosylated hemoglobin.

free —NH$_2$ of the beta chain of hemoglobin reacts with the carbonyl group of glucose to form a Schiff base (Section 6.4C). Step 1 is reversible. In a slower, irreversible second step, the Schiff base undergoes a type of keto-enol tautomerism (Section 6.7B) to form a glycosylated hemoglobin. A glycosylated hemoglobin molecule is shown schematically in Figure VIII-5.

Because this slow, reversible second step occurs continuously throughout the 120-day life span of a typical red blood cell, levels of glycosylated hemoglobin with a red blood cell pop-

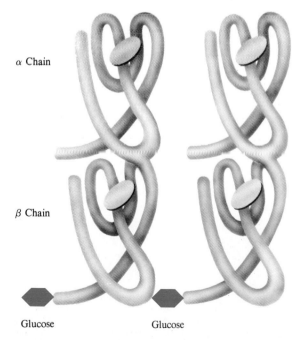

α Chain

β Chain

Glucose   Glucose

Figure VIII-5 Schematic drawing of a glycosylated hemoglobin.

reflect the average blood glucose levels during that period.

Normal levels of glycosylated hemoglobins usually fall within the range 4.5–8.5% of total hemoglobin. In cases of uncontrolled or poorly controlled diabetes, the percentage of glycosylated hemoglobins may rise to two or three times these values. Conversely, once long-term blood-glucose levels have been achieved, glycosylated hemoglobins gradually fall to within normal ranges. Thus, the level of glycosylated hemoglobin can be used to give a picture of the average blood glucose level over the previous 8 to 10 weeks.

References

Bunn, H. F. 1981. Evaluation of glycosylated hemoglobin in diabetic patients. *Diabetes* 30:613.

Garet, M. C., Blouquit, Y., Molko, F., and Rosa, J. 1979. HbA_1—A review of its structure, biosynthesis, clinical significance, and methods of assay. *Biomedicine* 30:234.

Henry, R. J., Cannon, D. O., and Winkleman, J. W., eds. 1974. *Clinical Chemistry, Principles and Techniques*. 2d ed. New York: Harper and Row.

Tietz, N., ed. 1976. *Fundamentals of Clinical Chemistry*. Philadelphia: W. B. Saunders.

12

Lipids

Lipids are a heterogeneous class of natural organic compounds, grouped together not by the presence of a distinguishing functional group or structural feature, but rather on the basis of common solubility properties. Lipids are all insoluble in water and very soluble in one or more organic solvents, including diethyl ether, chloroform, benzene, and acetone. In fact, these four solvents are often referred to as lipid-solvents, or fat-solvents. Proteins, carbohydrates, and nucleic acids are largely insoluble in these solvents.

In this chapter we will describe the structure and biological function of representative members of the five major types of lipids: fats and oils, phospholipids, the fat-soluble vitamins, steroids, and waxes. In addition, we will describe the structure of biological membranes.

12.1 Fats and Oils

You certainly are familiar with fats and oils because you encounter them every day in such things as milk, butter, oleomargarine, and corn oil and other liquid vegetable oils, and also in many other foods. Fats and oils are triesters of glycerol and are called **triglycerides**. Triglycerides are the most abundant lipids. Complete hydrolysis of triglyceride yields one molecule of glycerol and three molecules of fatty acid:

$$
\begin{array}{c}
\underset{\substack{\text{a triglyceride}\\ \text{(a fat or oil)}}}{
\begin{array}{c}
\qquad\qquad\qquad \overset{\text{O}}{\underset{\|}{}} \\
\text{CH}_2-\text{O}-\text{C}-\text{R} \\
\overset{\text{O}}{\underset{\|}{}}\qquad | \\
\text{R}-\text{C}-\text{O}-\text{CH} \\
| \\
\text{CH}_2-\text{O}-\text{C}-\text{R} \\
\qquad\qquad\underset{\|}{\overset{}{}}\;\text{O}
\end{array}}
\;+\;3\text{H}_2\text{O}\;\xrightarrow{\text{hydrolysis}}\;
\underset{\text{glycerol}}{
\begin{array}{c}
\text{CH}_2-\text{OH} \\
| \\
\text{CH}-\text{OH} \\
| \\
\text{CH}_2-\text{OH}
\end{array}}
\;+\;3\underset{\text{fatty acids}}{\text{R}-\overset{\overset{\text{O}}{\|}}{\text{C}}-\text{OH}}
\end{array}
$$

A. Fatty Acids

Fatty acids are monocarboxylic acids obtained from the hydrolysis of triglycerides. Over 70 different fatty acids have been isolated from various cells and tissues. Given in Table 12.1 are common names and structural formulas for several of the most abundant fatty acids.

Table 12.1 Some natural fatty acids.

Carbon Atoms	Structural Formula	Common Name	mp (°C)
Saturated fatty acids			
12	$CH_3(CH_2)_{10}CO_2H$	lauric acid	44
14	$CH_3(CH_2)_{12}CO_2H$	myristic acid	58
16	$CH_3(CH_2)_{14}CO_2H$	palmitic acid	63
18	$CH_3(CH_2)_{16}CO_2H$	stearic acid	70
20	$CH_3(CH_2)_{18}CO_2H$	arachidic acid	77
Unsaturated fatty acids			
16	$CH_3(CH_2)_5CH{=}CH(CH_2)_7CO_2H$	palmitoleic acid	−1
18	$CH_3(CH_2)_7CH{=}CH(CH_2)_7CO_2H$	oleic acid	16
18	$CH_3(CH_2)_4(CH{=}CHCH_2)_2(CH_2)_6CO_2H$	linoleic acid	−5
18	$CH_3CH_2(CH{=}CHCH_2)_3(CH_2)_6CO_2H$	linolenic acid	−11
20	$CH_3(CH_2)_4(CH{=}CHCH_2)_4(CH_2)_2CO_2H$	arachidonic acid	−49

We can generalize as follows about the most abundant fatty acid components of higher plants and animals.

1. Nearly all fatty acids have an even number of carbon atoms, most between 12 and 20 carbon atoms, in an unbranched chain. Those having 16 or 18 carbon atoms are by far the most abundant in nature.
2. The three most abundant fatty acids are palmitic, stearic, and oleic acids.
3. As the number of carbon atoms in a saturated fatty acid increases, its melting point increases.
4. Unsaturated fatty acids have lower melting points than their saturated counterparts.
5. The greater the degree of unsaturation in a fatty acid, the lower its melting point.
6. In most unsaturated fatty acids, cis isomers predominate; trans isomers are rare.

Because fatty acids have long hydrocarbon chains, they are insoluble in water. They do interact with water in a particular way, however. If a drop of fatty acid is placed on the surface of water, it spreads out to form a thin film one molecule thick (a monomolecular layer), with the polar carboxyl groups dissolved in the water and the nonpolar hydrocarbon chains forming a hydrocarbon layer on the surface of the water (Figure 12.1).

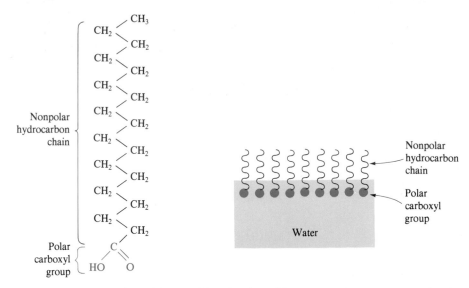

Figure 12.1 Interaction of fatty acid molecules with water to form a monomolecular layer.

B. Structure of Triglycerides

A triglyceride in which all three fatty acids are identical is called a simple triglyceride. Simple triglycerides are rare in nature; mixed triglycerides are much more common. Following is the structural formula of a mixed triglyceride formed from glycerol and molecules of stearic, oleic, and palmitic acids, the three most abundant fatty acids:

$$CH_3(CH_2)_7CH{=}CH(CH_2)_7C{-}OCH$$

An ester of oleic acid

An ester of stearic acid
$$CH_2O{-}C(CH_2)_{16}CH_3$$

$$CH_2O{-}C(CH_2)_{14}CH_3$$
An ester of palmitic acid

C. Physical Properties of Triglycerides

The physical properties of a triglyceride depend on its fatty acid components. In general, the melting point of a triglyceride increases as the number of carbons in its hydrocarbon chains increases and decreases as the degree of unsaturation increases. Triglycerides rich in oleic acid, linoleic acid, and linolenic and other unsaturated fatty acids are generally liquid at room temperature and are called **oils**. Triglycerides rich in palmitic, stearic, and other saturated fatty acids are generally semisolids or solids at room temperature and are called **fats**. Table

Table 12.2
Distribution of
saturated and
unsaturated fatty
acids in some
triglycerides.
Percentages are
given for the
most abundant
fatty acids; others
are present in
lesser amounts.

Source	% Triglyceride in Edible Portion	% Fatty Acids by Weight		
		Saturated	Oleic	Linoleic
Animal fat				
beef	5–37	41.5	49.6	2.5
butter	81	54.2	27.7	3.6
fish (tuna)	4	24	25	0.5
milk (whole)	4	57	33	3
pork	52	37	42	10
Vegetable oils				
coconut oil	100	76.2	7.5	0.5
corn oil	100	14.6	49.6	34.3
peanut oil	100	13.8	56.0	26
soybean oil	100	13.4	28.9	50.7
cottonseed oil	100	27.2	22.9	47.8

12.2 lists the percentage composition in grams of fatty acid per 100 grams of
triglyceride for several common fats and oils. Notice that beef tallow is approximately 41.5% saturated and 52.1% unsaturated fatty acids by weight. Vegetable
oils such as corn oil, soybean oil, and wheat germ oil are all approximately
80% by weight unsaturated fatty acids. Butter fat is distinctive in that it contains significant amounts of lower-molecular-weight fatty acids.

The lower melting points of triglycerides rich in unsaturated fatty acids
(vegetable oils, as compared with animal fats) are related to differences in three-
dimensional shape between the hydrocarbon chains of unsaturated and of
saturated fatty acid components. Shown in Figure 12.2 is the structural formula
of tripalmitin and a space filling model of this saturated triglyceride. Notice
that the three saturated hydrocarbon chains of tripalmitin lie parallel to each
other and that the molecule has an ordered, compact shape. Dispersion forces
between these hydrocarbon chains are strong. Because of this compact nature
and the interaction by dispersion forces, triglycerides rich in saturated fatty
acids have melting points above room temperature.

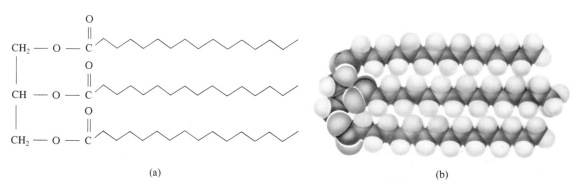

(a)   (b)

Figure 12.2 A saturated triglyceride, tripalmitin: (a) structural formula; (b) space-filling
model.

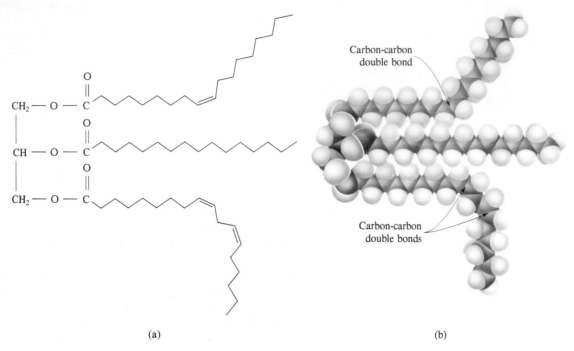

Figure 12.3 An unsaturated triglyceride: (a) structural formula; (b) space-filling model.

The three-dimensional shape of an unsaturated triglyceride is quite different from the shape of a saturated triglyceride. Figure 12.3 shows the structural formula of a triglyceride derived from one molecule each of palmitic acid, oleic acid, and linoleic acid. Notice the cis configuration about the double bonds in the hydrocarbon chains of oleic and linoleic acids. Also shown in Figure 12.3 is a space-filling model of this unsaturated triglyceride. An unsaturated triglyceride has a far less ordered structure than a saturated triglyceride, and because unsaturated triglycerides do not pack together so closely and compactly, dispersion forces between them are not so great as the dispersion forces between saturated triglycerides. Consequently, unsaturated triglycerides have lower melting points than the triglycerides that are more highly saturated.

Example 12.1

Following is the fatty acid composition by percentage of two triglycerides. Predict which triglyceride has the lower melting point.

	% Fatty Acid by Weight				
Triglyceride	**Palmitic**	**Stearic**	**Palmitoleic**	**Oleic**	**Linoleic**
A	24.0	8.4	5.0	46.9	10.2
B	9.8	2.4	0.4	28.9	50.7

☐ *Solution*

Triglyceride A is composed of approximately 32% saturated fatty acids and 62% unsaturated fatty acids. Triglyceride B is composed of 12% saturated fatty acids and 80% unsaturated fatty acids. Of the unsaturated fatty acids in B, more than 50% are linoleic acid, a fatty acid with two double bonds. Predict that triglyceride B has a lower melting point because of its higher degree of unsaturation. The fatty acid composition of triglyceride A is typical of human depot fat (mp 15°C). Triglyceride B is soybean oil (mp −16°C).

Problem 12.1

Why do beef tallow and corn oil, both composed of approximately 50% oleic acid, have such different melting points?

D. Rancidity

On exposure to air, most triglycerides develop an unpleasant odor and flavor, and are said to become rancid. In part, **rancidity** is the result of slight hydrolysis of the fat and oil, causing production of low-molecular-weight fatty acids. The odor of rancid butter is due largely to the presence of butanoic acid formed by the hydrolysis of butterfat. These same low-molecular-weight fatty acids can be formed by air oxidation of unsaturated fatty-acid side chains. The rate of rancidification varies with individual triglycerides, largely because of the presence of certain natural substances called antioxidants, which inhibit the process. One of the most common lipid antioxidants is vitamin E (Section 12.6C).

E. Reduction of Fats and Oils

For various reasons, partly convenience and partly dietary preference, conversion of oils to fats has become an important industry. The process is called **hardening** and involves reaction of an oil with hydrogen in the presence of a catalyst and **reduction** of some of the carbon-carbon double bonds of a triglyceride. If all double bonds are reduced (saturated with hydrogen), the resulting triglyceride is hard and brittle. In practice, the degree of hardening is carefully controlled to produce fat of a desired consistency. The resulting fats are sold for kitchen use (Crisco, Spry, and others). Oleomargarine and other butter substitutes are prepared by hydrogenation of cottonseed, soybean, corn, or peanut oils. The resulting product is often churned with milk and artificially colored to give it a flavor and a consistency resembling those of butter.

12.2 Soaps and Detergents

A. Structure and Preparation of Soaps

Soaps are potassium or sodium salts of fatty acids. Their preparation by boiling lard or other animal fat with potash is one of the most ancient organic reactions known. Potash (*pot* plus *ash*), so named because it is the solid residue obtained by extracting wood ashes with water and then evaporating the water in iron pots, is a mixture of potassium carbonate and potassium hydroxide. The reaction that takes place is **saponification** (hydrolysis in alkali) of the triglyceride, to give glycerol and potassium salts of fatty acids:

$$
\underset{\text{a triglyceride}}{
\begin{array}{c}
\text{O} \\
\parallel \\
\text{R}-\text{C}-\text{O}-\text{CH} \\
\end{array}
\begin{array}{c}
\text{O} \\
\parallel \\
\text{CH}_2-\text{O}-\text{C}-\text{R} \\
| \\
 \\
| \\
\text{CH}_2-\text{O}-\text{C}-\text{R} \\
\parallel \\
\text{O}
\end{array}}
\;+\;3\text{KOH}\;\xrightarrow[\text{(saponification)}]{\text{hydrolysis}}\;
\underset{\text{glycerol}}{
\begin{array}{c}
\text{CH}_2-\text{OH} \\
| \\
\text{CH}-\text{OH} \\
| \\
\text{CH}_2-\text{OH}
\end{array}}
\;+\;3\,\underset{\text{a potassium soap}}{\text{R}-\overset{\text{O}}{\overset{\parallel}{\text{C}}}-\text{O}^-\text{K}^-}
$$

In Europe, soap manufacture started in Marseilles in the Middle Ages but by no means was it a commonly available product. However, by the late 1700s, manufacture of soap was widespread throughout Europe and North America, and had become a big industry. As the soap as well as the glass and paper industries prospered, more and more wood had to be burned to provide the potash, and it seemed for a time that the forests of Europe might be threatened. Fortunately, through a new technology called the LeBlanc process, sodium carbonate (soda) became commercially available on a large scale and could in turn be used to produce sodium hydroxide. The change from potassium hydroxide to sodium hydroxide meant a change from potassium soaps to sodium soaps.

The most common triglycerides used today are from beef tallow (from meat-packing plants) and from coconut palm oil. After hydrolysis is complete, sodium chloride is added to precipitate the soap as thick curds. The water layer is then drawn off and glycerol is recovered by vacuum distillation. The crude soap contains sodium chloride, sodium hydroxide, and also other impurities. These are removed by boiling the curd in water and reprecipitating with more sodium chloride. After several such purifications, the soap can be used without further processing as an inexpensive industrial soap. Fillers such as pumice may be added to make a scouring soap. Other treatments transform the crude soap into pH-controlled cosmetic soaps, medicated soaps, and the like.

B. How Soaps Clean

Soap owes its remarkable cleansing properties to its ability to act as an emulsifying agent. Regarded from one end, the organic portion of a natural soap is a polar, negatively charged, hydrophilic carboxylate group that interacts with surrounding water molecules by ion-dipole interactions. Regarded from the other end, it is a long, nonpolar, hydrophobic hydrocarbon chain that does not interact at all with surrounding water molecules. Because the long hydrocarbon chains of natural soaps are insoluble in water, they tend to cluster in such a way as to minimize their contact with surrounding water molecules. The polar carboxylate groups, on the other hand, tend to remain in contact with the surrounding water molecules. The problem, then, is how to shield the hydrophobic hydrocarbon chains from contact with water but keep the hydrophilic carboxylate groups in contact with water. The solution is to cluster soap molecules into **micelles**. In soap micelles, the charged carboxylate groups form a negatively

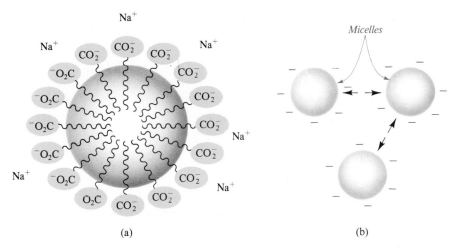

(a)   (b)

Figure 12.4 Soap micelles. (a) Diagram of a soap micelle showing nonpolar (hydrophobic) hydrocarbon chains clustered in the interior of the micelle and polar (hydrophilic) carboxylate groups spread on the surface of the micelle. (b) Soap micelles repel each other because of the negative charges on their surfaces.

charged surface and the nonpolar hydrocarbon chains lie buried within the center (Figure 12.4).

Most of the things we commonly think of as dirt, such as grease, oil, and fat stains are nonpolar and insoluble in water. When soap and this type of dirt are mixed together, as in a washing machine, the nonpolar hydrocarbon ends of soap micelles "dissolve" the nonpolar dirt molecules. In effect, new soap micelles are formed, this time with nonpolar dirt molecules in the center (Figure 12.5). In this way, nonpolar organic grease, oil, and dirt are dissolved and washed away in the polar wash water.

Soaps are not without their disadvantages. First, they are salts of weak acids, and in the presence of mineral acids, they are converted to free fatty acids. Whereas soaps are soluble in water as micelles, the free fatty acids are insoluble and form a scum. For this reason, soaps cannot be used in acidic solution.

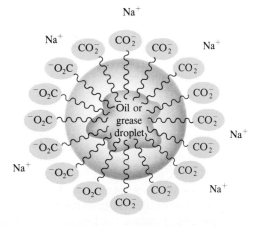

Figure 12.5 Soap micelles with a "dissolved" oil droplet.

$$CH_3(CH_2)_{16}CO_2^- \ Na^+ + HCl \longrightarrow CH_3(CH_2)_{16}CO_2H + NaCl$$

soluble in water   insoluble in
as micelles   water

Second, soaps form insoluble salts when used in water containing calcium, magnesium, or iron ions (hard water):

$$2CH_3(CH_2)_{16}CO_2^- \ Na^+ + Ca^{2+} \longrightarrow [CH_3(CH_2)_{16}CO_2^-]_2Ca^{2+} + 2Na^+$$

soluble in water   a water-insoluble
as micelles   calcium salt

These Ca(II), Mg(II), and Fe(III) fatty acid salts create problems, including rings around the bathtub, films that spoil the luster of hair, and grayness and roughness that build up on textiles after repeated washing.

C. Synthetic Detergents

Once the cleansing action of soaps was understood, the design criteria for a **synthetic detergent** could be established: a molecule with a long hydrocarbon chain, preferably 12 to 20 carbon atoms, and a polar group at one end of the molecule that does not form insoluble salts with Ca(II), Mg(II), or other ions present in hard water. Chemists recognized that the essential characteristics of a soap could be produced in a molecule containing a sulfate group instead of a carboxylate group. Calcium, magnesium, and iron salts of alkyl sulfate esters are much more soluble in water than comparable salts of fatty acids are.

The first synthetic detergent was made from 1-dodecanol (lauryl alcohol) by combining it with sulfuric acid to form a sulfate ester, followed by neutralization with sodium hydroxide to form sodium dodecyl sulfate (SDS):

$$CH_3(CH_2)_{10}CH_2OH + HO\overset{\overset{O}{\|}}{\underset{\underset{O}{\|}}{S}}OH \longrightarrow CH_3(CH_2)_{10}CH_2O\overset{\overset{O}{\|}}{\underset{\underset{O}{\|}}{S}}OH + H_2O$$

1-dodecanol   1-dodecyl hydrogen sulfate
(lauryl alcohol)   (lauryl hydrogen sulfate)

$$CH_3(CH_2)_{10}CH_2O\overset{\overset{O}{\|}}{\underset{\underset{O}{\|}}{S}}OH + NaOH \longrightarrow CH_3(CH_2)_{10}CH_2O\overset{\overset{O}{\|}}{\underset{\underset{O}{\|}}{S}}O^- \ Na^+ + H_2O$$

sodium dodecyl sulfate (SDS)
(sodium lauryl sulfate)

The physical resemblance between this synthetic detergent and natural soaps is obvious—a long nonpolar (hydrophobic) hydrocarbon chain and a highly polar (hydrophilic) end group. Large-scale commercial production of SDS was not possible because bulk quantities of 1-dodecanol were lacking. However, because of high foaming properties, SDS is used in numerous specialized applications, including many shampoos and cosmetics. It is also used in biochemistry to denature proteins and to disrupt biological membranes.

Currently, the most important synthetic detergent is dodecylbenzene sulfonate. Its preparation starts with dodecene, made by polymerizing four molecules of propylene (a byproduct of the petroleum refining industry). Next, benzene and the dodecene are reacted to form dodecylbenzene, and this is followed by sulfonation with sulfuric acid. The sulfonic acid is neutralized with NaOH and the product mixed with builders and spray-dried to give a smooth-flowing powder. The most common builders are sodium tripolyphosphate and sodium silicate.

$$4CH_3CH{=}CH_2 \xrightarrow{H_3PO_4} C_{12}H_{24} \xrightarrow[AlCl_3]{benzene} CH_3(CH_2)_{10}CH_2{-}\xrightarrow{H_2SO_4}$$

propylene   propylene   dodecylbenzene
                    tetramer

$$CH_3(CH_2)_{10}CH_2{-}{-}\overset{\displaystyle O}{\underset{\displaystyle O}{S}}OH \xrightarrow{NaOH} CH_3(CH_2)_{10}CH_2{-}{-}\overset{\displaystyle O}{\underset{\displaystyle O}{S}}O^-\,Na^+$$

dodecylbenzene sulfonic acid   sodium dodecylbenzene sulfonate

Alkylbenzene sulfonate detergents were introduced in the 1950s, and today they command close to 90% of the market once held by natural soaps.

Among common additives to most detergent preparations are foam stabilizers, bleaches, and optical brighteners. A common foam stabilizer added to liquid soaps but not laundry detergents (for obvious reasons—think of a top-loading washing machine with foam spewing up over the lid!) is the amide prepared from dodecanoic acid (lauric acid) and 2-aminoethanol (ethanolamine). The most common bleach is sodium perborate tetrahydrate, which decomposes at temperatures above 50°C to give hydrogen peroxide, the actual bleaching agent.

$$CH_3(CH_2)_{10}\overset{\displaystyle O}{\overset{\displaystyle \|}{C}}NHCH_2CH_2OH \qquad NaBO_3{\cdot}4H_2O$$

N-hydroxyethyldodecanamide   sodium perborate
(a foam stabilizer)   (a bleach)

Also added to laundry detergents are optical brighteners, also called optical bleaches, that are absorbed onto fabrics and fluoresce with a blue color, offsetting any yellowing due to aging of the fabric. Quite literally, these optical brighteners produce a "whiter-than-white" appearance. You most certainly have observed the effects of optical brighteners if you have ever been in the presence of black lights and seen the glow of "white" T-shirts, blouses, and the like.

12.3 Phospholipids

Phospholipids are the second most abundant kind of natural lipids. They are found almost exclusively in plant and animal membranes, which typically consist of about 40–50% phospholipids and 50–60% protein.

The most abundant phospholipids contain glycerol and fatty acids, as the simple fats do. They also contain phosphoric acid and a low-molecular-weight alcohol. The most common of these low-molecular-weight alcohols are choline, ethanolamine, serine, and inositol. In the following formulas, the —OH involved in formation of the phosphate ester is shown in color. Further, each molecule is shown as it would be ionized at pH 7.4.

$$\text{HOCH}_2\text{CH}_2\overset{+}{\text{N}}(\text{CH}_3)_3 \qquad \text{HOCH}_2\text{CH}_2\text{NH}_2 \qquad \text{HOCH}_2\text{CHCO}_2^- \;(\text{NH}_3^+) $$

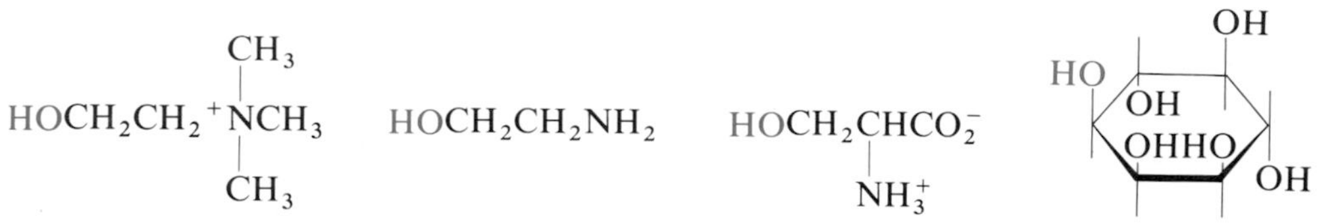

choline   2-aminoethanol   serine   inositol
               (ethanolamine)

The most abundant phospholipids in higher plants and animals are the **lecithins** and the **cephalins**.

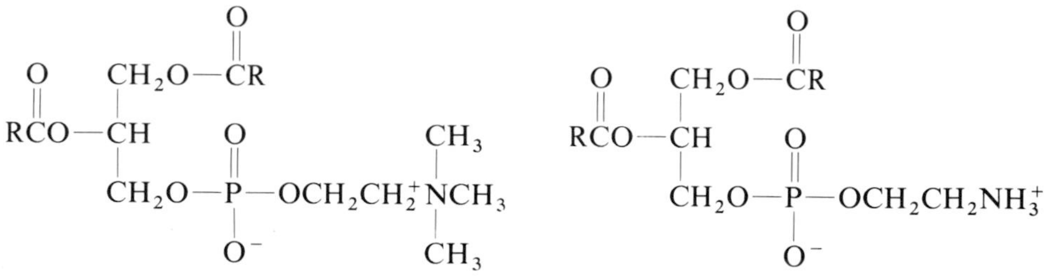

a lecithin
(a phospholipid
containing choline)

a cephalin
(a phospholipid
containing ethanolamine)

Lecithins are phosphate esters of choline and cephalins are phosphate esters of ethanolamine. Lecithin and cephalin are shown as they would be ionized at pH 7.4. At this pH, each of these molecules has a net charge of zero. The fatty acids most common in these membrane phospholipids are palmitic and stearic acids (both fully saturated) and oleic acid (one double bond in the hydrocarbon chain).

12.4 Biological Membranes

A. Composition

Membranes are an important feature of cell structure and are vital for all living organisms. Some of the most important functions of membranes can be illustrated by considering the **cell membrane**.

1. Cell membranes are mechanical barriers that separate the contents of cells from their environment.
2. Cell membranes control the passage of molecules and ions into and out of cells. For example, essential nutrients are transported into cells and metabolic wastes out of cells through their membranes. Cell membranes also help to regulate the concentrations of molecules and ions within cells.
3. Cell membranes provide structural support for certain proteins. Some of these proteins are "receptors" for hormone-carried "messages"; others are specific enzyme complexes.

The cell membrane is but one of the membranes possessed by most cells. Several other types of membranes found within a typical cell are shown in Figure 12.6.

The subcellular membranes shown in Figure 12.6 have functions similar to those of the cell membrane itself. For example, the nuclear membrane separates the nucleus from the rest of the cell. The inner membranes of the mitochondria contain the enzymes that catalyze the reactions of the final states of respiration (Chapter 16). The endoplasmic reticulum contains enzymes that carry out numerous synthetic reactions. The rough endoplasmic reticulum supports ribosomes and the enzymes that catalyze the synthesis of proteins from amino acids. The smooth endoplasmic reticulum contains hydroxylation enzymes, steroid synthesis enzymes, and enzymes for drug metabolism. Lysosomes contain enzymes that digest substances brought into the cell. Membranes are more than impervious, mechanical barriers separating the cell and its organelles from the environment. They are highly specialized structures that perform many tasks with great precision and accuracy.

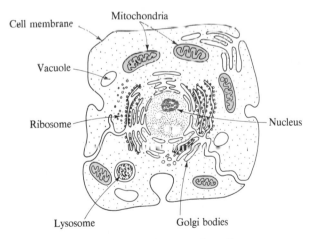

Figure 12.6 Diagram of a cell.

B. Structure

The determination of the detailed molecular structure of membranes is one of the most challenging problems in biochemistry today. Despite intensive research, many aspects of membrane structure and activity still are not understood. Before we discuss a model for membrane structure, let us first consider the shapes of phospholipid molecules and the organization of phospholipid molecules in aqueous solution.

Shown in Figure 12.7 is a structural formula and space-filling model of a lecithin, an important membrane phospholipid. Lecithin and other phospholipids are elongated, almost rodlike, molecules, with the nonpolar (hydrophobic) hydrocarbon chains lying essentially parallel to one another and with the polar (hydrophilic) phosphate ester group pointing in the opposite direction.

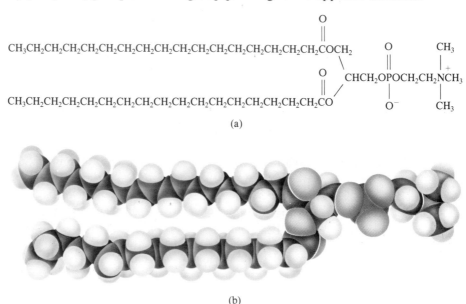

(a)

(b)

Figure 12.7 A lecithin: (a) structural formula; (b) space-filling model.

To understand what happens when phospholipid molecules are placed in aqueous medium, recall from Section 12.2B that soap molecules placed in water form micelles in which polar head groups interact with water molecules and nonpolar hydrocarbon tails cluster within the micelle and are removed from contact with water. One possible arrangement for phospholipids in water also is micelle formation (Figure 12.8).

Figure 12.8 Proposed micelle formation of phospholipids in an aqueous medium.

Another arrangement that satisfies the requirement that polar groups interact with water and nonpolar groups cluster together to exclude water is a **lipid bilayer**. A schematic diagram of a lipid bilayer is shown in Figure 12.9. The

Figure 12.9 A section of lipid bilayer (lower part). Enlarged (upper part) is a section of six phospholipid molecules in the bilayer. Note in the enlargement that 50% of the hydrocarbon chains are unsaturated.

favored structure for phospholipids in aqueous solution is a lipid bilayer instead of a micelle, because micelles can grow only to a limited size before holes begin to appear in the outer polar surface. Lipid bilayers can grow to an almost infinite extent and provide a boundary surface for a cell or organelle, whatever its size.

It is important to realize that self-assembly of phospholipid molecules into a bilayer is a spontaneous process driven by two types of noncovalent forces: (1) hydrophobic interactions that result when nonpolar hydrocarbon chains cluster together and exclude water molecules, and (2) electrostatic interactions and hydrogen bonding that result when polar head groups interact with water molecules.

Lipid bilayers are highly impermeable to ions and most polar molecules because of their structural characteristics and because it takes a great deal of energy to transport an ion or a polar molecule through the nonpolar interior of the bilayer. Water, however, passes readily in and out of a lipid bilayer. Glucose passes through lipid bilayers 10^4 times more slowly than water, and sodium ion 10^9 times more slowly than water.

The most satisfactory current model for the arrangement of proteins and phospholipids in plant and animal membranes is the **fluid-mosaic model**. According to this model, membrane phospholipids form a lipid bilayer and membrane proteins are embedded in this bilayer. Some proteins are exposed to the aqueous environment on the outer surface of the membrane; others provide channels that penetrate the membrane from the outer to the inner surface; and still others are embedded within the lipid bilayer. Four possible protein arrangements are shown schematically in Figure 12.10.

The fluid-mosaic model is consistent with the evidence provided by chemical analysis and electron-microscope pictures of cell membranes. However, this model does not explain just how membrane proteins act as pumps and gates for the transport of ions and molecules across the membrane, or how they act as receptors for hormone-borne messages and communications between one cell and another. Nor does it explain how enzymes bound on membrane surfaces catalyze reactions. All these questions are active areas of research today.

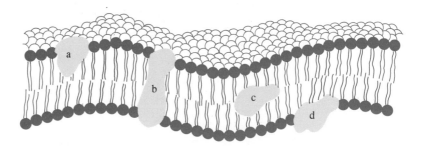

Figure 12.10 Fluid-mosaic model of a biological membrane, showing the lipid bilayer and membrane proteins oriented (a) on the outer surface of the membrane, (b) penetrating the entire thickness of the membrane, (c) embedded within the membrane, and (d) on the inner surface of the membrane.

12.5 Steroids

Steroids are a group of lipids that contain as a characteristic structural feature four fused carbon rings: 3 six-member rings designated A, B, and C; and 1 five-member ring designated D. Further, the parent compound from which all steroids are derived possesses methyl groups at the A/B and C/D ring junctions and a side chain of eight carbon atoms attached to ring D (Figure 12.11).

The steroid nucleus is found in many extremely important biomolecules, all of which are derived from cholesterol. In a sense, then, cholesterol can be called the parent steroid.

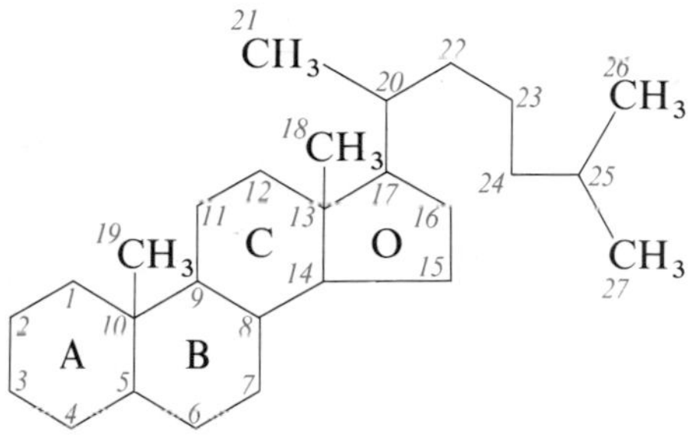

Figure 12.11 Letter designations and numbering system for steroids.

A. Cholesterol

Cholesterol is a white, water-insoluble compound found in varying amounts in practically all living organisms except bacteria. It is synthesized in the liver, intestine, and other tissues. Of these, synthesis in the liver is quantitatively the most important. Cholesterol contains multiple centers of chirality and possibilities for cis-trans isomerism about the five- and six-membered rings. The isomer that nature produces and the one that is involved in animal metabolism has somewhat the character of a staircase with risers and treads. It can be described as roughly planar with both angular methyl groups above the plane of the molecule.

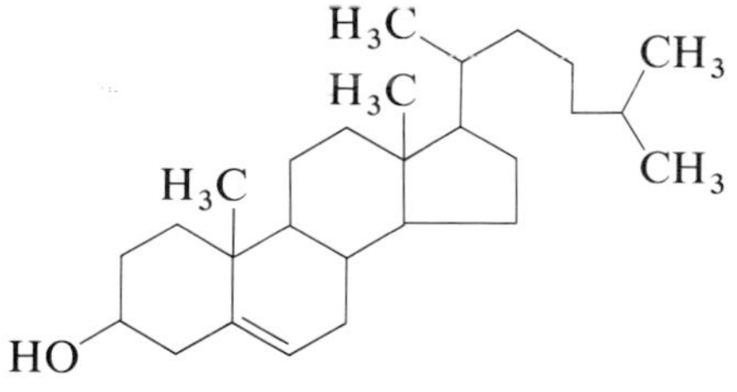

cholesterol

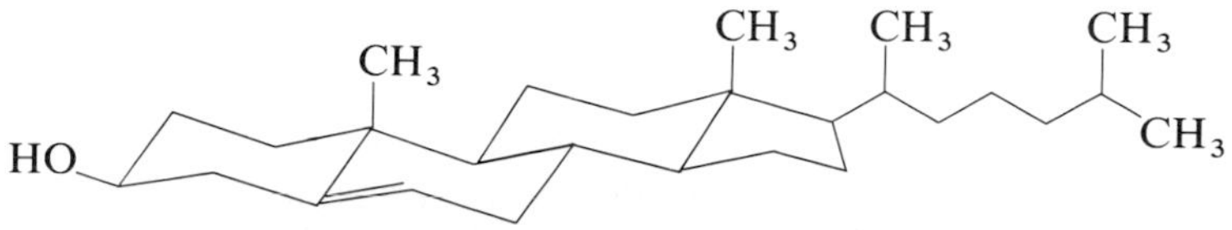

cholesterol is roughly planar with both angular
methyl groups above the plane of the molecule

Cholesterol is a vital part of human metabolism because it is:

1. An essential component of biological membranes. The total body content of cholesterol in a healthy adult is about 140 g, of which 120 g is present in biological membranes. In humans, for example, membranes of the central and peripheral nervous systems contain about 10% cholesterol by dry weight.
2. The precursor for the biosynthesis of bile acids, steroid hormones, and vitamin D.

Because measuring the concentration of cholesterol in serum is easy, much information has been collected in attempts to correlate serum cholesterol levels with various diseases. Human blood plasma contains an average of 50 mg of free cholesterol per 100 mL and about 200 mg of cholesterol esterified with fatty acids. One of the diseases associated with increased concentrations of cholesterol is arteriosclerosis, or hardening of the arteries, among the most common of the diseases of aging. With increasing age, humans normally develop decreased capacity to metabolize fat, and therefore, cholesterol concentration in membranes increases. Because of the roughly planar shape and hydrophobic character of cholesterol, it is able to fit between the hydrophobic fatty acid chains of phospholipids, with its polar —OH group on the surface of the membrane in contact with polar parts of the phospholipids and with the aqueous environment. Because cholesterol has a very ordered and rigid structure, its increased concentration in membranes decreases their fluidity and makes them more susceptible to stress and rupture. This is particularly evident in membranes of the coronary artery. The correlation between elevated blood cholesterol levels and coronary artery disease is due, at least partly, to increased incorporation of cholesterol in coronary membranes and their resulting increased fragility.

Arteriosclerosis accompanied by a buildup of cholesterol and other lipids on the inner surfaces of arteries is known as **atherosclerosis**; the condition causes a decrease in the diameter of the channels through which blood must flow. This decreased diameter, together with increased turbulence, leads to a greater probability of clot formation. If a blood vessel is blocked by a clot, cells may be deprived of oxygen and die. Death of tissue in this way is called infarction.

Infarction can occur in any tissue, and the clinical symptoms depend on which vessels and tissues are involved. Myocardial infarction involves the myocardium, or heart muscle tissue. (For a discussion of an important laboratory method for diagnosis of myocardial infarction, see Mini-Essay X, "Clinical Enzymology—The Search for Specificity.")

B. Adrenocorticoid Hormones

The cortex of the adrenal gland synthesizes two main classes of hormones, called **adrenocorticoid hormones**: (1) mineralocorticoid hormones, which affect water and electrolyte balance, and (2) glucocorticoid hormones, which affect carbohydrate and protein metabolism.

Aldosterone, the most effective mineralocorticoid hormone secreted by the adrenal cortex, acts on kidney tubules to stimulate resorption of sodium ions, thus regulating water and electrolyte metabolism. An adult on a diet with normal sodium content produces about 0.1 mg of aldosterone per day.

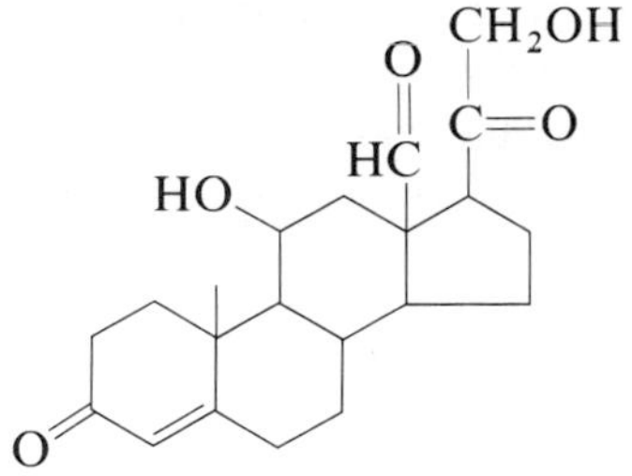

aldosterone
(a mineralocorticoid hormone
secreted by the adrenal cortex)

Cortisol is the principal glucocorticoid hormone of the adrenal cortex, which secretes about 25 mg of this substance per day. Cortisol affects the metabolism of carbohydrates, proteins, and fats; water and electrolyte balance; and inflammatory processes within the body. In the presence of cortisol, the synthesis of protein in muscle tissue is depressed, protein degradation is increased, and the supply of free amino acids is increased in both muscle cells and blood plasma. The liver, in turn, is stimulated to use the carbon skeletons of certain amino acids for the synthesis of glucose and glycogen. Thus, cortisol and other glucocorticoid hormones act to increase the supply of glucose and liver glycogen at the expense of body protein. Cortisol also has some mineralocorticoid action; it promotes resorption of sodium ions and water retention by the tubules of the kidney. However, it is far less potent as a mineralocorticoid than aldosterone is.

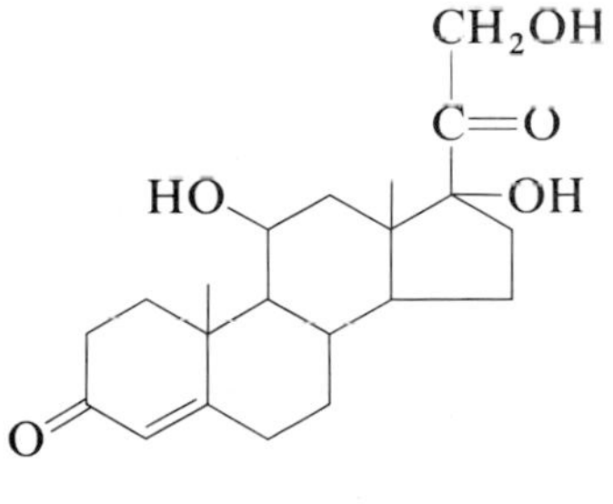

cortisol

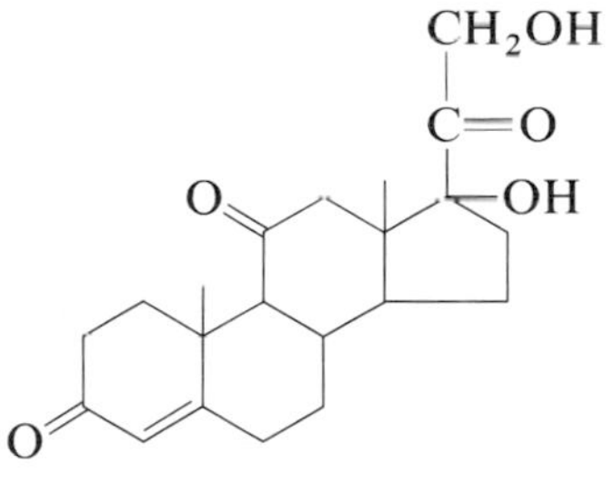

cortisone

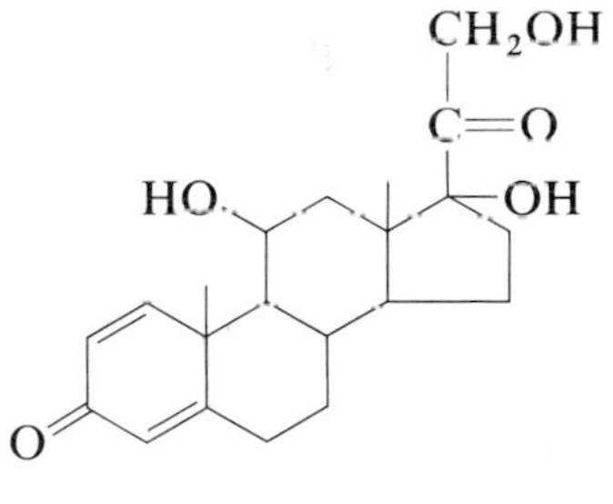

prednisolone

Cortisol and its oxidation product, cortisone, are probably best known for their use in clinical medicine as remarkably effective anti-inflammatory agents. They are used in the treatment of a host of inflammatory diseases, including acute attacks of rheumatoid arthritis and bronchial asthma, and inflammation of the eye, colon, and other organs. Laboratory research has produced a series of semisynthetic steroid hormones, including prednisolone, that are even more potent than cortisone in treating inflammatory diseases. Many of these semisynthetic hormones have the additional advantage over cortisone in not stimulating sodium retention and fluid accumulation.

C. Sex Hormones

Of the male sex hormones, or androgens, testosterone is the most important. It is produced in the testes from cholesterol. The chief function of testosterone is to promote normal growth of the male reproductive organs and development of the characteristic deep voice, pattern of facial and body hair, and musculature.

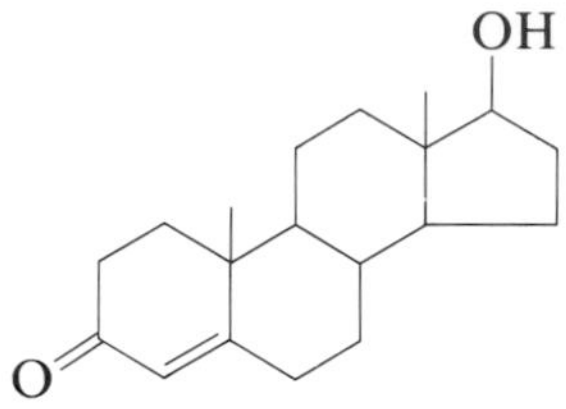

testosterone

The use of anabolic steroids among certain types of athletes to build muscle mass and strength, particularly for explosive sports, is common knowledge. Testosterone itself produces these effects but it is not active when taken orally because it is metabolized to an inactive steroid in the liver. Two laboratory modifications of testosterone have produced a synthetic steroid that does have these effects and at the same time has little of the virilizing effects of testosterone. Dianabol is synthesized by introduction of a methyl group on carbon 20 of ring D and a second carbon-carbon double bond in ring A.

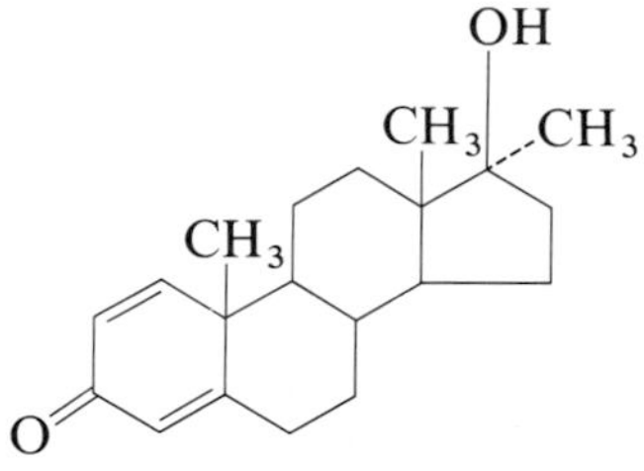

dianabol
(a synthetic anabolic steroid used to
increase muscle mass and strength)

It is thought that many male and female athletes take as much as 100 mg per day of dianabol while in training. Like any other drug, dianabol has its side effects, among which are increased sterility, impotence, and the risk of diabetes, coronary heart disease, and cancer. Increased awareness of these side effects coupled with routine urinalyses of athletes at major competitions, it is hoped, will help to prevent abuse of dianabol and other anabolic steroids.

Two types of female sex hormones are particularly important—progesterone, and a group of hormones known as estrogens. Changing rates of secretion of these hormones cause the periodic change in the ovaries and uterus known as the menstrual cycle. Immediately following menstrual flow, increased estrogen secretion causes growth of the uterus lining and ripening of ova. Estradiol is one of the most important of the estrogens, which are responsible for development of the female secondary sex characteristics.

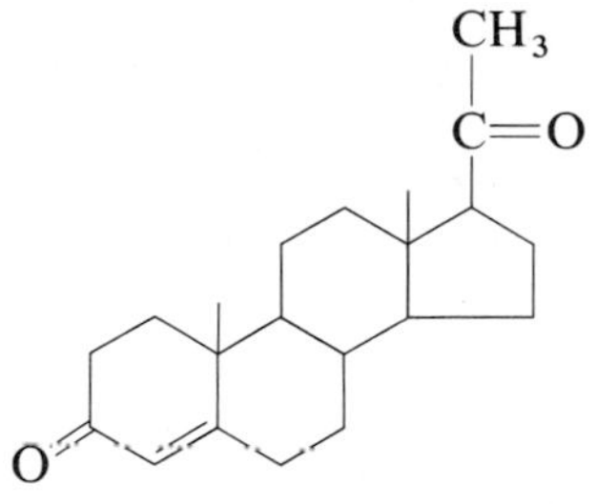

progesterone 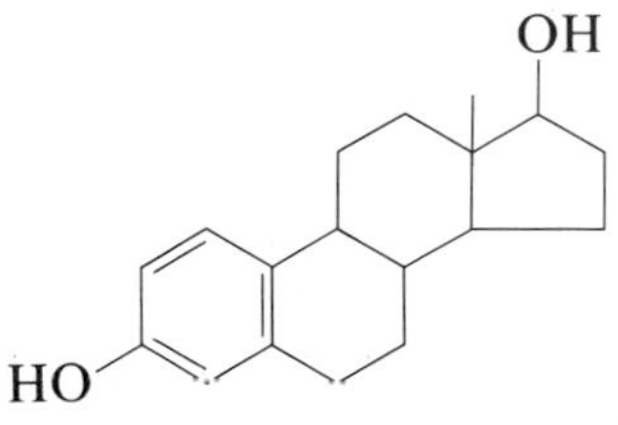   estradiol

Progesterone is synthesized in the ovaries from cholesterol. Its secretion just before ovulation keeps ova from ripening and also prepares the uterus for implantation and maintenance of a fertilized egg. If conception does not occur, progesterone production decreases and menstruation occurs. If fertilization and implantation do occur, production of progesterone continues, helping to maintain the pregnancy. One of the consequences of continued progesterone production is prevention of ovulation during pregnancy

Once the role of progesterone in inhibiting ovulation was understood, its potential as a possible contraceptive drug was realized. Unfortunately, progesterone itself is relatively ineffective when taken orally, and injection often produces local irritation. As a result of massive research programs, many synthetic steroids that could be administered orally became available in the early 1960s. When taken regularly, these drugs prevent ovulation, yet allow most women a normal menstrual cycle. Some of the most effective contain a progesteronelike analog, such as ethynodiol diacetate, combined with a smaller amount of an estrogenlike material. The small amount of estrogen prevents irregular menstrual flow ("breakthrough bleeding") during prolonged use of contraceptive pills.

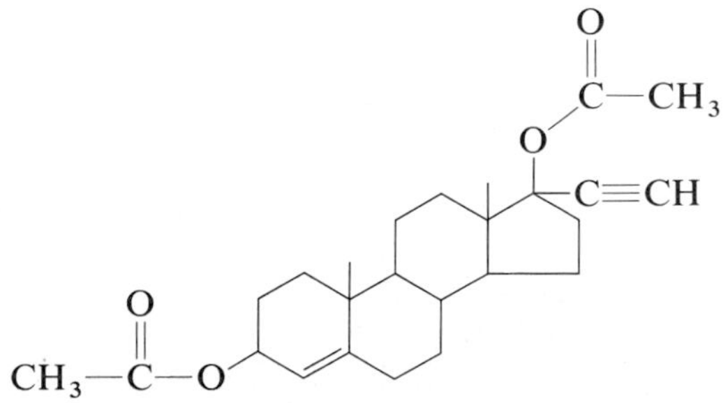

ethynodiol diacetate
(progesterone analog widely
used in oral contraceptives)

D. Bile Acids

Bile acids are synthesized in the liver from cholesterol and then stored in the gallbladder. During digestion, the gallbladder contracts and supplies bile to the small intestine by way of the bile duct. The primary bile acid in humans is cholic acid:

cholic acid
(an important constituent of human bile)

Bile acids have several important functions. First, they are products of the breakdown of cholesterol and thus are a principal pathway for the elimination of cholesterol from the body via the feces. Second, because bile acids can emulsify fats in the intestine, they aid in the digestion and absorption of dietary fats. Third, they can dissolve cholesterol by the formation of cholesterol–bile salt micelles or cholesterol–lecithin–bile salt micelles. In this way cholesterol, whether it is from the diet, synthesized in the liver, or removed from circulation by the liver, can be made soluble.

12.6 Fat-Soluble Vitamins

Vitamins are organic molecules required in trace amounts for normal metabolism but not able to be synthesized either at all or in amounts adequate for healthy growth. Therefore, because they cannot be synthesized, they must be supplied in the diet. This implies that they must be synthesized somewhere in the biological world, which of course they are. The main sources of vitamins for humans and all other animals are plants and microorganisms. We will concentrate on those vitamins that humans need. With one exception, all vitamins required by humans are also required by other mammals. The exception is vitamin C, which is needed only by primates and guinea pigs.

Vitamins are divided into two classes, depending on solubility characteristics; those that are fat-soluble and those that are water-soluble. Most of the twelve recognized water-soluble vitamins are components of coenzymes whose structure and functions we shall examine subsequently (Section 14.2C, and the following chapters on metabolism). Names and sources of the four **fat-soluble vitamins** and the serious clinical conditions that result from a deficiency of each are summarized in Table 12.3.

Table 12.3
Fat-soluble
vitamins required
by humans and
serious clinical
symptoms that
result from their
deficiency.

Vitamin	Source	Clinical Conditions Resulting from Deficiency
A	egg yolk, green and yellow vegetables, fruits, liver, and dairy products	night blindness, keratinization of mucous membranes
D	dairy products, action of sunlight on skin	rickets
E	green leafy vegetables	fragile red blood cells
K	leafy vegetables, intestinal bacteria	failure of blood clotting

A. Vitamin A

Vitamin A, or retinol, is a primary alcohol of molecular formula $C_{20}H_{30}O$. Vitamin A alcohol occurs only in the animal world, where the best sources are cod-liver oil and other fish-liver oils, animal livers, and dairy products. Vitamin A in the form of a precursor, or provitamin, is found in the plant world in pigments called carotenes. The most common of these, β-carotene ($C_{40}H_{56}$), has an orange-red color and is used as a food coloring. The carotenes have no vitamin A activity but are cleaved by enzymes in the intestine at the central carbon-carbon double bond, to give two molecules of retinol, which is then stored in the liver.

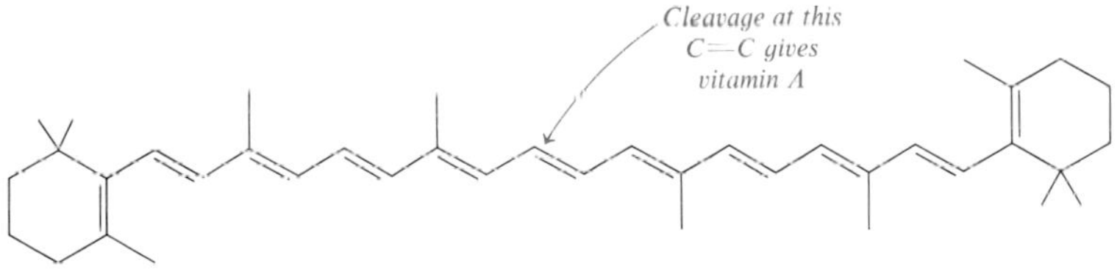

Cleavage at this C=C gives vitamin A

β-carotene

Enzyme-catalyzed cleavage at the middle—CH=CH—in the intestine

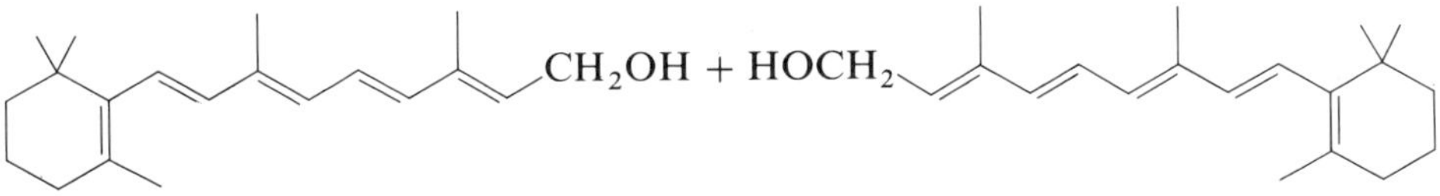

CH_2OH + $HOCH_2$

all *trans*-retinol
(vitamin A)

Probably the most understood role of Vitamin A is its participation in the **visual cycle** in rod cells. The eye contains two types of photoreceptors: rod cells, the agents for black and white vision in dim light (night vision); and cone cells, agents for color vision in brighter light. Each rod cell contains several million molecules of a protein called opsin. When the primary amine on the side chain of a lysine (one of the 20 amino acid components of proteins) in opsin reacts with the aldehyde group of 11-cis-retinal, an imine (Schiff base) called rhodopsin is formed. Neither opsin nor 11-cis-retinal absorbs visible light, but rhodopsin absorbs very strongly in the visible region of the spectrum.

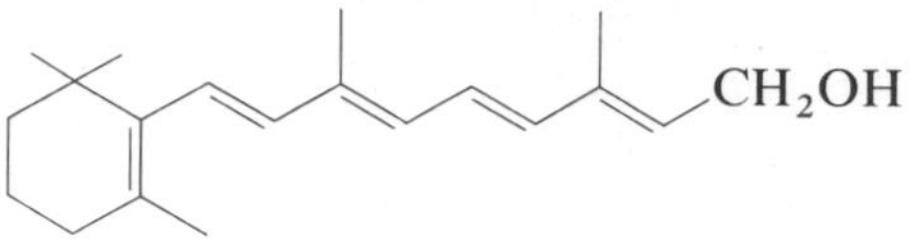

all-*trans*-retinol

oxidation of the primary
alcohol to an aldehyde

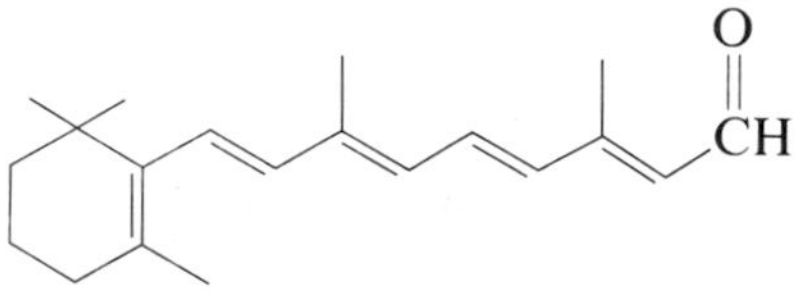

all-*trans* retinal

isomerization of the 11-trans double
bond to the cis configuration

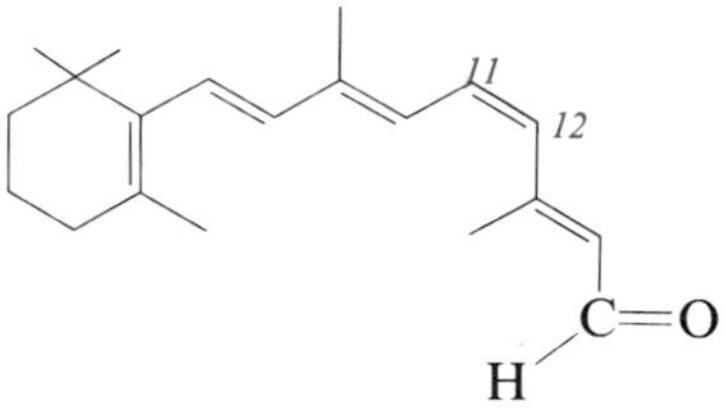

11-*cis* retinal

formation of an imine (a Schiff
base) with the protein opsin

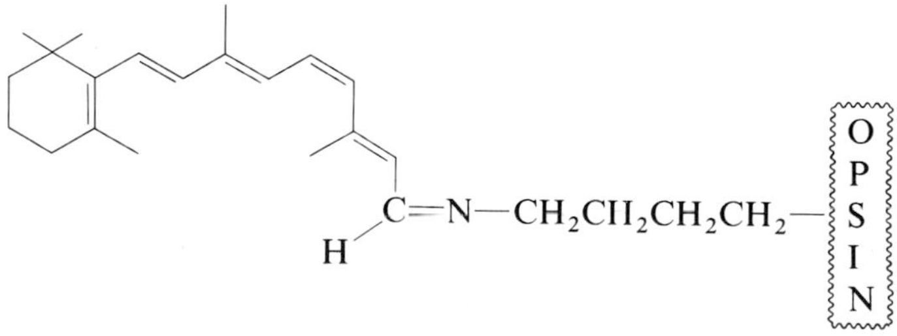

rhodopsin

The primary event in dark, or night, vision is interaction of a photon of light with a molecule of rhodopsin (Figure 12.12). Absorption of light causes photo-isomerization of 11-*cis*-retinal to all-*trans*-retinal. This change in shape in the retinal portion of rhodopsin brings about a change in conformation of opsin, which in some way not yet understood, generates an impulse in the optic nerve and sends a signal to the visual cortex in the brain. This entire process takes place in about one-millionth of a second and is accompanied by hydrolysis of rhodopsin to free all-*trans*-retinal and opsin. The visual cycle is completed by the following sequence of enzyme-catalyzed reactions: reduction of all-*trans*-retinal to all-*trans*-retinol, isomerization to 11-*cis*-retinol, oxidation to 11-*cis*-retinal, and reaction with opsin to regenerate rhodopsin.

Although the first symptom of vitamin A deficiency is night blindness, other clinical conditions develop too, indicating that vitamin A has other roles in the body besides photoreception. Probably its most important action is on epithelial cells, particularly those of the mucous membranes of the eye, respiratory tract,

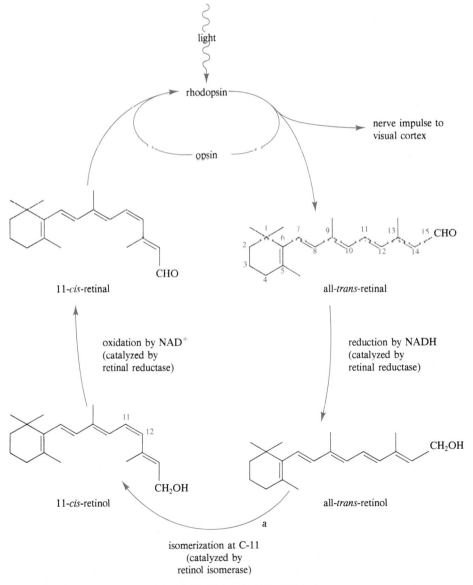

Figure 12.12 The visual cycle in rod cells. Similar cycles occur in cone cells.

and genitourinary tract. Without adequate supplies of vitamin A, these mucous membranes become hard and dry, a process known as keratinization. If untreated, keratinization of the cornea can lead to blindness. The mucous membranes of the respiratory, digestive, and urinary tracts also become keratinized in vitamin A deficiency, and become susceptible to infection.

B. Vitamin D

The term vitamin D is a generic name for a group of structurally related compounds produced by the action of ultraviolet light on certain provitamins. Vitamin D_3 (cholecalciferol) is produced in the skin of mammals by the action of sunlight on 7-dehydrocholesterol. When the skin has normal exposure to sunlight, enough 7-dehydrocholesterol is converted to vitamin D_3 so no dietary vitamin D is necessary. Only when the skin does not manufacture enough vitamin D_3 is there a need to supplement the diet with artificially fortified foods or multivitamins.

Vitamin D_3 has little or no biological activity, but must be metabolically activated before it can function in its target tissues. In the liver, vitamin D_3 is oxidized at carbon 25 of the side chain to form 25-hydroxyvitamin D_3. Although 25-hydroxyvitamin D_3 is the most abundant form in the circulatory system, it has only modest biological activity and undergoes further oxidation in the kidneys to form 1,25-dihydroxyvitamin D_3, the active form of the vitamin.

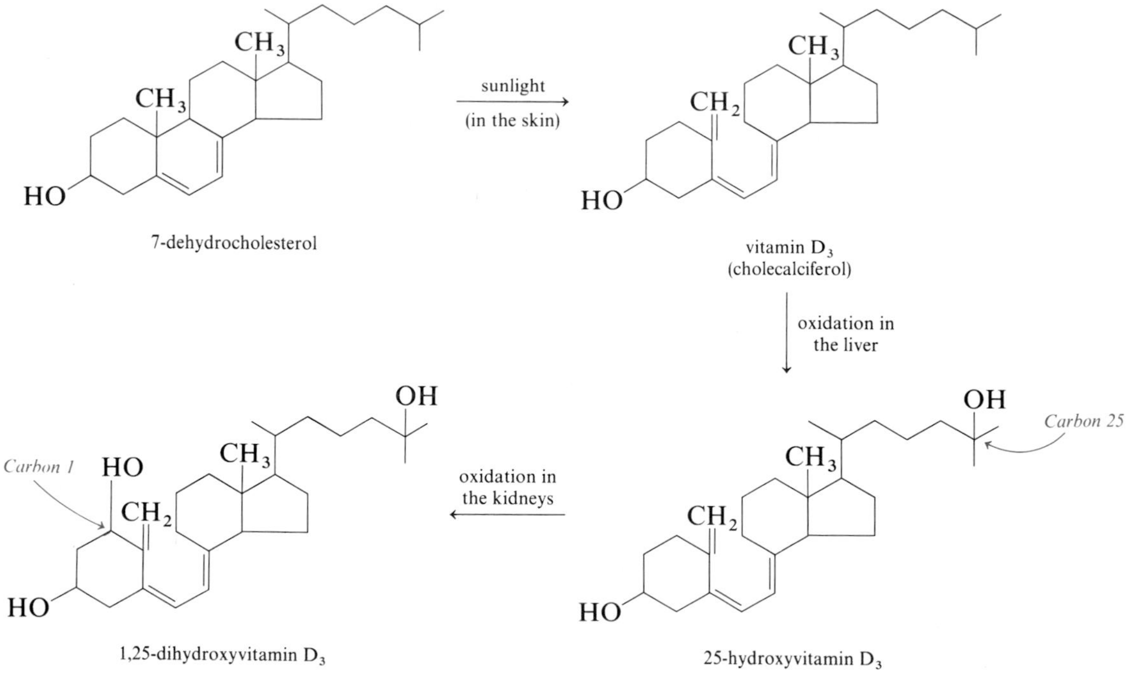

7-dehydrocholesterol

vitamin D_3
(cholecalciferol)

1,25-dihydroxyvitamin D_3

25-hydroxyvitamin D_3

The principal function of vitamin D metabolites is to regulate calcium metabolism. 1,25-Dihydroxyvitamin D_3 acts in the small intestine to facilitate absorption of calcium and phosphate ions; it acts in the kidneys to stimulate reabsorption of filtered calcium ions; and it acts in bone to stimulate demineralization and release calcium and phosphate ions into the bloodstream. A deficiency of vitamin D in childhood is associated with rickets, a mineral-metabolism disease that leads to bowlegs, knock-knees, and enlarged joints.

C. Vitamin E

Vitamin E is a group of eight structurally related compounds called tocopherols. Of these, α-tocopherol has the greatest potency. Vitamin E occurs in fish oil, in other oils such as cottonseed and peanut oil, and in green, leafy vegetables. The richest source of vitamin E is wheat germ oil.

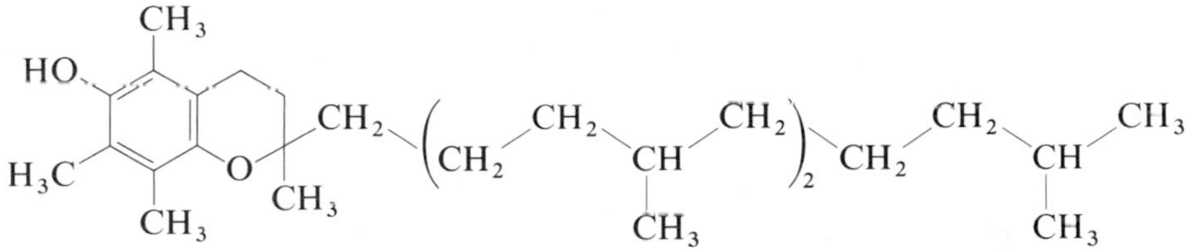

vitamin E (α-tocopherol)

Deficiency of vitamin E in laboratory animals causes infertility, hence the derivation of the name of this vitamin from the Greek, *tocopherol*, promoter of child birth. Deficiency in humans leads to premature destruction of erythrocytes and anemia. However, because vitamin E is present in green, leafy vegetables, rice, and so on, deficiency in humans is rare. Vitamin E is also an antioxidant in that it inhibits the oxidation of vitamin A, unsaturated fatty acids, phospholipids, and other unsaturated compounds by molecular oxygen. There is speculation that increased dosages of vitamin E may decrease the rate of aging, possibly by decreasing the rate of oxidation of susceptible biomolecules by molecular oxygen.

D. Vitamin K

Vitamin K exists in two principal forms, each consisting of two fused rings and a long, branched side chain of four or five isoprene units. The K_1 vitamins are synthesized in plants, particularly green, leafy vegetables, and they have a double bond only in the isoprene unit closest to the aromatic rings. The K_2 vitamins are synthesized by microorganisms in the large intestine and have double bonds in each of the isoprene units. Because of the combination of dietary intake and microbiological synthesis, deficiency of vitamin K in humans is rare.

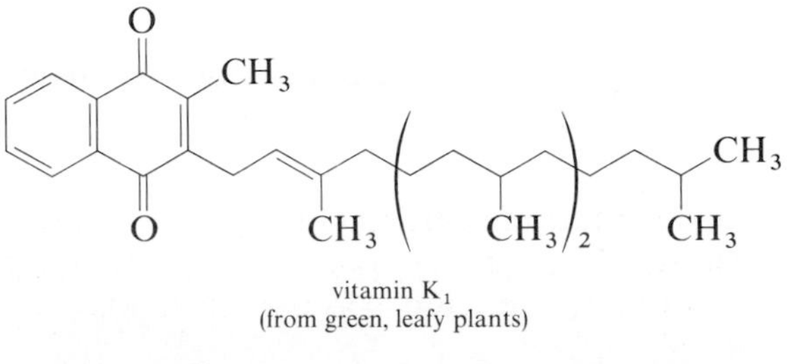

vitamin K_1
(from green, leafy plants)

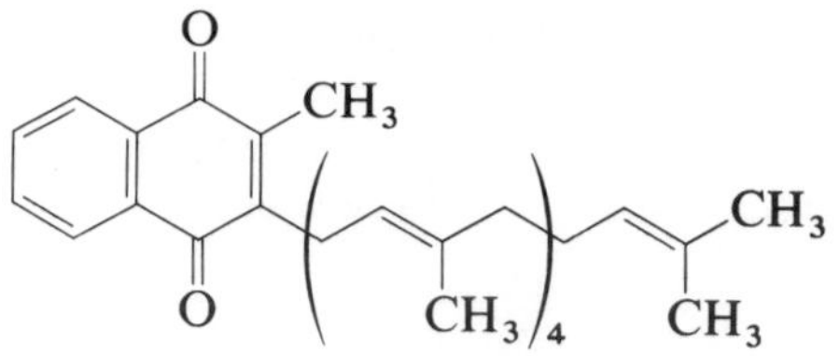

vitamin K$_2$
(from microorganisms in the large intestine)

The only known role of vitamin K is in blood-clotting; a deficiency of vitamin K leads to a slowing of clot formation.

12.7 Waxes

Waxes are esters of fatty acids and alcohols, each having from 16 to 34 carbon atoms. Carnauba wax coats the leaves of the carnauba palm, a native of Brazil; it is largely myricyl cerotate. Beeswax, secreted from the wax glands of the bee, is largely myricyl palmitate.

$$CH_3(CH_2)_{24}\overset{O}{\overset{\|}{C}}O(CH_2)_{29}CH_3 \qquad CH_3(CH_2)_{14}\overset{O}{\overset{\|}{C}}O(CH_2)_{29}CH_3$$

myricyl cerotate
(major component of carnauba wax)

myricyl palmitate
(major component of beeswax)

Waxes are harder, more brittle, and less greasy to the touch than fats. Applications are found in polishes, cosmetics, ointments, and other pharmaceutical preparations.

Key Terms and Concepts

adrenocorticoid hormones (12.5B)
atherosclerosis (12.5A)
bile acid (12.5D)
biological membrane (12.4)
cell membrane (12.4A)
cephalin (12.3)
detergents (12.2C)
fat (12.1C)
fat-soluble vitamins (12.6)
fatty acid (12.1A)
fluid-mosaic model (12.4B)
hardening of oils (12.1E)

lechithin (12.3)
lipid (introduction)
lipid bilayer (12.4B)
micelle (12.2B)
oil (12.1C)
phospholipid (12.3)
rancidity (12.1D)
reduction of triglycerides (12.1E)
saponification (12.2A)
sex hormone (12.5C)
soap (12.2A)
steroid (12.5)

synthetic detergent (12.2C)   visual cycle (12.6A)

triglyceride (12.1B)   waxes (12.7)

Key Reactions

1. Saponification of triglycerides: preparation of natural soaps (Section 12.2A).
2. Reaction of natural soaps in "hard water" to form water-insoluble salts (Section 12.2B).

Problems

Fats and oils
(Section 12.1)

12.2 List six important functions of lipids in the human body. Name a lipid representing each function.

12.3 How many isomers (including stereoisomers) are possible for a triglyceride containing one molecule each of palmitic, stearic, and oleic acids?

12.4 What is meant by the term *hardening* as applied to fats and oils?

12.5 Saponification number is defined as the number of milligrams of potassium hydroxide required to saponify 1 g of a fat or oil. Calculate the saponification number of tristearin, of molecular weight 890.

12.6 The saponification number of butter is approximately 230; of oleomargarine, approximately 195. Calculate the average molecular weight of butter fat and of oleomargarine.

Fatty acids
(Section 12.1)

12.7 Examine the structural formulas for lauric, palmitic, stearic, oleic, linoleic, and arachidonic acids. For each that shows cis-trans isomerism, state the total number of such isomers possible.

12.8 Compare saturated fatty acids and monosaccharides with respect to their solubility in water; in ether.

12.9 By using structural formulas, illustrate how fatty-acid molecules interact with water to form a monomolecular layer on the surface of water.

12.10 Draw structural formulas for the products formed by reaction of 9-octadecenoic acid (oleic acid) with the following:
  a. Br_2 **b.** $H_2/Pt/3$ atm pressure   **c.** $NaOH/H_2O$
  d. $NaHCO_3/H_2O$ **e.** $LiAlH_4$, followed by H_2O

Soaps and
detergents
(Section 12.2)

12.11 By using structural formulas, show how a soap "dissolves" fats, oils, and grease.

12.12 Show by balanced equations the reaction of a soap with (a) hard water, and (b) acidic solution.

12.13 Characterize the structural features necessary to make a good detergent. Illustrate by structural formulas two different classes of synthetic detergents. Name each example.

12.14 Following are structural formulas for a cationic detergent and a nonionic detergent. How do you account for the detergent properties of each?

$$CH_3$$
$$|$$
$$C_6H_5CH_2N^+\!\!-CH_3Cl^-$$
$$|$$
$$C_8H_{17}$$

benzyldimethyloctylammonium
chloride
(a cationic detergent)

$$\overset{\displaystyle O}{\parallel}\quad CH_2OH$$
$$CH_3(CH_2)_{14}COCH_2CCH_2OH$$
$$|$$
$$CH_2OH$$

pentaerythrityl palmitate
(a nonionic detergent)

Phospholipids
(Section 12.3)

12.15 Draw structural formulas for the products of complete hydrolysis of a lecithin; a cephalin.

Steroids
(Section 12.5)

12.16 Draw the structural formula of cholesterol; label all chiral carbons and state the total number of stereoisomers possible for cholesterol.

12.17 Esters of cholesterol and fatty acids are normal constituents of blood plasma. The fatty acids esterified with cholesterol are generally unsaturated. Draw the structural formula for cholesteryl oleate.

12.18 Cholesterol is an important component of the lipid fraction of cell membranes. How do you think a cholesterol molecule might be oriented in a biological membrane?

12.19 Examine the structural formulas of testosterone, a male sex hormone; of progesterone, a female sex hormone. What are the similarities in structure between the two? What are the differences?

12.20 Describe how a combination of progesterone and estrogen analogs functions as an oral contraceptive.

12.21 Examine the structural formula of cholic acid and account for the ability of this and other bile acids to emulsify fats and oils.

Fat-soluble
vitamins
(Section 12.6)

12.22 Examine the structural formula of vitamin A and state the number of cis-trans isomers possible for this molecule.

12.23 In fish-liver oils, vitamin A is present as esters of fatty acids. The most common of these esters is vitamin A palmitate. Draw its structural formula.

12.24 Describe the symptoms of severe vitamin A deficiency.

12.25 Examine the structural formulas of vitamins A, D_3, E, and K_2. From their structural formulas, would you expect them to be more soluble in water or in olive oil? Would you expect them to be soluble in blood plasma?

12.26 Explain why vitamin E is added to some processed foods.

Biological
membranes
(Section 12.4)

12.27 Two of the chief noncovalent forces directing the organization of biomolecules in aqueous solution are the tendencies to (1) arrange polar groups so that they interact with water by hydrogen bonding, and (2) arrange nonpolar groups so that they are shielded from water. Show how these forces direct micelle formation by soap molecules and lipid bilayer formation of phospholipids.

12.28 Describe the main features of the fluid-mosaic model of the structure of biological membranes.

13

Amino Acids and Proteins

Amino acids are compounds whose chemistry is built on that of amines and carboxylic acids. That amino acids are difunctional presents a special challenge to chemists, because in dealing with a reaction of the carboxyl group, we must also be aware of reactions the amino group might undergo, and vice versa.

Proteins are derived from amino acids. In studying these substances, we observe at first hand one of the principles of the molecular logic of living systems, namely, that in constructing macromolecules (biopolymers), living systems begin with small, readily available subunits (monomers). The monomer units from which proteins are derived are amino acids.

13.1 Amino Acids

A. Structure

Amino acids are compounds that contain both a carboxyl group and an amino group. While many types of amino acids are known, the α-amino acids are the most significant in the biological world because they are the units from which proteins are constructed. A general structural formula for an α-amino acid is shown in Figure 13.1. Although Figure 13.1(a) is a common way of writing structural formulas for amino acids, it is not accurate, since it shows an acid ($-CO_2H$) and a base ($-NH_2$) within the same molecule. These acidic and basic groups react with each other to form a dipolar ion or internal salt [Figure 13.1(b)]. The internal salt of an amino acid is given the special name of **zwitterion**. A zwitterion has no net charge; it contains one positive charge and one negative charge.

$$
\underset{\text{(a)}}{R-\underset{\underset{NH_2}{|}}{CH}-\overset{\overset{O}{\|}}{C}-OH} \qquad \underset{\text{(b)}}{R-\underset{\underset{NH_3^+}{|}}{CH}-\overset{\overset{O}{\|}}{C}-O^-}
$$

Figure 13.1 General formula for an α-amino acid: (a) un-ionized form; (b) dipolar ion.

B. Chirality of Amino Acids

With the exception of glycine, all protein-derived amino acids have at least one chiral carbon atom and show enantiomerism. Figure 13.2 shows stereorepresentations and Fischer projection formulas for the enantiomers of serine. Also shown for reference are a stereorepresentation and a Fischer projection of D-glyceraldehyde. Using D-glyceraldehyde as a reference, scientists have established that all α-amino acids in proteins have an L-configuration about the chiral α-carbon.

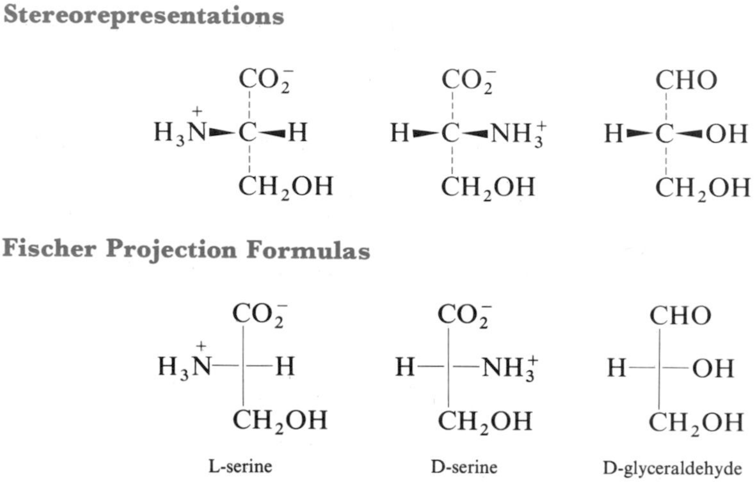

Stereorepresentations

$$
\begin{array}{ccc}
CO_2^- & CO_2^- & CHO \\
H_3\overset{+}{N}\!-\!C\!-\!H & H\!-\!C\!-\!NH_3^+ & H\!-\!C\!-\!OH \\
CH_2OH & CH_2OH & CH_2OH
\end{array}
$$

Fischer Projection Formulas

$$
\begin{array}{ccc}
CO_2^- & CO_2^- & CHO \\
H_3\overset{+}{N}\!-\!\!\!\mid\!\!\!-H & H\!-\!\!\!\mid\!\!\!-NH_3^+ & H\!-\!\!\!\mid\!\!\!-OH \\
CH_2OH & CH_2OH & CH_2OH \\
\text{L-serine} & \text{D-serine} & \text{D-glyceraldehyde}
\end{array}
$$

Figure 13.2 The enantiomers of serine.

C. Protein-Derived Amino Acids

Given in Table 13.1 are names, structural formulas, and standard three-letter abbreviations for the 20 common L-amino acids found in proteins. The amino acids in this table could be listed in several ways, for example, alphabetically, or in order of increasing molecular weight. A more useful classification, however, and one that will be of great value when we come to discuss the three-dimensional shapes of proteins, is according to the polarity of their side chains. For this reason, the amino acids in Table 13.1 are divided into the following three categories: nonpolar (hydrophobic) side chains, polar but uncharged side chains, and polar, charged side chains. Because our concern is for the behavior of amino acids in biological systems, we show them in Table 13.1 as they would be ionized at pH 7.0—all amino groups are protonated and bear a positive charge; all carboxyl groups are ionized and bear a negative charge.

1. *Nonpolar side chains.* Eight amino acids have nonpolar hydrophobic side chains. Of these, glycine, alanine, and proline have small side chains and are weakly hydrophobic. The other five amino acids in this category (phenylalanine, valine, leucine, isoleucine, and methionine) have larger side chains and are more strongly hydrophobic.

Table 13.1 The 20 common L-amino acids found in protein. Each is shown as it would be at pH 7.0.

Nonpolar side chains

$$\underset{\text{glycine (gly)}}{H-\underset{\overset{|}{NH_3^+}}{CH}-CO_2^-}$$

$$\underset{\text{L-leucine (leu)}}{\underset{H_3C}{\overset{H_3C}{\diagdown}}CH-CH_2-\underset{\overset{|}{NH_3^+}}{CH}-CO_2^-}$$

$$\underset{\text{L-alanine (ala)}}{CH_3-\underset{\overset{|}{NH_3^+}}{CH}-CO_2^-}$$

$$\underset{\text{L-isoleucine (ile)}}{\underset{CH_3-CH_2}{\overset{H_3C}{\diagdown}}CH-\underset{\overset{|}{NH_3^+}}{CH}-CO_2^-}$$

$$\underset{\text{L-valine (val)}}{\underset{H_3C}{\overset{H_3C}{\diagdown}}CH-\underset{\overset{|}{NH_3^+}}{CH}-CO_2^-}$$

L-proline (pro)

$$\underset{\text{L-phenylalanine (phe)}}{\diagesub-CH_2-\underset{\overset{|}{NH_3^+}}{CH}-CO_2^-}$$

$$\underset{\text{L-methionine (met)}}{CH_3-S-CH_2-CH_2-\underset{\overset{|}{NH_3^+}}{CH}-CO_2^-}$$

Polar uncharged side chains

$$\underset{\text{L-serine (ser)}}{HO-CH_2-\underset{\overset{|}{NH_3^+}}{CH}-CO_2^-}$$

$$\underset{\text{L-glutamine (gln)}}{H_2N-\overset{\overset{O}{\|}}{C}-CH_2-CH_2-\underset{\overset{|}{NH_3^+}}{CH}-CO_2^-}$$

$$\underset{\text{L-threonine (thr)}}{CH_3-\underset{\overset{|}{OH}}{CH}-\underset{\overset{|}{NH_3^+}}{CH}-CO_2^-}$$

$$\underset{\text{L-cysteine (cys)}}{HS-CH_2-\underset{\overset{|}{NH_3^+}}{CH}-CO_2^-}$$

$$\underset{\text{L-asparagine (asn)}}{H_2N-\overset{\overset{O}{\|}}{C}-CH_2-\underset{\overset{|}{NH_3^+}}{CH}-CO_2^-}$$

$$\underset{\text{L-tyrosine (tyr)}}{HO-\diagesub-CH_2-\underset{\overset{|}{NH_3^+}}{CH}-CO_2^-}$$

L-histidine (his)

L-tryptophan (trp)

Polar charged side chains

$$\underset{\text{L-aspartic acid (asp)}}{^-O-\overset{\overset{O}{\|}}{C}-CH_2-\underset{\overset{|}{NH_3^+}}{CH}-CO_2^-}$$

$$\underset{\text{L-glutamic acid (glu)}}{^-O-\overset{\overset{O}{\|}}{C}-CH_2-CH_2-\underset{\overset{|}{NH_3^+}}{CH}-CO_2^-}$$

$$\underset{\text{L-lysine (lys)}}{\overset{+}{H_3N}-CH_2-CH_2-CH_2-CH_2-\underset{\overset{|}{NH_3^+}}{CH}-CO_2^-}$$

$$\underset{\text{L-arginine (arg)}}{H_2N-\overset{\overset{NH_2^+}{\|}}{C}-NH-CH_2-CH_2-CH_2-\underset{\overset{|}{NH_3^+}}{CH}-CO_2^-}$$

2. *Polar but uncharged side chains.* Eight amino acids have polar but uncharged side chains. Of these, the side chains of serine and threonine have hydroxyl groups; asparagine and glutamine have amide groups; tyrosine has a phenol; tryptophan and histidine have heterocyclic aromatic amines; and cysteine has a sulfhydryl group. The sulfhydryl group of cysteine, the imidazole group of histidine, and the phenolic hydroxyl of tyrosine show some degree of ionization, depending on the pH.

3. *Charged side chains.* Four amino acids have charged side chains: aspartic acid, glutamic acid, lysine, and arginine. Aspartic and glutamic acids have carboxyl groups on their side chains that are negatively charged at pH 7.0. The side chain of lysine contains a primary amine and the side chain of arginine contains a guanidine group, each of which is protonated and therefore positively charged at pH 7.0.

D. Some Other Common L-Amino Acids

Thus far, we have studied only the 20 L-amino acids from which proteins are synthesized. While almost all plant and animal proteins are constructed from just these 20, a few other amino acids are also found in proteins. For example, L-4-hydroxyproline and L-5-hydroxylysine are important components of col-(Section 13.7B), but found in few other proteins. These and other special L-amino acids are synthesized in living systems by modification of one of the 20 common amino acids after it is incorporated into a protein.

L-4-hydroxyproline   L-5-hydroxylysine

Some important amino acids are either metabolic intermediates or parts of nonprotein biomolecules (Table 13.2). Ornithine and citrulline are found predominantly in the liver and are an integral part of the urea cycle (Section 19.3C), the metabolic pathway that converts ammonia to urea. 4-Aminobutanoic acid (gamma-amino butyric acid, GABA) is a neurotransmitter found in high concentration (0.8 mM) in the brain, but in no significant amounts in any other mammalian tissue. This amino acid is synthesized in neural tissue by decarboxylation of the alpha-carboxyl group of glutamic acid in a reaction catalyzed by pyridoxal phosphate:

glutamic acid
(an excitatory
neurotransmitter)

4-aminobutanoic acid
(an inhibitory
neurotransmitter)

Table 13.2 Several L-amino acids not found in proteins.

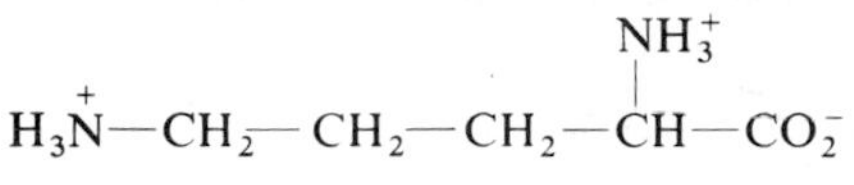

ornithine (orn)

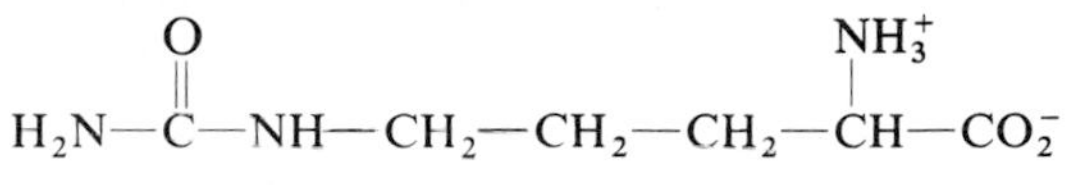

citrulline

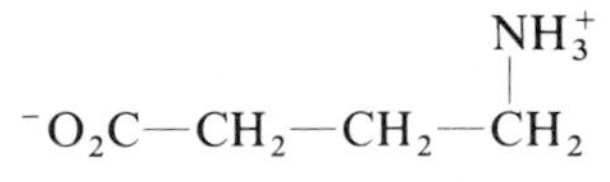

γ-aminobutyric acid
(GABA)

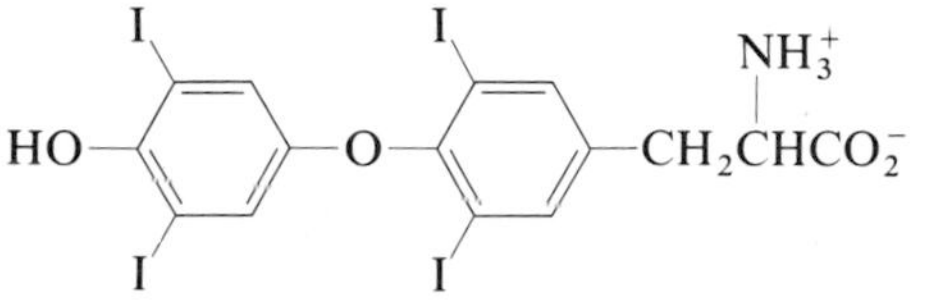

thyroxine
tetraiodothyronine, or T$_4$

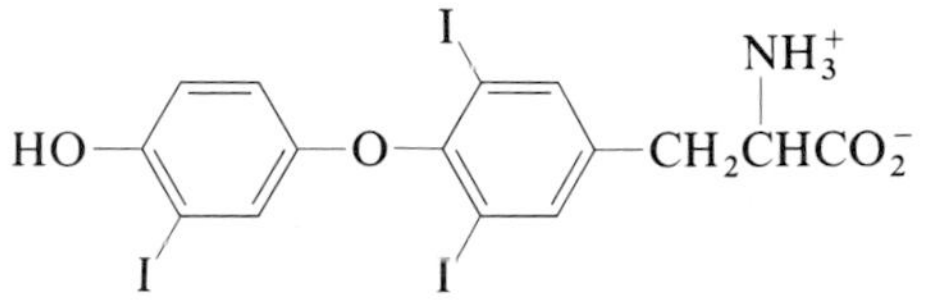

triiodothyronine, or T$_3$

Glutamic acid is one of the most important excitatory transmitters in the central nervous system of invertebrates and possibly in humans also. In contrast, 4-aminobutanoic acid is an inhibitory transmitter. Whereas an excitatory transmitter brings about depolarization (firing) of a postsynaptic membrane, an inhibitory transmitter leads to hyperpolarization and makes it more difficult to excite a postsynaptic membrane.

Thyroxine, known alternatively as T$_4$ because it contains 4 atoms of iodine, is one of several hormones derived from the amino acid tyrosine and was first isolated from thyroid tissue in 1914. In 1952, triiodothyronine, known alternatively as T$_3$ because it contains only three atoms of iodine, was also discovered in thyroid tissue. The principal function of these hormones is to stimulate metabolism in other cells and tissues. The levorotatory isomer of each is significantly more active than the dextrorotatory isomer.

D-amino acids are not found in proteins and are not a part of the metabolism of higher organisms. However, several D-amino acids along with their L-enantiomers are found in structural components and the metabolism of lower forms of life. As examples, both D-alanine and D-glutamic acid are structural

Table 13.3
Some D-amino acids found in peptide antibiotics.

Antibiotic	D-amino Acid	Agent of Production
Actinomycin D	D-valine	*Streptomyces parralus* and others
Bacitracin A	D-asparagine D-glutamic acid D-ornithine D-phenylalanine	*Bacillus subtilis*
Fungisporin	D-phenylalanine D-valine	*Penicillium* species
Gramicidin S	D-phenylalanine	*Bacillus brevis*

components of cell walls of certain bacteria. A variety of D-amino acids have been found in peptide antibiotics, a few of which are listed in Table 13.3.

13.2 Acid-Base Properties of Amino Acids

A. Ionization of Amino Acids

Given in Table 13.4 are pK_a values for ionizable groups of the 20 protein-derived amino acids. There are several things to notice from the data in this table.

Table 13.4 pK_a values for ionizable groups of amino acids.

Amino Acid	α-CO_2H Group	α-NH_3^+ Group	Side Chain
alanine	2.35	9.87	
arginine	2.01	9.04	12.48
asparagine	2.02	8.80	
aspartic acid	2.10	9.82	3.86
cysteine	1.86	10.25	8.00
glutamic acid	2.10	9.47	4.07
glutamine	2.17	9.13	
glycine	2.35	9.78	
histidine	1.77	9.10	6.10
isoleucine	2.32	9.76	
leucine	2.33	9.74	
lysine	2.18	8.95	10.53
methionine	2.28	9.20	
phenylalanine	2.58	9.24	
proline	2.00	10.60	
serine	2.21	9.15	
threonine	2.09	9.10	
tryptophan	2.38	9.39	
tyrosine	2.20	9.11	10.07
valine	2.29	9.72	

1. *Acidity of α-carboxyl groups.* The average pK_a value for an α-carboxyl group is 2.16. If you compare this value with that for acetic acid ($pK_a = 4.74$) and other simple carboxylic acids (Section 7.5), you will see that the α-carboxyl group of an amino acid is a considerably stronger acid.

$$\alpha\text{---}\overset{\overset{\displaystyle O}{\|}}{C}OH \;\rightleftharpoons\; \alpha\text{---}\overset{\overset{\displaystyle O}{\|}}{C}O^- + H^+ \qquad K_a = 6.9 \times 10^{-3} \qquad pK_a = 2.16$$

an α-carboxyl
group
(stronger acids than
simple carboxylic acids)

2. *Acidity of side chain carboxyl groups.* The side chain carboxyl groups of aspartic and glutamic acids are only slightly stronger acids than acetic acid.

3. *Basicity of α-amino groups.* The average pK_a value for an α-amino group is 9.56, a value very close to that for primary aliphatic amines (Section 5.3A).

$$\alpha\text{-NH}_3^+ \;\rightleftharpoons\; \alpha\text{-NH}_2 + \text{H}^+ \qquad K_a = 2.75 \times 10^{-10} \qquad pK_a = 9.56$$

an α-amino
group
(comparable in
strength to simple
amine salts)

4. *Basicity of the guanidine group of arginine.* The nitrogen-containing side chain of arginine is called a guanidine group and is a much stronger base than the amino group of lysine or other primary aliphatic amines. Consequently, the guanidine group of arginine is 100% protonated, with a charge of +1, at physiological pH.

$$\underset{\substack{\text{the guanidine}\\\text{group on the side}\\\text{chain of arginine}\\\text{(a weaker acid than}\\\text{simple amine salts)}}}{\text{R}-\text{NH}-\overset{\overset{\text{NH}_2^+}{\|}}{\text{C}}-\text{NH}_2} \rightleftharpoons \underset{\substack{\text{(a stronger base}\\\text{than simple amines)}}}{\text{R}-\text{NH}-\overset{\overset{\text{NH}}{\|}}{\text{C}}-\text{NH}_2} + \text{H}^+ \qquad K_a = 3.3 \times 10^{-13} \qquad pK_a = 12.48$$

5. *Basicity of the imidazole group of histidine.* The five-membered nitrogen-containing ring on the side chain of histidine is called an imidazole group (Section 5.1B). The imidazole group is a weaker base than simple aliphatic amines. Alternatively, the protonated form of the imidazole group is a stronger acid than the protonated form of a simple amine.

$$\underset{\substack{\text{the imidazole}\\\text{ring of histidine}\\\text{(a stronger acid than}\\\text{simple amine salts)}}}{} \rightleftharpoons \quad + \text{H}^+ \qquad K_a = 7.9 \times 10^{-7} \qquad pK_a = 6.10$$

B. Ionization of Amino Acids as a Function of pH

We have just considered the principal ionizable groups of amino acids separately. Let us now consider how the interaction of ionizable groups within a particular amino acid affects the properties of that amino acid. Each weak acid dissociates to give its conjugate base and H^+, and has its own acid dissociation constant, K_a.

$$\text{HA} \rightleftharpoons \text{A}^- + \text{H}^+ \qquad K_a = \frac{[\text{conjugate base}][\text{H}^+]}{[\text{weak base}]}$$

weak   conjugate
acid   base

Taking the logarithm of the acid dissociation equation, and then rearranging it gives the Henderson-Hasselbalch equation.

Henderson-Hasselbalch Equation

$$pH = pK_a + \log \frac{[\text{conjugate base}]}{[\text{weak acid}]}$$

The Henderson-Hasselbalch equation provides a particularly direct way of calculating the ratio of conjugate base to weak acid at any pH. Shown in Table 13.5 are values of this ratio for a weak acid with a dissociation constant pK_a. Also shown is the percentage of molecules present as undissociated weak acid at pH values higher and lower than the value of pK_a by one and two units.

Table 13.5
Ratio of
conjugate base
to weak acid as a
function of pH.

pH	$\dfrac{[\text{Conj Base}]}{[\text{Weak Acid}]}$	% Present as Weak Acid
$pK_a + 2$	100/1	0.990%
$pK_a + 1$	10/1	9.09%
pK_a	1/1	50.0%
$pK_a - 1$	1/10	90.9%
$pK_a - 2$	1/100	99.0%

Given the data in Table 13.5, we can generalize as follows:

1. If the pH of a solution is 2.0 or more units higher (more basic) than the pK_a of an ionizable group, then the group is more than 99% in the conjugate base form.
2. If the pH of a solution is 1.0 unit higher (more basic) than the pK_a of an ionizable group, then the group is approximately 90% in the conjugate base form and 10% in the acid form.
3. If the pH of a solution is equal to the pK_a of an ionizable group, then the group is 50% in the acid form and 50% in the conjugate base form.
4. If the pH of a solution is 1.0 unit lower (more acidic) than the pK_a of an ionizable group, then the group is approximately 10% in the conjugate base form and 90% in the acid form.
5. If the pH of a solution is 2.0 or more units lower (more acidic) than the pK_a of an ionizable group, then the group is more than 99% in the acid form.

Example 13.1

Draw a structural formula for L-serine and estimate what the net charge on this amino acid is at pH 3.0, 7.0, and 10.0.

☐ *Solution*

The pK_a of the α-carboxyl group of serine is 2.21. At pH of 3.0, which is 0.79 unit higher than its pK_a, the carboxyl group is approximately 90% in the ionized form and bears a negative charge. The pK_a of the α-amino group is 9.15. At pH 3.0, which is approximately 6 units lower than the pK_a of the ionizable group, the α-amino group is completely in the protonated form and bears a positive charge. The same type of calculations can be repeated at pH 7.0 and 10.0. Results are shown on these structural formulas:

| pH = 3.0 | pH = 7.0 | pH = 10.0 |
| net charge = + | net charge = 0 | net charge = − |

Problem 13.1

Draw a structural formula for lysine and estimate what the net charge is on each functional group at pH 3.0, 7.0, and 10.0.

C. Titration of Amino Acids

Values of pK_a for the ionizable groups of amino acids are usually obtained by acid-base **titration** and measuring the pH of the solution as a function of added base (or added acid, depending on how the titration is done). In an example of this experimental procedure, consider a solution containing 1.0 mol of glycine to which enough strong acid has been added so that both the carboxyl and the amino groups are fully protonated. Next, this solution is titrated with 1.0 M NaOH; the volume of base added and the pH of the resulting solution are recorded and then plotted as shown in Figure 13.3. The most acidic group and the one to react first with added sodium hydroxide is the carboxyl group. When exactly 0.5 mol of NaOH has been added, the carboxyl group is half-neutralized. At this point, the dipolar ion has a concentration equal to that of the positively charged ion, and the pH equals 2.35, the pK_a of the carboxyl group.

$$[H_3\overset{+}{N}-CH_2-CO_2H] = [H_3\overset{+}{N}-CH_2-CO_2^-] \quad \text{where pH} = pK_{\alpha\text{-CO}_2\text{H}}$$

positive ion (cation)　　　　　dipolar ion (no net charge)

The endpoint of the first part of the titration is reached when 1.0 mol of sodium hydroxide has been added. At this point, the predominant species in solution is the dipolar ion, and the observed pH of the solution is 6.07. The next section of the curve represents titration of the $-NH_3^+$ group. When another 0.5 mol of sodium hydroxide has been added (bringing the total to 1.5 mol), half the $-NH_3^+$ groups are neutralized and converted to $-NH_2$. At this point, the concentrations of the dipolar ion and the negatively charged ion are equal, and the observed pH is 9.78, the pK_a of the amino group of glycine.

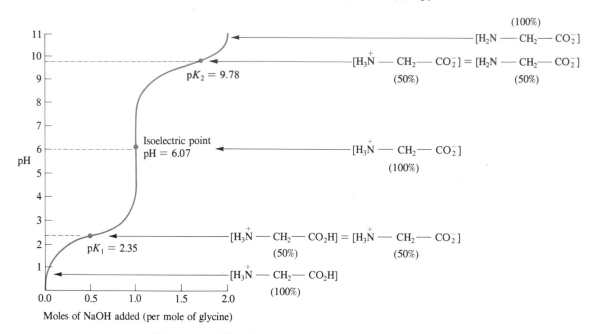

Figure 13.3 Titration of glycine with sodium hydroxide.

$$[\overset{+}{H_3N}-CH_2-CO_2^-] = [H_2N-CH_2-CO_2^-] \quad \text{where pH} = pK_{\alpha\text{-CO}_2H}$$

dipolar ion
(no net charge)

negative ion
(anion)

The second endpoint of the titration is reached when a total of 2.0 mol of sodium hydroxide has been added and glycine is converted entirely to an anion.

D. Isoelectric Point

Titration curves such as that for glycine permit us to determine pK_a values for the ionizable groups of an amino acid. They also permit us to determine another important property, namely, isoelectric point, abbreviated pI. **Isoelectric point** for an amino acid is defined as the pH at which the majority of the molecules in solution have no net charge, that is, a net charge of zero. By examining the titration curve, you can see that the isoelectric point for glycine is 6.07, halfway between the pK_a values for the α-carboxyl and α-amino groups.

$$\text{pI} = \tfrac{1}{2}(pK_{\alpha\text{-CO}_2H} + pK_{\alpha\text{-NH}_3^+})$$
$$= \tfrac{1}{2}(2.35 + 9.78) = 6.07 \quad \text{(isoelectric point for glycine)}$$

Calculation of the isoelectric points for amino acids such as aspartic and glutamic acids (with carboxyl side chains) or lysine and arginine (with basic side chains) can be done similarly. For aspartic and glutamic acids, the isoelectric point occurs at a pH at which the net charge on the two carboxyl groups is -1

and balances the charge of $+1$ on the α-amino group. For glutamic acid, the isoelectric point is

$$pI = \tfrac{1}{2}(pK_{\alpha\text{-CO}_2\text{H}} + pK_{\text{chain-CO}_2\text{H}})$$
$$= \tfrac{1}{2}(2.10 + 4.07) = 3.09 \quad \text{(isoelectric point for glutamic acid)}$$

For lysine and arginine, the isoelectric point occurs at a pH at which the net charge on the α-amino group and the amino side chain is $+1$, balancing the charge of -1 on the α-carboxyl group. For lysine, the isoelectric point is

$$pI = \tfrac{1}{2}(pK_{\alpha\text{-NH}_3^+} + pK_{\text{chain-NH}_3^+})$$
$$= \tfrac{1}{2}(8.95 + 10.53) = 9.74 \quad \text{(isoelectric point for lysine)}$$

Isoelectric points for amino acids with ionizable side chains are given in Table 13.6.

Table 13.6
Isoelectric points for amino acids with ionizable side chains

Amino Acid	Ionizable Side-Chain Group	pI (Isoelectric Point)
aspartic acid	carboxyl	2.98
glutamic acid	carboxyl	3.09
histidine	imidazole	7.60
cysteine	sulfhydryl	5.10
tyrosine	phenol	6.16
lysine	amine	9.74
arginine	guanine	10.76

Given a value for the isoelectric point of an amino acid, one can estimate the charge on that amino acid at any pH. For example, the charge on tyrosine is zero at pH 6.16 (its isoelectric point). A small fraction of tyrosine molecules are positively charged at pH 6.0 (0.16 unit lower than its pI) and virtually all are positively charged at pH 4.16 (2 units lower than its pI). As another example, at pH 9.74, the net charge on lysine is zero. At pH values lower than 9.74, an increasing fraction of lysine molecules are positively charged.

E. Electrophoresis

Electrophoresis is a process of separating compounds on the basis of their electrical charges. Electrophoretic separations can be carried out using paper, starch, agar, certain plastics, and cellulose acetate as solid supports. In paper electrophoresis, a paper strip saturated with an aqueous buffer of predetermined pH serves as a bridge between two electrode vessels. Next, a sample of amino acids is applied as a spot. When an electric potential is then applied to the electrode vessels, amino acids migrate toward the electrode carrying the charge opposite to their own. Molecules having a high charge density move more rapidly than those with a lower charge density. Any molecule already at its isoelectric point remains at the origin.

After electrophoretic separation is complete, the strip is dried and sprayed with a dye to make the separated components visible. The most common dye for amino acids is ninhydrin, a triketone with a benzene ring fused to a five-membered ring. In aqueous solution, the middle carbonyl group of the five-membered ring is almost fully hydrated, to give a compound called ninhydrin hydrate:

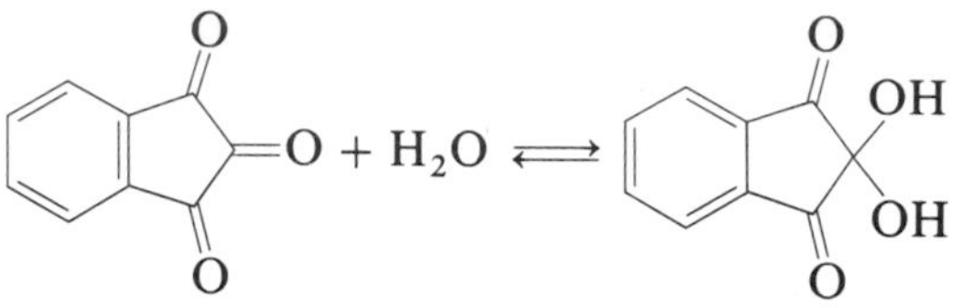

ninhydrin   ninhydrin hydrate

Ninhydrin reacts with α-amino acids to produce an aldehyde, carbon dioxide, and a purple anion with an absorption maximum at 580 nm:

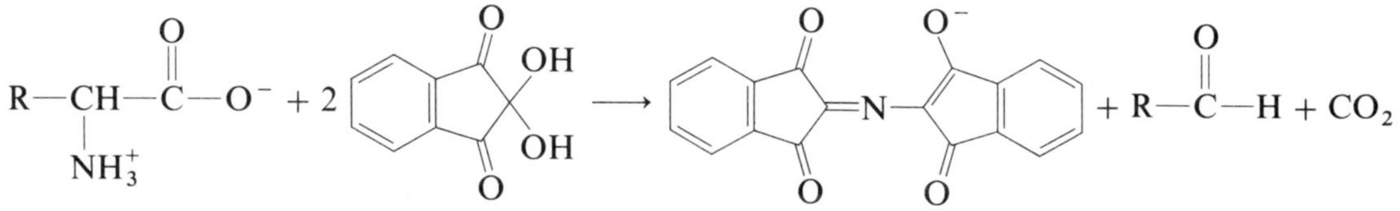

an α-amino   ninhydrin   anion
acid   hydrate   (purple)

Example 13.2

The isoelectric point of tyrosine is 6.16. Toward which electrode will tyrosine migrate during paper electrophoresis at pH 7.0?

Solution

When the pH of the solution equals 6.16, the isoelectric pH of tyrosine, tyrosine molecules bear no net charge and will remain where spotted on the paper. At pH 7.0, which is slightly more basic than tyrosine's isoelectric pH, tyrosine molecules will bear a small net negative charge and will move toward the positive electrode.

Problem 13.2

The isoelectric point of histidine is 7.60. Toward which electrode will histidine migrate on paper electrophoresis at pH 7.0?

Example 13.3

Electrophoresis of a mixture of lysine, histidine, and cysteine is carried out at pH 7.60. Describe the behavior of each amino acid under these conditions.

Solution

The isoelectric point of histidine is 7.60. At this pH, histidine has a net charge of zero and does not move from the origin. The pI of cysteine is 5.10; at pH

7.60, cysteine has a net negative charge and moves toward the positive electrode. The pI of lysine is 9.74; at pH 7.60 lysine has a net positive charge and moves toward the negative electrode (Figure 13.4).

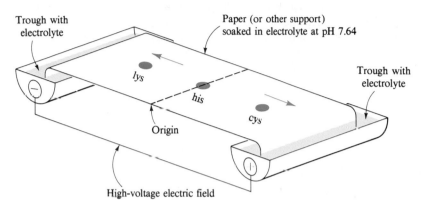

Trough with electrolyte

Paper (or other support) soaked in electrolyte at pH 7.64

Trough with electrolyte

lys

his

cys

Origin

High-voltage electric field

Figure 13.4 Electrophoresis of a mixture of histidine, lysine, and cysteine at pH 7.60.

Problem 13.3

Describe the behavior of a mixture of glutamic acid, arginine, and valine on paper electrophoresis at pH 6.0.

13.3 Polypeptides and Proteins

In 1902, Emil Fischer proposed that proteins are long chains of amino acids joined together by amide bonds between the α-carboxyl group of one amino acid and the α-amino group of another. For these amide bonds, Fischer proposed the special name **peptide bond**. Figure 13.5 shows the peptide bond formed between glycine and alanine in the dipeptide glycylalanine.

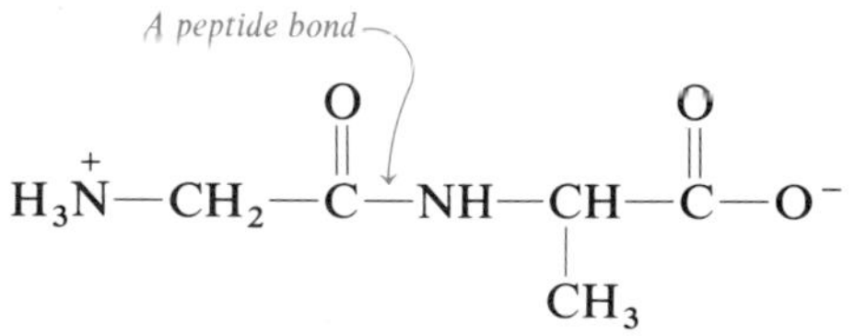

A peptide bond

$$H_3\overset{+}{N}-CH_2-\overset{\overset{\displaystyle O}{\|}}{C}-NH-CH-\overset{\overset{\displaystyle O}{\|}}{C}-O^-$$
$$|$$
$$CH_3$$

Figure 13.5 The peptide bond in glycylalanine.

A molecule containing two amino acids joined by an amide bond is called a dipeptide. Those containing larger numbers of amino acids are called tripeptides, tetrapeptides, pentapeptides, and so on. Molecules containing 10 or more amino acids are generally called polypeptides. Proteins are biological macromolecules of molecular weight 5000 or greater, consisting of one or more polypeptide chains.

By convention, polypeptides are written from the left, beginning with the amino acid having the free H_3N^+— group and proceeding to the right toward the amino acid with the free $-CO_2^-$ group. The amino acid with the free H_3N^+— group is called the **N-terminal amino acid** and that with the free $-CO_2^-$ group is called the **C-terminal amino acid**. The structural formula for a polypeptide sequence may be written out in full, or the sequence of amino acids may be indicated using the standard abbreviation for each.

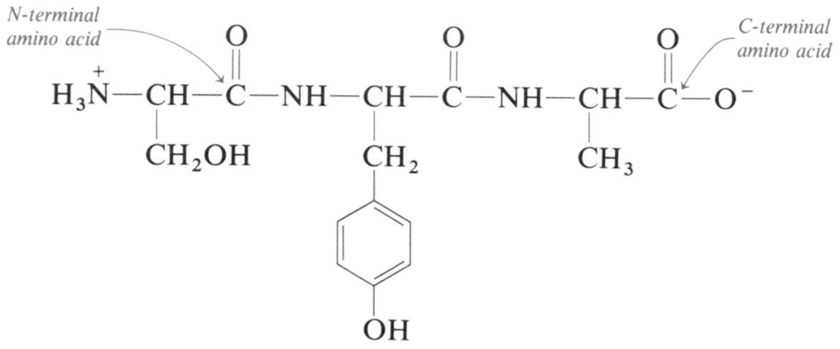

ser-tyr-ala
(seryltyrosylalanine)

Polypeptides are named by listing each amino acid in order, from the N-terminal end of the chain to the C-terminal end. The name of the C-terminal amino acid is given in full. The name of each other amino acid in the chain is derived by dropping the suffix -ine or -ic acid and adding -yl. For example, if the order of amino acids from the N-terminal end is serine-tyrosine-alanine (ser-tyr-ala), the name of the tripeptide is seryltyrosylalanine.

Example 13.4

Name and draw a structural formula for the tripeptide gly-ser-asp. Label the N-terminal amino acid and the C-terminal amino acid. What is the net charge on this tripeptide at pH 6.0?

☐ *Solution*

In writing the formula for this tripeptide, begin with glycine on the left. Then connect the α-carbonyl of glycine to the α-amino group of serine by a peptide bond. Finally, connect the α-carbonyl of serine to the α-amino group of aspartic acid by another peptide bond. If you have done this correctly, the backbone of the peptide chain should be a repeating sequence of nitrogen– alpha carbon– carbonyl. The net charge on this tripeptide at pH 6.0 is -1.

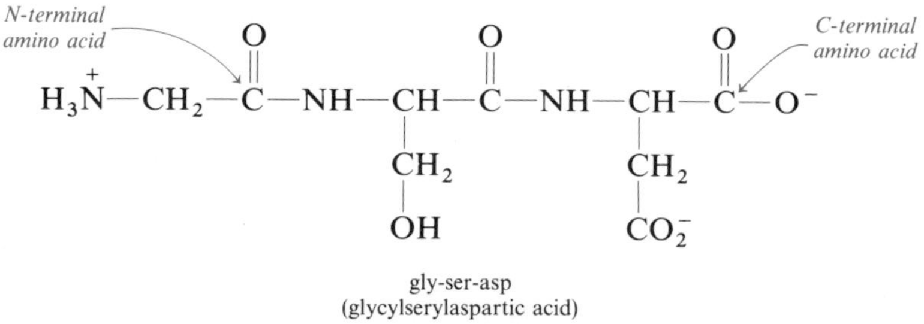

gly-ser-asp
(glycylserylaspartic acid)

**Problem
13.4**

Name and draw a structural formula for lys-phe-ala. Label the N-terminal
amino acid and the C-terminal amino acid. What is the net charge on this tripeptide at pH 6.0?

13.4 Amino Acid Analysis

The first step in analyzing a polypeptide is hydrolysis and quantitative determination of its amino acid composition. Recall that amide bonds are very resistant to hydrolysis (Section 8.8). Hydrolysis of polypeptides requires heating
in 6M HCl at 110°C for 24–70 hr, or heating in 2–4M NaOH at comparable
temperatures and for comparable times. Once the polypeptide is hydrolyzed, the
resulting mixture of amino acids is analyzed by ion-exchange chromatography.
Amino acids are detected as they emerge from the column by reaction with
ninhydrin (Section 13.2E). Current procedures for hydrolysis of polypeptides
and analysis of amino acid mixtures have been refined to the point where it is
possible to obtain amino acid composition from as little as 50 nanomoles
(50×10^{-9} mol) of polypeptide. Figure 13.6 shows the analysis of a polypeptide

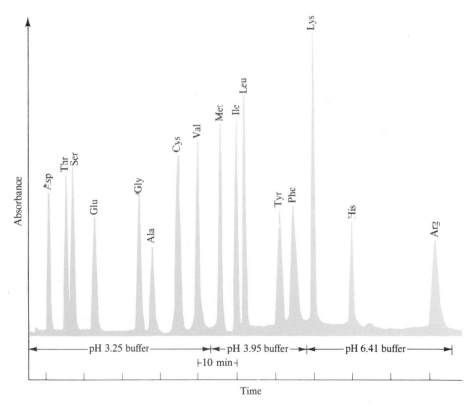

Figure 13.6 Analysis of a mixture of amino acids by ion-exchange chromatography.

hydrolysate by ion exchange chromatography. Note that during hydrolysis, the side-chain amides of asparagine and glutamine are hydrolyzed, and these amino acids are detected as glutamic acid and aspartic acid. For each glutamine or asparagine hydrolyzed, there is an equivalent amount of ammonia formed.

13.5 Primary Structure of Polypeptides and Proteins

Primary (1°) structure of polypeptides and proteins refers to the sequence of amino acids in a polypeptide chain. In this sense, primary structure is a complete description of all covalent bonding in a polypeptide or protein.

It is difficult to appreciate the incredibly large number of different polypeptides that can be constructed from the 20 amino acids, where the number of amino acids in a polypeptide can range from under ten to well over a hundred. With only three amino acids, there are 27 different tripeptides possible. For glycine, alanine, and serine, the 27 tripeptides are:

gly-gly-gly	ser-ser-ser	ala-ala-ala
gly-gly-ser	ser-ser-gly	ala-ala-gly
gly-gly-ala	ser-ser-ala	ala-ala-ser
gly-ser-gly	ser-gly-ser	ala-gly-ala
gly-ala-gly	ser-ala-ser	ala-ser-ala
gly-ser-ala	ser-gly-ala	ala-gly-ser
gly-ala-ser	ser-ala-gly	ala-ser-gly
gly-ser-ser	ser-gly-gly	ala-gly-gly
gly-ala-ala	ser-ala-ala	ala-ser-ser

For a polypeptide containing one each of the 20 different amino acids, the number of possible polypeptides is $20 \times 19 \times 18 \times \cdots \times 2 \times 1$, or about 2×10^{18}. With larger polypeptides and proteins, the possible combinations become truly countless!

13.6 Three-Dimensional Shapes of Polypeptides and Proteins

A. Geometry of a Peptide Bond

In the late 1930s, Linus Pauling began a series of studies designed to learn more about the three-dimensional shapes of proteins. One of his first discoveries was that a peptide bond itself is planar. As shown in Figure 13.7, the four atoms of a peptide bond and the two alpha-carbons joined to it all lie in the same plane.

Had you been asked earlier to predict the **geometry of a peptide bond**, you probably would have reasoned in the following way. There are three bundles of electron density around the carbonyl carbon; therefore predict bond angles of 120° about the carbonyl carbon. There are four bundles of electron density around the amide nitrogen; therefore predict bond angles of 109.5° about this atom. These predictions agree with the observed bond angles of approximately

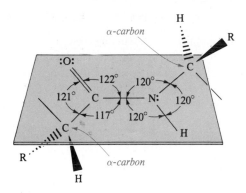

Figure 13.7 Planarity of a peptide bond. Bond angles about the carbonyl carbon and the amide nitrogen are approximately 120°.

120° about the carbonyl carbon. However, a bond angle of 120° about the amide nitrogen is unexpected. To account for this observed geometry, Pauling proposed that a peptide bond is more accurately represented as a resonance hybrid of two important contributing structures:

I $\longleftrightarrow$ II

Contributing structure I shows C=O double bond and a C—N single bond. Structure II shows a C—O single bond and a C=N double bond. If structure I is the major contributor to the hybrid, the C—N—C bond angle would be nearer 109.5°. If, on the other hand, structure II is the major contributor, the C—N—C bond angle would be nearer 120°. The fact, first observed by Pauling, is that the C—N—C bond angle is very near 120°, which means that the peptide bond is planar and structure II is the major contributor to the resonance hybrid.

Two configurations are possible for the atoms of a planar peptide bond. In one configuration, the two α-carbons are cis to each other; in the other, they are trans to each other:

cis   trans

The trans configuration is more favorable because the bulky α-carbons are farther from each other than they are in the cis configuration. Virtually all peptide bonds in natural proteins have the trans configuration.

B. Secondary Structure

Secondary (2°) structure refers to ordered arrangements (conformations) of amino acids in localized regions of a polypeptide, or protein, molecule. The first studies of polypeptide conformations were also carried out by Linus Pauling and Robert Corey, beginning in 1939. They assumed that in conformations of greatest stability, (1) all atoms in a peptide bond lie in the same plane, and (2) each amide group is hydrogen-bonded between the N—H of one peptide bond and the C=O of another, as shown in Figure 13.8.

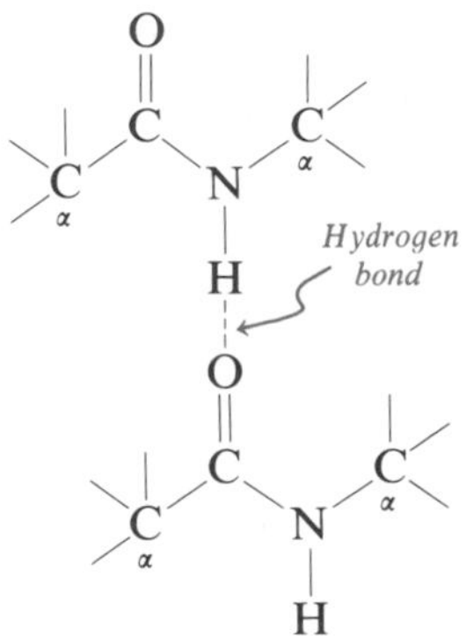

Figure 13.8 Hydrogen bonding between amide groups.

On the basis of model-building, Pauling and Corey proposed that two folding patterns should be particularly stable: the **α-helix** and the antiparallel **β-pleated sheet**. In the α-helix pattern shown in Figure 13.9, a polypeptide chain is coiled in a spiral. As you study the α-helix in Figure 13.9(c), note the following:

1. The helix is coiled in a clockwise, or right-handed, manner. Right-handed means that if you turn the helix clockwise, it twists away from you. In this sense, a right-handed helix is analogous to the right-hand thread of a common wood or machine screw.
2. There are 3.6 amino acids per turn of the helix.
3. Each peptide bond is trans and planar.
4. The N—H group of each peptide bond points roughly upward, parallel to the axis of the helix; and the C=O of each peptide bond points roughly downward, also parallel to the axis of the helix.
5. The carbonyl group of each peptide bond is hydrogen-bonded to the N—H group of the peptide bond four amino acid units away from it. Hydrogen bonds are shown as dotted lines.
6. All R— groups point outward from the helix.

Almost immediately after Pauling proposed the α-helix structure, other researchers proved the presence of α-helix in keratin, the protein of hair and wool.

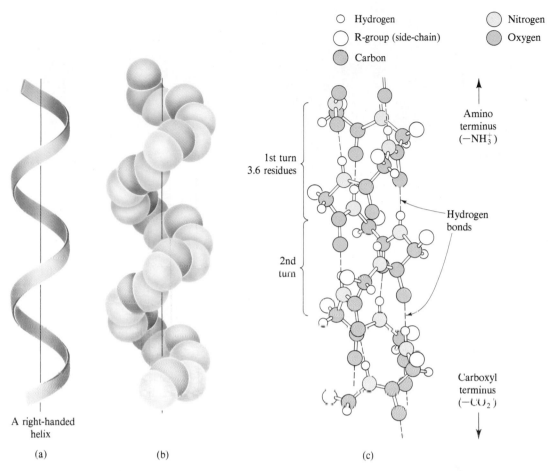

Figure 13.9 (a) A right-handed α-helix. (b) Space-filling model of the carbon-nitrogen backbone of an α-helix. (c) Ball-and-stick model of an α-helix showing intrachain hydrogen bonding. There are 3.6 amino acid residues per turn.

It soon became obvious that the α-helix is one of the fundamental folding patterns of polypeptide chains.

β-Pleated sheets consist of extended polypeptide chains with neighboring chains running in opposite (antiparallel) directions. Unlike the α-helix arrangement, N—H and C=O groups lie in the plane of the sheet and are roughly perpendicular to the long axis of the sheet. The C=O group of each peptide bond is hydrogen-bonded to the N—H group of a peptide bond of a neighboring chain.

As you study the section of β-pleated sheet shown in Figure 13.10, note the following:

1. The two polypeptide chains lie adjacent to each other and run in opposite (antiparallel) directions.

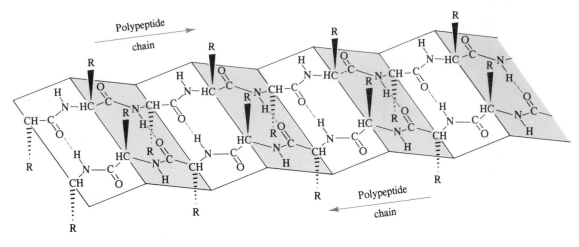

Figure 13.10 β-Pleated sheet conformation with two polypeptide chains running in opposite (antiparallel) directions. Hydrogen bonding between chains is indicated by dotted lines.

2. Each peptide bond is trans and planar.
3. The polypeptide is a chain of flat or planar sections connected at amino acid α-carbons.
4. The C=O and N—H groups of peptide bonds from adjacent chains point at each other and are in the same plane, so hydrogen bonding is possible between adjacent polypeptide chains.
5. The R— groups on any one chain alternate, first above the plane of the sheet and then below the plane of the sheet.

The pleated sheet conformation is stabilized by hydrogen bonding between N—H groups of one chain and C=O groups of an adjacent chain. By comparison, the α-helix is stabilized by hydrogen bonding between N—H and C=O groups within the same polypeptide chain.

The term secondary structure is used to describe α-helix, β-pleated sheet, and other types of periodic conformations in localized regions of polypeptide or protein molecules.

C. Tertiary Structure

Tertiary (3°) structure refers to the overall folding pattern and arrangement in space of all atoms in a single polypeptide chain. Actually, there is no sharp dividing line between secondary and tertiary structure. Secondary structure refers to the spatial arrangement of amino acids close to one another on a polypeptide chain, and tertiary structure refers to the three-dimensional arrangement of all atoms of a polypeptide chain.

Disulfide bonds (Section 4.8E) are important in maintaining tertiary structure. Disulfide bonds are formed between side chains of cysteine by oxidation of two thiol groups (—SH) to form a disulfide bond (—S—S—), as shown:

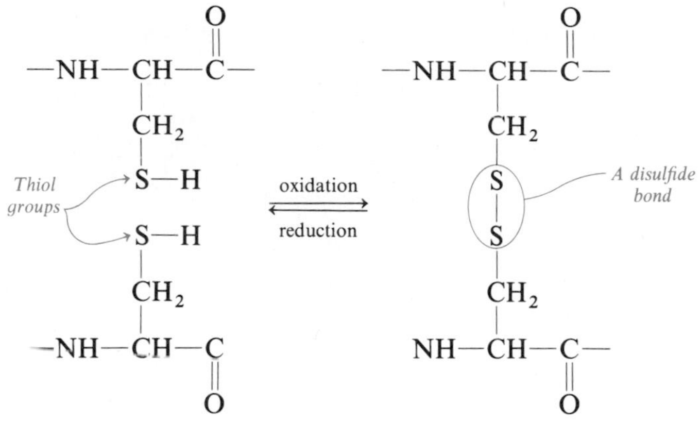

$$
\begin{array}{c}
\overset{\displaystyle O}{\underset{\displaystyle \|}{}} \\
-NH-CH-C- \\
\quad\quad\; | \\
\quad\quad CH_2 \\
\end{array}
\qquad
\begin{array}{c}
\overset{\displaystyle O}{\underset{\displaystyle \|}{}} \\
-NH-CH-C- \\
\quad\quad\; | \\
\quad\quad CH_2 \\
\end{array}
$$

Thiol groups →S—H   oxidation ⇌ reduction   S — *A disulfide bond*

→S—H   S

CH₂   CH₂

—NH—CH—C   NH—CH—C—

O   O

Treatment of a disulfide bond with a reducing agent regenerates the thiol groups.

Scientists have now determined the primary structure for several hundred polypeptides and proteins, and the secondary and tertiary structure of scores of these are also known. Let us look at the three-dimensional structure of myoglobin, for example, a protein found in skeletal muscle and particularly abundant in diving mammals such as seals, whales, and porpoises. Myoglobin and its structural relative, hemoglobin (Figure 13.12), are the oxygen transport and storage molecules of vertebrates. Hemoglobin binds molecular oxygen in the lungs and transports it to myoglobin in muscles. Myoglobin stores molecular oxygen until it is needed for metabolic oxidation.

Myoglobin consists of a single polypeptide chain of 153 amino acids. The complete amino acid sequence (primary structure) of the chain is known. Myoglobin also contains a single heme unit. Determination of the three-dimensional structure of myoglobin represented a milestone in the study of molecular architecture. J. C. Kendrew, for his contribution to this research, shared the Nobel Prize in chemistry in 1963. The secondary and tertiary structure of myoglobin are shown in Figure 13.11. The single polypeptide chain is folded into a complex, almost boxlike shape.

A more detailed analysis has revealed the exact location of all atoms of the peptide backbone and also the location of all side chains. The important structural features of myoglobin are:

1. The backbone consists of eight relatively straight sections of α-helix, each separated by a bend in the polypeptide chain. The longest section of α-helix has 23 amino acids, the shortest has 7. Some 75% of the amino acids are found in these eight regions of α-helix.

2. Hydrophobic side chains, such as those of phenylalanine, alanine, valine, leucine, isoleucine, and methionine, are clustered in the interior of the molecule, where they are shielded from contact with water. Hydrophobic interactions between nonpolar side chains are important in directing the folding of the polypeptide chain of myoglobin into this compact, three-dimensional shape.

3. The outer surface of myoglobin is coated with hydrophilic side chains, such as those of lysine, arginine, serine, glutamic acid, histidine, and glutamine, which

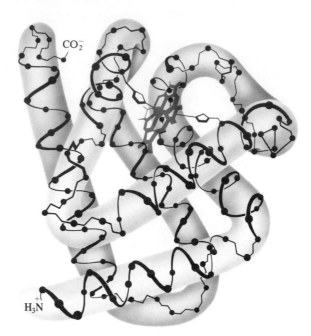

CO_2^-

$\overset{+}{H_3N}$

Figure 13.11 The three-dimensional structure of myoglobin. The heme group is shown in color. The N-terminal amino acid (indicated by $-NH_3^+$) is at the lower left and the C-terminal amino acid (indicated by $-CO_2^-$) is at the upper left.

interact with the aqueous environment by hydrogen bonding. The only polar side chains that point to the interior of the myoglobin molecule are those of two histidines. These side chains can be seen in Figure 13.11 as five-membered rings pointing inward toward the heme group.

4. Oppositely charged amino acids close to each other in the three-dimensional structure interact by electrostatic attractions called salt linkages. An example of a salt linkage is the attraction of the side chains of lysine ($-NH_3^+$) and glutamic acid ($-CO_2^-$).

The tertiary structures of several other globular proteins have also been determined. It is clear that globular proteins contain α-helix and β-pleated sheet structures, and also that the relative amounts of each vary widely. Lysozyme, with 129 amino acids in a single polypeptide chain, has only 25% of its amino acids in α-helix regions. Cytochrome, with 104 amino acids in a single polypeptide chain, has no α-helix structure but does contain several regions of β-pleated sheet. Yet, whatever the proportions of α-helix, β-pleated sheet, or other periodic structure, virtually all nonpolar side chains of globular proteins are directed toward the interior of the molecule, while polar side chains are on the surface of the molecule and are in contact with the aqueous environment. Thus the same type of hydrophobic/hydrophilic interactions that are responsible for formation of soap micelles (Section 12.2B) and phospholipid bilayers (Section 12.4B) are responsible for the three-dimensional shapes of globular proteins.

Example 13.5

With which of the following amino acid side chains can the side chain of threonine form hydrogen bonds?

a. valine   b. asparagine   c. phenylalanine
d. histidine   e. tyrosine   f. alanine

☐ *Solution*

The side chain of threonine contains a hydroxyl group that can participate in hydrogen bonding in two ways: oxygen has a partial negative charge and can function as a hydrogen bond acceptor; hydrogen has a partial positive charge and can function as a hydrogen bond donor. Therefore, the side chain of threonine can function as a hydrogen bond acceptor for the side chains of tyrosine, asparagine, and histidine. The side chain of threonine can also function as a hydrogen bond donor for the side chains of tyrosine, asparagine, and histidine.

Problem 13.5

At pH 7.4, with what amino acid side chains can the side chain of lysine form salt linkages?

D. Quaternary Structure

Most proteins of molecular weight greater than 50,000 consist of two or more noncovalently linked polypeptide chains. The arrangement of protein monomers in an aggregation is known as a **quaternary (4°) structure**. A good example is hemoglobin, a protein that consists of four separate protein monomers: two α-chains of 141 amino acids each and two β-chains of 146 amino acids each. The quaternary structure of hemoglobin is shown in Figure 13.12.

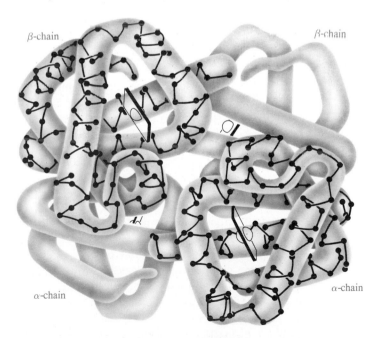

Figure 13.12 The quaternary structure of hemoglobin, showing the four subunits packed together. The flat disks represent four heme units.

The chief factor stabilizing the aggregation of protein subunits is hydrophobic interaction. When separate monomers fold into compact three-dimensional shapes to expose polar side chains to the aqueous environment and shield non-polar side chains from water, there are still hydrophobic "patches" on the surface, in contact with water. These patches can be shielded from water if two or more monomers assemble so their hydrophobic patches are in contact. The molecular weights, numbers of subunits, and biological functions of several proteins with quaternary structure are shown in Table 13.7.

Table 13.7 Quaternary structure of selected proteins.

Protein	Mol. Wt.	Number of Subunits	Subunit Mol. Wt.	Biological Function
insulin	11,466	2	5,733	a hormone regulating glucose metabolism
hemoglobin	64,500	4	16,100	oxygen transport in blood plasma
alcohol dehydrogenase	80,000	4	20,000	an enzyme of alcoholic fermentation
lactate dehydrogenase	134,000	4	33,500	an enzyme of anaerobic glycolysis
aldolase	150,000	4	37,500	an enzyme of anaerobic glycolysis
fumarase	194,000	4	48,500	an enzyme of the tricarboxylic acid cycle
tobacco mosaic virus	40,000,000	2200	17,500	plant virus coat

E. Denaturation

Globular proteins found in living organisms are remarkably sensitive to changes in environment. Relatively small changes in pH, temperature, or solvent composition, even for only a short period, may cause them to become denatured. **Denaturation** causes physical change, the most observable result of which is loss of biological activity. Except for cleavage of disulfide bonds, denaturation stems from changes in secondary, tertiary, or quaternary structure through disruption of noncovalent interactions, such as hydrogen bonds, salt linkages, and hydrophobic interactions. Common denaturing agents include the following:

1. Heat. Most globular proteins become denatured when heated above 50–60°C. For example, boiling or frying an egg causes egg-white protein to become denatured, forming an insoluble mass.

2. Large changes in pH. Adding concentrated acid or alkali to a protein in aqueous solution causes changes in the charged character of ionizable side chains and interferes with salt linkages. For example, in certain clinical chemistry tests where it is necessary first to remove any protein material, trichloroacetic acid (a strong organic acid) is added to denature and precipitate any protein present.

3. Detergents. Treating a protein with sodium dodecylsulfate (SDS, Section 12.2C), a detergent, causes the native conformation to unfold and exposes the nonpolar protein side chains to the aqueous environment. These side chains are then stabilized by hydrophobic interaction with hydrocarbon chains of the detergent.

4. Organic solvents such as alcohols, acetone, or ether.

5. Mechanical treatment. Most globular proteins are denatured in aqueous solution if they are stirred or shaken vigorously. An example is whipped egg whites.

6. Urea and guanidine hydrochloride cause disruption of protein hydrogen bonding and hydrophobic interactions. Because urea is a small molecule with a high degree of polarity, it is very soluble in water. A solution of 8M urea (480 g urea/L of water) is commonly used to denature proteins. Guanidine is a derivative of urea in which C=O is replaced by =NH. Guanidine is a strong base and reacts with HCl and to form the salt guanidine hydrochloride.

$$
\underset{\text{urea}}{H_2N-\overset{\overset{\displaystyle O}{\|}}{C}-NH_2}
\qquad
\underset{\text{guanidine}}{H_2N-\overset{\overset{\displaystyle NH}{\|}}{C}-NH_2}
\qquad
\underset{\substack{\text{guanidine}\\\text{hydrochloride}}}{H_2N-\overset{\overset{\displaystyle NH_2^+\ Cl^-}{\|}}{C}-NH_2}
$$

Denaturation can be partial or complete. It can also be reversible or irreversible. For example, the hormone insulin can be denatured with 8M urea and then the three disulfide bonds reduced to —SH groups. If urea is then removed and the disulfide bonds re-formed by oxidation, the resulting molecule has less than 1% of its former biological activity. In this case, denaturation is both complete and irreversible. Consider another example, ribonuclease, an enzyme that consists of a single polypeptide chain of 124 amino acids folded into a compact, three-dimensional structure partly stabilized by four disulfide bonds. Treatment of ribonuclease with urea causes the molecule to unfold, and the disulfide bonds can then be reduced to thiol groups. At this point, the protein is completely denatured—it has no biological activity. If urea is removed from solution and the thiol groups reoxidized to disulfide bonds, the protein regains its full biological activity. In this instance, denaturation has been complete but reversible.

F. 1° Structure Determines 2°, 3°, and 4° Structure

The primary structure of a protein is determined by information coded within genes. Once the primary structure of a polypeptide is established, the structure itself directs the folding of the polypeptide chain into a three-dimensional structure. In other words, information inherent in the primary structure of a protein determines its secondary, tertiary, and quaternary structures.

If the three-dimensional shape of a polypeptide or protein is determined by its primary structure, how can we account for the observation that denaturation is reversible for some proteins and not others?

The reason for this difference in behavior from one protein to another is that some proteins, like ribonuclease, are synthesized as single polypeptide chains, which then fold into unique three-dimensional structures with full biological activity. Others, like insulin, are synthesized as larger molecules that are not biologically active at first but "activated" later by specific enzyme-catalyzed peptide-bond cleavage. Insulin is synthesized in the beta cells of the pancreas as a single polypeptide chain of 84 amino acids. This molecule, called proinsulin, has no biological activity. When insulin is needed, a section of 33 amino acids is hydrolyzed from proinsulin in an enzyme-catalyzed reaction to produce the active hormone (Figure 13.13). Bovine insulin contains 51 amino acids in two polypeptide chains. The A chain contains 21 amino acids and has glycine (gly) at the $-NH_3^+$ terminus and asparagine (asn) at the $-CO_2^-$ terminus. The B chain contains 30 amino acids with phenylalanine (phe) at the $-NH_3^+$ terminus and alanine (ala) at the $-CO_2^-$ terminus.

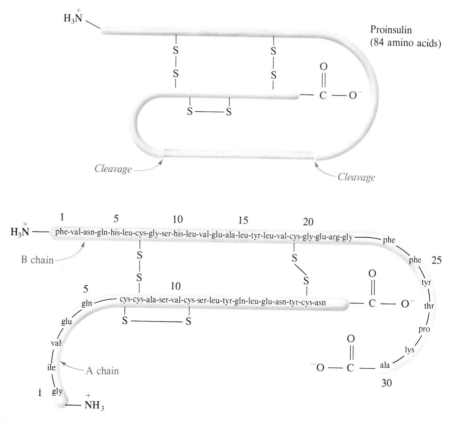

Figure 13.13 (Top) A schematic diagram of proinsulin, a single polypeptide chain of 84 amino acids. (Bottom) The amino acid sequence of bovine insulin.

Because the information directing the original folding of the single polypeptide chain of proinsulin is not present in the A and B chains of the active hormone, refolding of the denatured protein is irregular and denaturation is irreversible.

Zymogens are enzymes produced as inactive proteins, which are then activated by cleavage of one or more of the polypeptide bonds. The process of producing a protein in an inactive, storage form is common. For example, the digestive enzymes trypsin and chymotrypsin are produced in the pancreas as inactive proteins, named trypsinogen and chymotrypsinogen. There is a logical and simple reason for the synthesis of zymogens. In the case of trypsin and chymotrypsin, their function is to catalyze the hydrolysis of dietary proteins reaching the intestinal track. Proteins there are hydrolyzed to their component amino acids and then absorbed through the wall of the intestine into the bloodstream. If trypsin and chymotrypsin were produced as active enzymes, they might well catalyze their own hydrolysis as well as that of other proteins in the pancreas—in effect, a "self-destruct" system. But nature has protected against this happening by synthesizing and storing zymogens instead.

13.7   Fibrous Proteins

Fibrous proteins are stringy, physically tough macromolecules composed of rodlike polypeptide chains joined together by several types of cross-linkages to form stable, insoluble structures. The two main classes of fibrous proteins are the keratins of skin, wool, claws, horn, scales, and feathers, and the collagens of tendons and hides.

A.   The Alpha-Keratins

Hair and wool are very flexible and also elastic, so when tension is released, the fibers revert to their original length. At the molecular level, the fundamental structural unit of hair is a polypeptide wound into an α-helix conformation (Figure 13.14). Several levels of structural organization are built from the simple α-helix. First, three strands of α-helix are twisted together to form a larger cable called a protofibril. Protofibrils are then wound into bundles to form an 11-strand cable called a microfibril. These, in turn, are embedded in a larger matrix that ultimately forms a hair fiber. When hair is stretched, hydrogen bonds along turns of each α-helix are elongated. The main force causing stretched hair fibers to return to their original length is re-formation of hydrogen bonds in the α-helices.

Microfibril   Protofibril   α-Helix

Figure 13.14 Detailed structure of a hair fiber.

The α-keratins of horns and claws have essentially the same structure as hair but with a much higher concentration of cysteine and a greater degree of disulfide cross-linking between individual helices. These additional disulfide bonds greatly increase resistance to stretching and produce the hard keratins of horn and claw.

B. Collagen Triple Helix

Collagens are constituents of skin, bone, teeth, blood vessels, tendons, cartilage, and connective tissue. They are the most abundant protein in higher vertebrates and make up almost 30% of total body mass in humans. Table 13.8 lists the collagen content of several tissues. Note that bone, the Achilles tendon, skin, and the cornea of the eye are largely collagen.

Table 13.8
Collagen content of some body tissues.

Tissue	Collagen (% Dry Weight)
bone, mineral-free	88
Achilles tendon	86
skin	72
cornea	68
cartilage	46–63
ligament	17
aorta	12–24

Because collagen is abundant and widely distributed in vertebrates and because it is associated with a variety of diseases and problems of aging, more is known about this fibrous protein than about any other. Collagen molecules are very large and have a distinctive amino acid composition. One-third of all amino acids in collagen are glycine, and another 21% are either hydroxylysine or hydroxyproline (Section 13.1D). Both hydroxylated amino acids are formed after their parent amino acids (L-proline and L-lysine) and incorporated in collagen molecules. Because cysteine is almost entirely absent, there are no disulfide cross-links in collagen. When collagen fibers are boiled in water, they are converted to insoluble gelatins.

The polypeptide chains of collagen fold into a conformation that is particularly stable and unique to collagen. In this conformation, three protein strands wrap around each other to form a left-handed superhelix called the **collagen triple helix**. This unit, called tropocollagen, looks much like a three-stranded rope (Figure 13.15).

The hydroxyl groups of hydroxyproline and hydroxylysine residues help to maintain the triple helix structure by forming hydrogen bonds between adjacent chains. Fibers in which proline and lysine groups have not been hydroxylated are far less stable than fibers in which these groups have been hydroxylated. One of the important functions of vitamin C is in hydroxylation of collagen. Without adequate supplies of vitamin C, collagen metabolism is impaired, giving

Figure 13.15 Collagen triple helix.

rise to scurvy, a condition in which tropocollagen fibers do not form stable, physically tough fibers. Scurvy produces skin lesions, fragile blood vessels, and bleeding gums.

Collagen fibers are formed when many tropocollagen molecules line up side by side in a regular pattern and are then cross-linked by newly formed covalent bonds. Most covalent cross-linking involves the side chains of lysines. The extent and type of cross-linking vary with age and physiological condition. For example, the collagen of rat Achilles tendon is highly cross-linked, and collagen of the more flexible tendon of rat tail is less highly cross-linked. Further, it is not clear when, if ever, the process of cross-linking is completed. Some believe it continues throughout life, producing increasingly stiffer skin, blood vessels, and other tissues, which then contribute to the medical problems of aging and the aged.

13.8 Plasma Proteins: Examples of Globular Proteins

Human blood consists of a fluid portion (plasma) and cellular components. The cellular components, which make up 40–45% of the volume of whole blood, consist of red blood cells (erythrocytes), white blood cells (leukocytes), and blood platelets. Human plasma consists largely of water (90–92%) in which are dissolved various inorganic ions and a heterogeneous mixture of organic molecules, the largest groups of which are the **plasma proteins**. The earliest method of separating plasma proteins into fractions used ammonium sulfate to "salt out" different types of proteins. The fraction precipitated from plasma 50% saturated with ammonium sulfate was called **globulin**. The fraction not precipitated at this salt concentration but precipitated from plasma saturated with ammonium sulfate was called **albumin**. Today, electrophoresis is the most common method of separating proteins of biological fluids into fractions, especially in the clinical laboratory, where it is used routinely to measure proteins in human plasma, urine, and cerebrospinal fluid. It is estimated that between 15 and 20 million plasma-protein electrophoretic analyses are carried out each year in the United States and Canada.

In plasma-protein electrophoresis, a sample of plasma is applied as a narrow line to a cellulose acetate strip. The ends of the strip are then immersed in a buffer of pH 8.8, and a voltage is applied to the strip. At pH 8.8, plasma proteins have net negative charges and migrate toward the positive electrode. After a predetermined time, the cellulose acetate strip is removed, dried, and sprayed with a dye that selectively stains proteins. The separated protein fractions then appear on the developed strip as spots [Figure 13.16(a)]. The amount of protein in each spot is determined using a densitometer to measure intensity versus width of each spot. The concentration of protein in each spot is proportional to the area under each peak.

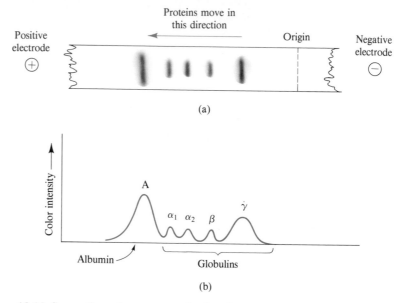

Figure 13.16 Separation of serum proteins by electrophoresis. (a) A sample is applied as a narrow line at the origin. After electrophoresis at pH 8.8, the paper is dried and stained. (b) A plot of color intensity of each spot.

Electrophoresis on cellulose acetate separates serum proteins into five large fractions: one albumin fraction and four globulin fractions. The four globulin fractions are arbitrarily designated α_1, α_2, β, and γ according to their electrophoretic mobilities. Serum albumin has an isoelectric point of about 4.9 and migrates farthest toward the positive electrode. Gamma-globulin has an isoelectric point of about 7.4 and migrates the shortest distance. Shown in Table 13.9 are the concentrations of the five large protein fractions of human serum.

Table 13.9
Concentrations of the important human serum proteins as determined by electrophoresis.

Fraction	(g/100 mL)	Total Protein (%)
albumin	3.5–5.0	52–67
globulins		
α_1	0.1–0.4	2.5–4.5
α_2	0.5–1.1	6.6–13.6
β	0.6–1.2	9.1–14.7
γ	0.5–1.5	9.0–21.6

The primary function of albumins is to regulate the osmotic pressure of blood. In addition, albumins are important in transporting fatty acids and certain drugs such as aspirin and digitalis. The α_1 and α_2 fractions transport other biomolecules, such as fats, steroids, and phospholipids and various other lipids. The α_1 fraction also contains antitrypsin, a protein that inhibits the protein-digesting enzyme trypsin. The α_2 fraction contains haptoglobulin, which binds any hemoglobin released from destroyed red blood cells, and ceruloplasmin, the principal copper-containing protein of the body. The α_2 fraction also contains prothrombin, an inactive form of the blood-clotting enzyme thrombin. The β fraction contains a variety of specific transport proteins, as well as substances involved in blood clotting.

The γ-globulin fraction consists primarily of **antibodies (immunoglobulins)**, whose function is to combat **antigens** (foreign proteins) introduced into the body. Specific antibodies are formed by the immune system in response to specific antigens. The response is the basis for immunization against such infectious diseases as polio, tetanus, and diphtheria. An antibody consists of a combination of heavy (high-molecular-weight) and light (low-molecular-weight) polypeptide chains held together by disulfide bonds (Figure 13.17). Each antibody has two identical binding sites that react with specific antigens to form an insoluble complex called precipitin (Figure 13.18). Formation of precipitin deactivates the antigen and permits its removal and breakdown by white blood cells.

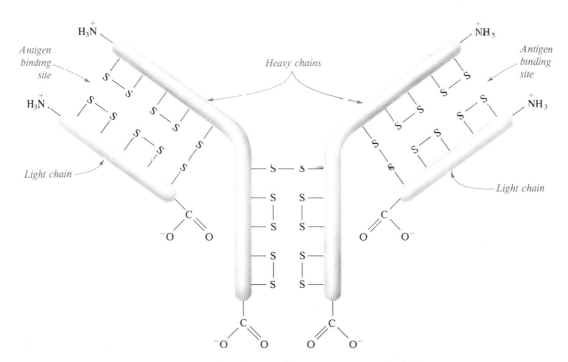

Figure 13.17 The three-dimensional shape of an antibody.

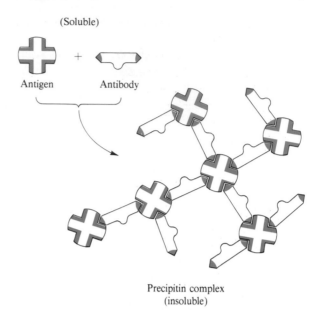

(Soluble)

Antigen   Antibody

Precipitin complex
(insoluble)

Figure 13.18 The action of an antibody and its specific antigen to form an inactive precipitin complex. The precipitated antigen-antibody complex is then ingested and broken down by white blood cells.

Key Terms and Concepts

albumin (13.8)	immunoglobulin (13.8)
amino acid (13.1A)	ionization of amino acids (13.2A)
antibody (13.8)	isoelectric point (13.2D)
antigen (13.8)	α-keratin (13.7A)
chirality of amino acids (13.1B)	peptide bond (13.3)
collagen (13.7B)	plasma protein (13.8)
collagen triple helix (13.7B)	β-pleated sheet (13.6B)
denaturation (13.6E)	primary structure (13.5)
electrophoresis (13.2E)	quaternary structure (13.6D)
fibrous protein (13.7)	secondary structure (13.6B)
geometry of a peptide bond (13.6A)	C-terminal amino acid (13.3)
globular protein (13.8)	N-terminal amino acid (13.3)
globulins (13.8)	tertiary structure (13.6C)
α-helix (13.6B)	titration of amino acids (13.2C)
hydrophobic side chain (13.1C)	zwitterion (13.1A)

Problems

Amino acids
(Section 13.1)

13.6   Explain the meaning of the designation L as it is used to indicate the stereochemistry (configuration) of amino acids found in proteins.

13.7 How do L-serine and D-glyceraldehyde compare in configuration about the chiral carbon? L-serine and L-glyceraldehyde?

13.8 **a.** Which amino acid found in proteins has no chiral carbon?
      b. Which amino acids found in proteins have two chiral carbons?

13.9 How many stereoisomers are possible for:
      a. L-4-hydroxyproline   **b.** L-5-hydroxylysine   **c.** ornithine
      d. citrulline   **e.** thyroxine

Acid-base properties of amino acids (Section 13.2)

13.10 For amino acids with nonionizable side chains, the value of pI (the isoelectric point) can be calculated from the equation

$$pI = \tfrac{1}{2}(pK_a \text{ of } \alpha\text{-CO}_2\text{H} + pK_a \text{ of } \alpha\text{-NH}_3^+)$$

Given the following values of pK_a, calculate the isoelectric point of each amino acid listed.

	pK_a of α-CO$_2$H	pK_a of α-NH$_3^+$
Glycine	2.35	9.78
Serine	2.21	9.15

13.11 For amino acids with a side-chain —CO$_2$H group, the value of pI can be calculated from the equation

$$pI = \tfrac{1}{2}(pK_a \text{ of } \alpha\text{-CO}_2\text{H} + pK_a \text{ of side-chain } -\text{CO}_2\text{H})$$

Given the following values of pK_a, calculate the pI of the following amino acids. Compare these values with those given in Table 13.6.

	pK_a of α-CO$_2$H	pK_a of α-NH$_3^+$	pK_a of Side-Chain —CO$_2$H
Asparatic acid	2.10	9.82	3.86
Glutamic acid	2.10	9.47	4.07

13.12 For amino acids with side-chain amino groups, the value of pI can be calculated from the equation

$$pI = \tfrac{1}{2}(pK_a \text{ of } \alpha\text{-NH}_3^+ + pK_a \text{ of side chain amino group})$$

Given the following values of pK_a, calculate the pI of each amino acid. Compare these values with those given in Table 13.6.

	pK_a of α-CO$_2$H	pK_a of α-NH$_3^+$	pK_a of Side-Chain Amino Group
Lysine	2.18	8.95	10.53
Arginine	2.01	9.04	12.48

13.13 For the following amino acids, draw a structural formula for the form you expect would predominate at pH 1.0, 6.0, and 12.0. Refer to Section 13.2D and Problems 13.10–13.12 for values of pI.
       a. alanine   **b.** tyrosine   **c.** aspartic acid   **d.** arginine

13.14   **a.** Estimate the pH at which the solubility of alanine in water is a minimum.
b. Why does the solubility of alanine increase as the pH is increased?
c. Why does the solubility of alanine increase as the pH is decreased?

13.15   Which of the following amino acids will migrate toward the (+) electrode and which will migrate toward the (−) electrode on electrophoresis at pH 6.0? at pH 8.6?

a. tyrosine   **b.** arginine   **c.** cysteine
d. aspartic acid   **e.** asparagine   **f.** histidine

Primary structure of polypeptides and proteins (Section 13.3)

13.16   What is the characteristic structural feature of a peptide bond?

13.17   Write a structural formula for the tripeptide glycylserylaspartic acid. Write a structural formula for an isomeric tripeptide. Calculate the net charge on each tripeptide at pH 6.0.

13.18   Write the structural formula for the tripeptide lys-asp-val. Calculate the net charge on this tripeptide at pH 6.0.

13.19   How many tetrapeptides can be constructed from the 20 amino acids
a. if each of the amino acids is used only once in the tetrapeptide?
b. if each amino acid can be used up to four times in the tetrapeptide?

13.20   Write an equation for the oxidation of two molecules of cysteine by O_2 to form a disulfide bond.

13.21   Following is the structural formula of glutathione (GSH), one of the most common small peptides in animals, plants, and bacteria.

$$\overset{+}{H_3N}-CH-CH_2-CH_2-\overset{\displaystyle O}{\overset{\|}{C}}-NH-CH-\overset{\displaystyle O}{\overset{\|}{C}}-NH-CH_2-\overset{\displaystyle O}{\overset{\|}{C}}-O^-$$

with CO_2^- below the first CH, and CH_2 / SH below the middle CH.

glutathione (GSH)

a. Name the amino acids in this tripeptide.
b. What is unusual about the peptide bond formed between the first two amino acids in this tripeptide?
c. Write an equation for the reaction of two molecules of glutathione with O_2 to form a disulfide bond.

Three-dimensional shapes of polypeptides and proteins (Section 13.6)

13.22   Using structural formulas, show how the theory of resonance accounts for the fact that a peptide bond is planar.

13.23   In constructing models of polypeptide chains, Linus Pauling assumed that for maximum stability, (1) all amide bonds are trans and coplanar, and (2) there is a maximum of hydrogen bonding between amide groups. Examine the alpha-helix (Figure 13.9) and the beta-pleated sheet (Figure 13.10) and convince yourself that in each conformation, amide bonds are planar and each carbonyl oxygen is hydrogen-bonded to an amide hydrogen.

13.24 Examine the alpha-helix conformation. Are the amino acid side chains arranged all inside the helix or all outside the helix, or are they randomly oriented?

13.25 Draw structural formulas to illustrate the noncovalent interactions indicated:
a. Hydrogen bonding between the side chains of thr and asn.
b. Salt linkage between the side chains of lys and glu.
c. Hydrophobic interactions between the side chains of two phenylalanines.

13.26 Consider a typical globular protein in an aqueous medium of pH 6.0. Which of the following amino acid side chains would you expect to find on the outside and in contact with water? which on the inside and shielded from contact with water?
a. glutamic acid   **b.** glutamine   **c.** arginine
d. serine   **e.** valine   **f.** phenylalanine
g. lysine   **h.** isoleucine   **i.** threonine

13.27 Which of the following peptides and proteins migrate to the (+) electrode and which migrate to the (−) electrode on electrophoresis at pH 6.0? at pH 8.6?
a. ala-glu-ile   **b.** gly-asp-lys   **c.** val-ala-leu
d. tyr-trp-arg   **e.** thyroglobulin (pI 4.6)   **f.** hemoglobin (pI 6.8)

13.28 The following proteins have approximately the same molecular weight and size. The pI of each is given.

carboxypeptidase   pI 6.0
pepsin   pI 1.0
human growth hormone   pI 6.9
ovalbumin   pI 4.6

a. State whether each is positively charged, negatively charged, or uncharged at pH 6.0.
b. Draw a diagram showing the results of electrophoresis of a mixture of the four at pH 6.0.

13.29 Examine the primary structure of bovine insulin (Figure 13.13) and list all asp, glu, lys, arg, and his in the molecule. Predict whether insulin has an isoelectric point nearer that of the acidic amino acids (pI 2.0–3.0), the neutral amino acids (pI 5.5–6.5), or the basic amino acids (pI 9.5–11.0).

13.30 Following is the primary structure of glucagon, a polypeptide hormone that helps to regulate glycogen metabolism. Glucagon is secreted by the alpha cells of the pancreas during the fasting state, when blood glucose levels are decreasing. This hormone stimulates the enzymes that catalyze

1   *5*   *10*   *15*
his-ser-glu-gly-thr-phe-thr-ser-asp-tyr-ser-lys-tyr-leu-asp-ser-arg-arg-

20   *25*   *29*
ala-gln-asp-phe-val-gln-trp-leu-met-asn-thr

glucagon

the hydrolysis of glycogen to glucose and thus helps to maintain blood glucose levels within a normal concentration range. Glucagon contains 29 amino acids and has a molecular weight of approximately 3500.

a. Estimate the net charge on glucagon at pH 6.0.

b. Predict whether the isoelectric point of glucagon would be nearer the isoelectric point of the acidic amino acids (pI 2.0–3.0), the neutral amino acids (pI 5.5–6.5), or the basic amino acids (pI 9.5–11.0).

13.31 Myoglobin and hemoglobin are globular proteins. Myoglobin consists of a single polypeptide chain of 153 amino acids. Hemoglobin is composed of four polypeptide chains, two of 141 amino acids and two of 146 amino acids. The three-dimensional structures of myoglobin and hemoglobin polypeptide chains are similar, yet myoglobin exists as a monomer in aqueous solution, while the four polypeptide chains of hemoglobin self-assemble to form a tetramer. Which polypeptide chains, those of myoglobin or hemoglobin, would have a higher percentage of nonpolar amino acids?

13.32 What is irreversible denaturation and how does it differ from reversible denaturation?

13.33 After water, proteins are the chief constituents of most tissues. Often in the analysis of lesser constituents, it is necessary first to remove all proteins. The most common reagents for this purpose are 0.5M trichloroacetic acid, ethanol, or acetone. Explain the basis for using these reagents to deproteinize a solution.

13.34 Is the following statement true or false? Explain your answer. "The principal factor directing the folding of globular proteins in an aqueous environment is hydrogen bonding between polar side chains and water molecules."

13.35 Suppose proinsulin is treated with a disulfide reducing agent in the presence of 8M urea. These reagents are then removed from solution and the denatured protein is allowed to refold in the presence of an oxidizing agent that converts thiols to disulfides. Do you expect that proinsulin after this treatment would have the same conformation as the original molecule or a different conformation? Explain your answer.

Fibrous proteins (Section 13.7)

13.36 What is the most characteristic type of secondary structure in the protein of (a) hair; (b) hooves; (c) collagen.

13.37 What is meant by the following statement? "An alpha-helix is flexible and it is also elastic."

13.38 What is the function of collagen? Describe (a) the macroscopic physical properties of collagen, and (b) the molecular structure of collagen.

Plasma proteins (Section 13.8)

13.39 Of the five main types of serum proteins, which has the highest isoelectric point? Which has the lowest isoelectric point?

13.40 Explain the process of "salting out."

13.41 What is the primary function of serum albumin?

13.42 What is the principal function of the proteins of the gamma-globulin fraction?

Abnormal Human Hemoglobins

There are an estimated 30 billion red blood cells (erythrocytes) in the bloodstream of an adult, each packed with about 270 million molecules of hemoglobin. For sheer numbers, hemoglobin is one of the most plentiful proteins in the body. Hemoglobin's function is to pick up molecular oxygen in the lungs and deliver it to all parts of the body for metabolic oxidation. Normal adult hemoglobin (HbA) is composed of four polypeptide chains: two alpha chains, each of 141 amino acids; and two beta chains, each of 146 amino acids. Each polypeptide chain surrounds one heme group that binds oxygen reversibly. The tetrameric structure of hemoglobin is stabilized principally by hydrophobic interactions. Shown in Figure IX-1 is the three-dimensional shape of a single beta chain. The N-terminal amino acid is indicated by $-NH_3^+$ and the C-terminal amino acid by $-CO_2^-$. The three-dimensional structure of hemoglobin A was determined by Max Perutz. For this pioneering work he shared in the Nobel Prize in 1963.

It is so-called abnormal human hemoglobins that have attracted particular attention because of the diseases associated with them. The best known of these diseases is sickle-cell anemia, a name derived from the characteristic sickle shape of affected red blood cells when they are deoxygenated. When combined with oxygen, red blood cells of persons with sickle-cell anemia have the flat, disclike conformation of normal erythrocytes. However, when oxygen pressure is reduced, affected cells become distorted and considerably more rigid and inflexible than normal cells (Figure IX-2).

Because sickled cells are larger than some of the blood channels through which they must pass, they tend to become wedged in capillaries, thereby blocking flow of blood. Surprisingly little is known about why some organs and tissues are affected more than others by the disease, the normal age at which the disease starts, and male versus female susceptibility. Some persons afflicted with sickle-cell anemia die at an early age, often due to childhood infections complicated by the disease. Others lead long, productive lives.

In 1949, Linus Pauling made a discovery that opened the way to an understanding of this disease at the molecular level. He observed that normal adult hemoglobin (HbA) differs significantly from sickle-cell hemoglobin (HbS). At pH 6.9, HbA has a net negative charge and HbS has a net positive charge. On paper electrophoresis at this pH, HbA moves toward the positive

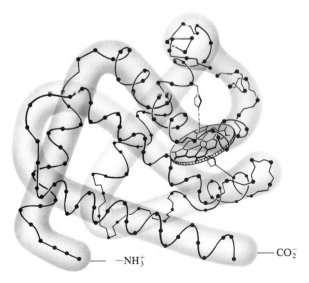

$-NH_3^+$ $-CO_2^-$

Figure IX-1 The beta-chain of hemoglobin.

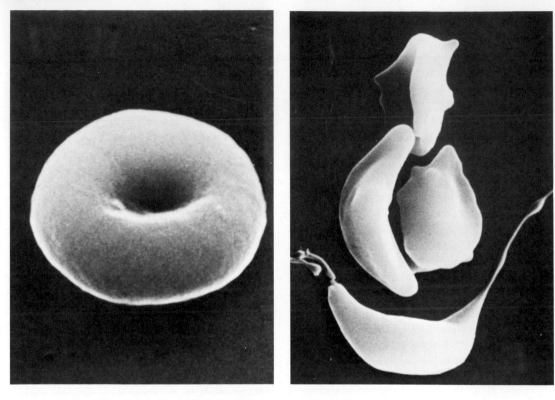

Figure IX-2 Normal red blood cell (left) magnified × 6750 and cells that have sickled (right) after discharging oxygen. Magnified × 8700.

electrode and HbS toward the negative electrode. Vernon Ingram pursued this discovery, and in 1956 showed that sickle-cell hemoglobin differs from normal hemoglobin only in the amino acid at the sixth position of the beta chain. Alpha chains of both are identical, but glutamic acid at position 6 of each beta chain of HbA is replaced by valine in HbS. As a result of the valine–glutamic acid substitution, two negatively charged, hydrophilic side chains are replaced by two uncharged, hydrophobic side chains.

How is the substitution of HbS for HbA in red blood cells related to the process of sickling? We know that HbS functions perfectly normally in transporting molecular oxygen from the lungs to cells. In this regard it is indistinguishable from HbA. However, when it gives up its oxygen, HbS tends to form polymers that separate from solution in crystalline form. There is now good evidence that the basic unit of crystalline HbS polymer is a double-stranded fiber stabilized by hydrophobic interactions, including that between valine at position 6 of one beta chain and a hydrophobic patch on another HbS molecule. Double-stranded HbS polymer molecules then interact to form multistranded cablelike structures. This is a remarkable phenomenon: polymerization of HbS is facilitated by the presence of valine at beta-6, but comparable polymerization of HbA is prevented by the presence of glutamic acid at beta-6.

Now that scientists understand sickle-cell anemia at the molecular level, the challenge is to devise specific medical treatments to prevent, or at least inhibit, the sickling process. One strategy being actively pursued is the search for substances that inhibit the polymerization of HbS by disrupting or preventing hydrophobic interactions of beta-6 valines.

Sickle-cell anemia is a genetic disease. Persons with an HbS gene from only one parent are said to have sickle-cell trait. About 40% of the hemoglobin in these individuals is HbS. Generally no ill effects are associated with sickle-cell trait except under extreme conditions. Persons with HbS genes from both parents are said to have sickle-cell disease, and all their hemoglobin is HbS. The mutant gene coding for HbS occurs in about 10% of black Americans and in about 20% of African blacks. The gene is also present in significant numbers of the populations of

HbS cells sickle at any one time, but the approximately 40% that do sufficiently reduce the severity of the malaria and prevent death.

The dramatic success in discovering the genetic and molecular basis for sickle-cell anemia spurred interest in searching for other abnormal hemoglobins. To date, several hundred have been isolated and the changes in primary structure have been determined. In most, there is but a single amino acid residue change in either the alpha or the beta chain, and each substitution is consistent with the change of a single nucleotide in one DNA codon (Section 15.11A). Several abnormal hemoglobins are listed in Table IX-1.

Table IX-1
Abnormal human hemoglobins. Many of these names are derived from the location of their discovery.

Hemoglobin Variant	Amino Acid Substitution		
	Position	From	To
alpha-chain			
J-Paris	12	ala	asp
G-Philadelphia	68	asn	lys
M-Boston	58	his	tyr
Dakar	112	his	gln
beta-chain			
S	6	glu	val
J-Trinidad	16	gly	asp
F	26	glu	lys
M-Hamburg	63	his	tyr

countries bordering the Mediterranean Sea and parts of India.

That there seems to be so much natural selection pressure against the HbS gene raises questions of why it has persisted so long in the gene pool and why sickle-cell trait is so common in populations of specific parts of the world. Of several explanations offered, the most likely, first advanced in 1949, is that sickle-cell trait provides some protection against *Plasmodium falciparum*, the parasite responsible for the most severe form of malaria. The falciparum parasite lives part of its life cycle in red blood cells and grows equally well in oxygenated cells containing either HbS or HbA. However, when infected cells containing HbS are deoxygenated and they sickle, the parasites living in them are killed. Not all infected

Although most of the abnormal hemoglobins differ from HbA by only a single amino acid substitution, several have been discovered in which there are either insertions or deletions of amino acids. For example, in hemoglobin-Leiden, discovered in 1968, glutamic acid at position 6 in each beta chain is missing altogether. In hemoglobin–Gun Hill (Figure IX-3), discovered in 1967 in a 41-year-old man and one of his three daughters, there is a deletion of five amino acids in each beta chain. Thus, each beta chain is shortened to 141 amino acids. In hemoglobin-Grady (Figure IX-3), discovered in 1974 in a 25-year-old woman and her father, there is insertion of three amino acids in each alpha chain. Thus, each alpha chain is elongated to 144 amino acid residues.

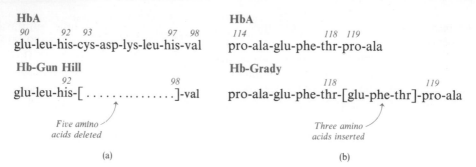

Figure IX-3 Some abnormal human hemoglobins; (a) a deletion mutation and (b) an insertion mutation.

Sources

Dayhoff, M. O. 1972. *Atlas of Protein Sequence and Structure.* Vol. 5. Washington, D. C.; National Biomedical Foundation.

Ingram, B. 1957. Gene mutations in human hemoglobin: The chemical difference between normal and sickle-cell hemoglobin. *Nature* 180:326.

Maugh, T. H. 1981. A new understanding of sickle cell emerges. *Science* 211:265–267.

Morimoto, H.; Lehmann, H.; and Perutz, M. F. 1971. Molecular pathology of human hemoglobins. *Nature* 232:408.

Enzymes

Living cells are unique in their ability to carry out complex reactions with remarkable specificity and remarkable speed. The agents responsible for this property of living matter are a group of protein biocatalysts called **enzymes**, each designed to catalyze a specific reaction or type of reaction. James Sumner, in 1926, was the first to isolate an enzyme in pure crystalline form. The enzyme was urease, which catalyzes the hydrolysis of urea to ammonia and carbon dioxide:

$$\underset{\text{urea}}{H_2N-\overset{\overset{\textstyle O}{\|}}{C}-NH_2} + H_2O \xrightarrow{\text{urease}} 2NH_3 + CO_2$$

All enzymes, it is now clear, are proteins, and the one feature that distinguishes them from other proteins is that they are catalysts.

Practically speaking, enzyme technology has been with us for centuries. The use of enzymes for fermentation of fruit juices and grains to make alcoholic beverages is a long-practiced art. Cheese was and still is made by treating milk with rennin, an enzyme obtained from the lining of calf stomachs. The active ingredient in commercially available meat tenderizers is an enzyme extracted from papaya plants.

Understanding enzymes and their properties is essential to understanding metabolism, metabolic diseases, and therapies for treating metabolic diseases, as well as many of the enzyme-based bioassays currently in use in the clinical chemistry laboratory.

14.1 Nomenclature and Classification of Enzymes

In the early work on metabolism, the numerous enzymes discovered were generally named according to their function; the suffix-*ase* was added to the name of the substrate whose degradation the enzyme catalyzed. Examples are urease, which catalyzes the hydrolysis of urea to carbon dioxide and ammonia, and arginase, an enzyme of the urea cycle (Section 19.2), which catalyzes the hydrolysis of arginine to urea and ornithine.

$$
\underset{\text{arginine}}{\overset{\overset{\displaystyle NH_2^+}{\|}}{H_2NCNHCH_2CH_2CH_2}\underset{\underset{\displaystyle NH_3^+}{|}}{CH}\overset{\overset{\displaystyle O}{\|}}{C}O^-} + H_2O \xrightarrow{\text{arginase}} \underset{\text{urea}}{H_2N\overset{\overset{\displaystyle O}{\|}}{C}NH_2} + \underset{\text{ornithine}}{H_3\overset{+}{N}CH_2CH_2CH_2\underset{\underset{\displaystyle NH_3^+}{|}}{CH}\overset{\overset{\displaystyle O}{\|}}{C}O^-}
$$

The practice of naming enzymes according to their function was soon extended to enzymes involved in nondegradative reactions and also to the type of reaction catalyzed. For example, enzymes catalyzing oxidations became known as oxidases (for instance, glucose oxidase). Others became classified as reductases, isomerases, synthetases (for instance, triose phosphate isomerase, citrate synthetase), and so on. Growth of this unsystematic nomenclature presented many problems, including more than one name for a single enzyme.

To overcome these difficulties, the International Union of Biochemistry in 1961 adopted the recommendations of its **Enzyme Commission** (EC) for a systematic nomenclature and classification of enzymes. For each enzyme, the EC proposed a systematic name and a unique numerical designation. In the Enzyme Commission classification, enzymes are divided into the following six large classes:

1. Oxidoreductases, which catalyze oxidation-reduction reactions.
2. Transferases, which catalyze the transfer of a group of atoms from one substrate to another or from one part of a substrate to another.
3. Hydrolases, which catalyze hydrolysis of esters, anhydrides, amides, imines, and the like.
4. Lyases, which catalyze nonhydrolytic removal of groups to form a double bond (for example, removal of HOH from an alcohol to form an alkene) or addition of groups to a double bond (for example, addition of HOH to an alkene to form an alcohol).
5. Isomerases, which catalyze isomerizations.
6. Ligases, which catalyze covalent bond formation coupled with hydrolysis of adenosine triphosphate (ATP) or similar nucleoside triphosphate.

However, because systematic names and numerical designations are often cumbersome, the Enzyme Commission also recommends a single, unique common name for each enzyme. Many of these common names are derived by naming the substrate and the type of reaction, and including the suffix -*ase*. An example is the common name of the enzyme that catalyzes the oxidation of lactate to pyruvate. Oxidations of this type, in which hydrogen atoms are removed from adjacent atoms, are often called dehydrogenations. In this enzyme-catalyzed reaction, the oxidizing agent is nicotinamide adenine dinucleotide (NAD$^+$).

$$\overset{\overset{\displaystyle OH}{|}}{CH_3CHCO_2^-} + NAD^+ \xrightarrow{\text{LDH}} \overset{\overset{\displaystyle O}{\|}}{CH_3CCO_2^-} + NADH + H^+$$

lactate   pyruvate

Substrate: lactate
Reaction type: dehydrogenation
Common name: lactate dehydrogen(ation) + *ase* = lactate dehydrogenase
Abbreviation: LDH

Because of the simplicity and wide use of these Enzyme Commission–recommended common names, we shall use them throughout the text.

There is a further problem to be dealt with in enzyme nomenclature and classification—namely, that a single organism or even a single cell may produce two or more enzymes that catalyze the same reaction. Such enzymes are termed isoenzymes, or isozymes. For example, lactate dehydrogenase (LDH) occurs as five isoenzymes, designated LDH_1, LDH_2, LDH_3, LDH_4, and LDH_5. (For further discussion of the isoenzymes of LDH and their use in diagnosis of myocardial infarction, see Mini-Essay X, "Clinical Enzymology—The Search for Specificity".) Although isoenzymes are similar in structure and physical properties, they do show certain differences. The property by which they are most commonly detected, separated, and identified is their behavior on electrophoresis. Because most isoenzymes have slightly different sizes, shapes, and net charges, they migrate at different rates. The Enzyme Commission has recommended that isoenzymes be designated by arabic numerals 1, 2, 3, . . . , with the lowest number given to the form that migrates most rapidly toward the anode during electrophoresis.

14.2 Cofactors, Prosthetic Groups, and Coenzymes

The enzymes that act as biocatalysts show considerable diversity of structure. Many enzymes are simple proteins, which means that the protein itself is the true catalyst. Other enzymes catalyze reactions of their substrates only when specific nonprotein molecules or metal ions are present. A cofactor is any nonprotein molecule or ion that is essential for the enzyme-catalyzed reaction. A prosthetic group can be defined similarly. The distinction between a cofactor and a prosthetic group is one of degree. A nonprotein molecule or ion that is very tightly bound to an enzyme is called a **prosthetic group**; one that is more loosely bound is called a **cofactor**. Since this is not an exact definition, it is entirely possible that what is a cofactor for one enzyme may be a prosthetic group for another. In the following discussion, we will concentrate on the types of nonprotein molecules and ions required for enzyme activity rather than the precise terminology for each.

A coenzyme is a true substrate for an enzyme-catalyzed reaction, and because it is a substrate, it undergoes a chemical change during the particular enzyme-catalyzed reaction. Unlike other substrates, a modified coenzyme is recycled by another metabolic pathway, and therefore can be used over and over for the same reaction or type of reaction.

A. Metal Ions

Metal ions function primarily by forming complexes with an enzyme proper or with other nonprotein groups required by the enzyme for catalytic activity. In some cases, metal ions appear to be only loosely associated with an active enzyme and can be removed easily from it. In other instances, metal ions are integral parts of the enzyme structure and are retained throughout isolation and purification procedures. As an example, virtually all reactions that involve adenosine triphosphate (ATP) and the hydrolysis of phosphate anhydride bonds require Mg^{2+} as a cofactor. In these reactions, the positively charged cation coordinates with the negatively charged oxygens of the phosphate groups. As another example, carbonic anhydrase requires one Zn^{2+} per molecule of enzyme for activity.

In recent years, we have come to realize that many other metals, some of them required only in trace amounts, are also essential for proper enzyme function in humans; hence, they are required for good health. In many instances, we have little or no understanding of how these metal ions function or why they are essential.

B. Prosthetic Groups

One important class of prosthetic groups comprises the hemes. The structure of heme consists of four substituted pyrrole rings joined by one-carbon bridges into a larger ring called porphyrin [Figure 14.1(a)]. Shown in Figure 14.1(b) is the heme group found in hemoglobin and myoglobin. Note that there is a metal atom in the center of the heme group. In hemoglobin and myoglobin, an iron atom occurs as Fe^{2+}. A magnesium ion, Mg^{2+}, embedded in a porphyrin ring is an integral part of chlorophyll.

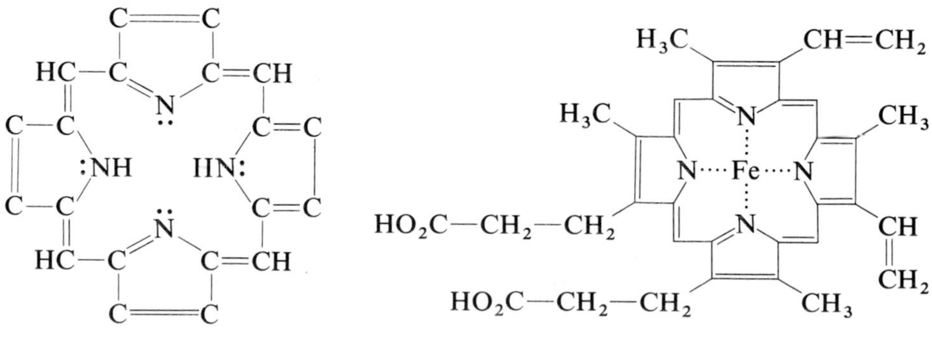

(a)   (b)

Figure 14.1 Heme coenzymes: (a) the porphyrin ring system; (b) the heme prosthetic group of hemoglobin, myoglobin, and certain enzymes.

C. Coenzymes and Vitamins

A **coenzyme** is a small organic molecule that binds reversibly to an enyme and functions as a second substrate for the enzyme. For example, the enzyme lactate dehydrogenase (LDH) requires nicotinamide adenine dinucleotide (NAD^+) for activity:

$$\underset{\text{lactate}}{CH_3\overset{\overset{\displaystyle OH}{|}}{C}HCO_2^-} + NAD^+ \xrightarrow{\underset{\text{dehydrogenase}}{\text{lactate}}} \underset{\text{pyruvate}}{CH_3\overset{\overset{\displaystyle O}{||}}{C}CO_2^-} + NADH + H^+$$

It should be obvious why NAD^+ is required—the organic molecule NAD^+ oxidizes lactate to pyruvate and in the process is reduced to NADH. Other metabolic pathways, chief among them electron transport and oxidative phosphorylation reoxidize NADH to NAD^+, so this vital coenzyme can be used over and over for further substrate oxidations.

Humans and many other organisms cannot synthesize certain coenzymes, and therefore, they must obtain from their diet either the coenzyme itself or a substance from which the coenzyme can be synthesized. These so-called essential coenzymes or coenzyme precursors are vitamins. Vitamins are divided into two classes, based on physical properties: those that are water-soluble and those that are fat-soluble. As might be expected, water-soluble vitamins are highly polar, hydrophilic substances, while fat-soluble vitamins are nonpolar, hydrophobic substances. Most water-soluble vitamins are either coenzymes themselves or small molecules from which coenzymes are synthesized within the body.

Table 14.1 lists the nine water-soluble vitamins required in human diets, the coenzyme derived from each, and the function of each coenzyme.

Table 14.1
Nine water-soluble vitamins, the coenzymes derived from each, and their biological function.

Vitamin	Coenzyme	Function of Coenzyme
D_3, nicotinic acid	nicotinamide adenine dinucleotide	oxidation-reduction
B_2, riboflavin	flavin adenine dinucleotide flavin mononucleotide	oxidation-reduction
C, ascorbic acid	————	oxidation-reduction
B_1, thiamine	thiamine pyrophosphate	decarboxylation
B_6, pyridoxine	pyridoxal phosphate	transfer of $-NH_2$ groups
pantothenic acid	coenzyme A	transfer of CH_3CO- groups
folic acid	tetrahydrofolic acid	transfer of $-CH_3$, $-CH_2OH$, and $-CHO$ groups
biotin	biocytin	transfer of $-CO_2H$ groups
B_{12}, cobalamin	coenzyme B_{12}	transfer of $-CO-SCoA$ groups

Table 14.2
Chief nutritional
sources of the
water-soluble
vitamins and
clinical symptoms
that result from
their deficiency.

Vitamin	Source	Effects of Deficiency
B_3, nicotinic acid	liver, lean meats, cereals, and legumes	pellagra; dermatitis; nervous and digestive problems
B_2, riboflavin	most foods; milk and meat products are especially rich	cornea vascularization; lesions of the skin, especially on the face
C, ascorbic acid	plants, especially rapidly growing fruits and vegetables	scurvy; failure to form connective tissue
B_1, thiamine	fresh vegetables, husks of cereal grains, liver and other organ meats	beriberi; neuritis; heart failure; mental disturbances
B_6, pyridoxine	meats, cereal grains, lentils	nervous disorders; dermatitis; lesions on the face
pantothenic acid	most foods, especially liver, meat, cereals, milk, fresh vegetables	vomiting; abdominal distress; insomnia
folic acid	organ meats, fresh green vegetables	anemia
biotin	yeast, meats, dairy products, grains, fruits and vegetables; intestinal bacteria	scaly skin; muscle pains; weakness; depression
B_{12}, cobalamin	meat and milk products	pernicious anemia

Table 14.2 lists the chief nutritional sources of each water-soluble vitamin and the clinical symptoms that result from their deficiency.

14.3 Mechanism of Enzyme Catalysis

A. Formation of Enzyme-Substrate Complex

Enzymes function as catalysts in much the same way as common inorganic or organic laboratory catalysts. A catalyst, whatever the kind, combines with a reactant to "activate" it. In enzyme-catalyzed reactions, reactants are referred to as substrates (S). In the first and critical step in all enzyme catalysis, enzyme and substrate combine to form an activated complex called an **enzyme-substrate (ES) complex**. This complex then undergoes a chemical change to form an enzyme-product complex from which one product or more then dissociate and regenerate the enzyme.

$$E + S \rightleftharpoons ES$$
(enzyme-substrate complex)

$$ES \longrightarrow EP$$
(enzyme-product complex)

$$EP \longrightarrow E + P$$

Interactions between an enzyme and a substrate generally involve noncovalent forces such as ion-ion or ion-dipole interactions, hydrogen bonding, and dispersion forces. In some cases, however, actual covalent bonds are formed between the enzyme and substrate. Whatever the types of interaction between enzyme and substrate and between enzyme and product, they must be sufficiently weak that the enzyme-product complex can break apart to liberate product and regenerate the enzyme.

B. Active Site

Virtually all enzymes are globular proteins, and even the simplest have molecular weights ranging from 12,000 to 40,000, meaning that they consist of 100–400 amino acids. It has been proposed that, because enzymes are so large compared with molecules whose reactions they catalyze, substrate and enzyme interact over only a small region of the enzyme surface, called the **active site**. It is at this site that substrate is bound, the reaction catalyzed, and product or products released.

To date, numerous enzymes have been isolated in pure crystalline form and studied by X-ray crystallography. In all cases in which the three-dimensional structure of an enzyme has been determined and the interactions between it and its substrate studied, the active site has been found to be a portion of the enzyme surface with a unique arrangement of amino acid side chains. X-ray crystallography studies suggest that as few as 10 amino acids may be involved at the active site.

In 1890, Emil Fischer likened the binding of an enzyme and its substrate to the interaction of a lock and key. According to this lock-and-key model, shown schematically in Figure 14.2, enzyme (the lock) and substrate (the key) have complementary shapes and fit together. Recall from our discussion of the significance of chirality in the biological world (Section 10.7) that we accounted for the remarkable ability of enzymes to distinguish between enantiomers by proposing that an enzyme and its substrate must interact through at least three specific binding sites on the surface of the enzyme. Groups on the enzyme surface that participate in binding enzyme and substrate to form an enzyme-substrate complex are called binding groups. In Figure 14.2, these three binding sites are labeled a, b, and c. The complementary regions on the substrate are labeled a', b', and c'.

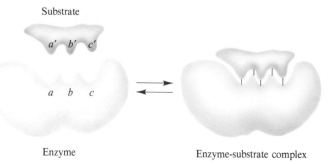

Figure 14.2 Lock-and-key model of the interaction between enzyme and substrate.

Once a substrate molecule is recognized and bound to the active site of the enzyme, certain functional groups in the active site called catalytic groups participate directly in the making and breaking of chemical bonds. In this sense, the active site on an enzyme is a unique combination of binding groups and catalytic groups.

C. Catalytic Efficiency

Enzymes can effect enormous increases in the rates of chemical reactions. An example is peroxidase, an enzyme that catalyzes the decomposition of hydrogen peroxide to oxygen and water:

$$2H_2O_2 \xrightarrow{\text{peroxidase}} 2H_2O + O_2$$

hydrogen peroxide

Many biological oxidations use molecular oxygen as the oxidizing agent, and in some of these, oxygen is reduced to hydrogen peroxide, a substance that is very toxic and must be decomposed rapidly to prevent damage to cellular compo-

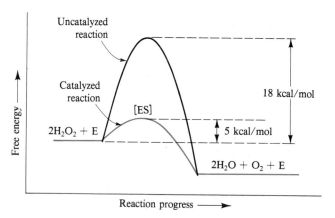

Figure 14.3 Energy changes for the uncatalyzed and peroxidase-catalyzed decomposition of hydrogen peroxide.

nents. As shown in Figure 14.3, the energy of activation for the uncatalyzed decomposition of hydrogen peroxide is 18 kcal/mol. Peroxidase provides an alternative pathway with an energy of activation of only 5.0 kcal/mol. Since the rate of a chemical reaction depends on the energy of activation, the lower the energy of activation for a reaction, the higher the rate. This seemingly small decrease in energy of activation (from 18 kcal/mol to 5 kcal/mol) corresponds to an increase in rate of reaction of almost 10^{10}!

A second example is carbonic anhydrase, an enzyme present in most tissues but in especially high concentration in erythrocytes; it catalyzes the reversible reaction between carbon dioxide and water to produce carbonic acid. The rate of hydration of carbon dioxide is normally very low but is increased almost 10^7 times by carbonic anhydrase:

$$CO_2 + H_2O \overset{\text{carbonic}}{\underset{}{\xrightleftharpoons{\text{anhydrase}}}} H_2CO_3$$

carbon   carbonic
dioxide   acid

D. How Enzymes Increase Rates of Biochemical Reactions

It is helpful, in thinking about rates of chemical reactions, to think about the following factors. To react, molecules must:

1. Collide.
2. Collide so the groups participating in the reaction are properly oriented to each other.
3. Collide with enough energy to allow bond breaking and bond making.

There are four main effects by which enzymes increase the rates of chemical reactions, all of which are related to the preceding requirements. Even though all four effects are involved in virtually every enzyme-catalyzed reaction, the balance between them may be entirely different in one reaction over another. Therefore, the order in which they are described should not be taken as a ranking of their importance.

1. Proximity

We can best appreciate the effects of proximity by referring to an example, namely the oxidation of lactate by NAD^+ catalyzed by lactate dehydrogenase (LDH). In this oxidation, the two different substrates must be brought close enough together to react. Because of the relatively low concentrations of each in a cell or other biological medium, the chance of their coming together in solution and reacting is very small. However, since LDH can bind both lactate and NAD^+, the chance that these two molecules are close enough to react (in effect, collide) becomes very high. Thus, an enzyme "collects" substrates from solution, thereby increasing the effective concentration of one relative to the other and in this way making reactions more likely.

2. Orientation

Not only do substrates have to come together (collide) to react, but they must do so with the proper orientation. For complex molecules in solution, the chances of a collision in which the reactive portions of each substrate are properly aligned may be very slight. With enzymes, however, substrates are oriented with great precision at the active site and therefore properly positioned for reaction.

3. Strain and Induced Fit

The lock-and-key model for enzyme catalysis that Emil Fischer proposed in 1890 is an overly simplified and undoubtedly inaccurate model for the interaction between enzyme and substrate. Much experimental evidence, including X-ray crystallographic pictures of actual enzyme-substrate complexes, suggests that when enzyme and substrate react to form an enzyme-substrate complex, the conformation of both enzyme and substrate change. The change in conformation of the enzyme in response to binding of substrate is called induced fit (see Figure 14.4). As the conformation of the enzyme changes in response to binding substrate, strain (most commonly distortion of bond angles and bond lengths) is induced in the substrate, thus reducing the energy required to convert substrate to product. In other words, inducing strain in the substrate makes it less stable and thereby reduces the energy of activation required for conversion of substrate to product.

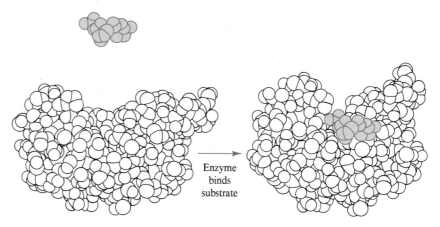

Enzyme
binds
substrate

Figure 14.4 As an enzyme and substrate interact, the conformation of the enzyme changes to bind the substrate tightly.

4. Catalytic Functional Groups at the Active Site

Many of the laboratory organic reactions we studied earlier were catalyzed by either acid (H^+) or base (OH^-). In most biological media, the concentration of these ions is very low. One big exception, of course, is gastric secretions, in which the concentration of hydrogen ion may be as high as 0.1M (pH = 1.0). Enzymes at their active sites contain functional groups such as the $—CO_2H$ side chains of aspartate and glutamate, which can function very effectively as hydrogen ion donors. Similarly, they contain functional groups such as the $—NH_2$ of lysine and the $—CO_2^-$ groups of aspartate and glutamate, which can function very effectively as hydrogen ion acceptors. Thus, although the concentration of acid or base may not be high in solution, the effective concentration of proton acceptors or proton donors at the active site of an enzyme may be very high.

In summary, enzymes are remarkably effective catalysts because they can bring molecules together so that reactive groups are properly positioned to interact, and so that functional groups at the active site can provide both proton donors and proton acceptors, also properly positioned to bring about reaction.

14.4 Factors That Affect Rate of Enzyme-Catalyzed Reactions

The rate of an enzyme-catalyzed reaction depends on many factors, the most important of which are concentration of the enzyme, concentration of the substrate, pH, temperature, and presence of enzyme inhibitors. Let us look at each of these factors in detail.

A. Concentration of Enzyme

In an enzyme-catalyzed reaction, concentration of the enzyme is very low compared with concentration of the substrate; under these conditions, rate of reaction is directly proportional to concentration of enzyme. For example, if the concentration of the enzyme is doubled, the rate of conversion of substrate to product is also doubled. The effect of enzyme concentration on rate of reaction is shown in Figure 14.5.

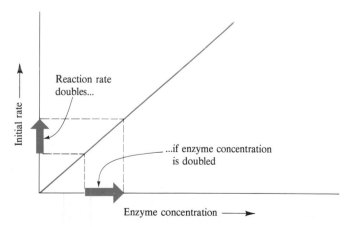

Figure 14.5 The effect of enzyme concentration on the rate of conversion of substrate to product.

B. Concentration of Substrate

To understand the effect of substrate concentration on the rate of an enzyme-catalyzed reaction, let us consider a series of experiments, each using the same concentration of enzyme but a different initial concentration of substrate.

Suppose we mix a given concentration of enzyme and substrate and then measure the initial rate of the reaction. In a second experiment, we mix the same concentration of enzyme but this time increase the concentration of substrate and again measure the initial reaction rate. This process is repeated, each time with the same enzyme concentration but a different substrate concentration. Figure 14.6 shows the relation between initial reaction rate and concentration of substrate. In region (a), an increase in initial substrate concentration causes a 1:1 increase in reaction rate; doubling the concentration of substrate doubles the reaction rate. Beyond region (a), an increase in initial substrate concentration also increases the reaction rate, but the effect becomes progressively smaller. Finally, in region (b), an increase in initial substrate concentration has no effect whatsoever on the reaction rate.

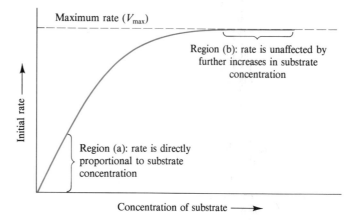

Figure 14.6 Dependence of initial rate of an enzyme-catalyzed reaction on concentration of the substrate.

C. Michaelis-Menton Equation

The shape of the curve shown in Figure 14.6 for the dependence of reaction rate on substrate concentration is one long familiar to mathematicians, namely, a hyperbola. The first to recognize this relation for enzyme-catalyzed reactions and put it in mathematical form were Lenore Michaelis and Maude Menton. In 1913, they proposed the following equation (that of a hyperbola) for enzyme kinetics. Although others have since refined and extended their work, the equation remains known as the **Michaelis-Menton equation**.

The Michaelis-Menton Equation

$$\text{Rate} = \frac{V_{max} \times [S]}{[S] + K_m}$$

The two constants in the equation are determined experimentally (Figure 14.7). The rate V_{max} is the maximum possible for a given enzyme concentration. The value K_m is the substrate concentration when the rate of the enzyme-catalyzed reaction is one-half the maximum rate.

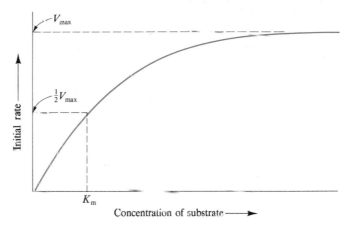

Figure 14.7 The value K_m is the concentration of substrate required to give an initial reaction rate of $\frac{1}{2}V_{max}$.

1. Significance of V_{max}

At V_{max}, all enzyme molecules have formed enzyme-substrate complexes and are operating continuously, catalyzing the conversion of substrate to product. We say that at this point the enzyme is saturated. Thus, values of V_{max} can be used to compare the maximum activity of one enzyme with that of another enzyme that might catalyze the same reaction.

2. Significance of K_m

The substrate concentration K_m required to convert one-half of all enzyme molecules to enzyme-substrate complexes is a measure of the affinity of enzyme for substrate. A small value for K_m means that a relatively low concentration of substrate is needed to saturate the enzyme; the enzyme has a high affinity for its substrate. On the contrary, a large value for K_m means that a relatively high concentration of substrate is needed to saturate the enzyme; the enzyme has a low affinity for its substrate.

Example 14.1

Following are data for five experiments, each using the same concentration of enzyme but a different initial concentration of substrate. Also given is the initial reaction rate for each substrate concentration. In these reactions, rate is measured as milligrams of substrate per milliliter of solution reacting per minute.

Initial Substrate Concentration (mol/L)	Initial Reaction Rate (mg/mL/min)
15.0×10^{-5}	0.38
10.0×10^{-5}	0.33
5.0×10^{-5}	0.25
3.3×10^{-5}	0.20
2.5×10^{-5}	0.166

a. Prepare a graph of initial rate versus substrate concentration and estimate V_{max}, $\frac{1}{2}V_{max}$, and K_m.

b. For this enzyme-catalyzed reaction, the initial rate is 0.166 mg/mL/min when the substrate concentration is 2.5×10^{-5}M. What will be the initial rate at this substrate concentration if the enzyme concentration is doubled?

Solution

a. In the accompanying graph, substrate concentration is plotted on the horizontal axis and initial reaction rate on the vertical axis. From this graph, estimate that V_{max} is approximately 0.41 mg of substrate reacting per milliliter of solution per minute. With this value, $\frac{1}{2}V_{max}$ is approximately 0.22 mg/mL/min and K_m is approximately 3.5×10^{-5}M. For this enzyme, a substrate concentration of 3.5×10^{-5} mol/L is required to convert half of all enzyme molecules to enzyme-substrate complexes.

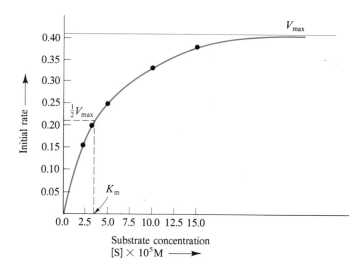

b. The rate of an enzyme-catalyzed reaction is directly proportional to the enzyme concentration. If the enzyme concentration is doubled, the initial rate doubles and the new initial rate becomes 0.332 mg/mL/min.

Problem 14.1

The initial rates of an enzyme-catalyzed reaction at five different substrate concentrations follow. Rate is given in units of micromoles of substrate reacting per minute.

Initial Substrate Concentration (mol/L)	Initial Reaction Rate (μmol/min)
2.0×10^{-3}	14
3.0×10^{-3}	18
4.0×10^{-3}	22
10.0×10^{-3}	31
12.0×10^{-3}	33

a. Prepare a graph of initial rate versus initial substrate concentration, and estimate V_{max}, $\frac{1}{2}V_{max}$, and K_m.
b. From your graph, estimate what the initial rate is when the substrate concentration is 6×10^{-3}M.
c. If the enzyme concentration is doubled, what will the new initial rate be when $[S] = 6 \times 10^{-3}$M?

D. pH

The catalytic activity of all enzymes is affected by the pH of the solution in which the reaction occurs. Often, small changes in pH cause large changes in the ability of a given enzyme to function as a biocatalyst. Shown in Figure 14.8 is a typical plot of enzyme activity versus pH. Notice that enzyme activity is a maximum in a narrow pH range and decreases at both higher and lower pHs.

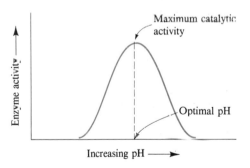

Figure 14.8 The dependence of enzyme activity on pH.

The pH corresponding to maximum enzyme activity is called the optimal pH. Most enzymes have maximum catalytic activity around pH 7, the pH of most biological fluids. Many enzymes, however, have maximum activity at considerably higher or lower pHs. For example, pepsin, a digestive enzyme of the stomach, has maximum activity around pH 1.5, the pH of gastric fluids. Table 14.3 lists optimal pH values for several enzymes.

Variations in enzyme activity with changes in pH depend on several factors, some or all of which may exist at the same time. First, changes in pH may cause partial denaturation of the enzyme protein. Second, catalytic activity may

Table 14.3
Optimal pH values for several enzymes.

Enzyme	Optimal pH
pepsin	1.5
acid phosphatase	4.7
α-glucosidase	5.4
urease	6.7
α-amylase (pancreatic)	7.0
carboxypeptidase	7.5
succinate dehydrogenase	7.6
trypsin	7.8
alkaline phosphatase	9.5
arginase	9.7

be possible only when certain amino acid side chains at the active site are in the correct states of ionization. Suppose, for example that for catalysis by a particular enzyme, the side chains of both lysine and a glutamic acid at the active site must be ionized (lys-NH_3^+ and glu-CO_2^-), as shown in Figure 14.9(b). This combination of ionization states is possible only in a particular pH range. At lower pH values (more acidic), catalytic activity decreases because the carboxyl of glutamate is protonated to $-CO_2H$ [Figure 14.9(a)]. At higher pH values (more basic), catalytic activity decreases because the side chain of lysine is deprotonated to $-NH_2$ [Figure 14.9(c)]. The effect of pH on catalytic activity may be much more complex than what is shown in Figure 14.9, particularly if the states of ionization of several amino acid side chains and of the substrate itself are important.

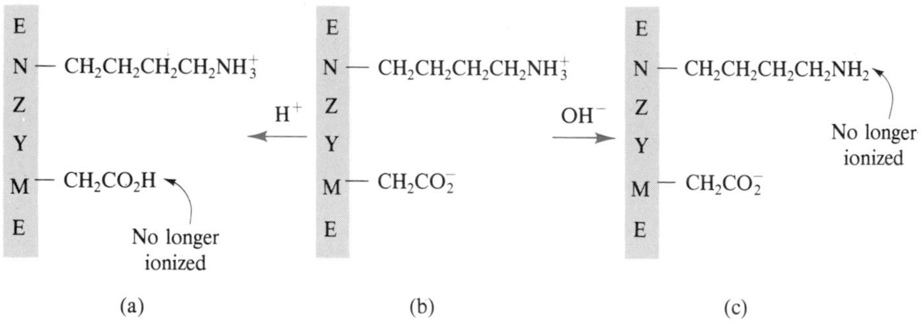

Figure 14.9 Dependence of catalytic activity on the ionization of amino acid side chains. (a) Below the optimal pH there is decreased or no catalytic activity. (b) At the optimal pH there is maximum catalytic activity. (c) Above the optimal pH there is decreased or no catalytic activity.

Example 14.2

Assume that one of the interactions binding a particular substrate to an enzyme is hydrogen bonding between a carbonyl group of the substrate and the —OH group of tyrosine.

$$(\text{enzyme})\text{---}\bigcirc\text{---}O\text{---}\overset{\delta+}{H}\text{------}\overset{\delta-}{O}\text{=}C\overset{CH_2\cdots}{\underset{CH_2\cdots}{\Big\langle}}\quad(\text{substrate})$$

What is the reason that the rate of this enzyme-catalyzed reaction is greatest below pH 8.1 but decreases at pHs greater (more basic) than 8.1? (The pK_a of the side chain of tyrosine is 10.1.)

Solution

For binding to occur between a carbonyl group of the substrate and the —OH group of tyrosine, it is essential to have the side chain of tyrosine in the un-ionized form. At pH 8.1 (two pH units more acidic than the pK_a of this group), the side chain is completely in the acid (—OH) form, and the rate is a maximum. As pH increases, the side chain becomes partially ionized, and at pH 10.1, the side chain is 50% ionized. The ionized form cannot participate in hydrogen bonding to bind substrate, and therefore the rate of reaction decreases.

Problem 14.2

For a particular enzyme-catalyzed reaction involving the side chain of histidine, the optimal pH is 4.0. Activity of the enzyme decreases above pH 4.0. Does the active form of this enzyme require the side chain of histidine to be in the acid form or in the conjugate base form? Explain.

E. Temperature

A fourth factor affecting the rate of an enzyme-catalyzed reaction is temperature. Just as there is an optimal pH for an enzyme-catalyzed reaction, so too is there an optimal temperature. Most enzymes have optimal temperatures in the range 25–37°C. Figure 14.10 shows a typical plot of enzyme activity as a function of temperature. Enzyme activity first increases with temperature because of an increase in the number of collisions between enzyme and substrate and an increase in the energy of these collisions. At higher temperatures, enzyme activity

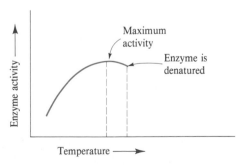

Figure 14.10 Dependence of enzyme activity on temperature.

decreases rapidly due to heat denaturation of the enzyme protein. The activity of most enzymes can be destroyed by heat treatment. There are, of course, exceptions, among which are the enzymes of bacteria living in hot springs at temperatures of 60–80°C.

F. Competitive Inhibition of Enzyme Activity

An **inhibitor** is any compound that can decrease the rate of an enzyme-catalyzed reaction. In fact, it is remarkably easy to inhibit enzymes. Certainly the denaturing agents like urea, guanidine hydrochloride, or detergents decrease the activity of enzymes because they bring about massive changes in the three-dimensional structure of the enzyme protein itself (Section 13.6E). Most often, these changes are irreversible and inhibition is irreversible. Some compounds, however, inhibit enzyme activity in such a way that the inhibition can be reversed, either by removing the offending compound or increasing the concentration of the substrate of the enzyme.

Understanding reversible enzyme inhibition is important for those in the health sciences. Many medicines act at the molecular level by inhibiting one or more enzymes, thereby decreasing the rates of the reactions these enzymes catalyze. For discussion of one particularly well understood example of how a drug works at the molecular level to inhibit a particular enzyme-catalyzed reaction, see Mini-Essay XI, "The Penicillins." The penicillins and also many other classes of drugs are administered to patients to decrease the flow of molecules through a particular metabolic pathway.

From a study of the effects of inhibitors on the rates of enzyme-catalyzed reactions and on both V_{max} and K_m, several types of reversible inhibition have been distinguished. **Competitive inhibition** is the most common type.

A competitive inhibitor binds to the active site of an enzyme and thus "competes" with substrate for the active site. Most often, competitive inhibitors are very closely related in structure to the enzyme substrate. Consider, as an example, the enzyme fumarase, which catalyzes the hydration of fumarate to malate:

fumarate   L-malate

Shown in Figure 14.11 are structural formulas for several competitive inhibitors of fumarase. Note the structural similarities between these inhibitors and fumarate.

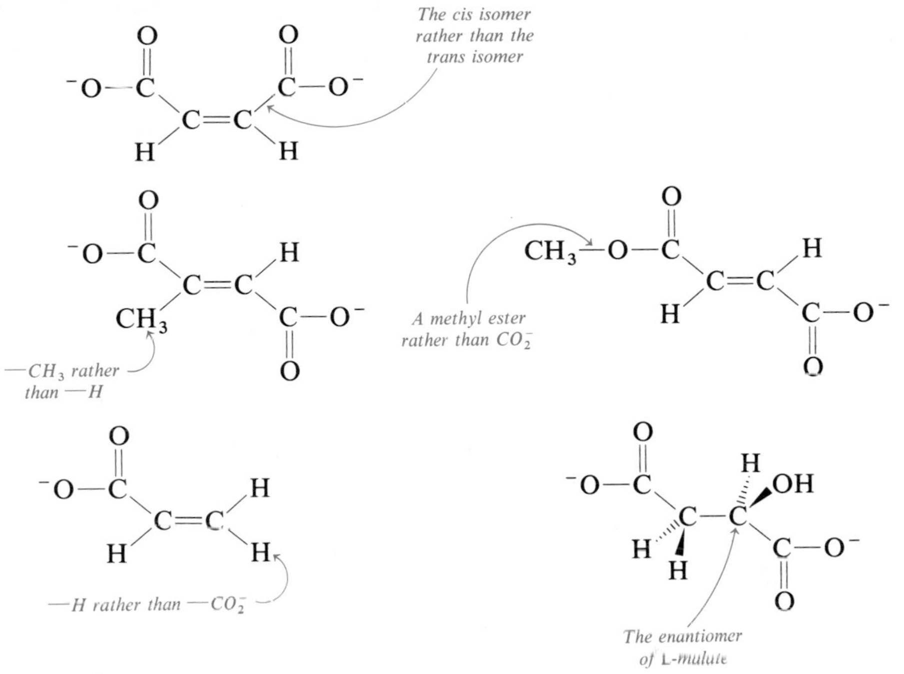

Figure 14.11 Several competitive inhibitors of the enzyme fumarase.

The nature of competitive inhibition is shown schematically in Figure 14.12. Part (a) of this figure shows an enzyme in the presence of both inhibitor and substrate. The two compete for the active site, and the inhibitor is shown forming an enzyme-inhibitor complex that blocks the active site for further catalytic activity.

A characteristic feature of competitive inhibition is that it can be reversed by increasing the concentration of substrate. We can account for this by looking at the two equilibria involved in a solution containing enzyme, substrate, and inhibitor. The first [Figure 14.12(a)] is between enzyme plus inhibitor, to form an enzyme-inhibitor complex. The second [Figure 14.12(b)] is between enzyme plus substrate, to form an enzyme-substrate complex. If we increase the concentration of substrate, then according to Le Chatelier's principle we also increase the concentration of enzyme-substrate complex at the expense of enzyme-inhibitor complex. The reversal of competitive inhibition is shown schematically in Figure 14.12(b).

Competitive inhibition can be detected by studying the rate of an enzyme-catalyzed reaction in the presence and in the absence of a competitive inhibitor. As shown in Figure 14.13, V_{max} is unchanged in the presence of a competitive inhibitor. However, K_m is increased, because a greater concentration of substrate is required to achieve one-half enzyme saturation in the presence of the inhibitor.

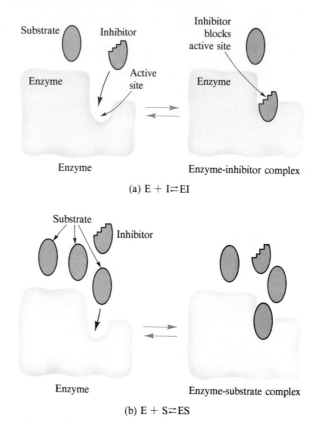

Figure 14.12 Competitive inhibition. (a) Formation of an enzyme-inhibitor complex. (b) When substrate is increased, the concentration of ES is increased and the concentration of EI is decreased, thus reversing the effect of the inhibitor.

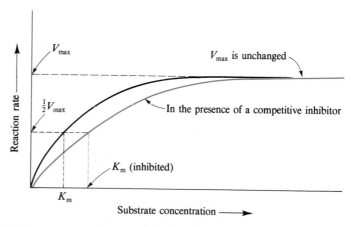

Figure 14.13 An example of competitive inhibition. When a competitive inhibitor is present, V_{max} is unchanged but K_m is increased.

14.5 Regulation of Enzyme Activity

Enzymes bring about enormous increases in the rates of chemical reactions, thus making possible life as we know it. But this is only part of the story of their importance for living systems. Not only can they increase rates of chemical reactions, but their activity can be regulated. As a consequence, rates of individual reactions can be controlled and integrated into separate metabolic pathways, which in turn are integrated into an overall metabolic system for the entire organism.

Regulation of enzyme activity is accomplished by two biological mechanisms. In the first, **allosteric regulation** (feedback control), the activity of key enzymes is altered by interaction of the enzyme with molecules produced within the cell itself. The result of this interaction may be either inhibition of activity or stimulation of activity. In the second biological mechanism, **genetic control**, the concentration of key enzymes is regulated by the rate of enzyme (protein) synthesis.

A. Allosteric Regulation—Feedback Control

The molecules responsible for regulating cellular metabolism are a special group of enzymes called **allosteric enzymes** (Greek, *allos*; other, plus *steros*, space). Regulation of metabolism through control of the activity of allosteric enzymes is immensely beneficial to an organism because through this control, the concentration of metabolites can be maintained within very narrow limits. Thus cells prevent unnecessary accumulation not only of the final product of a metabolic sequence but also of intermediates along the pathway. The evolution of **regulatory enzymes** was an essential step in achieving efficient use of cellular resources and in the evolution from single cells to multicellular organisms.

Following are the characteristics of an allosteric enzyme.

1. All allosteric enzymes have quaternary structure; that is, they are composed of two or more polypeptide chains.
2. All allosteric enzymes have two or more binding sites: an active site specific for substrate, and a regulatory site specific for a regulator molecule.
3. The regulatory site is distinct from the active site both in location on the enzyme and in shape. Because it is different from the active site, a regulatory site can bind molecules quite different in size and shape from the natural substrate for the enzyme.
4. Binding of a molecule at the regulatory site brings about a change in the three-dimensional shape of the enzyme protein, and in particular, of the active site. Molecules that increase enzyme activity are called allosteric activators; those that bring about a decrease in enzyme activity are called allosteric inhibitors.

To better understand the significance of an allosteric enzyme, suppose a cell needs a constant supply of molecule E, which is synthesized in a series of enzyme-catalyzed steps starting with molecule A. The simplest way for the concentration of E to be regulated within the cell is for E to be synthesized in a process that its own concentration regulates as shown here.

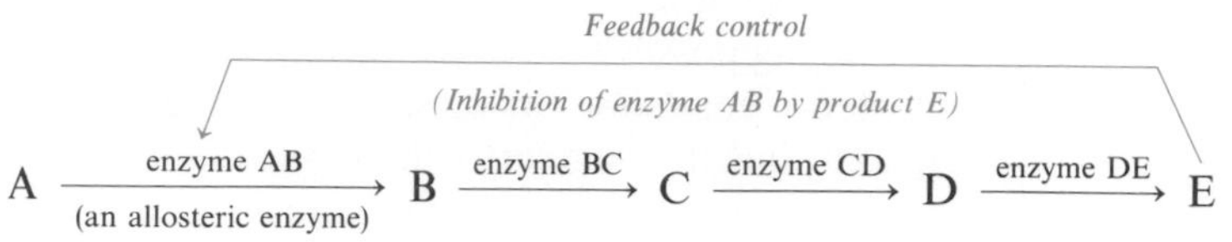

Feedback control

(Inhibition of enzyme AB by product E)

$$A \xrightarrow[\text{(an allosteric enzyme)}]{\text{enzyme AB}} B \xrightarrow{\text{enzyme BC}} C \xrightarrow{\text{enzyme CD}} D \xrightarrow{\text{enzyme DE}} E$$

If the concentration of E rises above that needed by the cell, it inhibits enzyme AB, the first enzyme in the sequence of steps that leads to the formation of E. On the other hand, if the concentration of E falls below that needed by the cell, inhibition of enzyme AB is decreased, thereby allowing increase in the rate of synthesis of E.

The first example of regulation of enzyme activity by **feedback control** was discovered in 1957 and involves the synthesis of isoleucine in the bacterium *E. coli.* This synthesis begins with threonine, and in a series of five sequential steps, each catalyzed by a different enzyme (E1, E2, . . . , E5), gives isoleucine. The concentration of isoleucine within the cell is regulated by the activity of isoleucine as an inhibitor of threonine deaminase, the first enzyme in this multi-step synthesis.

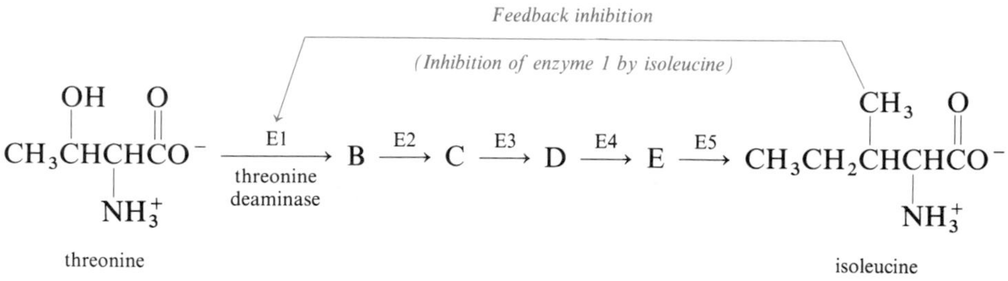

Feedback inhibition

(Inhibition of enzyme 1 by isoleucine)

$$\underset{\text{threonine}}{\overset{\displaystyle \underset{\overset{|}{NH_3^+}}{\overset{OH \quad O}{CH_3CHCHCO^-}}}{}} \xrightarrow[\text{threonine}]{\text{E1}}_{\text{deaminase}} B \xrightarrow{\text{E2}} C \xrightarrow{\text{E3}} D \xrightarrow{\text{E4}} E \xrightarrow{\text{E5}} \underset{\text{isoleucine}}{\overset{\displaystyle \underset{\overset{|}{NH_3^+}}{\overset{CH_3 \quad O}{CH_3CH_2CHCHCO^-}}}{}}$$

Because many of the examples in which allosteric regulation has been studied involve inhibition of a first, or at least an early, step in a metabolic pathway, allosteric regulation has become associated with inhibition. This is not the total picture, however. There are many examples in which allosteric regulation brings about activation of a metabolic pathway.

Allosteric modification of enzyme activity is significant because regulatory and catalytic sites have evolved separately. One important consequence is that allosteric regulators of a particular metabolic pathway may be products of entirely different pathways within the cell or even compounds produced outside the cell (like hormones). This special feature of allosteric regulation has permitted development of remarkably sensitive mechanisms for controlling metabolism within cells and within the organism as a whole.

B. Genetic Control

Genetic control of enzyme activity involves regulation of the rate of protein biosynthesis by induction and repression. The first demonstrated example of **enzyme induction** grew out of studies in the 1950s on the metabolism of the bacterium *E. coli.* Beta-galactosidase, an enzyme required for the utilization of lactose, catalyzes hydrolysis of lactose to D-galactose and D-glucose.

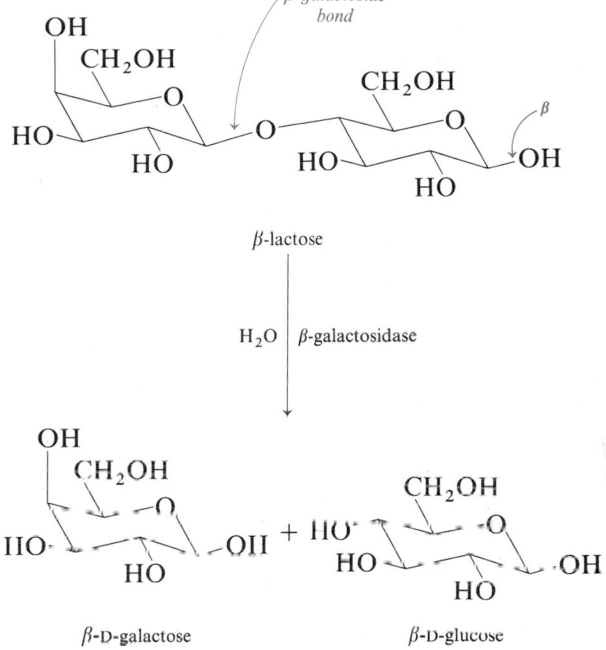

β-lactose

H_2O | *β*-galactosidase

β-D-galactose   *β*-D-glucose

When lactose is absent from the growth medium, no *β*-galactosidase is present in *E. coli.* If, however, lactose is added to the medium, within minutes the bacterium begins to produce this enzyme (Figure 14.14); lactose induces the synthesis of *β*-galactosidase. If lactose is then removed from the growth medium, production of *β*-galactosidase stops. Genetic control of enzyme activity is biologically significant because it allows an organism to adapt quickly to changes in its environment.

14.6 Enzymes and the Health Sciences

Our study of the structure and properties of enzymes so far has been descriptive, and in the case of enzyme kinetics, mathematical. This background has an important relation to the health sciences. The fact is, enzymology is an essential part of the everyday life of clinicians.

A. Enzymes in the Diagnosis of Diseases

Enzymes can be used as diagnostic tools because certain enzymes, such as those involved in blood coagulation, are normal constituents of plasma; their concentration in plasma is high compared with their concentration in cells. Other enzymes are normally present almost exclusively in cells and are released into the blood and other biological fluids only as a result of routine destruction of

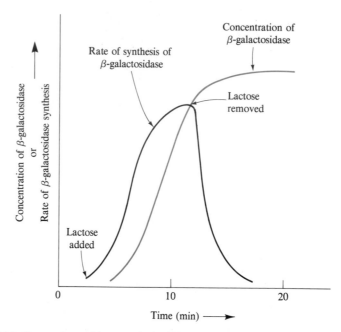

Figure 14.14 Beta-galactosidase, an inducible enzyme. Synthesis of β-galactosidase begins within minutes after lactose is added to the growth medium and ceases shortly after lactose is removed from the growth medium.

cells. Plasma levels of these enzymes are a million times or more lower than their cell levels. However, plasma concentrations of cellular enzymes may be elevated significantly in cases of cell injury and destruction (a damaged heart or skeletal muscle), or uncontrolled growth (as in cancer). Changes in plasma concentrations of appropriately chosen enzymes can be examined and used to detect cell damage; thus the site of damage or uncontrolled cell proliferation is suggested. Further, the degree of elevation of plasma concentration can often be used to determine the extent of cellular damage. Measurement of enzyme concentrations in blood plasma and other biological fluids has become a critical diagnostic tool, particularly for diseases of the heart, liver, pancreas, skeletal muscle, and bone, and for malignant diseases. In fact, certain enzyme deter-

Table 14.4
Some enzymes used in diagnostic enzymology.

Enzyme	Principal Clinical Condition in Which Enzyme Determination Is Used
lactate dehydrogenase (LDH)	heart or skeletal muscle damage
alkaline phosphatase	liver and bone disease
acid phosphatase	cancer of the prostate
serum glutamate oxaloacetate transaminase (SGOT)	heart and liver disease
creatine phosphokinase (CK)	myocardial infarction and muscle diseases
α-amylase	pancreatitis

minations are performed so often that they have become routine in the clinical chemistry laboratory. Several of these enzymes are listed in Table 14.4. For discussion of the use of assays for lactate dehydrogenase (LDH) and creatine phosphokinase (CK) as an invaluable tool for the diagnosis of myocardial infarction, see Mini-Essay X, "Clinical Enzymology—The Search for Specificity."

B. Enzymes as Laboratory Reagents

Many diseases are characterized by changes in the concentrations of specific compounds. The problem for the clinical chemistry laboratory is measuring the concentrations of these compounds in the presence of thousands of other compounds that might react similarly and therefore give either false positive or false negative results. Enzymes, because of their specificity, are ideal reagents for this type of determination. For example, the concentration of glucose in blood is normally 60–110 mg/100 mL. It is important for the diagnosis and effective treatment of patients with diabetes mellitus and other forms of hyperglycemia (abnormally high levels of blood glucose) to be able to determine blood glucose levels rapidly and accurately. It is likewise important to be able to determine blood glucose levels rapidly and accurately for those with hypoglycemia (abnormally low levels of blood glucose). The most common reagent in today's clinical chemistry laboratory for measuring glucose in blood and other biological fluids is the enzyme glucose oxidase. This enzyme is entirely specific for glucose and catalyzes the oxidation of glucose to gluconic acid and hydrogen peroxide:

$$\text{Glucose} + O_2 \xrightarrow{\text{glucose oxidase}} \text{gluconic acid} + H_2O_2$$

(For a more detailed discussion of the use of glucose oxidase as a reagent for the determination of glucose, see Mini-Essay VIII, "Clinical Chemistry—The Search for Specificity.")

C. Enzymes as Sites for the Action of Drugs

The pharmaceutical industry tries to produce drugs, either from natural sources or laboratory synthesis, that will counteract diseases (specific metabolic disorders). Because scientists have only an imperfect understanding of the underlying metabolic disorders responsible for most diseases, design of drugs is on a less rational basis than what would be desirable. In fact, many of today's most widely used drugs have been discovered quite by accident. Through research, scientists have been able to learn a great deal about their mechanisms of action at the molecular level. The sulfa drugs furnish a good example.

Sulfonic acids form amides in which the —OH is replaced by —NH_2. Following are structural formulas for benzenesulfonic acid and its amide, benzenesulfonamide. Also shown is the structural formula of *p*-aminobenzenesulfonamide, more familiar as sulfanilamide.

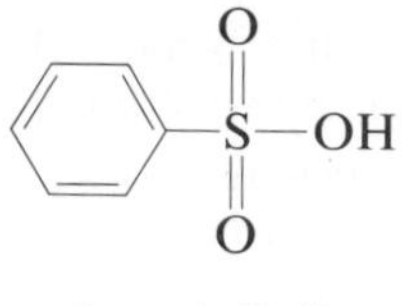

benzenesulfonic
acid

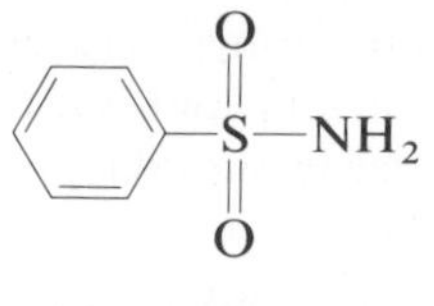

benzenesulfonamide

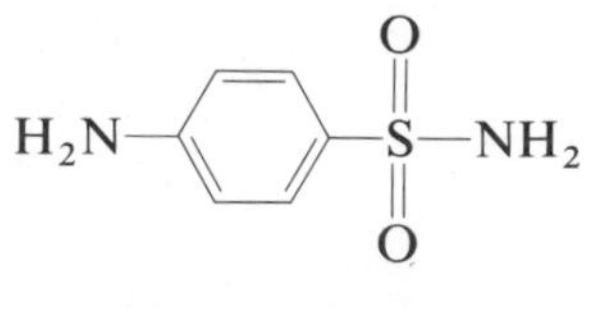

p-aminobenzenesulfonamide
(sulfanilamide)

The discovery of the medical uses of sulfanilamide and its derivatives was a milestone in the history of chemotherapy because it represents one of the first rational investigations of synthetic organic molecules as potential drugs to fight infection. Sulfanilamide was first prepared in 1908 in Germany, but not until 1932 was its possible therapeutic value realized. In that year, the dye Prontosil was prepared. During research over the next two years, the German scientist G. Domagk observed Prontosil's remarkable effectiveness in curing streptococcal and staphylococcal infections in mice and other experimental animals. Domagk further discovered that Prontosil is rapidly reduced in cells to sulfanilamide and that sulfanilamide, not Prontosil, is the actual antibiotic. His discoveries were honored in 1939 by a Nobel Prize in medicine.

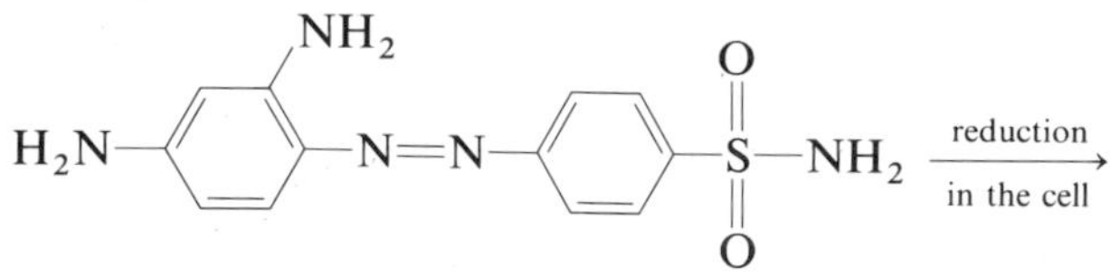

Prontosil

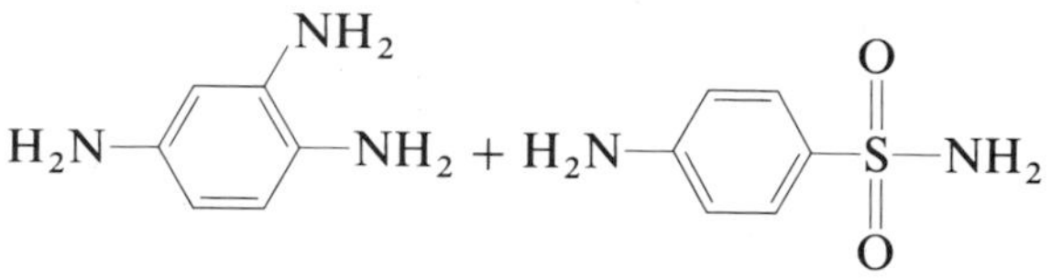

sulfanilamide

The key to understanding the action of sulfanilamide came in 1940 with the observation that inhibition of bacterial growth by sulfanilamide can be reversed by adding large amounts of p-aminobenzoic acid (PABA) to the growth medium. From this experiment, it was recognized that p-aminobenzoic acid is a growth factor for certain bacteria, and that in some way not then understood, sulfanilamide interferes with the bacteria's ability to use PABA. As you can see in the following structural formulas, obvious similarities in structure exist between p-aminobenzoic acid and sulfanilamide:

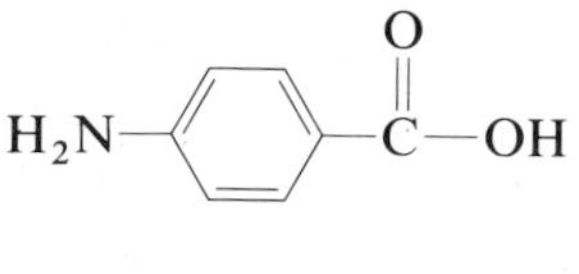

p-aminobenzoic acid
(PABA)

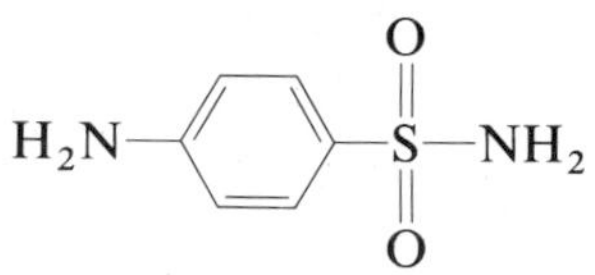

sulfanilamide

In the search for even better sulfa drugs, literally thousands of derivatives of sulfanilamide have been synthesized in the laboratory. Two of the most effective sulfa drugs are sulfathiazole and sulfadiazine:

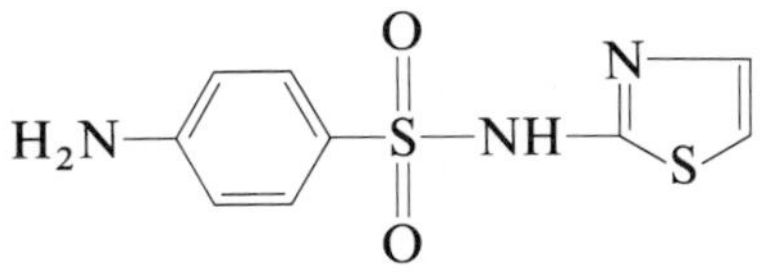

sulfathiazole 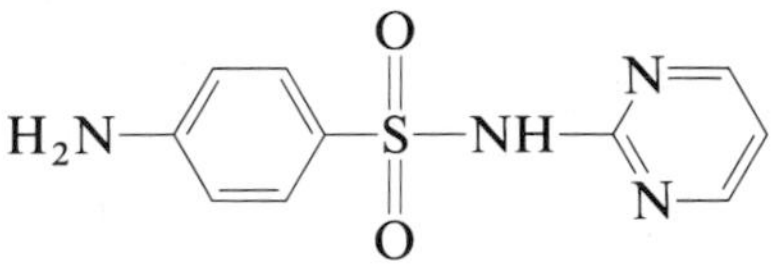   sulfadiazine

It now appears that sulfa drugs inhibit one or more of the enzyme-catalyzed steps in the synthesis of folic acid from *p*-aminobenzoic acid. Sulfanilamide can combat bacterial infections in humans without harming the patient because although humans also require folic acid, they do not make it from *p*-aminobenzoic acid. For humans, folic acid is a vitamin (Section 14.2C) and must be supplied in the diet.

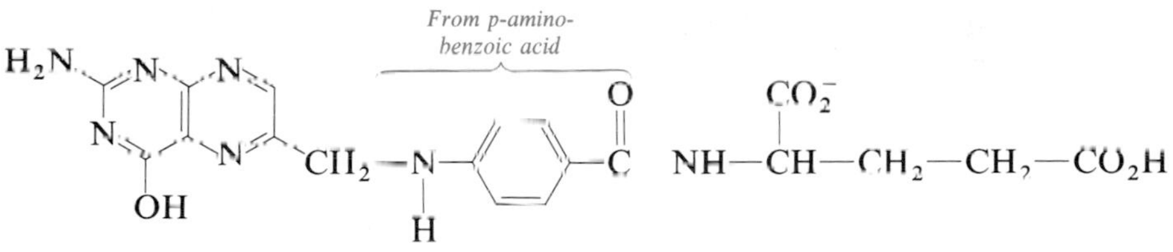

folic acid

Sulfa drugs, the first of the new "wonder drugs," were found to be effective in treating tuberculosis, pneumonia, and diphtheria, and they helped usher in a new era in public health in the United States in the 1930s. During World War II, they were routinely sprinkled on wounds to prevent infection. As a historical footnote, sulfa drugs were very soon eclipsed by an even newer class of wonder drugs for fighting bacterial infection, namely, the penicillins. (See Mini-Essay XI, "The Penicillins.")

Key Terms and Concepts

active site (14.3B)	feedback control (14.5A)
allosteric enzyme (14.5A)	genetic control (14.5B)
allosteric regulation (14.5A)	inhibitor (14.4F)
clinical enzymology (14.6)	Enzyme Commission (14.1)
coenzyme (14.2C)	K_m (14.4C2)
cofactor (14.2)	metal ion cofactor (14.2A)
competitive inhibition (14.4F)	Michaelis-Menton equation (14.4C)
enzyme (introduction)	prosthetic group (14.2)
enzyme induction (14.5B)	regulatory enzyme (14.5A)
enzyme-substrate complex (14.3A)	V_{max} (14.4C1)

Problems

14.3   Name three groups of enzyme cofactors.

14.4   What is the name given to cofactors that are permanently bound to an enzyme?

14.5   What is the relation between coenzymes and water-soluble vitamins?

14.6   Of the water-soluble vitamins, (a) which are coenzymes themselves? (b) which precursors from which coenzymes are synthesized in the body?

14.7   Is the following statement true or false? "Enzymes increase the rate at which a reaction reaches equilibrium but do not change the position of equilibrium for the reactions they catalyze." Explain.

14.8   For an uncatalyzed reaction and an enzyme-catalyzed reaction, compare (a) rate of formation of products; (b) the position of equilibrium (equilibrium constant) for the reaction.

14.9   List three characteristics of enzymes that make them superior catalysts compared with their nonbiological laboratory counterparts.

14.10   Binding of substrate to the surface of an enzyme can involve a combination of ionic interactions, hydrogen bonding, and dispersion forces. By what type of interaction or interactions might the side chains of the following amino acids bind a lecithin to form an enzyme-substrate complex?

    a. phenylalanine   **b.** serine   **c.** glutamic acid
    d. lysine   **e.** valine

14.11   Of the 20 protein-derived amino acids, the side chains of the following are most often involved as catalytic groups at the active site of an enzyme. What is the net charge on the side chain of each amino acid at pH 7.4?

    a. cys   **b.** his   **c.** ser   **d.** asp   **e.** glu   **f.** lys

14.12   The following precautions are commonly observed when storing and handling solutions of enzymes. Explain the importance of each.

    a. Enzyme solutions are stored at a low temperature, usually at 0°C or below.

    b. The pH of most enzyme solutions is kept near 7.0.

    c. Enzyme solutions are prepared by dissolving the enzyme in water distilled from all-glass apparatus; ordinary tap water is never used.

    d. When an enzyme is being dissolved in aqueous solution, it is dissolved with as little stirring as possible. Vigorous stirring or shaking is never used to hasten the dissolving.

14.13   Refer to Figure 14.6 and explain why in region (a), an increase in substrate concentration has a direct increase on reaction velocity. Also explain why in region (b), an increase in substrate concentration does not affect the reaction velocity.

14.14   What does it mean to say that an enzyme is saturated?

14.15   The value V_{max} for an enzyme-catalyzed reaction is 3 mg/mL/min. At what rate is product formed when only one-third of the enzyme molecules have substrate bound to them?

14.16 Following are the initial velocities of an enzyme-catalyzed reaction at six different substrate concentrations. Velocity is given in units of milligrams of product formed per milliliter of solution per minute.

[S] (mol/L)	Velocity (mg/mL/min)
0.5×10^{-3}	1.5×10^{-3}
1.0×10^{-3}	3.0×10^{-3}
1.5×10^{-3}	4.4×10^{-3}
2.0×10^{-3}	5.0×10^{-3}
3.0×10^{-3}	5.8×10^{-3}
4.0×10^{-3}	6.2×10^{-3}

   a. Prepare a graph of initial velocity versus substrate concentration, and from your graph, estimate V_{max}, $\frac{1}{2}V_{max}$, and K_m.
   b. At what substrate concentration will the initial velocity be 2.5×10^{-3} mg/mL/min?
   c. What percentage of enzyme is in the form of an enzyme-substrate complex when the initial velocity is 2.0×10^{-3} mg/mL/min?

14.17 Lysozyme catalyzes the hydrolysis of glycoside bonds of the polysaccharide components of certain types of bacterial cell walls. The catalytic activity of this enzyme is at a maximum at pH 5.0. The active site of lysozyme contains the side chains of asp and glu, and for maximum catalytic activity, the side chain of glu must be in the acid, or protonated, form and the side chain of asp in the conjugate base, or deprotonated, form. Explain why the velocity of lysozyme-catalyzed reactions decreases as the pH becomes more acidic than the optimal pH. Also explain why reaction velocity decreases as the pH becomes more basic than the optimal pH.

14.18 An enzyme isolated from yeast has an optimal temperature of 40°C. Explain why the velocity of the reaction catalyzed by this enzyme (a) decreases as the temperature is lowered to 0°C; (b) decreases as the temperature is increased above 40°C.

14.19 Following are equations for the ionization of the side chains of histidine and cysteine, along with the pK_a of each.

$-CH_2-S-H \rightleftharpoons -CH_2-S^- + H^+$ $pK_a = 8.0$

(This form required for maximum activity)

Assume that both histidine and cysteine are catalytic groups for a particular enzyme. Assume also that for maximum activity, the side chain of histidine must be in the protonated form and the side chain of cysteine must be in the deprotonated form. Estimate the pH at which the catalytic activity of this enzyme is the maximum, and sketch a pH-activity graph.

14.20 Various enzymes and their K_m values follow. Which enzyme has the highest affinity for substrate? Which has the lowest affinity for substrate?

Enzyme	K_m
sucrase	1.6×10^{-2}M
β-glucosidase	6×10^{-3}M
enolase	7×10^{-5}M
catalase	1.17M

14.21 Following is the initial velocity versus substrate concentration for an enzyme-catalyzed reaction and for the same reaction in the presence of an inhibitor. Inhibitor concentration is constant.

[S] (mol/L)	Velocity (mmol/min)	Velocity (Inhibited) (mmol/min)
1.0×10^{-4}	0.14	0.088
1.25×10^{-4}	0.16	0.105
1.67×10^{-4}	0.19	0.13
2.5×10^{-4}	0.24	0.18
5.0×10^{-4}	0.31	0.26
10.0×10^{-4}	0.38	0.36

a. Prepare a graph of initial velocity versus substrate concentration for the uninhibited reaction.
b. On the same graph, plot initial velocity versus substrate concentration for the inhibited reaction.
c. Is this inhibition competitive or noncompetitive?
d. Can the effects of this inhibitor be overcome? If so, how?

Regulation of enzyme activity (Section 14.5)

14.22 Name two biological mechanisms for regulating enzyme activity.
14.23 What is meant by the term regulatory enzyme?
14.24 What is meant by enzyme induction?
14.25 A metabolic pathway is shown, in which substance A is converted to B, and then B can be converted to substance D or substance F, depending on the needs of the cell at any particular time.

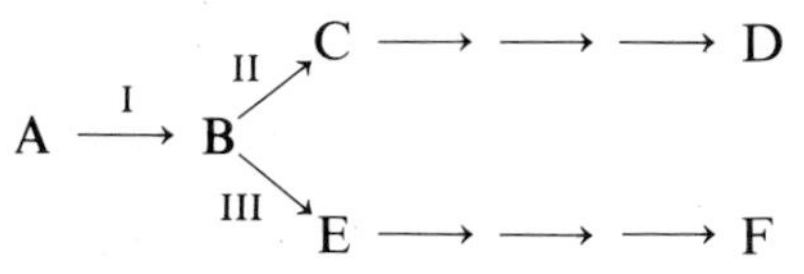

a. Which of these steps commits B to the synthesis of D?

b. Which step commits B to the synthesis of F?

c. Assume that D is an allosteric inhibitor of the enzyme catalyzing reaction B $\rightarrow$ C; explain how this relation can regulate the concentration of D.

d. Assume that both D and F are allosteric inhibitors of the enzyme catalyzing reaction A $\rightarrow$ B; show how this relation can regulate the concentration of B, C, and E.

Clinical Enzymology—The Search for Specificity

A 35-year-old, muscular man complains of severe chest pains and is admitted to the hospital. He lifts heavy objects all day in the course of his work. Is his chest pain due to overexertion that particular day, a temporary muscle spasm, a heart attack, or some other cause? The attending physician must rely on several types of information in making a correct diagnosis: patient history; the clinical pattern of the chest pains; electrocardiogram findings; and the rise and fall in the concentration of certain blood-serum enzymes. Today enzyme assays can be used to determine with virtually 100% certainty whether a patient has or has not had a heart attack and also the extent of damage to cardiac muscle tissue. To understand the basis for these enzyme tests, we must know what kinds of enzymes are normally found in serum and what isoenzymes are.

Serum Enzymes

Two kinds of enzymes are normally found in serum, functional enzymes and nonfunctional enzymes. Functional enzymes are secreted into the circulatory system, where they have clearly defined physiological roles. An example is

and blood only as a result of normal tissue breakdown. What makes the presence of nonfunctional enzymes significant is that their level in serum increases dramatically any time there is cellular injury or increased breakdown due to:

1. Localized trauma, such as a blow, a surgical procedure, or even an intramuscular injection.
2. Inadequate flow of blood to a particular tissue or area.
3. Any condition, such as cancer, in which there is an increase in cell growth accompanied by a corresponding increase in cell destruction.

Isoenzymes

Many enzymes exist in multiple forms, and while all forms of a particular enzyme catalyze the same reaction or reactions, they often do so at different rates. The term *isoenzyme* is used to distinguish between multiple forms of the same enzyme. Two of the most extensively studied sets of isoenzymes are those of lactate dehydrogenase and of creatine kinase.

Lactate dehydrogenase (LDH) catalyzes the following reversible reaction:

$$\underset{\text{lactate}}{CH_3\overset{\overset{\displaystyle OH}{|}}{C}HCO_2^-} + NAD^+ \underset{\text{dehydrogenase}}{\overset{\text{lactate}}{\rightleftharpoons}} \underset{\text{pyruvate}}{CH_3\overset{\overset{\displaystyle O}{\|}}{C}CO_2^-} + NADH + H^+$$

thrombin, an enzyme involved in blood clotting. Nonfunctional enzymes have no apparent role in serum. They are largely confined to cells and appear in the surrounding extracellular fluids

In the forward reaction, lactate is oxidized by NAD^+, and in the reverse reaction, pyruvate is reduced by NADH. High levels of LDH are found in the liver, the skeletal and heart muscles,

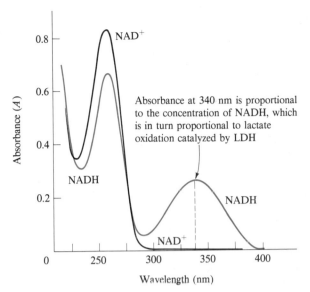

Absorbance at 340 nm is proportional to the concentration of NADH, which is in turn proportional to lactate oxidation catalyzed by LDH

Figure X-1 Absorbance versus wavelength for NADH and NAD$^+$.

trophoresis strip is incubated with lactate and NAD$^+$. At sites on the strip where LDH is present, NAD$^+$ is reduced to NADH. Figure X-1 shows a plot of absorbance versus wavelength for NAD$^+$ and NADH. Notice that only NADH absorbs radiation between 300 and 400 nm. In the clinical laboratory, absorbance at 340 nm is used to detect NADH.

Figure X-2(a) shows a typical LDH isoenzyme pattern. In the preparation of this strip, the sample is spotted at the right and the direction of migration is toward the positive electrode at the left.

The five LDH isoenzymes are all tetramers composed of different combinations of two polypeptide chains. One chain is designated H because it is found in highest concentration in heart muscle LDH. The other, found in highest concentration in skeletal muscle, is designated M. The fastest-moving LDH isoenzyme is a tetramer of four H chains and the slowest is a tetramer of four M chains. The other three are hybrids of H and M chains.

Table X-1 shows the distribution of LDH isoenzymes in several human tissues. Notice that each type of tissue has a distinct isoenzyme pattern. Liver and skeletal muscle, for example, contain particularly high percentages of LDH$_5$. Also notice that heart muscle and skeletal muscle have different ratios of LDH$_1$/LDH$_2$.

the kidneys, and erythrocytes. When extracts of these or other types of tissues are subjected to electrophoresis at pH 8.0, lactate dehydrogenase can be separated into five isoenzymes. By convention, the isoenzyme moving most rapidly toward the positive electrode is called LDH$_1$ and that moving most slowly is called LDH$_5$. The LDH isoenzymes are not visible to the eye. So that their locations can be determined, the elec-

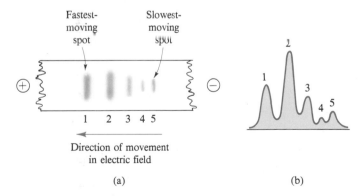

(a)   (b)

Figure X-2 An LDH isoenzyme assay. (a) The patterns of LDH isoenzymes in normal serum. (b) A densitometer plot showing the relative concentrations of each isoenzyme. The area under each peak is proportional to the concentration of isoenzyme.

Table X-1
Percentage distribution of LDH isoenzymes in several tissues.

Tissue	LDH$_1$ (H$_4$)	LDH$_2$ (H$_3$M)	LDH$_3$ (H$_2$M$_2$)	LDH$_4$ (HM$_3$)	LDH$_5$ (M$_5$)
serum	25	35	20	10	10
heart	40	35	20	5	0
kidney	35	30	25	10	0
liver	0	5	10	15	70
brain	25	35	30	10	0
skeletal muscle	0	0	10	30	60

Because LDH is widely distributed throughout the body, elevated serum levels of LDH are associated with a broad serum of diseases: anemias involving hemolysis of red blood cells, creatine kinase, until recently the preferred name was creatine phosphokinase (CPK). It is likely that you will encounter both names and both abbreviations in your readings.

$$^-O-\overset{\overset{O}{\|}}{\underset{\underset{O^-}{|}}{P}}-NH-\overset{\overset{NH_2^+}{\|}}{\underset{\underset{CH_3}{|}}{C}}-N-CH_2-CO_2^- + ADP \xrightarrow{\text{creatine kinase}} H_2N-\overset{\overset{NH_2^+}{\|}}{\underset{\underset{CH_3}{|}}{C}}-N-CH_2-CO_2^- + ATP$$

creatine phosphate creatine

acute liver diseases, congestive heart failure, pulmonary embolism, and muscular diseases, such as muscular dystrophy. This broad distribution makes an LDH assay a good initial test. If LDH activity is elevated, then an isoenzyme assay can be used to pinpoint the location and type of disease more accurately.

Creatine kinase (CK) catalyzes the transfer of a phosphate group from creatine phosphate to ADP (Section 16.6C). You should note that although the current name for this enzyme is

Creatine kinase consists of three isoenzymes, each of which is a dimer formed by combinations of B and M polypeptide chains. The B subunit is so named because it was first isolated from brain tissue. The M subunit was first isolated from skeletal muscle tissue. Table X-2 shows the percentage distribution of CK isoenzymes in several tissues. Notice that heart muscle is the only tissue containing a high percentage of CK-MB isoenzyme.

Table X-2
Percentage distribution of creatine kinase (CK) isoenzymes in several human tissues.

Tissue	CK$_1$ (B$_2$)	CK$_2$ (MB)	CK$_3$ (M$_2$)
serum	0	0	100
heart	0	40	60
lung	90	0	10
bladder	95	0	5
brain	90	0	10
skeletal muscle	0	0	100

Enzyme Profile of a Heart Attack

During a heart attack (myocardial infarction, or MI) a coronary artery is partially or completely blocked, reducing the flow of oxygen-rich blood to the heart muscle it serves. If the muscle is damaged because of oxygen starvation, cells die and release their contents into the surrounding extracellular fluid. Eventually, the cell contents find their way into the bloodstream. The levels of LDH begin to rise appproximately 6–24 hr following a heart attack and frequently reach two to three times normal serum levels. Peak LDH serum activity is usually reached within 2–3 days and may remain elevated for up to two weeks. Many types of cell and tissue damage lead to elevation in serum LDH activity. What is unique about the pattern that follows heart

damage is what happens to the ratio LDH_1/LDH_2. According to the data in Table X-1 normal serum contains approximately 25% LDH_1 and 35% LDH_2. Thus, in normal serum, the LDH_1/LDH_2 ratio is less than 1. In heart tissue, the LDH_1/LDH_2 ratio is greater than 1. Thus, following a heart attack, there is an elevation in LDH_1 activity in the serum and the LDH_1/LDH_2 ratio "flips"; that is, it changes from less than 1 to greater than 1. In 80% of patients with heart attacks, a flipped LDH_1/LDH_2 ratio appears within 24–48 hr following the attack. Although a flipped LDH_1/LDH_2 ratio is not always seen following a heart attack, there is almost invariably a significant increase in LDH_1 activity. For this reason, several companies have developed enzyme assay tests that are highly specific for LDH_1 only. Figure X-3 shows an LDH isoenzyme pattern after a mild heart

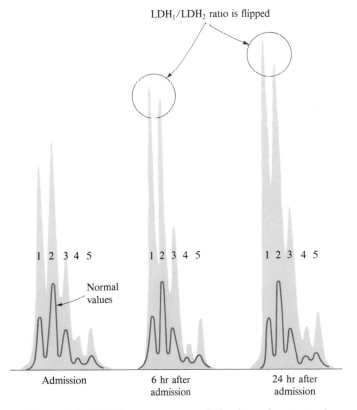

Figure X-3 LDH isoenzyme assay following a heart attack.

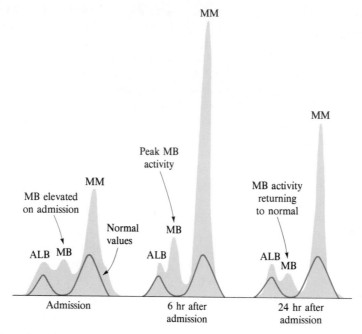

Figure X-4 Creatine kinase (CK) isoenzyme patterns following a heart attack. The peak labeled ALB is serum albumin and is used as a reference and calibration point.

attack. Note that the LDH_1/LDH_2 ratio is flipped at 12 and 24 hr following admission to the hospital.

Serum levels of creatine kinase also rise after many types of cell and tissue damage. As shown in Table X-2, heart muscle is the only human tissue with a high percentage of CK-MB isoenzyme, and for this reason, the serum CK isoenzyme pattern following heart damage is unique. Serum CK-MB begins to rise approximately 4–8 hr after myocardial infarction and reaches a peak at about 24 hr. Because CK iso-

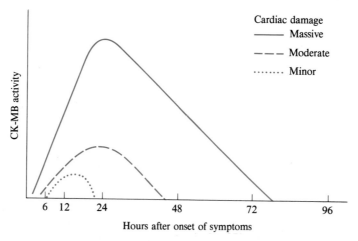

Figure X-5 Typical plots of serum creatine kinase–MB activity following heart damage.

enzymes are degraded rapidly, CK levels soon begin to drop and return to normal within a few days after the attack. Figure X-4 shows a creatine kinase isoenzyme pattern following a heart attack. Note that all samples, including the one on admission, are positive for CK-MB. The second sample shows greatest activity, and the third shows CK-MB returning to normal.

It is often possible to determine the extent of heart damage from the CK-MB pattern. Shown in Figure X-5 is a plot of CK-MB activity following minor, moderate, and massive heart damage. Notice that when the heart damage is massive, CK-MB appears in the serum sooner, its activity rises higher, and it returns more slowly to normal, compared with lesser attacks.

What about our 35-year-old man admitted to the hospital with chest pain? Enzyme assays over the next 48 hr showed increased levels of LDH and CK enzymes. Isoenzyme assays, however, showed no elevation in CK-MB and no significant change in the LDH_1/LDH_2 ratio. Therefore, his pains were associated with trauma to skeletal muscle and there was no indication of heart damage.

The Penicillins

The most successful of all antibiotics are the penicillins, the first of the so-called miracle drugs. These truly remarkable drugs are almost completely harmless to all living organisms except for certain classes of bacteria. As a result of extensive research on the structure, chemistry, and mechanism of antibacterial activity of penicillins, it is safe to say that we have a clearer understanding of the penicillins than of almost any other class of antibiotic.

The discovery of penicillin was purely accidental. In 1928, the Scottish bacteriologist Alexander Fleming (later to become Sir Alexander Fleming) reported:

> While working with staphylococcal variants, a number of culture plates were set aside on the laboratory bench and examined from time to time. In the examinations, these plates were necessarily exposed to the air and they became contaminated with various microorganisms. It was noticed that around a large colony of contaminating mold, the staphylococcal colonies became transparent and were obviously undergoing lysis . . . . I was sufficiently interested in the antibacterial substance produced by the mold to pursue the matter.

Because the contaminating mold was *Penicillium notatum*, Fleming named the antibacterial substance penicillin. Despite several attempts, he was unable to isolate and purify an active form of it. Nonetheless, he continued to maintain his cultures of the mold.

The outbreak of war in Europe in 1939 stimulated an intensive search for new drugs, and in Great Britain the potential of Fleming's penicillin was reinvestigated. Howard Florey, an Australian experimental pathologist, and Ernst Chain, a Jewish chemist who had fled Nazi Germany, worked together on the project. Using the newly discovered technique of freeze-drying (lyophilization), they succeeded in isolating penicillin from Fleming's cultures. Within a few months, larger quantities of purified penicillin were available, and many of its physical, chemical, and antibacterial properties had been determined. Florey and Chain published, in 1940, a report of treatment of bacterial infection in mice, and two years later, a report on treatment of humans. By 1943, pharmaceutical companies in Britain and the United States were producing penicillin on a large scale and it was authorized for use by the military. The following year, penicillin became available for civilian use, and in 1945 Fleming, Florey, and Chain were awarded the Noble Prize in medicine and physiology. Thus, in less than two decades, penicillin had progressed from a chance observation in a research laboratory to a drug that has been recognized as one of the greatest contributions of medical science in the service of humanity.

Preliminary investigations of the structure of penicillin presented a confusing picture until it was discovered that *P. notatum* produces different kinds of penicillin, depending on the composition of the medium in which the mold is grown. Initially, six different penicillins were recognized, all of which proved to be derivatives of 6-aminopenicillanic acid (Figure XI-1). Of the six, penicillin G (benzylpenicillin) became the most widely used and the standard against which others were judged.

The structural formula of penicillins consists of a five-member ring fused to a four-member ring. The four-member ring is a cyclic amide, the special name for which is lactam. Because the nitrogen atom of this lactam is on the carbon

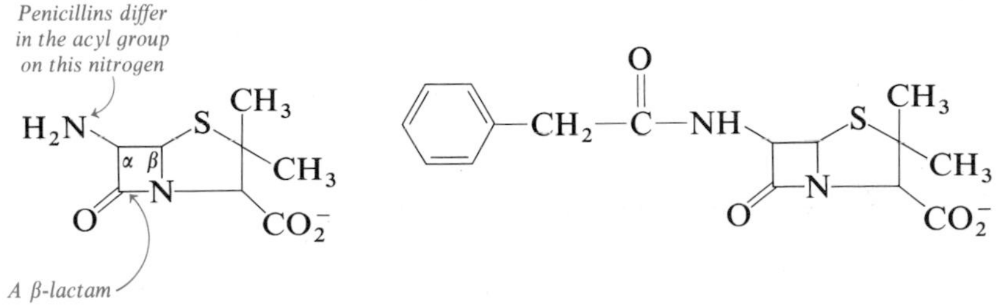

Penicillins differ
in the acyl group
on this nitrogen

A β-lactam

6-aminopenicillanic acid

penicillin G or
benzylpenicillin

Figure XI-1 The penicillins.

atom beta to the carbonyl group, the general name of this type of ring is beta-lactam.

The penicillins undergo a variety of chemical reactions, some of which are very complex. We will concentrate on just two of these, each chosen because it has important consequences for the medical use of these antibiotics. Treatment of penicillin with aqueous HCl brings about hydrolysis of the amide bond of the highly strained beta-lactam and cleavage of the five-member ring also (Figure XI-2). Both penaldic acid and penicillamine are devoid of antibacterial activity. Because penicillin G is rapidly inactivated by this type of hydrolysis in the acid conditions of the stomach, it cannot be given orally; it must be given by intramuscular injection.

The second important reaction of the penicillins is selective hydrolysis of the beta-lactam ring catalyzed by a group of enzymes called beta-lactamases (Figure XI-3). The product of this enzyme-catalyzed hydrolysis is penicilloic acid. Like penaldic acid and penicillamine, penicilloic acid has no antibacterial activity. The main basis for the natural resistance of certain bacteria to penicillins is their ability to synthesize beta-lactamases and thereby inactivate penicillins with which they come in contact.

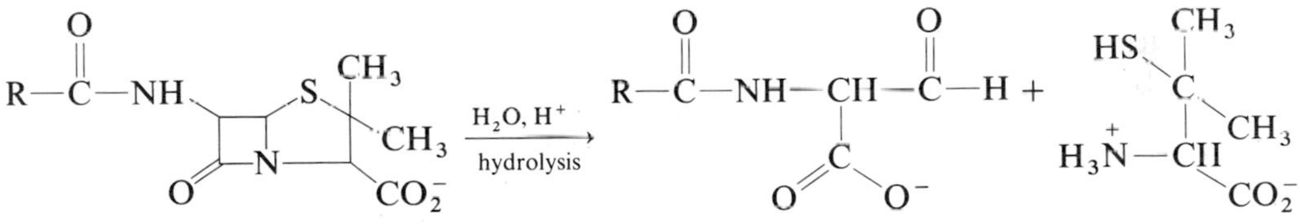

penicillin

penaldic acid

penicillamine

Figure XI-2 Acid-catalyzed hydrolysis of penicillin. Neither penaldic acid nor penicillamine has any antibacterial activity.

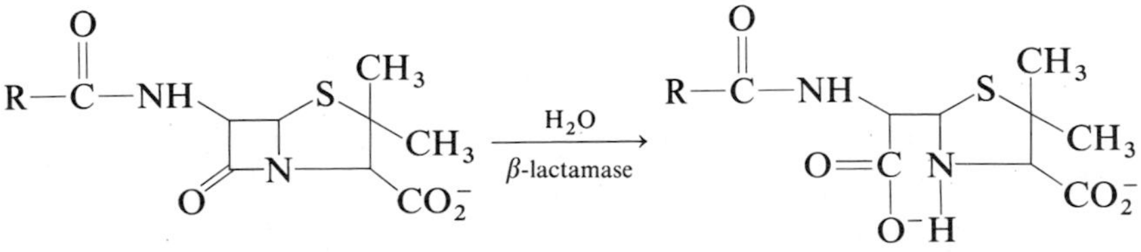

penicillin

penicilloic acid

Figure XI-3 Selective hydrolysis of the beta-lactam ring by beta-lactamase.

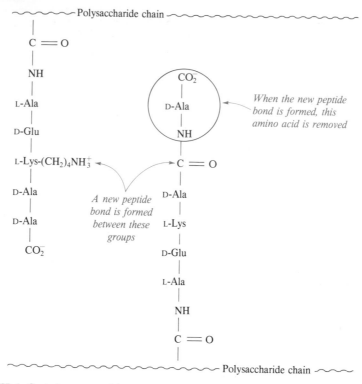

Figure XI-4 Certain types of bacterial cell walls are constructed of long polysaccharide chains to which are attached short polypeptide chains. In the final stage of cell-wall construction, cross-linking of the polypeptide chains creates an enormous bag-shaped macromolecule.

The penicillins owe their antibacterial activity to a common mechanism that inhibits the biosynthesis of a vital part of bacterial cell walls. Within certain types of bacterial cells, osmotic pressure is as high as 10–20 atm, and without the support of a cell wall, the bacterial cell would rupture. One of the simplest types of bacterial cell walls is contructed of long polysaccharide chains to which are attached short polypeptide chains (Figure XI-4).

The final reactions in construction of a bacterial cell wall are formation of amide bonds between adjacent polypeptide chains (Figure XI-4). Through this type of peptide bond interchange, polysaccharide chains are cross-linked, to form one enormous bag-shaped macromolecule.

D-alanyl-D-alanine

penicillin

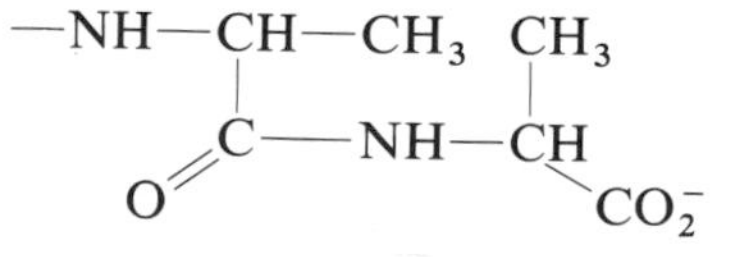

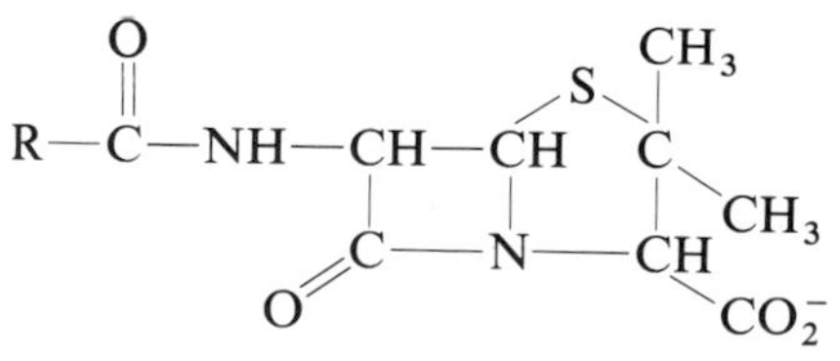

Figure XI-5 Structural formulas of D-alanyl-D-alanine and penicillin, drawn to suggest a structural similarity between the two.

It is formation of the final cross-linked macromolecule that penicillin inhibits. Numerous hypotheses have been proposed to explain just how this occurs. One hypothesis is that the penicillins have a structure like D-alanyl-D-alanine, the terminal amino acids of the short polypeptide chains that must be cross-linked. The similarity is illustrated in Figure XI-5. Penicillin is thought to bind selectively to the active site of the enzyme complex that catalyzes peptide bond interchange in the final step of cell-wall construction. Hence, at the molecular level, penicillin's antibacterial activity seems to be selective enzyme inhibition.

That this pattern of cell-wall construction is unique to certain forms of bacteria and not found in mammalian cells no doubt accounts for the lack of toxicity of penicillins to humans. However, the use of penicillins does have its problems. A significant percentage of the population have become hypersensitive to the drug and experience severe allergic reaction. The factor responsible for the allergic reaction is not penicillin itself but certain degradation products, particularly 6-aminopenicillanic acid (Figure XI-1).

The susceptibility of penicillin G to hydrolysis by acid and the emergence of beta-lactamase-producing strains of bacteria have provided incentive for researchers to develop more effective penicillins. As a result of such effort, by 1974 over 20,000 semisynthetic penicillins had been prepared. At present, the three most widely prescribed penicillins are ampicillin, penicillin VK, and penicillin G. The side chains of these three and of methicillin are shown in Figure XI-6.

Penicillin G is most effective against Gram-negative bacteria. Ampicillin is a antibiotic of broader spectrum and attacks both Gram-negative and Gram-positive bacteria. Penicillin VK is more resistant to acid hydrolysis than penicillin G and can be given orally. Methicillin is about 1000 times more resistant to the action of beta-lactamases than penicillin G and is used to treat infections caused by "penicillin-resistant" organisms.

Soon after the penicillins were introduced into medical practice, resistant strains began to appear, and they have proliferated ever since. Many argue that the widespread, unnecessary use of penicillins is the primary reason why resistant strains have emerged. One of the most serious is *Neisseria gonorrhoeae*; this strain causes a gonorrhea that is very difficult to treat. One approach to resistant strains is to synthesize newer, more effective beta-lactamase antibiotics, of which one class is the cephalosporins (Figure XI-7). The first cephalosporin was iso-

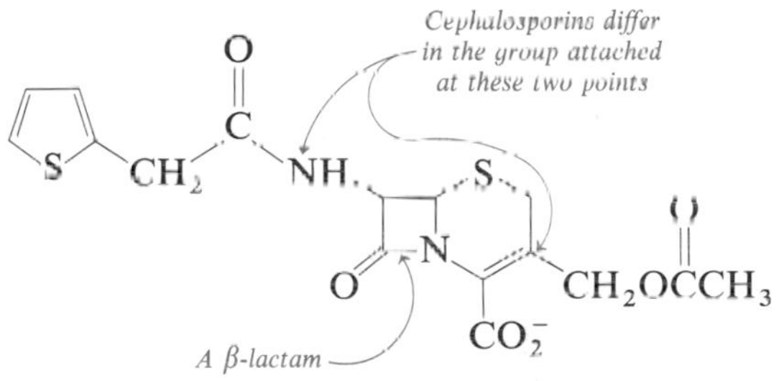

Figure XI-7 Cephalothin, one of the first of a new generation of beta-lactam antibiotics.

| penicillin G | ampicillin | penicillin VK | methicillin |

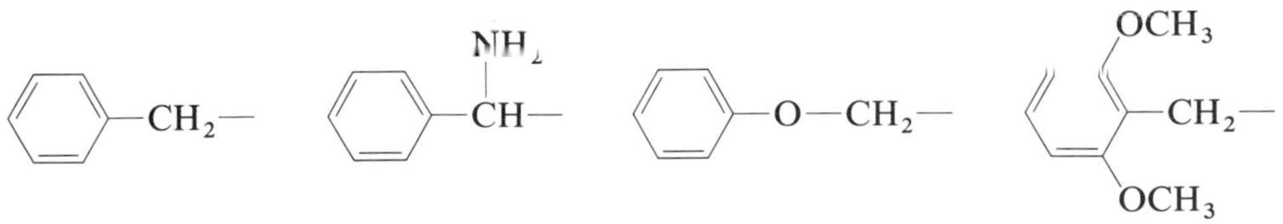

Figure XI-6 Structural relations between penicillin G and three widely used semisynthetic penicillins.

scribed penicillins are ampicillin, penicillin VK, and penicillin G. The side chains of these three and of methicillin are shown in Figure XI-6.

Penicillin G is most effective against Gram-negative bacteria. Ampicillin is a antibiotic of lated from the fungus *Cephalosporium acremonium*. Several cephalosporins have already been approved for clinical use, and approval of at least a dozen others is pending. Clinical studies indicate that this class of beta-lactam antibiotics has a broader spectrum of activity than the penicillins and at the same time a greater resistance to hydrolysis by beta-lactamases. Cephalosporins now account for approximately 35% of all antibiotic prescriptions.

The cephalosporins contain a beta-lactam ring fused to a six-member ring containing atoms of sulfur and nitrogen and a carbon-carbon double bond. As a family, cephalosporins differ from one another in the acyl group attached to the —NH$_2$ group of the beta-lactam ring and in the substituent on carbon 3 of the six-member ring.

An entirely different approach to the treatment of beta-lactam-resistant infections is to use either a penicillin or a cephalosporin combined with a compound that inhibits the activity of beta-lactamases. One such compound, clavulanic acid, has virtually no antibiotic activity by itself but is a powerful, irreversible inhibitor of beta-lactamases. Note that clavulanic acid itself is a beta-lactam (Figure XI-8). With this combination of drugs, it now appears possible to in-

A β-lactam

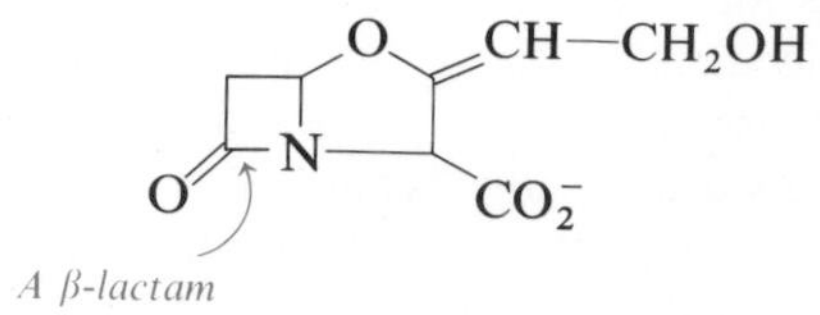

Figure XI-8 Clavulanic acid, a powerful inhibitor of beta-lactamases.

terfere not only with an infectious organism's essential biochemistry but also with its first line of defense against attack by drugs.

The penicillins and cephalosporins are among the most important anti-infective agents and almost certainly will continue to be so. Because the possibilities for substituting different groups on the essential ring structures of each are almost limitless, even more effective beta-lactam antibiotics are yet to be discovered. So, too, it is probable that in response to these drugs, even newer drug-resistant strains of bacteria will emerge.

15 Nucleic Acids and the Synthesis of Proteins

Nucleic acids are the third broad class of biopolymers that, like proteins and polysaccharides, are vital components of living materials. In this chapter we will look at the structure of nucleosides and nucleotides, and the manner in which these small building blocks are bonded together to form nucleic acids. Then, we will consider the three-dimensional structure of nucleic acids. Finally, we will examine the manner in which genetic information coded on deoxyribonucleic acids is expressed in protein biosynthesis.

15.1 Components of Deoxyribonucleic Acids (DNAs)

Controlled hydrolysis breaks DNA molecules into three components: (1) phosphoric acid; (2) 2-deoxy-D-ribose; and (3) heterocyclic aromatic amine bases. The bases fall into two classes: those derived from pyrimidine and those derived from purine. Following are structural formulas of the four bases most abundant in DNA, along with formulas for their parent bases.

pyrimidine   cytosine (C)   thymine (T)

purine   adenine (A)   guanine (G)

15.2 Nucleosides

A **nucleoside** is a glycoside in which nitrogen 9 of a purine base or nitrogen 1 of a pyrimidine base is bonded to 2-deoxy-D-ribose by a β-*N*-glycoside bond (Section 11.2A). In structural formulas for nucleosides and nucleotides, atoms of the purine and pyrimidine bases are designated by unprimed numbers; primed numbers are used to designate atoms of 2-deoxy-D-ribose. Two nucleosides, 2′-deoxyadenosine and 2′-deoxycytidine, are shown in Figure 15.1. The other two nucleosides found in DNA are 2′-deoxythymidine and 2′-deoxyguanosine.

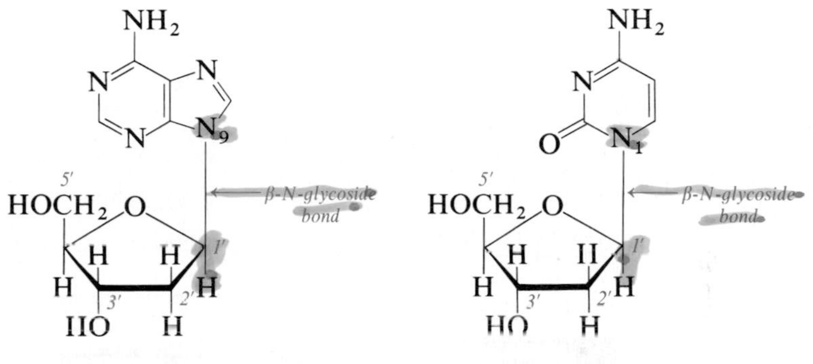

Figure 15.1 Nucleosides: 2′-deoxyadenosine and 2′-deoxycytidine.

15.3 Nucleotides

A **nucleotide** is a nucleoside monophosphate ester in which a molecule of phosphoric acid is esterified with a free hydroxyl group of 2-deoxy-D-ribose. Nucleoside monophosphates are illustrated in Figure 15.2 by the 5′-monophosphate

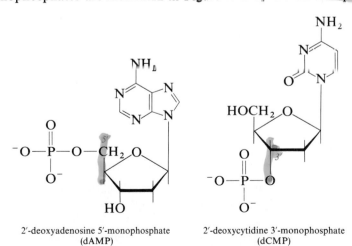

2′-deoxyadenosine 5′-monophosphate
(dAMP)

2′-deoxycytidine 3′-monophosphate
(dCMP)

Figure 15.2 Nucleotides (nucleoside monophosphate esters).

ester of 2′-deoxyadenosine and the 3′-monophosphate ester of 2′-deoxycytidine. Note that at pH 7.0, the two protons of a monophosphate ester are ionized, giving this group a net charge of −2.

Mononucleotides are commonly named as phosphate esters (for example, deoxyadenosine 5′-monophosphate) or as acids (for example, deoxyadenylic acid), or by four-letter abbreviations (for example, dAMP). In these four-letter abbreviations, the letter *d* indicates 2-deoxy-D-ribose, the second letter indicates the nucleoside, and the third and fourth letters indicate that the molecule is a monophosphate (MP) ester. Table 15.1 lists names for the principal mononucleotides derived from DNA.

Table 15.1
Names of the major mononucleotides derived from DNA.

As a Monophosphate	As an Acid	By a Four-Letter Abbreviation
deoxyadenosine monophosphate	deoxyadenylic acid	dAMP
deoxyguanosine monophosphate	deoxyguanidylic acid	dGMP
deoxycytidine monophosphate	deoxycytidylic acid	dCMP
deoxythymidine monophosphate	deoxythymidylic acid	dTMP

All nucleoside monophosphates can be further phosphorylated to form nucleoside diphosphates and nucleoside triphosphates. In diphosphates and triphosphates, the second and third phosphate groups are joined by anhydride bonds. At pH 7.0, all protons of diphosphate and triphosphate groups are fully ionized, giving them net charges of −3 and −4, respectively.

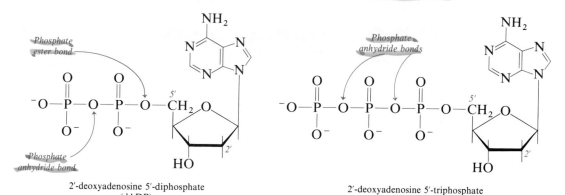

2′-deoxyadenosine 5′-diphosphate
(dADP)

2′-deoxyadenosine 5′-triphosphate
(dATP)

Example 15.1

Draw structural formulas for these mononucleotides.

a. 2′-deoxycytidine 5′-monophosphate (dCMP)
b. 2′-deoxyguanosine 5′-triphosphate (dGTP)

☐ *Solution*

a. Cytosine is joined by a β-N-glycoside bond between N-1 of cytosine and C-1 of the cyclic hemiacetal form of 2-deoxy-D-ribose. The 5′-hydroxyl of the pentose is bonded to phosphate by an ester bond.

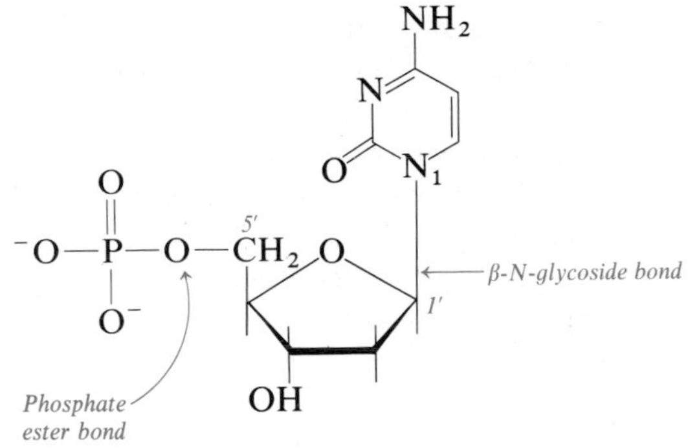

b. Guanine is joined by a β-N-glycoside bond between N-9 of guanine and C-1 of the cyclic hemiacetal form of 2-deoxy-D-ribose. The 5'-hydroxyl group of the pentose is joined to three phosphate groups by a combination of one ester bond and two anhydride bonds.

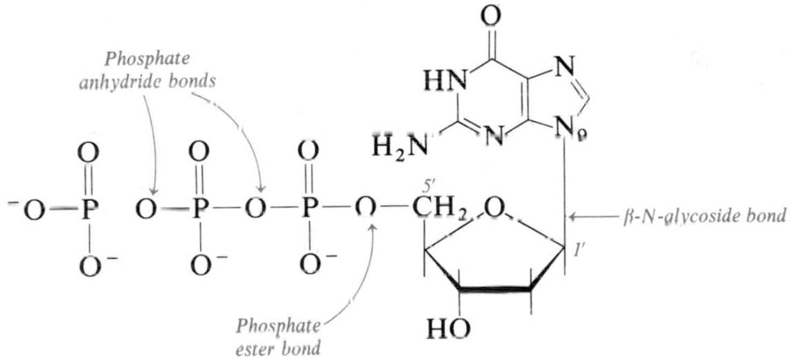

Problem 15.1   Draw structural formulas for (a) dTTP and (b) dGMP.

15.4 Structure of DNA

A. Primary Structure: The Covalent Backbone

Deoxyribonucleic acids (DNAs) consist of a backbone of alternating units of deoxyribose and phosphate, in which the 3'-hydroxyl of one deoxyribose is joined to the 5'-hydroxyl of the next deoxyribose by a phosphodiester bond (Figure 15.3). This pentose-phosphate backbone is constant throughout an entire DNA molecule. A heterocyclic base—adenine, guanine, thymine, or cytosine—is attached to each deoxyribose by a β-N-glycoside bond.

The sequence of bases in a DNA molecule is indicated by single-letter abbreviations for each base, beginning from the free 5'-hydroxyl end of the chain. According to this convention, the base sequence of the section of DNA shown

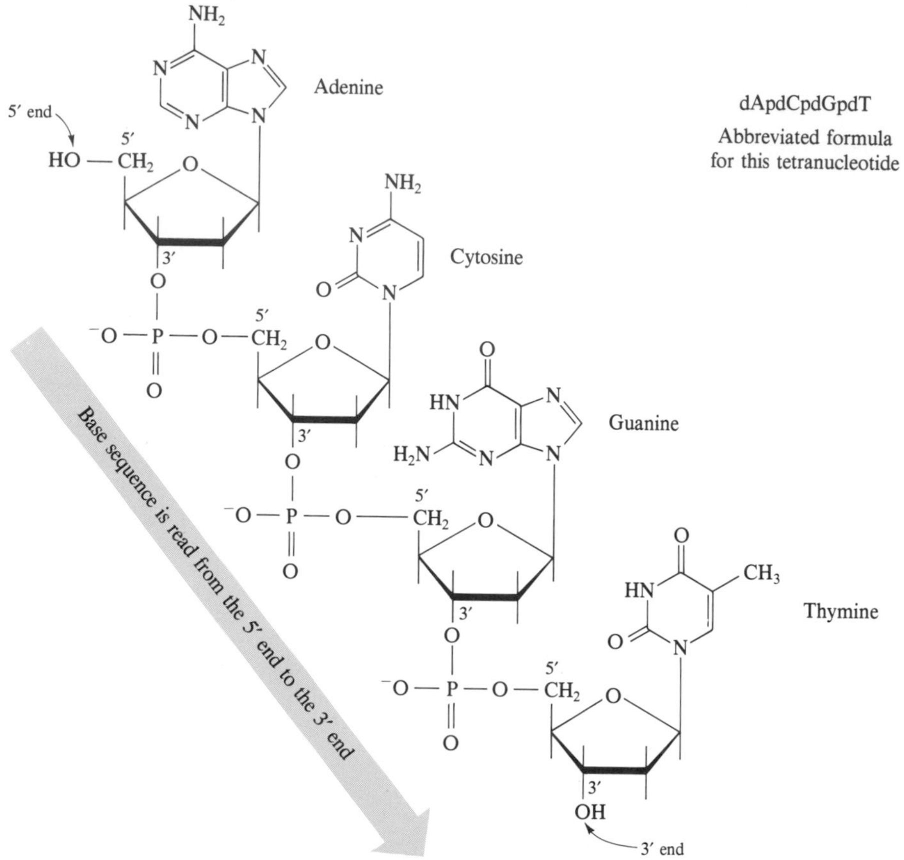

Figure 15.3 Partial structural formula of a deoxyribonucleic acid (DNA), showing a tetranucleotide sequence. In this abbreviated sequence, the bases of the tetranucleotide are read from the 5′ end of the chain to the 3′ end, as indicated by the arrow.

in Figure 15.3 is written dApdCpdGpdT, where the letter *d* indicates that each nucleoside monomer is derived from deoxyribose and the letter *p* indicates a phosphodiester bond in the backbone of the molecule. Alternatively, the base sequence can be written ACGT, a notation that emphasizes the order of the heterocyclic amine bases in the molecule.

Example 15.2

Draw a complete structural formula for a section of DNA containing the base sequence pdApdC.

☐ *Solution*

The first letter in the shorthand formula of this dinucleotide is *p*, indicating that the 5′-hydroxyl is bonded to phosphate by an ester bond. The last letter, *C*, shows that the 3′-hydroxyl is free, not esterified with phosphate.

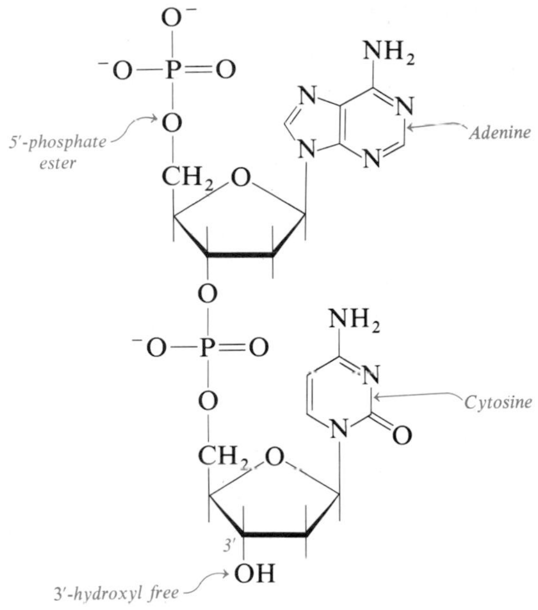

**Problem
15.2**

Draw a complete structural formula for a section of DNA containing the base sequence dCpdTpdGp.

B. Base Composition

By 1950, it was clear that DNA molecules consist of chains of alternating units of deoxyribose and phosphate linked by phosphodiester bonds, with a base attached to each deoxyribose by a β-N-glycoside bond. However, the precise sequence of bases along the chain of any particular DNA molecule was completely unknown. At one time, it was thought that the four principal bases occurred in equal ratios and perhaps repeated in a regular pattern along the pentose-phosphate backbone of the molecule. However, more precise determinations of base sequence by Erwin Chargaff revealed that the bases do not occur in equal ratios (Table 15.2).

Table 15.2 Comparison of base composition (in mole percentage) of DNA from several organisms.

Organism	A	G	C	T	A/T	G/C	Purines/Pyrimidines
human	30.9	19.9	19.8	29.4	1.05	1.00	1.04
sheep	29.3	21.4	21.0	28.3	1.03	1.03	1.03
sea urchin	32.8	17.7	17.3	32.1	1.02	1.02	1.03
marine crab	47.3	2.7	2.7	47.3	1.00	1.00	1.00
yeast	31.3	18.7	17.1	32.9	0.95	1.09	1.00
E. coli	24.7	26.0	26.0	23.6	1.04	1.01	1.03

From consideration of data such as shown in Table 15.2 the following conclusions emerged.

1. The mole-percentage base composition of DNA in any organism is the same in all cells and is characteristic of the organism.
2. The mole-percentage of adenine and of thymine are equal, and the mole-percentage of guanine and of cytosine are equal:

$$\%[\text{adenine}] = \%[\text{thymine}]$$
$$\%[\text{cytosine}] = \%[\text{guanine}]$$

3. The mole-percentage of purine bases (A + G) and of pyrimidine bases (C + T) are equal:

$$\%[\text{purines}] = \%[\text{pyrimidines}]$$

C. Molecular Dimensions of DNAs

Additional information on the structure of DNA emerged when X-ray diffraction photographs of DNA fibers taken by Rosalind Franklin and Maurice Wilkens were analyzed. These photographs showed that DNA molecules are long, fairly straight, and not more than a dozen atoms thick. Furthermore, even though the base composition of DNAs isolated from different organisms varies over a wide range, DNA molecules themselves are remarkably uniform in thickness. Herein lies one of the chief problems to be solved. How could the molecular dimensions of DNAs be so regular even though the relative percentages of the various bases differ so widely?

D. The Double Helix

With this accumulated information, the stage was set for the development of a hypothesis about DNA conformation. In 1953, F. H. C. Crick, a British physicist, and James D. Watson, an American biologist, postulated a precise model of the three-dimensional structure of DNA. The model not only accounted for many of the observed physical and chemical properties of DNA but also suggested a mechanism by which genetic information could be repeatedly and accurately replicated. Watson, Crick, and Wilkins shared the 1962 Nobel Prize in physiology and medicine for "their discoveries concerning the molecular structure of nucleic acids, and its significance for information transfer in living material."

The heart of the Watson-Crick model is the postulate that a molecule of DNA consists of two antiparallel polynucleotide strands coiled in a right-handed manner about the same axis to form a **double helix**. To account for the observed base ratios and the constant thickness of DNA, Watson and Crick postulated that purine and pyrimidine bases project inward toward the axis of the helix and are always paired in a very specific manner.

According to scale models, the dimensions of a thymine-adenine base pair are identical to the dimensions of a cytosine-guanine base pair, and the length of each pair is consistent with the thickness of a DNA strand (Figure 15.4). This fact gives rise to the principle of **complementarity**. In DNA, adenine is always

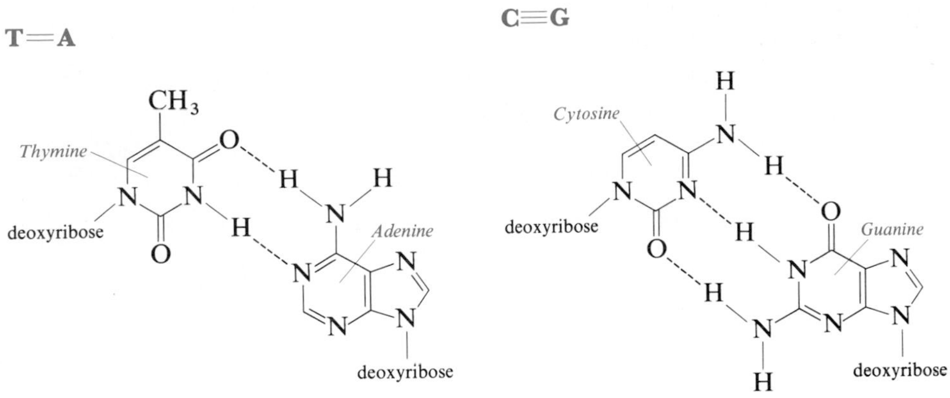

T=A

C≡G

Figure 15.4 Hydrogen-bonded interaction between thymine and adenine and between cytosine and guanine. The first pair is abbreviated T—A (showing two hydrogen bonds) and the second pair is abbreviated C≡G (showing three hydrogen bonds).

paired by hydrogen bonding with thymine; hence, adenine and thymine are complementary bases. Similarly, guanine and cytosine are complementary bases. A significant finding arising from Watson and Crick's model-building is that no other base pairing is consistent with the observed thickness of a DNA molecule. A pair of pyrimidine bases is too small to account for the observed thickness, while a pair of purine bases is too large. Thus, according to the Watson-Crick model, the repeating units in a double-stranded DNA molecule are not single bases of differing dimensions, but base pairs of identical dimensions.

To account for the periodicity observed from X-ray data, Watson and Crick postulated that base pairs are stacked one on top of the other, with a distance of 3.4×10^{-8} cm between base pairs. Exactly ten base pairs are stacked in one complete turn of the helix. There is one complete turn of the helix every 34×10^{-8} cm (Figure 15.5).

Example 15.3

One strand of a DNA molecule has a base sequence of 5′-ACTTGCCA-3′. Write the base sequence for the complementary strand.

☐ *Solution*

Remember that base sequence is always written from the 5′ end of the strand to the 3′ end, that A is always paired by hydrogen bonding with its complement T, and that G is always paired by hydrogen bonding with its complement C. In double-stranded DNA, the strands run in opposite (antiparallel) directions, so the 5′ end of one strand is associated with the 3′ end of the other strand. Hydrogen bonds between base pairs are shown by dashed lines.

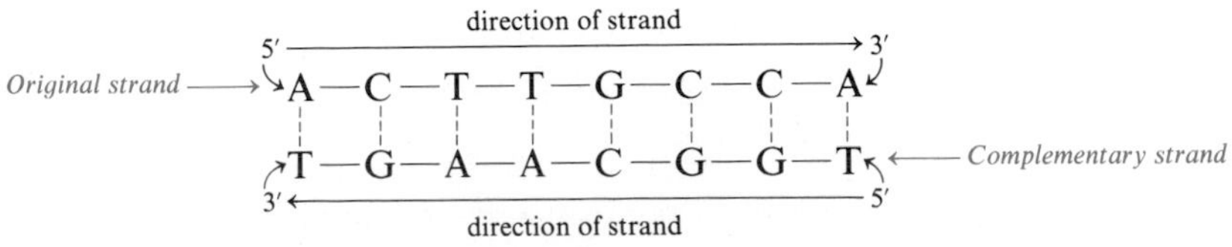

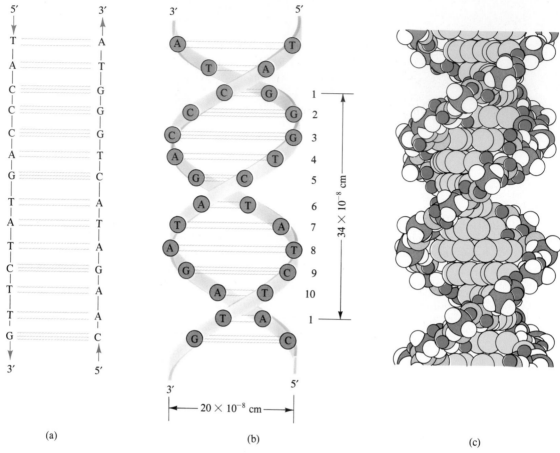

Figure 15.5 Abbreviated representation of the Watson-Crick double-helix model of DNA. (a) Two complementary antiparallel polynucleotide strands and the hydrogen bonds between complementary base pairs. (b) The strands are twisted in a double helix of thickness 20×10^{-8} cm and a repeat distance of 34×10^{-8} cm along the axis of the double helix. There are 10 base pairs per complete turn of the helix. (c) A space-filling model of a section of DNA double helix.

The complement of 5'-ACTTGCCA-3' is shown under it in the solution. Writing this strand poses a communication problem. DNA strands are always written from 5' to 3' end. Therefore, if the original strand is 5'-ACTTGCCA-3', its complement is 5'-TGGCAAGT-3'.

Problem 15.3

Write the complementary base sequence for 5'-CCGTACGA-3'

15.5 DNA Replication

At the time Watson and Crick proposed their model for the conformation of DNA, biologists had already amassed much evidence that DNA is the hereditary, or genetic, material. Detailed studies had revealed that during cell division, there was exact duplication of DNA. The challenge posed to molecular biologists was, How does the genetic material duplicate itself with such unerring fidelity?

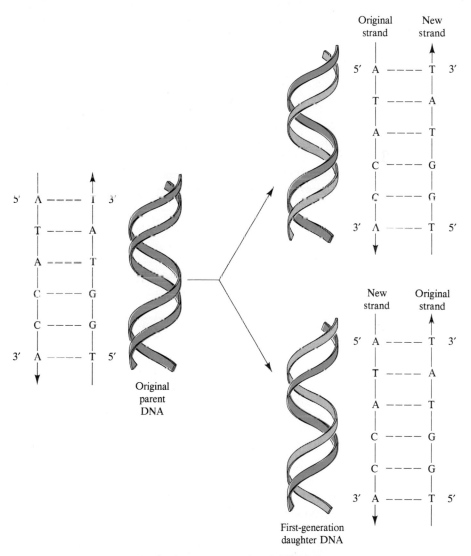

Figure 15.6 Schematic diagram of semiconservative replication. The double helix uncoils and each chain of the parent serves as a template for the synthesis of its complement. Each "daughter" DNA contains one strand from the original DNA and one newly synthesized strand.

One of the exciting things about the double-helix model is that it immediately suggested how DNA might produce an exact copy of itself. The double helix consists of two parts, one the complement of the other. If the two strands separate and each serves as a template for the construction of its own complement, then each new double strand will be an exact replica of the original DNA. Because each new double-stranded DNA molecule contains one strand from the parent molecule and one newly synthesized, **daughter strand**, the process is called **semiconservative replication** (Figure 15.6). Although Figure 15.6 shows the result of DNA replication, the actual process is much more complicated than shown. In *E. coli*, replication is thought to proceed by four main steps. While there are variations from species to species, replication in other organisms follows a similar process.

A. Initiation of Replication

Replication starts at a specific point on a chromosome, where unwinding proteins catalyze uncoiling of the DNA helix. During the unwinding, hydrogen bonds between complementary base pairs are broken and purine and pyrimidine bases in the center of double-stranded DNA are exposed (Figure 15.7). The point of unwinding is called a **replication fork.**

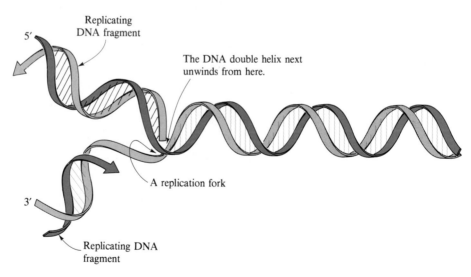

Replicating
DNA fragment

5′

The DNA double helix next
unwinds from here.

A replication fork

3′

Replicating DNA
fragment

Figure 15.7 Unwinding of DNA, creating a replication fork.

B. Formation of DNA Segments

In step 2, DNA replication proceeds along both branches of the exposed DNA template from the 3′ end toward the 5′ end. Because the two unwound DNA strands run in opposite directions, DNA synthesis proceeds toward the replication fork on one strand and away from the replication fork on the other. Addition of mononucleotides to growing DNA daughter strands is catalyzed by DNA polymerase, an enzyme that recognizes the 3'-hydroxyl end of the growing daughter strand and positions the proper complementary deoxynucleoside triphosphate to pair by hydrogen bonding with the next base pair on the template. Finally, the 3'-hydroxyl of the daughter strand displaces pyrophosphate from the 5' end of the next nucleoside triphosphate, to form a phosphodiester bond, and the daughter strand becomes elongated by one nucleotide (Figure 15.8).

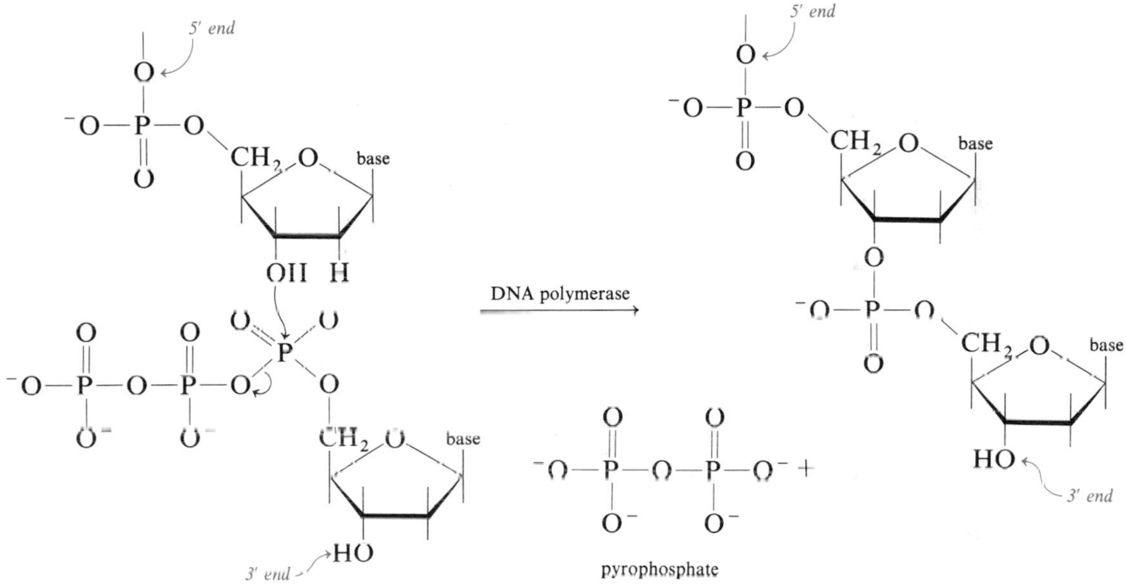

Figure 15.8 Formation of a phosphodiester bond and elongation of a daughter strand by one nucleotide.

C. Creation of a New Replication Fork and Continuation of DNA Synthesis

In step 3, a new section of DNA is unwound, creating a second replication fork, and replication is repeated. According to this mechanism, one daughter strand is synthesized as a continuous strand from the 3' end of the DNA template toward the first replication fork and then toward each newly created replication fork as further sections of double helix are unwound. The second daughter strand is synthesized as a series of fragments, each as long as the distance from one replication fork to the next. Breaks created at replication forks are called **nicks** (Figure 15.9) and the DNA fragments separated by nicks are called **Okazaki fragments**, after the biochemist who first discovered them. The isolation of Okazaki fragments is evidence that replication of DNA is not a continuous process.

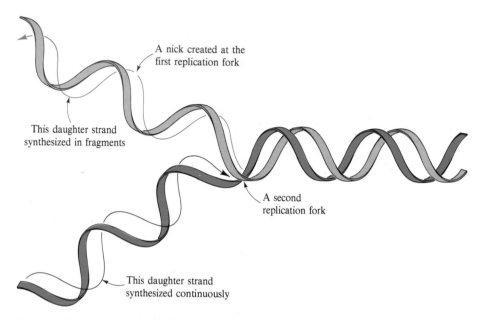

A nick created at the first replication fork

This daughter strand synthesized in fragments

A second replication fork

This daughter strand synthesized continuously

Figure 15.9 Synthesis of DNA daughter strands toward and away from a replication fork.

D. Completion of the DNA Strand

In step 4, an enzyme called DNA ligase closes nicks, to form a completed daughter strand.

15.6 Ribonucleic Acids (RNAs)

Ribonucleic acids (RNAs) are similar to deoxyribonucleic acids in that they, too, consist of long, unbranched chains of nucleotides joined by phosphodiester bonds between the 3'-hydroxyl of one pentose and the 5'-hydroxyl of the next. Thus, they have much the same structure as DNA (Figure 15.3). However, three main differences in structure exist between RNA and DNA:

1. The pentose unit of RNA is D-ribose rather than 2-deoxy-D-ribose.
2. The pyrimidine bases in RNA are uracil and cytosine rather than thymine and cytosine.
3. RNA is single-stranded rather than double-stranded.

Following are structural formulas of β-D-ribose and uracil.

β-D-ribose

uracil
(U)

Ribonucleic acids are distributed throughout the cell; they are present in the nucleus, in the cytoplasm, and in subcellular particles called mitochondria. Furthermore, cells contain three types of RNA: ribosomal RNA, transfer RNA, and messenger RNA. These three types of RNA differ in molecular weight and, as their names imply, they perform different functions within the cell.

A. Ribosomal RNAs

Ribosomal RNAs (rRNAs) have molecular weights of 0.5–1.0 million and comprise up to 85–90% of total cellular ribonucleic acid. The bulk of rRNAs are found in the cytoplasm in subcellular particles called **ribosomes**, which contain about 60% RNA and 40% protein. Complete ribosomes (referred to as 70S ribosomes) can be dissociated into two subunits of unequal size, known as 50S subunits and 30S subunits (Figure 15.10). The designation S stands for Svedberg units. Values of S are derived from rates of sedimentation during centrifugation and are used to estimate molecular weight and compactness of ribosomal particles. A large value of S indicates a high molecular weight; a small value indicates a low molecular weight. The 50S ribosomal subunit is about twice the size of the 30S subunit and further dissociates into 23S and 5S subunits and approximately 30 different polypeptides (Figure 15.10). The smaller 30S subunit dissociates into a single 16S subunit and about 20 different polypeptides.

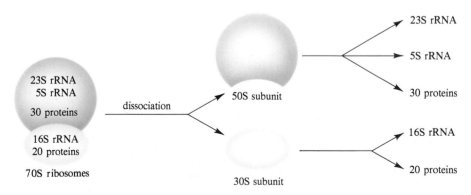

Figure 15.10 Dissociation of a complete ribosome into subunits.

Many of the proteins bound to ribosomes have a high precentage of lysine and arginine, and at the pH of cells, the side chains of these amino acids have net positive charges. It is likely that interactions between positively charged amino acid side chains and negatively charged phosphate groups of RNA are an important factor in stabilizing larger ribosomal particles.

B. Transfer RNA

Transfer RNA (tRNA) molecules have the lowest molecular weight of all nucleic acids. They consist of 75–80 nucleotides in a single chain; this chain is folded into a three-dimensional structure, stabilized by hydrogen bonding between complementary base pairs. Nearly all tRNA chains have G at the 5′ end and CCA at the 3′ end. The three-dimensional shapes of a number of tRNAs have been determined, typical of which is that of yeast phenylalanine tRNA (Figure 15.11).

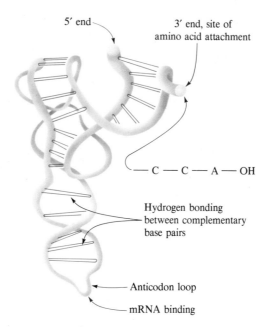

Figure 15.11 The three-dimensional shape of yeast phenylalanine tRNA.

The function of tRNA is to carry amino acids to the sites of protein synthesis on ribosomes. For this purpose, an amino acid is joined to the 3′ end of its specific tRNA by an ester bond, formed between the α-carboxyl group of the amino acid and the 3′-hydroxyl group of ribose. An amino acid thus bound to tRNA is said to be activated, because it is prepared for the synthesis of a peptide bond.

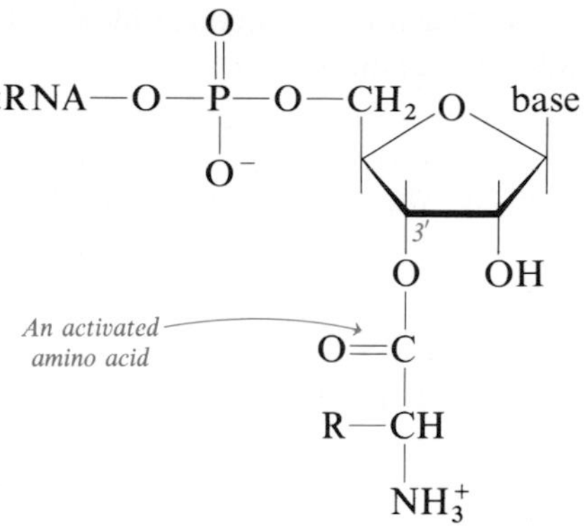

$$\text{tRNA}-\text{O}-\overset{\overset{\displaystyle\text{O}}{\|}}{\underset{\underset{\displaystyle\text{O}^-}{|}}{\text{P}}}-\text{O}-\text{CH}_2$$

An activated amino acid

$$\text{O}=\text{C}$$

$$\text{R}-\text{CH}$$

$$\text{NH}_3^+$$

C. Messenger RNA

Messenger RNA (mRNA) is present in cells in relatively small amounts and is very short-lived. It is single-stranded, has an average molecular weight of several hundred thousand, and has a base composition much like that of the DNA of the organism from which it is isolated. The name messenger RNA derives from the function of this type of RNA, which is made in cell nuclei on a DNA template and carries coded genetic information to the ribosomes for the synthesis of new proteins (Section 15.9).

15.7 Transcription of Genetic Information: mRNA Biosynthesis

Messenger RNA is synthesized from DNA in a manner similar to the replication of DNA. Double-stranded DNA is unwound, and a complementary strand of mRNA is synthesized along one strand of the DNA template, beginning from the 3' end. The synthesis of mRNA from a DNA template is called **transcription**, because genetic information contained in a sequence of bases of DNA is transcribed into a complementary sequence of bases in mRNA.

■

Example 15.4

☐ *Solution*

Following is a base sequence from a portion of DNA. Write the sequence of bases of the mRNA synthesized, using this section of DNA as a template.

3'-A-G-C-C-A-T-G-T-G-A-C-C-5'

RNA synthesis begins at the 3' end of the DNA template and proceeds toward the 5' end. The complementary mRNA strand is formed using the bases C, G, A, and U. Uracil (U) is the complement of adenine (A) on the DNA template.

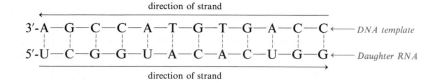

To write the base sequence of the DNA template, start from the 5′ end. If the DNA template is

$$5'\text{-C-C-A-G-T-G-T-A-C-C-G-A-}3'$$

then the complementary mRNA strand is

$$5'\text{-U-C-G-G-U-A-C-A-C-U-G-G-}3'$$

Problem 15.4

Following is a base sequence from a portion of DNA. Write the sequence of bases in the mRNA synthesized using this section of DNA as a template.

$$5'\text{-T-C-G-G-T-A-C-A-C-T-G-G-}3'$$

15.8 Genetic Code

A. Triplet Nature of the Code

It was clear by the early 1950s that the sequence of bases in DNA molecules constitutes a store of genetic information, and that the sequence of bases directs the synthesis of mRNA and of proteins. However, the statement that the sequence of bases in DNA directs the synthesis of proteins presented the following problem: How can a molecule containing only four variable units (adenine, cytosine, guanine, and thymine) direct the synthesis of molecules containing up to 20 units (the 20 common protein-derived amino acids)? How can an alphabet of 4 letters code for the order of letters in the 20-letter alphabet that occurs in proteins?

An obvious answer is that not one base but a combination of bases codes for each amino acid. If the code consists of nucleotide pairs, there are $4^2 = 16$ combinations, a more extensive code, but still not extensive enough to code for 20 amino acids. If the code consists of nucleotides in groups of three, $4^3 = 64$ combinations are possible, more than enough to code for the primary sequence of a protein. This appears to be a very simple solution to a system that must have taken eons of evolutionary trial and error to develop. Yet proof now exists, from comparison of gene (nucleic acid) and protein (amino acid) sequences, that nature does indeed use this simple 3-letter, or triplet, code to store genetic information. A triplet of nucleotides is called a **codon**.

B. Deciphering the Genetic Code

The next question is, Which of the 64 triplets code for which amino acids? In 1961, Marshall Nirenberg provided a simple experimental approach to the problem, based on the observation that synthetic polynucleotides direct polypeptide synthesis in much the same manner as natural mRNAs do. Nirenberg incubated ribosomes, amino acids, tRNAs, and appropriate protein-synthesizing enzymes. With only these components, there was no polypeptide synthesis. However, when he added synthetic polyuridylic acid (poly U), a polypeptide of high molecular weight was synthesized. What was more important, the synthetic poly-nucleotide contained only phenylalanine. With this discovery, the first element of the **genetic code** was deciphered: the triplet UUU codes for phenylalanine.

Similar experiments were carried out with different synthetic polyribonucleo-tides. It was found, for example, that polyadenylic acid (poly A) leads to the synthesis of polylysine, and that polycytidylic acid (poly C) leads to the synthesis of polyproline.

Codon on mRNA	Amino Acid
UUU	phenylalanine
AAA	lysine
CCC	proline

By 1966, all 64 codons had been deciphered (Table 15.3).

Table 15.3 The genetic code: mRNA codons and the amino acid whose incorporation each codon directs.

UUU	Phe	UCU	Ser	UAU	Tyr	UGU	Cys
UUC	Phe	UCC	Ser	UAC	Tyr	UGC	Cys
UUA	Leu	UCA	Ser	UAA	Stop	UGA	Stop
UUG	Leu	UCG	Ser	UAG	Stop	UGG	Trp
CUU	Leu	CCU	Pro	CAU	His	CGU	Arg
CUC	Leu	CCC	Pro	CAC	His	CGC	Arg
CUA	Leu	CCA	Pro	CAA	Gln	CGA	Arg
CUG	Leu	CCG	Pro	CAG	Gln	CGG	Arg
AUU	Ile	ACU	Thr	AAU	Asn	AGU	Ser
AUC	Ile	ACC	Thr	AAC	Asn	AGC	Ser
AUA	Ile	ACA	Thr	AAA	Lys	AGA	Arg
AUG	Met	ACG	Thr	AAG	Lys	AGG	Arg
GUU	Val	GCU	Ala	GAU	Asp	GGU	Gly
GUC	Val	GCC	Ala	GAC	Asp	GGC	Gly
GUA	Val	GCA	Ala	GAA	Glu	GGA	Gly
GUG	Val	GCG	Ala	GAG	Glu	GGG	Gly

C. Properties of the Genetic Code

Several features of the genetic code are evident from a study of Table 15.3.

1. Only 61 triplets code for amino acids. The remaining three (UAA, UAG, and UGA) are signals for chain terminations; that is, they signal to the protein-synthesizing machinery of the cell that the primary sequence of the protein is complete. The three-chain termination triplets are indicated in Table 15.3 by Stop.

2. The code is degenerate, which means that several amino acids are coded for by more than one triplet. If you count the number of triplets coding for each

amino acid, you will find that only methionine and tryptophan are coded for by just one triplet. Leucine, serine, and arginine are coded for by six triplets, and the remaining amino acids are coded for by two, three, or four triplets.

3. For the 15 amino acids coded for by two, three, or four triplets, the degeneracy is only in the last base of the triplet. For example, glycine is coded for by the triplets GGA, GGG, GGC, and GGU. In the codons for these 15 amino acids, it is only the third letter of the codon that varies.

4. There is no ambiguity in the code, meaning that each triplet codes for one and only one amino acid.

We must ask one last question about the genetic code: Is the code universal—that is, is it the same for all organisms? Every bit of experimental evidence available today from the study of viruses, bacteria, and higher animals, including humans, indicates that the code is universal. Furthermore, that it is the same for all these different organisms means that it has been the same over billions of years of evolution.

Example 15.5

During transcription, a portion of mRNA is synthesized with the following base sequence:

$$5'\text{-AUG-GUA-CCA-CAU-UUG-UGA-}3'$$

a. Write the base sequence of the DNA from which this portion of mRNA was synthesized.

b. Write the primary structure of the polypeptide coded for by this section of mRNA.

☐ Solution

a. During transcription, mRNA is synthesized from a DNA strand, beginning from the 3' end of the DNA template. The DNA strand must be complementary to the newly synthesized mRNA strand.

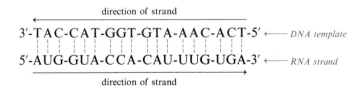

direction of strand

3'-TAC-CAT-GGT-GTA-AAC-ACT-5' ←— *DNA template*

5'-AUG-GUA-CCA-CAU-UUG-UGA-3' ←— *RNA strand*

direction of strand

b. The sequence of amino acids is shown below the mRNA strand.

$$5'\text{-AUG-GUA-CCA-CAU-UUG-UGA-}3'$$
      met   val   pro   his   leu   stop

The codon UGA codes for termination of a growing polypeptide chain; therefore, the sequence given in this problem codes for a pentapeptide only.

Problem 15.5

The following section of DNA codes for oxytocin, a polypeptide hormone.

mRA synthesis begins here

$$3'\text{-ACG-ATA-TAA-GTT-TTA-ACG-GGA-GAA-CCA-ACT-}5'$$

a. Write the base sequence of the mRNA synthesized from this section of DNA.

b. Given the sequence of bases in part (a), write the amino acid sequence of oxytocin.

■

15.9 Translation of Genetic Information: Biosynthesis of Polypeptides

The biosynthesis of polypeptides is usually described in three main processes: initiation of the polypeptide chain, elongation, and termination of the completed chain. These processes, along with the substances required for each, are summarized in Table 15.4.

Table 15.4
Main processes in polypeptide synthesis.

Process	Substances Required
initiation	tRNA carrying *N*-formylmethionine, mRNA, 30S and 50S subunits, GTP, protein-initiating factors
elongation	amino acyl tRNAs, protein elongation factors, GTP
termination	termination codon on mRNA, protein termination factors

A. Initiation of Polypeptide Synthesis

In bacteria, all polypeptide chains are initiated with the amino acid *N*-formylmethionine (fMet). *N*-formylmethionine is bound to a specific tRNA molecule and given the symbol tRNA$_{fMet}$.

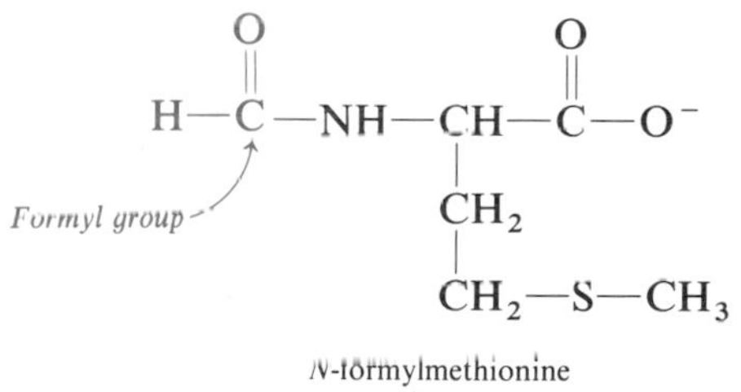

$$H-\overset{\overset{O}{\|}}{C}-NH-CH-\overset{\overset{O}{\|}}{C}-O^-$$

Formyl group

$$CH_2$$
$$CH_2-S-CH_3$$

N-formylmethionine
(fMet)

Many bacterial polypeptides do have *N*-formylmethionine as the *N*-terminal amino acid. However, for most bacterial proteins, *N*-formylmethionine, either by itself or with several other amino acids at the *N*-terminal end of the chain, is cleaved to give the native protein.

The first step in initiation is alignment of mRNA on a 30S ribosomal subunit so that the initiating codon is located at a specific site on the ribosome, called the P site. The initiating codon is most commonly AUG, the one for methionine. Next, tRNA carrying *N*-formylmethionine (fMet) binds to the initiating codon, and this complex, in turn, binds a 50S ribosomal subunit, to give a unit called an initiation complex (Figure 15.12).

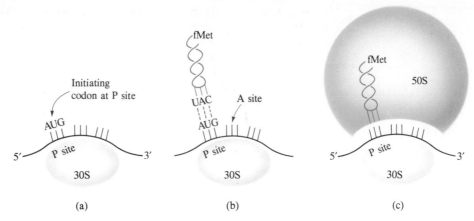

Figure 15.12 Formation of an initiation complex. (a) Alignment of mRNA on a 30S ribosomal subunit so that AUG, the initiating codon, is located at the P site. (b) Binding of tRNA carrying *N*-formylmethionine (fMet) to the initiating codon. (c) Association of the 50S ribosomal subunit to give an initiation complex.

B. Elongation of the Polypeptide Chain

Elongation of a polypeptide chain consists of three steps that are repeated over and over until the entire polypeptide chain is synthesized. In the first step, a "charged" tRNA (one carrying an amino acid esterified at the 3′ end of a tRNA chain) binds to the A site of an initiating complex (Figure 15.13).

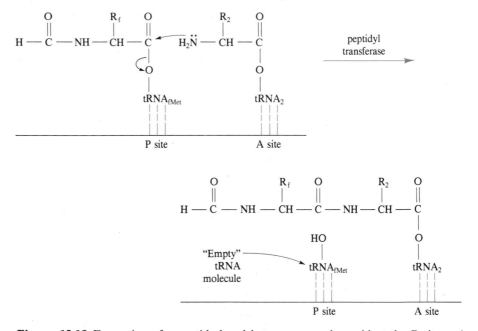

Figure 15.13 Formation of a peptide bond between an amino acid at the P site and another at the A site.

The second step is formation of a peptide bond between the carboxyl group of the tRNA-bound amino acid at the P site and the amino group of the tRNA-bound amino acid at the A site. Peptide bond formation is catalyzed by the enzyme peptidyl transferase. After a new peptide bond is formed, the tRNA attached to the P site is "empty" and the growing polypeptide chain is now attached to the tRNA bound to the A site.

The third step in the elongation cycle involves release of the empty tRNA from the P site and translocation of the growing polypeptide chain from the A site to the P site.

The three steps in the elongation cycle are shown schematically in Figure 15.14 for the synthesis of the tripeptide fMet-arg-phe from fMet-arg.

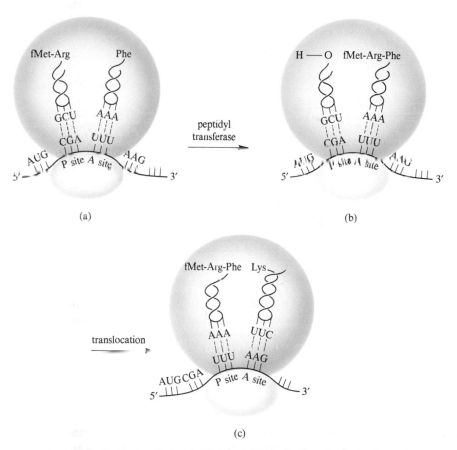

Figure 15.14 Chain elongation. (a) The growing polypeptide chain bound to arg-tRNA is aligned at the P site, and phe-tRNA is aligned at the A site. (b) Peptidyl transferase catalyzes peptide bond formation between the carbonyl group of arginine and the amino group of phenylalanine, and the growing polypeptide chain is transferred to phe-tRNA. (c) As a result of translocation, phe-tRNA is moved to the P site, and the next amino acid in the primary sequence, lys-tRNA, is aligned at the A site.

C. Termination of Polypeptide Synthesis

Polypeptide synthesis continues through the chain elongation cycle until the ribosome complex reaches a stop codon (UAA, UAG, or UGA) on mRNA. There, a specific protein called a termination factor binds to the stop codon and catalyzes hydrolysis of the completed polypeptide chain from tRNA. The "empty" ribosome then dissociates, ready to bind to another strand of mRNA and fMet-tRNA to form another initiation complex.

Figure 15.15 shows several ribosome complexes moving along a single strand of mRNA and illustrates that several identical polypeptide chains can be synthesized simultaneously from a single mRNA molecule. Figure 15.15 also shows that as a polypeptide chain grows, it folds spontaneously into its native three-dimensional conformation.

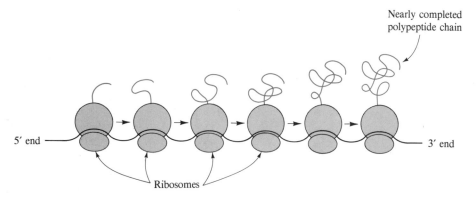

Figure 15.15 Simultaneous elongation of several identical polypeptide chains on a single strand of mRNA. The growing polypeptide chains spontaneously assume their natural three-dimensional conformation.

15.10 Inhibition of Protein Synthesis and the Action of Antibiotics

Several widely used antibiotics, including tetracycline, streptomycin, chloramphenicol, and puromycin (Figure 15.16), act by inhibiting protein synthesis in bacteria at the ribosomal level. Although the general process of protein synthesis described in Section 15.9 operates universally, some details of the processes in bacteria and animals are different. Because of these differences, many antibiotics inhibit protein synthesis in bacteria while having little or no effect on host cells.

The tetracyclines are a family of antibiotics of which all have the tetracyclic ring structure shown in Figure 15.16 but different groups attached to this ring pattern. Tetracyclines, because of their broad activity as antimicrobials, are among the most widely used antibiotics. Tetracyclines prevent binding of charged tRNAs to 30S ribosomal subunits and thereby disrupt protein synthesis.

Chloramphenicol binds specifically to the A site of a 50S ribosomal subunit and thereby prevents charged tRNAs from binding to it. Chloramphenicol is also a broad-spectrum antibiotic. However, in some persons it causes serious,

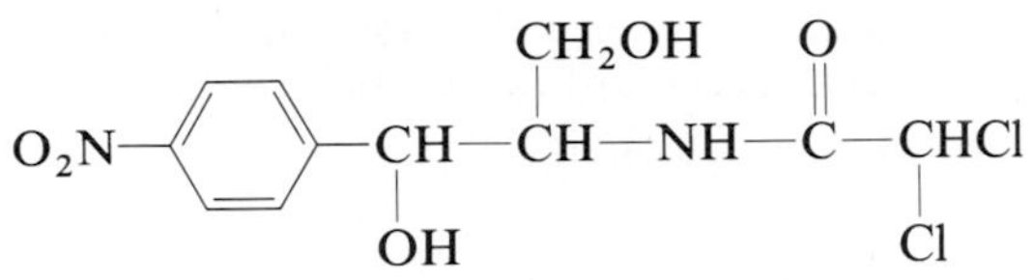

chloramphenicol (binds to the A site and inhibits binding of charged tRNAs)

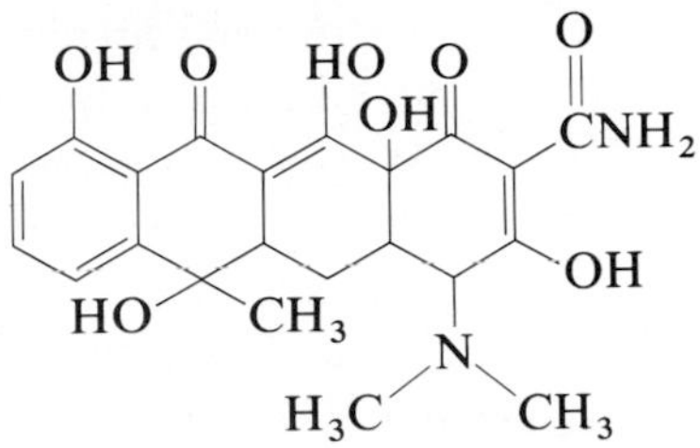

tetracycline (inhibits binding of charged tRNAs to the 30S ribosomal subunit)

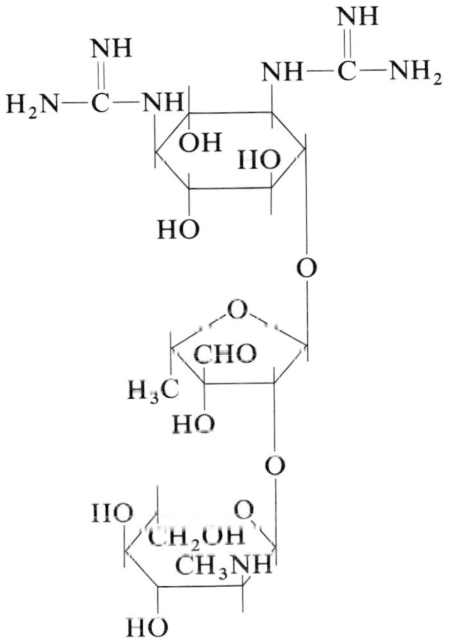

streptomycin (binds to proteins of the 30S ribosomal subunit and causes misreading of mRNA code)

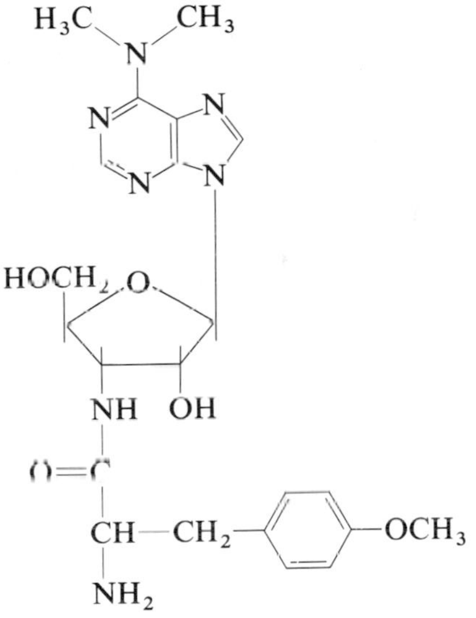

puromycin (is inserted in the growing polypeptide chain and causes premature termination of polypeptide synthesis)

Figure 15.16 Structural formulas for four antibiotics and their effects on protein synthesis in bacteria.

often toxic, side effects. For this reason, its use has been restricted largely to treatment of acute infections for which other antibiotics are ineffective or to cases in which, for medical reasons, other antibiotics cannot be used.

Puromycin is a structural analog of a charged tRNA molecule (one bearing an amino acid esterified to the 3′-hydroxyl of the terminal nucleotide) and binds to the A site during chain elongation. There, the enzyme peptidyl transferase catalyzes formation of a peptide bond between a growing polypeptide chain and the amino group of puromycin; at this point, further chain elongation ceases. Thus, puromycin causes premature termination of polypeptide synthesis.

Streptomycin binds with proteins of the 30S ribosomal subunit and interferes with interactions between mRNA codons and tRNAs. This interference gives rise to errors in reading the mRNA code and causes incorrect amino acids to be inserted into the growing polypeptide chain.

15.11 Mutations

A **mutation** is any change or alteration in the sequence of heterocyclic aromatic amine bases on DNA molecules. At the molecular level, mutation in its simplest form is a change in a single base pair in a DNA molecule. Such mutations are called **point mutations** and can be divided into three groups: (1) base-pair substitutions, (2) base-pair insertions, and (3) base-pair deletions. A point mutation in DNA is transmitted to mRNA during transcription and is ultimately expressed as a protein with an altered primary structure.

A. Base-Pair Substitutions

A point mutation leading to the substitution of one base for another in a section of double-stranded DNA affects only one codon, as illustrated in Figure 15.17.

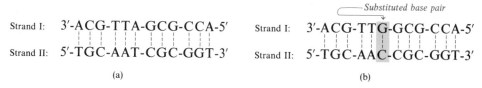

Strand I: 3′-ACG-TTA-GCG-CCA-5′   Strand I: 3′-ACG-TTG-GCG-CCA-5′

Strand II: 5′-TGC-AAT-CGC-GGT-3′   Strand II: 5′-TGC-AAC-CGC-GGT-3′

(a)   —*Substituted base pair*

                                          (b)

Figure 15.17 A portion of DNA (a) before a base-pair substitution, and (b) after.

Example 15.6

Refer to the DNA sequence in Figure 15.17.

a. Write the sequence of bases in the mRNA transcribed from the 3′ end of strand I of the original DNA. Also write the sequence of amino acids coded for by this section of mRNA. Remember that protein synthesis begins at the 5′ end of mRNA.

b. Do the same for strand I as it is after mutation and compare the two amino acid sequences.

☐ Solution

a. Original DNA: 3′-ACG-TTA-GCG-CCA-5′

          mRNA: 5′-UGC-AAU-CGC-GGU-3′

       Amino acids: cys   asn   arg   gly

b. Mutant DNA: 3′-ACG-TTG-GCG-CCA-5′

          mRNA: 5′-UGC-AAC-CGC-GGU-3′

       Amino acids: cys   asn   arg   gly

This point mutation does not cause any change in the amino acid sequence. Both AAU and AAC code for asparagine.

Following is a segment of DNA showing five triplets.

3'-GAC-TCC-GAT-CGC-GAT-5'

Problem 15.6

a. Write the sequence of bases in the mRNA transcribed from the 3' end of this section of DNA; write the amino acid sequence coded for by the complementary mRNA.

b. Assume that a point mutation changes the fifth base from the 3' end of the DNA strand from C to T. Write the mRNA sequence transcribed after this mutation and the amino acid sequence it codes for.

B. Base-Pair Insertion and Deletion

An insertion mutation involves addition of one or more base pairs to a DNA strand. A deletion mutation, on the contrary, involves removal of one or more base pairs. Both types of mutations change the reading frame of all bases after the mutation point; thus, all amino acids in the primary structure following the mutation are affected. Shown in Figure 15.18 is a section of double-stranded DNA and the same section after insertion of one base pair.

Strand I: 5'-ACG-TTA-GCG-CCA-3'

Strand II: 3'-TGC-AAT-CGC-GGT-5'

(a)

Inserted base pair

Strand I: 5'-ACG-TTA-CGC-GCC-A-3'

Strand II: 3'-TGC-AAT-GCG-CGG-T-3'

(b)

Figure 15.18 Insertion mutation. A section of double-stranded DNA (a) before insertion of a base pair, and (b) after.

Example 15.7

Following is a section of a single strand of DNA showing four triplets, the section of mRNA transcribed from the 3' end of this DNA template, and the amino acid sequence coded for by this section of mRNA. Insert A after TTA in the DNA template, write the sequence of bases in the mRNA produced by transcription from the mutant DNA, and write the amino acid sequence the new mRNA codes for.

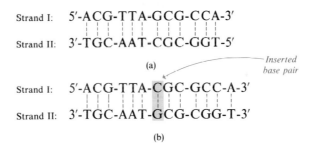

DNA template: 3'-TTA-GGT-TGT-TGG-5'

mRNA: 5'-AAU-CCA-ACA-ACC-3'

Amino acids: asn   pro   thr   thr

☐ *Solution*

Inserting A at the position indicated gives the following DNA template, mRNA, and amino acid sequence. Except for the first amino acid, the entire sequence is modified.

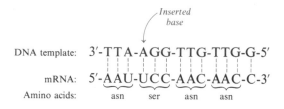

Inserted base

DNA template:	3'-TTA-AGG-TTG-TTG-G-5'
mRNA:	5'-AAU-UCC-AAC-AAC-C-3'
Amino acids:	asn ser asn asn

■ **Problem 15.7**

Consider the DNA strand sequence 3'-GAT-GGG-ATG-TCT-5'.

a. Write the base sequence of the complementary strand of mRNA and the amino acid sequence it codes for.

b. Delete T from GAT in the DNA strand and write the new mRNA sequence and the amino acid sequence it codes for.

■

C. Mutations and Mutagens

Replication of DNA occurs with astounding fidelity! From time to time, however, mutations do occur. Mutations that occur without any external environmental influence are termed spontaneous mutations. From studies with bacteria, it has been estimated that the frequency of spontaneous mutation is one error for every 10^9 to 10^{10} base pairs copied—a remarkably low number, but one important for evolution.

A second and much more prevalent cause for mutations involves environmental factors. Any environmental factor that brings about a mutation is called a **mutagen**. Common mutagens are ultraviolet light, X-rays, and chemicals. Consider, as an example of a chemical mutagen, 5-bromouracil, a compound used to treat certain types of skin cancer. Because its bromine atom is about the same size as the methyl group of thymine and in the same position, 5-bromouracil can substitute for thymine during replication.

thymine	5-bromouracil (keto form)	5-bromouracil (enol form)

The mutagenicity of 5-bromouracil arises because the bromine atom makes the enol form of this compound (which base-pairs with guanine) more stable than the keto form (which instead base-pairs with adenine).

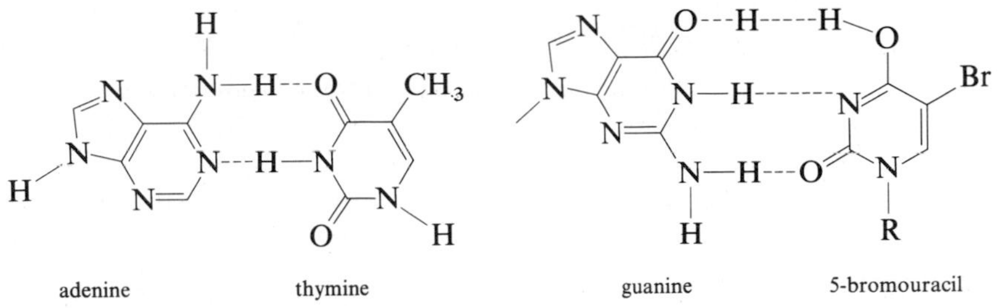

adenine   thymine   guanine   5-bromouracil

Thus, the chemical mutagenicity of 5-bromouracil can be laid to its incorporation in place of thymine and base-pairing with guanine instead of adenine.

Other chemical mutagens react with the heterocyclic aromatic amine bases and alter them chemically in ways that disrupt normal base pairing. Among the most potent chemical mutagenic agents are nitrous acid and compounds derived from it by reaction with secondary amines. Following are structural formulas for nitrous acid and N,N-dimethylnitrosamine, the product formed by reaction of dimethylamine and nitrous acid:

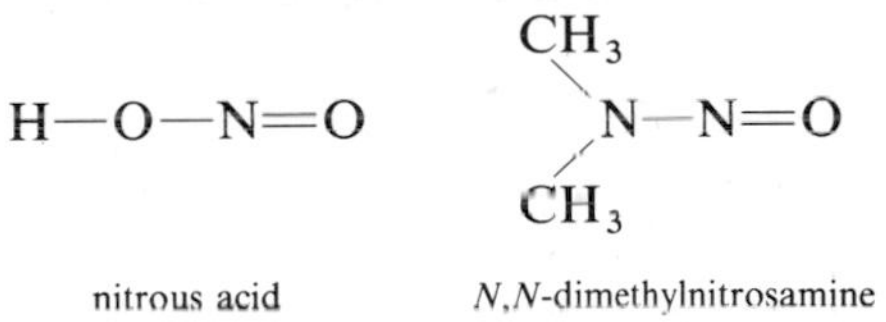

nitrous acid   N,N-dimethylnitrosamine

Nitrous acid and nitrosamines react with heterocyclic aromatic amines of DNA to convert a $C-NH_2$ group to a $C=O$ group, which in turn changes hydrogen bonding and base-pairing. Following is shown the reaction between cytosine and nitrous acid, to give uracil. Cytosine base-pairs with guanine, but uracil base-pairs with adenine.

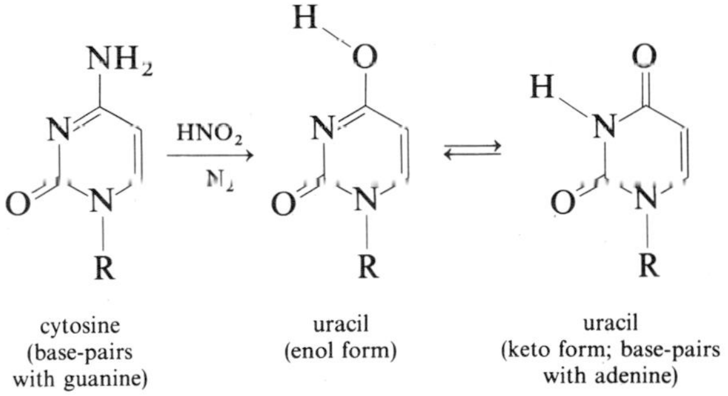

cytosine   uracil   uracil
(base-pairs   (enol form)   (keto form; base-pairs
with guanine)   with adenine)

5-bromouracil and nitrous acid are relatively specific in the mutations they cause. Others mutagens have broader and less specific properties. Fortunately, cells have several mechanisms available to repair altered (damaged) DNA, so the site of altered base-pairing can be repaired quickly and efficiently. Otherwise, an organism would soon be killed by the combination of spontaneous and environmentally caused mutations.

D. The Ames Test and Chemical Carcinogens

In view of the increasing number of chemicals being produced and released into our environment, it is important for a rapid and reliable test for mutagenicity to be available. The need for such a test is made even more critical because most chemical carcinogens (cancer-causing chemicals) are also mutagens. It must be emphasized that there is not a one-to-one correlation between mutagenicity and carcinogenicity; most carcinogens are mutagens, but not all mutagens are carcinogens.

The standard test for mutagenicity, developed by Bruce Ames and known as the **Ames test**, is based on the following observations:

1. A mutant of the bacterium *Salmonella typhimurium* lacks the ability to make histidine. Thus for it, histidine is an essential amino acid and must be supplied in the growth medium. Furthermore, this mutant has been made especially sensitive to mutagens by inactivation of several of its DNA repair mechanisms.
2. Mutagens cause this histidine-dependent strain to revert to its wild form, which can synthesize histidine from the growth medium; for the wild strain, histidine is not an essential amino acid.
3. Several compounds that are carcinogenic in animals are not carcinogenic in bacteria. The explanation for this observation is that these noncarcinogenic chemicals undergo chemical modification in the liver of animals and are there transformed to compounds that are carcinogenic.

In the Ames test, a histidine-requiring strain of *Salmonella typhimurium* is plated on agar with enough histidine in the growth medium to support a few rounds of cell division. Also present in the growth medium is a rat-liver microsomal fraction. A control plate has only growth medium, the histidine-dependent bacteria, and rat-liver microsomal fraction. The experimental plate has these same components plus the chemical to be tested. If no more wild colonies grow on the experimental plate than on the control plate (only equal numbers of spontaneous mutations), then the test chemical is not classified as a mutagen. If, however, significantly more wild colonies grow on the experimental plate, then the test chemical is reported to be a mutagen. By examining the relative growth rates stimulated by various test chemicals, one can establish their relative mutagenicities.

The Ames test has now been used for thousands of compounds, including industrial chemicals, pesticides, food additives, hair dyes, and cosmetics. Surprisingly, many compounds previously thought to be safe have been found to give positive Ames tests and have thus been identified as chemical mutagens.

At present, tests for mutagenicity are relatively simple and done on bacteria, whereas tests for carcinogenicity are complex and time-consuming and require testing with laboratory animals. Therefore the Ames test for mutagenicity has become a preliminary screening for potential carcinogenicity. It must be emphasized again that a positive Ames test does not demonstrate that the mutagen is also a carcinogen. However, the correlation is high between a positive Ames test and carcinogenicity.

Key Terms and Concepts

Ames test (15.11D)

base composition of DNA (15.4B)

codon (15.8A)

complementarity (15.4D)

daughter strand (15.5)

deoxyribonucleic acid, DNA (15.1)

double helix (15.4D)

genetic code (15.8)

messenger RNA (15.6C)

mutagen (15.11C)

mutation (15.11)

nicks (15.5C)

nucleoside (15.2)

nucleotide (15.3)

Okazaki fragment (15.5C)

point mutation (15.11)

replication (15.5)

replication fork (15.5A)

ribonucleic acids (RNAs) (15.6)

ribosomal RNA (15.6A)

ribosome (15.6A)

semiconservative replication (15.5)

transcription (15.7)

transfer RNA (15.6B)

translation (15.9)

Problems

Nucleosides and nucleotides (Sections 15.2 and 15.3)

15.8 Examine the structure of purine. Would this molecule be planar or puckered? Would it exist as several interconvertible conformations (like cyclohexane) or be rigid and inflexible? Explain the basis for your answer.

15.9 An important drug in the chemotherapy of leukemia is 6-mercaptopurine, a sulfur analog of adenine. Draw a structural formula for 6-mercaptopurine.

15.10 Explain the difference in structure between a nucleoside and a nucleotide.

15.11 Name and draw structural formulas for the following. In each label the *N*-glycoside bond:
a. a nucleoside composed of beta-D-ribose and adenine.
b. a nucleoside composed of beta-D-ribose and uracil.
c. a nucleoside composed of beta-2-deoxy-D-ribose and cytosine.

15.12 Name and draw structural formulas for the following. Label all *N*-glycoside bonds, ester bonds, and anhydride bonds.
a. ADP   **b.** dGMP   **c.** GTP

15.13 Calculate what the net charge on the following would be at pH 7.4.
a. ATP   **b.** 2'-deoxyadenosine   **c.** GMP

15.14 Cyclic-AMP (adenosine-3',5'-cyclic monophosphate), first isolated in 1959, is involved in many diverse biological processes as a regulator of metabolic and physiological activity. In it, a single phosphate group is esterified with both the 3'- and 5'-hydroxyls of adenosine. Draw the structural formula for this substance.

15.15   Following are sequences for several polynucleotides. Write structural formulas for each. Calculate what the net charge on each would be at pH 7.4.

    a. dApdGpdA   **b.** pppdCpdT   **c.** pdGpdCpdCpdTpdA

Structure of DNA (Section 15.4)

15.16   Compare the alpha-helix found in proteins with the double helix of DNA in regard to the following points.

    a. The units that repeat in the backbone of the chain.

    b. The projection in space of the backbone substituents (R groups in the case of amino acids; purine and pyrimidine bases in the case of DNA) relative to the axis of the helix.

15.17   List the postulates of the Watson-Crick model of DNA structure. This model is based on certain experimental observations of base composition and molecular dimensions. Describe these observations and show how the model accounts for each.

15.18   Explain the role of hydrophobic interaction in stabilizing (a) soap micelles; (b) lipid bilayers; (c) double-stranded DNA.

15.19   What type of bond or interaction holds monomers together in (a) proteins? (b) nucleic acids? (c) polysaccharides?

15.20   In terms of hydrogen bonding, which is more stable, an A—T base pair or a G—C base pair?

15.21   At high temperatures, nucleic acids become denatured; that is, they unwind into disordered single strands. Account for the fact that, the higher the content of G—C base pairs, the higher the temperature required to denature a given molecule of DNA.

15.22   Regarding the DNA triplet ATC, is its complement TAG or GAT? Explain.

DNA replication (Section 15.5)

15.23   What is the meaning of the adjective *semiconservative* in the term semiconservative replication?

15.24   From what direction is a DNA strand read during formation of its complement?

15.25   What is an Okazaki fragment? Explain how isolation and identification of Okazaki fragments provide evidence that the synthesis of DNA is not a continuous process.

Ribonucleic acids (RNAs) (Section 15.6)

15.26   Compare DNA and RNA on the following points:

    a. monosaccharide units   **b.** principal purine and pyrimidine bases

    c. primary structure   **d.** location in the cell

    e. function in the cell

15.27   Compare ribosomal RNA, messenger RNA, and transfer RNA on (a) molecular weight; (b) function in protein synthesis.

Transcription of genetic information: synthesis of RNA (Section 15.7)

15.28   Draw a diagram of an mRNA-ribosome initiation complex and label the following:

    a. 30S subunit   **b.** 50S subunit   **c.** 5′ and 3′ ends of mRNA

15.29   For the DNA strand sequence

$$5'\text{-ACC-GTT-GCC-AAT-G-}3'$$

(a) write the sequence of its DNA complement, and (b) of its mRNA complement.

The genetic code (Section 15.8)

15.30 Consider the mRNA sequence 5'-AGG-UCC-CAG-3'.
  a. What tripeptide is synthesized if the code is read from the 5' end to the 3' end?
  b. What tripeptide is synthesized if the code is read from the 3' end to the 5' end?
  c. Calculate what the net charge on each tripeptide would be at pH 7.4.
  d. Which way is the code read in the cell and which tripeptide is synthesized?

15.31 What peptide sequences are coded for by the following mRNA sequences? (Each is written in the 5' ⟶ 3' direction.)
  a. GCU-GAA-UGG   b. UCA-GCA-AUC
  c. GUC-GAG-GUG   d. GCU-UCU-UAA

15.32 Complete the table.

DNA	DNA Complement	mRNA Complement	Amino Acid Coded for
TGC	——	——	——
CAG	——	——	——
——	ACG	——	——
——	GTA	——	——
——	——	GUC	——
——	——	UGC	——
——	——	CAC	——

15.33 The alpha-chain of human hemoglobin has 141 amino acids in a single polypeptide chain.
  a. Calculate the minimum number of bases on DNA necessary to code for the alpha-chain. Include in your calculation the bases necessary for specifying termination of polypeptide synthesis.
  b. Calculate the length in centimeters of DNA containing this number of bases.

Translation of genetic information: biosynthesis of polypeptides (Section 15.9)

15.34 a. Draw the structural formula of *N*-formylmethionine.
  b. Draw structural formulas for the products of hydrolysis of the amide bond in *N*-formylmethionine. Show each product as it would be ionized at pH 7.0.

15.35 Each of the following reactions involves ammonolysis of an ester. Draw the structural formula of the amide produced in each reaction.

a.

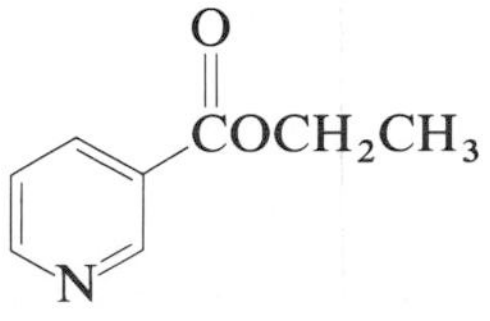

$$\text{(pyridine-}COCH_2CH_3) + NH_3 \longrightarrow \text{nicotinamide} + CH_3CH_2OH$$

b. $CH_3CH_2O\overset{\overset{O}{\|}}{C}OCH_2CH_3 + 2NH_3 \longrightarrow$ urea $+ 2CH_3CH_2OH$

c. $H_2C \overset{\overset{\displaystyle COCH_2CH_3}{\|}}{\underset{\underset{\displaystyle O}{\|}}{COCH_2CH_3}} + H_2N\overset{\overset{O}{\|}}{C}NH_2 \longrightarrow$ barbituric acid $+ 2CH_3CH_2OH$

15.36 Show that the reaction catalyzed by peptidyl transferase is an example of ammonolysis of an ester.

15.37 Are polypeptide chains synthesized from the *N*-terminal amino acid toward the C-terminal amino acid or vice versa?

Mutations
(Sections 15.11)

15.38 Following in (a) is an mRNA sequence written from the 5′ end to the 3′ end. Below it are three substitution mutations, one insertion mutation, and one deletion mutation. For what polypeptide sequence does the normal mRNA code, and what is the effect of each mutation on the resulting polypeptide?
a. Normal: 5′-UCC-CAG-GCU-UAC-AAA-GUA-3′
b. Substitution of A for C: 5′-UCC-AAG-GCU-UAC-AAA-GUA-3′
c. Substitution of A for C: 5′-UCC-CAG-GCU-UAA-AAA-GUA-3′
d. Substitution of A for C: 5′-UCA-CAG-GCU-UAC-AAA-GUA-3′
e. Insertion of A: 5′-UCC-CAG-GCU-AUA-CAA-AGU-A-3′
f. Deletion of A: 5′-UCC-CAG-GCU-UCA-AAG-UA-3′

15.39 In HbS, the abnormal human hemoglobin found in individuals with sickle-cell anemia, glutamic acid at position 6 of the beta chain, is replaced by valine.
a. List the two codons for glutamic acid and the four codons for valine.
b. Show that a glutamic acid codon can be converted to a valine codon by a single substitution mutation.

16 Flow of Energy in the Biological World

All living organisms need a constant supply of energy to support maintenance of cell structure and growth; in this sense, energy is the key to life itself. We will learn how cells extract energy from foodstuffs and how this energy is used in chemical, mechanical, and osmotic work.

16.1 Metabolism and Metabolic Pathways

Metabolism is defined as the sum of all chemical reactions an organism uses to grow, feed, move, excrete wastes, and communicate. Metabolism has two main components, catabolism and anabolism. **Catabolism** includes all reactions leading to the breakdown of biomolecules. **Anabolism** includes all reactions leading to the synthesis of biomolecules. In general, catabolism produces energy and anabolism consumes energy.

All reactions of a cell or organism are organized into orderly, carefully regulated sequences known as **metabolic pathways**. Each metabolic pathway consists of a series of consecutive steps that convert a starting material to an end product. The range of metabolic pathways in even a one-celled organism such as the bacterium *E. coli* is enormous. To support growth, a culture medium for *E. coli* needs to contain only glucose as a source of carbon atoms and energy, inorganic salts as sources of nitrogen and phosphorus, and a few other simple substances. Growth of *E. coli* under these conditions means that each cell has the metabolic pathways needed to extract energy from the culture medium and use it to synthesize all the carbohydrates, lipids, proteins, enzymes, coenzymes, nucleic acids, and other biomolecules necessary for maintenance and development. The ability of *E. coli* to grow under these conditions is truly remarkable, especially when you consider the complexity of some of the biomolecules found in living systems.

Fortunately for those who study the biochemistry of living systems, many similarities are found among the chief metabolic pathways in humans, *E. coli*, and for that matter, most other organisms. The number of individual reactions is large, but the number of different kinds of reactions is small. For example, the basic features of how different cells extract energy from foodstuffs and use it to synthesize other biomolecules, and even the means of self-regulation, are surprisingly similar. Because of these similarities, scientists can study the metabolism of simple organisms and then use these results to help understand the corresponding metabolic pathways in more complex organisms, including humans. We shall concentrate on human biochemistry, but much of what we say about human metabolism can be applied equally well to the metabolism of most other organisms.

16.2 Flow of Energy in the Biosphere

The uniqueness of living systems rests in their ability to capture energy from the environment, to store it, at least temporarily, and to use it to power the vast number of vital biological processes on which life depnds. The energy for all biological processes comes ultimately from the sun, whose enormous energy is derived from the fusion of hydrogen atoms, converting them to helium:

$$2H\cdot \xrightarrow{\text{nuclear fusion}} He\colon + \text{energy}$$

A portion of this energy streams toward us as sunlight and is absorbed by chlorophyll pigments in plants. There, this energy drives **photosynthesis**, the im-

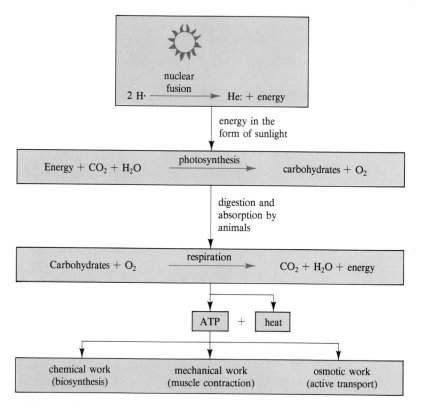

Figure 16.1 The flow of energy in the biosphere. (Adapted from David S. Page, *Principles of Biological Chemistry*, 2d ed. Boston: Willard Grant Press, 1981.)

mediate product of which is glucose:

$$6CO_2 + 6H_2O + \text{energy} \xrightarrow{\text{photosynthesis}} \underset{\text{glucose}}{C_6H_{12}O_6} + 6O_2$$

In secondary steps, plants convert glucose to other carbohydrates, triglycerides, and proteins, all chemical storage forms of energy. Animals get these energy-rich molecules either directly or indirectly from plants.

During **respiration**, both plants and animals oxidize these energy-rich compounds to carbon dioxide and water. Respiration is accompanied by the release of energy:

$$\underset{\text{energy}}{\overset{\text{glucose and other}}{\text{storage forms of}}} + O_2 \xrightarrow{\text{respiration}} CO_2 + H_2O + \text{energy}$$

A portion of the energy derived from respiration is transformed to adenosine triphosphate (ATP), a carrier of energy that can be used directly in performing biological work. The remainder of the energy of respiration is liberated as heat. Steps in the flow of energy in the biosphere are summarized in Figure 16.1.

16.3 ATP: The Central Carrier of Energy

A. Concept of Free Energy

Energy changes for reactions taking place in biological systems are commonly reported as changes in **free energy** ΔG^0. A change in free energy measures the maximum work that can be obtained from a given reaction or process. The symbol ΔG^0 stands for the change in free energy per mole of reactant when the reaction is carried out at standard temperature and 1 atm pressure.

Reactions that result in a decrease in free energy are said to be **exergonic**. Exergonic reactions include the oxidation of carbohydrates, fats, and proteins. The following equation shows that oxidation of glucose to carbon dioxide and water results in a decrease in free energy of 686,000 cal/mol of glucose:

$$\underset{\text{glucose}}{C_6H_{12}O_6} + 6O_2 \longrightarrow 6CO_2 + 6H_2O \qquad \Delta G^0 = -686,000 \text{ cal/mol}$$

Because exergonic reactions bring about a decrease in free energy, they are said to be spontaneous, which means that they proceed to the right as written; at equilibrium, the concentration of products is greater than concentration of reactants. The more negative the value of ΔG^0, the greater the concentration of products relative to reactants.

It is important to remember that although a negative value of ΔG^0 means that a reaction is spontaneous as written, it gives us no indication of the rate at which the reaction occurs. For example, ΔG^0 is negative for conversion of glucose and oxygen to carbon dioxide and water; this tells us that the reaction is spontaneous as written. Yet we know that in the absence of heat or appropriate catalysts, no reaction occurs.

Reactions that occur with an increase in free energy are said to be **endergonic**, which means that the reaction proceeds to the left as written; at equilibrium, the concentration of reactants is greater than concentration of products. The more positive the value of ΔG^0, the greater the concentration of reactants relative to products. The following equation shows that photosynthesis occurs with an increase in free energy.

$$6CO_2 + 6H_2O \longrightarrow \underset{\text{glucose}}{C_6H_{12}O_6} + 6O_2 \qquad \Delta G^0 = +686,000 \text{ cal/mol}$$

Because endergonic reactions occur with an increase in free energy, they are not spontaneous; at equilibrium very little product is formed unless energy is supplied to drive the reaction to the right. As shown in Figure 16.1, the energy to drive photosynthesis is supplied by sunlight. The relations between the sign of ΔG^0 and spontaneity are summarized in Figure 16.2.

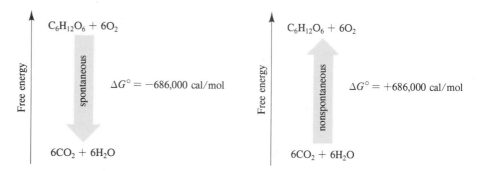

Figure 16.2 Conversion of glucose and oxygen to carbon dioxide and water occurs with a decrease in free energy; it is a spontaneous reaction. Conversion of carbon dioxide and water to glucose and oxygen occurs with an increase in free energy; it is not a spontaneous reaction.

B. ATP, a High-Energy Compound

The central role of **adenosine triphosphate (ATP)** in the transfer of energy in the biological world depends on the triphosphate end of the molecule. The structure of ATP is shown in Figure 16.3. At pH 7.4, all protons of the triphosphate group are ionized, giving ATP a charge of -4. In cells, ATP is most often present with Mg^{2+} in a 1:1 complex with a charge of -2. Figure 16.4 shows abbreviated structural formulas of ATP, adenosine diphosphate (ADP), and adenosine monophosphate (AMP). In these formulas, only phosphate ester and anhydride bonds are shown.

The key to understanding how ATP affects the flow of energy in the biological world is knowing that it can transfer a phosphoryl group, $-PO_3^{2-}$, to another molecule. For example, during hydrolysis of ATP in water, a phosphoryl group is transferred from ATP to water. The products of this hydrolysis are ADP and phosphate ion.

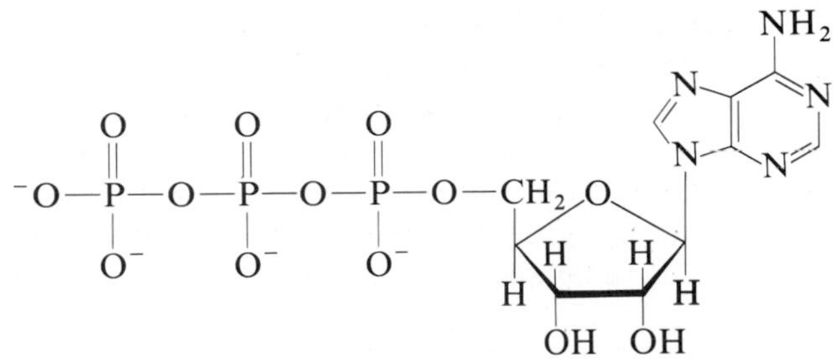

Figure 16.3 The structure of adenosine triphosphate (ATP).

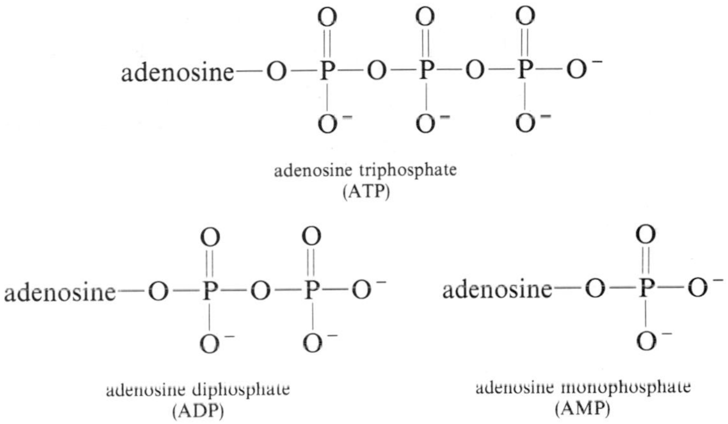

Figure 16.4 Abbreviated structural formulas for adenosine triphosphate (ATP), adenosine diphosphate (ADP), and adenosine monophosphate (AMP).

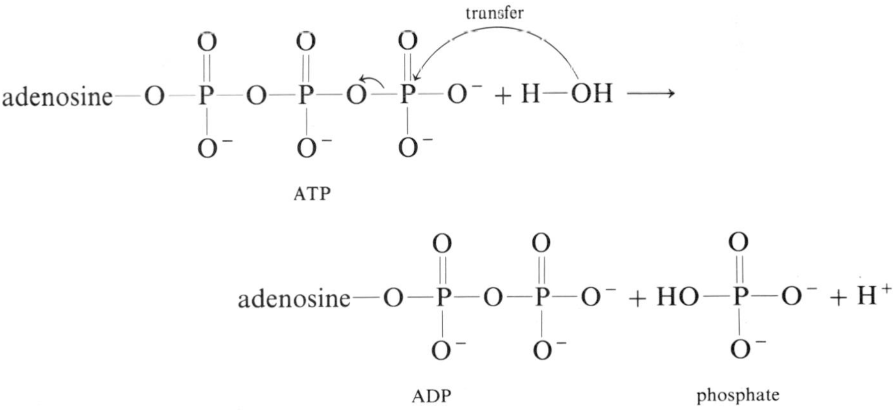

Transfer of a phosphoryl group from ATP to water is accompanied by a decrease in free energy, as shown in the following equation:

$$ATP^{4-} + HOH \longrightarrow ADP^{3-} + HPO_4^{2-} + H^+ \qquad \Delta G^0 = -7{,}300 \text{ cal/mol}$$

Adenosine triphosphate is but one of the many phosphate-containing compounds common in biological systems. Several of these, along with the **free**

energy of hydrolysis of each, are listed in Table 16.1. Notice that ATP has a free energy of hydrolysis larger than that of simple phosphate esters such as glucose 6-phosphate. Because of the size of its free energy of hydrolysis, ATP is called a high-energy compound. **High-energy compounds** have free energies of hydrolysis of $-7,000$ cal/mol or greater; **low-energy compounds** have free energies of hydrolysis of less than $-7,000$ cal/mol. As you can see from Table 16.1, the line between high-energy and low-energy compounds is not sharp.

Table 16.1
Free energy of hydrolysis of some phosphate-containing compounds present in biological systems.

Compound	Products of Hydrolysis	ΔG^0 (cal/mol)
phosphoenolpyruvate + $H_2O \longrightarrow$	pyruvate + phosphate	$-14,800$
1,3-diphosphoglycerate + $H_2O \longrightarrow$	3-phosphoglycerate + phosphate	$-11,800$
ATP + $H_2O \longrightarrow$	ADP + phosphate	$-7,300$
glucose 1-phosphate + $H_2O \longrightarrow$	glucose + phosphate	$-5,000$
fructose 6-phosphate + $H_2O \longrightarrow$	fructose + phosphate	$-3,800$
glucose 6-phosphate + $H_2O \longrightarrow$	glucose + phosphate	$-3,300$

Why does the hydrolysis of the phosphate anhydride bond of ATP have a ΔG^0 so much larger than that of the hydrolysis of a phosphate ester bond, say the bond in glucose 6-phosphate? The reason lies in the structure of ATP itself. At the pH of cells, the phosphate groups of ATP are fully ionized, giving ATP a net charge of -4. These negative charges are very close to each other and create an electrostatic strain within the molecule. Hydrolysis of the terminal phosphate anhydride gives inorganic phosphate and ADP, an ion with a net charge of -3. Thus, hydrolysis of ATP relieves some electrostatic strain. Be-

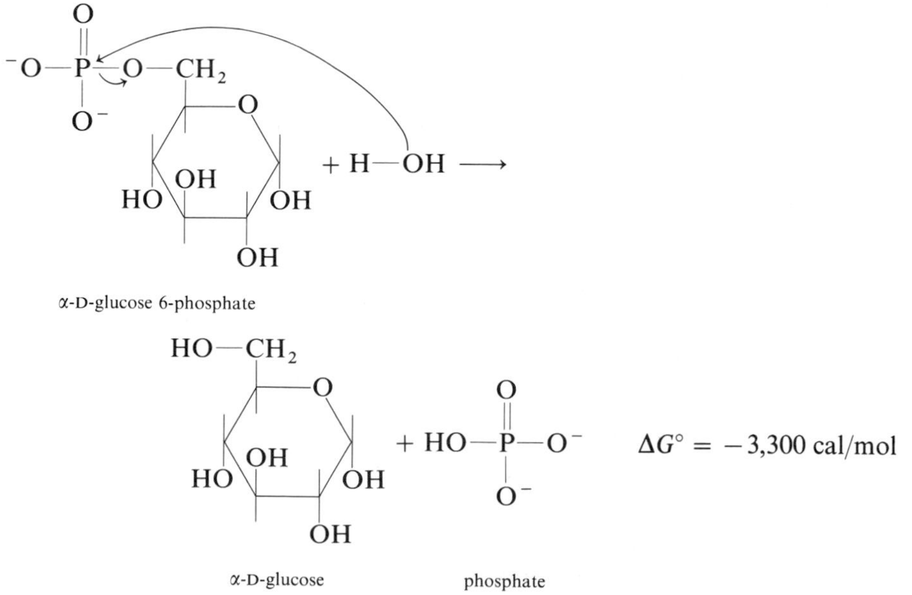

α-D-glucose 6-phosphate

α-D-glucose   phosphate

cause there is no such electrostatic strain in phosphate esters such as glucose 6-phosphate or glucose 1-phosphate, their hydrolysis releases comparatively less energy.

C. Other High-Energy Compounds

Two other high-energy compounds are also listed in Table 16.1. Both are intermediates in glycolysis, the metabolic pathway by which glucose is converted to pyruvate. Hydrolysis of phosphoenolpyruvate, the final step in glycolysis, gives the enol form of pyruvate and inorganic phosphate. (For a review of keto and enol forms, see Section 6.7B.)

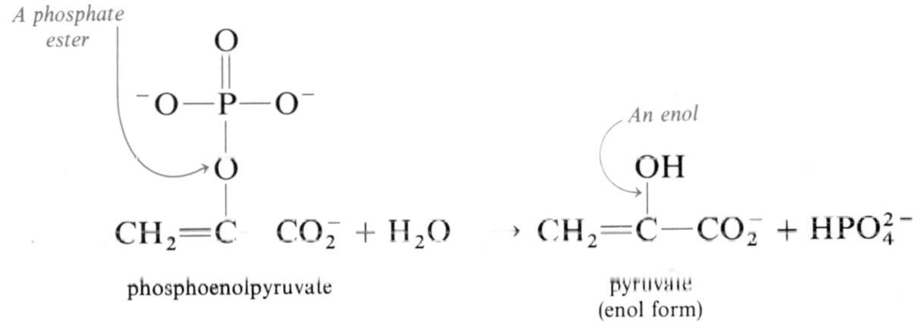

phosphoenolpyruvate   pyruvate (enol form)

The equilibrium between the keto and enol forms of pyruvate lies almost completely on the side of the keto form; accordingly, conversion to the keto form is accompanied by a large decrease in free energy.

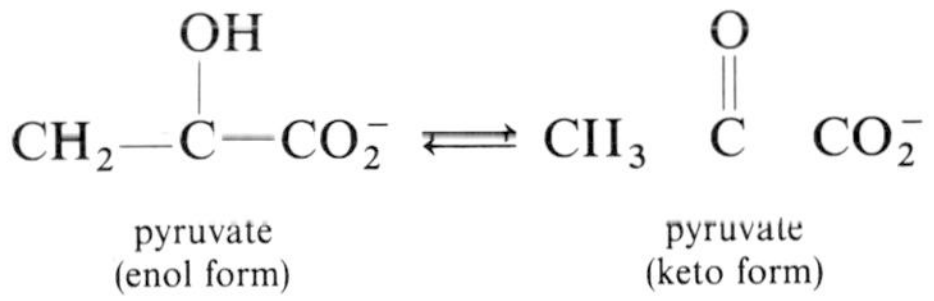

pyruvate (enol form)   pyruvate (keto form)

The other high-energy compound listed in Table 16.1 is 1,3-diphosphoglycerate.

1,3-diphosphoglycerate   3-phosphoglycerate

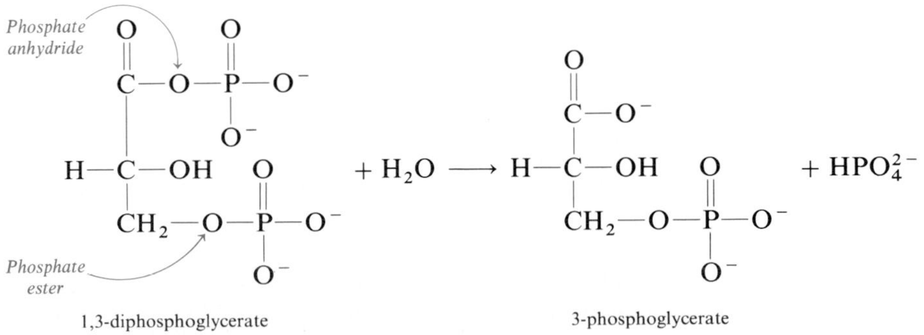

1,3-diphosphoglycerate contains a phosphate ester and a phosphate anhydride. Hydrolysis of the phosphate anhydride yields phosphate, 3-phosphoglycerate, and energy.

D. Central Role of ATP in Cellular Energetics

We have examined reactions involving transfer of a phosphoryl group to water. The same phosphate-containing compounds can, at least in principle, transfer a phosphoryl group to compounds of the type H—OR. Following is an equation for the transfer of a phosphoryl group from phosphoenolpyruvate to α-D-glucose to form pyruvate and α-D-glucose 6-phosphate. The flow of electrons in this reaction is shown by curved arrows.

phosphoenol-   α-D-glucose   pyruvate   α-D-glucose 6-phosphate
pyruvate

The change in free energy for this reaction can be calculated by (1) dividing the reaction into two separate equations for which changes in free energy are known; and (2) adding the separate equations and the free energy change for each.

phosphoenolpyruvate + H—OH $\longrightarrow$ pyruvate + HPO_4^{2-} $\Delta G^0 = -14,800$ cal/mol
glucose + HPO_4^{2-} $\longrightarrow$ glucose 6-phosphate + H—OH $\Delta G^0 = +\ 3,300$ cal/mol

glucose + phosphoenol- $\longrightarrow$ pyruvate + glucose 6- $\Delta G^0 = -11,500$ cal/mol
    pyruvate   phosphate

This calculation shows that transfer of a phosphoryl group from pyruvate to glucose occurs with a large decrease in free energy; it is a spontaneous reaction and proceeds to the right as written. Yet, although it is spontaneous, direct transfer of a phosphoryl group from phosphoenolpyruvate to glucose has not been observed in living systems. Rather, ATP is a common intermediate, or "medium of exchange," that links this and other high-energy phosphate donors to phosphate acceptors.

phosphoenolpyruvate + ADP $\longrightarrow$ pyruvate + ATP
glucose + ATP $\longrightarrow$ glucose 6-phosphate + ADP

So that you can appreciate how this means of phosphate transfer is valuable to cells, consider that virtually all reactions in living systems are enzyme-catalyzed.

Figure 16.5(a) shows the number of enzymes necessary to catalyze the transfer of phosphate from phosphoenolpyruvate (PEP), or 1,3-diphosphoglycerate (1,3-DPG), or ATP to five different phosphate acceptors. For the reactions in Figure 16.5(a), fifteen different enzymes are required. If ATP is used as a collector and a common donor of phosphate groups to other low-energy acceptors, only seven different enzymes are required [Figure 16.5(b)].

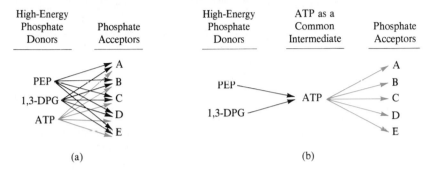

(a)   (b)

Figure 16.5 An illustration of the efficiency of using ATP as a common phosphate acceptor/donor. In (a), 15 different enzymes are required while in (b) only 7 are required.

16.4 Stages in Oxidation of Foodstuffs and Generation of ATP

The basic strategy used by all cells to extract energy from their surroundings is to oxidize foodstuffs and use a portion of the free energy released to convert ADP and HPO_4^{2-} to ATP. Oxidation of foodstuffs and the generation of ATP is accomplished in four stages.

A. Stage 1: Digestion and Absorption of Fuel Molecules

Stage 1, digestion of foods, involves hydrolysis of carbohydrates to monosaccharides, proteins to amino acids, and fats and oils to fatty acids and glycerol (Figure 16.6):

$$\text{polysaccharides} + H_2O \xrightarrow{\text{hydrolysis}} \text{monosaccharides}$$

$$\text{fats and oils} + H_2O \xrightarrow{\text{hydrolysis}} \text{fatty acids} + \text{glycerol}$$

$$\text{proteins} + H_2O \xrightarrow{\text{hydrolysis}} \text{amino acids}$$

Figure 16.6 Stage 1 in the oxidation of foodstuffs and generation of ATP: hydrolysis of complex fuel molecules to monosaccharides, fatty acids plus glycerol, and amino acids.

As a result of hydrolysis, the hundreds of thousands of different proteins, fats, oils, and carbohydrates ingested in the diet are converted to fewer than 30 lower-molecular-weight compounds, the most common of which are listed in Table 16.2.

Table 16.2
The 27 most common low-molecular-weight molecules derived from hydrolysis of carbohydrates, fats and oils, and proteins.

From Carbohydrates	From Fats and Oils	From Proteins
D-glucose	palmitic acid	20 amino acids
D-fructose	stearic acid	
D-galactose	oleic acid	
	glycerol	

B. Stage 2: Degradation of Fuel Molecules to Acetyl CoA

In stage 2 (Figure 16.7), the carbon skeletons of glucose, fructose, and galactose along with those of fatty acids, glycerol and several amino acids are converted to acetate in the form of a thioester named **acetyl coenzyme A**, or more commonly, acetyl CoA. The carbon skeletons of other amino acids are degraded to different small molecules, but eventually all go through the reactions of stage 3.

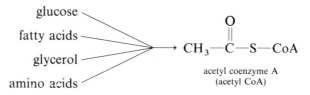

Figure 16.7 Stage 2 of the oxidation of foodstuffs and generation of ATP. The carbon skeletons of glucose, fatty acids, glycerol, and certain amino acids are degraded to the acetyl group of acetyl CoA.

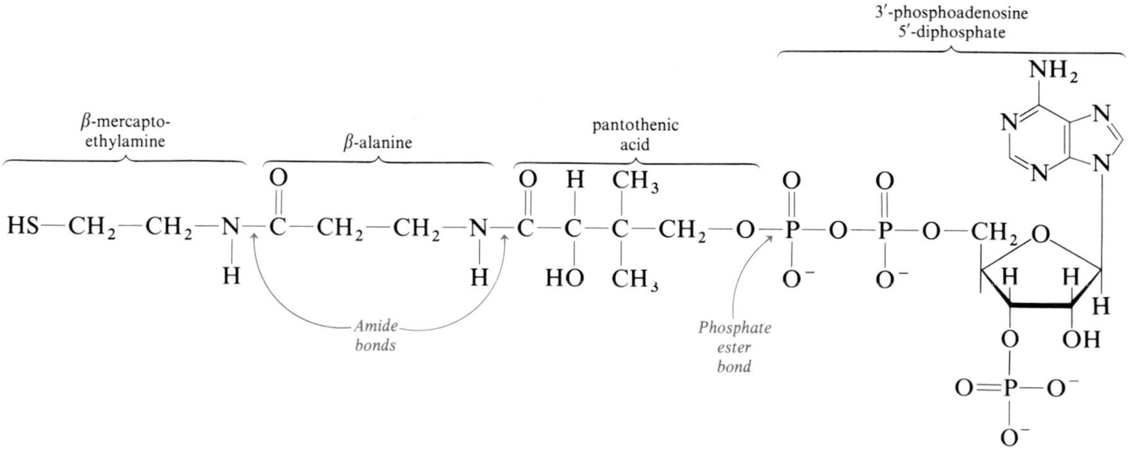

Figure 16.8 Coenzyme A. The acetylated form of this coenzyme, designated acetyl coenzyme A, or acetyl CoA, is the thioester of acetic acid and the terminal sulfhydryl group. Pantothenic acid is one of the vitamins of the B group.

Coenzyme A (Figure 16.8) is derived from four subunits. On the left is a two-carbon unit derived from beta-mercaptoethylamine, which is joined by an amide bond to the carboxyl group of beta-alanine. The amino group of beta-alanine is, in turn, joined by another amide bond to the carboxyl group of pantothenic acid, a vitamin of the B group (Section 14.2C). Finally, the —OH group of pantothenic acid is joined by an ester bond to the terminal phosphate of ADP. A key feature in the structure of coenzyme A is the presence of the terminal sulfhydryl group (—SH). Acetyl CoA is a thioester derived from the carboxyl group of acetic acid and the thiol group of coenzyme A.

C. Stage 3: The Krebs, or Tricarboxylic Acid, Cycle

Stage 3 consists of a series of reactions known alternatively as the **tricarboxylic acid (TCA) cycle**, the citric acid cycle, or the **Krebs cycle** (Figure 16.9). An important function of the tricarboxylic acid cycle is oxidation of the two-carbon acetyl group of acetyl CoA to two molecules of carbon dioxide.

$$CH_3-\overset{\overset{\displaystyle O}{\|}}{C}-S-CoA \xrightarrow{\text{tricarboxylic acid cycle}} 2CO_2 + CoA-SH$$

acetyl coenzyme A   coenzyme A

Figure 16.9 Stage 3 in the oxidation of foodstuffs and generation of ATP: the tricarboxylic acid cycle. The carbon atoms derived from stages 1 and 2 are oxidized to carbon dioxide.

The biological oxidizing agents for stage 3 are nicotinamide adenine dinucleotide (NAD^+) and flavin adenine dinucleotide (FAD). The former substance, NAD^+ (Figure 16.10), is the principal acceptor of electrons in the oxidation of fuel molecules.

The reactive group of NAD^+ is a pyridine ring, which accepts two electrons and one proton to form the reduced coenzyme NADH:

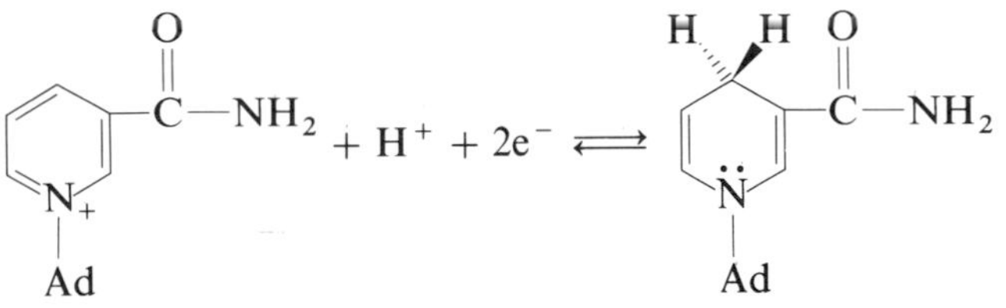

NAD$^+$ $+ H^+ + 2e^- \rightleftharpoons$ NADH

The second electron acceptor in the oxidation of fuel molecules is FAD (Figure 16.11). This molecule is composed of several subunits: a three-ring flavin group, a five-carbon group derived from D-ribose, and adenosine diphosphate.

The reactive group in FAD is a flavin group, which accepts two electrons and two protons to form the reduced coenzyme $FADH_2$.

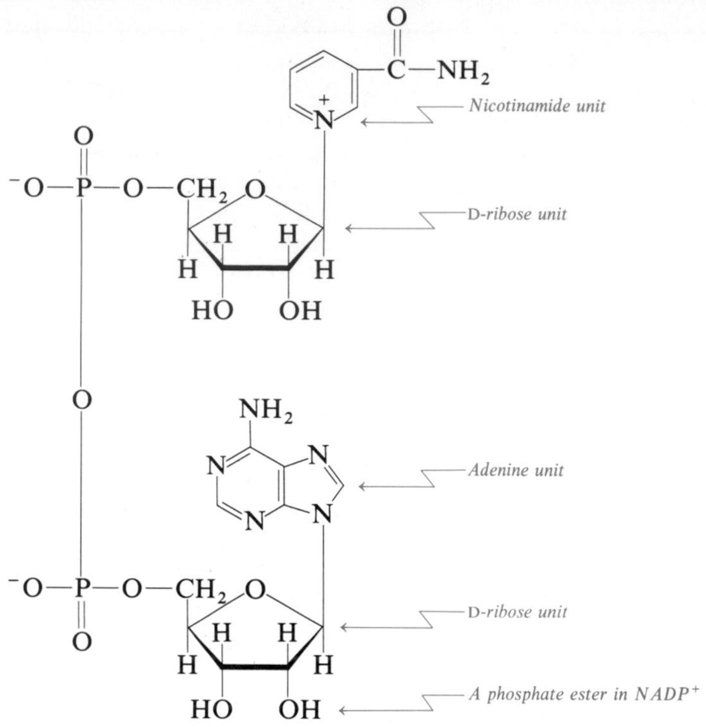

Figure 16.10 Nicotinamide adenine dinucleotide, NAD$^+$. Nicotinamide is one of the water-soluble vitamins. In nicotinamide adenine dinucleotide phosphate, NADP$^+$, the 2′ hydroxyl of D-ribose is esterified with phosphoric acid.

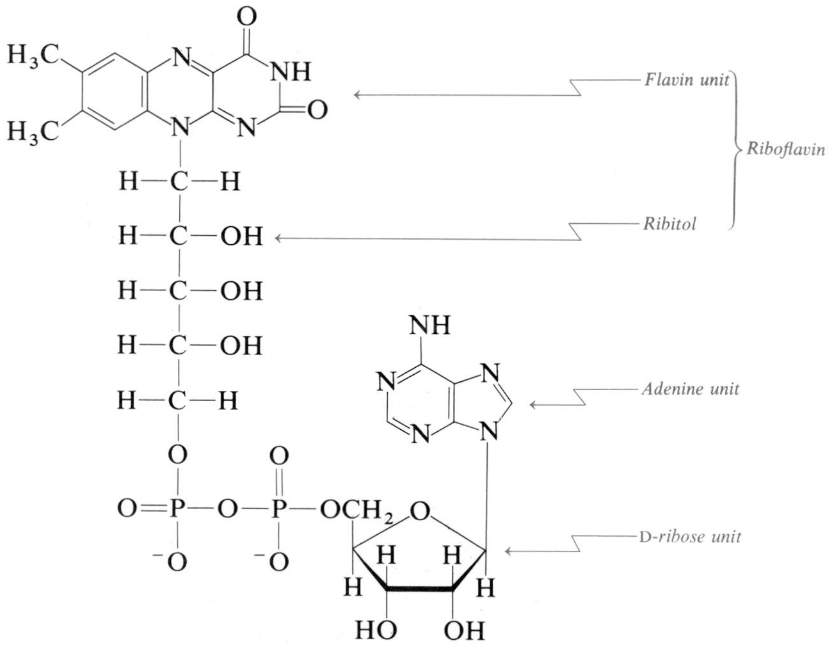

Figure 16.11 Flavin adenine dinucleotide, FAD. Riboflavin is one of the B vitamins.

$$+ 2H^+ + 2e^- \rightleftharpoons$$

FAD   FADH$_2$

All reactions of the tricarboxylic acid cycle (stage 3) and also of electron transport and oxidative phosphorylation (stage 4) take place within subcellular structures called **mitochondria** (singular, mitochondrion). To picture a mitochondrion (Figure 16.12), imagine two balloons, one larger than the other, and imagine that the larger balloon is extensively folded and stuffed inside the smaller balloon. Because the surface area of the inner membrane is so extensively and irregularly folded, it is approximately 10,000 times the area of the outer membrane. The folds of the inner membrane are called cristae, and the space that surrounds them is called the matrix. All enzymes required for catalysis of the tricarboxylic acid cycle are located within the mitochondria—some in the matrix, and others bound to the inner membrane.

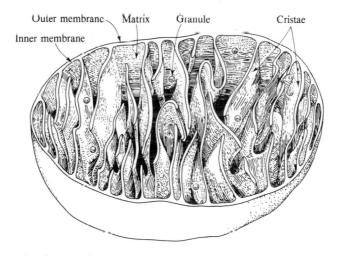

Figure 16.12 A mitochondrion. Both inner and outer membranes are phospholipid bilayers. The surface of the inner membrane is highly folded and several thousand times larger in surface area than the outer membrane.

D. Stage 4: Electron Transport and Oxidative Phosphorylation—A Central Pathway for Oxidation of Reduced Coenzymes and Generation of ATP

In stage 4, reduced coenzymes (NADH and FADH$_2$) accumulated from stages 2 and 3 are reoxidized by molecular oxygen; in effect, this is the aerobic phase of metabolism. Because reoxidation of NADH and FADH$_2$ is coupled with phosphorylation of ADP to ATP, stage 4 is called **oxidative phosphorylation**. The net reactions of stage 4 are shown in Figure 16.13.

$$2NADH + O_2 + 2H^+ \longrightarrow 2NAD^+ + 2H_2O$$
$$2FADH_2 + O_2 \longrightarrow 2FAD + 2H_2O$$
$$ADP + HPO_4^{2-} \longrightarrow ATP + H_2O$$

Figure 16.13 Stage 4 of the oxidation of foodstuffs and generation of ATP. Reoxidation of NADH and FADH$_2$ is coupled with phosphorylation of ADP to give ATP.

The four stages in the oxidation of foodstuffs and the generation of ATP are summarized in Figure 16.14.

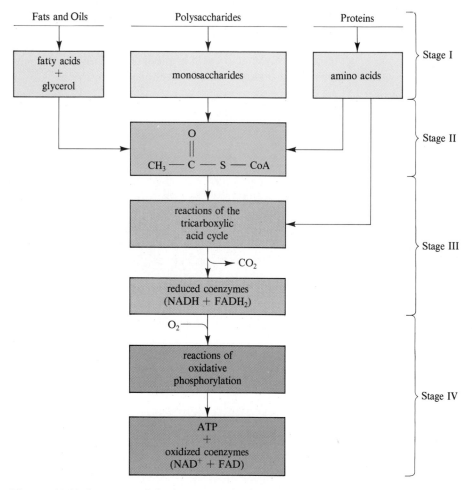

Figure 16.14 Summary of the four stages in the oxidation of foodstuffs and generation of ATP.

16.5 Electron Transport and Oxidative Phosphorylation: A Closer Look

A. Oxidation Part of Oxidative Phosphorylation

The final stage in oxidation of foodstuffs and generation of ATP involves re-oxidation of NADH and $FADH_2$ by molecular oxygen. As shown by the following equations, each oxidation is accompanied by a large decrease in free energy.

$$NADH + H^+ + \tfrac{1}{2}O_2 \longrightarrow NAD^+ + H_2O \qquad \Delta G^0 = -52,300 \text{ cal/mol}$$
$$FADH_2 + \tfrac{1}{2}O_2 \longrightarrow FAD + H_2O \qquad \Delta G^0 = -43,400 \text{ cal/mol}$$

We have written these equations as single reactions. Writing balanced half-reactions for the oxidation of NADH and $FADH_2$ and the reduction of oxygen to water allows a better appreciation of how cells bring about these oxidations.

Oxidation Half-Reactions

$$NADH \longrightarrow NAD^+ + H^+ + 2e^-$$
$$FADH_2 \longrightarrow FAD + 2H^+ + 2e^-$$

Reduction Half-Reaction

$$O_2 + 4H^+ + 4e^- \longrightarrow 2H_2O$$

Within mitochondria, the site of respiration, electrons are not passed directly from reduced coenzymes to molecular oxygen. Rather, they are passed from one acceptor to another and then to molecular oxygen by a pathway called the **electron transport** chain. All enzymes and cofactors required for electron transport are located on the inner membranes of mitochondria and are arranged in sequence so that electrons can be passed directly from one to the next. As illustrated in Figure 16.15, there are six intermediate carriers of electrons between NADH and molecular oxygen.

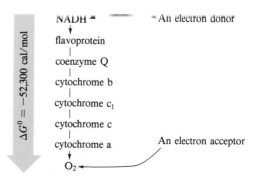

Figure 16.15 Six carriers of electrons separate NADH (an electron donor) from molecular oxygen (an electron acceptor) in the respiratory chain.

In the first step of the **respiratory chain**, a pair of electrons is transferred from NADH to a flavomononucleotide (FMN), a molecule similar to structure to riboflavin and FAD. This flavoprotein can exist in both oxidized (FMN) and reduced (FMNH$_2$) forms.

$$\text{NADH} + \text{H}^+ + \text{FMN} \longrightarrow \text{NAD}^+ + \text{FMNH}_2$$

(reduced form)   (oxidized form)   (oxidized form)   (reduced form)

The second carrier of electrons in the respiratory chain is coenzyme Q (Figure 16.16). This molecule has a long hydrocarbon chain of 6–10 isoprene units, which anchors it firmly in the nonpolar environment of the inner membrane of the mitochondrion. As you can see from the balanced half-reaction in Figure 16.16, the oxidized form of coenzyme Q is a two-electron oxidizing agent. In the second step of the respiratory chain, two electrons are transferred from the flavoprotein reduced in step 1 to the oxidized form of coenzyme Q.

$$\text{flavoprotein} + \text{coenzyme Q} \longrightarrow \text{flavoprotein} + \text{coenzyme Q}$$

(reduced form)   (oxidized form)   (oxidized form)   (reduced form)

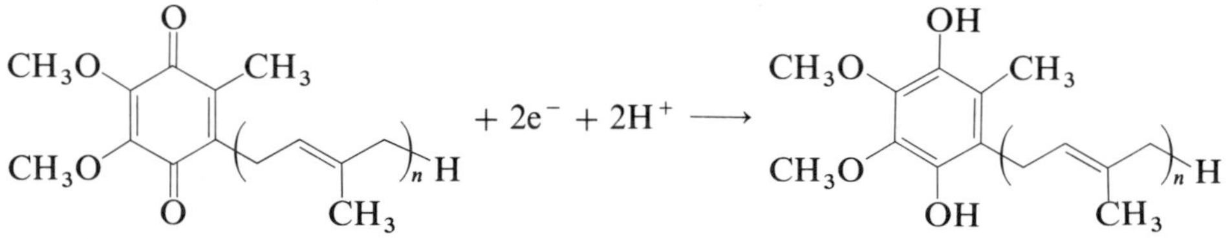

coenzyme Q
(oxidized form)

coenzyme Q
(reduced form)

Figure 16.16 Coenzyme Q. The nonpolar side chain of this molecule consists of six to ten ($n = 6$–10) isoprene units.

The remaining carriers of electrons in the respiratory chain are four structurally related proteins known as cytochromes. Cytochrome c, the most thoroughly studied of these electrons carriers, is a globular protein of molecular weight 12,400; it consists of a single polypeptide chain of 104 amino acids folded around a single heme group. The iron atom of all four cytochromes can exist in either Fe(II) or Fe(III) oxidation states. Thus, an atom of Fe(III) in a cytochrome molecule can accept an electron and be reduced to Fe(II), which in turn gives up an electron to reduce the next cytochrome in the chain.

In the final step of the respiratory chain, electrons are transferred from cytochrome a to a molecule of oxygen. In the following equation, cytochrome a is abbreviated as Cyt a.

$$2\text{Cyt a}(\text{Fe}^{2+}) + \tfrac{1}{2}\text{O}_2 + 2\text{H}^+ \longrightarrow 2\text{Cyt a}(\text{Fe}^{3+}) + \text{H}_2\text{O}$$

(reduced form)   (oxidized form)

The seven steps in the transfer of electrons from NADH to O_2 are summarized in Figure 16.17. Notice that the free-energy decrease in four of these steps is larger than that required for phosphorylation of ADP ($\Delta G^0 = +7,300$ cal/mol).

Electrons from $FADH_2$ are also transported via the intermediates of the respiratory chain to molecular oxygen. Electrons from $FADH_2$, however, enter the chain at coenzyme Q (Figure 16.17). There are only five intermediates in the transport of electrons from coenzyme Q to molecular oxygen.

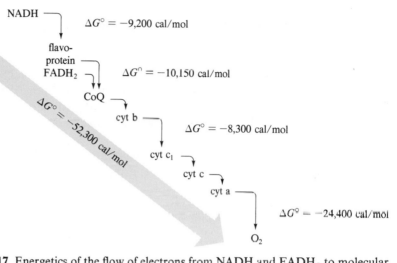

Figure 16.17 Energetics of the flow of electrons from NADH and $FADH_2$ to molecular oxygen in the respiratory chain.

B. Phosphorylation Part of Oxidative Phosphorylation

Cells have evolved a mechanism that couples the energy-releasing oxidation of reduced coenzymes with the energy-requiring phosphorylation of ADP. For each mole of NADH entering the respiratory chain, three moles of ATP are formed. The overall equation for oxidation of NADH and phosphorylation of ADP can be written as the sum of the exergonic oxidation of NADH and the endergonic phosphorylation of ADP:

$$\frac{\begin{array}{l} NADH + H^+ + \tfrac{1}{2}O_2 \longrightarrow NAD^+ + H_2O \\ 3H^+ + 3ADP + 3HPO_4^{2-} \longrightarrow 3ATP + 3H_2O \end{array}}{NADH + 4H^+ + \tfrac{1}{2}O_2 + 3ADP + 3HPO_4^{2-} \longrightarrow NAD^+ + 3ATP + 4H_2O}$$

Coupling the oxidation and phosphorylation reactions conserves $\frac{22}{52}$, or approximately 42%, of the decrease in free energy during the reoxidation of NADH.

Reoxidation of $FADH_2$ is coupled with phosphorylation of two moles of ADP, and approximately 34% of the decrease in free energy is conserved as ATP.

$$\frac{\begin{array}{l} FADH_2 + \tfrac{1}{2}O_2 \longrightarrow FAD + H_2O \\ 2ADP + 2HPO_4^{2-} + 2H^+ \longrightarrow 2ATP + 2H_2O \end{array}}{FADH_2 + \tfrac{1}{2}O_2 + 2ADP + 2HPO_4^{2-} + 2H^+ \longrightarrow FAD + 2ATP + 3H_2O}$$

Although we have written equations for oxidation of reduced coenzymes and phosphorylation as separate reactions, it is more accurate to consider them a single (but very complex), coupled reaction. *Coupled* means that one is tightly linked to the other. If electron transport is prevented (for example, by lack of oxygen as a terminal acceptor of electrons), then no ATP is produced. On the other hand, if there is a shortage of ADP or inorganic phosphate, so phosphorylation cannot take place, then electron transport does not occur.

C. Coupling of Oxidation and Phosphorylation: The Chemiosmotic Theory

While it has been possible to learn a great deal about how both oxidation and phosphorylation proceed during respiration, it has proved difficult to explain how these two processes are coupled. Biochemists reasoned that they must share a common intermediate that, in effect, couples them. By analogy, a phosphoryl group cannot be transferred directly from phosphoenolpyruvate to glucose. Rather, ATP is a common intermediate that couples reactions between phosphoenolpyruvate and glucose, to form glucose 6-phosphate and pyruvate (Section 16.3D).

$$
\begin{array}{l}
\text{phosphoenolpyruvate} + \text{ADP} \longrightarrow \text{pyruvate} + \text{ATP} \\
\underline{\qquad\quad \text{glucose} + \text{ATP} \longrightarrow \text{glucose 6-phosphate} + \text{ADP}} \\
\text{phosphoenolpyruvate} + \text{glucose} \longrightarrow \text{glucose 6-phosphate} + \text{pyruvate}
\end{array}
$$

It was assumed for many years that the common intermediate would be a chemical compound involved in both electron transport and ATP synthesis. But none could be discovered. More recently, the search for a common intermediate broadened with the realization that the "intermediate" need not be a covalent compound after all but some other form of stored chemical energy instead. The most widely accepted current model for coupling oxidation and phosphorylation was put forward in 1960 by Peter Mitchell, an English chemist. According to his **chemiosmotic theory**, as electrons are transferred along the respiratory chain during oxidation of reduced coenzymes, protons are transferred from inside the inner mitochondrial membrane to outside the membrane. It is the protons pumped outside that provide the driving force for ATP synthesis. As protons flow from the side of higher concentration outside the membrane toward the inside, they pass through a macromolecular structure called ATP synthase complex, where actual synthesis of ATP occurs.

Mitchell's theory stirred much controversy and was by no means widely accepted, even into the mid-1970s. Acceptance did develop, however, and in 1978 he was awarded the Nobel Prize in chemistry.

1. How the Proton Gradient Is Established

At the heart of the chemiosmotic theory is the concept of a "proton pump." Several electron-transport reactions in the respiratory chain involve protons as either reactants or products. Following are three such reactions, each written for simplicity as a balanced half-reaction:

$$\text{NADH} \longrightarrow \text{NAD}^+ + \text{H}^+ + 2e^-$$
$$\text{FADH}_2 \longrightarrow \text{FAD} + 2\text{H}^+ + 2e^-$$
$$\text{coenzyme Q} \longrightarrow \text{coenzyme Q} + 2\text{H}^+ + 2e^-$$
$$\text{(reduced form)} \qquad\qquad \text{(oxidized form)}$$

Mitchell proposes that sites of these reactions are arranged on the inner surface of the mitochondrial membrane in such a way that protons as reactants are taken from inside the membrane and those given off as products are released outside the membrane. Equally important, only protons are pumped outside the membrane; negative ions are not pumped out simultaneously. Through operation of the proton pump, two types of gradients are created: (1) a concentration gradient, because there are more protons outside than inside the membrane; and (2) a charge gradient, with the outside of the membrane positively charged compared with the inside (Figure 16.18). The energy involved

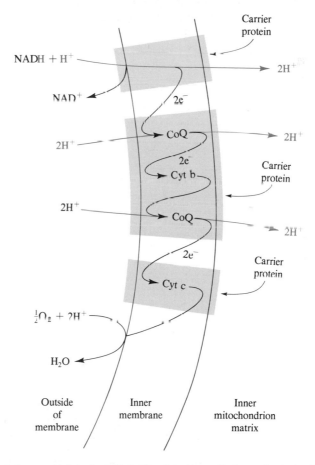

Figure 16.18 A key postulate in Mitchell's chemiosmotic hypothesis is that for each pair of electrons transferred from NADH to molecular oxygen, six hydrogen ions are pumped from the mitochondrial matrix through the inner membrane to the outside. Thus, there is established both a proton and a charge gradient between the inner mitochondrial matrix and the outer intermembrane space.

in both the concentration gradient and the charge gradient is involved in generating ATP.

2. How the Proton Gradient Is Coupled with ATP Synthesis

Mitchell proposes that ATP synthesis takes place at ATPase synthetase complexes located inside the inner mitochondrial membrane. Each complex consists of an F_1 system of five different proteins, which is joined to the inner membrane by an F_0 protein. One proposal is that as protons are propelled by the concentration and charge gradients back into mitochondria through the F_0 channel and into the F_1 system, they somehow cause a conformational change in the F_1 system, which in turn affects the enzyme ATP synthase in such a way that ADP and inorganic phosphate are converted to ATP (Figure 16.19). We know almost nothing about how a proton flux might be connected to a conformational change in the F_0–F_1 system or how that in turn might be connected with ATP synthesis.

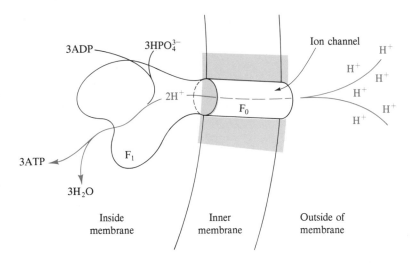

Figure 16.19 The chemiosmotic theory. Movement of protons through the F_0 channel, creation of a conformational change in the F_1 system, and synthesis of ATP.

D. Inhibitors of Electron Transport and Oxidative Phosphorylation

1. Inhibition of Electron Transport in the Respiratory Chain

Numerous chemicals interfere with specific steps in the transfer of electrons in the respiratory chain. Because they inhibit electron transport, they also inhibit phosphorylation and production of ATP. One of the best-known of these chemicals is cyanide ion, a powerful inhibitor of cytochrome oxidase, the enzyme that catalyzes the transfer of electrons from cytochrome c to cytochrome a. Cyanide ion also complexes with the iron atom of cytochrome a, to form a complex that is unable to function as a carrier of electrons. The result of cyanide ion poisoning is a block in the flow of electrons from NADH and $FADH_2$ to molecular oxygen. Cyanide ion has the same effect on the cell as lack of oxygen; death is by asphyxiation.

Another inhibitor of electron transport is rotenone, a powerful inhibitor of NADH-dehydrogenase, the enzyme that catalyzes the transfer of electrons from NADH to a flavoprotein in the first step of respiration.

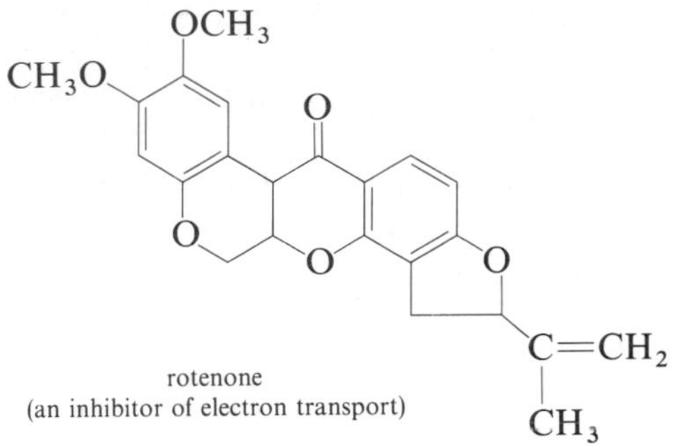

rotenone
(an inhibitor of electron transport)

Because rotenone passes readily into the breathing tubes of insects and is intensely toxic to these organisms, it is widely used as an insecticide. It is also toxic to fish, because it passes readily into their gills. Rotenone is not readily absorbed through the skin and therefore has a relatively low toxicity for humans and other vertebrates.

2. Inhibitors of Oxidative Phosphorylation

Other compounds have no effect on the transport of electrons but act to uncouple it from phosphorylation. Thus, reduced coenzymes are reoxidized by molecular oxygen, but there is no accompanying synthesis of ATP. One such uncoupling agent is 2,4-dinitrophenol, which appears to have its effect by increasing the permeability of the inner mitochondrial membrane to protons. With increased membrane permeability, both proton and charge gradients are reduced, thus decreasing the driving force for synthesis of ATP.

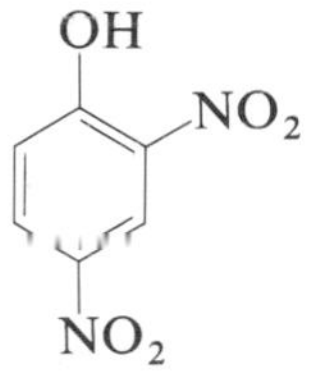

2,4-dinitrophenol
(uncouples electron
transport and phosphorylation)

16.6 Utilization of ATP for Cellular Work

As we have seen (Section 16.3), hydrolysis of the terminal phosphate anhydride bond of ATP occurs with a decrease in free energy. Cells are able to use a portion of this free energy to do three important types of work: **chemical work**, **mechanical work**, and **osmotic work**. Let us look in more detail at each type of work and see how it depends on ATP.

A. Chemical Work

The formation of peptide, glycoside, and ester bonds requires energy. For example, formation of a glycoside bond between glucose and fructose to form sucrose requires 5,500 cal for each mole of sucrose formed:

$$\text{glucose} + \text{fructose} \longrightarrow \text{sucrose} + H_2O \qquad \Delta G^0 = +5500 \text{ cal/mol}$$

On the other hand, hydrolysis of ATP decreases free energy:

$$\text{ATP} + H_2O \longrightarrow \text{ADP} + HPO_4^{2-} \qquad \Delta G^0 = -7300 \text{ cal/mol}$$

Adding these reactions gives a net reaction that occurs with a decrease in free energy and is spontaneous in the direction written:

$$
\begin{array}{ll}
\text{glucose} + \text{fructose} \longrightarrow \text{sucrose} + H_2O & \Delta G^0 = +5{,}500 \text{ cal/mol} \\
\text{ATP} + H_2O \longrightarrow \text{ADP} + HPO_4^{2-} & \Delta G^0 = -7{,}300 \text{ cal/mol} \\
\hline
\text{glucose} + \text{fructose} + \text{ATP} \longrightarrow \text{sucrose} + \text{ADP} + HPO_4^{2-} & \Delta G^0 = -1{,}800 \text{ cal/mol}
\end{array}
$$

If it were possible to capture a part of the free energy of the phosphate anhydride bond in ATP and channel it into glycoside bond formation, glucose and fructose could be converted to sucrose. Cells accomplish this by two sequential enzyme-catalyzed reactions involving a common intermediate:

$$
\begin{array}{l}
\text{glucose} + \text{ATP} \longrightarrow \text{glucose 1-phosphate} + \text{ADP} \\
\text{glucose 1-phosphate} + \text{fructose} \longrightarrow \text{sucrose} + HPO_4^{2-} \\
\hline
\text{glucose} + \text{fructose} + \text{ATP} \longrightarrow \text{sucrose} + \text{ADP} + HPO_4^{2-}
\end{array}
$$

Glucose 1-phosphate is the common intermediate. Together, these sequential reactions have a net free-energy change of -1800 cal/mol. Thus, a portion of the energy stored in ATP is captured in the form of a common intermediate and is then used to form a glycoside bond.

B. Osmotic Work: Transport across Membranes

The movement of molecules and ions across a membrane is called transport, and it is an essential process in all living organisms. Transport is of two types: passive and active. In **passive transport**, a molecule or an ion moves across a membrane from the side of high concentration to the side of lower concentration. Passive transport is spontaneous and requires no energy. In **active transport**, a molecule or an ion is moved across a membrane from the side of low concentration to the side of higher concentration. Active transport is nonspontaneous and requires energy. Figure 16.20 illustrates transport with and against a concentration gradient.

For an example of the results of active transport, compare the relative concentrations of ions and molecules in the intracellular and extracellular fluids of skeletal muscle tissue (Figure 16.21). Note that the concentrations of K^+, Mg^{2+},

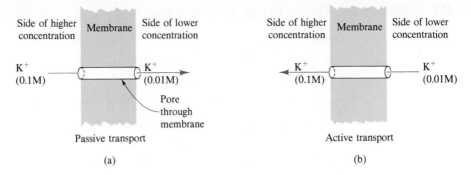

Passive transport

Active transport

(a)

(b)

Figure 16.20 Transport across membranes. (a) In passive transport, molecules and ions flow with a concentration gradient. (b) In active transport, molecules and ions flow against a concentration gradient. Active transport is nonspontaneous and requires energy.

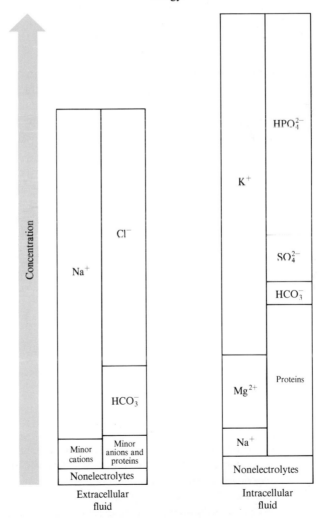

Figure 16.21 Relative concentrations of some molecules and ions in the intracellular and extracellular fluids of human skeletal muscle.

HPO_4^{2-}, and SO_4^{2-} are all much higher inside the cells of skeletal muscle than they are in the surrounding fluid. It requires energy to concentrate these ions within skeletal muscle cells.

At present, little is known about the mechanism of active transport. It is known, however, that active transport requires energy and that it is linked to the hydrolysis of high-energy phosphate bonds in ATP to give ADP and HPO_4^{2-}.

C. Mechanical Work: Muscle Contraction

Figure 16.22 is a schematic diagram of a section of skeletal muscle fiber. A fiber consists of two types of protein-containing filaments. One type of filament, containing the protein actin, consists of thin rods connected to a protein plate, or disc. Actin filaments connected to one plate do not make contact with those from an adjacent plate. A second type of filament, containing the protein myosin, consists of thicker rods that overlap actin filaments from adjacent plates.

Our best current model of muscle contraction is called the sliding filament model. During contraction, according to this model, actin filaments slide past myosin filaments, and in the process, the free ends of actin filaments are pulled close together. As actin filaments slide toward each other, they in turn pull the protein plates closer together, and the entire muscle fiber contracts. Contraction of muscle fibers is coupled with the hydrolysis of ATP to ADP and phosphate, but how these two processes are coupled is almost totally unknown.

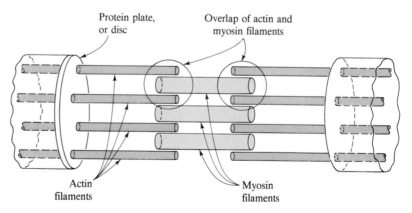

Figure 16.22 A schematic drawing of a skeletal muscle fiber.

Although ATP is the immediate source of energy to power contraction of skeletal muscle, it is not the form in which energy for muscle contraction is stored. In resting muscle, energy is stored as creatine phosphate, a high-energy compound containing a phosphate amide bond. Following is an equation for the hydrolysis of creatine phosphate. The free-energy change for this reaction is $-10,300$ cal/mol.

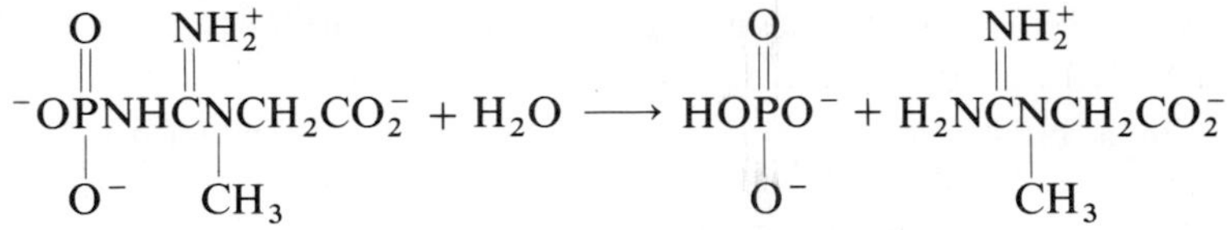

$$\underset{\substack{\text{creatine}\\\text{phosphate}}}{\overset{\begin{array}{cc}\text{O} & \text{NH}_2^+\\ \| & \|\end{array}}{{}^-\text{OPNHCNCH}_2\text{CO}_2^-}} + \text{H}_2\text{O} \longrightarrow \underset{\substack{\text{O}^-\\ \\ \text{creatine}}}{\overset{\text{O}}{\underset{}{\text{HOPO}^-}}} + \underset{\text{CH}_3}{\overset{\begin{array}{c}\text{NH}_2^+\\ \|\end{array}}{\text{H}_2\text{NCNCH}_2\text{CO}_2^-}} \qquad \Delta G^0 = -10{,}300 \text{ cal/mol}$$

The immediate chemical change on muscle contraction is hydrolysis of ATP and an increase in the concentration of ADP. In response, a phosphoryl group is transferred from creatine phosphate to ADP, and more ATP becomes available for muscle contraction. This reaction is catalyzed by the enzyme creatine kinase (CK):

$$\text{creatine phosphate} + \text{ADP} \underset{}{\overset{\substack{\text{creatine}\\\text{kinase}}}{\rightleftharpoons}} \text{creatine} + \text{ATP}$$

⎯⎯⎯ active muscle ⎯⎯⎯→

←⎯⎯⎯ resting muscle ⎯⎯⎯

During rest, the supply of ATP is regenerated, and in turn, used to regenerate the supply of creatine phosphate.

Key Terms and Concepts

acetyl coenzyme (16.4B)	Krebs cycle (16.4C)
active transport (16.6B)	low-energy compound (16.4B)
adenosine triphosphate (ATP) (16.3B)	mechanical work (16.6C)
anabolism (16.1)	metabolic pathway (16.1)
catabolism (16.1)	metabolism (16.1)
chemical work (16.6A)	mitochondrion (16.4C)
chemiosmotic theory (16.5C)	osmotic work (16.6B)
electron transport (16.5D)	oxidative phosphorylation (16.4D)
endergonic (16.3A)	passive transport (16.6B)
exergonic (16.3A)	photosynthesis (16.2)
free energy ΔG^0 (16.3A)	respiration (16.2)
free energy of hydrolysis (16.3B)	respiratory chain (16.5A)
high-energy compound (16.3B)	tricarboxylic acid (TCA) cycle (16.4C)

Key Reactions

1. Hydrolysis of high-energy compounds: ATP, ADP, phosphoenolpyruvate, and 1,3-diphosphoglycerate (Sections 16.3B and C).

2. NAD^+ as a biological oxidizing agent: reduction of NAD^+ to NADH (Section 16.4C).

3. NADH as a biological reducing agent: oxidation of NADH to NAD^+ (Section 16.4C).

4. FAD as a biological oxidizing agent: reduction of FAD to $FADH_2$ (Section 16.4C).

5. $FADH_2$ as a biological reducing agent: oxidation of $FADH_2$ to FAD (Section 16.4C).

Problems

Concept of free energy (Sections 16.1–16.3)

16.1   What is meant by the term *free energy*? by ΔG^0?

16.2   How is the sign of ΔG^0 related to:
a. the rate of a chemical reaction?
b. the position of equilibrium of a chemical reaction?
c. the spontaneity of a chemical reaction?

16.3   Define the term *high-energy compound* as it is used in biochemistry.

16.4   Adenosine triphosphate (ATP) is a phosphorylating agent. Explain what change in structural formula takes place when a molecule is phosphorylated.

16.5   Write equations for phosphorylation of the following compounds. Assume that ATP is the phosphorylating agent and that it is converted to ADP. Write structural formulas for each phosphorylated product.
a. phosphorylation of glucose to give alpha-D-glucose 6-phosphate
b. phosphorylation of glycerol to give glycerol 1-phosphate
c. phosphorylation of fructose 6-phosphate to give alpha-D-fructose 1,6-diphosphate

16.6   Calculate ΔG^0 for the following reactions. Which are spontaneous as written? Which are not spontaneous as written?
a. phosphoenolpyruvate + ADP $\longrightarrow$ pyruvate + ATP
b. 1,3-diphosphoglycerate + ADP $\longrightarrow$ 3-phosphoglycerate + ATP
c. glucose + ATP $\longrightarrow$ glucose 6-phosphate + ADP
d. glucose 1-phosphate + ADP $\longrightarrow$ glucose + ATP
e. glucose 1-phosphate $\longrightarrow$ glucose 6-phosphate

16.7   The change in free energy for complete oxidation of glucose to carbon dioxide and water is −686,000 cal/mol of glucose.

$$C_6H_{12}O_6 + 6O_2 \longrightarrow 6CO_2 + 6H_2O \qquad \Delta G^0 = -686,000 \text{ cal/mol}$$

Stages in the oxidation of foodstuffs and generation of ATP (Section 16.4)

If all this decrease in free energy could be channeled by a cell into conversion of ADP and HPO_4^{2-} to ATP, how many moles of ATP could be formed per mole of glucose oxidized?

16.8   Outline the four stages by which cells extract energy from foodstuffs. Of these four stages, which are concerned primarily with each of these processes?

a. degradation of fuel molecules   **b.** generation of ATP
c. generation of NADH and $FADH_2$ **d.** consumption of O_2

16.9   Name the separate units from which pantothenic acid is constructed.

16.10   **a.** Write an abbreviated structural formula for NAD^+, showing the portion of the molecule that functions as an oxidizing agent.
     b. Write an abbreviated structural formula for NADH, showing the portion of the molecule that functions as a reducing agent.
     c. Which water-soluble vitamin is an essential part of NAD^+?
     d. Complete and balance the following half-reaction:

$$NAD^+ + H^+ \longrightarrow$$

16.11   Write balanced equations for the oxidation of the following by NAD^+. Note that for each oxidation, the organic product is also given.

a. $CH_3CH_2OH \xrightarrow{\text{oxidation}} CH_3\overset{\displaystyle O}{\overset{\|}{C}}H$

b. $CH_3\underset{\underset{\textstyle OH}{|}}{C}HCO_2^- \xrightarrow{\text{oxidation}} CH_3\overset{\displaystyle O}{\overset{\|}{C}}CO_2^-$

c. $H\!-\!\underset{\underset{\textstyle CHCO_2^-}{|}}{\underset{\underset{\textstyle CH_2CO_2^-}{|}}{\overset{\overset{\textstyle OH}{|}}{C}}}\!-\!CO_2^- \xrightarrow{\text{oxidation}} \underset{\underset{\textstyle CHCO_2^-}{|}}{\underset{\underset{\textstyle CH_2CO_2^-}{|}}{\overset{\displaystyle O}{\overset{\|}{C}}}}\; CO_2^-$

d. $CH_3\overset{\displaystyle O}{\overset{\|}{C}}H \xrightarrow{\text{oxidation}} CH_3\overset{\displaystyle O}{\overset{\|}{C}}O^-$

16.12   Write the standard abbreviations for the oxidized and reduced forms of flavin adenine dinucleotide. Which water-soluble vitamin is an essential precursor for this molecule?

16.13   Write a balanced equation for the oxidation of succinate by FAD. The organic product is fumarate.

$$^-O_2CCH_2CH_2CO_2^- \xrightarrow{\text{oxidation}} \begin{array}{c} ^-O_2C \\ \\ H \end{array}\!\!C\!\!=\!\!C\!\!\begin{array}{c} H \\ \\ CO_2^- \end{array}$$

succinate   fumarate

Electron transport and oxidative phosphorylation (Section 16.5)

16.14   What is the function of the respiratory chain?

16.15   The final stage in aerobic metabolism involves oxidative phosphorylation. What is oxidized? What is phosphorylated?

16.16   Four of the carriers of electrons in the electron transport chain are structurally related proteins. Name the prosthetic group associated with each of these proteins. What metal is associated with each prosthetic group?

16.17 Explain why cyanide poisoning has the same effect on a cell as a lack of oxygen.

Utilization of ATP for cellular work (Section 16.6)

16.18 Name the three important types of cellular work that require ATP.

16.19 What is meant by the term *coupled reaction*? Give an example of the coupling of two biochemical reactions by a common intermediate.

16.20 What is the role of creatine phosphate in skeletal muscle?

17

Metabolism of Carbohydrates

Glucose is the key food molecule for most organisms, and virtually all organisms catabolize glucose by the same set of metabolic pathways. This fact suggests that glucose metabolism became a central feature at an early stage in the evolution of living systems. We shall concentrate on the metabolic pathways by which cells extract energy from glucose.

17.1 Digestion and Absorption of Carbohydrates

A. Digestion and Absorption

The main function of dietary carbohydrate is as a source of energy. In a typical American diet, carbohydrates provide about 50–60% of daily energy needs. The remainder is supplied by fats and proteins. During **digestion of carbohydrates**, disaccharides and polysaccharides are hydrolyzed to monosaccharides, chiefly glucose, fructose, and galactose.

$$\text{polysaccharides} + n\text{H}_2\text{O} \xrightarrow[\substack{\text{(mouth and} \\ \text{intestine)}}]{\alpha\text{-amylases}} n \text{ maltose}$$

$$\text{maltose} + \text{H}_2\text{O} \xrightarrow[\text{(intestine)}]{\text{maltase}} \text{glucose} + \text{glucose}$$

$$\text{sucrose} + \text{H}_2\text{O} \xrightarrow[\text{(intestine)}]{\text{sucrase}} \text{glucose} + \text{fructose}$$

$$\text{lactose} + \text{H}_2\text{O} \xrightarrow[\text{(intestine)}]{\text{lactase}} \text{glucose} + \text{galactose}$$

Hydrolysis of starch begins in the mouth, catalyzed by the enzyme α-amylase, a component of saliva. There, starch is broken down to smaller polysaccharides and the disaccharide maltose (Section 11.4A). Hydrolysis of sucrose, lactose, maltose, and the remaining polysaccharides is completed in the small intestine, catalyzed by the enzymes maltase, sucrase, lactase, and intestinal α-amylase.

Lactose (Section 11.4B) is the principal carbohydrate in milk; human and cow's milk is about 5% lactose by weight. Human babies are born with the digestive enzymes necessary to hydrolyze lactose to glucose and galactose. Many individuals lose the ability to hydrolyze lactose, a condition known as lactose intolerance. For them, lactose passes through the digestive system to the large intestine. There it increases the osmotic pressure of intestinal fluids, which in turn interferes with reabsorption of water and leads to diarrhea. Further, intestinal bacteria ferment lactose to gases, chiefly carbon dioxide, methane, and hydrogen, which further irritate the intestinal lining and lead to nausea and vomiting. Lactose intolerance quite predictably develops around the age of four, especially in African, Asian, Middle Eastern, Mediterranean, and American Indian peoples.

B. Normal Blood Glucose Levels

Under normal conditions, the concentration of glucose in blood is between 60 and 110 mg per 100 mL. This level rises following a meal and then falls to fasting level, a point that usually is associated with the onset of hunger. If blood glucose falls below about 60 mg per 100 mL, the condition is known as **hypoglycemia**. In hypoglycemia, there is danger that cells of the central nervous system and other tissues that depend on glucose for nourishment may not receive adequate supplies of glucose. When blood glucose levels rise above about 160 mg per 100 mL, the condition is known as **hyperglycemia**.

The liver is the key organ for regulating the concentration of glucose in the blood. As glucose is absorbed after a meal, the liver counters this increase by removing glucose from the bloodstream. Glucose removed from blood is used by the liver in two ways: (1) it can be converted to glycogen or triglycerides and stored in the liver; or (2) it can be catabolized to generate ATP and heat. Thus, the concentration of glucose in the bloodstream represents a balance between cellular intake, storage, and catabolism.

C. Glucose Tolerance Test

Any defect in the regulation of blood glucose levels can be detected by a **glucose tolerance test**, which measures the ability of tissues to absorb glucose from the blood. One part of the test depends on the limited ability of the kidneys to reabsorb glucose as they filter and purify the blood. When blood glucose levels are lower than approximately 160–180 mg/100 mL, virtually all glucose is reabsorbed by the kidneys and returned to the bloodstream. However, when blood glucose levels exceed 160–180 mg/100 mL, the kidneys can no longer absorb the excess and it is passed into the urine. The condition in which glucose appears in the urine is called glycosuria and the blood glucose level at which this occurs is called the renal threshold.

A glucose tolerance test is done in the following way. After an overnight fast, the patient is given a single dose of glucose, typically 50–100 g in a fruit-flavored drink. Specimens of blood and urine are taken before the glucose is administered, and then at regular intervals for 3–4 hr after the test dose is taken. In normal individuals, the blood glucose level increases within the first hour from 80 mg/100 mL to approximately 130 mg/100 mL; at the end of 2–3 hr, it returns to normal levels. For persons with diabetes, blood glucose begins at an elevated level and rises much higher after ingestion of the glucose test solution. Furthermore, the return to pretest levels is much slower than that observed in normal individuals. Figure 17.1 illustrates typical glucose tolerance curves for a normal individual and one with mild diabetes.

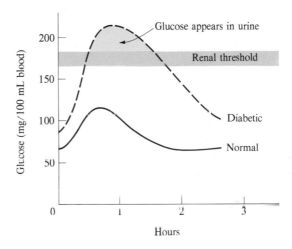

Figure 17.1 A typical glucose tolerance curve for a normal individual and one with mild diabetes mellitus.

17.2 Central Role of Glucose in Carbohydrate Metabolism

Because of the glucose requirements of cells, especially cells of the central nervous system, the body has developed a set of interrelated metabolic pathways designed to use glucose efficiently and to ensure an adequate supply of it in the bloodstream. Several of these pathways oxidize glucose to carbon dioxide and water and conserve a portion of the energy stored in glucose as ATP and other high-energy compounds. Other pathways "buffer" the concentration of glucose in the blood; that is, their job is to maintain blood-glucose levels within a narrow range. After an overview of the most important pathways of glucose metabolism, we shall study four of them (glycolysis, lactate fermentation, alcoholic fermentation, and the tricarboxylic acid cycle) in detail.

A. Glycolysis

Glycolysis is a series of ten consecutive reactions by which glucose is oxidized to two molecules of pyruvate. The oxidizing agent is NAD^+. Furthermore, two molecules of ATP are produced for each molecule of glucose oxidized to pyruvate. Following is the net reaction for glycolysis.

$$C_6H_{12}O_6 + 2NAD^+ + 2HPO_4^{2-} + 2ADP \xrightarrow{\text{glycolysis}} 2CH_3\overset{\displaystyle O}{\overset{\|}{C}}CO_2^- + 2NADH + 2ATP$$

glucose   pyruvate

B. Oxidation and Decarboxylation of Pyruvate

Following glycolysis, the carboxylate group of pyruvate is converted to carbon dioxide, and the remaining two carbons are converted to an acetyl group in the form of a thioester with coenzyme A. Several coenzymes including NAD^+ and coenzyme A are required for this metabolic pathway.

$$CH_3\overset{\displaystyle O}{\overset{\|}{C}}CO_2^- + NAD^+ + CoA-SH \xrightarrow[\text{of pyruvate}]{\substack{\text{oxidative} \\ \text{decarboxylation}}} CH_3\overset{\displaystyle O}{\overset{\|}{C}}-SCoA + CO_2 + NADH$$

pyruvate   acetyl CoA

C. The Tricarboxylic Acid Cycle

In the reactions of the tricarboxylic acid cycle, the two-carbon acetyl group of acetyl CoA is oxidized to two molecules of carbon dioxide:

$$CH_3\overset{\displaystyle O}{\overset{\|}{C}}-SCoA + 3NAD^+ + FAD + HPO_4^{2-} + ADP \xrightarrow{\text{TCA cycle}}$$

$$2CO_2 + 3NADH + FADH_2 + ATP + CoA-SH$$

The combination of glycolysis, oxidation of pyruvate to acetyl CoA, and the tricarboxylic acid cycle brings about complete oxidation of glucose to carbon dioxide and water and generates 2 moles of $FADH_2$, 10 moles of NADH, and 4 moles of ATP for each mole of glucose oxidized.

D. Oxidative Phosphorylation

Glycolysis, oxidation of pyruvate to acetyl coenyme A, and the tricarboxylic acid cycle are completely anaerobic, meaning that they do not involve molecular oxygen. Rather, there is a buildup of reduced coenzymes. Oxidation of the accumulated NADH and $FADH_2$ is coupled with phosphorylation of ADP during electron transport and oxidative phosphorylation (Section 16.5). It is **oxidative phosphorylation** that generates the major share of the ATP produced during glucose catabolism:

$$10NADH + 2FADH_2 + 6O_2 + 32ADP + 32HPO_4^{2-} + 10H^+ \xrightarrow[\text{phosphorylation}]{\text{oxidative}}$$

$$10NAD^+ + 2FAD + 32ATP + 44H_2O$$

Example 17.1

Glucose is oxidized to carbon dioxide and water by a combination of three metabolic pathways. How many molecules of CO_2 are produced in each pathway?

☐ *Solution*

No CO_2 is produced during glycolysis. Two molecules of CO_2 are produced in the oxidation and decarboxylation of pyruvate to acetyl coenzyme A. The remaining four molecules of CO_2 are produced through the reactions of the tricarboxylic acid cycle.

Problem 17.1

a. During glycolysis, how many moles of NADH and $FADH_2$ are produced per mole of glucose converted to pyruvate?

b. During the conversion of pyruvate to acetyl CoA, how many moles of NADH are produced per mole of pyruvate?

c. During the tricarboxylic acid cycle, how many moles of NADH and $FADH_2$ are produced per mole of acetyl CoA entering the cycle?

E. Pentose Phosphate Pathway

The **pentose phosphate pathway** is an alternative pathway for the oxidation of glucose to carbon dioxide and water:

$$C_6H_{12}O_6 + 12NADP^+ + 6H_2O \xrightarrow{\text{pentose phosphate pathway}} 6CO_2 + 12NADPH + 12H^+$$
glucose

At first glance, the pentose phosphate pathway appears to accomplish the same thing as a combination of glycolysis, oxidation of pyruvate to acetyl CoA, and the tricarboxylic acid cycle, namely, oxidation of glucose to carbon dioxide and water. While it is true that both sets of pathways bring about oxidation of glucose, there are important differences between them. The following reactions of the pentose phosphate pathway have been chosen to illustrate two of the most important differences. The first reaction of this pathway is oxidation of the aldehyde group of glucose 6-phosphate to a carboxylate group. Oxidation requires $NADP^+$ (Figure 16.10), which is a phosphorylated form of NAD^+, and the process is catalyzed by glucose 6-phosphate dehydrogenase. Next, oxidation of the secondary alcohol on carbon 3 of 6-phosphogluconate by a second molecule of $NADP^+$ gives a beta-ketoacid, which undergoes decarboxylation (Section 7.5D) to form ribulose 5-phosphate. In one of several reactions that follow, ribulose 5-phosphate is isomerized to ribose 5-phosphate.

These reactions are shown below using a convention widely used in biochemistry whenever there is a need to show reactants and products in a particularly compact manner. In this convention, a reactant may be shown at the tail of a curved arrow merging with the main arrow and a product may be shown at the head of an arrow branching off the main arrow. Curved arrows are used in the first and second equations to show that $NADP^+$ is the oxidizing agent and that it is reduced to NADPH. In the third reaction, a curved arrow is used to show that carbon dioxide is a product.

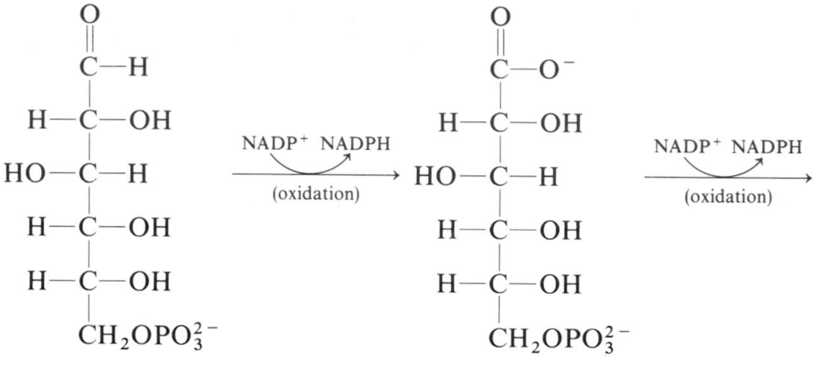

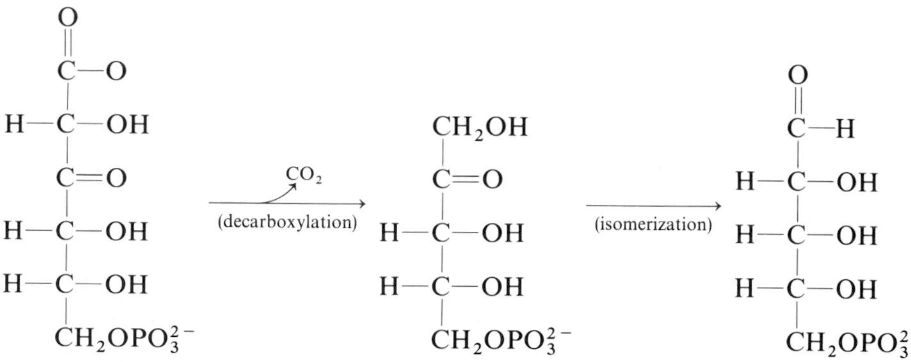

These four reactions of the pentose phosphate pathway illustrate two of its most important features. First, this metabolic pathway provides a pool of pentoses for the synthesis of nucleic acids. It also provides a pool of tetroses, not illustrated here. Second, it uses $NADP^+$ as an oxidizing agent and generates the reduced coenzyme NADPH. The major function of NADPH is as a reducing agent in the biosynthesis of other molecules. For example, adipose tissue, which has a high demand for reducing power to support the synthesis of fatty acids, is rich in $NADP^+/NADPH$. By comparison, glycolysis, oxidation of pyruvate, and the tricarboxylic acid cycle require NAD^+ and FAD as oxidizing agents, coenzymes that are reduced to NADH and $FADH_2$. These are in turn used for the generation of ATP through electron transport and oxidative phosphorylation (Section 16.5).

The pentose phosphate pathway is especially important for the normal functioning of red blood cells that depend on this pathway for a supply of NADPH—needed as a reducing agent to maintain iron atoms of hemoglobin in the Fe^{2+} state.

The activity of the pentose phosphate pathway is controlled by allosteric regulation of the first enzyme in the pathway, glucose 6-phosphate dehydrogenase. There is a genetic disease, **glucose 6-phosphate dehydrogenase deficiency**, that affects over 100 million people, mostly in Mediterranean and tropical areas. In this disease, operation of the pentose phosphate pathway is decreased due to a defect in glucose 6-phosphate dehydrogenase. As a result, concentrations of NADPH are lower than normal, erythrocyte membranes are fragile and rupture more easily than normal, and the average life span of erythrocytes is reduced. The clinical condition that results from glucose 6-phosphate dehydrogenase deficiency is called hemolytic anemia. Like sickle-cell anemia (see Mini-Essay IX, "Abnormal Human Hemoglobins"), this disease appears to make individuals more resistant to certain malaria parasites.

F. Glycogenesis and Glycogenolysis

Glycogenesis and glycogenolysis are probably the most important metabolic pathways contributing to a relatively constant blood glucose level. When dietary intake of glucose exceeds immediate needs, humans and other animals convert the excess to glycogen (Section 11.5B), which is stored in liver and muscle tissue. In normal adults, the liver can store about 110 g of glycogen, and muscles about 255 g. The pathway that converts glucose to glycogen is called **glycogenesis**:

$$(C_6H_{10}O_5)_n + nH_2O \underset{\text{glycogenesis}}{\overset{\text{glycogenolysis}}{\rightleftharpoons}} nC_6H_{12}O_6$$

$$\text{glycogen} \hspace{5cm} \text{glucose}$$

Liver and muscle glycogen are storage forms of glucose. When there is need for additional blood glucose, glycogen is hydrolyzed and glucose released into the bloodstream. The pathway that hydrolyzes glycogen to glucose is called **glycogenolysis**. This process is stimulated by the pancreatic hormone glucagon (Problem 13.30). The counterbalancing actions of glucagon and insulin in regulating normal, resting levels of blood glucose are shown schematically in Figure 17.2.

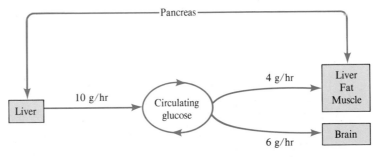

Figure 17.2 Under normal circumstances, the rate of glucagon-stimulated hydrolysis of glycogen and release of glucose into the bloodstream is balanced by insulin-stimulated uptake and metabolism of glucose by the brain and central nervous system, and muscle, adipose, and liver tissue.

G. Synthesis of Fatty Acids and Triglycerides

When carbohydrate intake is greater than the body's immediate needs for energy and its capacity to store glycogen, the excess is converted to fatty acids, which can be stored in almost unlimited quantities as triglycerides. To be stored as triglycerides, glucose is first catabolized to acetyl CoA, whose acetyl group provides the carbon atoms for the synthesis of fatty acids:

$$C_6H_{12}O_6 \longrightarrow \underset{\text{acetyl CoA}}{CH_3\overset{\overset{\displaystyle O}{\|}}{C}\!-\!SCoA} + 2CO_2$$

$$\updownarrow$$

$$\text{fatty acids}$$

Fatty acids are then combined with glycerol to form triglycerides. The synthesis of fatty acids from acetyl CoA represents a link between the metabolism of glucose and that of fatty acids. (We shall discuss the biochemistry of fatty acid synthesis and degradation in Chapter 18.)

H. Gluconeogenesis

The total supply of glucose in the form of liver and muscle glycogen and blood glucose can be depleted after about 12–18 hr of fasting. In fact, these stores of glucose often are not sufficient for the duration of an overnight fast between dinner and breakfast. Further, they also can be depleted in a short time during work or strenuous exercise. Without any way for additional supplies to be provided, nerve tissue, including the brain, would soon be deprived of glucose. Fortunately, the body has developed a metabolic pathway to overcome this problem.

Gluconeogenesis is the synthesis of glucose from noncarbohydrate molecules. During periods of low carbohydrate intake and when carbohydrate stores are being depleted rapidly, the carbon skeletons of lactate, glycerol (derived from the hydrolysis of fats), and certain amino acids are channeled into the synthesis of glucose.

$$\begin{matrix} \text{lactate} \\ \text{or} \\ \text{certain amino acids} \\ \text{or} \\ \text{glycerol} \end{matrix} \quad \xrightarrow{\text{gluconeogenesis}} \quad \text{glucose}$$

The major pathways in the metabolism of glucose are summarized in Figure 17.3.

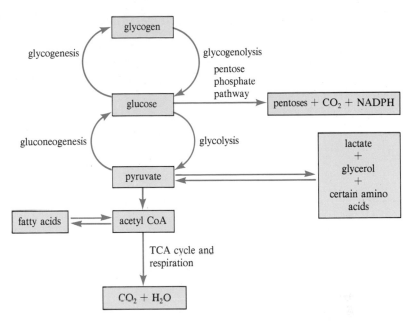

Figure 17.3 The flow of carbon atoms in the major metabolic pathways of glucose metabolism. The flow of energy (ATP generation and consumption) is not shown.

17.3 Glycolysis

A. Reactions of Glycolysis

Although writing the net reaction of **glycolysis** is simple (Section 17.2A), it took several decades of patient, intensive research by scores of scientists to discover the separate steps by which glucose is catabolized to pyruvate and to understand how this metabolic pathway is coupled with the production of ATP. By 1940, all the steps in glycolysis had been worked out. Glycolysis is frequently called the **Embden-Meyerhof pathway**, in honor of the two German biochemists, Gustav Embden and Otto Meyerhof, who contributed so greatly to our present knowledge of it.

All the reactions of glycolysis occur in the cytosol, and in a sense, the first reaction of glycolysis is transport of glucose across the cell membrane and into the cytosol. It is thought that transport involves combination of glucose with a carrier protein on the outer surface of the membrane, movement of this complex through the membrane, and release of glucose at the inner surface of the membrane. Because transport is from a high concentration outside the cell membrane toward a lower concentration inside the cell, the process is thought to be passive (Section 16.6B).

1. Phosphorylation of Glucose

The first step of glycolysis is phosphorylation of glucose by ATP to yield glucose 6-phosphate. Transfer of a phosphate group from ATP to an organic molecule is one of the basic reaction types in living systems, and any enzyme that catalyzes this type of reaction is called a kinase. The enzyme that catalyzes the transfer of a phosphate group from ATP to glucose (a hexose) is called hexokinase.

$$
\begin{array}{ccc}
\text{CHO} & & \text{CHO} \\
\text{H}\!-\!\!-\text{OH} & & \text{H}\!-\!\!-\text{OH} \\
\text{HO}\!-\!\!-\text{H} & \xrightarrow{\text{hexokinase}} & \text{HO}\!-\!\!-\text{H} \\
\text{H}\!-\!\!-\text{OH} & +\text{ATP} & \text{H}\!-\!\!-\text{OH} \quad +\text{ADP}\\
\text{H}\!-\!\!-\text{OH} & & \text{H}\!-\!\!-\text{OH} \\
\text{CH}_2\text{OH} & & \text{CH}_2\text{OPO}_3^{2-}
\end{array}
$$

glucose   glucose 6-phosphate

Phosphorylation of glucose at this stage serves a very important function. Whereas glucose passes freely through membranes, phosphorylated intermediates normally do not pass through either cell or mitochondrial membranes. Thus, phosphorylation of glucose at this early stage confines it and all subsequent phosphorylated intermediates to the cytosol, where all glycolytic enzymes are located.

2. Isomerization of Glucose 6-Phosphate to Fructose 6-Phosphate

The second step of glycolysis, isomerization of glucose 6-phosphate to fructose 6-phosphate, is catalyzed by the enzyme phosphoglucoisomerase. This isomerization involves formation of an enediol intermediate (see Problem 6.21), which then forms the carbonyl group of the ketone in fructose 6-phosphate.

$$
\begin{array}{ccc}
\text{CHO} & \left[\begin{array}{c}\text{H}\!-\!\text{C}\!-\!\text{OH}\\ \text{C}\!-\!\text{OH}\end{array}\right] & \text{CH}_2\text{OH} \\
\text{H}\!-\!\text{C}\!-\!\text{OH} & & \text{C}\!=\!\text{O} \\
\text{HO}\!-\!\!-\text{H} & \text{HO}\!-\!\!-\text{H} & \text{HO}\!-\!\!-\text{H} \\
\text{H}\!-\!\!-\text{OH} & \text{H}\!-\!\!-\text{OH} & \text{H}\!-\!\!-\text{OH} \\
\text{H}\!-\!\!-\text{OH} & \text{H}\!-\!\!-\text{OH} & \text{H}\!-\!\!-\text{OH} \\
\text{CH}_2\text{OPO}_3^{2-} & \text{CH}_2\text{OPO}_3^{2-} & \text{CH}_2\text{OPO}_3^{2-}
\end{array}
$$

glucose 6-phosphate   (an enediol)   fructose 6-phosphate
(an aldohexose)   (a ketohexose)

3. Phosphorylation of Fructose 6-Phosphate

In the third step of glycolysis, a second mole of ATP is used to convert fructose 6-phosphate to fructose 1,6-diphosphate. This phosphorylation is catalyzed by phosphofructokinase, a allosteric enzyme whose activity is a key control point in the regulation of glycolysis. As will be discussed in Section 17.3C, the catalytic activity of this enzyme is regulated (increased as well as decreased) by a number of metabolites that are indicators of energy balance within the cell. Thus, through modulation of the catalytic activity of phosphofructokinase, glycolysis can be increased when there is demand for more ATP, or decreased when there is an adequate supply of ATP.

$$
\begin{array}{ccc}
\text{CH}_2\text{OH} & & \text{CH}_2\text{OPO}_3^{2-} \\
| & & | \\
\text{C}=\text{O} & & \text{C}=\text{O} \\
| & \xrightarrow[\text{fructokinase}]{\text{phospho-}} & | \\
\text{HO}-\!\!-\text{H} & +\text{ATP} & \text{HO}-\!\!-\text{H} \quad +\text{ADP} \\
| & & | \\
\text{H}-\!\!-\text{OH} & & \text{H}-\!\!-\text{OH} \\
| & & | \\
\text{H}-\!\!-\text{OH} & & \text{H}-\!\!-\text{OH} \\
| & & | \\
\text{CH}_2\text{OPO}_3^{2-} & & \text{CH}_2\text{OPO}_3^{2-}
\end{array}
$$

fructose
6-phosphate

fructose
1,6-diphosphate

4. Cleavage of Fructose 1,6-Diphosphate into Two Triose Phosphates

In the fourth step of glycolysis, fructose 1,6-diphosphate is cleaved to dihydroxy-acetone phosphate and glyceraldehyde 3-phosphate by a reaction that is the reverse of an aldol condensation (Section 6.8). Recall that an aldol condensation involves addition of the alpha carbon of one carbonyl-containing compound to the carbonyl group of another to form a beta-hydroxyaldehyde or a beta-hydroxyketone. Among the functional groups in fructose 1,6-diphosphate is a ketone, and beta to it is a secondary alcohol. Because the cleavage of fructose 1,6-diphosphate is like the reverse of an aldol condensation, the enzyme that catalyzes this reaction is named aldolase.

$$
\begin{array}{ccc}
\text{CH}_2\text{OPO}_3^{2-} & & \text{CH}_2\text{OPO}_3^{2-} \\
| & & | \\
\text{C}=\text{O} & & \text{C}=\text{O} \qquad \text{dihydroxyacetone} \\
| & & | \qquad\qquad \text{phosphate} \\
\text{HO}-\!\!-\text{H} & \xrightarrow{\text{aldolase}} & \text{CH}_2\text{OH} \\
| & & + \\
\text{H}-\!\!-\text{OH} & & \text{H}-\text{C}=\text{O} \\
| & & | \\
\text{H}-\!\!-\text{OH} & & \text{H}-\text{C}-\text{OH} \quad \text{glyceraldehyde} \\
| & & | \qquad\qquad 3\text{-phosphate} \\
\text{CH}_2\text{OPO}_3^{2-} & & \text{CH}_2\text{OPO}_3^{2-}
\end{array}
$$

fructose 1,6-diphosphate

5. Isomerization of Dihydroxyacetone Phosphate to Glyceraldehyde 3-Phosphate

In the fifth step of glycolysis, dihydroxyacetone phosphate is converted to gly-ceraldehyde 3-phosphate by the same type of enediol intermediate we have already seen in the isomerization of glucose 6-phosphate to fructose 6-phosphate.

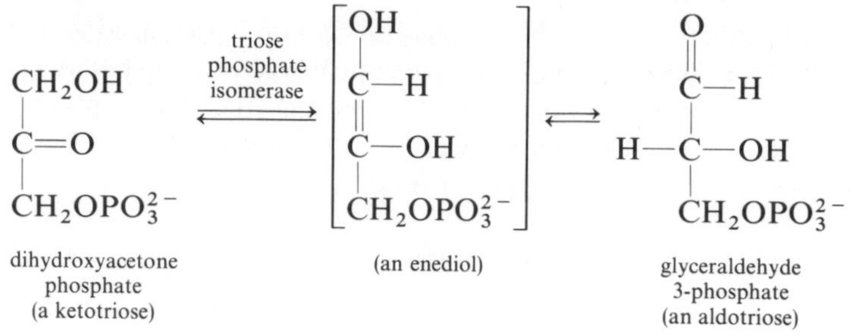

dihydroxyacetone
phosphate
(a ketotriose)

(an enediol)

glyceraldehyde
3-phosphate
(an aldotriose)

6. Oxidation of the Aldehyde of Glyceraldehyde 3-Phosphate

In the sixth step, glyceraldehyde 3-phosphate is oxidized by NAD^+. In this reaction, catalyzed by glyceraldehyde 3-phosphate dehydrogenase, one phosphate ion is required. The immediate product of the aldehyde oxidation is the mixed anhydride, 1,3-diphosphoglycerate.

glyceraldehyde
3-phosphate

1,3-diphosphoglycerate

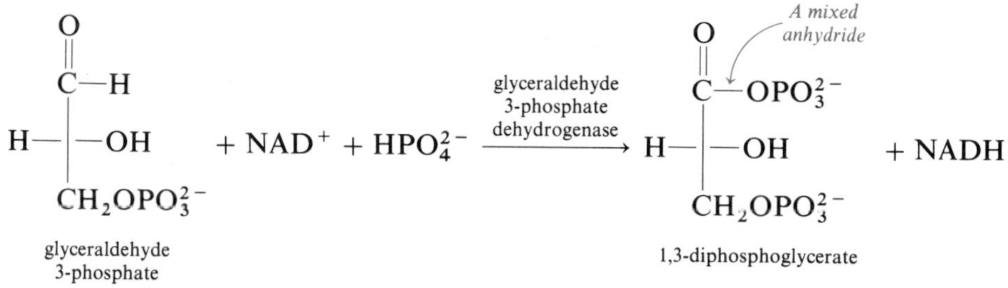

7. Transfer of a Phosphoryl Group to ADP to Form ATP

Transfer of a phosphoryl group from 1,3-diphosphoglycerate to ADP in the seventh step produces the first ATP generated in glycolysis. In this reaction, catalyzed by phosphoglycerate kinase, the anhydride bond of 1,3-diphosphoglycerate is exchanged for the terminal phosphate anhydride bond in ATP. This synthesis of ATP is known as substrate phosphorylation, to distinguish it from the oxidative phosphorylation of ADP that occurs during electron transport and oxidative phosphorylation in mitochondria.

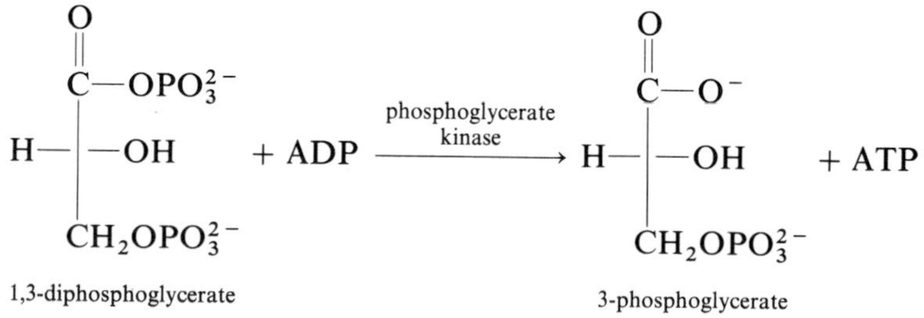

1,3-diphosphoglycerate

3-phosphoglycerate

Let us stop and look at the energy balance to this point. Two molecules of ATP were consumed in the conversion of glucose to fructose 1,6-diphosphate.

Now, with the oxidation of two molecules of glyceraldehyde 3-phosphate to 3-phosphoglycerate (remember that the original glucose molecule has been split into 2 three-carbon fragments), two molecules of ATP have been generated. Thus, through the first seven steps of glycolysis, the energy debit and credit are balanced; there is neither profit nor loss.

8. Isomerization of 3-Phosphoglycerate to 2-Phosphoglycerate

In step 8, a phosphate group is transferred from the hydroxyl group of carbon 3 to the hydroxyl group on carbon 2. This isomerization is an exchange of phosphate esters between hydroxyl groups on glycerate. The reaction proceeds by enzyme-catalyzed transfer of a phosphate group from a phosphorylated enzyme, to form 2,3-diphosphoglycerate. The 3-phospho group of this intermediate is then transferred back to the enzyme, leaving 2-phosphoglycerate.

$$
\begin{array}{c}
CO_2^- \\
| \\
H—C—OH \\
| \\
CH_2OPO_3^{2-}
\end{array}
\; + \; E—OPO_3^{2-}
\xrightarrow{\text{phospho-glyceromutase}}
\begin{array}{c}
CO_2^- \\
| \\
H—C—OPO_3^{2-} \\
| \\
CH_2OPO_3^{2-}
\end{array}
\; + \; E—OH
$$

3-phosphoglycerate ⟶ 2,3-diphosphoglycerate

$$
\begin{array}{c}
CO_2^- \\
| \\
H—C—OPO_3^{2-} \\
| \\
CH_2OPO_3^{2-}
\end{array}
\; + \; E—OH
\xrightarrow{\text{phospho-glyceromutase}}
\begin{array}{c}
CO_2^- \\
| \\
H—C—OPO_3^{2-} \\
| \\
CH_2OH
\end{array}
\; + \; E—OPO_3^{2-}
$$

2,3-diphosphoglycerate ⟶ 2-phosphoglycerate

9. Dehydration of 2-Phosphoglycerate

In step 9, 2-phosphoglycerate is dehydrated by removal of H and OH from adjacent carbons, to form phosphoenolpyruvate. Because the product of this reaction contains an alcohol (as a phosphate ester) on a carbon-carbon double bond and is therefore an enol, the enzyme catalyzing the reaction is named enolase.

$$
\begin{array}{c}
CO_2^- \\
| \\
H—C—OPO_3^{2-} \\
| \\
CH_2OH
\end{array}
\xrightarrow{\text{enolase}}
\begin{array}{c}
CO_2^- \\
| \\
C—OPO_3^{2-} \\
\| \\
CH_2
\end{array}
\; + \; H_2O
$$

2-phosphoglycerate ⟶ phosphoenolpyruvate

10. Transfer of a Phosphate Group from Phosphoenolpyruvate to ADP

The tenth and final step of glycolysis is what gives the energy profit. Phosphoenolpyruvate is a high-energy compound (Section 16.3C), and in this step, transfers its phosphate group to ADP, to form ATP. In the following equation, the terminal phospate group of ADP is written out to show how phosphate is transferred from phosphoenolpyruvate to ADP. Note that the name of the enzyme catalyzing this reaction, like all enzymes involving ADP and ATP, includes the word *kinase*.

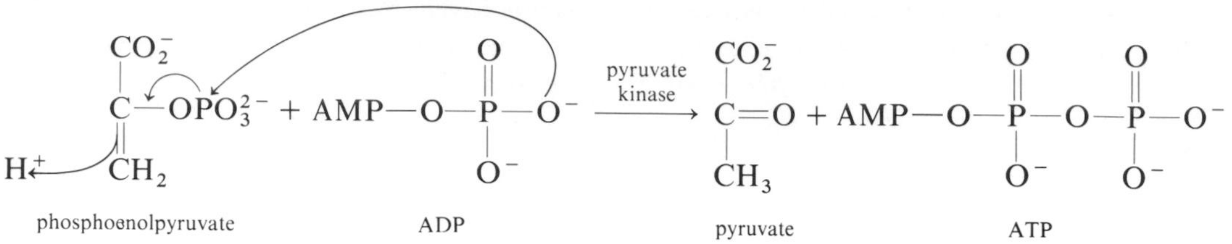

$$\text{phosphoenolpyruvate} + \text{ADP} \xrightarrow{\text{pyruvate kinase}} \text{pyruvate} + \text{ATP}$$

Because the carbon skeleton of glucose entering glycolysis is split into two triose phosphates, eventually giving two molecules of phosphoenolpyruvate, step 10 of glycolysis produces two molecules of ATP for each molecule of glucose entering the pathway.

The ten steps in the conversion of glucose to pyruvate, including those that consume NAD^+ and ATP as well as those that generate ATP, are summarized in Figure 17.4.

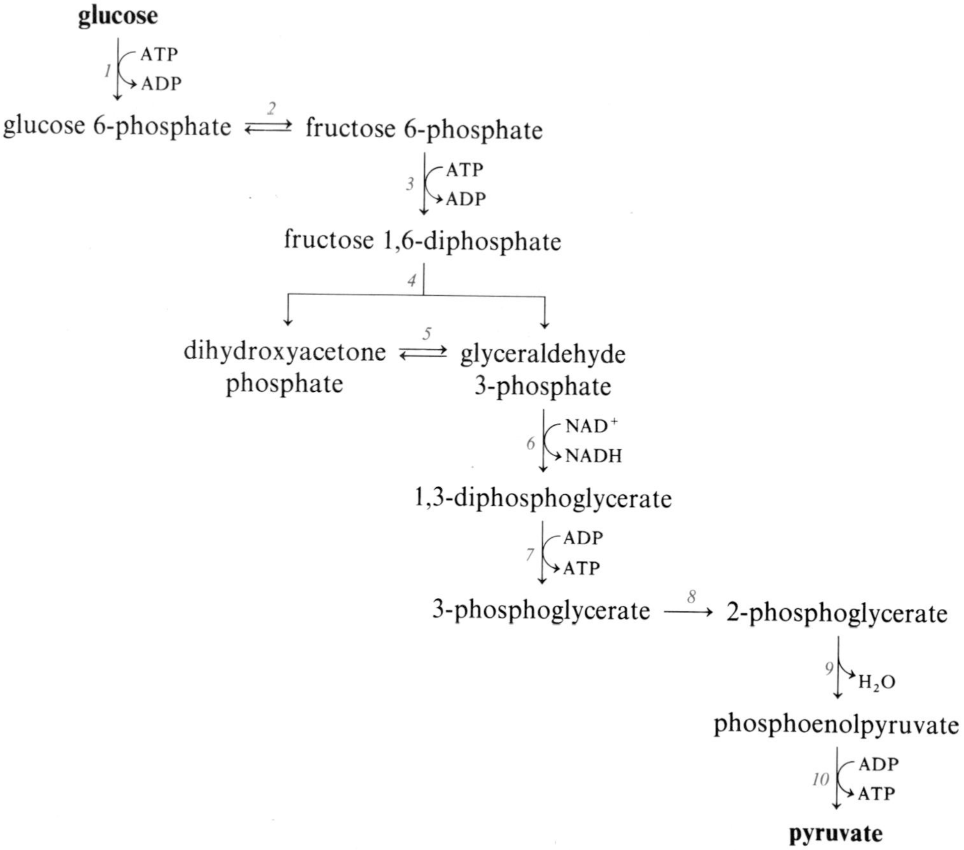

Figure 17.4 The ten steps of glycolysis.

Summing the energy balance for glycolysis, on the debit side two moles of ATP are required (steps 1 and 3) to convert glucose to fructose 1,6-diphosphate. On the credit side, two moles of ATP are produced in step 7 from reaction between 1,3-diphosphoglycerate and ADP, and another two moles in step 10 from reaction between phosphoenolpyruvate and ADP. Thus, there is a net profit of two moles of ATP per mole of glucose entering glycolysis.

Example 17.2

List all reactions of glycolysis that involve isomerization.

☐ *Solution*

An isomerization is a reaction in which a reactant and product are isomers of each other. There are three isomerizations in glycolysis:

Step 2: glucose 6-phosphate $\longrightarrow$ fructose 6-phosphate
Step 5: dihydroxyacetone phosphate $\longrightarrow$ glyceraldehyde 3-phosphate
Step 8: 3-phosphoglycerate $\longrightarrow$ 2-phosphoglycerate

Problem 17.2

List all reactions of glycolysis that involve (a) oxidation; (b) cleavage of carbon-carbon bonds.

B. Entry of Fructose and Galactose into Glycolysis

Transformation of fructose into glycolytic intermediates takes place in the liver, and begins with phosphorylation of fructose by ATP, catalyzed by the enzyme fructokinase. Fructose 1-phosphate, the product, is then split into two trioses by the same type of reaction we have already seen in step 4 of glycolysis—in which fructose 1,6-diphosphate is cleaved into two trioses.

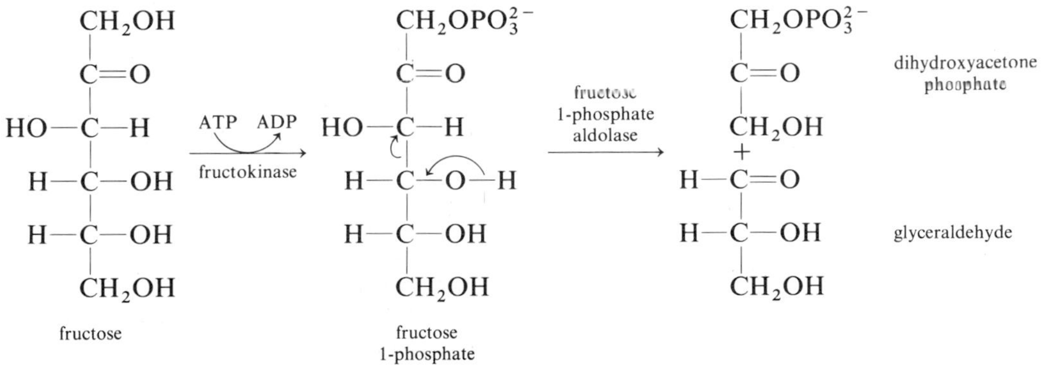

Dihydroxyacetone phosphate is a glycolytic intermediate and enters glycolysis at step 5. Glyceraldehyde is metabolized in several ways. It can be phosphorylated to glyceraldehyde 3-phosphate and then enter glycolysis. Alternatively, depending on the needs of the organism, it can be reduced to glycerol and then phosphorylated to glycerol phosphate, a metabolic intermediate required for the synthesis of phospholipids.

$$
\begin{array}{ccc}
\text{CHO} & & \text{CH}_2\text{OH} & & \text{CH}_2\text{OH} \\
| & \xrightarrow{\text{NADH}\ \ \text{NAD}^+} & | & \xrightarrow{\text{ATP}\ \ \text{ADP}} & | \\
\text{H}-\text{C}-\text{OH} & & \text{H}-\text{C}-\text{OH} & & \text{H}-\text{C}-\text{OH} \\
| & & | & & | \\
\text{CH}_2\text{OH} & & \text{CH}_2\text{OH} & & \text{CH}_2\text{OPO}_3^{2-}
\end{array}
$$

glyceraldehyde   glycerol   glycerol phosphate

As a result of a genetic disease, some humans lack the enzyme fructokinase and cannot metabolize fructose in the normal way. Fructose is instead excreted in the urine. Those who lack the enzyme fructose 1-phosphate aldolase have a much more serious problem. Because fructose 1-phosphate cannot be broken down to trioses, it accumulates in the liver and interferes with the activity of several enzyme systems, including those of gluconeogenesis. Kidney function is also disturbed. This condition is known as fructose intolerance. Vomiting and loss of appetite are early symptoms of **fructose intolerance**.

Galactose enters glycolysis by way of a series of reactions that convert it to glucose 6-phosphate:

$$
\begin{array}{ccc}
\text{CHO} & & \text{CHO} \\
\text{H}-\!\!\!-\text{OH} & & \text{H}-\!\!\!-\text{OH} \\
\text{HO}-\!\!\!-\text{H} & \xrightarrow[\text{(several steps)}]{\text{ATP}\ \ \text{ADP}} & \text{HO}-\!\!\!-\text{H} \\
\text{HO}-\!\!\!-\text{H} & & \text{H}-\!\!\!-\text{OH} \\
\text{H}-\!\!\!-\text{OH} & & \text{H}-\!\!\!-\text{OH} \\
\text{CH}_2\text{OH} & & \text{CH}_2\text{OPO}_3^{2-}
\end{array}
$$

D-galactose   D-glucose 6-phosphate

The result of these steps is inversion of configuration at carbon 4 and phosphorylation of the hydroxyl group at carbon 6. Among humans, there is an inherited disease, **galactosemia**, which manifests itself by an inability to metabolize galactose. The genetic defect leading to galactosemia is the liver's failure to produce a key enzyme involved in the inversion of configuration at carbon 4 that converts galactose to glucose. In persons with this disease, galactose accumulates in the blood and various tissues, including those of the central nervous system, and causes damage to cells. Early symptoms are similar to those of lactose and fructose intolerance, namely diarrhea, nausea, and vomiting. Without treatment, an infant with galactosemia is likely to suffer irreversible brain damage and even death. If recognized in time, however, it can be treated simply.

Because milk is the only source of galactose, it is excluded from the infant's diet.

C. Regulation of Glycolysis

Within cells, the rate of glycolysis is controlled by the activity of two allosteric enzymes, hexokinase and phosphofructokinase. Hexokinase catalyzes the phosphorylation of glucose to glucose 6-phosphate:

$$\text{glucose} + \text{ATP} \xrightarrow{\text{hexokinase}} \text{glucose 6-phosphate} + \text{ADP}$$

Inhibited by glucose 6-phosphate

Hexokinase represents a control point not only for glycolysis but also for other pathways in the metabolism of glucose. Unless glucose is first phosphorylated, it cannot enter glycolysis, the pentose phosphate pathway, or glycogenesis. Hexokinase is inhibited by a high concentration of glucose 6-phosphate, the end product of the reaction it catalyzes. Thus, phosphorylation of glucose is under self-control by feedback inhibition.

The second and more important control point for glycolysis is phosphofructokinase. Once glucose is converted to fructose 1,6-diphosphate, it is committed irreversibly to glycolysis. Because phosphorylation of fructose 6-phosphate represents a committed step, the enzyme that catalyzes it is ideally suited for being a regulatory enzyme. Phosphofructokinase is inhibited by high concentrations of ATP and citrate (an intermediate in the TCA) and it is activated by high concentrations of ADP and AMP.

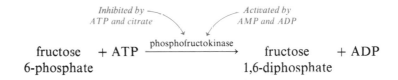

Inhibited by ATP and citrate   *Activated by AMP and ADP*

$$\text{fructose 6-phosphate} + \text{ATP} \xrightarrow{\text{phosphofructokinase}} \text{fructose 1,6-diphosphate} + \text{ADP}$$

To understand the molecular logic behind these methods of enzyme regulation, remember that glycolysis provides fuel (in the form of acetyl CoA) for the tricarboxylic acid cycle, which in turn provides fuel (in the form of NADH and $FADH_2$) for respiration and oxidative phosphorylation. When a cell or organism is in a state of low energy demand (supplies of ATP are adequate for immediate energy needs), there is no need to commit glucose to carbohydrate degradation. Hence, the inhibition of phosphofructokinase by ATP. On the other hand, when a cell or organism is using a great deal of energy and the concentration of ATP decreases and of ADP and AMP increase, ADP and AMP become metabolic signals to speed the degradation of carbohydrates so that more ATP is produced.

17.4 Fates of Pyruvate

A key to understanding the fates of pyruvate is to recognize that it is produced by oxidation of glucose through the reactions of glycolysis. NAD$^+$ is the oxidizing agent and is reduced to NADH:

$$C_6H_{12}O_6 + 2NAD^+ \xrightarrow{\text{glycolysis}} 2CH_3\overset{\displaystyle O}{\overset{\|}{C}}CO_2^- + 2NADH$$

$$\text{glucose} \qquad\qquad\qquad \text{pyruvate}$$

A continuing supply of NAD$^+$ is necessary for continued operation of glycolysis. Therefore, pyruvate is metabolized in ways that regenerate NAD$^+$.

A. Oxidation to Acetyl CoA

In most mammalian cells operating with a good supply of oxygen, O_2 is the terminal acceptor of electrons from NADH. During respiration, NADH is oxidized to NAD$^+$, oxygen is reduced to H_2O, and these processes are coupled with phosphorylation of ADP:

$$NADH + H^+ + \tfrac{1}{2}O_2 \xrightarrow{\text{respiration}} NAD^+ + H_2O$$

Thus, under aerobic conditions, where supplies of ATP and NAD$^+$ are furnished by electron transport and oxidative phosphorylation, pyruvate is transported into the mitochondria by a specific carrier protein and there oxidized to acetyl coenzyme A, a fuel for the tricarboxylic acid cycle.

$$CH_3\overset{\displaystyle O}{\overset{\|}{C}}CO_2^- + NAD^+ + CoA\!-\!SH \xrightarrow[\text{dehydrogenase}]{\text{pyruvate}} CH_3\overset{\displaystyle O}{\overset{\|}{C}}\!-\!SCoA + NADH + CO_2$$

$$\text{pyruvate} \qquad\qquad\qquad\qquad\qquad\qquad\qquad \text{acetyl CoA}$$

Oxidation is catalyzed by pyruvate dehydrogenase, a multienzyme complex attached to the inner wall of the mitochondrion. Coenzymes required for oxidation of pyruvate to acetyl CoA are NAD$^+$, coenzyme A, lipoic acid, FAD, and thiamine pyrophosphate.

B. Reduction to Lactate: Lactate Fermentation

Under anaerobic conditions, when oxygen is not available to accept electrons, the electron carriers of the electron transport system (Section 16.5) become almost totally reduced. Consequently, even if NADH produced in the cytosol by glycolysis were transported into the mitochondria, there would be no way for it to be reoxidized to NAD$^+$. Yet, glycolysis must proceed, because it is the only way to generate ATP under anaerobic conditions. In vertebrates, the most important pathway for regeneration of NAD$^+$ under anaerobic conditions is reduction of pyruvate to lactate catalyzed by lactate dehydrogenase:

$$\underset{\text{pyruvate}}{CH_3\overset{O}{\overset{\|}{C}}CO_2^-} + NADH + H^+ \xrightarrow{\underset{\text{dehydrogenase}}{\text{lactate}}} \underset{\text{lactate}}{CH_3\overset{OH}{\overset{|}{C}}HCO_2^-} + NAD^+$$

Adding the reduction of lactate to the net reaction of glycolysis gives an overall reaction for a metabolic pathway called **lactate fermentation**:

$$\underset{\text{glucose}}{C_6H_{12}O_6} + 2ADP + HPO_4^{2-} \xrightarrow{\underset{\text{fermentation}}{\text{lactate}}} \underset{\text{lactate}}{2CH_3\overset{OH}{\overset{|}{C}}HCO_2^-} + 2ATP + 2H^+$$

While lactate fermentation allows glycolysis to continue in the absence of oxygen and generates some ATP, it also brings about an increase in the concentration of lactate, and perhaps more important, protons, in muscle tissue and in the bloodstream. This buildup of lactate and protons is associated with fatigue. When blood lactate reaches a concentration of about 0.4 mg/100 mL, muscle tissue becomes almost completely exhausted.

Most of the lactate formed in active skeletal muscle is transported by the bloodstream to the liver, where it is converted to glucose by the reactions of gluconeogenesis. This newly synthesized glucose is then returned to skeletal muscles for further anaerobic glycolysis and generation of ATP. In this way, a part of the metabolic burden of active skeletal muscle is shifted, at least temporarily, to the liver. Transport of lactate from muscle to the liver, resynthesis of glucose by gluconeogenesis, and return of glucose to muscle tissue is called the **Cori cycle** (Figure 17.5).

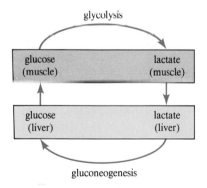

Figure 17.5 The Cori cycle.

C. Reduction to Ethanol: Alcoholic Fermentation

Yeast and several other organisms have developed an alternative pathway to regenerate NAD^+ under anaerobic conditions. In the first step of this pathway, pyruvate is decarboxylated to acetaldehyde:

$$CH_3\overset{\overset{\displaystyle O}{\|}}{C}CO_2^- + H^+ \xrightarrow[\text{decarboxylase}]{\text{pyruvate}} CH_3\overset{\overset{\displaystyle O}{\|}}{C}H + CO_2$$

pyruvate   acetaldehyde

The carbon dioxide produced in this reaction is responsible for the foam on beer and the carbonation of naturally fermented wines and champagnes. In a second step, acetaldehyde is reduced by NADH to ethanol:

$$CH_3\overset{\overset{\displaystyle O}{\|}}{C}H + NADH + H^+ \xrightarrow[\text{dehydrogenase}]{\text{alcohol}} CH_3CH_2OH + NAD^+$$

acetaldehyde   ethanol

Adding reactions for the decarboxylation of pyruvate and reduction of acetaldehyde to the net reaction of glycolysis gives the overall reaction for **alcoholic fermentation**:

$$C_6H_{12}O_6 + 2ADP + 2HPO_4^{2-} \xrightarrow[\text{fermentation}]{\text{alcoholic}} 2CH_3CH_2OH + 2CO_2 + 2ATP$$

glucose   ethanol

Note that both alcoholic fermentation and lactate fermentation represent ways in which cells can continue glycolysis under anaerobic conditions, that is, under conditions where NADH cannot be reoxidized by O_2.

Example 17.3

Under conditions of high oxygen concentration in muscle cells, which compound in each set would be present in higher concentration?

a. NAD^+ or NADH   **b.** acetyl CoA or coenzyme A

☐ *Solution*

a. If there is an adequate supply of oxygen to cells, NADH can be oxidized to NAD^+, and therefore, NAD^+ will be present in higher concentration than NADH.
b. If NAD^+ is present in high concentrations, glycolysis can continue and pyruvate is produced. Pyruvate is, in turn, converted to acetyl CoA, and therefore, the concentration of acetyl CoA increases and of free coenzyme A decreases.

Problem 17.3

Under conditions of low oxygen concentration in muscle cells, which of the following compounds, do you predict, would be present in higher concentration?

a. acetyl CoA or lactate   **b.** lactate or pyruvate

17.5 The Tricarboxylic Acid Cycle

Under aerobic conditions, the central metabolic pathway for the oxidation of the carbon skeletons of not only carbohydrates but also of fatty acids and amino acids to carbon dioxide is the **tricarboxylic acid cycle (TCA)**, known alternatively as the citric acid cycle, or **Krebs cycle**. The last-mentioned name is in honor of Sir Adolph Krebs, the biochemist who in 1937 first proposed the cyclic nature of this pathway. The name *citric acid cycle* is little used today, because it gives undue importance to what is but one of the intermediates in this metabolic pathway. All enzymes of the TCA cycle are located within the mitochondria, most within the mitochondrial matrix.

Through the reactions of the TCA cycle, the carbon atoms of the acetyl group of acetyl CoA are oxidized to carbon dioxide. As you can see from the balanced half-reaction, this is an eight-electron oxidation.

$$CH_3\overset{O}{\overset{\|}{C}}-SCoA + 3H_2O \longrightarrow 2CO_2 + CoA-SH + 8H^+ + 8e^-$$

This oxidation is brought about by three molecules of NAD^+ and one molecule of FAD. Following are balanced half-reactions for the reduction of NAD^+ to NADH and FAD to $FADH_2$.

$$\begin{aligned} 3NAD^+ + 3H^+ + 6e^- &\longrightarrow 3NADH \\ FAD + 2H^+ + 2e^- &\longrightarrow FADH_2 \\ \hline 3NAD^+ + FAD + 5H^+ + 8e^- &\longrightarrow 3NADH + FADH_2 \end{aligned}$$

Adding the balanced half-reactions for the oxidation of the two-carbon acetyl group of acetyl CoA and the reduction of three moles of NAD^+ and one mole of FAD gives the net reaction of the tricarboxylic acid cycle:

$$CH_3\overset{O}{\overset{\|}{C}}SCoA + 3NAD^+ + FAD + 3H_2O \xrightarrow{\text{tricarboxylic acid cycle}}$$

$$2CO_2 + CoA-SH + 3NADH + FADH_2 + 3H^+$$

As we study the individual reactions of the TCA cycle, we shall concentrate on the four reactions that involve oxidations and produce reduced coenzymes, and the two that produce carbon dioxide.

A. Steps in the Tricarboxylic Acid Cycle

1. Formation of Citrate

The two-carbon acetyl group of acetyl coenzyme A enters the TCA cycle by carbonyl condensation between the alpha carbon of acetyl CoA and the carbonyl group of oxaloacetate. The product of this reaction is citrate, the tricarboxylic acid from which the cycle derives one of its names. In this reaction, catalyzed by citrate synthase, carbonyl condensation is coupled with hydrolysis of the thioester to form free coenzyme A:

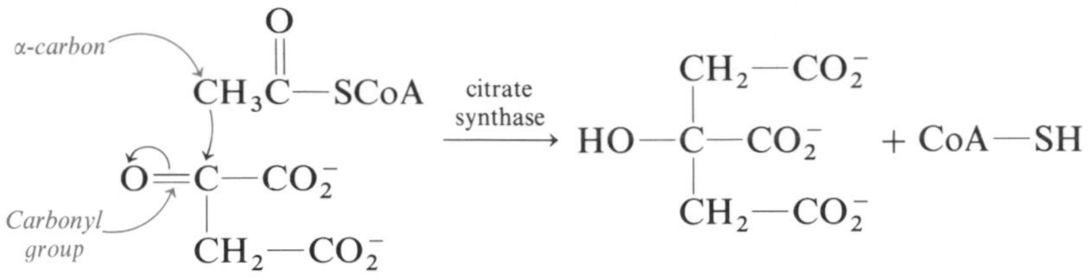

oxaloacetate   citrate

Oxaloacetate is in a sense a starting point of the TCA cycle, and it is also an endpoint. As we shall see, subsequent reactions of the cycle regenerate oxaloacetate, thus providing for entry of further acetyl groups into the cycle.

2. Isomerization of Citrate to Isocitrate

In the second reaction of the cycle, catalyzed by aconitase, citrate is converted to an isomer, isocitrate. It is thought that this isomerization is accomplished in two reactions, both catalyzed by aconitase. First, in a reaction analogous to acid-catalyzed dehydration of an alcohol (Section 4.4A), citrate undergoes enzyme-catalyzed dehydration to aconitate. Then, in a reaction analogous to acid-catalyzed hydration of an alkene (Section 3.5C), aconitate undergoes enzyme-catalyzed hydration to form isocitrate.

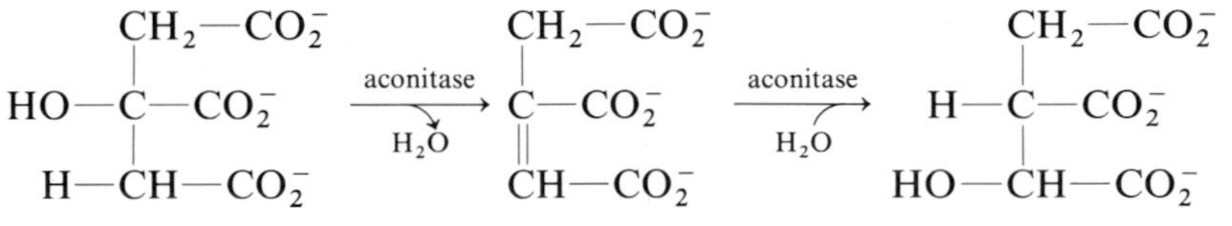

citrate   aconitate   isocitrate

3. Oxidation and Decarboxylation of Isocitrate

In step 3, the secondary alcohol of isocitrate is oxidized to a ketone by NAD^+ in a reaction catalyzed by isocitrate dehydrogenase. The product, oxalosuccinate, is a beta-ketoacid and undergoes decarboxylation (Section 7.5D) to produce alpha-ketoglutarate.

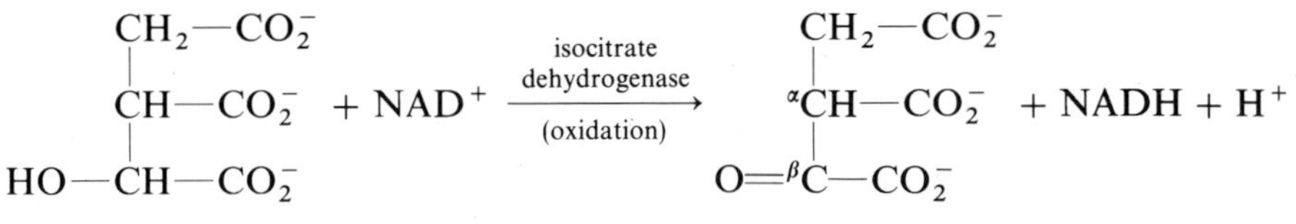

isocitrate   oxalosuccinate
                                                        (a β-ketoacid)

$$\underset{\text{oxalosuccinate}}{\begin{matrix} CH_2-CO_2^- \\ | \\ CH-CO_2^- \\ | \\ O=C-CO_2^- \end{matrix}} + H^+ \xrightarrow[\text{(decarboxylation)}]{\substack{\text{isocitrate} \\ \text{dehydrogenase}}} \underset{\alpha\text{-ketoglutarate}}{\begin{matrix} CH_2-CO_2^- \\ | \\ CH_2 \\ | \\ O=C-CO_2^- \end{matrix}} + CO_2$$

4. Oxidation and Decarboxylation of Alpha-Ketoglutarate

The second molecule of carbon dioxide is generated by the TCA cycle in the same type of oxidative decarboxylation as that for the conversion of pyruvate to acetyl CoA and carbon dioxide (Section 17.4A). In oxidative decarboxylation of alpha-ketoglutarate, the carboxyl group is converted to carbon dioxide and the adjacent ketone is oxidized to a carboxyl group in the form of a thioester with coenzyme A.

$$\underset{\alpha\text{-ketoglutarate}}{\begin{matrix} CH_2-CO_2^- \\ | \\ CH_2 \\ | \\ O=C-CO_2^- \end{matrix}} + NAD^+ + CoA-SH \xrightarrow{\substack{\alpha\text{-ketoglutarate} \\ \text{dehydrogenase}}} \underset{\substack{\text{succinyl coenzyme A} \\ \text{(succinyl CoA)}}}{\begin{matrix} CH_2-CO_2^- \\ | \\ CH_2 \\ | \\ O=C-SCoA \end{matrix}} + NADH + CO_2$$

Then, in coupled reactions catalyzed by succinyl CoA synthetase, succinyl coenzyme A, HPO_4^{2-}, and guanosine diphosphate (GDP) react, to form succinate, guanosine triphosphate (GTP), and coenzyme A:

$$\underset{\text{succinyl CoA}}{\begin{matrix} CH_2-CO_2^- \\ | \\ CH_2-C-SCoA \\ \| \\ O \end{matrix}} + GDP + HPO_4^{2-} \xrightarrow{\substack{\text{succinyl CoA} \\ \text{synthetase}}} \underset{\text{succinate}}{\begin{matrix} CH_2-CO_2^- \\ | \\ CH_2-CO_2^- \end{matrix}} + GTP + CoA-SH$$

The terminal phosphate group of GTP can be transferred to ADP according to the reaction

$$GTP + ADP \rightleftharpoons GDP + ATP$$

Thus, one molecule of a high-energy compound (either GTP or ATP) is produced for each molecule of acetyl CoA entering the tricarboxylic acid cycle. This is the only reaction of the TCA cycle that conserves energy as ATP.

5. Oxidation of Succinate

In the third oxidation of the cycle, catalyzed by succinate dehydrogenase, succinate is oxidized to fumarate. The oxidizing agent is FAD, which is reduced to $FADH_2$. Succinate dehydrogenase is inhibited by such compounds as malonate (Section 14.4F).

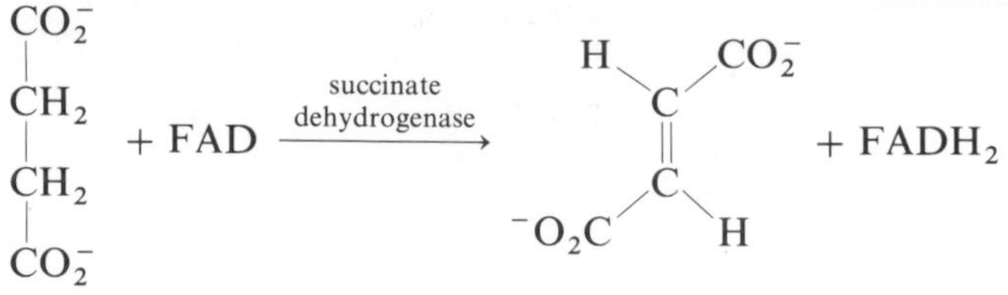

succinate + FAD $\xrightarrow{\text{succinate dehydrogenase}}$ fumarate + FADH$_2$

6. Hydration of Fumarate

In the second hydration of the tricarboxylic acid cycle, fumarate is converted to L-malate in a reaction catalyzed by the enzyme fumarase:

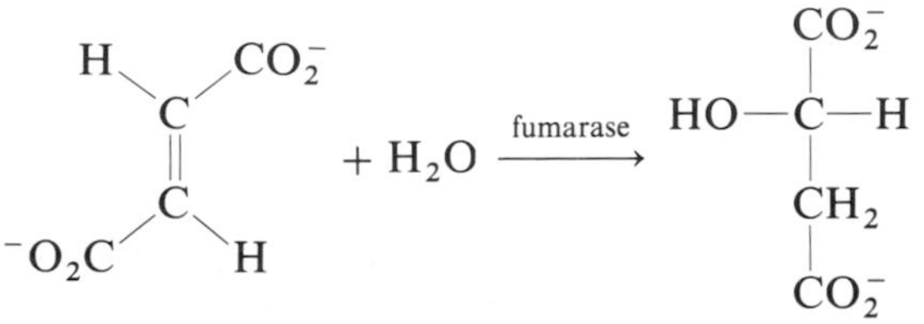

fumarate + H$_2$O $\xrightarrow{\text{fumarase}}$ L-malate

Fumarase shows a high degree of specificity; it recognizes only fumarate (a trans isomer) as a substrate and gives only L-malate (one member of a pair of enantiomers) as the product.

7. Oxidation of L-Malate

In the fourth and final oxidation of the TCA cycle, L-malate is oxidized by NAD$^+$ to oxaloacetate:

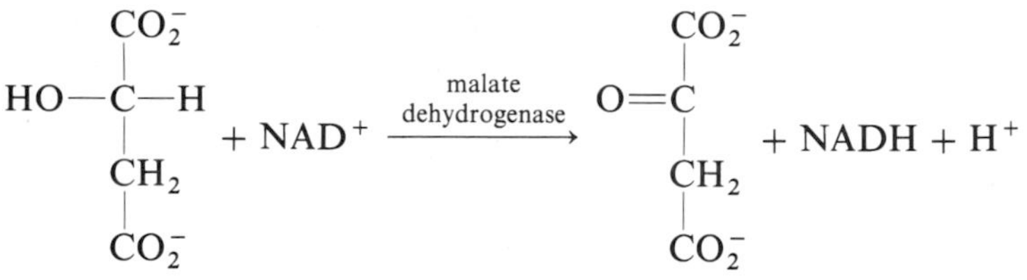

L-malate + NAD$^+$ $\xrightarrow{\text{malate dehydrogenase}}$ oxaloacetate + NADH + H$^+$

With production of oxaloacetate, the reactions of the tricarboxylic acid cycle are complete. Continued operation of the cycle requires two things: (1) a supply of carbon atoms in the form of acetyl groups from acetyl CoA, and (2) a supply of oxidizing agents in the form of NAD$^+$ and FAD. The TCA cycle is linked to glycolysis and the oxidation of pyruvate to get acetyl CoA as a fuel, and it is also linked to the breakdown of fatty acids and amino acids. For a supply of NAD$^+$ and FAD, the cycle depends on reactions of the electron transport system and oxidative phosphorylation. Recall (Section 16.5) that since oxygen is the final acceptor of electrons in the electron transport system, continued operation of the TCA cycle depends ultimately on an adequate supply of oxygen.

Another important feature of the cycle is best seen by returning to the balanced equation for the cycle (Section 17.2C):

$$\underset{\text{O}}{\overset{\text{O}}{\text{CH}_3\overset{\|}{\text{C}}\text{—SCoA}}} + 3\text{NAD}^+ + \text{FAD} + \text{HPO}_4^{2-} + \text{ADP} \xrightarrow{\text{TCA cycle}}$$

$$2\text{CO}_2 + 3\text{NADH} + \text{FADH}_2 + \text{ATP} + \text{CoA—SH}$$

The TCA cycle is truly catalytic; its intermediates do not enter into the balanced equation for this pathway, since they are neither destroyed nor synthesized in the net reaction. The only function of the TCA cycle is to accept acetyl groups from acetyl CoA, oxidize them to carbon dioxide, and at the same time produce a supply of reduced coenzymes as fuel for electron transport and oxidative phosphorylation. In fact, if any of the intermediates of the cycle is removed, then operation of the cycle ceases, because there is no way to regenerate oxaloacetate. As we shall see, however, the cycle is connected through several of its intermediates to other metabolic pathways. In practice, certain intermediates of the cycle can be used for the synthesis of other biomolecules, provided that another intermediate is supplied that in turn can be converted to oxaloacetate, making up for the intermediate withdrawn.

The reactions of the tricarboxylic acid cycle, including those that generate carbon dioxide, reduced coenzymes, and high-energy phosphates, are summarized in Figure 17.6. This figure also shows the central role of this cycle and its linkage to other metabolic pathways.

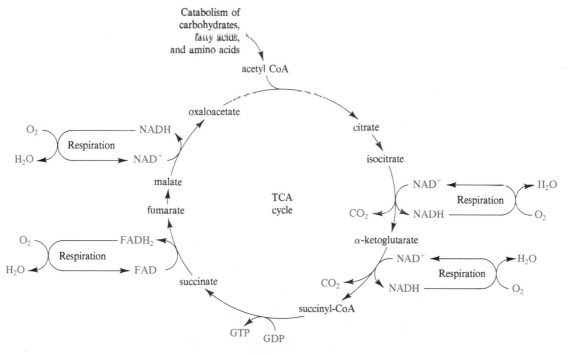

Figure 17.6 The tricarboxylic acid cycle. Fuel for the cycle is derived from catabolism of carbohydrates, fatty acids, and amino acids. For a continuing supply of NAD$^+$ and FAD, the TCA cycle depends on respiration and oxygen.

B. Control of the Tricarboxylic Acid Cycle

The chief points of control of the TCA cycle are two regulatory enzymes.

Enzyme Name	How Regulated
citrate synthetase	inhibited by ATP and NADH
isocitrate dehydrogenase	inhibited by ATP and NADH, activated by ADP

To appreciate the importance of these regulatory enzymes, remember that the main function of the TCA cycle is to provide fuel in the form of reduced coenzymes for electron transport and oxidative phosphorylation (Section 16.5). Under conditions where supplies of ATP are adequate for the immediate needs of cells, ATP interacts with citrate synthetase to reduce its affinity for acetyl coenzyme A. Thus, ATP acts as a negative modifier of citrate synthetase and inhibits entry of acetyl CoA into the TCA cycle. In this case, acetyl CoA is channeled into the synthesis of fatty acids and triglycerides (Section 18.5). Similarly, NADH, a product of the TCA cycle, acts as a negative modifier of citrate synthetase. Isocitrate dehydrogenase, the primary control point in the cycle, is also inhibited by ATP and NADH. Because the activity of this enzyme is increased by interaction with ADP, this substance is said to be a positive modifier of isocitrate dehydrogenase.

17.6 Energy Balance for Oxidation of Glucose to Carbon Dioxide and Water

Now that we have examined the biochemical pathways by which the carbon atoms of glucose are oxidized to carbon dioxide, let us look at the energy changes in these transformations. Complete oxidation of glucose to carbon dioxide occurs with a large decrease in free energy:

$$C_6H_{12}O_6 + 6O_2 \longrightarrow 6CO_2 + 6H_2O \qquad \Delta G^0 = -686,000 \text{ cal/mol}$$

The number of moles of ATP derived from aerobic catabolism of glucose are summarized in Table 17.1. Note that glycolysis takes place in the cytoplasm. For glucose catabolized in skeletal muscle, one ATP is required to transport each NADH produced by glycolysis from the cytosol to the mitochondrion, site of the TCA cycle, thus yielding 36 mol ATP per mole of glucose. In the liver and heart muscle, however, NADH is transported from the cytoplasm to the mitochondria by a different mechanism, which does not require expenditure of ATP. Thus, in the liver and heart muscle, 38 mol ATP is produced per mole of glucose.

We can write the net reaction and the associated energy changes for the complete oxidation of glucose as the sum of an exergonic oxidation of glucose to carbon dioxide and water and an endergonic phosphorylation of 36 moles of ADP.

Table 17.1 Yield of ATP from the complete oxidation of glucose to carbon dioxide and water in skeletal muscle.

Reaction	Process	Yield of ATP (moles)
Glycolysis		
glucose $\longrightarrow$ glucose 6-phosphate	phosphorylation	-1
fructose 6-phosphate $\longrightarrow$ fructose 1,6-diphosphate	phosphorylation	-1
glyceraldehyde 3-phosphate $\longrightarrow$ 1,3-diphosphoglycerate	oxidation by NAD^+	$+6$
transport of NADH from cytosol to mitochondrion	active transport	-2
1,3-diphosphoglycerate $\longrightarrow$ 3-phosphoglycerate	phosphorylation	$+2$
phosphoenolpyruvate $\longrightarrow$ pyruvate	phosphorylation	$+2$
Oxidation of pyruvate		
pyruvate $\longrightarrow$ acetyl CoA + CO_2	oxidation by NAD^+	$+6$
Tricarboxylic acid cycle		
isocitrate $\longrightarrow$ α-ketoglutarate + CO_2	oxidation by NAD^+	$+6$
α-ketoglutarate $\longrightarrow$ succinyl CoA + CO_2	oxidation by NAD^+	$+6$
succinyl CoA $\longrightarrow$ succinate	phosphorylation	$+2$
succinate $\longrightarrow$ fumarate	oxidation by FAD	$+4$
malate $\longrightarrow$ oxaloacetate	oxidation by NAD^+	$+6$
Net yield of ATP (skeletal muscle) per mole of glucose		$+36$

Exergonic Reaction

$$C_6H_{12}O_6 + 6O_2 \longrightarrow 6CO_2 + 6H_2O \qquad \Delta G^0 = -686,000 \text{ cal/mol}$$

Endergonic Reaction

$$36ADP + 36HPO_4^{2-} \longrightarrow 36ATP + 36H_2O \qquad \Delta G^0 = +263,000 \text{ cal/mol}$$

Overall Reaction

$$C_6H_{12}O_6 + 6O_2 + 36ADP + 36HPO_4^{2-} \longrightarrow$$
$$6CO_2 + 36ATP + 42H_2O \qquad \Delta G^0 = -423,000 \text{ cal/mol}$$

Can be used for biochemical work   *Liberated as heat*

The total energy conserved as a result of aerobic oxidation of 1 mol of glucose is 36 mol of ATP, or 263,000 cal/mol. The efficiency of energy conservation during glucose metabolism is

$$\frac{263,000}{686,000} \times 100 = 38\% \text{ free energy conserved as ATP}$$

It is an impressive feat for living cells to trap this amount of energy as ATP!

The decrease in free energy for lactate fermentation (glucose $\longrightarrow$ lactate) is 47,000 cal/mol, a much smaller value than that for aerobic oxidation of glucose to carbon dioxide. We can write the net reaction and associated energy changes for lactate fermentation as the sum of the exergonic conversion of glucose to lactate and the endergonic phosphorylation of ADP.

Exergonic Reaction

$$\text{Glucose} \longrightarrow 2 \text{ lactate} \qquad \Delta G^0 = -47{,}000 \text{ cal/mol}$$

Endergonic Reaction

$$2\text{ADP} + 2\text{HPO}_4^{2-} \longrightarrow 2\text{ATP} + 2\text{H}_2\text{O} \qquad \Delta G^0 = +14{,}600 \text{ cal/mol}$$

Overall Reaction

$$\text{Glucose} + 2\text{ADP} + 2\text{HPO}_4^{2-} \longrightarrow 2 \text{ lactate} + 2\text{ATP} + 2\text{H}_2\text{O} \qquad \Delta G^0 = -32{,}400 \text{ cal/mol}$$

Thus, the total energy conserved through lactate fermentation of 1 mol of glucose is 2 mol of ATP or 14,600 cal/mol.

$$\frac{14{,}600}{686{,}000} \times 100 = 2\% \text{ free energy conserved as ATP}$$

Lactate fermentation keeps glycolysis going, but at a cost. Only 2 mol of ATP are produced per mole of glucose by lactate fermentation, compared to 36 mol of ATP produced by complete oxidation of glucose to carbon dioxide and water. The same is true for alcoholic fermentation. Thus, in the use of fuel molecules for producing heat and ATP, aerobic catabolism of glucose is 18 times more efficient than either lactate or alcoholic fermentation.

Key Terms and Concepts

alcoholic fermentation (17.4C)

blood glucose levels (17.1B)

Cori cycle (17.4B)

digestion of carbohydrates (17.1A)

Embden-Meyerhoff pathway (17.3A)

fructose intolerance (17.3B)

galactosemia (17.3B)

gluconeogenesis (17.2H)

glucose 6-phosphate dehydrogenase deficiency (17.2E)

glucose tolerance test (17.1C)

glycogenesis (17.2F)

glycogenolysis (17.2F)

glycolysis (17.3)

hyperglycemia (17.1B)

hypoglycemia (17.1B)

Krebs cycle (17.5)

lactate fermentation (17.4B)

oxidative phosphorylation (17.2D)

pentose phosphate pathway (17.2E)

tricarboxylic acid cycle (17.5)

Key Reactions

1. Hydrolysis of disaccharides and polysaccharides to monosaccharides (Section 17.1A).
2. Net reaction of glycolysis (Section 17.2A).
3. Net reaction for oxidative decarboxylation of pyruvate (Section 17.2B).
4. Net reaction for the tricarboxylic acid cycle (Section 17.2C).
5. Net reaction for the pentose phosphate pathway (Section 17.2E).
6. Net reaction for glycogenolysis (Section 17.2F).
7. Net reaction for glycogenesis (Section 17.2F).
8. The 10 steps in glycolysis (Section 17.3A).
9. Entry of fructose and galactose into glycolysis (Section 17.3B).
10. Oxidative decarboxylation of pyruvate to acetyl CoA during aerobic respiration (Section 17.4A).
11. Reduction of pyruvate to lactate during lactate fermentation (Section 17.4B).
12. Reduction of pyruvate to ethanol during alcoholic fermentation (Section 17.4C).
13. The 7 steps of the tricarboxylic acid cycle (Section 17.5A).

Problems

Central role of glucose in carbohydrate metabolism (Section 17.2)

17.4 Match the names with the processes listed.
    a. glycolysis   **b.** gluconeogenesis
    c. glycogenolysis   **d.** glycogenesis
    _____ synthesis of glucose from noncarbohydrate molecules
    _____ breakdown of glucose to pyruvate
    _____ hydrolysis of glycogen to glucose
    _____ conversion of glucose to glycogen

17.5 How many moles of ATP are produced either directly or by oxidation of reduced coenzymes and phosphorylation of ADP when:
    a. 2 mol glucose is oxidized to CO_2?
    b. 2 mol glucose is oxidized to pyruvate?
    c. 2 mol glucose 6-phosphate is oxidized to 2 mol ribose 5-phosphate?
    d. 2 mol acetyl CoA is oxidized to fumarate?

17.6 When liver stores of glycogen are very low (in the morning or during vigorous exercise) and blood glucose levels are low, how can glucose be produced and energy supplied?

17.7 Name two important functions of the pentose phosphate pathway.

17.8 What is the difference in structural formula between NAD^+ and $NADP^+$?

17.9 The degradation of carbohydrates provides the cell with three things: energy; NADPH as a reducing agent for biosynthesis; and a pool of

intermediates for the biosynthesis of other molecules. Which of these three are produced by the following pathways?
a. conversion of glucose to lactate
b. tricarboxylic acid cycle
c. pentose phosphate pathway
d. conversion of glucose to ethanol

Glycolysis
(Section 17.3)

17.10   Name one coenzyme required for glycolysis. From what vitamin is this coenzyme derived?

17.11   Number the carbon atoms of glucose 1 through 6 and show the fate of each atom in glycolysis.

17.12   Write equations for the two reactions of glycolysis that consume ATP.

17.13   Write equations for the two reactions of glycolysis that produce ATP.

17.14   Although glucose is the principal source of carbohydrates for glycolysis and other pathways, fructose and galactose are also metabolized for energy.
a. What is the main dietary source of fructose? of galactose?
b. Explain how the carbon skeleton of fructose enters glycolysis.
c. Explain how the carbon skeleton of galactose enters glycolysis.

17.15   Describe the genetic defect leading to (a) fructose intolerance; (b) galactosemia.

17.16   The feedback effects of ATP, ADP, and AMP are important in regulating both glycolysis and the tricarboxylic acid cycle. Explain the effect of ATP and ADP on (a) isocitrate dehydrogenase, a regulatory enzyme of the tricarboxylic acid cycle; (b) phosphofructokinase, a regulatory enzyme of glycolysis.

Fates of
pyruvate
(Section 17.4)

17.17   Number the carbon atoms of glucose 1 through 6 and show the fate of each in (a) alcoholic fermentation; (b) lactate fermentation.

17.18   In what ways are alcoholic fermentation and lactate fermentation similar? In what ways do they differ?

17.19   Write balanced half-reactions for the following conversions:
a. glucose $\longrightarrow$ pyruvate
b. glucose $\longrightarrow$ lactate
c. lactate $\longrightarrow$ pyruvate
d. pyruvate $\longrightarrow$ ethanol + carbon dioxide

17.20   What is the principal function of the Cori cycle?

17.21   From your knowledge of glycolysis, the fates of pyruvate, and the tricarboxylic acid cycle, propose a series of steps for the following biochemical conversions.
a. glycerol $\longrightarrow$ lactate
b. 3-phosphoglycerol $\longrightarrow$ ethanol + carbon dioxide
c. 3-phosphoglyceraldehyde $\longrightarrow$ glucose 6-phosphate
d. glycerol $\longrightarrow$ acetyl CoA
e. ethanol $\longrightarrow$ carbon dioxide

The
tricarboxylic
acid cycle
(Section 17.5)

17.22   What is the main function of the TCA cycle?

17.23   Write equations for the step or steps in the TCA cycle that involve:

a. formation of a new carbon-carbon bond.
b. oxidation by NAD^+.
c. oxidation by FAD.
d. formation of a high-energy phosphate bond.

17.24 Why is GTP just as effective a high-energy compound as ATP?

17.25 What does it mean to say that the TCA cycle is catalytic? that it does not produce any new compounds?

17.26 The main control points of the TCA cycle are the regulatory enzymes, citrate synthetase, and isocitrate dehydrogenase.
a. Write an equation for the reaction catalyzed by each enzyme.
b. Each enzyme is inhibited by NADH and ATP. Explain the benefit to the cell of this means of regulation.

Energy balance for glucose metabolism (Section 17.6)

17.27 A maximum of 36 mol ATP can be formed as the result of complete metabolism of 1 mol glucose to carbon dioxide and water. How many of the 36 moles are formed in:
a. glycolysis?
b. the tricarboxylic acid cycle?
c. the electron transport system?

17.28 The total amount of energy that can be obtained from complete oxidation of glucose is 686,000 cal/mol. What fraction of this energy is conserved as ATP in alcoholic fermentation? (Note that although this fraction is small, it is enough for the survival of anaerobic cells.)

18

Metabolism of Fatty Acids

In this chapter, we shall discuss the metabolic pathways for the catabolism of fatty acids and show how these pathways are coupled with the generation of ATP. In addition, we shall discuss the biosynthesis of fatty acids and then compare and contrast the steps by which cells degrade and synthesize these vital molecules. Finally we will show some of the interrelationships between the metabolism of fatty acids and carbohydrates.

18.1 Fatty Acids as Sources of Energy

For available energy, fatty acids have the highest caloric value of any food. Following are balanced equations for the complete oxidation of glucose and palmitic acid, one of the most abundant fatty acids. As you can see by comparing the changes in free energy, complete oxidation of a gram of palmitic acid yields almost 2.5 times the energy obtained from a gram of glucose.

	ΔG^0 (cal/mol)	ΔG^0 (cal/g)
$C_6H_{12}O_6 + 6O_2 \longrightarrow 6CO_2 + 6H_2O$	$-686,000$	$-3,800$
glucose		
$CH_3(CH_2)_{14}CO_2H + 23O_2 \longrightarrow 16CO_2 + 16H_2O$	$-2,340,000$	$-9,300$
palmitic acid		

The yield of energy per gram is larger because the hydrocarbon chain of a fatty acid is more highly reduced than the oxygenated chain of a carbohydrate. This can be seen by comparing the number of moles of oxygen consumed per carbon atom. One mole of oxygen is consumed per carbon atom of glucose, while $\frac{23}{16}$, or 1.44, moles of oxygen are consumed per carbon atom of palmitic acid.

Fatty acids constitute about 40% of the calories in a typical American diet. Further, because they can be stored in large quantities, fatty acids as triglycerides are the most important storage form of energy. Adipose tissue contains specialized cells, called adipocytes, whose sole function is to store fats.

18.2 Hydrolysis of Triglycerides

The first phase of catabolism of fatty acids involves their release from triglycerides by **hydrolysis**, catalyzed by a group of enzymes called lipases:

$$
\underset{\text{a triglyceride}}{
\begin{array}{c}
\quad\quad\quad\overset{\displaystyle O}{\overset{\|}{C}} \\
\overset{O}{\overset{\|}{RC}}-O\overset{\displaystyle CH_2O-CR}{\underset{\displaystyle CH_2O-CR}{CH}} \\
\quad\quad\quad\underset{\displaystyle O}{\underset{\|}{}}
\end{array}}
\;+\;3H_2O\;\xrightarrow{\text{lipase}}\;
\underset{\text{glycerol}}{
\begin{array}{c}
CH_2OH \\
CHOH \\
CH_2OH
\end{array}}
\;+\;\underset{\text{fatty acids}}{3R\overset{O}{\overset{\|}{C}}O^-}\;+\;3H^+
$$

Fatty acids cannot be transported as such in the bloodstream since they are insoluble in water. Instead they are transported in combination with albumin, the most abundant of the serum proteins (Section 13.8).

Release of fatty acids from adipose tissue into the bloodstream is stimulated by several hormones, including epinephrine, adrenocorticotropic hormone, growth hormone, and thyroxine. Accumulation of fatty acids in adipose tissue and storage as triglycerides is stimulated by high levels of glucose and insulin in the bloodstream.

18.3 Essential Fatty Acids

Somewhat more than 50% of the fatty acids in human triglycerides are unsaturated. These unsaturated fatty acids have several vital functions. Their presence lowers the melting points of triglycerides (Section 12.1C) and ensures that triglyceride droplets stored in the body remain liquid at body temperature. For comparison, tripalmitin, a saturated triglyceride, has a melting point of 65°C. Similarly, the fluidity of biological membranes depends on the presence of unsaturated fatty acids in membrane phospholipids. Further, prostaglandins and related compounds are synthesized from unsaturated fatty acids, chiefly arachidonic acid (Table 12.1). (For discussion of the biosyntheses of several prostaglandins and their functions in the body, see Mini-Essay VI, "The Prostaglandins.")

It was discovered in 1929 that if fatty acids (in the form of fats) were withheld from the diet of rats, they soon began to suffer from retarded growth, scaly skin, kidney damage, and premature death, even though they were fed adequate supplies of energy in the form of carbohydrates and proteins. Adding unsaturated fatty acids (specifically linoleic, linolenic, and arachidonic acids) to their diet prevented these conditions in control animals. In humans, signs of unsaturated-fatty-acid deficiency include scaly and thickened skin and decrease in growth rate. Because of the natural occurrence of fatty acids in normal diets, deficiency diseases are generally seen only in infants fed on a formula that lacks these unsaturated fatty acids, and in hospital patients who are fed intravenously for prolonged periods.

Humans and other higher animals produce the enzymes necessary to catalyze the synthesis of saturated fatty acids and of oleic acid, an unsaturated fatty acid with a double bond between carbons 9 and 10. However, they cannot synthesize linoleic acid, an unsaturated fatty acid with double bonds between carbons 9 and 10, and 12 and 13. Because this unsaturated fatty acid must be obtained from the diet to have normal growth and well-being, it is classified as an **essential fatty acid**.

$$CH_3(CH_2)_7\overset{10}{C}H{=}\overset{9}{C}H(CH_2)_7CO_2H$$

oleic acid
(can be synthesized in the
liver by humans and higher animals)

$$CH_3(CH_2)_4\overset{13}{C}H{=}\overset{12}{C}HCH_2\overset{10}{C}H{=}\overset{9}{C}H(CH_2)_7CO_2H$$

linoleic acid
(cannot be synthesized by
humans and higher animals)

Linolenic and arachidonic acids are also required for normal growth and development. However, these unsaturated fatty acids can be synthesized in the liver from linoleic acid, and therefore are not classified as essential fatty acids.

$$CH_3CH_2(CH{=}CHCH_2)_3(CH_2)_6CO_2H$$

linolenic acid

$$CH_3(CH_2)_4(CH{=}CHCH_2)_4(CH_2)_2CO_2H$$

arachidonic acid
(prostaglandins are synthesized)
from this unsaturated fatty acid)

Although no minimum daily requirement for linoleic acid and the other polyunsaturated fatty acids has been established, the Food and Nutrition Board suggests an intake of about 6 g per day for adults. For infants and premature babies, the requirements are higher.

Linoleic acid and other unsaturated fatty acids are especially abundant in plant oils. As little as one teaspoon of corn oil a day supplies all necessary unsaturated fatty acids.

18.4 Oxidation of Fatty Acids

The three major stages in **oxidation of fatty acids** are activation of free fatty acids in the cytoplasm by formation of a thioester with coenzyme A; transport across the inner mitochondrial membrane; and oxidation within mitochondria to carbon dioxide and water. Let us look at each stage separately.

A. Activation of Free Fatty Acids

As the first step in catabolism, a free fatty acid in the cytoplasm is converted to a thioester formed with coenzyme A. The product is called a fatty acyl CoA. Thioester formation is an endergonic reaction, and the energy to drive it is derived by coupling thioester formation with hydrolysis of ATP. In this coupled hydrolysis, ATP is converted to AMP and pyrophosphate. Within the cytoplasm, pyrophosphate is further hydrolyzed to two phosphate ions.

ΔG^0
(cal/mol)

Endergonic Reaction

$$\underset{\displaystyle \text{RCO}^-}{\overset{\displaystyle O}{\overset{\displaystyle \|}{}}} + \text{HS—CoA} \longrightarrow \underset{\displaystyle \text{RC—SCoA}}{\overset{\displaystyle O}{\overset{\displaystyle \|}{}}} + \text{OH}^- \qquad +7{,}300$$

Exergonic Reactions

$$\text{ATP}^{4-} + \text{H}_2\text{O} \longrightarrow \text{AMP}^{2-} + {}^-\text{O—}\underset{\displaystyle \text{O}^-}{\overset{\displaystyle O}{\overset{\displaystyle \|}{\text{P}}}}\text{—O—}\underset{\displaystyle \text{O}^-}{\overset{\displaystyle O}{\overset{\displaystyle \|}{\text{P}}}}\text{—O}^- + 2\text{H}^+ \qquad -7{,}600$$

pyrophosphate

$${}^-\text{O—}\underset{\displaystyle \text{O}^-}{\overset{\displaystyle O}{\overset{\displaystyle \|}{\text{P}}}}\text{—O—}\underset{\displaystyle \text{O}^-}{\overset{\displaystyle O}{\overset{\displaystyle \|}{\text{P}}}}\text{—O} + \text{H}_2\text{O} \longrightarrow 2\text{HPO}_4^{2-} \qquad -8{,}300$$

pyrophosphate

Adding the one endergonic reaction and two exergonic reactions gives the overall equation for activation of a fatty acid:

$$\underset{\displaystyle \text{RCO}^-}{\overset{\displaystyle O}{\overset{\displaystyle \|}{}}} + \text{HS—CoA} + \text{ATP}^{4-} + 2\text{H}_2\text{O} \longrightarrow$$

$$\underset{\displaystyle \text{RC—SCoA}}{\overset{\displaystyle O}{\overset{\displaystyle \|}{}}} + 2\text{HPO}_4^{2-} + \text{AMP}^{2-} + \text{H}^+ \qquad \Delta G^0 = -8{,}600 \text{ cal/mol}$$

Because activation of fatty acids is coupled with hydrolysis of two high-energy phosphate anhydride bonds, the initial investment by a cell in fatty acid oxidation is equivalent to 2 moles of ATP for each mole of fatty acid oxidized.

B. Transport of Activated Fatty Acids into Mitochondria

Mitochondrial membranes do not contain a system for transporting fatty acid thioesters of coenzyme A. They do, however, contain a system for transporting fatty acids in the form of esters with the molecule carnitine. Functional groups in carnitine are a carboxylate group, a secondary alcohol, and a quaternary ammonium ion. Fatty acyl CoA and carnitine undergo a reaction in which the fatty acyl group is transferred from the sulfur atom of coenzyme A to the oxygen atom of the secondary alcohol of carnitine. This reaction is an example of exchange of ester groups between molecules.

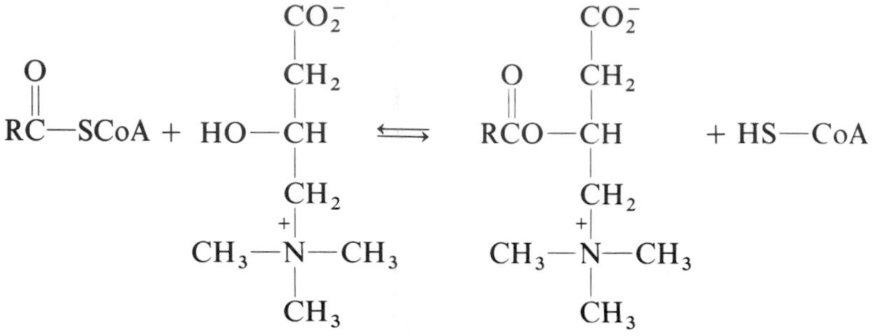

$$
\begin{array}{ccc}
\underset{\substack{\parallel \\ O}}{RC}\!-\!SCoA + HO\!-\!CH & \rightleftharpoons & RCO\!-\!CH + HS\!-\!CoA
\end{array}
$$

carnitine   fatty acid ester
of carnitine

The fatty acid ester of carnitine is transported through the inner mitochondrial membrane, and there the reaction is reversed; the fatty acid ester and a molecule of coenzyme A react, to form a fatty acyl CoA and regenerate carnitine. The freed carnitine is then returned to the cytoplasm to repeat the cycle. The effect of these two reactions, one in the cytoplasm and the other in the mitochondria, is to transfer fatty acyl CoA from the cytoplasm into mitochondria.

C. Beta-Oxidation Spiral

Once an activated fatty acid is in a mitochondrion, its carbon chain is degraded two carbons at a time. The metabolic pathway by which this is accomplished is called beta-oxidation because in two separate steps, a beta carbon is oxidized.

1. Oxidation

In the first step of beta-oxidation, the carbon chain is oxidized and a double bond is formed between the alpha and beta carbons (carbons 2 and 3) of the hydrocarbon chain. The oxidizing agent, FAD, is reduced to $FADH_2$, which is subsequently oxidized in the respiratory chain (Section 16.5):

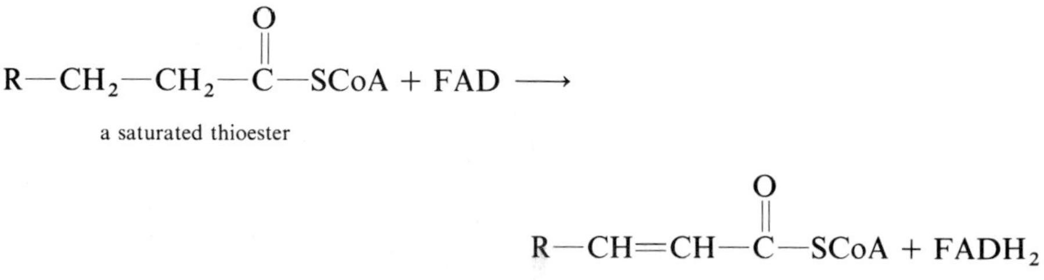

$$
R\!-\!CH_2\!-\!CH_2\!-\!\underset{\substack{\parallel \\ O}}{C}\!-\!SCoA + FAD \longrightarrow
$$

a saturated thioester

$$
R\!-\!CH\!=\!CH\!-\!\underset{\substack{\parallel \\ O}}{C}\!-\!SCoA + FADH_2
$$

an α,β-unsaturated thioester

2. *Hydration*

Next, in a reaction analogous to acid-catalyzed hydration of an alkene (Section 3.5C), water is added to the carbon-carbon double bond, to form a beta-hydroxythioester:

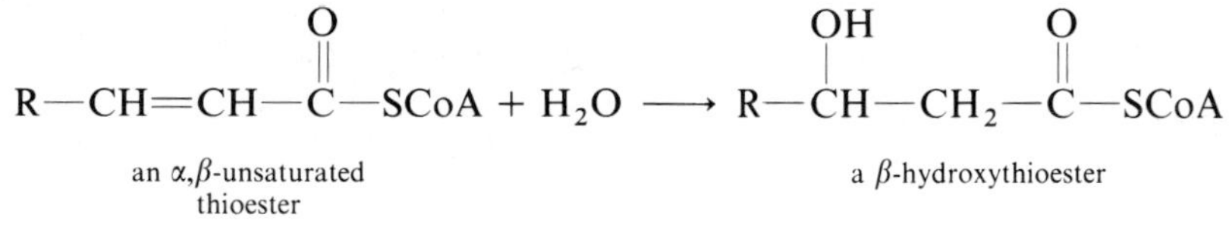

$$R—CH=CH—\overset{\overset{\displaystyle O}{\|}}{C}—SCoA + H_2O \longrightarrow R—\overset{\overset{\displaystyle OH}{|}}{CH}—CH_2—\overset{\overset{\displaystyle O}{\|}}{C}—SCoA$$

an α,β-unsaturated   a β-hydroxythioester
thioester

The effect of the first two reactions of beta-oxidation is conversion of a —CH$_2$— group on carbon 3 of the hydrocarbon chain to a —CHOH— group.

3. *Oxidation*

In the second oxidation of beta-oxidation, the secondary alcohol of the β-hydroxythioester is oxidized to a ketone. The oxidizing agent is NAD$^+$, which is reduced to NADH:

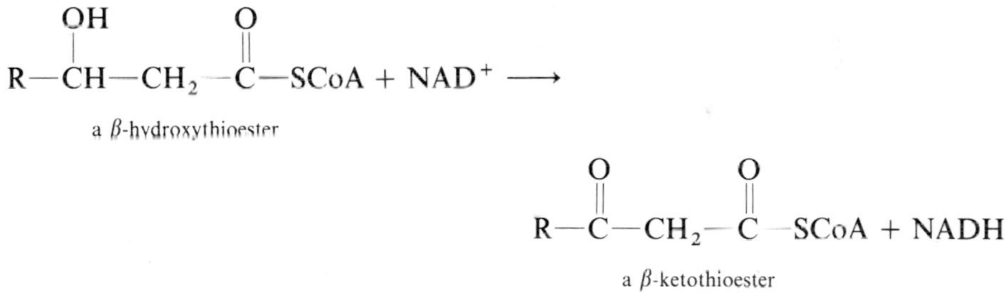

$$R—\overset{\overset{\displaystyle OH}{|}}{CH}—CH_2—\overset{\overset{\displaystyle O}{\|}}{C}—SCoA + NAD^+ \longrightarrow$$

a β-hydroxythioester

$$R—\overset{\overset{\displaystyle O}{\|}}{C}—CH_2—\overset{\overset{\displaystyle O}{\|}}{C}—SCoA + NADH$$

a β-ketothioester

4. *Cleavage of Acetyl Coenzyme A*

In the final step of beta-oxidation, reaction of the beta-ketothioester with a molecule of coenzyme A brings about cleavage of a carbon-carbon bond and gives a molecule of acetyl CoA and a fatty acyl CoA molecule now shortened by two carbon atoms:

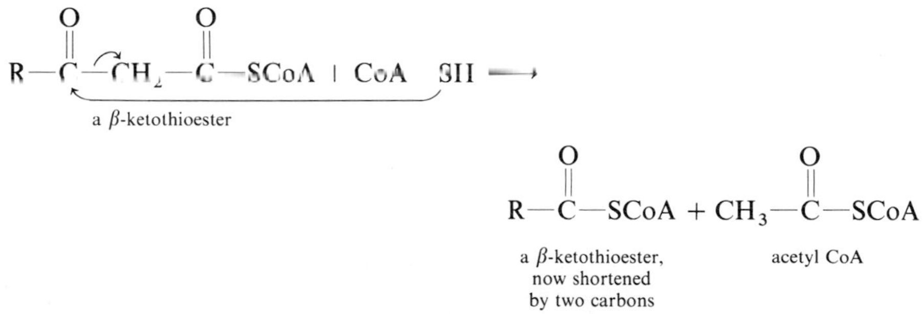

$$R—\overset{\overset{\displaystyle O}{\|}}{C}—CH_2—\overset{\overset{\displaystyle O}{\|}}{C}—SCoA + CoA—SH \longrightarrow$$

a β-ketothioester

$$R—\overset{\overset{\displaystyle O}{\|}}{C}—SCoA + CH_3—\overset{\overset{\displaystyle O}{\|}}{C}—SCoA$$

a β-ketothioester,   acetyl CoA
now shortened
by two carbons

The same series of steps is then repeated on the shortened fatty acyl chain and another molecule of acetyl CoA is cleaved. This series of steps is continued until the entire chain is degraded to acetyl CoA. The steps of beta-oxidation are called a **spiral**, because after each series of four reactions, the carbon chain is shortened by two carbon atoms. Figure 18.1 illustrates beta-oxidation of the carbon chain of palmitic acid to eight molecules of acetyl CoA.

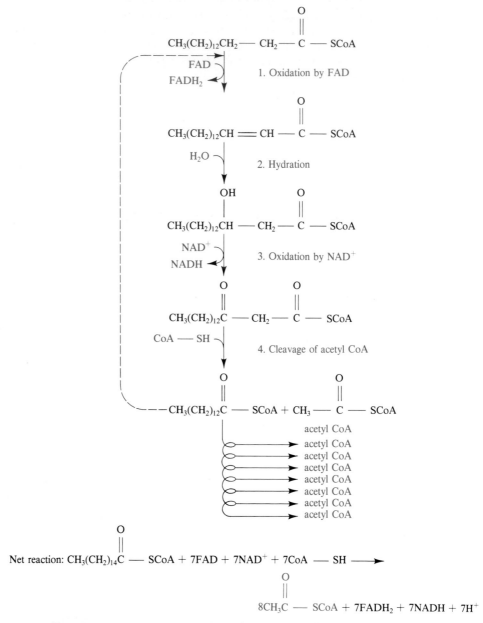

Figure 18.1 Beta-oxidation of palmitic acid to eight molecules of acetyl CoA.

D. Energetics of Fatty Acid Oxidation

Now that we have examined the steps of beta-oxidation, let us calculate how much of the free energy available from complete oxidation of a fatty acid to carbon dioxide and water is conserved as ATP. Let us take palmitic acid as a specific example. Seven turns of beta-oxidation converts 1 mole of palmitic acid to 8 moles of acetyl CoA and generates 7 moles of $FADH_2$ and 7 moles of NADH. Reoxidation of each $FADH_2$ is coupled with formation of 2 ATP and reoxidation of each NADH is coupled with formation of 3 ATP (Section 16.5). Furthermore, oxidation of each acetyl CoA in the tricarboxylic acid cycle, followed by oxidation of all reduced coenzymes, generates another 12 ATP. Because 2 phosphate anhydride bonds (equivalent to 2 moles of ATP) are required to activate each mole of palmitic acid, 2 ATP must be subtracted. The ATP balance for the oxidation of 1 mole of palmitic acid is

$$
\begin{array}{ll}
14\ \text{ATP} & \text{from oxidation of 7 } FADH_2 \\
21\ \text{ATP} & \text{from oxidation of 7 NADH} \\
96\ \text{ATP} & \text{from oxidation of 8 acetyl CoA} \\
-2\ \text{ATP} & \text{from activation of palmitic acid} \\
\hline
129\ \text{ATP} &
\end{array}
$$

Coupling the exergonic oxidation of palmitic acid with the endergonic phosphorylation of ADP gives:

Exergonic Reaction

$$CH_3(CH_2)_{14}CO_2H + 23O_2 \longrightarrow 16CO_2 + 16H_2O \qquad \Delta G^0 = -2{,}340\ \text{kcal/mol}$$

Endergonic Reaction

$$129ADP + 129HPO_4^{2-} \longrightarrow 129ATP + 129H_2O \qquad \Delta G^0 = +940\ \text{kcal/mol}$$

Net Reaction

$$CH_3(CH_2)_{14}CO_2H + 23O_2 + 129ADP + 129HPO_4^{2-} \longrightarrow$$
$$16CO_2 + 145H_2O + 129ATP \qquad \Delta G^\circ = -1{,}400\ \text{kcal/mol}$$

Thus we see that some $\frac{940}{2340}$, or 40%, of the standard free energy of oxidation of palmitate is conserved as ATP and can be used by cells for doing work. This fraction of energy conserved as ATP is comparable to that conserved in the complete oxidation of glucose to carbon dioxide and water (Section 17.6).

Example 18.1

Beta-oxidation of stearic acid, $CH_3(CH_2)_{16}CO_2H$, produces 9 moles of acetyl CoA, 8 moles of NADH, and 8 moles of $FADH_2$. Calculate the number of moles of ATP produced:

a. by oxidative phosphorylation of 8 moles of NADH.
b. by oxidative phosphorylation of 8 moles of $FADH_2$.
c. by oxidation of 9 moles of acetyl CoA through the reactions of the tricarboxylic acid cycle.
d. by oxidative phosphorylation of the NADH and $FADH_2$ produced during the oxidation of 9 moles of acetyl CoA in the tricarboxylic acid cycle.
e. in parts a–d combined.

□ *Solution*

a. Three moles of ATP are produced by oxidative phosphorylation of each mole of NADH. Therefore, oxidative phosphorylation of 8 moles of NADH gives 24 moles of ATP.

b. Two moles of ATP are produced per mole of $FADH_2$. Therefore, oxidative phosphorylation of 8 moles of $FADH_2$ gives 16 moles of ATP.

c. The reactions of the tricarboxylic acid cycle produce 1 mole of GTP (which can be converted to ATP) per mole of acetyl CoA entering the cycle. Therefore, oxidation of 9 moles of acetyl CoA in the TCA cycle gives 9 moles of ATP.

d. Each turn of the TCA cycle gives 3 moles of NADH and 1 mole of $FADH_2$ per mole of acetyl CoA entering the cycle. Therefore, 9 moles of acetyl CoA give 27 moles of NADH and 9 moles of $FADH_2$. Oxidative phosphorylation of these reduced coenzymes gives $81 + 18 = 99$ moles of ATP.

e. Complete oxidation of 1 mole of stearic acid and oxidative phosphorylation of the resulting NADH and $FADH_2$ give a total of $24 + 16 + 9 + 99 = 148$ moles of ATP. Because two high-energy phosphate anhydride bonds are required in the activation of a fatty acid for beta-oxidation, the net yield per mole of stearic acid is 146 moles of ATP. Note that almost 65% of this ATP is produced by oxidative phosphorylation of NADH and $FADH_2$ generated through the tricarboxylic acid cycle.

■
Problem 18.1

The structural formula of myristic acid is

$$CH_3(CH_2)_{12}CO_2H$$

myristic acid

In the complete beta-oxidation of myristic acid to acetyl CoA:
a. How many moles of ATP are required?
b. How many moles of NADH and $FADH_2$ are produced?
c. How many moles of acetyl CoA are produced?

■

E. Oxidation of Propanoate

Thus far we have dealt with beta-oxidation of fatty acids with an even number of carbon atoms. These acids are degraded completely to acetyl CoA. While fatty acids with odd numbers of carbon atoms are not nearly so common, they do occur in nature. The final beta-oxidation product of an odd-numbered carbon chain is the thioester of propanoic acid. The IUPAC name of this ester is propanoyl CoA. Its common name is propionyl CoA (derived from the common name propionic acid). In normal human metabolism, the most important source of propanoyl CoA is not odd-chain fatty acids but the carbon skeletons of the branched-chain amino acids isoleucine and valine.

$$CH_3CH_2\overset{\overset{\displaystyle O}{\|}}{C}{-}SCoA$$

propanoyl CoA
(propionyl CoA)

Propanoyl CoA is converted to succinyl CoA, an intermediate in the TCA cycle. In the first reaction of this conversion, propanoyl CoA is carboxylated to give methylmalonyl CoA:

propanoyl CoA   methylmalonyl CoA

This reaction requires ATP as a source of energy to form the new carbon-carbon bond. It also requires the coenzyme biotin. Next, methylmalonyl CoA is iso-merized to succinyl CoA in an unusual reaction that requires vitamin B_{12} as a cofactor. In vitamin B_{12} deficiency, both propanoate and methylmalonate appear in the urine.

methylmalonyl CoA   succinyl CoA
                                             (an intermediate
                                             in the TCA cycle)

Studies using radioisotopes have revealed that this isomerization involves migration of the entire thioester group to the methyl carbon and exchange of a hydrogen atom for it.

The importance of vitamin B_{12} was first realized when it was discovered that liver extracts could be used to treat patients with pernicious anemia; this condition was so named because it does not respond to treatment with iron. Symptoms of pernicious anemia include tiredness, anorexia, headaches, and neurological disorders. The principle in liver extract that reverses pernicious anemia was isolated, purified, and crystallized in 1948, and its complete three-dimensional structure was worked out by Dorothy Hodgkin in 1956. For her work in deter-mining the structure of this complex molecule, she received the Nobel Prize in chemistry in 1964.

Most cases of pernicious anemia are not due to lack of B_{12} in the diet, but to a deficiency of a substance called intrinsic factor, which is normally present in gastric juice. Intrinsic factor is necessary for absorption of B_{12} through the walls of the gastrointestinal tract and into the bloodstream. Because most B_{12}-deficiency diseases are due to a lack of intrinsic factor and reduced absorption of the vitamin, the most common means of administration of B_{12} is direct intramuscular injection.

18.5 Formation of Ketone Bodies

Acetoacetate, beta-hydroxybutyrate, and acetone are classed as ketone bodies. Note that the names we have given for the first two of these compounds are common names. Each also has an IUPAC name. However, as is so often the case in both organic chemistry and biochemistry, these and many other compounds are still known largely by their common names.

$$\underset{\substack{\text{3-hydroxybutanoate}\\(\beta\text{-hydroxybutyrate})}}{CH_3\overset{\overset{\text{OH}}{|}}{C}H CH_2\overset{\overset{\text{O}}{\|}}{C}O^-} \qquad \underset{\substack{\text{3-oxobutanoate}\\(\text{acetoacetate})}}{CH_3\overset{\overset{\text{O}}{\|}}{C}CH_2\overset{\overset{\text{O}}{\|}}{C}O^-} \qquad \underset{\text{acetone}}{CH_3\overset{\overset{\text{O}}{\|}}{C}CH_3}$$

Acetoacetate is so named because it is a derivative of acetate. The substituent group is CH_3CO-, which in the common system of nomenclature, is called an aceto group. Hence, the name acetoacetate.

Ketone bodies are products of human metabolism and are always present in blood plasma. However, under normal conditions, their concentration in plasma is low. In humans and most other animals, the liver is the only organ that produces any significant amounts of ketone bodies. Most tissues, with the notable exception of the brain, have the capacity to use them as energy sources.

Ketone bodies are synthesized from acetyl coenzyme A. In a series of three reactions, acetyl CoA is converted to acetoacetate:

$$\underset{\text{acetyl CoA}}{2 CH_3\overset{\overset{\text{O}}{\|}}{C}-SCoA} + H_2O \xrightarrow{\substack{\text{three}\\\text{steps}}} \underset{\text{acetoacetate}}{CH_3\overset{\overset{\text{O}}{\|}}{C}CH_2\overset{\overset{\text{O}}{\|}}{C}O^-} + 2CoA-SH + H^+$$

In one subsequent reaction, the ketone group of acetoacetate is reduced by NADH to a secondary alcohol. The product is beta-hydroxybutyrate.

$$\underset{\text{acetoacetate}}{CH_3\overset{\overset{\text{O}}{\|}}{C}CH_2\overset{\overset{\text{O}}{\|}}{C}O^-} + NADH + H^+ \xrightarrow{\text{(reduction)}} \underset{\beta\text{-hydroxybutyrate}}{CH_3\overset{\overset{\text{OH}}{|}}{C}H CH_2\overset{\overset{\text{O}}{\|}}{C}O^-} + NAD^+$$

In another reaction, acetoacetate loses carbon dioxide to give acetone. Recall that decarboxylation is a characteristic reaction of beta-ketoacids (Section 7.5D).

$$\underset{\text{acetoacetate}}{CH_3\overset{\overset{\text{O}}{\|}}{C}CH_2\overset{\overset{\text{O}}{\|}}{C}O^-} + H^+ \xrightarrow{\text{(decarboxylation)}} \underset{\text{acetone}}{CH_3\overset{\overset{\text{O}}{\|}}{C}CH_3} + CO_2$$

When the production of acetoacetate, beta-hydroxybutyrate, and acetone exceeds the capacity of the body to metabolize them, the condition is known as **ketosis**. Of the three ketone bodies, acetoacetic acid and beta-hydroxybutyric acid are the most significant because they are acids and must be buffered in the blood and most other body fluids to keep their accumulation from disrupting normal acid-base balance. The acidosis that results from the accumulation of ketone bodies is called **ketoacidosis**. During ketoacidosis, the effectiveness of hemoglobin in transporting oxygen is reduced. In the extreme, a deficiency in the supply of oxygen to the brain can produce a fatal coma. The presence of ketone bodies in the urine indicates an advanced state of ketoacidosis and provides a clear signal that immediate medical attention is essential.

Several abnormal conditions, including starvation, unusual diets, and diabetes mellitus lead to increased production of ketone bodies, to ketoacidosis, and to spilling of ketone bodies into the urine. After a short period of starvation, carbohydrate reserves become depleted and the cell or organism turns to beta-oxidation of fatty acids as a source of energy. Fatty acid degradation in the liver is increased, and in turn leads to an increase in the concentration of acetyl CoA. Under normal conditions, acetyl CoA in the liver can be channeled into three metabolic pathways: the tricarboxylic acid cycle, for oxidation to carbon dioxide and water; the synthesis of fatty acids and triglycerides, for storage in the liver; and finally, the synthesis of ketone bodies. During starvation and under conditions where carbohydrate metabolism is drastically reduced, the synthesis of pyruvate and other intermediates of the tricarboxylic acid cycle is also reduced. Therefore, the increased supply of acetyl CoA generated during starvation is channeled into the production of ketone bodies (Figure 18.2).

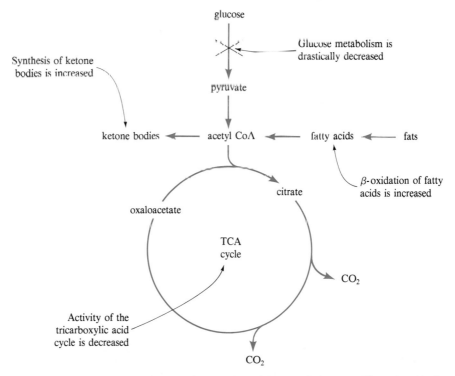

Figure 18.2 As a result of starvation, an abnormal diet, or diabetes mellitus, β-oxidation of fatty acids increases, the concentration of acetyl CoA increases, and more and more of it is channeled into the synthesis of ketone bodies.

18.6 Synthesis of Fatty Acids

In principle, fatty acids could be synthesized by reversal of beta-oxidation, that is, by addition of successive units of two-carbon acetyl groups to acetyl CoA and reduction of the resulting ketone groups to methylene groups. However, synthetic reactions are rarely the reverse of degradation reactions. As an example of this generalization, degradation and **synthesis of fatty acids** are quite different biochemical pathways both in mechanism and in location within the cell. Among the main differences are the following:

1. Synthesis of fatty acids takes place in the cytoplasm, whereas degradation takes place in mitochondria.
2. Synthesis involves two reductions, one of a ketone to a secondary alcohol and the other of a carbon-carbon double bond to a carbon-carbon single bond. In a sense, these two reductions are the reverse of two steps in the oxidation of fatty acid hydrocarbon chains. However, synthesis of fatty acids uses NADPH/NADP$^+$ for both reductions; by comparison, fatty acid oxidation uses FAD/FADH$_2$ for one oxidation and NAD$^+$/NADH for the other.
3. Fatty acid chains are built up by successive addition of two-carbon units derived not from acetyl CoA but from malonyl CoA.

$$\overset{\displaystyle O}{\overset{\displaystyle \|}{{}^-O-C-CH_2-}}\overset{\displaystyle O}{\overset{\displaystyle \|}{C-SCoA}}$$

malonyl CoA

4. Coenzyme A is involved in degradation of fatty acids but not in their synthesis.

A. Formation of Malonyl CoA

Malonyl CoA is formed by carboxylation of acetyl CoA in a reaction that requires biotin as a cofactor and acetyl CoA carboxylase as an enzyme catalyst. You might compare this reaction with the carboxylation of propanoyl CoA to give methylmalonyl CoA (Section 18.4E).

$$O{=}C{=}O + \overset{\displaystyle O}{\overset{\displaystyle \|}{CH_3-C-SCoA}} \xrightarrow[\substack{\text{acetyl CoA}\\ \text{carboxylase}}]{\text{biotin}} \overset{\displaystyle O}{\overset{\displaystyle \|}{{}^-O-C-CH_2-}}\overset{\displaystyle O}{\overset{\displaystyle \|}{C-SCoA}} + H^+$$

acetyl CoA   malonyl CoA

Acetyl CoA carboxylase is a regulatory enzyme, and the rate of fatty acid synthesis is controlled by modulation of its activity.

B. Synthesis of Fatty Acid Hydrocarbon Chains

Synthesis of fatty acid hydrocarbon chains is catalyzed by a multienzyme complex called fatty acid synthase. A key part of this enzyme complex is a low-molecular-weight protein called acyl carrier protein (ACP). Synthesis of a fatty acid chain begins with transfer of an acetyl group from the sulfur atom of coenzyme A to a sulfur atom of ACP. In effect, this reaction is the exchange of one sulfur ester for another. Next, the acetyl group is shifted to an adjacent sulfhydryl group on the enzyme complex. Then a malonyl group is transferred from the sulfur atom of coenzyme A to the first sulfhydryl group of ACP. The result of these three reactions is positioning of one acetyl group and one malonyl group as thioesters at adjacent sites on fatty acid synthetase (Figure 18.3).

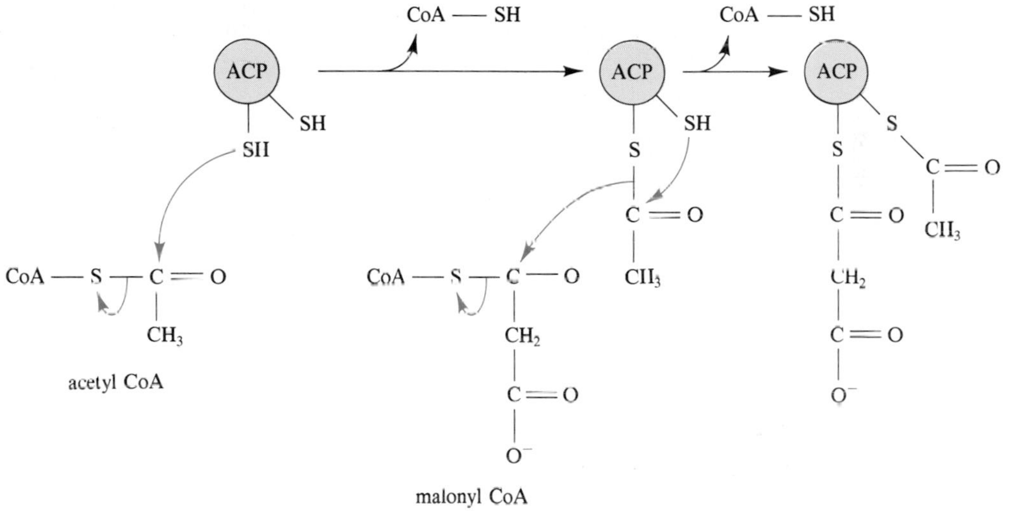

Figure 18.3 Binding of acetyl and malonyl groups to the acyl carrier protein (ACP).

At this point, a pair of two-carbon fragments are activated, one as acetyl-ACP and the other as malonyl-ACP. They next undergo a series of four reactions that cause elongation of the hydrocarbon chain by two atoms at a time. These four steps are condensation, reduction, dehydration, and reduction, each catalyzed by a separate enzyme component of the fatty acid synthase complex. To visualize the operation of this complex, think of ACP in the center of a circle, surrounded by the enzymes that catalyze each step of chain elongation. Further, think of ACP as turning within this enzyme complex, with the growing hydrocarbon chain as a flexible arm that gets longer and longer as the hydrocarbon chain is elongated. The relation of acyl carrier protein to the four enzyme systems of chain elongation is illustrated in Figure 18.4.

1. Condensation

The first step in chain elongation is formation of a carbon-carbon bond between the carbonyl group of acetyl-ACP and the beta-carbon of malonyl-ACP to give

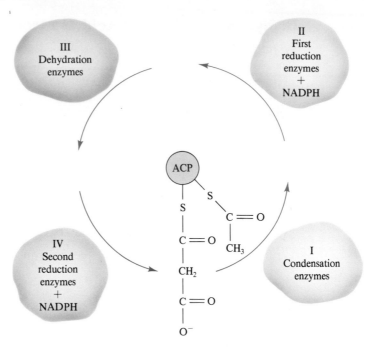

Figure 18.4 The relation between acyl carrier protein and the four enzyme systems responsible for chain elongation. In this arrangement, the growing chain is bound as a thioester to ACP and swings from one enzyme in the multienzyme complex to the next.

acetoacetyl-ACP:

$$CH_3C\overset{O}{\overset{\|}{C}}-S-ACP + CH_2C\overset{O}{\overset{\|}{C}}-S-ACP \longrightarrow$$

$$CO_2^-$$

acetyl-ACP   malonyl-ACP

$$CH_3C\overset{O}{\overset{\|}{C}}CH_2C\overset{O}{\overset{\|}{C}}-S-ACP + ACP-SH + CO_2$$

acetoacetyl-ACP
(a β-ketothioester)

This enzyme-catalyzed formation of a new carbon-carbon bond is analogous to a Claisen condensation (Section 8.9) and forms a beta-ketothioester. Note that this condensation is coupled with the loss of carbon dioxide by decarboxylation. Thus, although carbon dioxide (actually bicarbonate) is required for fatty acid synthesis, it does not appear in the newly synthesized fatty acid.

2. Reduction of a Ketone to a Secondary Alcohol

In step 2, the beta–keto group of the growing fatty acid chain is reduced to a secondary alcohol:

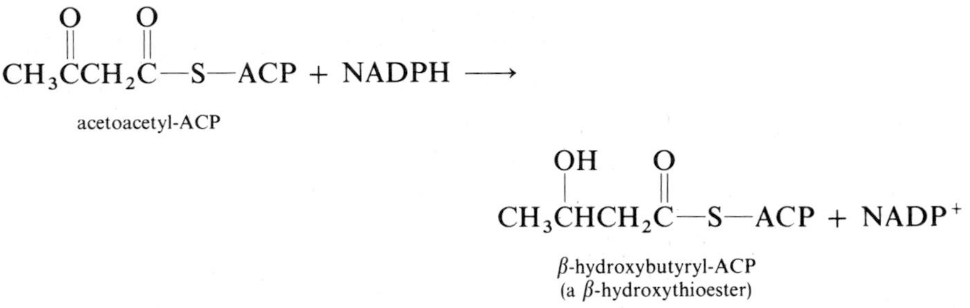

$$\underset{\text{acetoacetyl-ACP}}{CH_3\overset{O}{\overset{\|}{C}}CH_2\overset{O}{\overset{\|}{C}}-S-ACP} + NADPH \longrightarrow$$

$$\underset{\substack{\beta\text{-hydroxybutyryl-ACP} \\ (\text{a }\beta\text{-hydroxythioester})}}{CH_3\overset{OH}{\overset{|}{C}}HCH_2\overset{O}{\overset{\|}{C}}-S-ACP} + NADP^+$$

The reducing agent is NADPH, consistent with the principal use of this coenzyme as a reducing agent in biosynthesis (Section 17.2E). Note that the carbon bearing the —OH group is chiral. Experiments have shown that the configuration of this chiral carbon is the opposite of the comparable chiral carbon involved in fatty acid oxidation (that is, it is the enantiomer).

3. Dehydration

In step 3, the beta-hydroxythioester is dehydrated, to form an α,β-unsaturated thioester in a reaction analogous to acid-catalyzed dehydration of an alcohol:

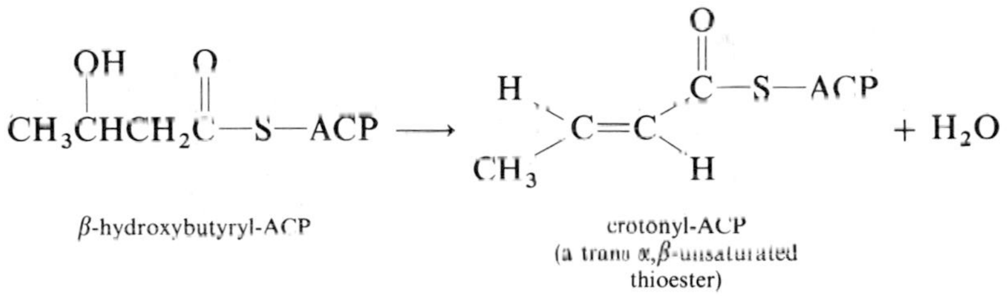

$$\underset{\beta\text{-hydroxybutyryl-ACP}}{CH_3\overset{OH}{\overset{|}{C}}HCH_2\overset{O}{\overset{\|}{C}}-S-ACP} \longrightarrow \underset{\substack{\text{crotonyl-ACP} \\ (\text{a trans }\alpha,\beta\text{-unsaturated} \\ \text{thioester})}}{\overset{\displaystyle H}{\underset{\displaystyle CH_3}{}}C=C\overset{\overset{O}{\overset{\|}{C}}-S-ACP}{\underset{\displaystyle H}{}}} + H_2O$$

The configuration about the double bond formed in step 3 is trans. By comparison, configurations in the unsaturated fatty acid components of triglycerides and phospholipids are entirely cis.

4. Reduction to a Carbon-Carbon Single Bond

In the final reaction of chain elongation, the carbon-carbon double bond of the growing chain is reduced to a carbon-carbon single bond. The reducing agent is another molecule of NADPH.

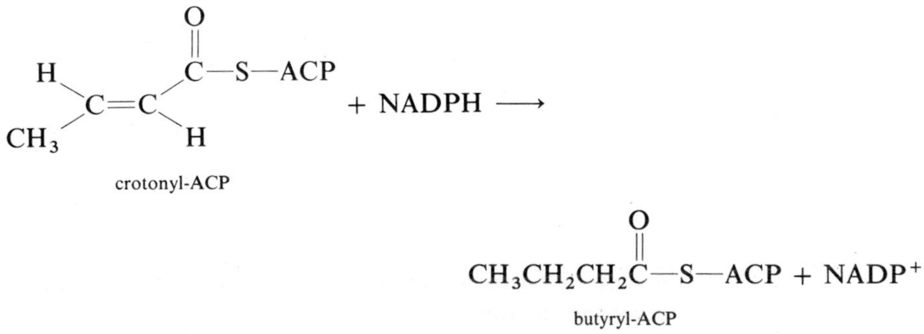

$$\underset{\text{crotonyl-ACP}}{\overset{\displaystyle H}{\underset{\displaystyle CH_3}{}}C=C\overset{\overset{O}{\overset{\|}{C}}-S-ACP}{\underset{\displaystyle H}{}}} + NADPH \longrightarrow$$

$$\underset{\text{butyryl-ACP}}{CH_3CH_2CH_2\overset{O}{\overset{\|}{C}}-S-ACP} + NADP^+$$

The thioester formed after steps 1–4 is shown in Figure 18.5. At this point, the acyl group of the four-carbon thioester is transferred to the adjacent —SH group and a second malonyl group is transferred to ACP.

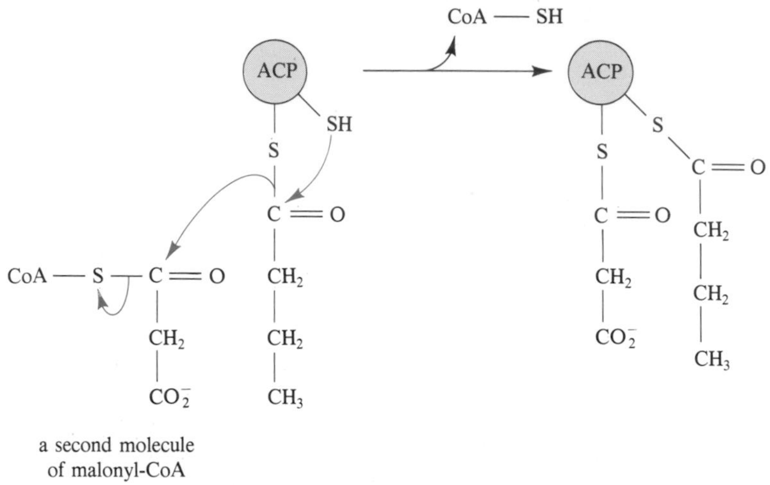

a second molecule
of malonyl-CoA

Figure 18.5 Preparation for a second cycle of chain elongation reactions.

The second cycle of chain elongation begins with formation of a carbon-carbon bond between the carbonyl carbon of the four-carbon thioester and the alpha carbon of malonyl-ACP (Figure 18.6).

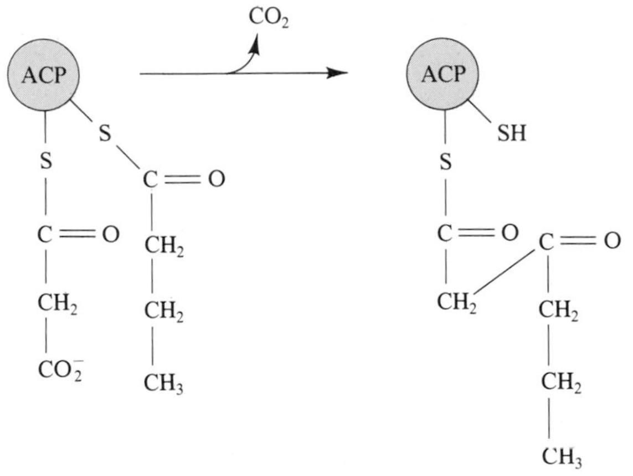

Figure 18.6 The beginning of the second cycle of chain elongation reactions.

The cycle of condensation, reduction, dehydration, and reduction continues until the 16-carbon chain of palmitoyl-ACP is formed. At this point, the thioester of palmitoyl-ACP is hydrolyzed and palmitic acid is released from the multienzyme complex.

$$CH_3(CH_2)_{14}\overset{O}{\underset{\|}{C}}-S-ACP + H_2O \longrightarrow CH_3(CH_2)_{14}\overset{O}{\underset{\|}{C}}OH + ACP-SH$$

palmityl-ACP   palmitic acid
(hexadecanoic acid)

Fatty acid synthesis catalyzed by fatty acid synthase yields only palmitic acid. Yet approximately 60% of the fatty acids of triglycerides and phospholipids contain fatty acids of 18 and 20 carbon atoms. The reactions of further chain elongation appear to be identical with those catalyzed by fatty acid synthase, but are catalyzed by other enzymes found associated with the endoplasmic reticulum.

C. Synthesis of Unsaturated Fatty Acids

A significant portion of fatty acids in human triglycerides are unsaturated and can be synthesized by oxidation of saturated fatty acids. Insertion of unsaturation is catalyzed by enzymes called fatty acid desaturase, as illustrated here by the conversion of stearic acid to oleic acid. Note that molecular oxygen, O_2, is the oxidizing agent and that in this reaction, both NADH and the saturated fatty acid chain are oxidized.

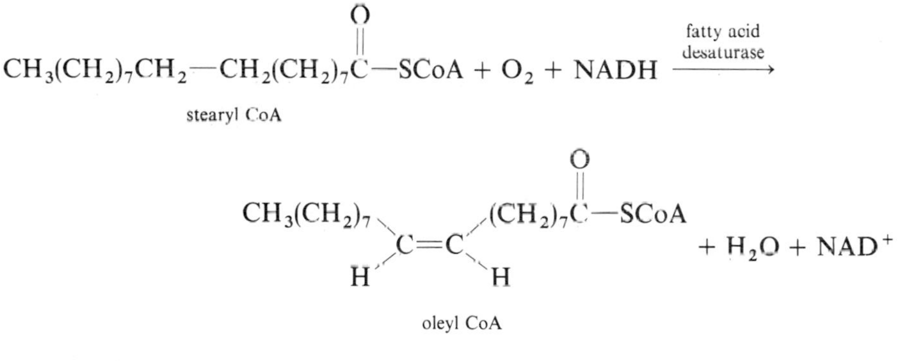

$$CH_3(CH_2)_7CH_2-CH_2(CH_2)_7\overset{O}{\underset{\|}{C}}-SCoA + O_2 + NADH \xrightarrow{\text{fatty acid desaturase}}$$

stearyl CoA

$$CH_3(CH_2)_7\diagdown \atop H \diagup C=C \diagup (CH_2)_7\overset{O}{\underset{\|}{C}}-SCoA \atop \diagdown H \quad + H_2O + NAD^+$$

oleyl CoA

Mammalian fatty acid desaturases have three important characteristics:

1. Double bonds introduced in fatty acid chains have a cis configuration.
2. The enzymes are active on hydrocarbon chains of only 18 or fewer carbon atoms.
3. Double bonds can be introduced between carbons 4–5, 5–6, 6–7, and 9–10. What is important to note is that these desaturases cannot introduce double bonds beyond carbons 9–10.

Linoleic, linolenic, and arachidonic acids are polyunsaturated. Linoleic acid has double bonds between carbons 9–10 and 12–13. In linolenic acid, there are double bonds between carbons 9–10, 12–13, and 15–16; in arachidonic acid, they are between carbons 5–6, 8–9, 11–12, and 14–15. Because mammalian fatty acid desaturases cannot introduce double bonds beyond carbons 9–10, linoleic acid must be supplied in the diet. It in turn can be used for the synthesis of the other polyunsaturated fatty acids.

D. Synthesis of Fatty Acids from the Carbon Atoms of Glucose

There is a close relationship between the metabolism of glucose and the metabolism of fatty acids (see Figure 16.14). Specifically, the body has only a limited capacity to store glucose. However, it has a very large capacity to store fatty acids in the form of triglycerides. Through the metabolic pathways we have already covered, the carbon atoms of glucose can be used for the synthesis of fatty acids. The flow of carbon atoms from glucose to palmitic acid is traced in Figure 18.7.

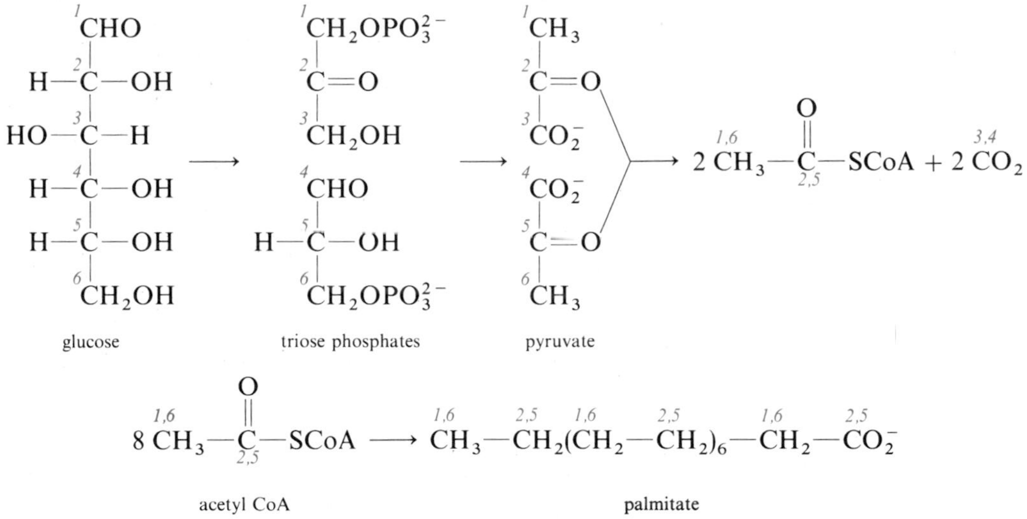

Figure 18.7 The synthesis of palmitate from glucose. The carbon atoms of glucose are numbered 1 through 6.

Cleavage of glucose during glycolysis produces one molecule each of 3-phosphoglyceraldehyde and dihydroxyacetone phosphate. Dihydroxyacetone phosphate is isomerized to 3-phosphoglyceraldehyde and both continue in glycolysis to form pyruvate. Oxidation and decarboxylation of pyruvate (Section 17.4A) give acetyl CoA and carbon dioxide. The carbonyl carbons of acetyl CoA are derived from carbons 2 and 5 of glucose, and the methyl carbons are derived from carbons 1 and 6 of glucose. Acetyl CoA in the form of malonyl CoA is the key building block for the synthesis of fatty acids including palmitic acid. As shown in Figure 18.7, carbon atoms 1, 2, 5, and 6 of glucose can become incorporated into the carbon skeleton of a fatty acid.

18.7 Synthesis of Cholesterol and Other Steroids

Cholesterol is a necessary component of biological membranes. Further, it is the molecule from which all other classes of steroids are derived, including bile acids and sex hormones. In Western society, cholesterol is usually obtained in the diet. If dietary sources are not sufficient, however, it can be synthesized. All nucleated cells have the ability to synthesize cholesterol, but in man, 80–95% of all cholesterol synthesis is done in the liver and intestine.

It was a challenge to biochemists to discover how this complex molecule is constructed. The availability of radioactive isotopes for research purposes, in particular carbon-14, gave scientists the probe they needed. They discovered that the carbon skeleton of cholesterol is derived entirely from acetate, and further, that the two-carbon acetate unit is not broken in the process. In other words, carbon atoms of cholesterol are derived in order from a methyl carbon and then the carbonyl carbon, then another methyl carbon and its carbonyl carbon, and so on. This pattern is but another demonstration that in putting together large and complex molecules, nature begins with small, readily available subunits and puts them together piece by piece. In the case of cholesterol and all other steroids, the readily available subunit is acetyl CoA, a molecule produced in the degradation of glucose and other monosaccharides, fatty acids, and, as we shall see in Chapter 19, also from amino acids.

The **synthesis of cholesterol** begins by condensation of two molecules of acetyl CoA to produce acetoacetyl CoA, in a reaction catalyzed by acetyl-CoA acetyl-transferase:

$$CH_3\overset{O}{\overset{\|}{C}}-SCoA + CH_3\overset{O}{\overset{\|}{C}}-SCoA \longrightarrow CH_3\overset{O}{\overset{\|}{C}}CH_2\overset{O}{\overset{\|}{C}}-SCoA + CoA-SH$$

acetyl CoA   acetyl CoA   acetoacetyl CoA

At this point, either one or another of two enzyme systems is activated. One enzyme system leads to ketone bodies; acetoacetyl CoA is converted to acetoacetate and in turn to acetone and β-hydroxybutyrate. A second enzyme system catalyzes condensation of acetoacetyl CoA with a third molecule of acetyl CoA, to form 3-hydroxy-3-methylglutaryl-CoA (HMG-CoA):

$$CH_3\overset{O}{\overset{\|}{C}}CH_2\overset{O}{\overset{\|}{C}}-SCoA + CH_3\overset{O}{\overset{\|}{C}}-SCoA \longrightarrow$$

acetoacetyl CoA   acetyl CoA

$$^-O\overset{O}{\overset{\|}{C}}CH_2\underset{\underset{CH_3}{|}}{\overset{OH}{\overset{|}{C}}}CH_2\overset{O}{\overset{\|}{C}}-SCoA + CoA-SH$$

3-hydroxy-3-methylglutaryl-CoA

Once formed, 3-hydroxy-3-methylglutaryl CoA is reduced to mevalonate (the anion of mevalonic acid), and then, in several subsequent steps, converted to 3,3-dimethylallylpyrophosphate.

$$^-OCCH_2CCH_2CH_2OH \longrightarrow CH_3C{=}CHCH_2O{-}P{-}O{-}P{-}O^-$$

mevalonate	3,3-dimethylallylpyrophosphate

Each of these compounds is particularly important. Mevalonate is important because it is the first step that specifically commits the carbon skeletons of acetate to the synthesis of cholesterol. The enzyme that catalyzes this reaction, HMG-CoA reductase, is a regulatory enzyme. The compound 3,3-dimethylallyl-pyrophosphate is important because it contains the carbon skeleton of an iso-prene unit and is the basic building block from which all terpenes are derived—including vitamin A, the carotenes, the side chains of vitamins E and K, and coenzyme Q. (For further discussion of terpenes, see Mini-Essay II, "Terpenes.") All the remaining steps in the synthesis of cholesterol are well understood. However, we will not discuss them here.

Before we leave this section, let us stop for a moment to emphasize the central importance of acetyl CoA in metabolism. As we have indicated, it is produced from the degradation of carbohydrates, fatty acids, and proteins. It is also the starting material for the synthesis of several classes of biomolecules including fatty acids, steroids, and terpenes.

Carbohydrates
Fatty acids $\longrightarrow$ $CH_3C{-}SCoA$ $\longrightarrow$ Steroids, Fatty acids, Terpenes
Certain amino acids

Key Terms and Concepts

beta-oxidation spiral (18.4C)	ketosis (18.5)
essential fatty acids (18.3)	oxidation of fatty acids (18.4)
hydrolysis of triglycerides (18.2)	fatty acid oxidation spiral (18.4C)
ketoacidosis (18.5)	synthesis of cholesterol (18.7)
ketone bodies (18.5)	synthesis of fatty acids (18.6)

Key Reactions

1. Activation of free fatty acids by conversion to a thioester of acetyl CoA (Section 18.4A).

2. The four reactions of the β-oxidation spiral (Section 18.4C).

3. Carboxylation of propanoyl CoA to methylmalonyl CoA (Section 18.4E).

4. Isomerization of methylmalonyl CoA to succinyl CoA (Section 18.4E).

5. Formation of ketone bodies from acetyl CoA (Section 18.5).

6. Carboxylation of acetyl CoA to form malonyl CoA (Section 18.6A).

7. The four steps in the cycle of synthesis of fatty acid hydrocarbon chains (Section 18.6B).

Problems

Fatty acids as sources of energy (Section 18.1)

18.2 Compare carbohydrates and fatty acids as energy sources. How do you account for the difference between the amount of energy released on complete oxidation of each?

18.3 Write structural formulas for palmitic, oleic, and stearic acids, the three most abundant fatty acids.

Oxidation of fatty acids (Section 18.4)

18.4 A fatty acid must be activated before it can be metabolized in cells. Write a balanced equation for the reaction that activates palmitic acid.

18.5 Name three coenzymes necessary for the catabolism of fatty acids to acetyl CoA. What vitamin precursor is associated with each coenzyme?

18.6 Outline the four steps in the fatty acid oxidation spiral.

18.7 How much energy in the form of ATP is produced directly in the oxidation of palmitic acid to acetyl CoA?

18.8 Review the oxidation reactions in the catabolism of glucose and fatty acids. Prepare a list of (a) types of functional groups oxidized, and (b) the oxidizing agent used for each type. Compare the types of functional groups oxidized by FAD with those oxidized by NAD^+.

18.9 Calculate the number of moles of ATP produced when:
  a. palmitoyl CoA is oxidized to acetyl CoA and all the reduced coenzymes produced in the process are reoxidized by molecular oxygen.
  b. palmitoyl CoA is oxidized to CO_2 and all the reduced coenzymes produced in the process are reoxidized by molecular oxygen.

18.10 In patients with pernicious anemia, up to 50–90 mg of a dicarboxylic acid of molecular formula $C_4H_6O_4$ appear in the urine daily. Draw a structural formula for this dicarboxylic acid. How do you account for its formation?

18.11 The respiratory quotient (RQ) is used in studies of energy metabolism and exercise physiology. It is defined as the ratio of the volume of carbon dioxide produced to the volume of oxygen used:

$$RQ = \frac{\text{volume of } CO_2}{\text{volume of } O_2}$$

  a. Show that the RQ for glucose is 1.00. (*Hint*: Look at the balanced equation for the complete oxidation of glucose.)
  b. Calculate the RQ for triolein, a triglyceride of molecular formula $C_{57}H_{104}O_6$.

    c. Calculate the RQ on the assumption that triolein and glucose are oxidized in equal molar amounts.

    d. For an individual on a normal diet, the RQ is approximately 0.85. Would this value increase or decrease if ethanol were to supply an appreciable portion of caloric needs?

Formation of ketone bodies (Section 18.5)

18.12 What is the only organ in humans that produces significant amounts of ketone bodies?

18.13 Explain why ketone-body formation increases markedly when excessive amounts of fatty acids are oxidized and carbohydrate availability is limited.

18.14 Explain why the accumulation of ketone bodies leads to acidosis.

Synthesis of fatty acids (Section 18.6)

18.15 Starting with butyryl-ACP, show all steps in the synthesis of hexanoic acid. Name the type of reaction involved in each step.

$$CH_3CH_2CH_2\overset{\displaystyle O}{\overset{\|}{C}}SACP$$

butyryl-ACP

18.16 For the synthesis of stearic acid from acetyl-ACP:

    a. How many moles of NADPH are required?

    b. How many moles of malonyl-ACP are required?

18.17 During fatty acid synthesis, NADPH is oxidized to $NADP^+$. Name the metabolic pathway primarily responsible for regeneration of NADPH.

18.18 If glucose were the only source of acetyl CoA, how many moles of glucose would be required for the synthesis of one mole of palmitate? How many grams of glucose are required for the synthesis of one mole of palmitate?

18.19 Explain why almost all fatty acids have an even number of carbon atoms in an unbranched chain.

19

Metabolism of Amino Acids

In the broadest sense, amino acids have three vital functions in the human body. Amino acids are (1) building blocks for the synthesis of proteins; (2) sources sources of carbon and nitrogen atoms for the synthesis of other biomolecules; and (3) sources of energy. Compared with the metabolism of carbohydrates and fatty acids, the metabolism of amino acids is extremely complex. Unlike hexoses and fatty acids, which share common metabolic pathways (for example, glycolysis for hexoses and beta-oxidation for fatty acids), each amino acid is degraded and synthesized by a separate pathway. We shall not go into the details of the metabolic pathways for the degradation and synthesis of each amino acid, but instead examine the overall process. Of specific interest is how the carbon skeletons of amino acids are used as sources of energy and how amino acid nitrogen atoms are collected, converted to urea, and excreted.

19.1 Amino Acid Metabolism—An Overview

A. Amino Acids Are Used for the Synthesis of Body Proteins

The most important function of amino acids, at least in terms of total amino acid use, is as building blocks for the synthesis of proteins. It is estimated that approximately 75% of amino acid metabolism in a normal, healthy adult is devoted to this purpose. The maintenance of body proteins is not a simple matter. Tissue proteins are being hydrolyzed constantly through normal "wear and tear." At the same time, we eat proteins that are hydrolyzed in the gut to amino acids. The amino acids from food and from hydrolysis of tissue proteins combine to create what is called the **amino acid pool**, from which new proteins are synthesized.

The use of radioisotopes has given us some idea of the extent of turnover of body proteins and the amino acid pool. The **half-life** of liver proteins, for example, is about 10 days. This means that over a 10-day period, half the proteins in the liver are hydrolyzed to amino acids and replaced by equivalent proteins. The half-life of plasma proteins is also about 10 days, hemoglobin about 30 days, and muscle protein about 180 days. The half-life of collagen is much longer. Some proteins, particularly enzymes and polypeptide hormones, have much shorter half-lives. That of insulin, once it is released from the pancreas, is estimated to be only 7–10 minutes.

Clearly, the stability of body proteins is more apparent than real and represents a dynamic balance between degradation and synthesis. In spite of this turnover, the total amount of body protein remains relatively constant in most adults. This means that the quantity of amino acids required for the synthesis of body proteins and the quantity obtained from hydrolysis of body proteins are roughly equivalent. Therefore, the amino acid pool of most adults contains a surplus approximately equal to the amino acids obtained in the diet. This surplus is used for the synthesis of nonprotein compounds and as a source of fuel, or it is excreted.

B. Amino Acids as Sources of Carbon and Nitrogen for the Synthesis of Other Biomolecules

Tissues constantly draw on the amino acid pool for the synthesis of nonprotein biomolecules. These molecules include nucleic acids; porphyrins, such as those in the prosthetic groups of hemoglobin and myoglobin; choline and ethanolamine, which are building blocks of phospholipids; glucosamine and other amino sugars; and neurotransmitters, such as acetylcholine, dopamine, norepinephrine, and serotonin. Like proteins, these compounds are also being broken down and replaced constantly.

C. Amino Acids as Fuel

Unlike carbohydrates and fatty acids, amino acids in excess of immediate needs cannot be stored for later use. Their nitrogen atoms are converted to ammonium ions, urea, or uric acid, depending on the organism, and excreted. Their carbon skeletons are degraded to pyruvate, acetyl CoA, or one of the intermediates of the tricarboxylic acid cycle. The various metabolic pathways of amino acid metabolism are summarized in Figure 19.1.

D. Essential Amino Acids

Most dietary proteins contain all the amino acids humans need for synthesizing their proteins. They are often present in widely different proportions, however. For protein synthesis to take place, all the required amino acids must be present at the time of synthesis and in the correct proportions. Studies of protein synthesis have led to the concept of essential and nonessential amino acids. Nonessential amino acids are synthesized in the body at a rate equal to the needs of protein biosynthesis. Essential amino acids cannot be synthesized within the body fast enough to support normal protein synthesis.

In humans, the most widely used experimental procedure for classifying amino acids as essential or nonessential is nitrogen balance. A normal, healthy

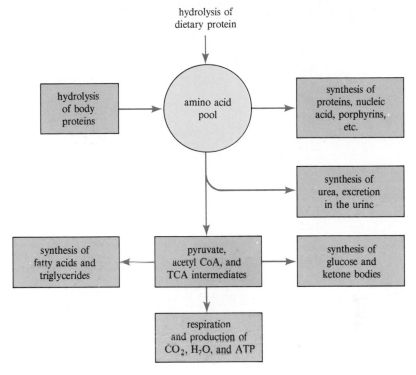

Figure 19.1 An overview of the metabolism of amino acids. Average daily turnover of amino acids is approximately 400 g.

adult, for whom the rate of nitrogen intake equals the rate of its loss in feces, urine, and sweat, is said to be in nitrogen balance. In periods of active growth or tissue repair, when more nitrogen is ingested than excreted, the individual is said to be in positive nitrogen balance. In the opposite condition, starvation or malnutrition, when body proteins are being used for fuel and more nitrogen is excreted than is taken in, the individual is said to be in negative nitrogen balance. Whether an amino acid is essential or nonessential can be determined by withholding it alone from the diet, supplying all other amino acids. If the absence of this amino acid from the diet brings about a negative nitrogen balance, the amino acid is classified as essential.

Of the 20 amino acids that the human body needs, adequate amounts of 12 can be synthesized by enzyme-catalyzed reactions starting from carbohydrates or lipids, and a source of nitrogen. For the remaining amino acids, either no biochemical pathways are available for their synthesis, or the available pathways do not provide adequate amounts for proper nutrition. Therefore, these eight amino acids must be supplied in the diet, and are called essential amino acids (Table 19.1).

Tyrosine is synthesized from phenylalanine in the body. Therefore, in Table 19.1, the requirements for tyrosine and phenylalanine are combined. Similarly, the sulfur-containing amino acids methionine and cysteine are combined.

Table 19.1
Essential amino acids required for humans.

	Daily Requirement, mg/kg Body Weight		
Essential Amino Acid	**Infant (4–6 mo)**	**Child (10–12 yr)**	**Adult**
histidine	33	—	—
isoleucine	83	28	12
leucine	135	42	16
lysine	99	44	12
total S-containing amino acids (methionine and cysteine)	49	22	10
total aromatic amino acids (phenylalanine and tyrosine)	141	22	16
threonine	68	28	8
tryptophan	21	4	3
valine	92	25	14

Histidine is essential for growth in infants and may be needed by adults also. Arginine is synthesized in the body, but the rate of internal synthesis is not adequate to meet the needs of the body during periods of rapid growth and protein synthesis. Therefore, 8, 9, or 10 amino acids are essential for humans, depending on age and state of health.

E. Biological Value of Dietary Proteins

It should now be clear that not all proteins are equivalent, at least for human nutrition. Some contain a more appropriate blend of essential amino acids than others do. One measure of the relative blend of essential amino acids for human nutrition is a protein's biological value, that is, the percentage that is absorbed and used to build body tissue. Some of the first information on the biological value of dietary proteins came from studies on rats. In one series of experiments, young rats were fed diets containing protein in the form of casein (a milk protein), gliadin (a wheat protein), or zein (a corn protein). With casein as the sole dietary source of protein, the rats remained healthy and grew normally. Those fed gliadin maintained their weight but did not grow much. Those fed zein not only failed to grow but lost weight, and if kept on this diet, eventually died. Because casein evidently supplies all required amino acids in the correct proportions needed for growth, it is called a complete protein. Analysis revealed that gliadin contains too little lysine, and that zein is low in both lysine and tryptophan. When a gliadin diet is supplemented with lysine, or the zein diet with lysine and tryptophan, test animals grew normally.

Table 19.2 shows the biological value for rats of some common dietary sources of protein. The proteins in egg are the best-quality natural protein. The proteins of milk have a biological value of 84, and meat and soybeans, about 74. The legumes, vegetables, and cereal grains are in the range 50–70.

Animal proteins generally contain a blend of essential amino acids similar to those that humans need. Many plant proteins, however, are low in one or more

Table 19.2
Biological value for rats of some common sources of dietary proteins.

Food	Protein as % of Dry Solid	Biological Value of Protein, %
hen's egg, whole	48	94
cow's milk, whole	27	84
fish	72	83
beef	45	74
soybeans	41	73
rice, brown	9	73
potato, white	6	67
wheat, whole grain	14	65
corn, whole grain	11	59
dry beans, common	25	58

essential amino acids. Fortunately, not all plant proteins are deficient in the same amino acids. For example, beans and other legumes are low in methionine but adequate in lysine. Wheat and other cereal grains have just the opposite pattern; they are low in lysine but adequate in methionine. Thus it is possible, by eating wheat and beans together, to increase by 33% the usable protein you would get by eating either of these foods alone.

Estimates of recommended daily intake of protein have varied over the years as our understanding of the relation between diet and nutrition has advanced. In 1980, the Food and Nutrition Board of the National Academy of Sciences stated that a generous protein allowance for a healthy adult is 0.8 g of high-quality protein per kilogram of "ideal" body weight per day. For example, for a healthy 68-kg (150-lb) male of medium frame, the RDA is 0.8 × 68, or 54, g protein per day.

The provision of a diet adequate in protein and essential amino acids is a grave problem in the world today, especially in areas of Asia, Africa, and South America. The overriding dimension of this problem is poverty and an inability to select foods of adequate protein and caloric content. The best overall foods are the cereal grains, which provide not only calories but protein also. When these are supplemented with animal protein or a proper selection of plant protein, the diet is adequate for even the most vulnerable. However, as income decreases, there is less animal protein in the diet, and even cereal grains are often replaced by cheaper sources of calories, such as starches or tubers—foods that have either very little or no protein. The poorest 25% of the world's population consumes diets with caloric and protein content that fall dangerously below the calculated minimum daily requirements.

Those most apt to show symptoms of too little food, too little protein, or both, are young children in the years immediately following weaning. They fail to grow properly, and their tissues become wasted. This sickness is called marasmus, a name derived from a Greek word meaning "to waste away." Muscles become atrophied and the face develops a wizened "old man's" look. Another disease, kwashiorkor, leads to tragically high death rates among children. As long as a child is breast-fed, it is healthy. At weaning (often forced when a second child is born), the first child's diet is switched to starch or other inadequate sources of protein. Symptoms of the disease are edema, cessation of growth, severe body wasting, diarrhea, mental apathy, and a peeling of the skin that leaves areas of raw flesh. Those afflicted with the disease are doomed to short lives. Kwashiorkor was unknown in the medical literature until 1933, when it was described by Cicely Williams, who studied the condition while she was working among the tribes of West Africa. She gave it the name by which it was known in the African Ga tribe.

One solution to the problem of quantity and quality of protein has been to breed new varieties of cereal grains with higher protein content or better-quality protein, or both. Alternatively, cereals and their derived products can be supplemented (fortified) with amino acids in which they are deficient: principally lysine for wheat; lysine or lysine plus threonine for rice; and lysine plus tryptophan for corn. New methods of synthesis and fermentation now provide cheap sources of these amino acids, thus making the economics of food fortification entirely practical. In another attack on the problem of protein malnutrition, nutritionists have developed several high-protein, low-cost infant foods. Clearly, advances in food chemistry and technology have provided the means to eradicate hunger and malnutrition. What remains is for the world's political and social systems to put this knowledge into practice.

19.2 Hydrolysis (Digestion) of Dietary Proteins

Proteins ingested in the diet cannot be absorbed as such through the epithelial cells of the intestine and passed into the bloodstream. First, they must be hydrolyzed (digested) into amino acids, dipeptides, and tripeptides. Digestion of proteins begins in the stomach, where the highly acid gastric juices (pH approximately 1.0) cause denaturation of polypeptide chains, thus making them more susceptible to the catalytic activity of proteases (protein-hydrolyzing enzymes). The early stages of protein hydrolysis are catalyzed by four specific enzymes: pepsin, trypsin, chymotrypsin, and elastase. These and all other enzymes involved in the hydrolysis of dietary proteins are synthesized as inactive pro-enzymes, namely, pepsinogen, trypsinogen, chymotrypsinogen, and pro-elastase. Each pro-enzyme is larger than the active enzyme by one or more short sections of polypeptide, which effectively prevent the pro-enzyme from folding into a biologically active conformation. Pepsinogen is secreted directly into the stomach. There its extra polypeptide section, 42 amino acids long, is cleaved by hydrolysis of one peptide bond:

$$\text{Pepsinogen} + H_2O \xrightarrow[\text{(stomach)}]{H^+} \text{pepsin} + \text{polypeptide of 42 amino acids}$$

Trypsinogen, chymotrypsinogen, and pro-elastase are secreted by the pancreas into the small intestine. There trypsinogen is hydrolyzed to trypsin, which in turn catalyzes hydrolysis of remaining trypsinogen, chymotrypsinogen, and pro-elastase to their active forms.

$$\text{Trypsinogen} + H_2O \xrightarrow{\text{trypsin}} \text{trypsin} + \text{polypeptide}$$

$$\text{Chymotrypsinogen} + H_2O \xrightarrow{\text{trypsin}} \text{chymotrypsin} + \text{polypeptide}$$

$$\text{Pro-elastase} + H_2O \xrightarrow{\text{trypsin}} \text{elastase} + \text{polypeptide}$$

These four proteases hydrolyze larger polypeptides into a series of smaller polypeptides. Digestion is then continued by several other classes of enzymes, including the aminopeptidases and carboxypeptidases. Aminopeptidases catalyze the hydrolysis of amino acids from the *N*-terminal end of a polypeptide chain; carboxypeptidases do the same but from the *C*-terminal end of a polypeptide chain.

In the final stage of digestion, the water-soluble amino acids and di- and tripeptides are absorbed through the wall of the intestine and passed directly into the bloodstream.

19.3 Catabolism of Amino Acids

Since learning the different sequences of reactions by which each amino acid is synthesized and degraded, and covering pathways for even a few amino acids, would be an advanced undertaking, we shall instead concentrate on the general principles of amino acid degradation, without referring to specific amino acids. As we shall see, the final stage in metabolism of the carbon skeletons of amino acids is the tricarboxylic acid cycle, which is why the metabolism of amino acids is very closely integrated with the metabolism of both carbohydrates and fatty acids.

A. Transamination

The loss of the alpha-amino group is the first stage in the metabolism of most amino acids. It is accomplished by two different reactions: transamination and oxidative deamination.

In **transamination**, an alpha-amino group is transferred from a donor alpha-amino acid to an acceptor alpha-ketoacid. In the process, the acceptor alpha-ketoacid is transformed into a new alpha-amino acid. While several different alpha-ketoacids participate in transamination reactions, the most important are pyruvate and alpha-ketoglutarate. During the transamination phase of amino acid catabolism, all amino groups (except possibly those of lysine and threonine) are channeled into either alanine or glutamate.

Alanine Transaminase

$$\overset{\overset{\displaystyle NH_3^+}{|}}{R CHCO_2^-} + \overset{\overset{\displaystyle O}{\|}}{CH_3 C CO_2^-} \rightleftharpoons \overset{\overset{\displaystyle O}{\|}}{R C CO_2^-} + \overset{\overset{\displaystyle NH_3^+}{|}}{CH_3 CHCO_2^-}$$

(an α-amino acid)	pyruvate (a ketoacid)	(a new α-ketoacid)	L-alanine

Glutamate Transaminase

$$\underset{\substack{\text{(an }\alpha\text{-amino} \\ \text{acid)}}}{\overset{\overset{+}{\text{NH}_3}}{\text{RCHCO}_2^-}} + \underset{\substack{\alpha\text{-ketoglutarate} \\ \text{(an }\alpha\text{-ketoacid)}}}{\overset{\overset{\text{O}}{\|}}{^-\text{O}_2\text{CCH}_2\text{CH}_2\text{CCO}_2^-}} \;\rightleftharpoons\; \underset{\substack{\text{(a new} \\ \alpha\text{-ketoacid)}}}{\overset{\overset{\text{O}}{\|}}{\text{RCCO}_2^-}} + \underset{\substack{\text{L-glutamate} \\ \text{(a new }\alpha\text{-amino acid)}}}{\overset{\overset{+}{\text{NH}_3}}{^-\text{O}_2\text{CCH}_2\text{CH}_2\text{CHCO}_2^-}}$$

Example 19.1

Draw structural formulas for the starting materials and products of these transamination reactions:

a. tyrosine + alpha-ketoglutarate $\longrightarrow$ **b.** valine + pyruvate $\longrightarrow$

☐ *Solution*

a. $\text{HO}-\!\!\!\underset{\text{L-tyrosine}}{\bigcirc}\!\!\!-\overset{\overset{+}{\text{NH}_3}}{\text{CH}_2\text{CHCO}_2^-} + \underset{\alpha\text{-ketoglutarate}}{\overset{\overset{\text{O}}{\|}}{^-\text{O}_2\text{CCH}_2\text{CH}_2\text{CCO}_2^-}} \longrightarrow$

$\text{HO}-\!\!\!\underset{p\text{-hydroxyphenylpyruvate}}{\bigcirc}\!\!\!-\overset{\overset{\text{O}}{\|}}{\text{CH}_2\text{CCO}_2^-} + \underset{\text{L-glutamate}}{\overset{\overset{+}{\text{NH}_3}}{^-\text{O}_2\text{CCH}_2\text{CH}_2\text{CHCO}_2^-}}$

b. $\underset{\text{L-valine}}{\overset{\overset{+}{\text{NH}_3}}{\underset{\underset{\text{CH}_3}{|}}{\text{CH}_3\text{CHCHCO}_2^-}}} + \underset{\text{pyruvate}}{\overset{\overset{\text{O}}{\|}}{\text{CH}_3\text{CCO}_2^-}} \longrightarrow \underset{\substack{\text{2-oxo-3-methyl-} \\ \text{butanoate} \\ (\alpha\text{-ketoisovalerate})}}{\overset{\overset{\text{O}}{\|}}{\underset{\underset{\text{CH}_3}{|}}{\text{CH}_3\text{CHCCO}_2^-}}} + \underset{\text{L-alanine}}{\overset{\overset{+}{\text{NH}_3}}{\text{CH}_3\text{CHCO}_2^-}}$

Problem 19.1

Draw structural formulas for the starting materials and products of these transamination reactions:

a. phe + pyruvate $\longrightarrow$ **b.** his + α-ketoglutarate $\longrightarrow$

Transaminations are catalyzed by a specific group of enzymes called amino-transferases, or more commonly, transaminases. Though transaminases are found in all cells, their concentrations are particularly high in heart and liver tissues. Damage to either heart or liver leads to release of transaminases into the blood, and determination of serum levels of these enzymes can provide the clinician with valuable information about the extent of heart or liver damage. The two transaminases most commonly assayed for this purpose are serum glutamate oxaloacetate transaminase (SGOT) and serum glutamate pyruvate transaminase (SGPT).

For catalytic activity, all transaminases require pyridoxal phosphate, a coenzyme derived from pyridoxine, vitamin B_6 (Figure 19.2). In its role as a catalyst,

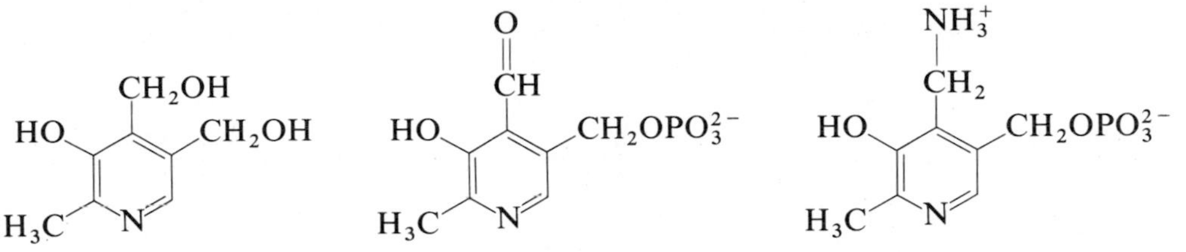

pyridoxine (vitamin B_6) — pyridoxal phosphate (PLP) — pyridoxamine phosphate (PMP)

Figure 19.2 Pyridoxine, or vitamin B_6. Pyridoxal phosphate (PLP) and pyridoxamine phosphate (PMP) are coenzymes derived from pyridoxine.

this coenzyme undergoes reversible transformations between an aldehyde (pyridoxal phosphate) and a primary amine (pyridoxamine phosphate). It first reacts as pyridoxal phosphate with an α-amino group, to form a Schiff base (Section 6.4C), which by rearrangement of two hydrogen atoms and the double bond, forms an isomeric Schiff base. This isomeric Schiff base then undergoes hydrolysis to give pyridoxamine and an α-ketoacid. The amino group of pyridoxamine then reacts with the keto group of a different α-ketoacid, to form another Schiff base, which after isomerization and hydrolysis, regenerates pyridoxal and a new α-amino acid. These reversible transformations are illustrated in Figure 19.3. To simplify the drawings, pyridoxal phosphate is abbreviated PP—CHO and pyridoxamine is abbreviated PP—CH_2NH_2.

Transaminations have two vital functions. First, they provide a means of readjusting the relative proportions of a number of amino acids to meet the particular needs of the organism; in most diets, the amino acid blend does not correspond precisely to what is needed. Second, transaminations collect the nitrogen atoms of all amino acids as glutamate, the primary source of nitrogen atoms for synthesis.

B. Oxidative Deamination

The second major pathway by which amino groups are removed from amino acids is **oxidative deamination** of glutamate in the liver, to form ammonium ion and alpha-ketoglutarate. The oxidation requires either NAD^+ or $NADP^+$ and is catalyzed by the enzyme glutamate dehydrogenase:

$$\underset{\text{glutamate}}{^{-}O_2CCH_2CH_2\overset{\overset{\displaystyle NH_3^+}{|}}{C}HCO_2^-} + NAD^+ + H_2O \xrightarrow[\text{dehydrogenase}]{\text{glutamate}}$$

$$\underset{\alpha\text{-ketoglutarate}}{^{-}O_2CCH_2CH_2\overset{\overset{\displaystyle O}{\|}}{C}CO_2^-} + NADH + NH_4^+ + H^+$$

In this way, amino groups collected from other amino acids are converted to ammonium ion.

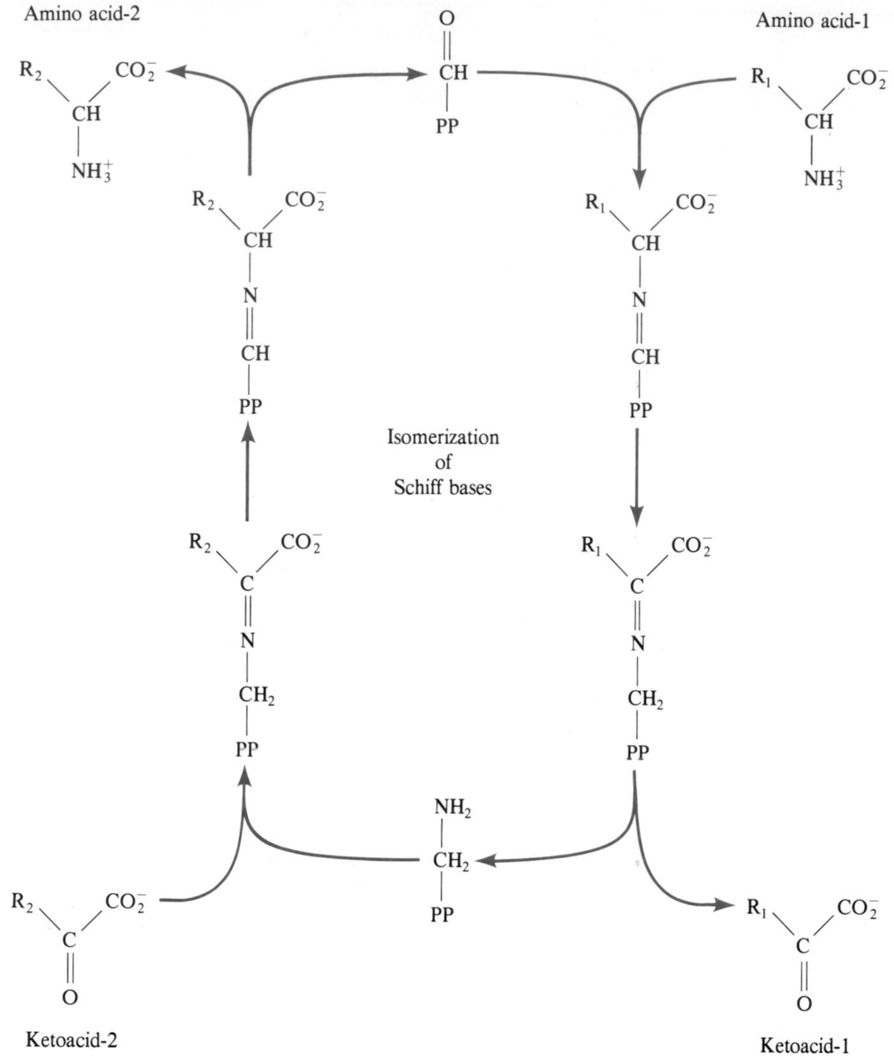

Figure 19.3 Mechanism of transamination catalyzed by pyridoxal/pyridoxamine phosphate.

Normal concentrations of ammonium ion in plasma are 0.025 to 0.04 mg/L. Ammonium ion is extremely toxic and must be eliminated. The main pathway by which ammonium ion is detoxified and eliminated in humans is formation of urea, a neutral nontoxic compound, which is then excreted in the urine.

C. Synthesis of Urea

Urea synthesis in mammals occurs exclusively in the liver. The metabolic pathway that catalyzes the formation of urea is called the urea cycle, or the Krebs-

Henseleit cycle, after Hans Krebs and Kurt Henseleit, who proposed it in 1932. This cycle accepts one carbon atom in the form of bicarbonate (or carbon dioxide) and two nitrogen atoms, one from ammonium ion and the other from aspartate, and in a five-step cycle, generates urea and fumarate. The net reaction of the **urea cycle** is shown in Figure 19.4.

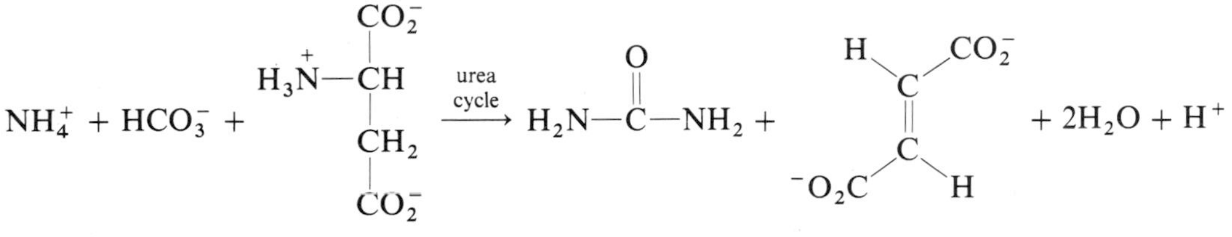

Figure 19.4 The net reaction of the urea cycle.

The five reactions of the urea cycle are shown in Figure 19.5. The first step in the formation of urea is synthesis of carbamoyl phosphate from bicarbonate, ammonium ion, and inorganic phosphate. This reaction is catalyzed by carbamoyl phosphate synthetase and is coupled with hydrolysis of two molecules of ATP to ADP. In step 2, carbamoyl phosphate reacts with ornithine, to form citrulline. The third step, condensation of citrulline with aspartate, is coupled with hydrolysis of a third molecule of ATP and incorporates a second nitrogen atom into the cycle. Cleavage of argininosuccinate produces arginine and fumarate (step 4). Finally, hydrolysis of arginine (step 5) yields one molecule of urea and regenerates ornithine.

Operation of the cycle requires a continuous supply of carbamoyl phosphate and aspartate. Carbamoyl phosphate is supplied by reaction of ammonium ion and bicarbonate. Aspartate is supplied from fumarate via reactions of the tricarboxylic acid cycle, followed by transamination. The conversion of fumarate to aspartate via oxaloacetate demonstrates the interrelation between the urea cycle and the tricarboxylic acid cycle (Figure 19.6).

Virtually all tissues produce NH_4^+, and ammonium ion is highly toxic, especially to the nervous system. In humans, the synthesis of urea in the liver is the only important route for detoxification and elimination of NH_4^+. Failure of the urea-synthesizing pathway for any reason, including liver malfunction or inherited defects in any of the five enzymes of the urea cycle, causes an increase of ammonium ion in the blood, liver, and urine, a condition that produces **ammonia intoxication**. Symptoms of ammonia intoxication are protein-induced vomiting, blurred vision, tremors, slurred speech, and ultimately, coma and death. In treating ammonia intoxication, it is essential to decrease the intake of dietary protein in order to decrease ammonium ion formation. Genetic defects in each of the five enzymes of the urea cycle do exist, but fortunately these inborn errors of metabolism are rare.

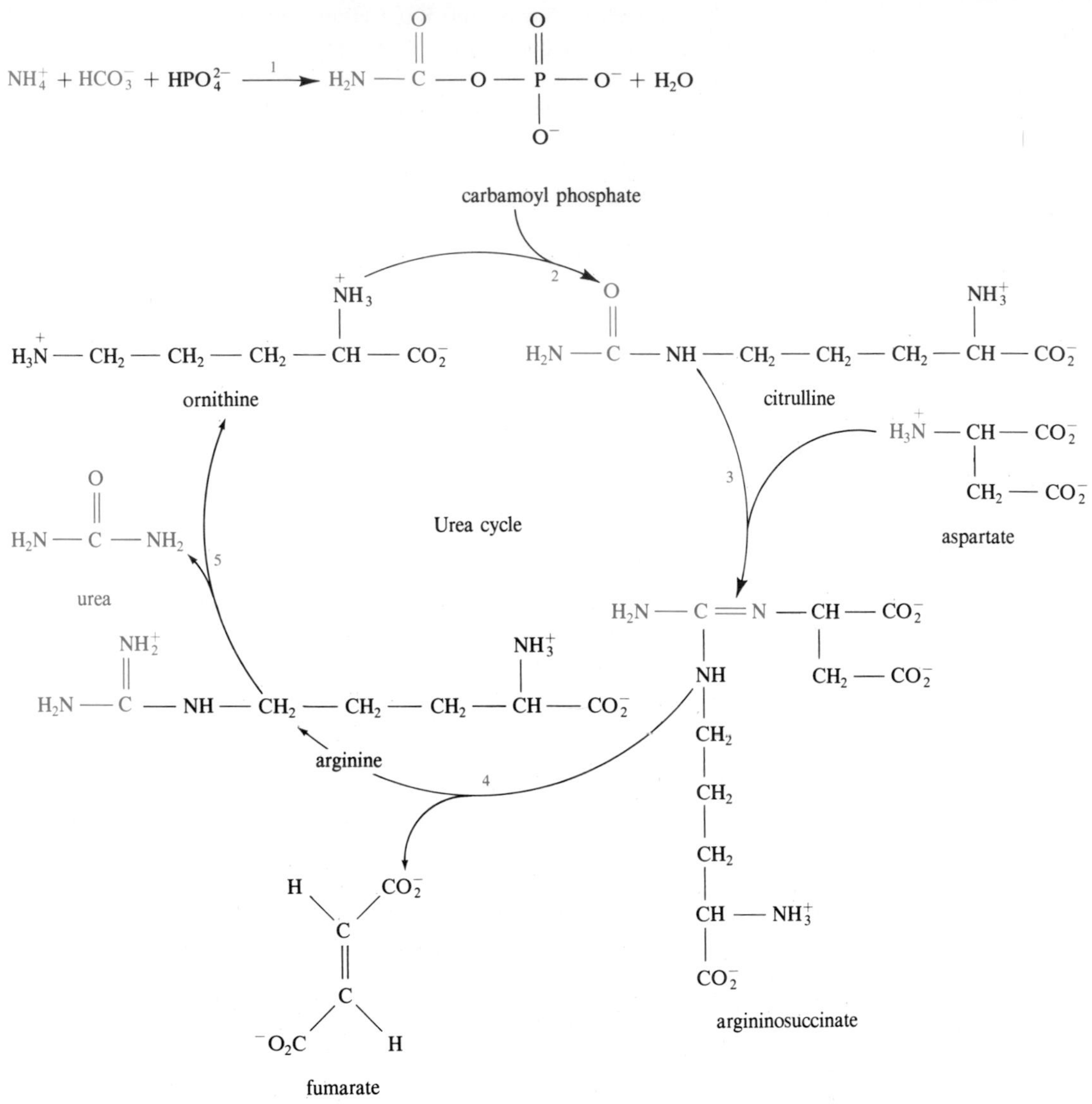

Figure 19.5 The urea cycle. The enzyme-catalyzed steps are numbered 1–5.

D. Fates of Carbon Skeletons of Amino Acids

By transamination, oxidative deamination, and a few other reactions, the carbon skeletons of amino acids are converted to one of six common intermediates, all of which we have seen at one time or another (Chapters 17 and 18, and this chapter). These common intermediates and the amino acids from which they are derived are summarized in Table 19.3. The central pathway for the oxidation of these common intermediates to carbon dioxide and water is the

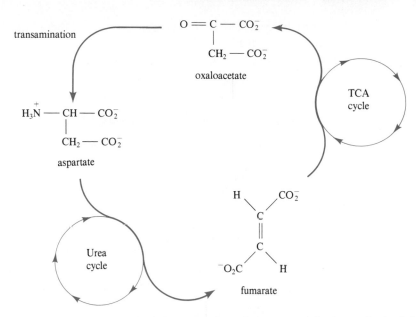

Figure 19.6 Aspartate, required for operation of the urea cycle, is synthesized from fumarate, a product of the urea cycle, via the tricarboxylic acid cycle, followed by transamination.

	Common Intermediate	Name	Amino Acid Source
Table 19.3 Common intermediates derived from amino acids.	$CH_3\overset{\overset{\text{O}}{\|}}{C}CO_2^-$	pyruvate	alanine, glycine, serine, cysteine, tryptophan
	$^-O_2CCH_2CH_2\overset{\overset{\text{O}}{\|}}{C}CO_2^-$	α-ketoglutarate	arginine, histidine, proline, glutamine, glutamate
	$^-O_2CCH_2CH_2\overset{\overset{\text{O}}{\|}}{C}{-}SCoA$	succinyl CoA	valine, isoleucine, methionine, threonine
	fumarate structure	fumarate	phenylalanine, tyrosine
	$^-O_2CCH_2\overset{\overset{\text{O}}{\|}}{C}CO_2^-$	oxaloacetate	asparagine, aspartate
	$CH_3\overset{\overset{\text{O}}{\|}}{C}{-}SCoA$	acetyl CoA	leucine, phenylalanine, tyrosine, lysine, tryptophan, isoleucine

tricarboxylic acid cycle. The points at which these intermediates enter the cycle are shown in Figure 19.7.

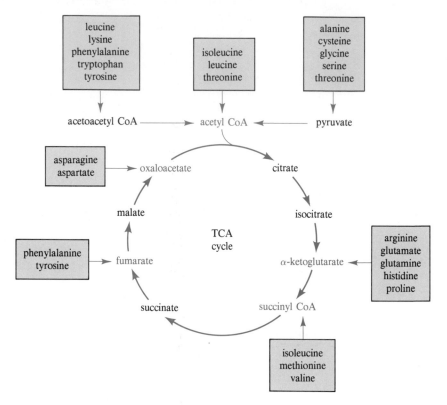

Figure 19.7 Pathways by which carbon skeletons from amino acid degradation enter the tricarboxylic acid.

E. Conversion to Glucose and Ketone Bodies

Amino acids degraded to pyruvate, alpha-ketoglutarate, succinyl CoA, fumarate, and oxaloacetate can be converted to phosphoenolpyruvate and then to glucose. These amino acids are said to be glycogenic, and the synthesis of glucose from them is called gluconeogenesis (Section 17.2H). **Glycogenic amino acids**, along with glycerol, provide alternative sources of glucose during periods of low carbohydrate intake or when stores are being rapidly depleted.

Amino acids degraded directly to acetyl CoA cannot be converted to glucose, since humans and other animals have no biochemical pathways for the synthesis of glucose from acetyl CoA. It is for this reason that fatty acids cannot serve as sources of carbon atoms for the synthesis of glucose. Acetyl CoA derived from the degradation of amino acids can be transformed to ketone bodies (Section 18.5). Amino acids that are degraded to acetyl CoA and then transformed to ketone bodies are said to be **ketogenic**.

The first experimental attempts to classify amino acids as glycogenic or ketogenic were carried out on test animals when the metabolic pathways for the degradation of individual amino acids were only poorly understood. In the earliest studies, laboratory dogs were made diabetic, either by selective chemical destruction of the insulin-producing ability of the pancreas, or by removal of the pancreas itself. Blood-glucose levels were then controlled by injecting insulin. These diabetic dogs excreted glucose in the urine even when glycogen and fat stores had been depleted and when they were fed a diet containing protein as the sole source of metabolic fuel. They also excreted urea, the means by which amino acid-derived nitrogen atoms are detoxified and eliminated. The molar ratio of glucose to urea indicates the extent to which the carbon skeletons of amino acids can be used for the synthesis of glucose. These studies revealed that in diabetic dogs, a maximum of 58 g glucose can be derived from 100 g protein. In other words, 58% of protein is glycogenic.

To determine which of the 20 amino acids were glycogenic, diabetic test animals were fed pure amino acids, one at a time. If glucose was excreted in the urine following such a feeding, the amino acid was classified as glycogenic. If acetoacetate, beta-hydroxybutyrate, or acetone was excreted, the amino acid was classified as ketogenic. Of the 20 protein-derived amino acids, only leucine is purely ketogenic. Six amino acids are both glycogenic and ketogenic, and the remaining thirteen amino acids are purely glycogenic. The glycogenic and ketogenic amino acids are:

Glycogenic		Glycogenic and Ketogenic	Ketogenic
alanine	glycine	isoleucine	leucine
arginine	histidine	lysine	
asparagine	methionine	phenylalanine	
aspartic acid	proline	threonine	
cysteine	serine	tryptophan	
glutamic acid	valine	tyrosine	
glutamine			

Key Terms and Concepts

amino acid pool (19.1A)

ammonia intoxication (19.3C)

glycogenic amino acids (19.3E)

ketogenic amino acids (19.3E)

oxidative deamination (19.3B)

protein half-life (19.1A)

transamination (19.3A)

urea cycle (19.3C)

Key Reactions

1. Reactions of pyridoxal phosphate/pyridoxamine phosphate in transamination (Section 19.3A).
2. Oxidative deamination of glutamate by NAD^+ (Section 19.3B).
3. Net reaction of the urea cycle (Section 19.3C).
4. The five reactions of the urea cycle (Section 19.3C).

Problems

An overview of amino acid metabolism (Section 19.1)

19.2 List the three vital functions served by amino acids in the body.

19.3 Define the term half-life as it is applied in this chapter to tissue and plasma proteins.

19.4 Compare the degree to which carbohydrates, fats, and proteins can be stored in the body for later use.

19.5 What percentage of the total energy requirement of the average adult is supplied by carbohydrates? by fats? by proteins?

Catabolism of amino acids (Section 19.3)

19.6 Complete the following reactions. Show structural formulas for products and reactants.

a. leucine + oxaloacetate $\xrightarrow{\text{transaminase}}$

b. alanine + alpha-ketoglutarate $\xrightarrow{\text{transaminase}}$

c. glycine + pyruvate $\xrightarrow{\text{transaminase}}$

d. glycine + pyridoxal phosphate $\xrightarrow{\text{transaminase}}$

e. pyridoxamine phosphate + pyruvate $\xrightarrow{\text{transaminase}}$

19.7
a. Write an equation for the reaction catalyzed by the enzyme serum glutamate oxaloacetate transaminase (SGOT); for the reaction catalyzed by serum glutamate pyruvate transaminase (SGPT).
b. Describe how an assay for the presence of these enzymes in blood can provide information about possible heart and liver damage.

19.8 In nutritional studies on rats, it has been found that certain alpha-ketoacids may substitute for essential amino acids. Shown here are structural formulas for three such alpha-ketoacids.

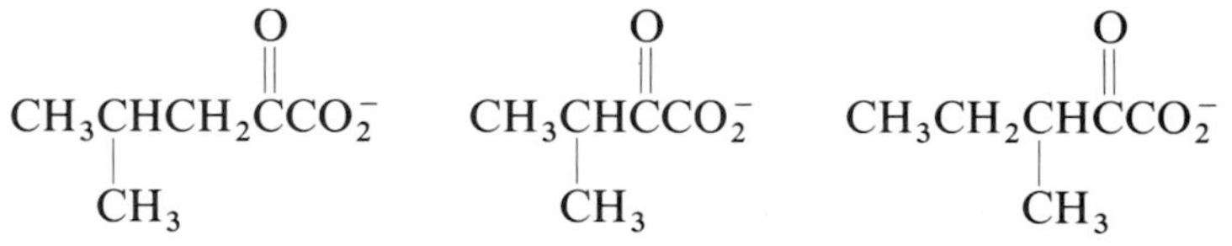

$$\underset{\text{(a)}}{\overset{\displaystyle \quad}{CH_3CHCH_2\overset{\displaystyle O}{\overset{\displaystyle \|}{C}}CO_2^-}} \qquad \underset{\text{(b)}}{\overset{\displaystyle \quad}{CH_3CH\overset{\displaystyle O}{\overset{\displaystyle \|}{C}}CO_2^-}} \qquad \underset{\text{(c)}}{\overset{\displaystyle \quad}{CH_3CH_2CH\overset{\displaystyle O}{\overset{\displaystyle \|}{C}}CO_2^-}}$$

with CH_3 substituents below each structure

a. Account for substitution, in certain instances, of these alpha-ketoacids for essential amino acids.
b. For which essential amino acid might each substitute?

19.9 The following reaction is the first step in the degradation of ornithine. Propose a metabolic pathway to account for this transformation.

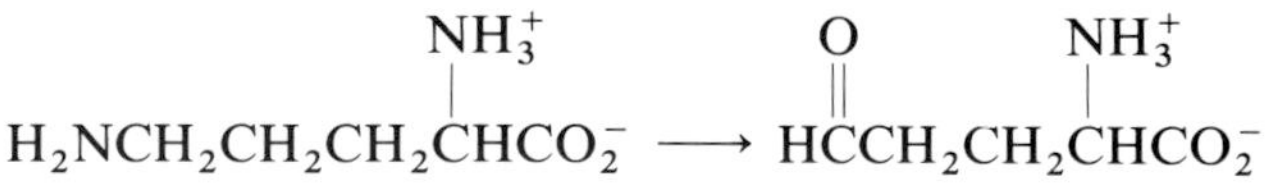

$$\underset{\text{ornithine}}{H_2NCH_2CH_2CH_2\overset{NH_3^+}{CHCO_2^-}} \longrightarrow \underset{\text{glutamate semialdehyde}}{\overset{O}{\overset{\|}{H}CCH_2CH_2}\overset{NH_3^+}{CHCO_2^-}}$$

19.10 Write an equation for the oxidative deamination of glutamate.

19.11 What is the function of the urea cycle?

19.12 Using structural formulas, write an equation for the hydrolysis of arginine to ornithine and urea.

19.13 In what organ does the synthesis of urea take place?

19.14 Urea has two nitrogen atoms. What is the source of each? What is the source of the single carbon atom in urea?

19.15 Write a balanced equation for the net reaction of the urea cycle.

19.16 What is meant by ammonia intoxication?

19.17 List the five points at which carbon skeletons derived from amino acids enter the tricarboxylic acid cycle.

19.18 **a.** What does it mean to say that an amino acid is glycogenic?
   b. What does it mean to say that an amino acid is ketogenic?
   c. Is it possible for an amino acid to be both glycogenic and ketogenic? Explain.

19.19 Propose biochemical pathways to explain how the cell might carry out the following transformations. Name each type of reaction (for example, hydration, dehydration, oxidation, hydrolysis).
   a. phenylalanine $\longrightarrow$ phenylacetate
   b. 3-phosphoglycerate $\longrightarrow$ serine
   c. citrate $\longrightarrow$ glutamate
   d. ornithine $\longrightarrow$ glutamate
   e. methylmalonyl CoA $\longrightarrow$ oxaloacetate

19.20 Following is the metabolic pathway for the conversion of isoleucine to acetyl CoA and propionyl CoA. You have already studied each type of reaction shown here, though not necessarily in this chapter. For each step in this sequence, name the type of reaction and specify any coenzymes involved.

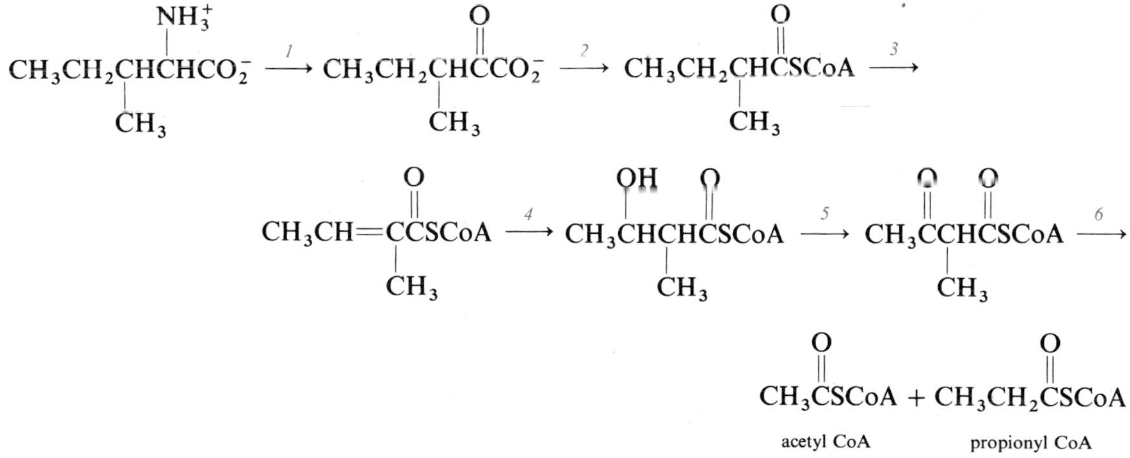

$$\underset{\underset{CH_3}{|}}{CH_3CH_2\overset{\overset{NH_3^+}{|}}{CH}CHCO_2^-} \xrightarrow{\text{\textit{1}}} \underset{\underset{CH_3}{|}}{CH_3CH_2\overset{\overset{O}{\|}}{CH}CCO_2^-} \xrightarrow{\text{\textit{2}}} \underset{\underset{CH_3}{|}}{CH_3CH_2\overset{\overset{O}{\|}}{CH}CSCoA} \xrightarrow{\text{\textit{3}}}$$

$$\underset{\underset{CH_3}{|}}{CH_3CH=\overset{\overset{O}{\|}}{C}CSCoA} \xrightarrow{\text{\textit{4}}} \underset{\underset{CH_3}{|}}{CH_3\overset{OH}{CH}CH\overset{O}{\|}CSCoA} \xrightarrow{\text{\textit{5}}} \underset{\underset{CH_3}{|}}{CH_3\overset{O}{\|}CCH\overset{O}{\|}CSCoA} \xrightarrow{\text{\textit{6}}}$$

$$\underset{\text{acetyl CoA}}{CH_3\overset{\overset{O}{\|}}{C}SCoA} + \underset{\text{propionyl CoA}}{CH_3CH_2\overset{\overset{O}{\|}}{C}SCoA}$$

19.21 Following is the metabolic pathway for the conversion of proline to glutamate. Name the type of reaction involved in each step of this transformation and specify any coenzymes you think might be involved.

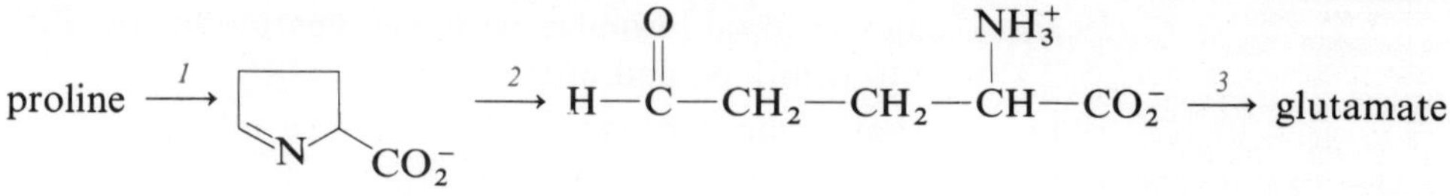

$$\text{proline} \xrightarrow{\;1\;} \underset{\underset{\displaystyle N}{}}{\bigsqcup}_{CO_2^-} \xrightarrow{\;2\;} \underset{\displaystyle O}{\overset{\displaystyle O}{H-C}}-CH_2-CH_2-\underset{\underset{\displaystyle }{}}{\overset{\displaystyle NH_3^+}{CH}}-CO_2^- \xrightarrow{\;3\;} \text{glutamate}$$

19.22 Most substances in the biological world are built from just thirty or so smaller molecules. Review the biochemistry we have discussed in this textbook and make your own list of these thirty or so fundamental building blocks of nature.

Index